SIGNAL TRANSDUCTION IN LUNG CELLS

LUNG BIOLOGY IN HEALTH AND DISEASE

Executive Editor

Claude Lenfant
Director, National Heart, Lung, and Blood Institute
National Institutes of Health
Bethesda, Maryland

1. Immunologic and Infectious Reactions in the Lung, *edited by Charles H. Kirkpatrick and Herbert Y. Reynolds*
2. The Biochemical Basis of Pulmonary Function, *edited by Ronald G. Crystal*
3. Bioengineering Aspects of the Lung, *edited by John B. West*
4. Metabolic Functions of the Lung, *edited by Y. S. Bakhle and John R. Vane*
5. Respiratory Defense Mechanisms (in two parts), *edited by Joseph D. Brain, Donald F. Proctor, and Lynne M. Reid*
6. Development of the Lung, *edited by W. Alan Hodson*
7. Lung Water and Solute Exchange, *edited by Norman C. Staub*
8. Extrapulmonary Manifestations of Respiratory Disease, *edited by Eugene Debs Robin*
9. Chronic Obstructive Pulmonary Disease, *edited by Thomas L. Petty*
10. Pathogenesis and Therapy of Lung Cancer, *edited by Curtis C. Harris*
11. Genetic Determinants of Pulmonary Disease, *edited by Stephen D. Litwin*
12. The Lung in the Transition Between Health and Disease, *edited by Peter T. Macklem and Solbert Permutt*
13. Evolution of Respiratory Processes: A Comparative Approach, *edited by Stephen C. Wood and Claude Lenfant*
14. Pulmonary Vascular Diseases, *edited by Kenneth M. Moser*
15. Physiology and Pharmacology of the Airways, *edited by Jay A. Nadel*
16. Diagnostic Techniques in Pulmonary Disease (in two parts), *edited by Marvin A. Sackner*
17. Regulation of Breathing (in two parts), *edited by Thomas F. Hornbein*
18. Occupational Lung Diseases: Research Approaches and Methods, *edited by Hans Weill and Margaret Turner-Warwick*
19. Immunopharmacology of the Lung, *edited by Harold H. Newball*
20. Sarcoidosis and Other Granulomatous Diseases of the Lung, *edited by Barry L. Fanburg*

21. Sleep and Breathing, *edited by Nicholas A. Saunders and Colin E. Sullivan*
22. *Pneumocystis carinii* Pneumonia: Pathogenesis, Diagnosis, and Treatment, *edited by Lowell S. Young*
23. Pulmonary Nuclear Medicine: Techniques in Diagnosis of Lung Disease, *edited by Harold L. Atkins*
24. Acute Respiratory Failure, *edited by Warren M. Zapol and Konrad J. Falke*
25. Gas Mixing and Distribution in the Lung, *edited by Ludwig A. Engel and Manuel Paiva*
26. High-Frequency Ventilation in Intensive Care and During Surgery, *edited by Graziano Carlon and William S. Howland*
27. Pulmonary Development: Transition from Intrauterine to Extrauterine Life, *edited by George H. Nelson*
28. Chronic Obstructive Pulmonary Disease: Second Edition, Revised and Expanded, *edited by Thomas L. Petty*
29. The Thorax (in two parts), *edited by Charis Roussos and Peter T. Macklem*
30. The Pleura in Health and Disease, *edited by Jacques Chrétien, Jean Bignon, and Albert Hirsch*
31. Drug Therapy for Asthma: Research and Clinical Practice, *edited by John W. Jenne and Shirley Murphy*
32. Pulmonary Endothelium in Health and Disease, *edited by Una S. Ryan*
33. The Airways: Neural Control in Health and Disease, *edited by Michael A. Kaliner and Peter J. Barnes*
34. Pathophysiology and Treatment of Inhalation Injuries, *edited by Jacob Loke*
35. Respiratory Function of the Upper Airway, *edited by Oommen P. Mathew and Giuseppe Sant'Ambrogio*
36. Chronic Obstructive Pulmonary Disease: A Behavioral Perspective, *edited by A. John McSweeny and Igor Grant*
37. Biology of Lung Cancer: Diagnosis and Treatment, *edited by Steven T. Rosen, James L. Mulshine, Frank Cuttitta, and Paul G. Abrams*
38. Pulmonary Vascular Physiology and Pathophysiology, *edited by E. Kenneth Weir and John T. Reeves*
39. Comparative Pulmonary Physiology: Current Concepts, *edited by Stephen C. Wood*
40. Respiratory Physiology: An Analytical Approach, *edited by H. K. Chang and Manuel Paiva*
41. Lung Cell Biology, *edited by Donald Massaro*
42. Heart–Lung Interactions in Health and Disease, *edited by Steven M. Scharf and Sharon S. Cassidy*
43. Clinical Epidemiology of Chronic Obstructive Pulmonary Disease, *edited by Michael J. Hensley and Nicholas A. Saunders*
44. Surgical Pathology of Lung Neoplasms, *edited by Alberto M. Marchevsky*

45. The Lung in Rheumatic Diseases, *edited by Grant W. Cannon and Guy A. Zimmerman*
46. Diagnostic Imaging of the Lung, *edited by Charles E. Putman*
47. Models of Lung Disease: Microscopy and Structural Methods, *edited by Joan Gil*
48. Electron Microscopy of the Lung, *edited by Dean E. Schraufnagel*
49. Asthma: Its Pathology and Treatment, *edited by Michael A. Kaliner, Peter J. Barnes, and Carl G. A. Persson*
50. Acute Respiratory Failure: Second Edition, *edited by Warren M. Zapol and Francois Lemaire*
51. Lung Disease in the Tropics, *edited by Om P. Sharma*
52. Exercise: Pulmonary Physiology and Pathophysiology, *edited by Brian J. Whipp and Karlman Wasserman*
53. Developmental Neurobiology of Breathing, *edited by Gabriel G. Haddad and Jay P. Farber*
54. Mediators of Pulmonary Inflammation, *edited by Michael A. Bray and Wayne H. Anderson*
55. The Airway Epithelium, *edited by Stephen G. Farmer and Douglas Hay*
56. Physiological Adaptations in Vertebrates: Respiration, Circulation, and Metabolism, *edited by Stephen C. Wood, Roy E. Weber, Alan R. Hargens, and Ronald W. Millard*
57. The Bronchial Circulation, *edited by John Butler*
58. Lung Cancer Differentiation: Implications for Diagnosis and Treatment, *edited by Samuel D. Bernal and Paul J. Hesketh*
59. Pulmonary Complications of Systemic Disease, *edited by John F. Murray*
60. Lung Vascular Injury: Molecular and Cellular Response, *edited by Arnold Johnson and Thomas J. Ferro*
61. Cytokines of the Lung, *edited by Jason Kelley*
62. The Mast Cell in Health and Disease, *edited by Michael A. Kaliner and Dean D. Metcalfe*
63. Pulmonary Disease in the Elderly Patient, *edited by Donald A. Mahler*
64. Cystic Fibrosis, *edited by Pamela B. Davis*
65. Signal Transduction in Lung Cells, *edited by Jerome S. Brody, David M. Center, and Vsevolod A. Tkachuk*

ADDITIONAL VOLUMES IN PREPARATION

Tuberculosis: A Comprehensive International Approach, *edited by Lee B. Reichman and Earl S. Hershfield*

Fluid and Solute Transport in the Airspaces of the Lungs, *edited by Richard M. Effros and H. K. Chang*

Pharmacology of the Respiratory Tract: Experimental and Clinical Research, *edited by K. Fan Chung and Peter J. Barnes*

Prevention of Respiratory Diseases, *edited by Albert Hirsch, Marcel Goldberg, and J. P. Martin*

Sleep and Breathing: Second Edition, Revised and Expanded, *edited by Nicholas A. Saunders and Colin E. Sullivan*

Airway Secretion: Biological Bases for the Control of Mucus Hypersecretion, *edited by Tamotsu Takishima and Sanae Shimura*

Sarcoidosis and Other Granulomatous Disorders, *edited by D. Geraint James*

Epidemiology of Lung Cancer, *edited by Jonathan Samet*

Pneumocystis carinii Pneumonia: Second Edition, Revised and Expanded, *edited by Peter D. Walzer*

SIGNAL TRANSDUCTION IN LUNG CELLS

Edited by

Jerome S. Brody
David M. Center

Boston University School of Medicine
Boston, Massachusetts

Vsevolod A. Tkachuk

Cardiology Research Center
Moscow, Russia

Marcel Dekker, Inc. **New York • Basel • Hong Kong**

Library of Congress Cataloging-in-Publication Data

Signal transduction in lung cells / edited by Jerome S. Brody, David M. Center, Vsevolod A. Tkachuk.
p. cm. — (Lung biology in health and disease ; v. 65)
Includes bibliographical references and indexes.
ISBN 0-8247-8813-3 (alk. paper)
1. Lungs—Cytology. 2. Cellular signal transduction. I. Brody, Jerome S. II. Center, David M. III. Tkachuk, V. A. (Vsevolod Arsen 'evich) IV. Series.
[DNLM: 1. Lung—cytology. 2. Signal Transduction. W1 LU62 v.65 1993 / WF 600 S5775 1993]
QP121.S54 1993
591.87-dc20
DNLM/DLC
for Library of Congress 92-48426
CIP

This book is printed on acid-free paper.

MARCEL DEKKER, INC.
270 Madison Avenue, New York, New York 10016

Current printing (last digit)
10 9 8 7 6 5 4 3 2 1

Printed in the United States of America

INTRODUCTION

"Transduction" is a relatively new word. It derives from a much older one, "transducer," which is defined as a "device that receives energy (of a signal) from one system and retransmits it, often in a different form, to another."

This volume, *Signal Transduction in Lung Cells*, is a first for the Lung Biology in Health and Disease series. No other volume addresses a similar topic, even though today's biology has recognized for quite a while that the communications between cells and the functions of different cells are as complex as the structure of the cells themselves. Thus, this volume is timely as well as unprecedented. Clinicians and investigators will find in one place the newest and most provocative information on the events that trigger the function of the many cells that are present in the lungs.

For generations, theories have existed to explain the functions of cells and the manner in which these functions are activated. Claude Bernard once said:

> However good a theory may be, it is never as good as the truth or an actual fact. . . . There is a discovery underneath a revolution, as they say; for science is revolutionary and does not advance by successive addition. . . .

For the last two or three decades, science has undergone an extraordinary, explosive upheaval fueled in great part by new approaches. This volume reports the discoveries that have resulted from many revolutions.

But this monograph represents another first for the Lung Biology series, as well. Other volumes have been acclaimed for their international authorship, generally from several countries and continents. Here, the international authorship involves only two countries: the United States and Russia. Its concept was developed years ago in an era when Russia was the USSR and it was difficult for the Soviet scientists to work with scientists from the United States. However, those involved in this enterprise persevered, believing as did Claude Bernard that:

> . . . in the most elevated region of human knowledge exists an atmosphere common to every intellectual and cultural enterprise, where artist, man of culture, and scientist must necessarily meet and understand each other.

And thus the concept of this book became a reality.

As the editor of the series, I take great pride in introducing this volume. All the editors and authors are to be commended for having persisted in its completion. I am particularly grateful to my two friends, Jerry Brody and Seva Tkachuk, for having been so enthusiastic about this project.

Claude Lenfant, M.D.
Bethesda, Maryland

PREFACE

This book results from the convergence of two unique and quite special circumstances. The first was the opportunity of U.S. scientists to meet researchers from the Cardiology Research Center of the (formerly) USSR Academy of Medical Sciences. This opportunity led to scientific interactions between investigators from the Cardiovascular Research Center in Moscow and pulmonary scientists in the United States. A number of investigators from the Research Center and in particular from Professor Tkachuk's laboratory have collaborated with U.S. investigators on cardiovascular and pulmonary topics related to signal transduction. These collaborations have involved U.S. scientists working in either Moscow in the Republic of Russia or Bishkek (formerly Frunze) in Kirghizistan, or have involved scientists from these institutes working in laboratories in Boston, San Francisco, Miami, Denver, or Hershey, Pennsylvania. This scientific-cultural interchange has resulted in not only a number of scientific publications, including this book, but cultural understanding and friendships that will continue long after the scientific work has concluded.

The second special program leading to this publication was the initiation in 1991 of an annual basic science symposium devoted to the cell and molecular biology of the lung. This symposium, which we have called "Lung Cell

Biology at Woods Hole," has been organized by the Pulmonary Center at Boston University School of Medicine to foster interchange between pulmonary scientists and to expose pulmonary scientists to other basic scientists whose work does not involve the lung but who are working in fields that have application to lung cell biology. The first of these symposia, on signal transduction in lung cells, was held at the Marine Biology laboratories in Woods Hole from October 6 to 9, 1991. It was sponsored by both the Pulmonary Center and Dr. Michael Fox, currently Senior Vice President at Astra Pharmaceuticals, and Smith Kline and Beecham, Inc.

We chose as the topic of our first meeting signal transduction in lung cells in part because of the above-noted scientific interaction between U.S. scientists and scientists in the then USSR and in part because of the increasing importance this topic plays in understanding the biology of the lung and virtually all its cells.

The plan for the meeting and the resulting book was to begin with a series of talks (chapters) covering basic aspects of signal transduction and then to consider special aspects for signal transduction as they apply to cells of the lung. Because almost every cellular process in the lung involves some sort of signal transduction, we had to be selective in choosing the content of this book. Since much of the collaborative science supported by the above-mentioned collaboration has involved G proteins, the book is weighted toward this aspect of cell signaling, and important topics such as signaling and gene regulation or nuclear signaling have not been dealt with in depth. It was our goal that the chapters provide a general overview of a topic and then specific new data.

The part devoted to general concepts includes chapters that range from G proteins and second messenger systems to membrane receptors, their kinases, and phosphatases. The lung-related chapters cover epithelial cells, smooth muscle cells, endothelial cells, and various cells of the immune system. While certainly not exhaustive, the general and lung-related chapters deal with most current aspects of cell signaling and provide a blueprint of the major directions of current research.

Jerome S. Brody
David M. Center
Vsevolod A. Tkachuk

CONTRIBUTORS

A. A. Aldashev, Ph.D. Laboratory of Molecular Cardiology, Kirghiz Institute of Cardiology, Bishkek, Kirghizstan, Russia

Pavel Avdonin, M.D. Department of Biochemistry, Laboratory of Molecular Endocrinology, Institute of Experimental Cardiology, Cardiology Research Center, Moscow, Russia

A. Bakhramov, Ph.D. Laboratory of Molecular Endocrinology, Cardiology Research Center, Moscow, Russia

Kim E. Barrett, Ph.D. Associate Professor of Medicine, Division of Allergic and Immunologic Diseases, Department of Medicine, School of Medicine, University of California, San Diego, San Diego, California

Dennis J. Beer, M.D. Associate Professor of Medicine, Pulmonary Center, Boston University School of Medicine, Boston, Massachusetts

Dale J. Benos, Ph.D. Professor, Department of Physiology and Biophysics, University of Alabama at Birmingham, Birmingham, Alabama

Norbert Berndt, Ph.D. Assistant Professor, Department of Pediatrics, University of Southern California School of Medicine, and Childrens Hospital Los Angeles, Los Angeles, California

Philippe Bertrand, Ph.D. Research Associate, Department of Cell Biology, Baylor College of Medicine, Houston, Texas

Gary St. John Bird, Ph.D. Laboratory of Cellular and Molecular Pharmacology, National Institute of Environmental Health Sciences, Research Triangle Park, North Carolina

Lutz Birnbaumer, Ph.D. Professor, Division of Neuroscience, Departments of Cell Biology, Physiology, and Molecular Biophysics, Baylor College of Medicine, Houston, Texas

Mariel Birnbaumer, Ph.D. Associate Professor, Department of Cell Biology, Baylor College of Medicine, Houston, Texas

James C. Bonner, Ph.D. Senior Staff Fellow, Pulmonary Pathology Group, Laboratory of Pulmonary Pathobiology, National Institute of Environmental Health Sciences, Research Triangle Park, North Carolina

Philippe Brabet, Ph.D. Research Associate, Department of Cell Biology, Baylor College of Medicine, Houston, Texas

P. Bregestovski, Ph.D. Professor of Biochemistry, Laboratory of Molecular Endocrinology, Cardiology Research Center, Moscow, Russia

Arnold R. Brody, Ph.D. Head, Pulmonary Pathology Group, Laboratory of Pulmonary Pathobiology, National Institute of Environmental Health Sciences, Research Triangle Park, North Carolina

Alexander G. Chuchanlin, Ph.D. Laboratory of Molecular Endocrinology, Cardiology Research Center, Moscow, Russia

Juan Codina, Ph.D. Research Associate Professor, Department of Cell Biology, Baylor College of Medicine, Houston, Texas

Deborah E. Cool, Ph.D. Research Associate, Department of Biochemistry, University of Washington, Seattle, Washington

Roberto Coria, Ph.D. Research Associate, Department of Cell Biology, Baylor College of Medicine, Houston, Texas

William W. Cruikshank Assistant Professor of Medicine, Pulmonary Center, Boston University School of Medicine, Boston, Massachusetts

Primal de Lanerolle, Ph.D. Associate Professor, Department of Physiology and Biophysics, University of Illinois at Chicago, Chicago, Illinois

Balazs Dezso, M.D. Research Associate, Department of Physiology and Biophysics, University of Illinois at Chicago, Chicago, Illinois

Burton F. Dickey, M.D. Chief, Pulmonary and Critical Care Section, Houston VA Medical Center, and Associate Professor, Departments of Medicine, Cell Biology, and Experimental Therapeutics, Baylor College of Medicine, Houston, Texas

George R. Dubyak, Ph.D. Associate Professor, Department of Physiology and Biophysics, Case Western Reserve University, Cleveland, Ohio

Harrison W. Farber, M.D. Associate Professor of Medicine, Pulmonary Center, Boston University School of Medicine, Boston, Massachusetts

Jacob N. Finkelstein, Ph.D. Associate Professor, Departments of Pediatrics, Radiation Oncology, and Environmental Medicine, University of Rochester School of Medicine and Dentistry, Rochester, New York

Lois J. Geist, M.D. Assistant Professor, Department of Internal Medicine, University of Iowa Hospitals and Clinics, Iowa City, Iowa

Major Clay Glennon Biological Laboratory Technician, Laboratory of Cellular and Molecular Pharmacology, National Institute of Environmental Health Sciences, Research Triangle Park, North Carolina

Gayle Gorgas Research Assistant, Department of Physiology and Biophysics, University of Illinois at Chicago, Chicago, Illinois

Rolf Graf, Ph.D. Research Associate, Department of Cell Biology, Baylor College of Medicine, Houston, Texas

Thomas Gudermann, M.D. Research Associate, Department of Cell Biology, Baylor College of Medicine, Houston, Texas

John R. Hadcock, Ph.D. Research Biologist, Department of Molecular and Cellular Biology, Agricultural Research Division, American Cyanamid Company, Princeton, New Jersey

Frederick L. Hall, Ph.D. Director and Assistant Professor, Orthopedic Surgery Research Laboratory, Departments of Surgery, Molecular Pharmacology, and Toxicology, University of Southern California School of Medicine, and School of Pharmacy, Childrens Hospital Los Angeles, Los Angeles, California

Brendan Hayes, Ph.D. Department of Surgery, Washington University School of Medicine, St. Louis, Missouri

Morley D. Hollenberg, D.Phil., M.D. Professor, Departments of Pharmacology, Therapeutics, and Medicine, Faculty of Medicine, University of Calgary, Calgary, Alberta, Canada

Gary W. Hunninghake, M.D. Professor and Director, Pulmonary Disease Division, Department of Internal Medicine, University of Iowa Hospitals and Clinics, Iowa City, Iowa

K. Chul Kim, Ph.D. Associate Professor, Department of Pharmacology and Toxicology, University of Maryland School of Pharmacy, Baltimore, Maryland

Andreas Klos, M.D. Research Associate, Department of Cell Biology, Baylor College of Medicine, Houston, Texas

Irna L. Korichneva Laboratory of Molecular Endocrinology, Cardiology Research Center, Moscow, Russia

Steven L. Kunkel, Ph.D. Professor, Department of Pathology, The University of Michigan Medical School, Ann Arbor, Michigan

Johngsoon Lee Department of Biochemistry, Boston University School of Medicine, Boston, Massachusetts

Finn Olav Levy, M.D. Research Associate, Department of Cell Biology, Baylor College of Medicine, Houston, Texas

Craig C. Malbon, Ph.D. Professor of Pharmacology, Director, Diabetes and Metabolic Diseases Research Program, School of Medicine, Health Sciences Center, State University of New York at Stony Brook, Stony Brook, New York

Diana L. Marquardt, M.D. Associate Professor, Department of Medicine, School of Medicine, University of California, San Diego, San Diego, California

Sadis Matalon, Ph.D. Professor of Anesthesiology, Physiology and Biophysics, and Pediatrics, Department of Anesthesiology, University of Alabama at Birmingham, Birmingham, Alabama

Frank S. Menniti, Ph.D. Pfizer Central Research, Groton, Connecticut

V. Mokh Laboratory of Molecular Endocrinology, Cardiology Research Center, Moscow, Russia

Kathleen G. Morgan, Ph.D. Associate Professor in Medicine, Department of Medicine, Harvard Medical School, Boston, Massachusetts

Kutsumi Nogimori, Ph.D. Kaken Pharmaceutical Company, Ltd., Kyoto, Japan

Johnny F. Obie Laboratory of Cellular and Molecular Pharmacology, National Institute of Environmental Health Sciences, Research Triangle Park, North Carolina

Youngsuk Oh Departments of Physiology and Biophysics, University of Alabama at Birmingham, Birmingham, Alabama

Timothy O'Hare, M.D., Ph.D. Department of Biochemistry, Boston University School of Medicine, Boston, Massachusetts

Kerry G. Oliver Senior Staff Fellow, Laboratory of Cellular and Molecular Pharmacology, National Institute of Environmental Health Sciences, Research Triangle Park, North Carolina

Mikhail P. Panchenko, Ph.D. Laboratory of Molecular Endocrinology, Cardiology Research Center, Moscow, Russia

Paul F. Pilch, Ph.D. Professor, Department of Biochemistry, Boston University School of Medicine, Boston, Massachusetts

James W. Putney, Jr., Ph.D. Chief, Laboratory of Cellular and Molecular Pharmacology, National Institute of Environmental Health Sciences, Research Triangle Park, North Carolina

Mark W. Rolfe, M.D. Pulmonary Fellow, Department of Internal Medicine, Division of Pulmonary and Critical Care Medicine, University of Michigan Medical Center, Ann Arbor, Michigan

Michel F. Rossier, Ph.D. Division of Endocrinology, University Hospital, Geneva, Switzerland

Jeffrey B. Rubins, M.D. Assistant Professor, Department of Medicine, Minneapolis Veterans Affairs Medical Center and University of Minnesota, Minneapolis, Minnesota

Sergey A. Rudchenko Laboratory of Molecular Endocrinology, Cardiology Research Center, Moscow, Russia

Uwe Rudolph, M.D. Instructor, Department of Cell Biology, Baylor College of Medicine, Houston, Texas

Rita M. Ryan, M.D. Assistant Professor, Department of Pediatrics (Neonatology), University of Rochester School of Medicine and Dentistry, Rochester, New York

Thomas C. Ryan, Ph.D. Research Associate, Pulmonary Center, Boston University School of Medicine, Boston, Massachusetts

Una S. Ryan, Ph.D. Monsanto Company and Washington University School of Medicine, St. Louis, Missouri

Vitalyi O. Rybin, Ph.D. Laboratory of Molecular Endocrinology, Cardiology Research Center, Moscow, Russia

Jack Sanford, Ph.D. Research Associate, Department of Cell Biology, Baylor College of Medicine, Houston, Texas

Kimberly Schluns Research Assistant, Department of Physiology and Biophysics, University of Illinois at Chicago, Chicago, Illinois

Theodore J. Standiford, M.D. Assistant Professor, Division of Pulmonary and Critical Care Medicine, Department of Internal Medicine, University of Michigan Medical Center, Ann Arbor, Michigan

Robert M. Strieter, M.D. Assistant Professor, Division of Pulmonary and Critical Care Medicine, Department of Internal Medicine, University of Michigan Medical Center, Ann Arbor, Michigan

Vsevolod A. Tkachuk, Ph.D. Professor of Biochemistry, Laboratory of Molecular Endocrinology, Cardiology Research Center, Moscow, Russia

Vladimir G. Tupykin Laboratory of Molecular Endocrinology, Cardiology Research Center, Moscow, Russia

David Warburton, B.Sc., M.B., M.R.C.P. Professor of Pediatrics, University of Southern California School of Medicine, and Division of Neonatology and Pediatric Pulmonology, Childrens Hospital Los Angeles, Los Angeles, California

Stephen I. Wasserman, M.D. Helen M. Ranney Professor and Chairman, Department of Medicine, University of California, San Diego, San Diego, California

Allison K. Wilson, Ph.D. Research Associate, Department of Physiology and Biophysics, University of Illinois at Chicago, Chicago, Illinois

CONTENTS

Part One

GENERAL CONCEPTS

1

Molecular Diversity and Functional Correlates in Transmembrane Signaling by G Proteins

LUTZ BIRNBAUMER, JUAN CODINA, ROLF GRAF, ROBERTO CORIA, JACK SANFORD, THOMAS GUDERMANN, FINN OLAV LEVY, PHILIPPE BERTRAND, PHILIPPE BRABET, ANDREAS KLOS, UWE RUDOLPH, and MARIEL BIRNBAUMER

Baylor College of Medicine
Houston, Texas

I. Historical Perspective

The first report that the receptor-sensitive signal-transducing adenylyl cyclase system discovered in 1957 by Sutherland and co-workers (Rall et al., 1956, 1957) is regulated not only by hormones but also by GTP was published in 1971 (Rodbell et al., 1971). In 1975, Pfeuffer and Helmreich presented preliminary evidence that the site of action of GTP was on a molecule separate from adenylyl cyclase (Pfeuffer and Helmreich, 1975). By 1977, this molecule had been resolved by Pfeuffer, who characterized it as a GTP-binding factor with a molecular weight of 42,000 and reconstituted it with adenylyl cyclase (Pfeuffer, 1977). Hormone receptors, which had been suggested in 1969 to be molecular entities separate from adenylyl cyclase (Birnbaumer et al., 1969), were proven to be independent molecules in cell-cell transfer experiments by Schramm's laboratory in 1977 (Orly and Schramm, 1976). Involvement of a GTP regulatory step in light perception was discovered in 1977 (Wheeler and Bitensky, 1977). By 1980 the light-activated GTP-binding component, now termed transducin, and the GTP-binding regulatory component of adenylyl cyclase, originally termed G/F and N and now designated as G_s, had both been purified (Kühn, 1980; Northup et al., 1980). The two GTP-binding proteins were quite similar. Both were activated by GTP under the influence of a receptor, the light receptor (rhodopsin) for transducin and a hormone receptor for G_s. Both were $\alpha\beta\gamma$ heterotrimers formed of homologous but distinct α subunits, and of an interchangeable $\beta\gamma$ dimer. The α subunits of both bound and hydrolyzed GTP, and the purified trimeric holoproteins dissociated upon interaction with nonhydrolyzable GTP analogs to give two products: a free α-G nucleotide complex and the $\beta\gamma$ dimer. Of these, it was shown that it is the α subunit that regulates the effector functions—a cGMP-specific phosphodiesterase in the case of transducin (Fung et al., 1981), and adenylyl cyclase in the case of G_s (Northup et al., 1983)—and $\beta\gamma$ dimers play a regulatory role in what turned out to be a shuttling of the α subunit between effector and receptor.

Parallel to these in-depth studies on the biochemical and molecular basis of phototransduction and activation of adenylyl cyclase, other studies led to the discovery, first, that a GTP-dependent step intervened not only in hormonal stimulation but also in hormonal inhibition of adenylyl cyclase (Londos et al., 1978), and then, that actions of hormones and neurotransmitters affecting cell metabolism through means other than cAMP were also dependent on a GTP-binding process. It is now recognized that hormonal regulation of phosphoinositide hydrolysis by a phospholipase of the C type, releasing inositol trisphosphate (IP_3) plus diacylglycerol (DAG) (Litosch et al., 1985; Cockroft and Gomperts, 1985), activation of a phospholipase A_2, with consequential release of arachidonic acid (AA) (Burch et al., 1986), stimulation of several types of K^+ channels (Pfaffinger et al., 1985; Breitwieser and Szabo, 1985; Yatani et al., 1987a; Ribalet et al., 1989; Kirsch et al., 1990; VanDongen et al., 1988), and at least one type of Ca^{2+} channel (Yatani et al., 1988a; Hamilton et al., 1991) all occur also with the intervening participation of G proteins. G proteins that regulate phospholipases and the K^+ channel are termed G_p and G_k, respectively. Three similar yet distinct G proteins, termed G_{i1}, G_{i2}, and G_{i3}, have G_k activity (Yatani et al., 1988b). A set of G proteins, termed G_q and G_{11}, have been identified as having G_p activity (Smrka et al., 1991; Taylor et al., 1991). Smell receptors in cilia of the olfactory neuroepithelium activate adenylyl cyclase through a unique "olfactory" G protein, G_{olf}, the α subunit of which is highly homologous to that of G_s but encoded in a separate gene expressed exclusively in these cells (Jones et al., 1990).

II. Signals Transduced by G Proteins

Table 1 presents a general diagram of the elements involved in G-protein-dependent signal transductions. Numerically, about 80% of all known hormones and neurotransmitters, as well as many neuromodulators and auto- and paracrine factors that regulate cellular interactions, collectively termed *pri-*

Table 1 Functional and Structural Elements That Mediate the Flow of Information Through G-Protein-Dependent Signal Transduction Systems in Vertebrates

	Input	Transduction			Output
Location					
	Extracellular milieu	Plasma membrane		Intracellular milieu or plasma membrane	
Functional elements		G-protein-coupled receptors		G-protein-regulated effectors	
	Primary Messengers ⇨	R ⇨	G Protein ⇨	E ⇨	Second messengers Membrane potential
Molecular diversity					
Known	ca. 65	91 (60 cloned)	16	10	cAMP, Ca^{2+} IP's, DAG,
Estimated	?	Ca. 150 (?)[a]	Up to 20 (?)	Over 20	AA

[a] Excludes smell and taste.

mary messengers, elicit cellular responses by combining to specific receptors which are coupled to effector functions by G proteins. To name but a few, the neurotransmitters and hormones acting through G-protein-coupled receptors include those for catecholamines, acetylcholine (muscarinic actions), serotonin, dopamine, neurokinins, angiotensin II, bradykinin, hypothalamic release-stimulating and release-inhibiting peptides and factors, gut hormones (cholecystokinin, gastrin, secretin, vasoactive intestinal peptide), pituitary hormones with the exception of prolactin and growth hormone, prostanoids, parathyroid hormone, calcitonin, and glucagon.

Receptors coupled to effectors by G proteins span the plasma membrane seven times. Except for the metabotropic glutamate receptor, all other G-protein-coupled receptors exhibit a significant degree of similarity in their transmembrane segments that indicates a common evolutionary origin. In spite of this, they do have very characteristic features that make them quite different and specific in the rest of their molecules. Figure 1 illustrates these features in a schematic form.

A summary of commonly accepted signals transduced by G proteins, as well as their receptors as they are known by pharmacological and/or molecular means, is presented in Table 2. It is clear from this table that even though the primary messengers are many, the number of distinct receptors that mediate their action is even larger. This is well exemplified by the existence of at least 10 types of adrenergic receptors (β_1, β_2, β_3, α_{1A}, α_{1B}, α_{1C}, α_{1D}, α_{2A}, α_{2B}, and

Table 2 Examples of Ligands Acting Through G-Protein-Coupled Receptors in Mammals and Molecular Diversity of These Receptors

Ligand(s)	Minimum molecular diversity of receptors: By pharmacological criteria	Minimum molecular diversity of receptors: Cloned
1. Neurotransmitters and autacoids		
Catecholamines	10 (α_{1A-1D}; $\alpha_{2A-2C}\beta_1$, β_2, β_3)	10
M-acetylcholine	3-4 (M1, M2, M3)	5 (M1-M5)
Dopamine	2 (D1, D2A, D2B)	5 (D1-5)
Serotonin	>8 ($5HT_1$-like; $5HT_{1A-E}$, $5HT_2$, $5HT_4$)	7(1A-C, 1Dα, 1Dβ, S31, 2)
Histamine	3 (H1, H2, H3)	2 (H1,H2)
GABA	1 (type B)	None
Glutamate	1 (metabotropic)	1
Purines	4 (P1 (Ado1, Ado2); P2x, P2y)	2 (Ado1, Ado2)
2. Peptide and glycoprotein hormones		
ACTH	1	None
Opioids (>5 ligands)	3 (, δ, k)	None
MSH, CRF, GRF, TRH, GnRH, SRIF	6 (1 each)	2 (TRH; SRIF)
Vasopressin	4 (V1a, V1b, V1c, V2)	3 (V1, V2, OT)
Oxytocin	1	None
Glucagon	2 (G1, G2)	None
Glucagon 19-29	1	1
PTH	2	1
Ang. II	1	1
Secretin	1	1
NPY	3 (Y_1-, Y_2-, Y_3-like)	None
VIP	2 (VIP-1, VIP-2)	1 (VIP-1)
PHI (PMI)	1	None
Bradykinin	2 (BK1, BK2)	1 (BK2)

LH, FSH, TSH	3 (1 each)	3
Endothelin	2 (ET-A, ET-B)	2
Neurokinins (Subst. P, subst. K, nmdn. K)	3 (NK1, NK2, NK3)	3
CCK, GRP, galanin, CGRP, calcitonin, kyotorphin	8 (1 each)	2 (GRP, CT)
3. Arachidonic acid metabolites		
Thromboxane	1 (>1?)	1
PGE1/2	2 (PGE-s, PGE-i)	None
PGD, PGI_2, $PGF_{2\alpha}$, LtB_4, LtD_4, LtE_4	6 (1 each)	None
4. Other		
Chemoattractant (fMet-Leu-Phe)	1	2 (fMPL-I and fMLP-II)
C5a	1	1
IL-8	1	1
C3a	1	None
Thrombin	1	1
PAF	1	1
Phosphatidic acid	1	None
5. Sensory		
Light	4 (low, red, green, blue)	4
Odor	?	>15
Taste (?)	?	None
6. Orphan receptors		
Alkylaminoimidazol	1	None
TCH[a]	1	1
B-cell X	?	1
GIP[b]	?	1

[a] THC = terahydrocannabinol.

[b] GIP = glucocorticoid-induced protein.

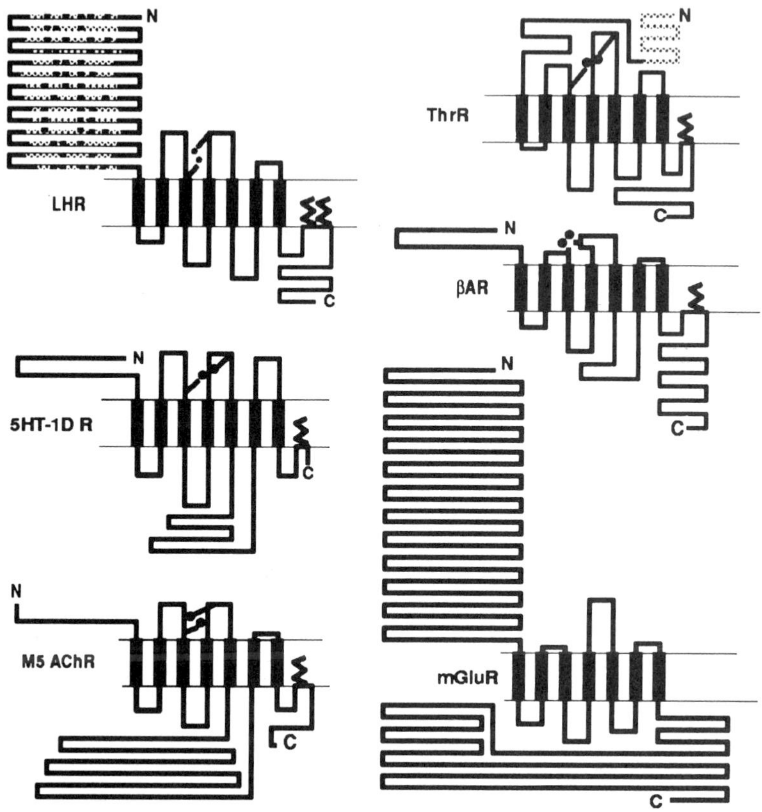

Figure 1 Schematic representation of the LH receptor (LHR), the self-activating thrombin receptor (ThR), a 5HT-1D receptor (5HT-1DR), the M5 acetylcholine receptor (M5R), and the metabotropic glutamate receptor (mGluR). Transmembrane domains, as deduced from hydropathicity plots, are presented as black boxes; lengths of N-terminal and C-terminal domains and of intracellular and extracellular loops are proportional to length in amino acid. All, except the mGLuR, appear to have one conserved cysteine each in the first and second extracellular loops, which are presented as bonded by a disulfide bridge on the basis of studies with rhodopsin. All, except the mGluR, have at least one cysteine in the C-terminal tail, which is depicted as a palmitoylated residue anchored in the membrane on the basis of studies with rhodopsin and the β-adrenergic receptor. The longest extracellular domain, next to that of the mGluR, is that of the N-terminus of the luteinizing receptor, which is depicted with its imperfect homology repeats. The longest intracellular domain is that of the third intracellular loop of the M5R. Primary sequence information for rat LHR is from McFarland et al. (1989); for the thrombin receptor is from Vu et al. (1991); for the rat M5 acetylcholine receptor is from Liao et al. (1989); for the hamster β_2-adrenergic receptor is from Dixon et al. (1986); for the human serotonin 5HT-1Dβ (S12) receptor is from Levy and Birnbaumer (unpublished); and for the rat metabotropic glutamate receptor is from Masu et al. (1991).

α_{2C}), which interact with but two catecholamines. There are five muscarinic acetylcholine receptors (M1 through M5), at least five dopamine receptor types (D1-D5), with one exhibiting "microheterogeneity" due to alternative MRNA splicing. There are also several types of histamine receptors, and many serotonin, purinergic, and light (opsin) receptors. Multiplicity of receptors is not restricted to neurotransmitters. There are at least four types of vasopressin receptors (V1a, V1b, V1c, and V2), possibly two types of receptor for each glucagon, bradykinin and endothelin. Angiotensin II and vasoactive intestinal peptide (VIP) are known to interact with receptors that stimulate either adenylyl cyclase or phosphoinositide hydrolysis. At this time a little over 100 distinct receptors can be identified that recognize about 60 hormones, neurotransmitters, and neuropeptides.

III. Diversity of Effector Systems

In contrast to receptors, the number of final effector functions regulated by these receptors appears to be much smaller, probably not much more than 20 to 30. The effectors include adenylyl cyclase, the "original" effector system, the cGMP-specific phosphodiesterase of photoreceptor cells, phospholipases of the C, D, and A_2 types, and various classes of ionic channels, including one specific for K^+ that conducts preferentially in the inward direction, another also specific for K^+, but less rectifying and sensitive to ATP and sulfonyureas, another specific for Ca^{2+}, strongly dependent on membrane potential for its activity and sensitive to dihydropyridines, and a group of neuronal monovalent cation channels either K^+ or both K^+ and Na^+ selective. There may be more types of effector functions, as well as more types of ion channels. As a consequence of ion channel modulation, the occupancy of G-protein-coupled receptors by primary messengers leads not only to changes of intracellular second messengers—cyclic nucleotides, inositol phosphates, diacylglycerol, arachidonic acid, and Ca^{2+}—but also to changes in the cell's membrane potential, which itself is a potent regulator of cellular function.

IV. G-Protein Diversity

Based on the identification of distinct α subunits (for general structure of G proteins and some structure/function relationships deduced thus far for α subunits, see Figs. 2 and 3), we know at this time of a minimum of 16 distinct α-subunit genes: α_s, α_{i1}, α_{i2}, α_{i3}, α_o, α_{t-r}, α_{t-c}, α_{olf}, α_z, α_q, α_{11} through α_{14}, $\alpha_{16/15}$ and α_{gust}. The number of distinct G proteins is higher than the number of Gα genes because of two reasons. One is that there are splice variants for some of the α subunits: four for α_s that differ in the presence/absence of the

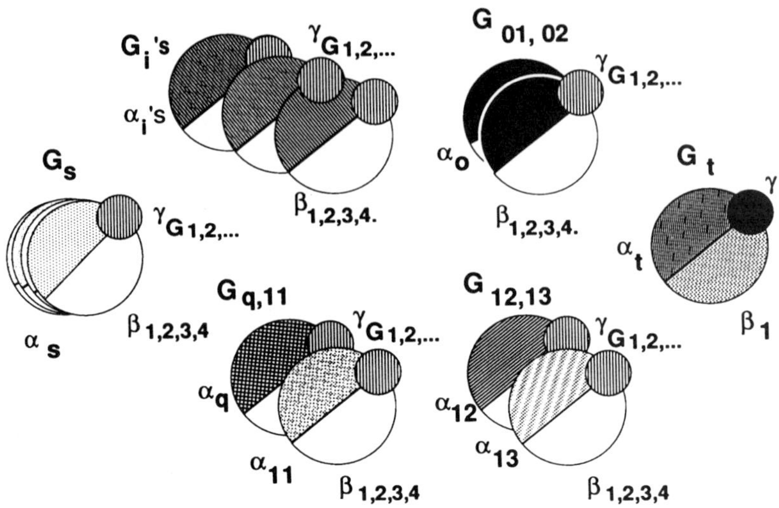

Figure 2 Schematic representation of subunit composition of typical vertebrate G proteins. G proteins have molecular weights (Mr) of 80,000 to 95,000. Except for transducin, which is easily washed off membranes with low salt solutions, all other known G proteins are tightly bound to membranes and can be extracted only with detergents. All G proteins involved in receptor effector coupling, including transducin, are formed of distinct Gα subunits (Mr 39,000 to 50,000), one of four types of β subunits) β_1 through β_4 (Mr = 35,000 to 36,000), and one of more than five types of γ subunits [γ_T and γ_{G1} through γ_{G3} (Mr = 7,000 to 10,000]. One γ (type T) is water soluble with a high proportion of charged amino acids. It is expressed only in retina and hence is found in transducin γ_T. The other γ's are somewhat more hydrophobic and are found in all other G proteins studied to date. They are thought to play a role in anchoring G proteins to the plasma membrane. β and γ subunits have not been purified separate from each other, only as dimers or complexed to α subunits as holo-G proteins. Relative abundance of the various β and γ subunits varies significantly from cell to cell. Extreme cases are retinal rod cells, which have exclusively β_1 and γ_T, and placenta, which do not express β_1. Within any given tissue, the different α subunits appear to interact with a common pool of $\beta\gamma$ dimers, and $\beta\gamma$ heterogeneity is the same for each holo-G protein. G_s is depicted as a group of $\beta\gamma$ dimers interacting with any one of four α_s splice variants. The two splice variants of G_o (Hsu et al., 1990; Strathmann et al., 1990) are represented in similar form. The family of highly homologous G_i proteins and those of highly homologous G_q/G_{11} (Strathmann and Simon, 1990) and G_{12}/G_{13} (Strathmann and Simon, 1991) G proteins are presented as clusters of independent G proteins.

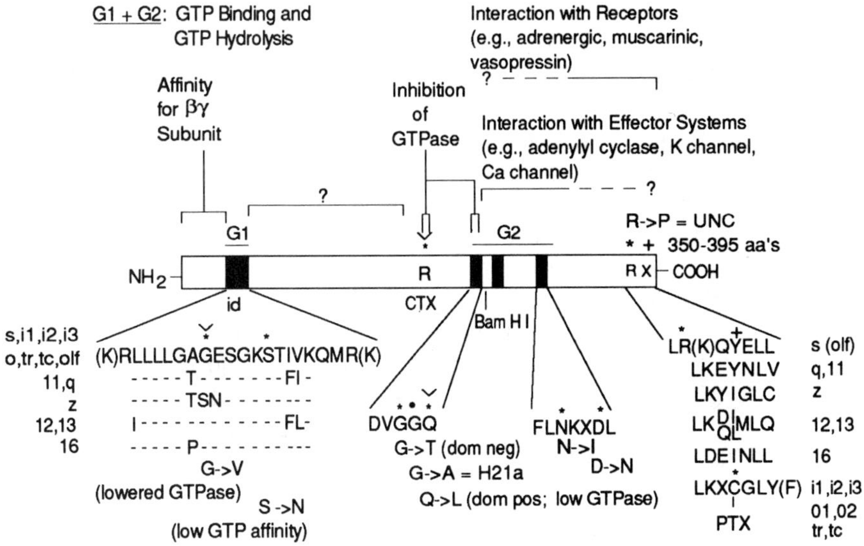

Figure 3 Summary of assignments of functional domains to structural domains of α subunits. A typical α subunit is represented in linearized form. Amino acid sequences of interest are highlighted. These include: (1) the identity box (id), conserved in α_s, α_i's, α_o, α_t's, and α_{olf} and the deviations found in $\alpha_{Z/X}$, α_q/α_{11} and higher-numbered α's [termini of id are either Arg (R) or Lys (K)]; (2) location of the Arg (R) ADP-ribosylated by CTX; location of the Gln (Q) which when mutated leads to loss of GTPase activity and spontaneous activation by GTP (Masters et al., 1989) and of Gly (G) residues which when mutated impede activation by guanine nucleotides (Osawa and Johnson, 1991; Miller et al., 1988); (3) location of a *Bam*HI restriction site used by Bourne and collaborators for contraction of an α_1/α_s chimera that retained α_s function; (4) location of another rather invariant squence, FLNKXDL, engaged in defining specificity for the guanine ring, where X may vary and in which mutation of the N or D leads to altered activity in some but not all GTP-binding proteins; and (5) the carboxyl termini of various α subunits, with the location of the Cys (C) ADP-ribosylated by PTX and an Arg (R) at position -6, which when mutated uncouples the G protein from receptor [the carboxyl terminal amino acid of the known PTX substrates is either Try (Y) or Phe (F)]. A truncated version of α_{13}, starting with Met[18], has reduced affinity of $\beta\gamma$ dimers (Graf, R., Codina J., and Birnbaumer, L., unpublished).

45-bp-long exon 3 and use of an alternate 5′ splice junction at the 3′ end of exon 4, and two for α_o, each encoded in eight exons of which there are alternate A and B versions of exons 7 and 8, which encode the carboxyterminal third of the protein. The other is that there are four separate Gβ genes and at least as many Gγ genes, the products of which can combine to give as many as 16 distinct isoforms of $\beta\gamma$ dimers, each of which is able to interact with any one of the 16 or so α subunits for a potential total of 16×16 G proteins. Not all of these genes are expressed in any single cell, some are expressed in single cells only, and the functional significance of this tremendous potential diversity is not fully understood as yet.

Based on the amino acid sequence of their α subunits, G proteins have been grouped by Simon and his collaborators into four major homology groups that include (Fig. 4): G_s (four isoforms) and G_{olf} (group 1); three G_i's, G_o's, two transduscins, T-r and T-c, expressed in rod and cone cells, respectively, which are all substrates for pertussin toxin, and G_z (group 2); the G_q class of proteins with unique differences in the G1 region comprising the closely related G_q, G_{11}, and G_{14} widely distributed and G_{16} and G_{15}, which appear to be expressed exclusively in hematopoietic cells (group 3), and G_{12} and G_{13} (group 4).

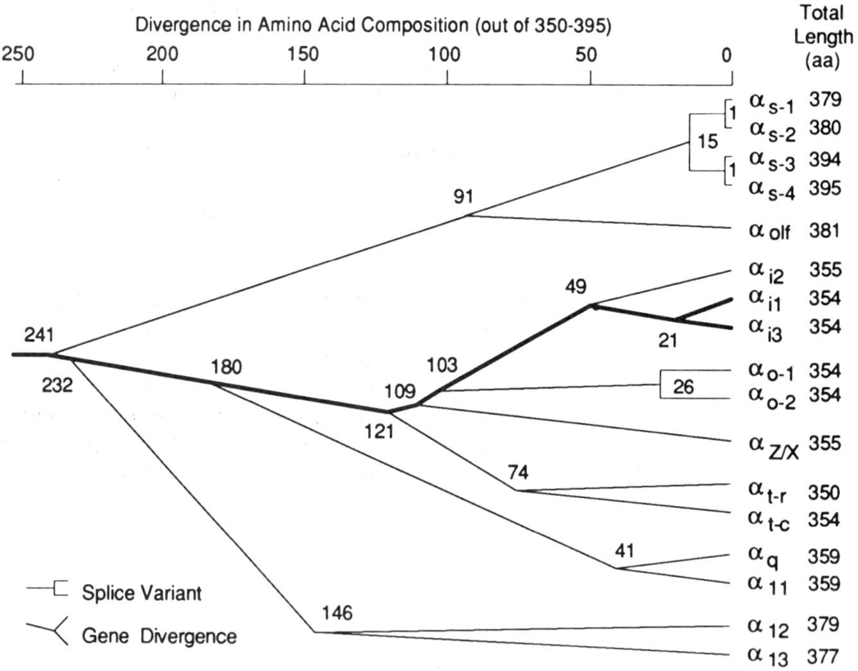

Figure 4 Similarity tree for G protein α subunits.

V. Mechanism of Action of a G Protein: Molecular Basis for Receptor-Effector Coupling

The mechanism by which binding of ligand to receptor leads to increased or decreased activity of an effector, with the intermediary participation of a G protein, has been unraveled at the molecular level. At the heart of receptor action lie two reactions of G proteins: (1) the binding of GTP followed by dissociation of the G protein to give a GTP-liganded α subunit (Gα*GTP) plus G$\beta\gamma$; and (2) in situ hydrolysis of GTP, which leads to reassociation of G-protein subunits (Fig. 5). It is now accepted that the role of the receptor is merely to catalyze the first of these events and that effector stimulation (or inhibition) is due to its interaction with the Gα*GTP complexes. Upon hydrolysis of GTP, the Gα changes its conformation and loses its affinity for effector. As a consequence, Gα dissociates from the effector. Because of this, the GTPase reaction is equivalent to an inactivation reaction. This is followed by reassociation with G$\beta\gamma$ to give a stable GDP-liganded trimeric G protein (G$\alpha\beta\gamma$-GDP), which requires the aid of receptor to reinitiate the cycle. Receptor reinitiates the cycle because, in the presence of GTP, it favors the formation of α*GTP · G$\beta\gamma$ through a series of equilibrium reactions involving GDP dissociation, GTP binding, and changes in protein conformation. The cycle would come to a rest here if it were not that the "starred" conformation of Gα has low affinity for G$\beta\gamma$, which therefore dissociates. This in turn leads to the release of the hormone-receptor complex. Thus protein-protein interactions between receptor and G protein and between G protein and effector are based not only on the conformation of the individual subunits, but also on the multimeric/polymeric state transition of the G protein: An effector interacts only with the monomeric Gα*GTP complex, while receptor interacts only with a trimeric form of the G protein (G$\alpha\beta\gamma$-GDP, G$\alpha\beta\gamma$, G$\alpha\beta\gamma$-GTP, and Gα*GTP · G$\beta\gamma$). For further reading, see Birnbaumer et al. (1985, 1990).

A. Consequences of the Catalytic Nature of Receptor Action

The receptor-stimulated cycle described above and shown in Fig. 5 allows the activated Gα*-GTP to be regulating the effector at the same time that the same receptor that led to its formation is free to mediate the activation of another G protein by GTP. Two important consequences of this mechanism are (1) that the receptor signal is amplified, and (2) that several receptors may be engaged simultaenously in activating the same G-protein pool. Amplification was well demonstrated for phototransduction, where a single photon can lead to activation of up to 10 rhodopsin molecules (Fung et al., 1981) and for β-adrenergic stimulation of adenylyl cyclase, where progressive chemical inactivation of the receptor results in a reduction in the rate at which the membrane pool of G_s is activated without a reduction in the extent to which G_s is activated (Tolkovsky

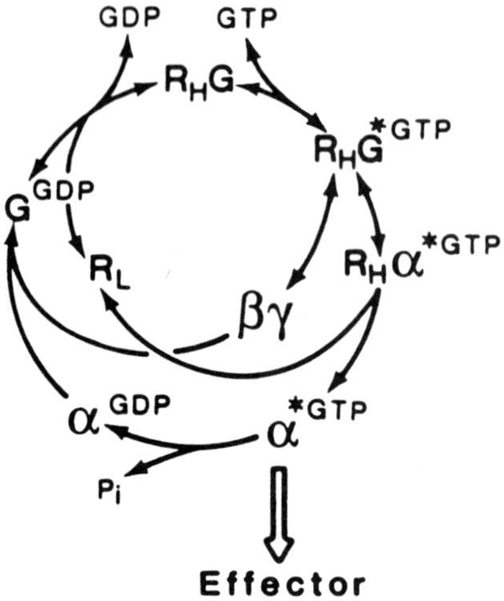

Figure 5 Catalysis of G-protein activation by hormone-receptor complex. The scheme incorporates three overlapping and mutually dependent regulatory cycles: (1) the G protein undergoes the cyclical dissociation-reassociation reaction, with receptor interacting only with the trimeric form of the G protein; (2) the G protein oscillates between GDP, nucleotide-free, and GTP states, driven energetically forward by its capacity to hydrolyze GTP; and (3) receptor intervenes as a catalyst by stabilizing an "activated" GTP-liganded form of the G protein, which, however, is no longer stable as a trimer and decomposes into free $G\beta\gamma$ dimer plus receptor-$G\alpha$*GTP, which, in turn, because of the absence of the $G\beta\tau$ portion of the G protein, loses its ability to stay associated with receptor and decomposes further into activated $G\alpha$*GTP plus the free form of the receptor. The receptor is shown in two forms or states. One has low (l) affinity for agonist and the other has high (h) affinity for agonist. These are forms that it adopts when it is free and associated with $G\alpha\beta\gamma$, respectively. Although the assumption is made that regulation of effector is possible only after $G\beta\tau$ and receptor have dissociated, it is in fact not yet known at which point the $G\alpha$ subunit becomes competent to associate with effector. Likewise, it is also not known at which point the $G\alpha$ subunit acquires the capacity to hydrolyze GTP. In spite of these uncertainties, note that without subunit dissociation, receptors could not act catalytically. Note also that since the $G\beta\gamma$ dimers are necessary for interaction of the G protein with the receptor, their overall effect is to stimulate $G\alpha$ activation. All G proteins involved in coupling receptors to effectors appear to conform to this scheme.

and Levitzki, 1978). The second consequence leads to synergism between hormones at low concentrations but lack of additivity at high concentrations, such as may happen in tissues with multiple receptors of different ligand specificities but the same cellular action. Examples of this kind are fat, liver, and both heart atrial and ventricular cells. In fact, ACTH, β-adrenergic, secretin, and glucagon receptors all potentiate each other to induce cAMP-mediated lipolysis by catalyzing the activation of G_s that stimulates adenylyl cyclase (Birnbaumer et al., 1969; Rodbell et al., 1970). In liver, α_1-adrenergic, type-1a vasopressin and angiotensin II receptors cooperatively induce Ca^{2+}-mediated glycogenolysis by catalyzing the activation of G_p, the G protein that stimulates membrane phospholipase C activity, leading to formation of the second messenger inositol trisphosphate (IP_3), which in turn causes release of Ca^{2+} from intracellular stores, which eventually activates the phosphorylase system (Exton, 1988). In heart atria, muscarinic M2, adenosine A1, and in some species neuropeptide Y receptors all have a bradycardic effect caused through activation of K^+ channels stimulated by G_k; and in the heart ventricle the adenylyl cyclase system is positively regulated by adrenergic β_1 and β_2 receptors as well as by serotonin receptors of the 5HT-4 type (e.g., Lemoine et al., 1988; Kaumann et al., 1991).

VI. Activities of G Proteins

A. Pertussis Toxin Substrates

Hormonal inhibition of adenylyl cyclase is blocked by pertussin toxin (PTX), and PTX substrates have been purified. Although it has been impossible to reconstitute biochemically, hormonal inhibition of adenylyl cyclase in vitro, cloned and mutationally activated G_{i1}, G_{i2}, and G_{i3} α subunits, but not G_o α subunits have recently been shown to have the expected adenylyl cyclase inhibitory effect (Wong et al., 1991; Bourne, H., personal communication). In addition, the three G_i's have G_k activity, being equipotent in stimulating the inwardly rectifying K^+ channel involved in the bradycardic effects of vagal stimulation (Yatani et al., 1988b), and in activating the ATP-sensitive K^+ channel (Kirsch et al., 1990). Since G_{i3} stimulates the analogous K^+ channel involved in inhibition of pituitary hormone secretion by somatostatin (Yatani et al., 1987b), it is likely that G_i's do the same in this tissue as in heart, and that one, the other, or all G_i-type PTX substrates are responsible for PTX-sensitive, slow inhibitory, postsynaptic potentials (s-ipsp's) in the central nervous system. Receptors mediating such s-ipsp's are $GABA_B$, muscarinic M2, serotonin, and dopamine D1 receptors. Like $G_i\alpha$'s, a purified $G_o\alpha$, structurally either α_{o1} or α_{o2}, also stimulates ionic channels in CNS neurons. But the channels are different and of more than one type. They include nonselective monovalent cation channels and a K^+-selective but nonrectifying channel, all insensitive to any of the $G_i\alpha$'s (VanDongen et al., 1988). In addition to its activity to stimulate monovalent cation channels, G_o is also directly or indirectly involved in the receptor-triggered inhibition of neuronal Ca^{2+} channels, such as are elicited by opioid (Hescheler et al., 1987) and neuropeptide Y receptors (Ewald et al., 1989), and in the inhibition of Ca^{2+} currents in endocrine cells, such as are elicited by somatostatin and carbachol in GH-type cells (Schlegel et al., 1985).

G_o is of interest. Both of its splice variants and the corresponding forms of the protein have been identified (Hsu et al., 1990; Codina et al., 1991). Both splice variants and both forms of the protein are found in brain, heart, and endocrine cells, but only one appears to be present in lung (Bertrand et al., 1990). In GH-type cells it has been shown, using α_{o1}- and α_{o2}-specific antisense oligonucleotides, that acetylcholine and somatostatin receptors each interact with a different G_o subtype: Anti-α_{o1} abolishes the effect of the acetylcholine receptor without interfering with that of the somatostatin receptor, and, vice versa, α_{o2} abolishes the effect of somatostatin on Ca^{2+} current without affecting that of the muscarinic receptor (Kleuss et al., 1991).

It is interesting to note that all of the receptors that stimulate K^+ channels via G_i and inhibit Ca^{2+} channels with involvement of G_o, also inhibit adenylyl cyclase. It may be that the PTX substrates of the G_i and G_o type all have multiple effects, providing for a coordinate cellular response in which cAMP levels are lowered (inhibition of adenylyl cyclase), Ca^{2+} levels are lowered (inhibition of Ca^{2+} channels), and the cell is hyperpolarized (K^+ channel activation). In contrast, using the *Xenopus* oocyte as a test system, Iyengar, Landau, and collaborators found that microinjection of purified and activated $G_o\alpha$ leads to an increase in intracellular Ca^{2+}, most likely due to activation of the oocyte's phospholipase C (Moriarty et al., 1990). This would suggest that G_o has G_p activity and points to the fact that uncertainty exists in both the assignment of function to G_o and the identification of G proteins that are physiological G_p's (see below).

B. G_p-type G Proteins: Hydrolysis of Membrane Phospholipids

Some of the G proteins that mediate activation of membrane phospholipases have recently been biochemically identified. They are referred to functionally as G_p (Cockroft and Gomperts, 1985) and, depending on whether the phospholipase is of the C, D, or A_2 type, also as Gplc, Gpld, or Gpla. Purified mixtures of G_q and G_{11} have Gplc-type G_p activity (Smrcka et al., 1991; Taylor et al., 1991). Studies with PTX indicate that there are at least two types of phospholipase C-specific G_p's: one sensitive to PTX and the other not. Thus, responses of vasopressin, angiotensin II, and α_1-adrenergic receptors in liver and to thyrotropin- and gonadotropin-releasing hormones in the pituitary involve stimulation of a phospholipase C by a G protein that is unaffected by PTX. On the other hand, responses to chemoattractants of neutrophils, which also involve the activation of a phospholipase C, are sensitive to PTX. Neither the α subunit of G_p nor that of G_{11} has the consensus sequence for ADP ribosylation by PTX (Strathmann and Simon, 1990; Fig. 3). Both are found in liver, and it stands to reason that these are the natural mediators of phospholipase C stimulation in liver. As mentioned earlier, G_o has G_p activity (Moriarty et al., 1990). Since G_o is a substrate for PTX, it is a candidate for the PTX-sensitive type of G_p in cells where receptor-mediated activation of this enzyme is inhibited by PTX. However, neutrophils in which PTX blocks receptor-mediated activation of phospholipase C do not express α_o. It follows that the G_p function in neutrophils has to reside in one or more G proteins that are neither G_o, G_q, G_{11}, any of the G_i's, G_{12}, G_{13}, or G_{16}. G_i's are excluded because they mediate inhibition of adenylyl cyclase in neutrophils as they do in other cells (Wong et al., 1991; Bourne, H., personal communication). Type-12, -13, and -16 G proteins are excluded because, like G_q and G_{11}, they also lack the PTX substrate consensus sequence (Fig. 3). Likewise, the α subunits identified by Melvin Simon's group as G_{14} and G_{15} lack a PTX ADP-ribosylation site. It follows that there should be additional G proteins whose primary structure is still unknown. Similar arguments can be made for the G_p acting on phospholipase D and A_2 activities. To date, all phospholipase A_2 activations, such as are seen upon stimulation of mast cells and macrophages by activating substances or upon stimulation of thyroid cells by α_1-adrenergic receptors, is due to activation of a type-A_2 phospholipase and occurs with the participation of a PTX-sensitive G_p. This rules out all non-PTX substrates as candidates. As for phospholipase D, one possibility is that G_q and/or G_{11} have multiple effectors as targets and that both have Gplc and Gpld activities. Future studies, involving G-protein purification, biochemical characterization of cloned G-protein α subunits, and possibly more molecular cloning of G-protein α subunits, are required to settle these interesting issues.

VII. Networks

Single G proteins are designed to interact with classes of receptors as opposed to single receptor subtypes. This conclusion stems from the discoveries in the late 1960s that up to five different hormone receptors can activate a single adenylyl cyclase system in an isolated membrane (Birnbaumer et al., 1969; Rodbell et al., 1970) and, in the early and mid-1970s, that receptors can be transferred from one cell to another (Orly and Schramm, 1976). Little, if any, species- and/or tissue-dependent restrictions as to the source of G_s were seen for reconstitution of hormonally stimulable adenylyl cyclase in *cyc*$^-$ membranes. The discovery that the same splice variant of G_s that activates adenylyl cyclase is also able to regulate Ca^{2+} channel activity (Yatani et al., 1988a; Mattera et al., 1989) indicates that one G protein can interact with more than one effector. The discovery that three different G_i proteins all activate the same K^+ channels (Yatani et al., 1988b; Kirsch et al., 1990) indicates that several G proteins may regulate a single effector. Ashkenazi et al., working with muscarinic receptors (Ashkenazi et al., 1987, 1989), as well as more recently our laboratory, working with the LH receptor (Gudermann, T., Birnbaumer, M., and Birnbaumer, L., unpublished results), showed clearly that single receptors may affect more than one G protein. It follows that transmembrane wiring diagrams are in fact combinations of the following four basic configurations:

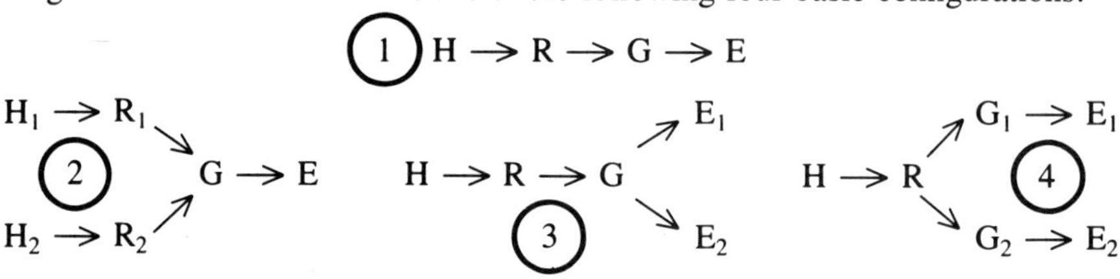

The complexity that may exist in the wiring of transmembrane signal transmission was well illustrated by the findings of Ewald et al. (1988, 1989) in PTX-treated rat sensory neurons. Upon studying the efficacy with which brain G_i and G_o reconstitute Ca^{2+} current regulation by NPY and bradykinin, they discovered that the effect of NPY could be fully constituted with G_o, with G_i being much less potent, while the effect of bradykinin could be only partially reconstituted by G_o, requiring G_i to achieve full reconstitution (Diagram 5).

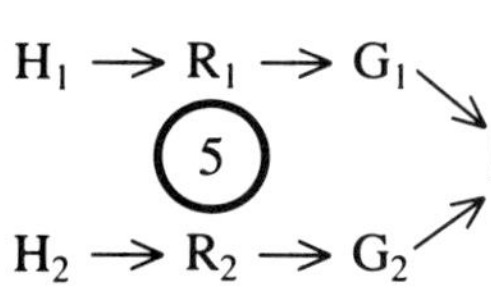

Studies by Schultz, Rosenthal, and colleagues showing splice variant specificity of receptors for separate G-protein splice variants, which both regulate the same effector, provide further proof for this type of wiring (Kleuss et al., 1991).

The important notion that emerges from these findings is that the wiring diagrams describing signal transduction by G proteins need to be determined individually and separately for each cell of interest. This includes not only the determination of which receptors are present but also which G proteins and which effectors.

VIII. Conclusion

Signal transduction by G proteins is a fundamental and widespread mechanism used by a wide variety of hormones, neurotransmitters, and auto- and paracrine factors to regulate cellular functions. G proteins modulate not only cAMP formation, but also intracellular Ca^{2+} mobilization, arachidonic acid release, and, very importantly, membrane potential. The latter is not only a trigger for neurotransmitter release and conduction of nerve impulses. In tissues such as secretory cells, it is the main regulator of Ca^{2+} entry. In heart, action potentials play the dual role of determining the frequency of contraction, and through modulation of the duration of the depolarized state, membrane potential determines Ca^{2+} entry and the force of contraction. More subtle changes in resting membrane potential alters the cell's predisposition to be stimulated by other factors and hormones. It is easy to imagine that persistent changes in membrane potential may affect acutely and chronically the cell's proliferative properties.

The mechanism by which G proteins are activated provides for amplification, reversal of action, and continued monitoring of hormone: for amplification, because few receptor molecules may act catalytically to activate many G proteins molecules; for reversal of action, because they have an internal turnoff mechanism whereby the Gα subunit hydrolyzes GTP to GDP; and for continued monitoring of the primary messenger level, because each activation cycle requires not only GTP but also occupied receptor.

Not all G proteins are known, and some are known of which their functions are still unknown. Both more G proteins and more effector functions affected by them will surely be found. Work is in progress to unravel a complicated network of interactions between receptors, G proteins, and effector systems, which affects not only regulation of metabolic activities of organs such as liver, heart, and fat, but also the integrative functions of the central nervous system.

Acknowledgments

This work was supported in part by National Institutes of Health research grants DK-19318, HD-09581, HL-45198, and GM-45198 to L.B., and DK-41233 to M.B.; by a fellowship from the Institute National de la Sante et de la Recerche Medicale (INSERM, France) to P.B.; by fellowships from the Deutsche Forschungsgemeinshaft (FRG) to T.G. and U.R.; by a fellowship from the National Institutes of Health Fogarty International Center (FOL), and by National Institutes of Health center grants HD-07495 and DK-27685; and by the Welch Foundation (Q1075).

References

Ashkenazi A, Peralta EG, Winslow JW, Ramachandran J, Capon DJ. Functionally distinct G proteins selectively couple different receptors to PI hydrolysis in the same cell. Cell 1989; 56:487–93.

Ashkenazi A, Winslow JW, Peralta EG, Peterson GL, Schimerlik MI, Capon DJ, Ramachandran J. An M2 muscarinic receptor subtype coupled to both adenylyl cyclase and phosphoinositide turnover. Science 1987; 238:672–5.

Bertrand P, Sanford J, Rudolph U, Codina J, Birnbaumer L. At least three alternatively spliced mRNAs encoding two α subunits of the G_o GTP-binding protein can be expressed in a single tissue. J Biol Chem 1990; 265:18576–80.

Birnbaumer L, Abramowitz J, Brown AM. Signal transduction by G proteins. Biochim Biophys Acta (Reviews in Biomembranes) 1990; 1031:163–224.

Birnbaumer L, Pohl SL, Rodbell M. Adenyl cyclase in fat cells. I. Properties and the effects of adenocorticotropin and fluoride. J Biol Chem 1969; 244:3468–76.

Birnbaumer L, Hildebrandt JD, Codina J, Mattera R, Cerione RA, Hildebrandt JD, Sunyer T, Rojas FJ, Caron MG, Lefkowitz RJ, Iyengar R. In: Cohen P, Houslay MD, eds., Molecular mechanisms of signal transduction. Amsterdam: Elsevier/North Holland Biomedical Press, 1985; 131–182.

Breitweiser GE, Szabo G. Uncoupling of cardiac muscarinic and beta-adrenergic receptors from ion channels by a guanine nucleotide analogue. Nature 1985; 317:538–40.

Burch RM, Luini A, Axelrod J. Phospholipase A_2 and phospholipase C are activated by distinct GTP-binding proteins in response to alpha$_1$-adrenergic stimulation in FRTL-5 cells. Proc Natl Acad Sci USA 1986; 83:7201–5.

Cockroft S, Gomperts BD. Role of guanine nucleotide binding protein in the activation of polyphosphoinositide phosphodiesterase. Nature 1985; 314:534–6.

Codina J, Grenet D, Chang K-J, Birnbaumer L. Urea gradient/SDS-PAGE: a useful tool in the investigation of signal transducing G proteins. J Receptor Res 1991; 11:587–601.

Dixon RAF, Kobilka BK, Strader DJ, Benovic JL, Dohlman HC, Frielle T, Bolanowski MA, Bennett CD, Rands E, Diehl RE, Mumford RA, Slater EE, Sigal IS, Caron MG, Lefkowitz RJ, Strader CD. Cloning of the gene and cDNA for mammalian beta-adrenergic receptor and homology with rhodopsin. Nature 1986; 321:75–9.

Ewald DA, Pang I-H, Sternweis PC, Miller RJ. Differential G protein-mediated coupling of neurotransmitter receptors to Ca^{2+} channels in rat dorsal root ganglion neurons in vitro. Neuron 1989; 2:1185–93.

Ewald DA, Sternweis PC, Miller RJ. Guanine nucleotide-binding protein G_o-induced coupling of neuropeptide Y receptors to Ca^{2+} channels in sensory neurons. Proc Natl Acad Sci USA 1988; 85:3633–7.

Exton JH. Role of phosphoinositides in regulation of liver function. Hepatology 1988; 8:152–66.

Fung BK-K, Hurley JB, Stryer L. Flow of information in the light-triggered cyclic nucleotide cascade of vision. Proc Natl Acad Sci USA 1981; 78:152–6.

Hamilton SL, Codina J, Hawkes MJ, Yatani A, Sawada T, Strickland FM, Froehner

SC, Spiegel AM, Toro L, Stefani E, Birnbaumer L, Brown AM. Evidence for direct interaction of $G_s\alpha$ with the Ca^{2+} channel of skeletal muscle. J Biol Chem 1991; 266:19528–35.

Hescheler J, Rosenthal W, Trautwein W, Schultz G. The GTP-binding protein, G_o, regulates neuronal calcium channels. Nature 1987; 325:445–7.

Hsu WH, Rudolph U, Sanford J, Bertrand P, Olate J, Nelson C, Moss LG, Boyd AE III, Codina J, Birnbaumer L. Molecular cloning of a novel splice Dr. variant of the α subunit of the mammalian G_o protein. J Biol Chem 1990; 265:11220–6.

Jones DT, Masters SB, Bourne, HR, Reed RR. Biochemical characterization of three stimulatory GTP-binding proteins. The large and small forms of G_s and the olfactory-specific G-protein, G_{olf}. J Biol Chem 1990; 265:2671–6.

Kaumann AJ, Sanders L, Brown AM, Murray KJ, Brown MJ. A 5-HT_4-like receptor in human right atrium. Naunyn-Schmiedeberg's Arch Pharmacol 1991; 344:150–9.

Kirsch G, Codina J, Birnbaumer L, Brown AM. Coupling of ATP-sensitive K^+ channels to purinergic receptors by G-proteins in rat ventricular myocytes. Am J Physiol 1990; 259:H820–6.

Kleuss C, Hescheler J, Ewel C, Rosenthal W, Schultz G, Wittig B. Assignment of G-protein subtypes to specific receptors inducing inhibition of calcium currents. Nature 1991; 353:43–8.

Kühn H. Light-induced, reversible binding of proteins to bovine photoreceptor membranes. Influence of nucleotides. Neurochemistry 1980; 1:269–85.

Lemoine H, Schönell H, Kaumann AJ. Contribution of B_1- and B_2-adrenoaline as assessed with (−) atenolol. Br J Pharm 1988; 95:55–66.

Liao C-F, Themmen APN, Joho R, Barberis C, Birnbaumer M, Birnbaumer L. Molecular cloning and expression of a fifth muscarinic acetylcholine receptor (M5-mAChR). J Biol Chem 1989; 264:7328–37.

Litosch I, Wallis C, Fain JN. 5-Hydroxytryptamine stimulates inositol phosphate production in a cell-free system from blowfly salivary glands. Evidence for a role of GTP in coupling receptor activation to phosphoinositide breakdown. J Biol Chem 1985; 260:5464–71.

Londos C, Cooper DMF, Schlegel W, Rodbell M. Adenosine analogs inhibit adipocyte adenylate cyclase by a GTP-dependent process: basis for actions of adenosine and methylxanthines on cyclic AMP production and lipolysis. Proc Natl Acad Sci USA 1978; 75:5362–6.

Masters SB, Miller RT, Chi M-H, Chang F-H, Beiderman B, Lopez NG, Bourne HR. Mutations in the GTP-binding site of $G_{s\alpha}$ alter stimulation of adenylyl cyclase. J Biol Chem 1989; 264:15467–74.

Masu M, Tanabe Y, Tsuchida K, Shigemoto R, Nakanishi S. Sequence and expression of a metabotropic glutamate receptor. Nature 1991; 349:760–5.

Mattera R, Graziano MP, Yatani A, Zhou Z, Graf R, Codina J, Birnbaumer L, Gilman AG, Brown AM. Individual splice variants of the α subunit of the G protein G_s activate both adenylyl cyclase and Ca^{2+} channels. Science 1989; 243:804–7.

McFarland KC, Sprengel R, Phillips HS, Köhler M, Rosemblit N, Nikolics K, Segaloff DL, Seeburg PH. Lutropin-choriogonadotropin receptor: an unusual member of the G protein-coupled receptor family. Science 1989; 245:494–528.

Miller RT, Masters SB, Sullivan KA, Beiderman B, Bourne HR. A mutation that prevents GTP-dependent activation of the α chain of G_s. Nature 1988; 334:712–5.

Moriarty TM, Padrell E, Corby DJ, Omri G, Landau EM, Iyengar R. G_o as the signal transducer in the pertussis toxin sensitive phosphatidylinositol pathway. Nature 1990; 343:79–82.

Northup JK, Smigel MD, Sternweis PC, Gilman AG. The subunits of the stimulatory regulatory component of adenylate cyclase. Resolution of the activated 45,000-dalton (alpha) subunit. J Biol Chem 1983; 258:11369–76.

Northup JK, Sternweis PC, Smigel MD, Schleifer LS, Ross EM, Gilman AG. Purification of the regulatory component of adenylate cyclase. Proc Natl Acad Sci USA 1980; 77:6516–20.

Orly J, Schramm M. Coupling of catecholamine receptor from one cell with adenylate cyclase from another cell by cell fusion. Proc Natl Acad Sci USA 1976; 73:4410–4.

Osawa S, Johnson GL. A dominant negative $G\alpha_s$ mutant is rescued by secondary mutation of the α chain amino terminus. J Biol Chem 1991; 266:4673–6.

Pfaffinger PJ, Martin JM, Hunter DD, Nathanson NM, Hille B. GTP-binding proteins couple cardiac muscarinic receptors to a K channel. Nature 1985; 317:536–8.

Pfeuffer T. GTP-binding proteins in membranes and the control of adenylate cyclase activity. J Biol Chem 1977; 252:7224–34.

Pfeuffer T, Helmreich EJM. Activation of pigeon erythrocyte membrane adenylate cyclase by guanylnucleotide analogs and separation of nucleotide binding protein. J Biol Chem 1975; 250:867–76.

Rall TW, Sutherland EW, Wosilait WD. The relationship of epinephrine and glucagon to liver phosphorylase. III. Reactivation of liver phosphorylase in slices and in extracts. J Biol Chem 1956; 218:483–95.

Rall TW, Sutherland EW, Berthet J. The relationship of epinephrine and glucagon to liver phosphorylase. IV. Effect of epinephrine and glucagon on the reactivation of phosphorylase in liver homogenates. J Biol Chem 1957; 224:463–75.

Rall TW, Sutherland EW, Berthet J. The relationship of epinephrine and glucagon to liver phosphorylase. J Biol Chem 1957; 224:463–75.

Ribalet B, Ciani S, Eddlestone GT. Modulation of ATP-sensitive K channels in RINm5F cells by phosphorylation and G proteins. Biophys J 1989; 55:587A.

Rodbell M, Birnbaumer L, Pohl SL. Adenyl cyclase in fat cells. III. Stimulation by secretin and the effects of trypsin on the receptors for lipolytic hormones. J Biol Chem 1970; 245:718–22.

Rodbell M, Krans HMJ, Pohl SL, Birnbaumer L. The glucagon-sensitive adenyl cyclase system in plasma membranes of rat liver. III. Binding of glucagon: Method of assay and specificity. J Biol Chem 1971; 246:1861–71.

Schlegel W, Wuarin F, Zbaren C, Wollheim CB, Zahnd GR. Pertussis toxin selectively abolishes hormone induced lowering of cytosolic calcium in GH3 cells. FEBS Lett 1985; 189:27–32.

Smrcka AV, Helper JR, Brown KO, Sternweis PC. Regulation of polyphosphoinositide-specific phospholipase C activity by purified G_q. Science 1991; 251:804–7.

Strathmann MP, Simon MI. G protein diversity: a distinct class of α subunits is present in vertebrates and invertebrates. Proc Natl Acad Sci USA 1990; 87:9113–7.

Strathmann MP, Simon MI. $G\alpha 12$ and $G\alpha 13$ subunits define a fourth class of G protein α subunits. Proc Natl Acad Sci USA 1991; 88:5582–6.

Strathmann MP, Wilkie T, Simon MI. Alternative splicing procedures transcripts encoding two forms of the α subunit of GTP-binding protein G_o. Proc Natl Acad Sci USA 1990; 87:6477–81.

Taylor SJ, Chae HZ, Rhee SG, Exton JH. Activation of the $\beta 1$ isozyme of phospholipase C by α subunits of the G_q class of G proteins. Nature 1991; 350:516–8.

Tolkovsky AM, Levitzki A. Mode of coupling between the β-adrenergic receptor and adenylate cyclase in turkey erythrocytes. Biochemistry 1978; 17:3795–810.

VanDongen A, Codina J, Olate J, Mattera R, Joho R, Birnbaumer L, Brown AM. Newly identified brain potassium channels gated by the guanine nucleotide binding (G) protein G_o. Science 1988; 242:1433–7.

Vu T-KH, Hung DT, Wheaton VI, Coughlin SR. Molecular cloning of a functional thrombin receptor reveals a novel proteolytic mechanism of receptor activation. Cell 1991; 64:1057–68.

Wheeler GL, Bitensky MW. A light-activated GTPase in vertebrate photoreceptors: regulation of light-activated cyclic GMP phosphodiesterase. Proc Natl Acad Sci USA 1977; 74:4238–42.

Wong UH, Federman A, Pace AM, Zachary I, Evanst T, Pouysségur, Bourne, HR. Mutant α subunits of G_{i2} inhibit cyclic AMP accumulation. Nature 1991; 351:63–5.

Yatani A, Codina J, Brown AM, Birnbaumer L. Direct activation of mammalian atrial muscarinic potassium channels by GTP regulatory protein G_k. Science 1987a; 235:207–11.

Yatani A, Codina J, Sekura RD, Birnbaumer L, Brown AM. Reconstitution of somatostatin and muscarinic receptor mediated stimulation of K^+ channels by isolated G_k protein in clonal rat anterior pituitary cell membranes. Mol Endocrinol 1987b; 1:283–9.

Yatani A, Imoto Y, Codina J, Hamilton SL, Brown AM, Birnbaumer L. The stimulatory G protein of adenylyl cyclase, G_s, directly stimulates dihydropyridine-sensitive skeletal muscle Ca^{2+} channels. Evidence for direct regulation independent of phosphorylation by cAMP-dependent protein kinase. J Biol Chem 1988a; 263:9887–95.

Yatani A, Mattera R, Codina J, Graf R, Okabe K, Padrell E, Iyengar R, Brown AM, Birnbaumer L. The G protein-gated atrial K^+ channel is stimulated by three distinct $G_i\alpha$-subunits. Nature 1988b; 336:680–2.

2

Regulation of G-Protein-Linked Receptors: Analysis from the Genome to the Cell Membrane

CRAIG C. MALBON

School of Medicine
Health Sciences Center
SUNY/Stony Brook
Stony Brook, New York

JOHN R. HADCOCK

American Cyanamid Co.
Princeton, New Jersey

I. Introduction

Cells respond to stimuli through a variety of distinct pathways (for references, see Hadcock and Malbon, 1991a). Three prominent pathways are those through which steroid hormones, growth factors, and catecholamines operate (Fig. 1). In the case of the steroid hormones, such as estrogen or glucocorticoids, the hormones traverse the cell membrane and combine with heterobifunctional receptors that bind both the steroid ligand as well as specific DNA sequences. Growth factors, such as insulin, PDGF, and EGF, bind to cell surface receptors that possess an intrinsic tyrosine kinase activity. The activity of this tyrosine kinase is ligand-regulated, and it is the phosphorylation of specific cellular targets that propagates the signal intracellularly. In addition to these two major pathways is a third pathway, typified by the ones employed in catecholamine hormone action. This system is composed of three membrane-bound elements, receptors, effectors, and GTP-binding regulatory proteins, which transduce information from activated receptors to effectors (Birnbaumer et al., 1987; Gilman, 1987).

G-protein-linked receptors include not only the catecholamine receptors, but more than 100 receptors that have been characterized via molecular clon-

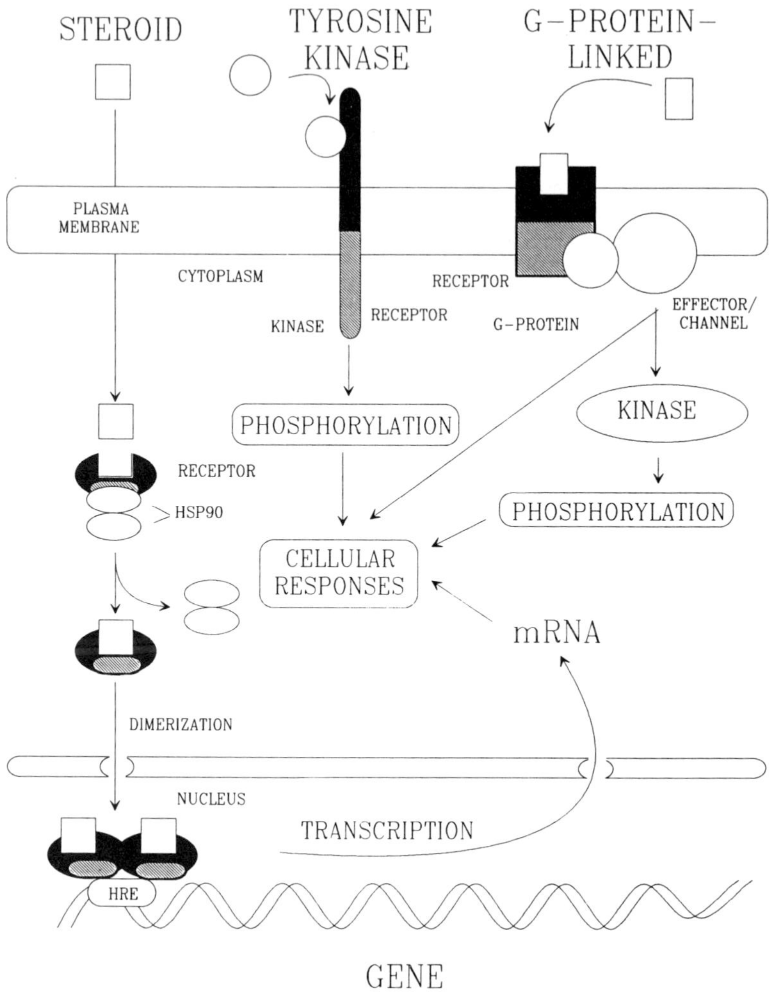

Figure 1 Agonist activation of steroid, tyrosine kinase, and G-protein-linked receptors. Steroid receptors, operating as trans-acting factors, undergo dimerization after hormonal activation. The scheme shows the first steps in the pathways mediating the actions of the three classes of receptors. Localization of nonactivated steroid receptors has been reported to be both in the cytoplasm and the nucleus. Activation by hormone, therefore, may promote a translocation from the cytoplasm to the nucleus or may already be localized in the nucleus prior to activation. Shading represents plasma membrane. See the text for details. (From Hadcock and Malbon, 1991.)

ing. Several families of G-protein-linked receptors exist. For the catecholamines epinephrine and norepinephrine, β-adrenergic (β_1, β_2, β_3) and α-adrenergic (α_1, α_2, and their many subtypes) receptors can be easily segregated into families. The muscarinic acetylcholine receptors (M1-M5), too, constitute a family of pharmacologically distinct but homologous gene products (Bonner et al., 1987). Transmembrane signaling in sensory physiology also makes use of the G-protein-linked receptor pathway for vision (in which the rhodopsin photopigments operate) and for olfaction. Protein phosphorylation plays a prominent role in transmembrane signaling via cell surface receptors. Many excellent reviews are available on the molecular biology and characterization of G-protein-linked receptors (Wang et al., 1991; Collins et al., 1989a).

In this work we seek to provide a snapshot of the current understanding of the regulation of G-protein-linked receptors. For this purpose we shall confine the analysis to the α- and β-adrenergic receptors and the pathways that they mediate. Highlighted will be the stimulatory and inhibitory G-protein-mediated control of adenylylcylase (Fig. 2). β_2-adrenergic receptors typify the stimulatory receptors (Hs) linked via $G_{s\alpha}$ to adenylyl cylcase, and α_2-adrenergic or A1-purinergic receptors typify the inhibitory receptors (Hi) linked via G_i to adenylylcyclase. Changes in intracellular cAMP alter the activity of the protein kinase A, the kinase that mediates the cellular effects mediated by cAMP. Although focused on regulation at the level of the receptor, this chapter will include discussion of physiological regulation exerted at the level of G-proteins and effectors, when necessary for continuity in developing basic themes of regulation.

II. Receptor Structure

G-protein-linked receptors share, with few exceptions, common properties. Most are expressed at steady-state levels characteristic of low-abundance membrane proteins. The level of expression is often a few thousand to tens of thousands copies per cell, with a prominent exception being the expression of the photopigment rhodopsin in the retina. These receptors are all glycoproteins, possessing N-linked complex oligosaccharides. The role of glycosylation in receptor biology and function appears to be dependent on the specific receptor as well as on the cell type in which the receptor is expressed. For the β-adrenergic receptors, blockade of glycosylation with inhibitors such as tunicamycins or through mutagenesis of the N-glycosylation recognition sequences appears to have little influence on receptor expression or function (George et al., 1986; Wang et al., 1991). For prostaglandin receptors, inhibition of glycosylation by tunicamycin results in a loss of the stimulatory adenylylcyclase response to this hormone (George et al., 1986).

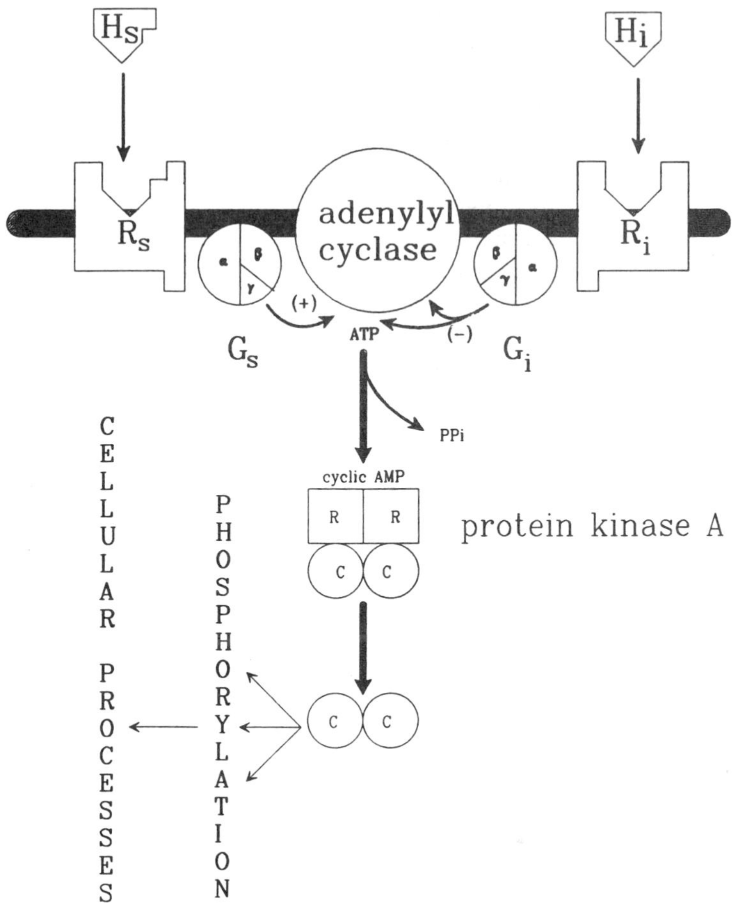

Figure 2 A schematic of the stimulatory and inhibitory adenylyl cyclase pathway in mammalian cells. For details, see text.

Analysis by polyacrylamide gel electrophoresis in the presence of sodium dodecyl sulfate reveals that most G-protein-linked receptors display a molecular weight in the range of 50,000 to 80,000 Da. β-Adrenergic receptors typically display a more narrow range of Mr, 64,000 to 67,000 Da (Graziano et al., 1984). The exception to the rule is a form of the human β-adrenergic receptor displaying an Mr of approximately 76,000 Da (Bahouth et al., 1986). The structural basis for this larger version of the receptor has not been identified, but more heavy N-glycosylation might be a likely possibility.

Many G-protein-linked receptors display a thiol-induced shift in their apparent Mr (Malbon et al., 1986). First described for the mammalian β_1-

adrenergic receptor (Moxham and Malbon, 1985), thiol-induced increases in the Mr of G-protein-linked receptors may best be envisioned as the result of cleavage of disulfide bonds that keep the native molecule in a more compact conformation. Demonstration that thiol treatment can activate purified β-adrenergic receptors in vitro when co-reconstituted with purified G_s (Pederson and Ross, 1985; Moxham et al., 1985) provided a functional insight into the possible role of disulfide-sulfhydryls in both receptor structure and operation. Molecular cloning has revealed a conservation of cysteinyl residues among G-protein-linked receptors and the presence of cysteinyl residues in multiple regions of the molecule including the hydrophobic, transmembrane-spanning domains. Much further structural analysis will be required if we are to elucidate the significance of disulfide-free sulfhydryl groups to the conformation and function of G-protein-linked receptors.

The receptors mediating their effects via G-proteins or via intrinsic tyrosine kinase activity (Fig. 1) share the common feature of being themselves substrates for protein phosphorylation. β-Adrenergic receptors, for example, are substrates for at least three major protein kinases, protein kinase A, protein kinase C, and the β-adrenergic receptor kinase (Clark et al., 1988; Hausdorf et al., 1990). Lefkowitz and co-workers have provided compelling evidence for the regulation of adrenergic receptors via phosphorylation. This topic will be discussed at greater length with reference to agonist-induced desensitization. The notion that receptors that mediate the activation of one or more protein kinases in their mode of action are also substrates for these same kinases is very appealing. How phosphorylation influences the activity of these membrane receptors remains largely unexplored.

Finally, the elucidation of the primary sequences of rhodopsin earlier as well as the mammalian β_2-adrenergic receptor identified a major motif for G-protein-linked receptors, seven hydrophobic, potentially membrane-spanning domains. The topography of the β_2-adrenergic receptor was first tested and established by the development of site-specific antipeptide antibodies to each of the hydrophilic regions that flank and interdigitate the seven hydrophobic domains. The results of these efforts (Fig. 3) established the orientation of each hydrophilic region with respect to the cytoplasmic versus extracellular faces of the lipid bilayer (Wang et al., 1989c). Thus, G-protein-linked receptors must undergo considerable protein folding to achieve proper organization in the membrane.

III. Physiological Regulation of G-Protein-Linked Receptors

The molecular cloning of more than 100 G-protein-linked receptors has provided a wealth of information on both general structural features (e.g., consensus sites for phosphorylation, N-linked glycosylation, fatty acylation) as

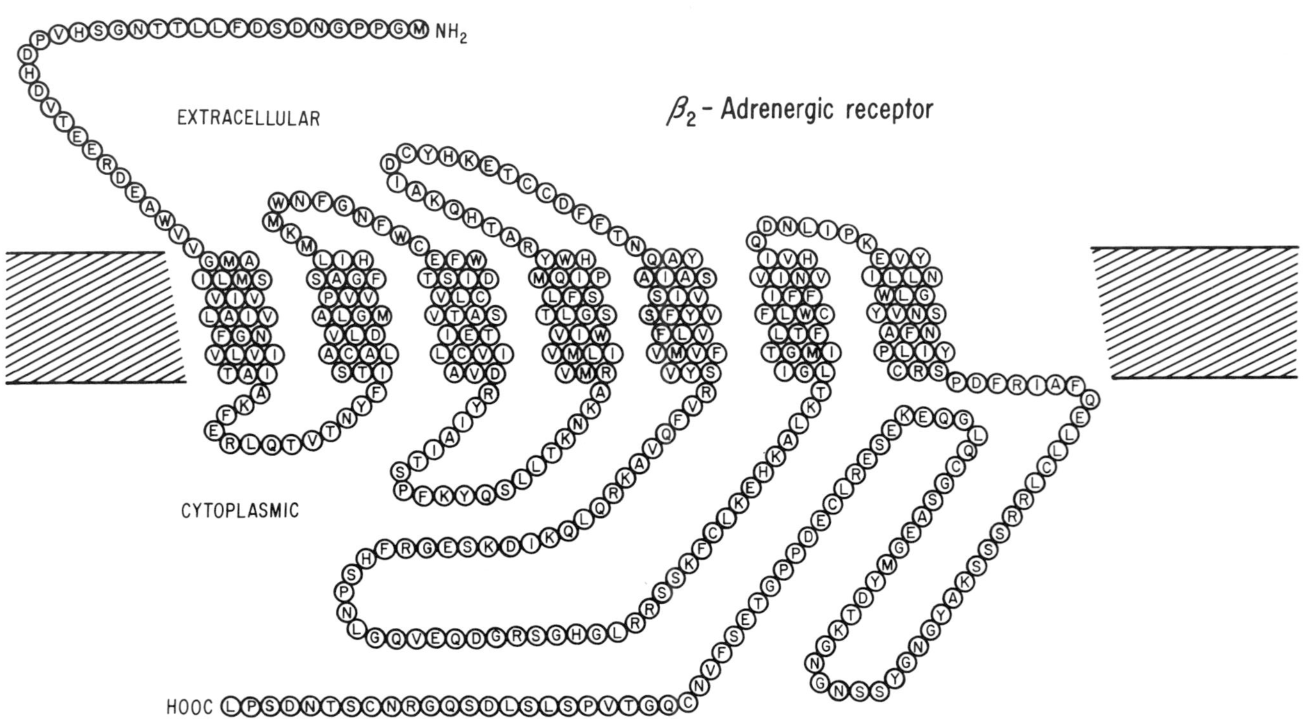

Figure 3 Topography of the β_2-adrenergic receptor in the lipid bilayer. The sequence information is derived from Dixon et al. (1986). The analysis of the topography was reported by Wang et al. (1989).

well as features unique to specific receptors (e.g., variable size of cytoplasmic and extracellular hydrophilic domains, N-terminal peptide, C-terminal peptide). Equally impressive as this wealth of information is our ignorance about the motifs and mechanisms by which the expression and function of these receptors are regulated physiologically. For the sake of discussion, physiological regulation of receptors has been divided into three modes: up-regulation of receptor expression; down-regulation of receptor expression; and cross-regulation among elements of G-protein-mediated pathways, including receptors and G-proteins.

A. Up-Regulation of Receptor Expression

A variety of stimuli are capable of modulating the expression of G-protein-linked receptors. Glucocorticoid regulation of β-adrenergic receptors will be employed as a template for the larger class of G-protein-linked receptors. Glucocorticoid excess and deficiency have profound effects on G-protein-mediated pathways (Haynes, 1990). The ability of adrenal steroids to modulate the response of target tissues to other hormones is referred to as the "permissive effects" of these steroids (Malbon et al., 1988). A detailed review of permissive hormone regulation of G-protein-mediated pathways is available elsewhere (Davies and Lefkowitz, 1984; Hadcock and Malbon, 1991) and will not be repeated here.

Regulation of β-adrenergic receptor-mediated responses by "permissive" hormones, such as glucocorticoids, is well documented (Collins et al., 1988; Hadcock and Malbon, 1988). Only recently, however, has the basis for the permissive effects of glucocorticoids on β-adrenergic receptor action been clarified. Glucocorticoids increase the number of β-adrenergic receptors in a variety of target tissues, an effect that has been duplicated in several cell lines in culture. Incubating DDT_1MF-2 hamster smooth muscle cells with the synthetic glucocorticoid dexamethasone promoted an increased expression of β-adrenergic receptors, as vividly displayed in Fig. 4. Indirect immunofluorescence microscopy of cells stained with antireceptor antibodies reveals increased epifluorescence in the steroid-treated cells. The receptors stain as punctate structures, suggestive of macromolecular clustering of the receptors in the cell membrane (Wang et al., 1989a, 1989c). Radioligand binding confirmed that receptor expression had increased several-fold within 72 h of exposure to the steroid. Further insight into the regulation was gained by an analysis of the receptor mRNA levels. For low-abundance proteins the analysis of mRNA levels if often problematic. By use of DNA-excess solution hybridization (Hadcock et al., 1989c) with a single-stranded probe corresponding to nucleotides +12 to 182 of the hamster receptor cDNA, it was possible to quantify the steady-state levels of receptor mRNA. Receptor mRNA levels were ~0.37 pg of mRNA per microgram of total cellular RNA, increasing to

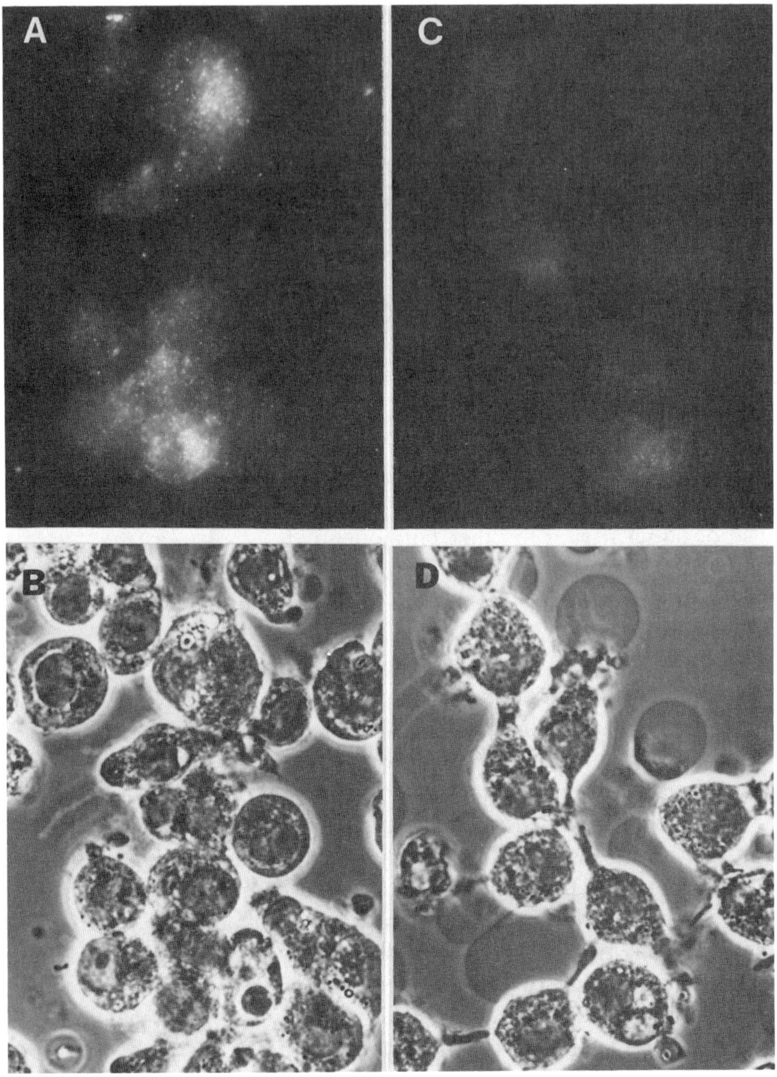

Figure 4 Agonist and glucocorticoid regulation of β-adrenergic receptor expression in hamster DDT_1MF-2 vas deferens cells: analysis in situ by indirect immunofluorescence. DDT_1MF-2 cells were incubated for 72 h in medium alone (A and B), medium plus 10 μM isoproterenol (C and D), medium plus 500 nM dexamethasone (E and F), or medium plus 500 nM dexamethasone and 10 μM isoproterenol (G and H). Phase-contrast (B, D, F, and H) and epifluorescence (A, C, E, and G) images are shown. Fixed, intact cells were stained with anti-β-adrenergic receptor antiserum (A–H). (Bar = 0.25 μM.) (From Hadcock et al., 1990.)

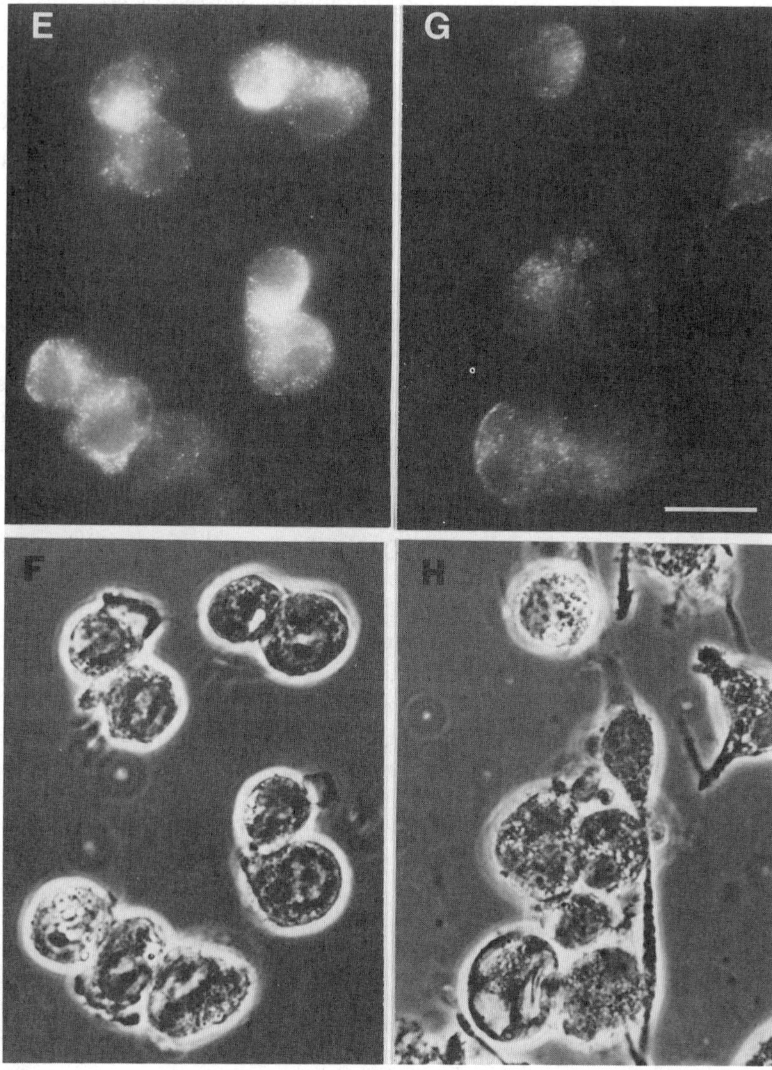
E
G
F
H

1.05 pg within 2 to 4 h of exposure to dexamethasone (500 nM). This sharp rise in mRNA levels was transient, declining within 4 to 6 h to a level onefold greater than that observed in the control, unstimulated cells (Hadcock and Malbon, 1988b).

Dexamethasone up-regulation of β_2-adrenergic receptor mRNA levels displayed half-maximal stimulation with 20 to 40 nM steroid. Neither estradiol nor testosterone (500 nM) influenced receptor mRNA levels. The steroid-induced up-regulation was sensitive to actinomycin D, and the half-life of the mRNA (12 h) was similar in control and dexamethasone-treated cells (Hadcock and Malbon, 1988b). Nuclear run-on measurements of the relative rates of transcription revealed a five- to sevenfold increase in the transcription rate in nuclei isolated from dexamethasone-treated as compared to untreated cells (Collins et al., 1988; Hadcock et al., 1989b). Thus, the basis for the glucocorticoid up-regulation of these G-protein-linked receptors was increased transcription of the β-adrenergic receptor gene.

Analysis of the structure of the gene encoding the human β_2-adrenergic receptor revealed the presence of several putative glucocorticoid-responsive elements (GREs) in the 5′-flanking, 3′-flanking, and open-reading frame (Emorine et al., 1987). Evidence has been provided that the GREs in the 5′-noncoding region of the hamster β_2-adrenergic receptor gene are obligate for glucocorticoid regulation of receptor mRNA levels (Malbon and Hadcock, 1988). Elimination of the GRE positioned at -1031 of the hamster β_2-adrenergic receptor gene resulted in the loss of glucocorticoid sensitivity, in spite of the fact that several other GREs remained within the modified gene. The question of whether or not the change in receptor expression induced by glucocorticoids is sufficient to explain the physiology was defined by examining β-adrenergic stimulation of cAMP accumulation in cells stably transfected with an expression vector harboring the receptor gene. As the level of receptor expression increased, so did the maximal hormonal response, while the ED_{50} for the response declined. Increasing receptor expression two- to threefold increased the sensitivity and the maximal response to a similar extent (George et al., 1988). Thus the up-regulation by glucocorticoids appears to reflect steroid-induced increases in the transcription rate of the gene, exerted at a GRE in the 5′-noncoding region.

Up-regulation of receptor expression is not confined to either glucocorticoids or steroids in general. For β-adrenergic receptors, thyroid hormones (Lazar-Wesley et al., 1991; Bahouth, 1991), retinoic acid (Galvin-Parton et al., 1990), and agonist (Collins et al., 1989b), in addition to glucocorticoids, have been shown to up-regulate expression of these G-protein-linked receptors. As greater information on the 5′-noncoding upstream and 3′-noncoding downstream regions accumulates, the structural basis for the regulation by thyroid hormones, retinoic acid, and other agents at the level of the gene may be elucidated.

B. Down-Regulation of Receptor Expression

Attenuation of a biological response upon a chronic challenge of agonist is a nearly universal phenomenon in biology. Tolerance to therapeutic agents and photoadaptation are examples in which "desensitization" of the biological response follows a sustained stimulatory insult. Chronic stimulation of cells with catecholamines results in an attenuated response to agonist with two temporal components. In the short term (minutes), agonists induce desensitization, a phenomenon discussed further below. On a longer term, agonists induce a down-regulation, most commonly observed as a reduction in the maximal binding capacity for radiolabeled antagonist ligands The molecular basis for agonist-induced down-regulation of G-protein-linked receptors remained a mystery until recently. Critical insights into agonist-induced down-regulation of receptor were obtained through analysis of the phenomenon at the level of the protein (using immunocytochemistry) and at the level of the mRNA (using DNA-excess solution hybridization).

Immunocytochemical analysis of receptor distribution revealed two important features about agonist-induced down-regulation of G-protein-linked receptors (Wand et al., 1989b). First, staining of A431 human epidermoid carcinoma cells treated with agonist for minutes to hours demonstrated that the receptor complement on the cell surface was similar to that observed in untreated cells. Thus, the loss in radioligand binding that occurs in the short term reflects only that, i.e., a loss of binding capacity, but not receptor protein. Cells fixed and stained in the absence of detergent permeabilization displayed equivalent amounts of cell surface receptor staining when treated with agonist, agonist plus antagonist, or with vehicle alone. Subsequently, chronic challenge of cells with agonist results in a loss in radioligand binding as well as a loss both in immunoreactive receptor content of membranes (as evidenced in immunoblots) and in the epifluorescence signal of cells stained with antireceptor antibodies (Wang et al., 1989a, 1989c).

At the level of receptor mRNA, a startling observation was made. Agonist treatment induced a down-regulation in the mRNA for the receptor to which the agonist binds. This phenomenon is referred to as an "agonist-induced down-regulation of receptor mRNA levels" (Hadcock and Malbon, 1988a). In DDT_1MF-2 smooth muscle cells in culture, mRNA levels for the β_2-adrenergic receptor were found to be ~0.38 pg of mRNA per microgram of total cellular RNA. Incubation with agonist for 12 to 16 h reduced receptor mRNA levels by more than 40%. The agonist-induced down-regulation of receptor mRNA displayed all of the pharmacological properties expected for a typical β_2-adrenergic receptor-mediated response. The response was reversible by wash-out of agonist or by addition of antagonist, returning to 90% of normal values within 12 h.

Although the phenomenology of the decline in receptor mRNA in response to agonist was clear, it was unclear whether the extent, duration, and time of onset of agonist-induced down-regulation of mRNA differed among cells or experimental conditions. Answers to these critical questions were provided by analysis of the rat C6 glioma cell line. Agonist treatment of C6 glioma cells resulted in an 80% reduction in β_2-adrenergic receptor mRNA levels within 2 h of challenge (Hough and Chuang, 1990). Thus, agonist-induced down-regulation of receptor mRNA displays the capacity to reduce mRNA levels from 40% to 80% on a time scale of hours (2 to 12 h). This loss in receptor mRNA provides the basis for the reduction in receptor expression observed earlier (Neve and Molinoff, 1986).

What factor(s) mediates the down-regulation of receptor mRNA in cells challenged with agonist? Since the output of the β-adrenergic stimulation of adenylyl cyclase is elevated intracellular cAMP levels, perhaps cAMP mediated this control of receptor mRNA. Elevation of intracellular cAMP with cholera toxin treatment or exposure of the cells to the diterpene forskolin was found to induce a down-regulation of the β_2-adrenergic receptor (Hadcock and Malbon, 1988a). The extent of this down-regulation induced by cholera toxin or forskolin was approximately half of that induced by challenge with agonist, even though the levels of intracellular cAMP accumulation generated in response to the three agents were comparable. These data suggested that there was a cAMP-dependent component and also some additional component induced by agonist, but not by agents that elevate intracellular cAMP to the same extent.

Analysis of agonist regulation of β-adrenergic receptor mRNA in S49 mouse lymphoma mutants provided a means to address how far the hormone signal had to be propagated to induce down-regulation of the mRNA (Hadcock et al., 1990). The analysis was performed utilizing wild-type cells, three variants with mutations in $G_{s\alpha}$ (UNC, cyc-H21A), and a variant (kin-) with a transdominant mutuation in the catalytic subunit of protein kinase-A (Hadcock et al., 1990). The UNC variant demonstrates uncoupling between the β-adrenergic receptor and the mutant $G_{s\alpha}$. In this variant, isoproterenol failed to increase cAMP levels or to induce down-regulation of receptor mRNA, whereas activation of the adenylylcyclase with forskolin promoted cAMP accumulation and the wild-type level of down-regulation. The cyc- variant lacks $G_{s\alpha}$ and phenotypically appeared just like the UNC variant with respect to down-regulation of receptor mRNA in response to isoproterenol and to forskolin. The H21A variant displays mutant $G_{s\alpha}$ that can interact productively with activated receptors, but fails to propagate the signal to the adenylyl cyclase. Like the other variant with mutations in $G_{s\alpha}$, H21A cells accumulate cAMP in response to forskolin, but not isoproterenol. Too, the H21A variant displayed forskolin-induced down-regulation of receptor mRNA. Most interesting and unexpected was the observation that isoproterenol was able to down-regulate receptor mRNA in H21A variant cells, though unable to stimuate elevated levels of intracellular cAMP. Like the dichotomy between agonist and cholera toxin (or forskolin) in DDT_1MF-2 cells, there appears to be not only a cAMP-dependent component to down-regulation of receptor mRNA, but also some other component in addition.

The analysis of receptor mRNA down-regulation in the S49 mouse lymphoma kin- variant was extraordinarily productive. With the wild-type pathway intact up to the level of protein kinase A, the kin- mutants displayed robust stimulation of cAMP levels in response to either isoproterenol or forskolin. What is remarkable is that neither isoproterenol nor forskolin were able to induce a down-regulation of receptor mRNA in the kin- cells. The role of cAMP and protein kinase A was established by these experiments; i.e., there is an obligatory role for some basal level of cAMP and protein kinase A activity for agonist-induced down-regulation of receptor mRNA. Elevated levels of intracellular cAMP are capable of inducing the down-regulation of mRNA, but are insufficient to achieve the level of down-regulation obtained in response to agonist.

The molecular basis for agonist-induced down-regulation remains to be established. The reduction in steady-state levels of receptor mRNA in response to agonist challenge has been shown, however, to be the result of destabilization of mRNA. The obligate role of cAMP in this process prompted us to explore the possibility that the cyclic nucleotide was altering the rate of transcription. Analysis of the β_2-adrenergic receptor gene identified two sequences with high homology to the consensus cAMP-response element (CRE) found in other genes (Hadcock and Malbon, 1988a). Suppressor rather than enhancer activity of a CRE would have been a novel observation. Nuclear run-on assays of relative rates of transcription of the receptor gene using nuclei isolated from control, untreated cells as compared to agonist-treated (4 h) cells suggested that the regulation was not at the level of transcription. Subsequent analysis of the half-life of the β_2-adrenergic receptor mRNA revealed that for control cells the $t_{1/2}$ was approximately 12 h. In contrast, the $t_{1/2}$ for receptor mRNA declined to 5 h in the agonist-treated cells. Together these data demonstrated that the reduction in receptor mRNA observed in cells following agonist treatment was the result of a destabilization of the mRNA and not due to an alteration in the rate of transcription (Hadcock and Malbon, 1988a; Hadcock et al., 1989b).

Analysis of the two opposing regulatory forces on receptor expression (i.e., up-regulation versus down-regulation) was performed utilizing the interplay between glucocorticoids and agonist treatment with respect to the expression of β_2-adrenergic receptors. At the level of mRNA, the balance and interplay between glucocorticoid-induced up-regulation and agonist-induced down-regulation were quite impressive (Hadcock et al., 1989b). For glucocorticoid regulation, concurrent challenge with agonist did not influence the early transient peak of receptor mRNA resulting from enhanced transcription. The sustained, elevated level of mRNA observed in glucocorticoid-treated cells following the transient peak was, however, abolished by challenge with agonist. Thus the transcriptional activation induced early by steroid operated without interference, but was later counterbalanced by destabilization of receptor mRNA under the influence of agonist. In cells challenged with agonist in which the level of receptor mRNA has declined by 40–50%, glucocorticoids were able to stimulate a robust, transient peak of receptor mRNA. Thus, the decline in receptor mRNA resulting from destabilization could be countered by a sharp but transient increase in transcription induced by glucocorticoid. The beauty of this counterregulation is self-evident (Fig. 5). Analysis of the destabilization at the molecular level, too, should provide greater insight into the interplay between up- and down-regulation.

C. Cross-Regulation Among G-Protein-Linked Pathways

Cross-Regulation from Stimulatory to Inhibitory Adenylyl Cyclase

Adenylyl cyclase activity is controlled by two opposing pathways, a stimulatory pathway activated through stimulatory receptors and mediated by G_s, and an inhibitory pathway activated through inhibitor receptors and mediated by G_i (Fig. 2). Although we tend to explore and think about single G-protein-linked pathways in isolation, in the cell information must be processed simultaneously from several independent G-protein-linked pathways the converge at several levels. One approach to examine the interplay among these pathways is to persistently activate a single pathway and explore to what extent, if any, the persistent activation of one pathway influences information being processed through a second G-protein-linked pathway. Persistent (24-h) activation of the stimulatory adenylyl cyclase pathway by either forskolin or by β-adrenergic agonist was examined in S49 mouse lymphoma cells and found to enhance dramatically the effects of somatostatin mediated via the opposing inhibitory adenylyl cyclase pathway (Hadcock et al., 1990). Remarkably, activating the stimulatory pathway resulted in a threefold increase in the steady-state levels of $G_{i\alpha 2}$ and a simultaneous 25% decline in the level of $G_{s\alpha}$, as quantified by immunoblotting $G_{i\alpha 2}$ mRNA levels, quantified by DNA-excess solution hybridization, were observed to increase by fourfold within 12 h of persistent activation of the stimulatory pathway. $G_{s\alpha}$ mRNA levels, in contrast, increased initially, but then declined to 75% of the control. This phenomenon of the activation of one pathway influencing the throughput and expression of the elements

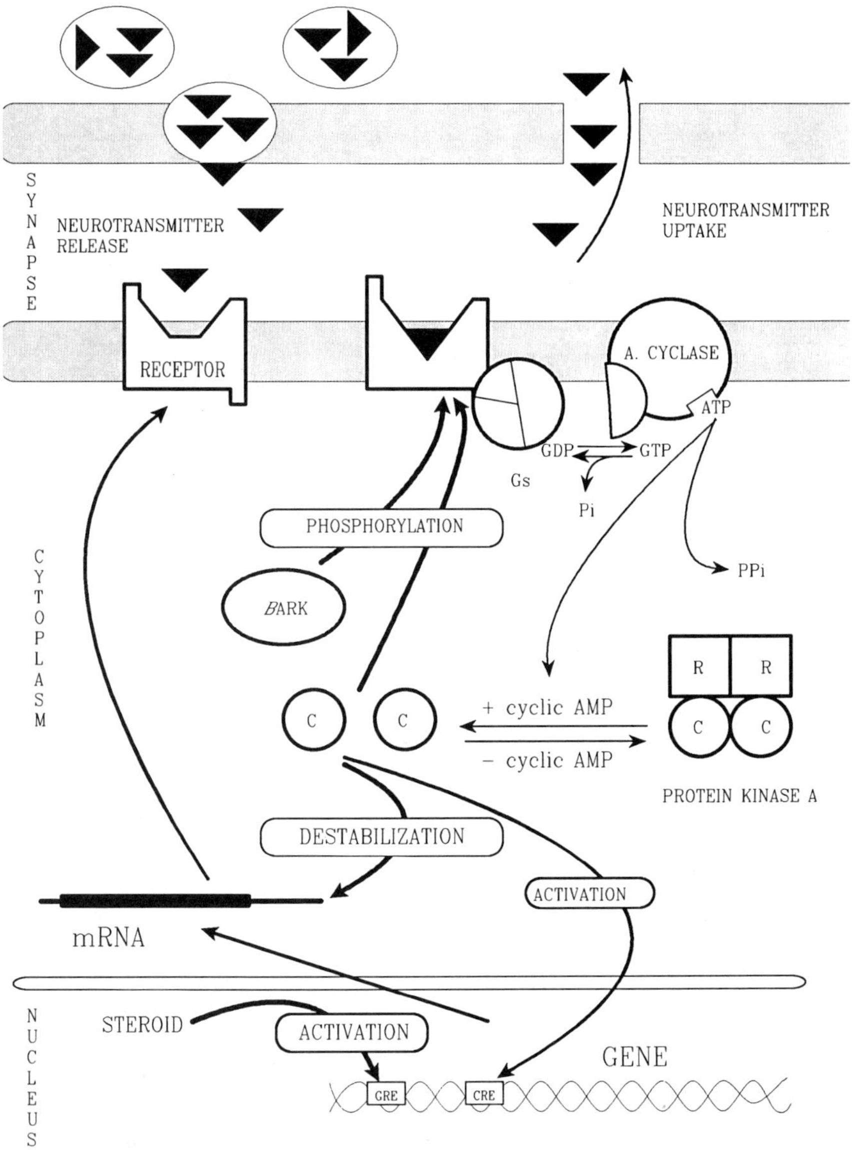

Figure 5 Agonist regulation of G-protein-linked receptor mRNA: transcriptional, posttranscriptional, and posttranslational controls. Agonist regulation of β-adrenergic receptors is divided into two distinct phases. In the early phase (0 to 4 h), uptake of the neurotransmitter and phosphorylation of the receptor play central roles in attenuating the response to catecholamines. An early, transient, agonist-induced increase in transcription of the β_2-adrenergic receptor gene has been reported, although its significance remains to be established. In the late phase of agonist regulation (>4 h), the rate of transcription of the receptor gene is unaffected by agonist. Agonist-induced down-regulation of receptor reflects down-regulation of receptor mRNA via destabilization of the mRNA.

comprising a second G-protein-linked pathway has been termed "cross-regulation" (Hadcock et al., 1990).

The availability of the kin- variant of S49 mouse lymphoma cells (the variant with a transdominant mutation in the catalytic subunit of protein kinase A) permitted analysis of cross-regulation from the stimulatory to inhibitory adenylyl cyclase in cells deficient in protein kinase A activity. In the kin- cells, persistent activation of the stimulatory pathway with isoproterenol failed to alter the expression of $G_{i\alpha 2}$ at either the mRNA or protein levels. Metabolic labeling of the cells with ^{35}S-methionine and immuneprecipitation of G-protein subunits permitted study of the relative synthesis and degradation rates for $G_{s\alpha}$ and $G_{i\alpha 2}$. For $G_{i\alpha 2}$, a threefold increase in the rate of synthesis and no change in rate of degradation was observed in cells in which the stimulatory adenylyl cyclase pathway was activated. The synthesis rate of $G_{s\alpha}$ increased modestly in response to activation of the stimulatory adenylyl cyclase, but this increase was counterbalanced by a decline in the $t_{1/2}$ of $G_{s\alpha}$ from 55 h in naive cells to 34 h in agonist-treated cells. Thus these two pathways controlling adenylyl cyclase display a complex "cross-regulation" that influences not only mRNA levels but also the turnover of elements in the pathways. Persistent activation of the stimulatory pathway increases $G_{i\alpha 2}$-mediated responses, $G_{i\alpha 2}$ mRNA levels, and $G_{i\alpha 2}$ expression. Transient elevation of $G_{s\alpha}$ mRNA are counterbalanced by a decline in the $t_{1/2}$ for the degradation rate of the subunit. Protein kinase A plays an obligate role in this cross-regulatory phenomenon (Hadcock et al., 1990).

Cross-Regulation from Inhibitory to Stimulatory Adenylyl Cyclase

The extent to which cross-regulation occurs from the inhibitory to the stimulatory adenylyl cyclase pathways was explored recently (Hadcock et al., 1991). The inhibitory adenylyl cyclase pathway of DDT_1MF-2 smooth muscle cells was persistently activated by the A1-adenosine receptor agonist (−)-N6(R-phenylisopropyl)adenosine (PIA). In response to the activation of the inhibitory pathway, the stimulatory adenylyl cyclase pathway was enhanced. PIA treatment (48 h) resulted in a doubling of the stimulatory response of the adenylyl cyclase to the β-adrenergic agonist isoproterenol. Furthermore, the ED_{50} for isoproterenol-stimulated adenylyl cyclase was reduced 50-fold in cells that were treated with PIA as compared to untreated, naive cells. The ED_{50} declined from 50 nM to ~1 nM isoproterenol in the cells activated for 48 h with PIA.

At the level of a stimulatory receptor, persistent activation of the inhibitory adenylyl cyclase resulted in an increased expression of β_2-adrenergic receptors (analyzed by both radioligand binding as well as immunoblotting) that was preceded by an increase in receptor mRNA levels. The $t_{1/2}$ of the β_2-

adrenergic receptor was unaffected by persistent activation of the inhibitory pathway. The cross-regulation from the inhibitory to the stimulatory adenylyl cyclase responses was also shown to be dependent on both the time and the dose of PIA employed. As one would expect from the data gleaned from studies of the β-adrenergic receptors, activation of the inhibitory A1-adenosine receptors with PIA were found to display down-regulation of radioligand binding by the A1-receptors. $G_{i\alpha 2}$ steady-state expression declined by half in the cells persistently activated by PIA. This decline occurred in spite of an increase in $G_{i\alpha 2}$ mRNA and synthesis rate for the subunit, reflecting the reduction in the $t_{1/2}$ for the degradation of $G_{i\alpha 2}$ from ~75 h in naive cells to ~40 h in cells in which the inhibitory pathway had been persistently activated. $G_{s\alpha}$ mRNA levels and expression were largely unaffected by persistent activation of the inhibitory pathway. Cross-regulation from the inhibitory to stimulatory adenylyl cyclase involves the following basic features: (1) up-regulation of β_2-adrenergic receptors; (2) enhancement of receptor-mediated activation of stimulatory pathway by a reduction in the ED_{50} of the stimulatory receptor-mediated response; (3) down-regulation of the A1-adenosine receptors; and (4) a down-regulation of $G_{i\alpha 2}$ at the level of protein turnover (Fig. 6). Thus, the basic motif appears to be that activation of a G-protein-linked pathway attenuates the response propagated via the pathway, while amplifying throughput propagated by opposing G-protein-linked pathways.

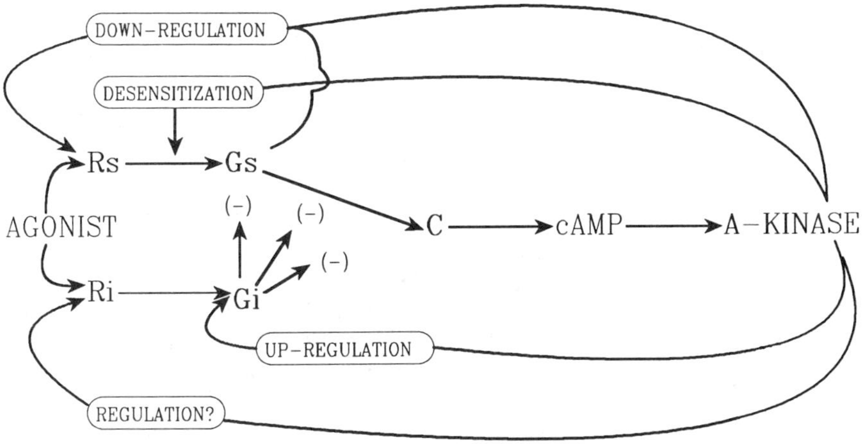

Figure 6 A schematic of the up- and down-regulation of the hormone-sensitive pathways controlling the inhibition and stimulation of adenylylcyclase. See text for details. (From Hadcock et al., 1990.)

Cross-Regulation from Stimulatory Adenylyl Cyclase to Stimulatory Phospholipase C Pathways

Cross-regulation between G-protein-coupled receptors is not confined to those coupled only to adenylyl cyclase. Recently, a novel form of cross-regulation from the stimulatory adenylyl cyclase to the stimulatory phospholipase C (PLC) pathways has been reported. Activation of the effector phospholipase C appears to be mediated by a novel G-protein termed G_q and results in the breakdown of phosphatidylinositol phospholipid to diacylglycerol (DAG) and water-soluble inositol phosphates. The DAG activates protein kinase C, and one of the inositol phosphates (inositol 1,4,5 trisphosphate) stimulates the release of stored intracellular calcium pools. Studies were initiated in DDT_1MF-2 smooth muscle cells in culture which display an α_1-adrenergic receptor-mediated activation of PLC (Morris et al., 1991). Stimulation of cells with epinephrine promoted a time-dependent and dose-dependent loss of α_1-adrenergic receptor-specific ligand binding. Cells displayed ~ 0.7 amol (10^{-18} mol) of α_1-adrenergic receptor mRNA in monolayer cultures. Stimulation by epinephrine, exerted at both α_1- and β_2-adrenergic receptors in these cells, promoted a short-term (2- to 8-h) increase in the abundance of α_1-receptor mRNA and expression. Interestingly, this increase in receptor mRNA was sensitive to competition by the β-adrenergic antagonist propranolol, but not by the α_1-antagonist ligand phentolamine. Phentolamine, however, did block a later phase characterized by the decrease in α_1-receptor-specific radioligand binding. The epinephrine-induced increase in α_1-adrenergic receptor mRNA was demonstrated to reflect a cross-regulation from the stimulatory adenylyl cyclase to the stimulatory PLC pathways. Via activation of the β-adrenergic receptors, epinephrine increased intracellular cAMP levels, which in turn appeared to increase the steady-state levels of α_1-receptor mRNA. Synthetic analogs of cAMP, or forskolin activation of the adenylyl cyclase itself, also promoted the increase in α_1-receptor mRNA. This study was the first to demonstrate a heterologous up-regulation of mRNA levels of adrenergic receptors (Morris et al., 1991).

The decline in α_1-adrenergic receptor radioligand binding and mRNA levels that follows the earlier phase of cross-regulation suggested that agonist-induced down-regulation of receptor mRNA levels for α_1-receptors might operate much as described above for β_2-receptors. Analysis of the half-life of the α_1-receptor mRNA, however, revealed no apparent change in cells in which the α_1-adrenergic receptor-mediated PLC response had been persistently activated. The $t_{1/2}$ for α_1-receptor mRNA was about 7 h for both the naive, untreated cells and for cells challenged with epinephrine for up to 12 h (Morris et al., 1991). These data suggest that, unlike the situation for agonist-induced down-regulation of β_2-adrenergic receptor mRNA promoted by destabilization of mRNA, the agonist-induced down-regulation of α_1-adrenergic receptor mRNA involves not mRNA stability, but perhaps transcriptional regulation. Much further work will be required to elucidate the molecular basis for the down-regulation of α_1-adrenergic receptor mRNA observed in epinephrine-treated cells. Thus, cross-regulation between α_1- and β_2-adrenergic receptor-mediated pathways at 4 h occurs at the level of receptor mRNA, whereas the later phase, characterized by a decline in α_1-receptor radioligand binding and mRNA levels, appears to be propagated via the α_1-receptors themselves (homologous regulation). A schematic representation of these relationships is provided in Fig. 7.

Short-Term Cross-Regulation from Inhibitory to Stimulatory Adenylyl Cyclase

The approach of persistently activating a single G-protein-linked receptor pathway to observe possible cross-regulation to other G-protein-mediated pathways has produced a wealth of new information. Use of persistent activation for these studies left unanswered the question concerning to what extent does

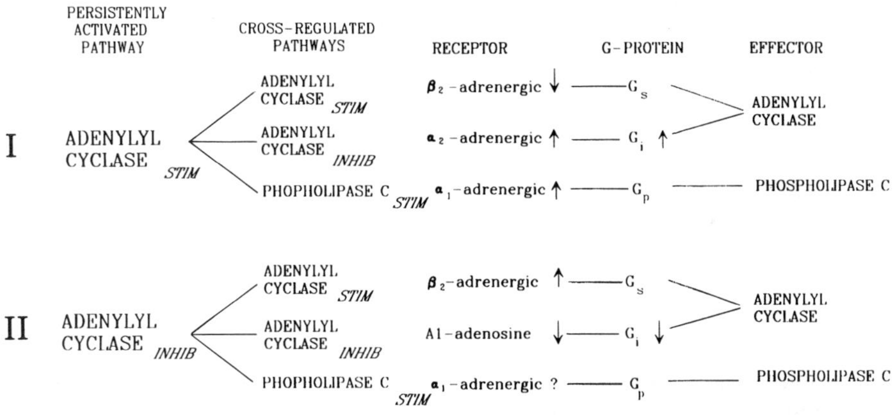

Figure 7 Cross-regulation between G-protein-mediated pathways: regulation of expression. For the stimulatory pathway of adenylyl cyclase (pathway I), cyclic AMP is the output, activating the cAMP-dependent protein kinase (protein kinase A). For the inhibitory adenylyl cyclase pathway (pathway II), attenuation of protein kinase A activity represents the major effector response. For the stimulatory pathway of phospholipase C, activation of protein kinase C and increases in intracellular Ca^{2+} mobilization are the two outputs. For details of the regulation of system elements in response to activations of the $G_{s\alpha}$-mediated and $G_{i\alpha}$-mediated pathways, see text. STIM = stimulatory; INHIB = inhibitory. (From Hadcock et al., 1991.)

cross-regulation operate on a short-term basis. This topic has been investigated, and the preliminary results indicate that there is much to be learned about cross-regulation in the short term.

Short-term (30- to 60-min) activation of the inhibitory adenylyl cyclase pathway of DDT_1MF-2 smooth muscle cells in culture by PIA was shown to enhance the stimulatory adenylyl cyclase pathway, as did the persistent activation by PIA (Port et al., 1992). β_2-adrenergic receptor-stimulated adenylyl cyclase activity was enhanced in the cells treated with PIA for the short term. The effects of the short-term regulation were confined to the level of the stimulatory receptor. Activation of the stimulatory adenylyl cyclase at the level of G_s by GTPγS was unaffected by short-term challenge with PIA, as was the activation of the adenylyl cyclase itself by forskolin. Although the β-adrenergic response was enhanced in the PIA-treated cells, the number and affinity of these receptors for the radiolabeled β-adrenergic antagonist ^{125}I-cyanopindolol were not altered. Metabolic labeling of the cells with ^{32}P-orthophosphate followed by immuneprecipitation from cell membrane extracts with anti-β_2-adrenergic receptor antibodies revealed that the phosphorylation state of the receptor was enhanced by epinephrine challenge (threefold) and reduced 75% by short-term activation of the inhibitory adenylyl cyclase pathway with PIA. Exposing the cells to a cAMP analog was found to abolish the PIA-induced enhancement of the β_2-adrenergic response controlling adenylyl cyclase. The results of this study suggest that short-term cross-regulation from the inhibitory to the stimulatory adenylyl cyclase pathways involves the following: (1) decreased intracellular cAMP levels and a decline in the activity of the protein kinase A; (2) decreased phosphorylation of the β_2-adrenergic receptor; and (3) enhanced receptor-mediated activation of G_s.

Cross-Regulation from Tyrosine Kinase to G-Protein-Linked Receptor-Mediated Pathways

Although little data have been accumulated about cross-regulation between G-protein-linked pathways and tyrosine-kinase receptor-mediated pathways (Fig. 1), it seems highly probable, and necessary, for integration and cross-regulation to operate between these two major templates for transmembrane signaling. There are many cases, such as the well-known counterregulatory effects of insulin and catecholamines, in which one might anticipate some cross-regulation to exist between the tyrosine-kinase receptors and G-protein-linked pathways. With the advent of new biochemical and molecular probes for the elements of both the tyrosine-kinase receptor pathways and G-protein-linked receptor pathways, many of these fundamental questions of physiological regulation will come under detailed examination.

IV. Receptor Activation

In spite of a vast increase in our knowledge of the nature of the elements that compose G-protein-linked pathways, our understanding of the process of "receptor activation" remains primitive. Receptor activation refers to the structural features induced by an agonist ligand in its receptor that provide the basis for transducing that signal to G-protein activation. The low abundance of these receptors, the inherent difficulties in purifying these receptors, and the problems associated with developing probes for their structural analysis (Malbon, 1991, Malbon et al., 1991) have severely hampered progress in this area. One of the most promising avenues appears to be the study of the role of receptor disulfides and free sulfhydryl groups in receptor structure as well as in receptor activation (Malbon et al., 1987). As discussed above, it is clear that disulfide bonds play a critical role in maintaining the conformation and binding domain of the β-adrenergic receptor and many other G-protein-linked receptors. Too, the demonstration that thiols activate the purified receptor and yield a receptor form that itself activates G_s in a manner similar to that of agonist-liganded receptors prompts us to examine cysteinyl residues and the role of disulfide bonds in greater detail.

Several important advances have expanded opportunities to approach the question of receptor structure from a biochemical and molecular viewpoint. The successful, high-efficiency expression of mammalian G-protein-linked receptors in baculovirus-infected insect cells is capable of producing adequate levels of receptor for analysis by conventional techniques of peptide chemistry (George et al., 1989). Milligram quantities of β-adrenergic receptor can be expressed by *Spodoptera frugiperda* cells (Sf9) infected with a baculovirus expression vector engineered to harbor the β_2-adrenergic receptor cDNA. The receptor is fully functional and displays the structural features expected for the receptor expressed in its normal mammalian context. The only exception appears to be an altered pattern of glycosylation by the insect cells that does not duplicate the mammalian N-glycosylation with fidelity. Previous work demonstrating the normal functioning and expression of the unglycosylated form of the β_2-adrenergic receptor (George et al., 1986) suggests that this shortcoming does not compromise analysis of the receptor structure. In combination with site-directed mutagenesis, baculovirus-infected Sf9 cell expression appears to provide the tools necessary for more detailed biochemical and biophysical analyses to be accomplished.

V. Concluding Remarks

Even just several years ago, it would have been difficult to envision the explosion of knowledge concerning the structure of elements of transmembrane signaling systems. G-protein-linked receptor pathways, no longer confined to the regulation of adenylyl cyclase, but rather a spectrum of effectors including phospholipases C and A2, cGMP phosphodiesterases, and ion channels, represent a major motif in information transduction across biological membranes. As the number of G-protein-coupled pathways increased, it became clear that cells process information transduced from multiple pathways simultaneously. Physiological regulation of transmembrane signaling must include not only knowledge of the operation of each individual pathway, but also of the intracellular networking that must integrate input signals. Although confined largely to adrenergic receptors and the G-proteins mediating the dual control of adenylyl cyclase, the chapter highlights the rich diversity that exists for mechanisms of physiological regulation. Our task is to extend our level of understanding even further to include a description of the molecular basis for up-regulation, down-regulation, and cross-regulation. Much will be learned when we expand the scope of analysis to the broader spectrum of G-protein-linked receptor systems. Clearly, a hybrid approach employing molecular and cellular biology approaches in a physiological context will yield new dimensions in our understanding of transmembrane signaling and more complex questions for us to ponder.

Acknowledgments

This work was supported by research grants provided with the generous support of the National Institutes of Health, American Heart Association, and Weight Watchers International Foundation.

References

Bahouth SW. Thyroid hormones transcriptionally regulate the β_1-adrenergic receptor gene in cultured ventricular myocytes. J Biol Chem 1991; 266:15863–69.

Bahouth SW, Kelley LK, Smith CH, Lautens L, Ruoho AE, Malbon CC. Identification of a novel Mr = 76,000 Da form of β-adrenergic receptors. Biochem Biophy Res Commun 1986; 141:411–7.

Benovic JL, DeBlasi A, Stone WC, Caron MG, Lefkowitz RJ. β-Adrenergic receptor kinase primary structures delineates a multigene family. Science 1989; 246:235–40.

Benovic JL, Pike LJ, Cerione RA, Staniszewski C, Yoshimasa T, Codina J, Caron MG, Lefkowitz RJ. Phosphorylation of the mammalian β-adrenergic receptor by cyclic AMP-dependent protein kinase. J Biol Chem 1985; 260:7094–101.

Benovic JL, Strasser RH, Caron MG, Lefkowitz RJ. β-Adrenergic receptor kinase: identification of a novel protein kinase that phosphorylates the agonist-occupied form of the receptor. Proc Natl Acad Sci USA 1986; 83:2797–801.

Birnbaumer L, Codina J, Mattera R, Yatani A, Scherer N, Toro M-J, Brown AM. Signal transduction by G-proteins. Kidney Int 1987; 32(suppl. 23):S14–37.

Bonner TI, Buckley NJ, Young AC, Brann MR. Identification of a family of muscarinic acetylcholine receptor genes. Science 1987; 237:527–32.

Bouvier M, Collins S, O'Dowd BF, Campbell PT, de Blasi A, Kobilka BK, MacGregor GP, Caron MG, Lefkowitz RJ. Two distinct pathways for cAMP-mediated down-regulation of the β_2-adrenergic receptor: phosphorylation of the receptor and regulation of its mRNA level. J Biol Chem 1989; 264:16786–92.

Bouvier M, Leeb-Lundberg LMF, Benovic JL, Caron MG, Lefkowitz RJ. Regulation of adrenergic receptor function by phosphorylation. II. Effects of agonist occupancy on phosphorylation of α_1 and β_2-adrenergic receptor by protein kinase C and the cyclic AMP-dependent protein kinase. J Biol Chem 1987; 262:3106–13.

Clark RB, Kunkel MW, Friedman J, Goka TJ, Johnson JA. Activation of cAMP-dependent protein kinase is required for heterologous desensitization of adenylyl cyclase in S49 wild-type lymphoma cells. Proc Natl Acad Sci USA 1988; 85:1442–6.

Collins S, Bolanowski MA, Caron MG, Lefkowitz RJ. Genetic regulation of β-adrenergic receptors. Ann Rev Physiol 1989a; 51:203–15.

Collins S, Bouvier M, Bolanowski MA, Caron MG, Lefkowitz RJ. cAMP stimulates transcription of the β_2-adrenergic receptor gene in response to short-term agonist exposure. Proc Natl Acad Sci USA 1989b; 86:4853–7.

Collins S, Caron MG, Lefkowitz RJ. β_2-Adrenergic receptors in hamster smooth muscle cells are transcriptionally regulated by glucocorticoids. J Biol Chem 1988; 263:9067–70.

Davies AO, Lefkowitz RJ. Regulation of β-adrenergic receptors by steroid hormones. Ann Rev Physiol 1984; 46:119–30.

Dixon RAF, Kobilka BK, Strader DJ, Benovic JL, Dohlman, HG, Frielle T, Bolanowski MA, Bennett CD, Rands E, Diehl RE, Mumford RA, Slater EE, Sigal IS, Caron MG, Lefkowitz RJ, Strader CD. Cloning of the gene and cDNA for mammalian βAR and homology with rhodopsin. Nature 1986; 321:75–9.

Dohlman HG, Caron MG, Lefkowitz RJ. A family of receptors coupled to guanine nucleotide regulatory proteins. Biochemistry 1987; 26:2657–64.

Emorine LJ, Marulo S, Delavier-Klutchko C, Kaveri SV, Durieu-Trautman O, Strosberg AD. Structure of the gene for human β_2-adrenergic receptor: expression and promoter characterization. Proc Natl Acad Sci USA 1987; 84:6995–9.

Galvin-Parton PA, Watkins DC, Malbon CC. Retinoic acid modulation of transmembrane signaling: analysis in F9 teratocarcinoma cells. J Biol Chem 1990; 265:17771–9.

George ST, Arbabaian MA, Ruoho AE, Kiely J, Malbon CC. High-efficiency expression of mammalian β-adrenergic receptors in baculovirus-infected insect cells. Biochem Biophys Res Commun 1989; 163:1265–9.

George ST, Berrios M, Hadcock JR, Wang H-Y, Malbon CC. Receptor density and cAMP accumulation: analysis in CHO cells exhibiting stable expression of a cDNA that encodes the β_2-adrenergic receptor. Biochem Biophys Res Commun 1988; 150:665–72.

George ST, Ruoho AE, Malbon CC. N-glycosylation in expression and function of β-adrenergic receptors. J Biol Chem 1986; 261:16559–64.

Gilman AG. G-proteins: transducers of receptor-generated signals. Ann Rev Biochem 1987; 56:615–49.

Graziano MP, Moxham CP, Malbon CC. Purified rat hepatic β_2-adrenergic receptor: Structural similarities to the rat fat cell β_1-adrenergic receptor. J Biol Chem 1985; 260:7665–74.

Guest SG, Hadcock JR, Watkins DC, Malbon CC. β_1- and β_2-adrenergic receptors: agonist-induced reduction in receptor mRNA levels. Proc Natl Acad Sci USA 1990; 85:5021–5.

Hadcock JR, Malbon CC. Down-regulation of β-adrenergic receptors: agonist-induced reduction in receptor mRNA levels. Proc Natl Acad Sci USA 1988a; 85:5021–5.

Hadcock JR, Malbon CC. Regulation of β-adrenergic receptors by "permissive" hormones: glucocorticoids increase steady-state levels of receptor mRNA. Proc Natl Acad Sci USA 1988b; 85:8415–9.

Hadcock JR, Malbon CC. Regulation of receptor expression by agonists: transcriptional and post-transcriptional controls. Trends Neuro Sci 1991a; 14:242–7.

Hadcock JR, Malbon CC. Adrenal dysfunction and G-protein-mediated pathways. In: Milligan G, Wakelam M, eds. G-proteins: signal transduction and disease states. London: Academic Press, 1991b; 109–27.

Hadcock JR, Port JD, Malbon CC. Cross-regulation between G-protein-mediated pathways: activation of the inhibitory pathway of adenylylcyclase increases the expression of β_2-adrenergic receptors. J Biol Chem 1991; 266:11915–22.

Hadcock JR, Ros M, Malbon CC. Agonist regulation of β-adrenergic receptor mRNA: analysis in S49 mouse lymphoma mutants. J Biol Chem 1989a; 264:13956–61.

Hadcock JR, Ros M, Watkins DC, Malbon CC. Cross-regulation between G-protein-mediated pathways: stimulation of adenylylcyclase increases expression of the inhibitory G-protein, $G_{i\alpha2}$. J Biol Chem 1990; 265:14784–90.

Hadcock JR, Wang H-Y, Malbon CC. Agonist-induced destabilization of β-adrenergic receptor mRNA: attenuation of glucocorticoid-induced up-regulation of β-adrenergic receptors. J Biol Chem 1989b; 264:19928–33.

Hadcock JR, Williams DL, Malbon CC. Physiological regulation at the level of mRNA: analysis of steady-state levels of specific mRNAs by DNA-excess solution hybridization. Am J Physiol 1989c; 257:C457–65.

Hausdorff WP, Caron MG, Lefkowitz R. Turning off the signal: Desensitization of β-adrenergic function. FASEB J 1990; 4:2881–89.

Haynes RC. Adrenocorticotropic hormone: Adrenocortical steriods and their synthetic analogs: Inhibitors of the synthesis and actions of adrenocortical hormones. In Gilman AG, Rall TW, Nies AS, Taylor P, eds. The pharmacological basis of therapeutics, 8th ed. New York: Pergamon Press, 1990; 1431–62.

Hough C, Chuang D-M. Biochem Biophys Res Commun 1990; 170:46–52.

Lazar-Wesley E, Hadcock JR, Malbon CC, Kunos G, Ishac EJN. Tissue-specific regulation of α_{1b}-, β_1-, and β_2-adrenergic receptor mRNAs by thyroid state in the rat. Endocrinology 1991; 129:1116–8.

Malbon CC. Purification of beta-adrenergic receptors: isolation of mammalian β_1- and β_2-subtypes. In Litwack G, ed., Receptor purification. Clifton, NJ: Humana Press, 1991; in press.

Malbon CC, George ST, Moxham CP. Intramolecular disulfide bridges: Avenues to receptor activation? Trends Biochem Sciences 1987; 12:172–6.

Malbon CC, Hadcock JR. Evidence that glucocorticoid response elements in the 5′ non-coding region of the hamster β_2-adrenergic receptor gene are obligate for glucocorticoid regulation of hamster mRNA levels. Biochem Biophys Res Commun 1988; 154:676–81.

Malbon CC, Moxham CP, Brandwein HJ. Antibodies to beta-adrenergic receptors. In Perkins J, ed. The β-adrenergic receptors. Clifton, NJ: Humana Press, 1991; 181–261.

Malbon CC, Rapiejko PJ, Watkins DC. Permissive hormone regulation of hormone-sensitive effector systems. Trends Pharmacol Sci 1988; 9:33–6.

Morris GM, Hadcock JR, Malbon CC. Cross-regulation between G-protein-coupled receptors: activation of β_2-adrenergic receptors increases α_1-adrenergic receptor mRNA levels. J Biol Chem 1991; 266:2233–8.

Moxham CP, Malbon CC. Fat cell β_1-adrenergic receptor: structural evidence for existence of disulfide bridges essential for ligand binding. Biochemistry 1985; 24:6072–7.

Moxham CP, Ross EM, George ST, Malbon CC. β-Adrenergic receptors display intramolecular disulfide bridges in situ: analysis by immunoblotting and functional reconstitution. Mol Pharmacol 1988; 33:486–92.

Neve KA, Molinoff PB. Turnover of β_1- and β_2-adrenergic receptors after down-regulation or irreversible blockade. Mol Pharmacol 1986; 30:104–11.

Pederson SE, Ross EM. Functional activation of beta-adrenergic receptors by thiols in the presence or absence of agonists. J Biol Chem 1985; 260:14150–7.

Port JD, Hadcock JR, Malbon CC. Cross-regulation between G-protein-mediated pathways: Acute activation of the inhibitory pathway of adenylylcyclase reduces β_2-adrenergic receptor phosphorylation and increases β-adrenergic responsiveness. J Biol Chem 1992; 267:8468–72.

Rhee SG, Suh P-G, Ryu SH, Lee SY. Studies of inositol phospholipid-specific phospholipase C. Science 1989; 244:546–50.

Sibley DR, Lefkowitz RJ. Molecular mechanisms of receptor desensitization using the β-adrenergic receptor coupled-adenylate cyclase system as a model. Nature 1985; 317:124–9.

Wang H-Y, Berrios M, Hadcock JR, Malbon CC. The biology of β-adrenergic receptors: analysis in human epidermoid carcinoma A431 cells. Int J Biochem 1991; 23:7–20.

Wang H-Y, Berrios M, Malbon CC. Indirect immunofluorescence localization of β-adrenergic receptors and G-proteins in human A431 cells. Biochem J 1989a; 263:519–32.

Wang, H-Y, Berrios M, Malbon CC. Localization of β-adrenergic receptors in A431 cells in situ: effect of chronic exposure to agonist. Biochem J 1989b; 263:533–8.

Wang H-Y, Lipfert L, Malbon CC, Bahouth SW. Site-directed anti-peptide antibodies define the topography of the β-adrenergic receptor. J Biol Chem 1989c; 264:14424–31.

3

G Proteins in Pulmonary and Cardiovascular Signal Transduction Systems

VSEVOLOD A. TKACHUK, VITALYI O. RYBIN, and MIKHAIL P. PANCHENKO

Cardiology Research Center
Moscow, Russia

I. Introduction

The receptors for catecholamines, hypothalamic and hypophyseal hormones, vasoactive agents, glucagon, light (photoreceptor rhodopsin), and odorants are coupled in membranes to the family of the regulatory GTP-binding proteins (G proteins) that modulate activity of the cellular effector systems, such as adenylyl cyclase, polyphosphoinositide-specific phospholipase C, phospholipase A_2, phospholipase D, K^+ channels, Ca^{2+} channels, cGMP-specific phosphodiesterase, etc. (Birnbaumer et al., 1987, 1990; Gilman, 1987; Freissmuth et al., 1989; Birnbaumer, 1990). Binding of the ligand (hormone or its agonist) to the specific receptor induces the GDP/GTP exchange reaction at the nucleotide-binding site of the G protein coupled to the receptor. Binding of the GTP to G protein induces the dissociation of the heterotrimeric G protein to the alpha GTP subunit and beta/gamma subunit complex. Free alpha GTP subunits interact with the effector(s), inducing the transition of the effector into a new functional state. Hydrolysis of the GTP to GDP and P_i, catalyzed by G protein alpha subunits, results in dissociation of alpha GDP from the effector and its reassociation with beta/gamma complex. It is in this fashion that G-protein-coupled transmembrane signal transduction occurs, resulting in a change of plasma membrane potential and/or in generation/degradation of the intracellular second messengers: cyclic AMP, cyclic GMP, Ca^{2+}, and polyphosphoinositide turnover products. Besides modulating effector activity, G proteins are able to regulate the functional state of the receptors: GTP-dependent dissociation of the G protein from the receptor induces its transition from the high- to the low-affinity state of the hormone (Birnbaumer et al., 1987, 1990; Gilman, 1987; Freissmuth et al., 1989; Birnbaumer, 1990).

II. Mechanisms of Adenylyl Cyclase Regulation by G Proteins and Calmodulin

The involvement of G proteins in the hormone-dependent regulation of adenylyl cyclase activity has been described in detail. In a wide variety of mammalian cells, the binding of catecholamines to beta-adrenergic receptors induces the isomerization of G_s protein(s) into an active state (one hormone/receptor complex catalytically stimulates several G_s protein molecules), followed by G_s-mediated stimulation of adenylyl cyclase activity. On the other hand, catecholamines (via alpha$_2$-adrenergic receptors), acetylcholine (via muscarinic cholinergic receptors), and many other hormones and neurotransmitters induce the inhibition of adenylyl cyclase activity via hormone/receptor-induced coupling of the enzyme to the G_i proteins: G_i1, G_i2 (dominant mediator of cyclase inhibition), and G_i3 (Birnbaumer et al., 1987, 1990; Gilman, 1987; Freissmuth et al., 1989; Simonds et al., 1989a, 1989b; Birnbaumer, 1990; Mckenzie and Milligan, 1990; Morris et al., 1990).

The subunit dissociation model of heterotrimeric G-protein activation developed by Gilman and colleagues, as well as others, has been well documented as the mechanism by which G_s or G_i proteins stimulate or inhibit adenylyl cyclase (Hildebrandt et al., 1984; Katada et al., 1984a, 1984b; Bokoch, 1987; Hildebrandt and Kohnken, 1990). It appears conclusive that, in the majority of cases, the effects of G_s and G_i on adenylyl cyclase activity are due to their alpha subunits, which are generated in their activated form (GTP-liganded form) upon the stimulation of receptors by ligands. The hormonal inhibition of adenylyl cyclase occurs in a simple noncompetitive manner by the hormone-induced stimulation of enzyme activity; i.e., the naturally existing G_i can inhibit the catalytic activity of adenylyl cyclase without having to affect the activity state of G_s, it does so in a manner that is kinetically noncompetitive with respect to the activity of G_s (Hildebrandt et al., 1984; Hildebrandt and Kohnken, 1990). The noncompetitive character of opposite G_s and G_i regulatory effects on adenylyl cyclase activity is illustrated by data presented on Fig. 1. The beta-adrenergic agonist isoproterenol accelerates the Gpp(NH)p-induced isomerization of G_s protein into an activated state evidenced by the

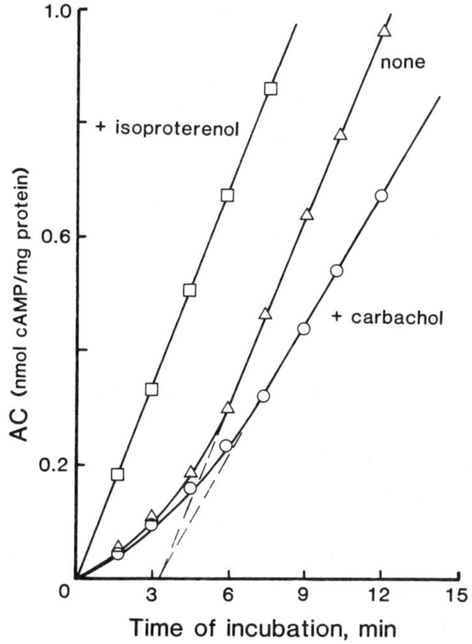

Figure 1 Kinetics of Gpp(NH)p-induced stimulation of adenylyl cyclase in rabbit heart sarcolemma in the absence and in the presence of isoproterenol or carbachol. Gpp(NH)p, isoproterenol, and carbachol concentrations were 10 μM.

disappearance of a lag phase in the stimulatory action of Gpp(NH)p on adenylyl cyclase activity, whereas the steady-state level of Gpp(NH)p-stimulated adenylyl cyclase activity remains unchanged in the presence of isoproterenol. Such phenomena reflect the fact that beta-agonists accelerate the isomerization of G_s proteins into a Gpp(NH)p-stimulated state without affecting the number of catalytically active adenylyl cyclase-$G_s \cdot$ Gpp(NH)p complexes formed under the action of Gpp(NH)p. Unlike hormone stimulators, the agonist of muscarinic cholinergic receptors, carbachol (via modulating of the Gpp(NH)p-induced G_i protein coupling to the adenylyl cyclase), inhibits pre-steady-state as well as the steady-state Gpp(NH)p-stimulated adenylyl cyclase activities and does not affect the lag phase in Gpp(NH)p action on adenylyl cyclase; i.e., Gpp(NH)p-stimulated G_i fails to alter the velocity of G_s isomerization into the Gpp(NH)p-stimulated state, as well as coupling the Gpp(NH)p-stimulated G_s to the adenylyl cyclase (Fig. 1).

The cholera toxin-induced ADP-ribosylation of G_s alpha subunits results in: (1) a GDP/GTP exchange reaction at G_s nucleotide-binding sites (Lad et al., 1980; Burns et al., 1982); (2) abrogation of G_s GTPase activity (Cassel and Selinger, 1977; Cassel and Pfeuffer, 1978); and (3) dissociation of $G_s\alpha^* \cdot$GTP from the beta/gamma complex (Kahn and Gilman, 1984a). The ADP-ribosylated G_s-protein alpha · GTP complex irreversibly binds to and stimulates adenylyl cyclase in a manner similar to that induced by G_s-protein alpha subunits activated by the nonhydrolazable GTP analog Gpp(NH)p (Table 1; Enomoto and Gill, 1980; Tamir and Gill, 1988).

On the other hand, the ADP-ribosylation of G_i alpha subunits by pertussis toxin uncouples G_i protein from the hormonal receptor, followed by the abolition of the hormone-induced inhibition of adenylyl cyclase and an increase of the effects of hormone-stimulators on enzyme activity (Table 2; Katada and Ui, 1982; Kurose and Ui, 1983).

Attenuation of hormone-induced inhibition of adenylyl cyclase activity and enhancement of effects of hormone-stimulators on adenylyl cyclase, as well as an increase of basal adenylyl cyclase activity, occur in cells treated with phorbol esters (Table 3; Katada et al., 1985; Bell and Brunton, 1986; Avdonin et al., 1989). The effects of phorobol esters (which are known to activate the kinase C of whole cells) on adenylyl cyclase regulatory properties can be mimicked by plasma membrane treatment with purified protein kinase C (Table 3; Naghshineh et al., 1986) and correlates with the protein kinase C-dependent phosphorylation of the alpha subunit of the G_i protein (Katada et al., 1985) or any other G-protein (probably G_z) alpha subunit (Carlson et al., 1989).

Table 1 Comparison of GTP and Gpp(NH)p Effects on Adenylyl Cyclase Activity in Control and Cholera Toxin-Treated Rabbit Heart Sarcolemma

	Adenylyl cyclase activity (pmol cAMP/min · mg protein)	
Nucleotide	Control sarcolemma	Cholera toxin-treated sarcolemma
No additions	81 ± 4	83 ± 3
GTP, 100 μM	109 ± 9	246 ± 5
Gpp(NH)p, 100 μM	297 ± 29	308 ± 27

Sarcolemma preparations (protein concentration 1 mg/mL) were incubated in ADP-ribosylation mixture in the absence (control) or in the presence of activated cholera toxin (100 μg/mL) for 20 min at 37°C. During the incubation time, complete ADP-ribosylation of sarcolemmal G_s alpha was achieved. Then sarcolemma preparations were washed with buffer by centrifugation and adenylyl cyclase activity was immediately measured. Data are means ±S.E. of values from triplicate assays.

Table 2 Effect of Pertussis Toxin on Hormone-Induced Inhibition of Adenylyl Cyclase Activity in Heart and Lung Membranes

Additions	Adenylyl cyclase activity (pmol cAMP/min · mg protein) Control membranes	Pertussis toxin-treated membranes
	Rabbit heart sarcolemma	
No additions	49.1 ± 0.1	66.2 ± 0.8
Isoproterenol, 10 μM	80.2 ± 0.1	107 ± 3
Isoproterenol, 10 μM, + carbachol, 10 μM	51.5 ± 0.4	108 ± 8
	Rat lung plasma membranes	
No additions	12.1 ± 0.2	37.6 ± 0.1
Isoproterenol, 10 μM	24.7 ± 3.7	78.4 ± 6.1
Isoproterenol, 10 μM, + atrial natriuretic peptide, 1 μM	14.8 ± 0.9	89.0 ± 3.3

Membranes (protein concentration 1 mg/mL) were incubated in ADP-ribosylation mixture in the absence (control) or in the presence of activated pertussis toxin (10 μg/mL) for 30 min at 37°C. During the incubation time, complete ADP-ribosylation of membrane-bound G_i alpha was achieved. Then membrane preparations were washed with buffer by centrifugation and adenylyl cyclase activity was immediately measured. Adenylyl cyclase assay mixture contained 100 μM GTP. Data are means ±S.E. of values from triplicate assays.

The pertussis toxin- or protein kinase C-induced changes in hormonal sensitivity of adenylyl cyclase are probably accomplished at the level of G_i and G_s protein subunit "cross-talk." The release of G_i beta/gamma complexes that occur as a result of the activation of G_i by hormonal receptors has been proposed as one of the mechanisms by which G_i-mediated attenuation of G_s alpha subunit-stimulated adenylyl cyclase activity occurs (Katada et al., 1984b; Bokoch, 1987; Linder et al., 1990). Since the beta/gamma complexes of G_s and G_i proteins appear to be functionally and physically similar (Amatruda et al., 1988; Freissmuth et al., 1989; Schmidt and Neer, 1991), the activation of G_i by its receptor would lead to the release of beta/gamma subunits which would be able to interact effectively with alpha subunits of G_s, tending to suppress their activation. On the other hand, the stabilization of G_i protein in its heterotrimeric form (which may take place in response to G_i-protein alpha subunit ADP-ribosylation by pertussis toxin or its phosphorylation by protein kinase C) would result in a decrease in the amount of free beta/gamma com-

Table 3 Effects of PMA on the Inhibition of Adenylyl Cyclase Activity by Platelet-Activating Factor and on the Activation of Adenylyl Cyclase by Prostaglandin E_1 in Human Platelets. Effects of Purified Protein Kinase C on Adenylyl Cyclase Activity and Its Sensitivity to Carbachol in Rabbit Heart Sarcolemma

	Adenylyl cyclase activity (pmol cAMP/min · mg protein)	
Additions	Control	PMA-treated
	Human platelet membranes	
No additions	22.4 ± 1.0	23.1 ± 0.4
Platelet-activating factor, 0.1 μM	15.6 ± 0.2	22.9 ± 0.6
Prostaglandin E_1, 10 μM	201 ± 10	439 ± 20
	Rabbit heart sarcolemma	
	Control	Protein kinase C-treated
No additions	122 ± 10	277 ± 14
Isoproterenol, 10 μM	285 ± 21	640 ± 17
Carbachol, 10 μM	85.1 ± 4.2	315 ± 10
Forskolin, 10 μM	1750 ± 90	4490 ± 21

Platelets were incubated for 2 min at 37°C with 1 μM 4-alpha-phorobol 12,13-didecanoate (control) or 1 μM 4-beta-phorobol-12-beta-myristate-13-alpha-acetate (PMA-treated) before preparation of membranes, followed by measurement of adenylyl cyclase activity in the presence of 10 μM GTP. Rabbit heart sarcolemma (protein concentration 1 mg/mL) was incubated for 5 min at 37°C with partially purified brain protein kinase C (5 μg/mL) in the presence of 100 μM ATP, 100 μM Ca^{2+}, and 5 mM Mg^{2+}. Control incubation was without protein kinase C. Then sarcolemma preparations were washed with buffer by centrifugation and adenylyl cyclase was immediately measured in the incubation mixture contained 100 μM GTP. Data are means ±S.E. of values from triplicate assays.

plexes and provide much more pronounced effects of G_s alpha-subunit mediation of hormone-stimulators on adenylyl cyclase. The effects of beta/gamma complex would depend on the relative affinities of G_s and G_i alpha subunits for this complex, the relative amounts of each G protein, the degree of subunit dissociation of G protein induced by the receptor, the degree of association of G_s alpha with adenylyl cyclase, the relative amounts of unbound beta/gamma subunits in membrane, etc. In addition, the direct interaction of the beta/gamma complex with calmodulin is likely to be involved in beta/gamma complex-induced inhibition of calmodulin-sensitive adenylyl cyclase (Katada et al., 1987).

The inhibitory effects of the G_i protein's alpha · GTP and beta/gamma subunits on adenylyl cyclase activity probably are a result of different functional states of the enzyme: G_i alpha · GTP is the primary mediator of hormonal inhibition of adenylyl cyclase when the enzyme is stimulated by coupling with G_s alpha · GTP, whereas the G_i beta/gamma complex may have a role as a mediator of inhibition of basal (nonhormone/receptor-mediated) and calmodulin-stimulated adenylyl cyclase activities.

Other known roles of beta/gamma complex in transduction of hormonal signals to adenylyl cyclase are (1) to promote interactions of G_s and G_i with receptors, which it does by reassociating with alpha · GDP; and (2) to allow for receptor dissociation from GTP-activated alpha, which it does by dissociating from activated h · R* beta/gamma · alpha · GTP. An additional role of the beta/gamma complex, which is hydrophobic, is to serve as an anchor for the hydrophilic alpha subunit, which remains localized with plasma membrane (Birnbaumer et al., 1987, 1990; Gilman, 1987; Freissmuth et al., 1989; Birnbaumer, 1990).

Most cells contain two distinguishable forms of G_s alpha subunits, 52 and 45 kDA in apparent molecular mass, which are produced by alternative splicing of primary transcripts of a single G_s-protein alpha gene (Robishaw et al., 1986). The ADP-ribosylation of G_s-protein alpha subunit by cholera toxin strongly depends on the presence of GTP or its nonhydrolizable analogs Gpp(NH)p or GTPgammaS (Fig. 2). Such strong dependence of cholera toxin-induced ADP-ribosylation of G_s protein on the presence of GTP, Gpp(NH)p, or GTPgammaS results from the potentiation of cholera toxin-catalyzed ADP-ribosylation by an endogenous low-molecular-weight G protein, ARF, or S protein (Enomoto and Gill, 1980; Gill and Meren, 1983; Kahn and Gilman, 1984b, 1986; Gill and Coburn, 1987). In GTP, Gpp(NH)p or the GTPgammaS-liganded form ARF interacts with cholera toxin ADP-ribosyltransferase and stimulates its activity (Tsai et al., 1988). The cholera toxin-dependent ADP-ribosylation of G_s 52- and 45-kDa alpha subunits also requires the beta/gamma complex (Graziano et al., 1987, 1989).

The pertussis toxin-dependent ADP-ribosylation of 40- to 41-kDa G_i alpha subunits is slightly potentiated in the presence of the GDP analog, GDPbetaS (Fig. 2), i.e., by the guanine nucleotide that stabilizes the heterotrimeric structure of the G protein, and is strongly decreased in the presence of GTPgammaS, i.e., by the nucleotide that induces the dissociation of the heterotrimeric G protein to free alpha subunit and beta/gamma complex (Gilman, 1987). Such dependence of pertussis toxin-induced ADP-ribosylation of G_i alpha on the presence of guanine nucleotides seems reasonable, since it is well known that G_i alpha subunits in complex with beta/gamma subunits

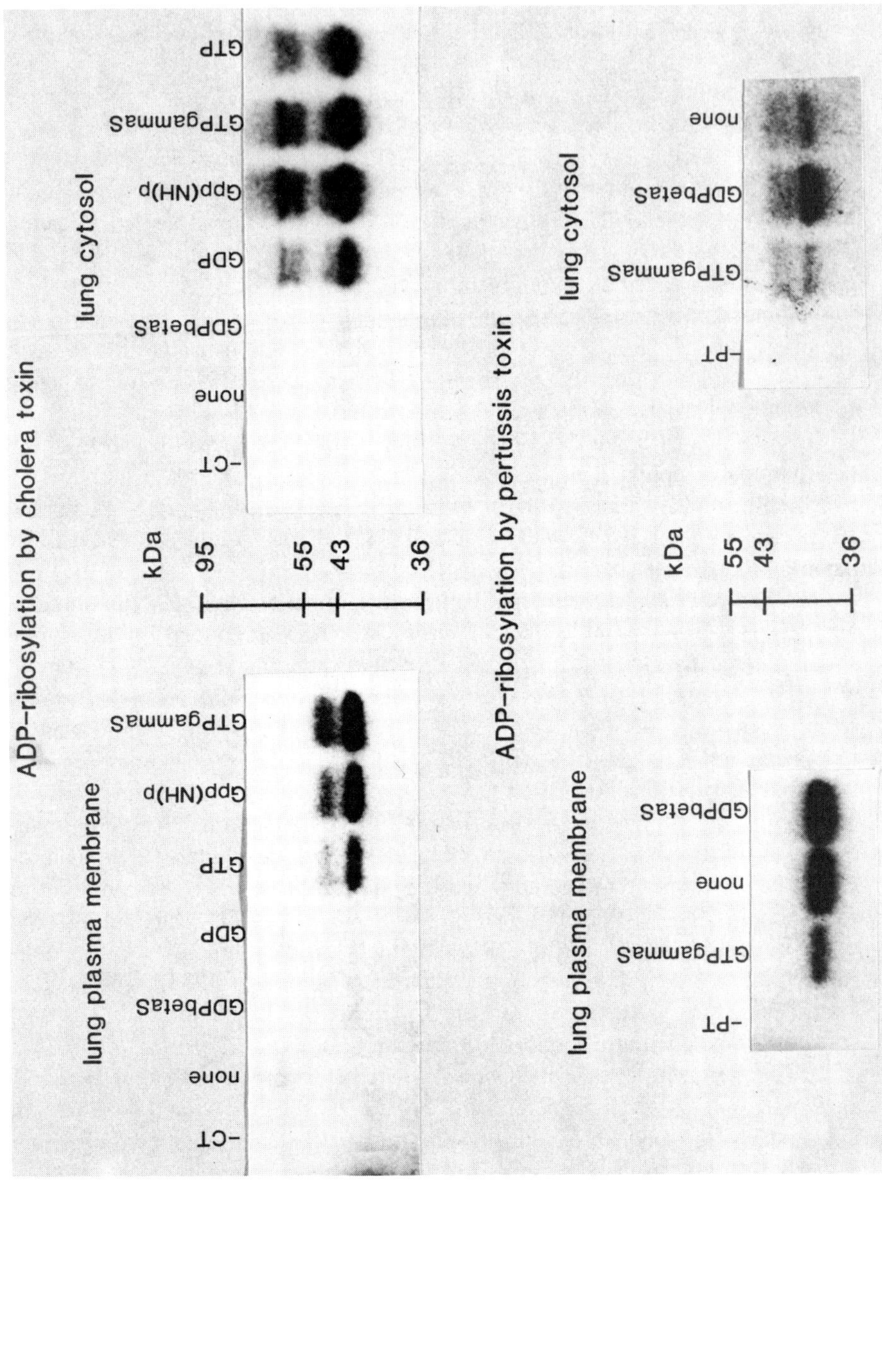
ADP-ribosylation by cholera toxin
lung plasma membrane
-CT
none
GDPbetaS
GDP
GTP
Gpp(NH)p
GTPgammaS
kDa
95
55
43
36
lung cytosol
-CT
none
GDPbetaS
GDP
Gpp(NH)p
GTPgammaS
GTP
ADP-ribosylation by pertussis toxin
lung plasma membrane
-PT
GTPgammaS
none
GDPbetaS
kDa
55
43
36
lung cytosol
-PT
GTPgammaS
GDPbetaS
none

represent a more preferred substrate for pertussis toxin than free alpha subunits (Neer et al., 1984; Katada et al., 1986).

Using cholera- and pertussis-induced ADP-ribosylations, we (in lung tissue) and others (in several other tissues and cell types) were able to identify cholera toxin-sensitive 52- and 45-kDa G_s alpha subunits and pertussis toxin-sensitive 40- to 41-kDa G_i alpha subunits, not only in membranes but also in the soluble cytosolic cellular (tissue) fraction (Fig. 2; Rosenthal et al., 1987; Tkachuk et al., 1989; Rudolph et al., 1989; Takahashi et al., 1991). About 7% to 10% of total ADP-ribosylation of pertussis toxin-sensitive G_i alpha subunit(s) in lung tissue homogenate is detected in the supernatant after centrifugation at 200,000 × g for 90 to 120 min. The strict dependence of cytosolic protein ADP-ribosylation by cholera toxin on the presence of GTP, Gpp(NH)p, or GTPgammaS suggests that this covalent modification of soluble 52- and 45-kDa G_s alpha subunits, as well as other, yet-to-be-identified soluble cholera toxin substrates, is stimulated by cytosolic ARF (Fig. 2). In fact, ARF can be identified in cytosolic as well as in plasma membrane lung fractions by Western blot analysis (Fig. 3). The clear dependence of pertussis toxin-dependent ADP-ribosylation of the cytosolic 40- to 41-kDa G_i alpha subunits on guanine nucleotide presence: GDPbetaS potentiates ADP-ribosylation, whereas GTPgammaS attenuates it (Fig. 2), may reflect the coupling of cytosolic G_i-protein alpha subunits with soluble beta/gamma complex and/or their existence in cytosol in a "pertussis toxin-sensitive conformation" even in the absence of beta/gamma complex. The molecular mass of the cytosolic pertussis toxin substrate(s) in the absence of the activating agents GTPgammaS and Mg^{2+} has been estimated to be about 75 to 80 kDa (Tkachuk et al., 1989; Rudolph et al., 1989; Takahashi et al., 1991), which is close to the molecular mass of G-protein heterotrimer, although the beta/gamma complex has not been detected in cytosolic fractions (Rudolph et al., 1989; Takahashi et al., 1991). The cytosolic G_i-protein alpha subunit has been reconstituted with the purified beta/gamma complex into pertussis toxin-sensitive heterotrimeric G-

---

Figure 2 Autoradiograms of the SDS-PAGE of pig lung plasma membranes and cytosol G_s and G_i proteins ^{32}P-ADP-ribosylated by pertussis and cholera toxins. Membrane or cytosol preparations were ADP-ribosylated for 30 min at 37°C in a final volume of 100 μl containing 3 μg of membrane and 50 μg of cytosolic protein (in the case of ADP-ribosylation by pertussis toxin) or 30 μg of membrane and 50 μg of cytosolic protein (in the case of ADP-ribosylation by cholera toxin), 10 μM NAD, 3 μCi ^{32}P-NAD, 50 mM Tris-Cl (pH 7.5), 2 mM $MgCl_2$, 1 mM EDTA, 10 mM dithiotreitol, 1 mM ATP, 1 mM NADP, 10 μM of guanine nucleotide (where indicated in the figure), 20 μg/mL pertussis toxin, or 100 μg/mL cholera toxin.

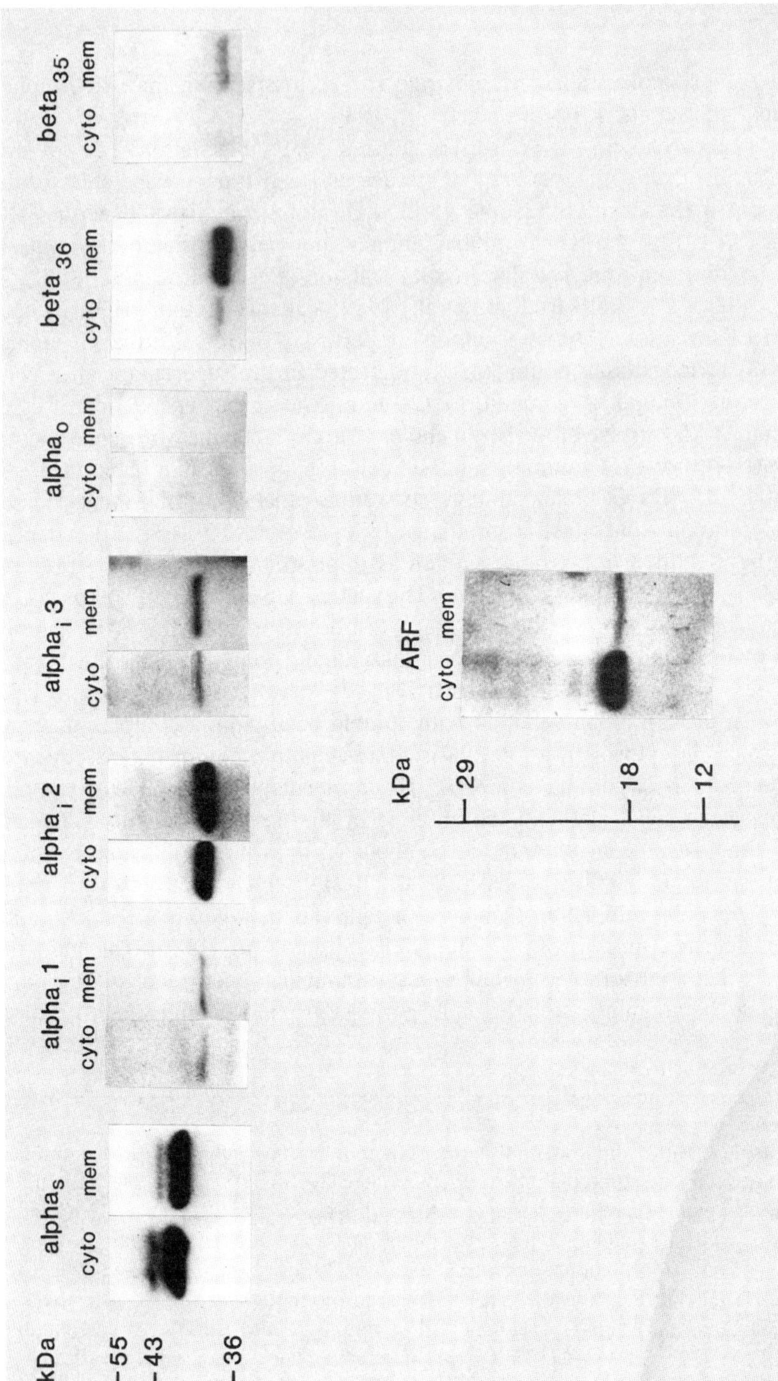
kDa
55
43
36
alphas
cyto mem
alphai 1
cyto mem
alphai 2
cyto mem
alphai 3
cyto mem
alphao
cyto mem
beta 36
cyto mem
beta 35
cyto mem
kDa
29
18
12
ARF
cyto mem

protein form (Rudolph et al., 1989; Takahashi et al., 1991). On the basis of immunoblots with antisera specific to 35- and 36-kDa beta subunits, we were able to show trace amounts of 36-kDa beta subunit but not 35-kDa beta subunit in lung cytosol, whereas both beta-subunit forms, 35 kDa (minor) and 36 kDa (major), can be identified in lung plasma membrane (Fig. 3). On the other hand, by immunoblotting with specific antisera to different G-protein alpha subunits, the 41-kDa G_i1 alpha (minor pertussis toxin-sensitive subunit), the 40-kDa G_i2 alpha (major pertussis toxin-sensitive subunit), the 41-kDA G_i3 alpha (moderate pertussis toxin-sensitive subunit), the 52-kDa G_s alpha (minor cholera toxin-sensitive subunit), and the 45-kDa G_s alpha (major cholera toxin-sensitive subunit) have been identified in both lung plasma membrane and cytosol fractions (Fig. 3). About 1% to 2% of the total membrane-bound amount of the 40- to 41-kDa G_i1, -2, -3 alpha subunits, the 52- and 45-kDa G_s alpha subunits, as well as the 36-kDa beta subunit, but not the 35-kDa beta subunit, was released from lung plasma membrane into the 200,000 × g supernatant with membrane treatment with GTPgammaS and Mg^{2+} at 37°C (Panchenko et al., 1992). The coexistence of $G_{s/i}$-protein alpha subunits and 36-kDa beta subunits in membrane-bound as well as in soluble (cytosolic) forms may reflect the in-situ distribution of G proteins throughout different cell compartments. The existence of G-protein subunits in a soluble form may be the result of their release from the plasma membrane under the receptor-mediated reorganization of G protein/cytoskeleton polydisperse structures (Huang and Devanney, 1986; Sarndahl et al., 1989; Bourguignon et al., 1990; Bengtsson et al., 1990).

Besides G_s and G_i proteins, another regulatory protein, calmodulin (in a Ca^{2+}-dependent manner) modulates activity of adenylyl cyclase in a wide

---

Figure 3 Immunochemical identification of G-protein alpha, beta subunits and ARF in lung plasma membrane and cytosol fractions. Membrane and cytosol proteins (15 μg and 100 μg per lane, respectively) were subjected to SDS-PAGE, gels were electroblotted onto nitrocellulose paper, then nitrocellulose strips were incubated with sequence-specific antibodies to different alpha, beta subunits or ARF followed by incubation with ^{125}I-labeled donkey anti-(rabbit IgG) antibodies. The ^{125}I-labeled bands were visualized by autoradiography. Antiserum specific to G_i1 alpha was obtained from J. M. Stadel (Department of Molecular Pharmacology, Smith Kline and French Laboratories, Philadelphia, PA). Antisera specific to G_i2, G_i3, G_o, and G_s alpha, 35-kDa and 36-kDa G-protein beta subunits, were provided by J. D. Robishaw (Weis Research Center, Geisinger Clinic, Danville, PA). Antiserum specific to ARF was a gift to D. M. Gill and J. Coburn (Department of Molecular Biology and Microbiology, Tufts University, Boston, MA).

variety of mammalian tissues (for review, see Bradham and Cheung, 1980). Calmodulin mediates the Ca^{2+}-dependent stimulation of adenylyl cyclase by interacting directly with the catalytic component of the enzyme (Smigel, 1986). Two isoforms of adenylyl cyclase, calmodulin-sensitive and calmodulin-insensitive, have been identified to date (Rosenberg and Storm, 1987). As is the case with the calmodulin-insensitive form, the activity of calmodulin-sensitive adenylyl cyclase is also regulated by hormones, guanine nucleotides, fluoride-aluminum complex, and forskolin. All these modulators change neither adenylyl cyclase affinity to calmodulin nor the biphasic character of Ca^{2+} action on adenylyl cyclase activity (for review, see Tkachuk and Panchenko, 1987). Magnesium ions (at millimolar concentrations) potentiate the activities of calmodulin-insensitive and calmodulin-sensitive adenylyl cyclases. Ca^{2+} at concentrations more than 10^{-5} M inhibit (not via calmodulin but in a manner competitive to Mg^{2+}-dependent stimulation of enzyme) both adenylyl cyclase isoforms but have no effect on the G_s-protein transition to the stimulated state (Tkachuk and Panchenko, 1987).

The Ca^{2+}-dependent effect of calmodulin on adenylyl cyclase activity develops rapidly (without a detectable lag phase) and does not have any effect on the G_s-protein transition into the Gpp(NH)p-stimulated state (Panchenko and Tkachuk, 1984a). Since calmodulin-sensitive adenylyl cyclase activity increases in proportion to the formation of catalytically active complexes between adenylyl cyclase and Gpp(NH)p-stimulated G_s, it was proposed that calmodulin potentiates catalytically active adenylyl cyclase holoenzyme activity by increasing its catalytic turnover numbers (Panchenko and Tkachuk, 1984a). The evidence for this mechanism was provided by reconstitution experiments. Partially purified adenylyl cyclase catalytic component (free from detectable amounts of G_s, G_i proteins, and calmodulin), as well as functionally active adenylyl cyclase holoenzyme adenylyl cyclase/Gpp(NH)p-stimulated G_s protein complex, possess a similar affinity for calmodulin (Panchenko and Tkachuk, 1984b). This speaks in favor of a suggestion that calmodulin interacts directly with the catalytic component of adenylyl cyclase and does not interfere with G proteins. Reconstitution of the catalytic component with activated G_s protein does not change the degree of the adenylyl cyclase stimulation by calmodulin; i.e., the calmodulin-dependent adenylyl cyclase activity is proportional to the amount of functionally active adenylyl cyclase holoenzyme molecules, the amount of complexes formed between the adenylyl cyclase and Gpp(NH)p-activated G_s (Fig. 4). The results obtained show that both Ca^{2+}-dependent stimulation of adenylyl cyclase (which occurs via coupling of calmodulin with the catalytic component of adenylyl cyclase) and Ca^{2+}-dependent inhibition of enzyme (which takes place via competion of Ca^{2+} with Mg^{2+} at the special regulatory site localized on the catalytic component of adenylyl cyclase) are kinetically revealed as increase or decrease,

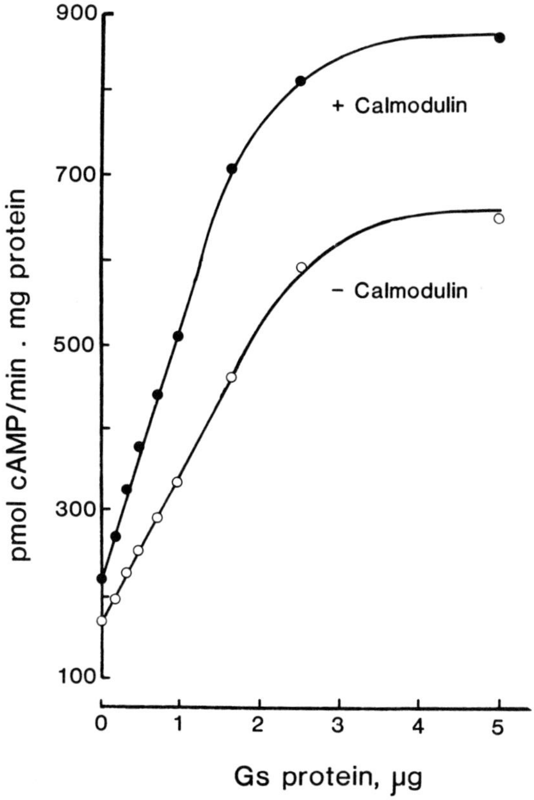

Figure 4 Calmodulin-dependent heart adenylyl cyclase activity on the formation of adenylyl cyclase holoenzyme:G_s/catalytic component complex. Prior to reconstitution with adenylyl cyclase catalytic component, the partially purified G_s protein was stimulated by incubation with 10 μM Gpp(NH)p and 5 mM Mg^{2+} for 15 min at 37°C. The protein content of the partially purified adenylyl cyclase catalytic component (in 50 μL of adenylyl cyclase assay mixture containing 10 μM of homogenous calmodulin and 1 μM of free Ca^{2+}) was 1 μg.

respectively, of the catalytic turnover numbers of the adenylyl cyclase molecules and do not alter the formation of the adenylyl cyclase holoenzyme molecules, which are due to the G_s/adenylyl cyclase complexes.

III. G-Protein Multiplicity and Regulation of Expression

G_i protein(s) not only modulate inhibition of adenylyl cyclase by hormones but also couple the receptors for Ca^{2+}-mobilizing hormones to phospholipase C, specifically hydrolyzing polyphosphoinositides. As a result, two second messengers are formed: diacylglycerol (protein kinase C activator) and inositoltrisphosphate (Ca^{2+}-mobilizing agent). In many types of cells, G_i protein(s) are also involved in phospholipase A_2 activation (Berridge, 1987; Birnbaumer et al., 1990).

Some types of membrane receptors stimulate phospholipase C or phospholipase A_2 through G proteins that are not sensitive to the pertussis toxin. These are the recently described G_q and G_{11} proteins and probably G_z protein as well (Fong et al., 1988; Smrcka et al., 1991). It is possible that low-molecular-weight GTP-binding proteins, revealed in both membrane and cytosol fractions of cells, also participate in the modulation of the regulatory effect of some hormone receptors on phospholipase C (Berridge, 1987). For example, in cultured endothelial cells, histamine and bradykinin activate polyphosphoinositide turnover in a GTP-dependent manner (Voyno-Yasenetskaya et al., 1989a, 1989b). The histamine effect on polyphosphoinositide metabolism can be blocked by pertussis toxin-dependent ADP-ribosylation of G_i2, 3 proteins, whereas that of bradykinin can be blocked by treatment with C_3 ADP-ribosyltransferase from *Clostridium botulinum.* The latter strongly correlates with ADP-ribosylation of 21- to 22-kDa low-molecular-weight GTP-binding proteins related to the Rho and/or Rac families (Voyno-Yasenetskaya et al., 1989a).

There also is a G_o protein, which resembles G_i proteins and serves as a substrate for pertussis toxin. Its function is not clear. G_o protein has 39-kDa alpha subunit with a beta/gamma complex similar to that of G_s and G_i proteins (Sternweis and Robishaw, 1984). On two-dimensional electrophoresis gels, from three to four species of the G_o protein alpha subunits have been revealed (Birnbaumer et al., 1990). This might reflect either the presence of G_o1 and G_o2 alpha subunit isoforms and/or their posttranslational modification. When reconstituted with muscarinic cholinoceptors, G_o proteins, like G_i proteins, increase receptor affinity to agonists and make the process of agonist binding to receptors sensitive to guanine nucleotides (Sternweis and Robishaw, 1984; Florio and Sternweis, 1985).

Another GTP-binding protein, transducin (G_t protein), which has molecular masses of alpha, beta, and gamma subunits similar to those of G_o protein, is the most thoroughly studied G protein. In retinal rod outer segments, transducin is responsible for transmitting the light signal from rhodopsin to cGMP-phosphodiesterase. Under experimental conditions, transducin may be reconstituted with adenylyl cyclase to activate it and also to make it sensitive to activated (illuminated) rhodopsin (Rasenick et al., 1981; Cerione et al., 1985). The alpha subunit of G_t may be ADP-ribosylated by both cholera and pertussis toxins (Abood et al., 1982; Rybin and Gureeva, 1985).

The definition of the primary structure of G proteins resulted in the discovery of the multiplicity of their alpha subunit isoforms (Freissmuth et al., 1989; Birnbaumer, 1990). Two forms of the G_s alpha subunit (52 and 45 kDa) are transcribed on one gene, but the differences in their molecular masses are the result of mRNA alternative splicing (Robishaw et al., 1986). Moreover, both "long" and "short" forms of G_s alpha subunits have been shown to exist in two subforms containing or not containing additional serine residue (Graziano et al., 1987).

Three different genes with different-sized mRNAs coding different G_i alpha subunits have been described to date (Birnbaumer et al., 1990). Three forms of G_i subunits are expressed by these genes: G_i1, G_i2, and G_i3 alpha subunits, respectively (Suki et al., 1987). Therefore it may be expected that the existing number of G_i-protein alpha subunit forms exceeds that of genes coding them. By studying of mammalian brain and sperm cDNA libraries, the complementary DNA coding of at least five more G-protein alpha subunits were discovered (Strathmann et al., 1989).

The important factor affecting the functional diversity of G proteins is the posttranslational modification of their subunits. The fact that the affinity of the recombinant 45- and 52-kDa G_s alpha subunits to the catalytic component of brain adenylyl cyclase is five to 10 times less than that of liver G_s alpha subunit suggests that the alpha subunits of G_s protein undergo posttranslational modification(s) which are crucial for their effective coupling to adenylyl cyclase (Graziano et al., 1987). Apparently, some of the posttranslational modifications of G-protein alpha subunits are under the control of sex hormones (see below).

The NAD-dependent mono-ADP-ribosylations of several G-protein alpha subunits catalyzed by ADP-ribosyltransferases of cholera or pertussis toxins are the most well established covalent modifications of G proteins (Birnbaumer et al., 1987; Gilman, 1987). In addition, the endogenous cellular ADP-ribosyltransferases which ADP-ribosylate G-protein alpha subunits have been described and purified (Moss and Vaughan, 1978; Tanuma and Endo, 1989).

As a result of ADP-ribosylation, important changes in G-protein structure occur that determine functional properties of G proteins. For example, cholera toxin-dependent ADP-ribosylation of G_s alpha subunit induces the dissociation of heterotrimeric G protein and consequent release of some of its alpha subunits from membranes (Kahn and Gilman, 1984a). As a result of cholera toxin-induced ADP-ribosylation of G_s or G_t alpha subunits, their guanine nucleotide-binding and GTP-hydrolyzing properties are also changed. In the ADP-ribosylated state, alpha subunits of G_s and G_t become "irreversible" GTP-dependent stimulators of their effectors: adenylyl cyclase and retinal cGMP phosphodiesterase, respectively (Table 1; Cassel and Pfeuffer, 1978;

Navon and Fung, 1984). In some cell types, such as adipocytes, erythrocytes, and cardiomyocytes, the existing endogenous ADP-ribosylation of G_s-protein alpha subunit modulates the basal activity of adenylyl cyclase and the degree of enzyme stimulation by agonists (Moss and Vaughan, 1978).

The pertussis toxin-catalyzed ADP-ribosylation of G_i1,2,3 G_o or G_t alpha subunits at the Cys residue in the C-terminus part of the alpha subunit uncouples G protein from the receptor, followed by the stabilization of G protein in heterotrimeric form which is inactive. In the pertussis toxin-induced ADP-ribosylated state, transducin cannot participate in rhodopsin-catalyzed GDP/Gpp(NH)p exchange (Figure 5: 83% of GDP remains tightly bound to protein), and this is followed by disappearance of its rhodopsin-dependent

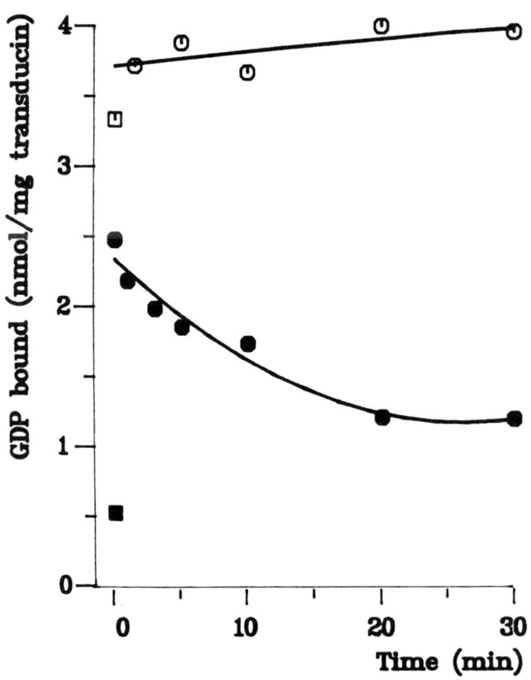

Figure 5 Time course of ^{3}H-GDP release from transducin by 1 mM Gpp(NH)p. Transducin (0.6 mg/mL) in the presence of rod outer segment membranes (3 mg/mL) and 30 μM ^{3}H-GDP was incubated for 80 min at 37°C with activated pertussis toxin without (open symbols) or with 0.2 mM NAD (closed symbols). Then, at 4°C, Gpp(NH)p was added to a final concentration of 1 mM, and a indicated times aliquots were withdrawn and assayed for ^{3}H-GDP bound to transducin (circles). The last aliquot was warmed at 37°C for 5 min and assayed for ^{3}H-GDP bound to transducin (squares).

GTPase activity and its ability to stimulate retinal cGMP phosphodiesterase (Rybin et al., 1989).

The other type of covalent modification that G proteins undergo is the phosphorylation of G-protein alpha subunits by protein kinase C as well as several tyrosyl kinases (Katada et al., 1985; O'Brien et al., 1987). It has been shown that the level of endogenous protein kinase C-dependent phosphorylation of G_i2-protein in rat hepatocytes is increased in experimentally induced diabetes. Apparently, this modification of G_i proteins as well as the reduction of G_i2 protein expression are the main reasons for the attenuation of the G_i-mediated pathway on hepatocyte adenylyl cyclase that occurs in diabetes (Bushfield et al., 1990).

Acylation and/or lipidation modifications of G-protein alpha and gamma subunits have also been established. For example, the myristoylations of G_i and G_o alpha subunits at the N terminus have been well characterized (Buss et al., 1987). On the other hand, the Cys residue of the gamma subunit located on the C-terminal region of the protein is able to undergo several step-by-step modifications. First, a specific endopeptidase splits off three amino acid residues from the C terminus of the gamma subunit. Then carboxymethylation of opened Cys occurs (Yamane et al., 1990). In addition, lipidation of the C-terminal Cys residue is also characteristic for several G-protein gamma subunits. For instance, modification with C_{20}-all-*trans*-geranylgeranyl residue of brain gamma subunit or modification with C_{15}-farnesyl residue of yeast gamma subunit has been demonstrated (Hancock et al., 1989; Yamane et al., 1990). These modifications may be essential in the association of the G-protein heterotrimer with membranes, since nonmodified gamma and alpha subunits are hydrophilic.

As has already been noted, the dissociation of G protein to a free alpha subunit and beta/gamma complex is crucial in G-protein-mediated signaling. The specificity of G protein (with respect to both the receptor as well as the effector) is defined predominantly by the type of alpha subunit, while the beta/gamma complex is relatively similar in all G proteins. At the same time, at least four types of G-protein beta subunits, differing in both their molecular mass (36 and 35 kDa) and their structure, have been defined (Boege et al., 1991). Also, not less than six or seven types of gamma subunits, differing in structure and properties, have been described (Gautam et al., 1990). Taking into account that more than a dozen of various forms of G-protein alpha subunits are involved in signal transduction, it would not be difficult to propose the existence at least 100 alpha/beta/gamma-heterotrimeric compositions that can be created in membranes. This number of G-protein heterotrimeric compositions is close to the number (about 90 to date) of membrane receptor types known to be coupled to G proteins. Therefore, it is possible to suggest that each membrane receptor type is coupled to its strictly defined heterotrimeric G protein, which ensures the "individuality" of each regulatory signal.

The receptors coupled to G proteins may affect, via different second messenger systems, not only cell metabolism, secretion, or contraction, but also cell proliferation, differentiation, morphology, etc. In turn, expression, posttranslational modification, and compartmentalization of membrane receptors, G proteins, and/or effectors depend on the functional state of the cell (stage of cell cycle, intensity of protein synthesis, cytoskeleton organization, etc.) and are controlled by steroid and thyroid hormones.

Thyroid and steroid hormones possess permissive effects on the action of hormones coupled to adenylyl cyclase. For example, under hypothyroid conditions, the beta-adrenergic agonist-dependent stimulation of adipocyte adenylyl cyclase is reduced, while the adenosine-dependent inhibition of the enzyme is potentiated (Ros et al., 1988; Mazurkiewicz and Saggerson, 1989). Glucocorticoids in a number of tissues also modulate the effects of different stimulators on adenylyl cyclase (Ros et al., 1989). During the past five years it has been shown that thyroid hormones in adipocytes (Ros et al., 1988; Rapiejko et al., 1989; Milligan and Saggerson, 1990), brain (Mazurkiewicz and Saggerson, 1989), and glucocorticoids in brain (Saito et al., 1989), hepatocytes (Garcia-Sainz et al., 1989), and adipocytes (Ros et al., 1989) modulate the expression of G proteins coupled to adenylyl cyclase.

One should underline species and tissue specificity of the effects of thyroid and steroid hormones. Thyroidectomy does not lead to essential changes in the contents of G protein subunits in rat hepatocytes (Garcia-Sainz et al., 1989) or in lung (Table 4), whereas adrenalectomy is associated in tissues and cells with different (sometimes contrasting) changes in G-protein level (Table 4; Garcia-Sainz et al., 1989; Ros et al., 1989). Probably the modulatory effects of thyroid and glucocorticoid hormones on the G-protein level are nondirect and are a result of tissue-specific transcription factors or tissue-specific post-transcription levels of G-protein synthesis.

According to our data, a change of estrogen status in rats results in a change in the number of the G_i2-protein alpha subunit subforms (Nickashin et al., 1992). In plasma membranes isolated from the lungs of ovariectomized rats, in addition to the well-known 40-kDa G_i2 alpha subunit, an additional 39-kDa protein is recognized by antiserum specific for G_i2 alpha subunit (Figure 6). This 39-kDa G_i2 alpha-like protein is not present in lung plasma membrane isolated from ovariectomized rats which have been injected with the estrogen analog, diethylstilbestrol. It is important that the 39-kDa G_i2 alpha-like protein is not ADP-ribosylated by pertussis toxin. This may be due to the absence (or blockage) in its primary structure of the "pertussis toxin-sensitive" Cys residue or/and to its low affinity to beta/gamma complex. The change of G_i2 alpha subunit number is very specific, since ovariectomy in rats does not

Table 4 Levels of G-Protein Subunits in Membranes from Several Tissues of Rats with Different Status of Thyroid or Steroid Hormones

G-protein subunit type	Lung		Heart		Kidney	
	Thyroidectomy	Adrenalectomy	Thyroidectomy	Adrenalectomy	Thyroidectomy	Adrenalectomy
$G\alpha_i2$	121 ± 6	116 ± 3	123 ± 26	174 ± 29	137 ± 2	131 ± 1
$G\alpha_i3$	93 ± 13	74 ± 4	n.d.	n.d.	41 ± 4	154 ± 4
$G\alpha_s45$	135 ± 10	122 ± 20	77 ± 15	67 ± 1	137 ± 8	112 ± 2
$G\alpha_s52$	103 ± 3	125 ± 2	n.d.	n.d.	159 ± 29	122 ± 9
$G\beta36$	104 ± 11	127 ± 4	122 ± 24	230 ± 20	116 ± 6	113 ± 1
$G\beta35$	107 ± 15	108 ± 5	n.d.	n.d.	n.d.	n.d.

Lung, heart, and kidney plasma membranes from thyroidectomized or adrenalectomized and control rats were isolated and then subjected to SDS/polyacrylamide gel electrophoresis. Immunoblottings of the resolved proteins were performed with an anti-G-protein antipeptide antisera, and a second ^{125}I-labeled goat anti-rabbit IgG. The level of each G-protein subunit in each tissue of control rats was considered to be 100%, and the results are expressed as the percentage of these values (means ± S.E.M. for three experiments). Anti-G_s antiserum 584 and anti-G 36 antiserum U49 were provided by Alfred G. Gilman (University of Texas Health Science Center, Dallas); anti-G_i2 antiserum A54, anti-G_i3 antiserum A56, and anti-G 35 antiserum GC2 were provided by Janet D. Robishaw (Weis Center for Research, Geisinger Clinic, Danville, PA).

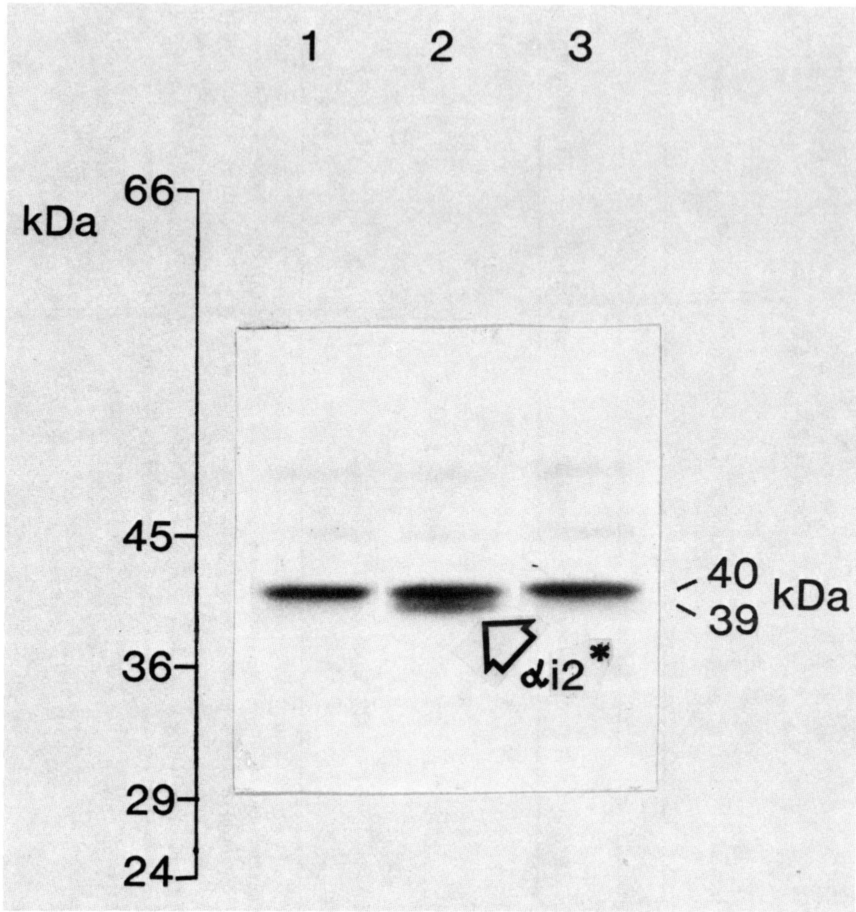

Figure 6 Anti-G alpha$_i$2 peptide antiserum immunoreactive proteins in lung membranes from rats with different status of estrogens. Membranes isolated from control (1), ovariectomized (2), and ovariectomized, diethylstilbestrol-treated (3) rats were subjected to SDS/polyacrylamide gel electrophoresis and then to immunoblotting with anti-G alpha$_i$2 antiserum and at second ^{125}I-labeled goat anti-rabbit IgG. Autoradiogram of the blot is shown. Anti-G alpha$_i$2 antiserum A54 was a gift of Janet D. Robishaw (Weis Center for Research, Geisinger Clinic, Danville, PA).

result in the appearance or disappearance of the other G-protein alpha subunits, G_i3 or G_s alpha. The castration of male rats also did not lead to the appearance of 39-kDa G_i2 alpha-like protein (Nickashin et al., 1992). It should be noted that the above-mentioned appearance of 39-kDa G_i2 alpha-like protein may be the result of specific covalent modification (limited proteolysis, acylation, lipidation of C terminus, etc.) of the "classical" 40-kDa G_i2 alpha subunit normally suppressed by estrogens or the result of the derepression in the absence of estrogens of the specific gene coding the special 39-kDa G_i2 alpha-like protein.

IV. Is The Na/K-ATPase One More Effector System Regulated by G Protein(s)?

Muscarinic cholinergic receptors are coupled to various cellular effector systems via the family of regulatory proteins. In cardiac sarcolemma, the G_i protein mediates muscarinic agonist-induced inhibition of adenylyl cyclase and opening of K^+ channels, whereas the G_o protein probably participates in the opening of Na^+ channels. The pertussis toxin-insensitive G protein transduces the activating signal from muscarinic cholinergic receptor to phospholipase C (specific for polyphosphoinositides) (for review, see Berridge, 1987; Gilman, 1987; Birnbaumer et al., 1990).

Our preliminary experiments have shown that, at micromolar concentrations, acetylcholine inhibits Na/K-ATPase activity, an enzymatic system of Na^+ pumping, in the sarcolemma of skeletal muscle, myocardium, and smooth muscle (Severin et al., 1974; Tkachuk et al., 1975). The cholinergic agonist-induced inhibition of sarcolemmal Na/K-ATPase is blocked by the receptor antagonist, atropine, and enhanced by micromolar concentrations of GTP. The stable analog of GTP, GTPgammaS, inhibits the enzyme in the absence of the agonist. Taken together, the above data allowed us to suggest that G protein(s) may be involved in the coupling between muscarinic cholinergic receptors and Na/K-ATPase. We have attempted to test this hypothesis by investigating the effect of carbachol, a stable acetylcholine analog, on oubain-sensitive K^+-activated 4-nitrophenylphosphatase (K^+-p-NPPase), which is a partial reaction of Na/K-ATPase. This enzymatic activity, which can be determined without ATP, was assayed to exclude (1) the effects of GTP contamination in commercial ATP preparations and/or nucleosidediphosphokinase-mediated phosphorylation of G protein-bound GDP to form GTP, and (2) possible modulatory effects of ATP-dependent endogenous sarcolemmal protein kinase-induced phosphorylations of individual components in the muscarinic cholinergic receptor/G-protein/Na/K-ATPase signal-transduction pathway.

Using highly purified light sarcolemma preparations isolated from rabbit and pig ventricular myocardium, we have established that carbachol (in a micromolar concentration range) inhibits K^+-p-NPPase activity only when GTP or GTPgammaS (in micromolar concentrations) is added to the incubation medium (Fig. 7; Danilenko et al., 1990). The carbachol-induced inhibition was blocked by the antagonist atropine. The GDP analog GDPbetaS was not able to promote this action of carbachol. GTPgammaS alone produces a concentration-dependent decrease in K^+-p-NPPase activity with a lag phase which is eliminated by the carbachol (Danilenko et al., 1990). The GTPgammaS-promoted effect of carbachol is abolished in sarcolemma treated with pertussis toxin, which catalyzes ADP-ribosylation of alpha subunits of G_i2 and G_i3 of sarcolemma (Fig. 8; Danilenko et al., 1990). In agreement with the currently accepted criteria for participation of G proteins in hormonal signal transduction (Gilman, 1987), our data provide evidence that the G_i protein may be involved directly in coupling between muscarinic cholinergic receptors and Na/K-ATPase in mammalian cardiac sarcolemma.

To prove the hypothesis stated above, we also tested the G_i-dependent cholinergic regulation of the overall Na/K-ATPase activity in cardiac sarcolemma. According to data obtained, the inhibitory action of carbachol on

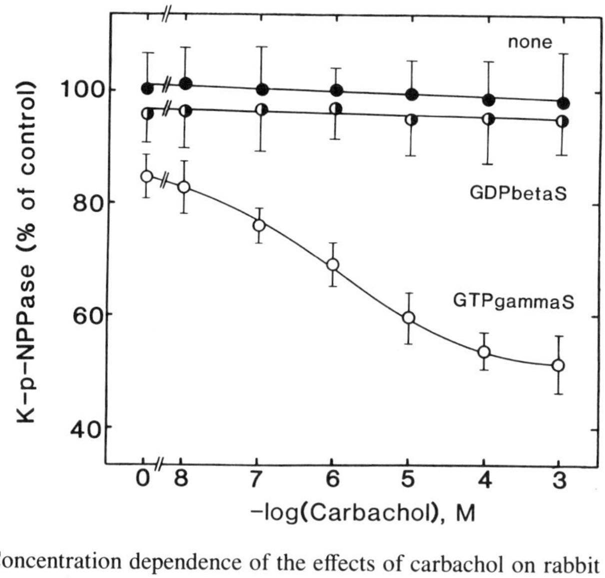

Figure 7 Concentration dependence of the effects of carbachol on rabbit sarcolemmal ouabain-sensitive K^+-p-NPPase activity in the absence of nucleotides and in the presence of 10 μM GDPbetaS or 1 μM GTPgammaS. Enzyme activity was measured as in Danilenko et al. (1990).

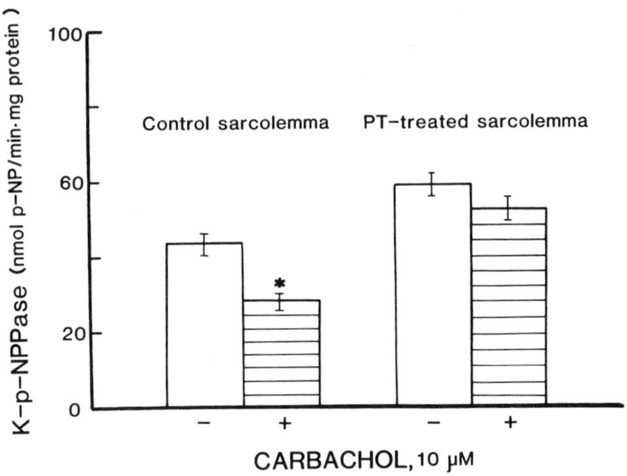

Figure 8 Effect of pertussis toxin on carbachol-induced GTPgammaS-dependent inhibition of K^+-p-NPPase activity in cardiac sarcolemma. Rabbit cardiac sarcolemma was treated with or without pertussis toxin and enzyme activity was measured as in Danilenko et al. (1991).

ouabain-sensitive Na/K-ATPase activity (measured with ATP free from GTP contamination) depended strictly on the presence of micromolar concentrations of GTPgammaS and was blocked in the presence of atropine. On the other hand, carbachol increased markedly (10 times) the apparent affinity of the Na/K-ATPase for GTPgammaS (Danilenko et al., 1991).

There is a growing body of evidence that G proteins participate in hormonal regulation of a number of plasma membrane ion-transporting systems, such as K^+ and Ca^{2+} channels in cardiac myocytes and the Ca^{2+}-pumping ATPase in hepatocytes and cardiac myocytes (for review, see Birnbaumer et al., 1990). Recently, it has been shown that the receptor-mediated inhibition of Na/K-ATPase activity in kidney proximal tubules by dopamine involves a pertussis toxin-sensitive G protein and activation of protein kinase C (Bertorello and Aperia, 1989). Also, it has been reported that the alpha$_1$-adrenergic agonist, phenylephrine, is able to stimulate Na^+ pump activity in cardiac Purkinje fibers in a pertussis toxin-sensitive manner (Shah et al., 1988). In order to avoid the modulatory action of endogenous sarcolemmal protein kinases in the inhibitory effect of muscarinic cholinergic agonist on Na/K-ATPase activity in cardiac sarcolemma, we have compared the modulation of K^+-p-NPPase activity by carbachol with different assay conditions. Since, in our hands, carbachol elicits a similar inhibitory effect on K^+-p-NPPase activity in the absence or in the presence of ATP as well as in the absence or in the presence of the Ca^{2+} chelator EGTA, it seems unlikely that the inhibition is mediated by phosphorylation of membrane proteins that are involved in the transduction of G_i-mediated signal from muscarinic cholinergic receptor to Na/K-ATPase. Moreover, data obtained strongly suggest that GTP or its analog, but not ATP, is essential for the inhibitory effect of carbachol on cardiac Na/K-ATPase activity. Since the cardiac G protein which couples muscarinic cholinergic receptors to phospholipase C is insensitive to pertussis toxin (Masters et al., 1985), it is tempting to speculate that the pertussis toxin-sensitive inhibitory effect of carbachol on Na/K-ATPase activity is not related to the activation of inositol phospholipid hydrolysis in sarcolemma.

In agreement with the hypothesis of direct coupling of Na/K-ATPase with the G_i protein are the results of Na/K-ATPase and G_i protein reconstitution experiments. We were able to separate the Na/K-ATPase and G_i protein activities by gel filtration from a detergent extract of cardiac sarcolemma (Danilenko et al., 1991). The separation of Na/K-ATPase from G_i protein abolished the inhibition of enzyme activity by GTP or its stable analogs, while the reconstitution of Na/K-ATPase with pertussis toxin-sensitive G_i in liposomes resulted in restoration of enzyme sensitivity to the inhibitory action of micromolar concentrations of guanine nucleotides (Table 5; Danilenko et al.,

Table 5 Reconstitution of Soluble Na/K-ATPase with Soluble G_i Protein in Liposomes

Nucleotides	Na/K-ATPase activity (μmol P_i/h mg protein) Na/K-ATPase without G_i protein	 Na/K-ATPase + G_i protein
No additions	67.7 ± 6.2	238 ± 12
GTP, 10 μM	70.9 ± 6.7	213 ± 19
GDP, 10 μM	69.1 ± 7.0	241 ± 10
Gpp(NH)p, 10 μM	74.8 ± 8.1	161 ± 14
Gpp(CH_2)p, 10 μM	69.3 ± 6.9	188 ± 10
GTPgammaS, 10 μM	76.2 ± 9.0	176 ± 12
GDPbetaS, 10 μM	61.5 ± 4.3	211 ± 17

Cardiac sarcolemma was solubilized with Chaps, then Na/K-ATPase and pertussis toxin-sensitive G_i-protein activities were separated into two different fractions by gel filtration, reconstituted into common liposomes, followed by measurement of the Na/K-ATPase activity (Danilenko et al., 1991). The threefold-higher specific activity of the Na/K-ATPase reconstituted into liposomes than the activity of the resolved delipidated Na/K-ATPase is due to the relipidation of the enzyme in liposomes but is not due to the Na/K-ATPase interaction with the G_i protein in liposomes.

1991). The strong dependency of the G_i-protein-mediated ATPase inhibition on nonhydrolizable GTP analogs indicates that the "active" GTP form (conformation) of the G_i alpha and/or its free beta/gamma complex affects the activity of NA/K-ATPase.

There is convincing evidence that G proteins in cardiac tissue mediate muscarinic cholinergic receptor agonist-induced inhibition of adenylyl cyclase, opening of K^+ and Na^+ channels, and activation of inositol-phospholipid-specific phospholipase C (for review, see Birnbaumer et al., 1990). Our results suggest that in ventricular myocardium the agonist-induced activation of muscarinic cholinergic receptor may also lead to the inhibition of Na/K-ATPase activity via the direct coupling of the enzyme to the G_i protein. There, Na/K-ATPase may be one more effector system involved in muscarinic cholinergic receptor signaling in cardiac myocytes.

Acknowledgments

We thank our collaborators, Ella E. Illarionova, Michail Y. Men'shikov, Grigory N. Baldenkov, Pavel V. Avdonin, Elena G. Chekneva, George Y. Grigorian, Alexei V. Nickashin, Irina V. Svitina-Ulitina, Marina G. Starikova, Eugenia V. Nupenko, Alexander V. Grishin, Simon I. Hoffenberg, and Tatyana A. Voyno-Yasenetskaya, without whose help many of the results from our laboratory would have not been obtained. From outside the Cardiology Research Center, we thank Mikhail P. Danilenko, Oleg V. Yesirev, Vera C. Turmukhambetova, Olga D. Lopina, Alexander A. Boldyrev, Sergei E. Severin (studies with K^+-p-NPPase and Na/K-ATPase); Nikolay B. Gusev and Alexander V. Vorotnikov (for supplying purified brain protein kinase C); Alexander I. Ephimenko, Klaus Aktories, and D. Michael Gill (for generous gifts of C_3 ADP-ribosyltransferase preparations); Therese J. Resink and Fritz R. Buhler (studies on ANF regulation of lung adenylyl cyclase); Victor B. Rosen (for supplying rats with different status of steroid and thyroid hormones); Alfred G. Gilman, Suzanne M. Mumby, Janet D. Robishaw, D. Michael Gill, and Jeffrey M. Stadel (for generous gifts of anti-peptide G-protein antisera).

References

Abood ME, Hurley JB, Pappone M-C, Bourne HR, Stryer L. Functional homology between signal-coupling proteins: cholera toxin inactivates the GTPase activity of transducin. J Biol Chem 1982; 257:10540–3.

Amatruda TT III, Gautam N, Fong HKW, Nothup JK, Simon MI. The 35- and 36-kDa beta subunits of GTP-binding proteins are products of separate genes. J Biol Chem 1988; 263:5008–11.

Avdonin PV, Svitina-Ulitina IV, Tkachuk VA. Selective inactivation by endogenous protein kinase C of human platelet high-affinity GTPase coupled with PAF receptors. J Mol Cell Cardiol 1989; 21 (suppl. I):139–43.
Bell JD, Brunton LL. Enhancement of adenylate cyclase activity in S49 lymphoma cells by phorbol esters. Withdrawal of GTP-dependent inhibition. J Biol Chem 1986; 261:12036–41.
Bengtsson T, Sarndahl E, Stendahl O, Andersson T. Involvement of GTP-binding proteins in actin polymerization in human neutrophils. Proc Natl Acad Sci USA 1990; 87:2921-5.
Berridge MJ. Inositol triphosphate and diacylglycerol: two interacting second messengers. Ann Rev Biochem 1987; 56:159–93.
Bertorello A, Aperia A. Am J Physiol 1989; 256:F370–3.
Birnbaumer L. G proteins in signal transduction. Ann Rev Pharmacol Toxicol 1990; 30:675–705.
Birnbaumer L, Abramowitz J, Brown AM. Receptor-effector coupling by G proteins. Biochim Biophys Acta 1990; 1031:163–224.
Birnbaumer L, Codina J, Mattera R, Yatani A, Scherer N, Toro, M-J, Brown AM. Signal transduction by G proteins. Kidney Int 1987; 32:S14–37.
Boege F, Neumann E, Helmreich EJM. Structural heterogeneity of membrane receptors and GTP-binding proteins and its functional consequences for signal transduction. Eur J Biochem 1991; 199:1–15.
Bokoch GM. The presence of free G protein beta/gamma subunits in human neutrophils results in suppression of adenylate cyclase activity. J Biol Chem 1987; 262:589–94.
Bourguignon LYW, Walker G, Huang HS. Interactions between a lymphoma membrane-associated guanosine 5′-triphosphate-binding protein and the cytoskeleton during receptor patching and capping. J Immunol 1990; 144:2242–52.
Bradham LS, Cheung WY. Calmodulin-dependent adenylate cyclase. In: Cheung WY, ed. Calcium and cell function. Calmodulin. Vol. 1. New York: Academic Press, 1980; 109–26.
Burns DL, Moss J, Vaughan M. Choleragen-stimulated release of guanyl nucleotides from turkey erythrocyte membranes. J Biol Chem 1982; 257:32–4.
Bushfield M, Griffiths SL, Murphy GJ, Pyne NJ, Knowler JT, Milligan G, Parker PJ, Mollner S, Houslay MD. Diabetes-induced alterations in the expression, functioning and phosphorylation state of the inhibitory guanine nucleotide regulatory protein G_I-2 in hepatocytes. Biochem J 1990; 271:365–72.
Buss JE, Mumby SE, Casey PJ, Gilman AG, Sefton BM. Myristoylated alpha subunits of guanine nucleotide-binding regulatory proteins. Proc Natl Acad Sci USA 1987; 84:7493–7.
Carlson KE, Brass LF, Manning DR. Thrombin and phorbol esters cause the selective phosphorylation of G protein other than G_i in human platelets. J Biol Chem 1989; 264:13298–305.
Cassel D, Pfeuffer T. Mechanism of cholera toxin action: covalent modification of the guanyl nucleotide-binding protein of the adenylate cyclase system. Proc Natl Acad Sci USA 1978; 75:2669–73.

Cassel D, Selinger Z. Mechanism of adenylate cyclase activation by cholera toxin: inhibition of GTP hydrolysis at the regulatory site. Proc Natl Acad Sci USA 1977; 74:3307–11.

Cerione RA, Staniszenski C, Benovic JL, Lefkowitz RJ, Caron MG, Gierschik P, Somers R, Spiegel AM, Codina J, Birnbaumer L. Specificity of the functional interactions of the beta-adrenergic receptor and rhodopsin with guanine nucleotide regulatory proteins reconstituted in phospholipid vesicles. J Biol Chem 1985; 260:1493–500.

Danilenko MP, Panchenko MP, Yesyrev OV, Tkachuk VA. Involvement of pertussis-toxin-sensitive G protein in muscarinic-receptor-mediated inhibition of K^+-activated 4-nitrophenylphosphatase activity of cardiac sarcolemma. Eur J Biochem 1990; 194:155–60.

Danilenko MP, Turmukhambetova VC, Yesyrev OV, Tkachuk VA, Panchenko MP. Na^+/K^+-ATPase-G protein coupling in myocardial sarcolemma: separation and reconstitution. Am J Physiol 1991; 261:87–91.

Enomoto K, Gill DM. Cholera toxin activation of adenylate cyclase. Roles of nucleoside triphosphates and a macromolecular factor in the ADP ribosylation of the GTP-dependent regulatory component. J Biol Chem 1980; 255:1252–8.

Florio VA, Sternweis PC. Reconstitution of resolved muscarinic cholinergic receptors with purified GTP-binding proteins. J Biol Chem 1985; 260:3477–83.

Fong HKW, Yoshimoto KK, Eversole-Cire P, Simon MI. Identification of a GTP-binding protein alpha subunit that lacks an apparent ADP-ribosylation site for pertussis toxin. Proc Natl Acad Sci USA 1988; 85:3066–70.

Freissmuth M, Casey PJ, Gilman AG. G proteins control diverse pathways of transmembrane signaling. FASEB J 1989; 3:2125–31.

Garcia-Sainz JA, Huerta-Bahena ME, Malbon CC. Hepatocyte beta-adrenergic responsiveness and guanine nucleotide-binding regulatory proteins. Am J Physiol 1989; 256:C384–9.

Gautam N, Northup J, Tamir H, Simon MI. G protein diversity is increased by associations with a variety of gamma subunits. Proc Natl Acad Sci USA 1990; 87:7973–7.

Gill DM, Coburn J. ADP-ribosylation by cholera toxin: functional analysis of a cellular system that stimulates the enzymic activity of cholera toxin fragment A_1. Biochemistry 1987; 26:6364–71.

Gill DM, Meren R. A second nucleotide-binding site associated with adenylate cyclase. Distinct nucleotides activate adenylate cyclase and permit ADP-ribosylation by cholera toxin. J Biol Chem 1983; 258:11908–14.

Gilman, AG. G proteins: transducers of receptor-generated signals. Ann Rev Biochem 1987; 56:615–49.

Graziano MP, Casey PJ, Gilman AG. Expression of cDNAs for G proteins in *Escherichia coli.* Two forms of $G_{s\,alpha}$ stimulate adenylate cyclase. J Biol Chem 1987; 262:11375–81.

Graziano MP, Freissmuth M, Gilman A. Expression of $G_{s\,alpha}$ in *Escherichia coli.* Purification and properties of two forms of the protein. J Biol Chem 1989; 264:409–18.

Hancock JF, Magee AI, Childs JE, Marshall CJ. All ras proteins are polyisoprenylated but only some are palmitoylated. Cell 1989; 57:1167–77.

Hildebrandt JD, Kohnken RE. Hormone inhibition of adenylyl cyclase. Differences in the mechanisms for inhibition by hormones and G protein beta/gamma. J Biol Chem 1990; 265:9825–30.

Hildebrandt JD, Codina J, Birnbaumer L. Interaction of the stimulatory and inhibitory regulatory proteins of the adenylyl cyclase system with the catalytic component of cyc$^-$ S49 cell membranes. J Biol Chem 1984; 259:13178–85.

Huang C-K, Devanney JF. GTPgammaS-induced solubilization of actin and myosin from rabbit peritoneal neutrophil membrane. FEBS Lett 1986; 202:41–4.

Kahn RA, Gilman AG. ADP-ribosylation of G_s promotes the dissociation of its alpha and beta subunits. J Biol Chem 1984a; 259:6235–40.

Kahn RA, Gilman AG. Purification of a protein cofactor required for ADP-ribosylation of the stimulatory regulatory component of adenylate cyclase by cholera toxin. J Biol Chem 1984b; 259:6228–34.

Kahn RA, Gilman AG. The protein cofactor necessary for ADP-ribosylation of G_S by cholera toxin is itself a GTP binding protein. J. Biol Chem 1986; 261:7906–11.

Katada T, Ui M. Direct modification of the membrane adenylate cyclase system by islet-activating protein due to ADP-ribosylation of a membrane protein. Proc Natl Acad Sci USA 1982; 79:3129–33.

Katada T, Bokoch GM, Smigel MD, Ui M, Gilman AG. The inhibitory guanine nucleotide-binding regulatory component of adenylate cyclase. Subunit dissociation and the inhibition of adenylate cyclase in S49 lymphoma cyc$^-$ and wild type membranes. J Biol Chem 1984a; 259:3586–95.

Katada T, Gilman AG, Watanabe Y, Bauer S, Jakobs KH. Protein kinase C phosphorylates the inhibitory guanine-nucleotide-binding regulatory component and apparently suppresses its function in hormonal inhibition of adenylate cyclase. Eur J Biochem 1985; 151:431–7.

Katada T, Kusakabe K, Oinuma M, Ui M. A novel mechanism for the inhibition of adenylate cyclase via inhibitory GTP-binding proteins. Calmodulin-dependent inhibition of the cyclase catalyst by the beta/gamma-subunits of GTP-binding proteins. J Biol Chem 1987; 262:11897–900.

Katada T, Northup JK, Bokoch GM, Ui M, Gilman AG. The inhibitory guanine nucleotide-binding regulatory component of adenylate cyclase. Subunit dissociation and guanine nucleotide-dependent hormonal inhibition. J Biol Chem 1984b; 259:3578–85.

Katada T, Oinuma M, Ui M. Two guanine nucleotide-binding proteins in rat brain serving as the specific substrate of islet-activating protein, pertussis toxin. Interaction of the alpha-subunits with beta/gamma-subunits in development of their biological activities. J Biol Chem 1986; 261:8182–91.

Kurose H, Ui M. Functional uncoupling of muscarinic receptors from adenylate cyclase in rat cardiac membranes by the active component of islet-activating protein, pertussis toxin. J Cyclic Nucleotide Protein Phosphoryl 1983; 9:305–18.

Lad PM, Nielsen TB, Preston MS, Rodbell M. The role of the guanine nucleotide exchange reaction in the regulation of the beta-adrenergic receptor and in the actions of catecholamines and cholera toxin on adenylate cyclase in turkey erythrocyte membranes. J Biol Chem 1980; 255:988–95.

Linder ME, Ewald DA, Miller RJ, Gilman AG. Purification and characterization of $G_{o\,alpha}$ and three types of $G_{i\,alpha}$ after expression in *Escherichia coli.* J Biol Chem 1990; 265:8243–51.

Masters SB, Martin MW, Harden TK, Brown JH. Pertussis toxin does not inhibit muscarinic-receptor-mediated phosphoinositide hydrolysis or calcium mobilization. Biochem J 1985; 227:933–7.

Mazurkiewicz D, Saggerson D. Inhibition of adenylate cyclase in rat brain synaptosomal membranes by GTP and phenylisopropyladenosine is enhanced in hypothyroidism. Biochem J 1989; 263:829–35.

McKenzie FR, Milligan G. Opioid-receptor-mediated inhibition of adenylate cyclase is transduced specifically by the guanine-nucleotide-binding protein G_i2. Biochem J 1990; 267:391–8.

Milligan G, Saggerson ED. Concurrent up-regulation of guanine-nucleotide-binding proteins G_{i1}alpha, G_{i2}alpha and G_{i3}alpha in adipocytes of hypothyroid rats. Biochem J 1990; 270:765–9.

Morris D, McHugh-Sutkowski E, Moos M, Simonds WF, Spiegel AM, Seamon KB. Immunoprecipitation of adenylate cyclase with an antibody to a carboxyl-terminal peptide from $G_{s\,alpha}$. Biochemistry 1990; 29:9079–84.

Moss J, Vaughan M. Isolation of an avian erythrocyte protein possessing ADP-ribosyltransferase activity and capable of activating adenylate cyclase. Proc Natl Acad Sci USA 1978; 75:3621–4.

Naghshineh S, Noguchi M, Huang K-P, Londos C. Activation of adipocyte cyclase by protein kinase C. J Biol Chem 1986; 261:14534–8.

Navon SE, Fung BK-K. Characterization of transducin from bovine retinal rod outer segments. Mechanism and effects of cholera toxin-catalyzed ADP-ribosylation. J Biol Chem 1984; 259:6686–93.

Neer EJ, Lok JM, Wolf LG. Purification and properties of the inhibitory guanine nucleotide regulatory unit of brain adenylate cyclase. J Biol Chem 1984; 259:14222–9.

Nickashin AV, Rybin VO, Tkachuk VA. 1992; in preparation.

O'Brien RM, Houslay MD, Milligan G, Siddle K. The insulin receptor tyrosyl kinase phosphorylates holomeric forms of the guanine nucleotide regulatory proteins G_i and G_o. FEBS Lett 1987; 212:281–8.

Panchenko MP, Tkachuk VA. Calmodulin activates adenylate cyclase from rabbit heart plasma membranes. FEBS Lett 1984a; 174:50–4.

Panchenko MP, Tkachuk VA. Participation of calmodulin and regulatory guanine nucleotide binding component in the regulation of cardiac adenylyl cyclase. Doklady Akademii Nauk SSSR 1984b; 275:1219–23 (in Russian).

Panchenko MP, Starikova MG, Grishin AV, Tkachuk VA. 1992; in preparation.

Rapiejko PJ, Watkins DC, Ros M, and Malbon CC. Thyroid hormones regulate G-protein beta-subunit mRNA expression in vivo. J Biol Chem 1989; 264:16183–9.

Rasenick MM, Stein PJ, Bitensky MW. Regulatory subunit of adenylate cyclase interacts with cytoskeletal components. Nature 1981; 294:560–2.

Robishaw JD, Smigel MD, Gilman AG. Molecular basis for two forms of the G protein that stimulates adenylate cyclase. J Biol Chem 1986; 261:9587–90.

Ros M, Northup JK, Malbon CC. Steady-state levels of G-proteins and beta-adrenergic receptors in rat fat cells. Permissive effects of thyroid hormones. J Biol Chem 1988; 263:4362–8.

Ros M, Northup JK, Malbon CC. Adipocyte G-proteins and adenylate cyclase. Effects of adrenylectomy. Biochem J 1989; 257:737–44.

Rosenberg GB, Storm DR. Immunological distinction between calmodulin-sensitive and calmodulin-insensitive adenylate cyclases. J Biol Chem 1987; 262:7623–8.

Rosenthal W, Binder T, Schultz G. NADP efficiently inhibits endogenous but not pertussis toxin-catalyzed covalent modification of membrane proteins incubated with NAD. FEBS Lett 1987; 211:137–43.

Rudolph U, Schultz G, Rosenthal W. The cytosolic G-protein alpha-subunit in human neutrophils responds to treatment with guanine nucleotides and magnesium. FEBS Lett 1989; 251:137–42.

Rybin VO, Gureeva AA. Influence of bacterial toxins on GTPase activity of transducin from bovine retinal rod outer segments. Biokhimia 1985; 50:1556–65 (in Russian).

Rybin VO, Gureeva AA, Tkachuk VA. Inhibition by pertussis toxin of guanyl nucleotides exchange on transducin in bovine rod cell membranes. Membrane Biochem 1989; 8:115–26.

Saito N, Guitart X, Hayward M, Tallman JF, Duman RS, Nestler EJ. Corticosterone differentially regulates the expression of $G_s$alpha and $G_i$alpha messenger RNA and protein in rat cerebral cortex. Proc Natl Acad Sci USA 1989; 86:3906–10.

Sarndahl E, Linddroth M, Bengtsson T, Fallman M, Gustavsson J, Stendahl O, Andersson T. Association of ligand-receptor complexes with actin filaments in human neutrophils: a possible role for a G-protein. J Cell Biol 1989; 109:2791–9.

Schmidt CJ, Neer EJ. In vitro synthesis of G protein beta/gamma dimers. J Biol Chem 1991; 266:4538–44.

Severin SE, Boldyrev AA, Tkachuk VA. Some properties of Na^+/K^+-stimulated ATPase from normal and denervated muscles of rabbit with a special reference to its sensitivity to acetylcholine. Comp Gen Pharmacol 1974; 5:181–5.

Shah A, Cohen IS, Rosen MR. Stimulation of cardiac alpha receptors increases Na/K pump current and decreases g_k via a pertussis toxin-sensitive pathway. Biophys J 1988; 54:219–25.

Simonds WF, Goldsmith PK, Codina J, Unson CG, Spiegel AM. G_{i2} mediates alpha$_2$-adrenergic inhibition of adenylyl cyclase in platelet membranes: in situ identification with G_{alpha} C-terminal antibodies. Proc Natl Acad Sci USA 1989a; 86:7809–13.

Simonds WF, Goldsmith PK, Woodard CJ, Unson CC, Spiegel AM. Receptor and effector interactions of G_s. Functional studies with antibodies ti the alpha$_s$ carboxyl-terminal decapeptide. FEBS Lett 1989b; 249:189–94.

Smigel MD. Purification of the catalyst of adenylate cyclase. J Biol Chem 1986; 261:1976–82.

Smrcka AV, Hepler JR, Brown KO, Sternweis PC. Regulation of polyphosphoinositide-specific phospholipase C activity by purified G_q. Science 1991; 251:804–7.

Sternweis PC, Robishaw JD. Isolation of two proteins with high affinity for guanine nucleotides from membranes of bovine brain. J Biol Chem 1984; 259:13806–13.

Strathmann M, Wilkie TM, Simon MI. Diversity of the G-protein family: sequences from five additional alpha subunits in the mouse. Proc Natl Acad Sci USA 1989; 86:7407–9.

Suki WN, Abramowitz J, Mattera R, Codina J, Birnbaumer L. The human genome encodes at least three non-allellic G proteins with $alpha_i$-type subunits. FEBS Lett 1987; 220:187–92.

Takahashi S, Hashida K, Yatsunami K, Fukui T, Negishi M, Katada T, Ui M, Kanaho Y, Asano T, Ichikawa A. Characterization of cytosolic pertussis toxin-sensitive GTP-binding protein in mastocytoma P-815 cells. Biochim Biophys Acta 1991; 1093:207–15.

Tamir A, Gill DM. ADP-ribosylation by cholera toxin of membranes derived from brain modifies the interaction of adenylate cyclase with guanine nucleotides and NaF. J Neurochem 1988; 50:1791–7.

Tanuma S-I, Endo H. Mono(ADP-ribosyl)ation of Gi by eukaryotic cysteine-specific mono(ADP-ribosyl)transferase attenuates inhibition of adenylate cyclase by epinephrine. Biochim Biophys Acta 1989; 1010:246–9.

Tkachuk VA, Panchenko MP. Heart adenylate cyclase is a calmodulin-dependent enzyme. In: Smirnov VN, Katz AM, eds. Myocardial metabolism. Harwood Academic Publishers, 1987; 240–268.

Tkachuk VA, Hoffenberg SI, Starikova MG, Panchenko MP. Signal-transducing GTP-binding proteins of mammalian heart and lungs. J Mol Cell Cardiol 1989; 21 (suppl. 1):91–5.

Tkachuk VA, Lopina OD, Boldyrev AA. About the mechanism of the inhibitory action of acetylcholine and analogs on the sarcolemmal Na/K-ATPase. Biokhimia 1975; 40:1032–8 (in Russian).

Tsai S-C, Noda M, Adamik R, Chang PP, Chen H-C, Moss J, Vaughan M. Stimulation of choleragen enzymatic activities by GTP and two soluble proteins purified from bovine brain. J Biol Chem 1988; 263:1768–72.

Voyno-Yasenetskaya TA, Panchenko MP, Nupenko EV, Rybin VO, Tkachuk VA. Histamine and bradykinin stimulate the phosphoinositide turnover in human umbilical vein endothelial cells via different G-proteins. FEBS Lett 1989a; 259:67–70.

Voyno-Yasenetskaya TA, Tkachuk VA, Cheknova EG, Panchenko MP, Grigorian GY, Vavrek RJ, Stewart JM, Ryan US. Guanine nucleotide-dependent, pertussis toxin-insensitive regulation of phosphoinositide turnover by bradykinin in bovine pulmonary artery endothelial cells. FASEB J 1989b; 3:44–51.

Yamane HK, Farnsworth CC, Xie H, Howald W, Fung BKK, Clarke S, Gelb MH, Glomset JA. Brain G protein gamma subunits contain an all-trans-geranylgeranyl-cysteine methyl ester at their carboxyl termini. Proc Natl Acad Sci USA 1990; 87:5868–72.

4

Small GTPases

JEFFREY B. RUBINS

Minneapolis VA Medical Center and
University of Minnesota
Minneapolis, Minnesota

BURTON F. DICKEY

Houston VA Medical Center
and Baylor College of Medicine
Houston, Texas

I. Overview of Small GTPases

GTP binding and hydrolyzing proteins (GTPases) are two-state "molecular switches" (Wittinghofer and Pai, 1991; Milburn et al., 1990). By employing a cycle of GTP binding and hydrolysis, followed by the dissociation of GDP and the initiation of another round of GTP binding and hydrolysis, GTPases shuttle between two activity states which are determined by the phosphorylation state of the bound guanine nucleotide (Fig. 1). These two-state switches appear to be variously employed to control the duration, direction, and accuracy of a wide variety of critical cellular processes. Protein families which exhibit the functional motif of GTP binding and subsequent hydrolysis include the cytoplasmic factors involved in protein synthesis (e.g., EF-Tu), the trimeric signal transducing G proteins, cytoskeletal structural and motor proteins (e.g., tubulin and dynamin), signal recognition particles and their receptors, and the family of small ras-related GTPases. The GTPase superfamily is thus a striking example of evolution's ability to create functional diversity from a conserved structure and mechanism (see Section II, below), and is the subject of several recent reviews (Bourne et al., 1990, 1991; Downward, 1990; Hall, 1990).

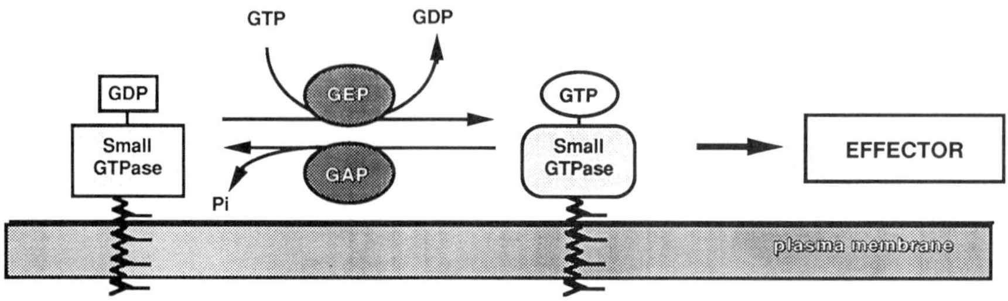

Figure 1 The basic GTPase cycle. GTPases shuttle between two activity states and two conformations (indicated by slight differences in shape), which are allosterically determined by the phosphorylation state of the bound guanine nucleotide. Nucleotide exchange is regulated by GEPs (guanine nucleotide exchange proteins) which may include both GDP dissociation stimulators and GDP dissociation inhibitors. GTP hydrolysis to GDP is regulated by protein-specific GAPs (GTPase accelerating proteins). The small GTPase shown here is anchored to the membrane by an isoprenoid.

Ras is historically the grandfather of the family of "small" GTPases with molecular masses of 20 to 35 kDa, and consequently lends its name to the entire family, as well as specifying one of three gene products described in detail in Section III. Ras genes were discovered in the 1960s as the transforming principle of rat sarcoma viruses, and were shown biochemically to code for GTP binding and hydrolyzing proteins in the 1970s (Barbacid, 1987). While the family of 39- to 52-kDa trimeric signal-transducing G proteins grew steadily through the 1980s, no additional members of the ras-related family were identified until the latter half of the decade. Since that time there has been an explosion in the identification of family members as a result of a variety of techniques including homology probing, protein purification, and recognition of conserved guanine nucleotide-binding sequences in yeast mutants and mammalian genes. The ras-related family of small GTPases now numbers over 50 proteins, and continues to grow at a rapid rate.

Based on degrees of sequence homology, the Ras-related family can be divided into several subfamilies (Valencia et al., 1991; Santos and Nebreda, 1989) (Fig. 2). The most distantly related members of the ras family share approximately 30% homology, usually in several well-defined GTP-binding and -hydrolyzing regions which are shared with other families of the GTPase superfamily (Fig. 4, below). At the other extreme, closely related family members may differ by only a few amino acids. In addition to quantitative assessment of degree of homology, certain primary structural motifs, described in greater detail in Section II, help define small GTPase subfamilies.

As might be anticipated from their discovery as transforming mutants, ras protooncogenes clearly play a role in cellular growth and differentiation. The related Rap and Ral proteins, which share approximately 70% and 50%

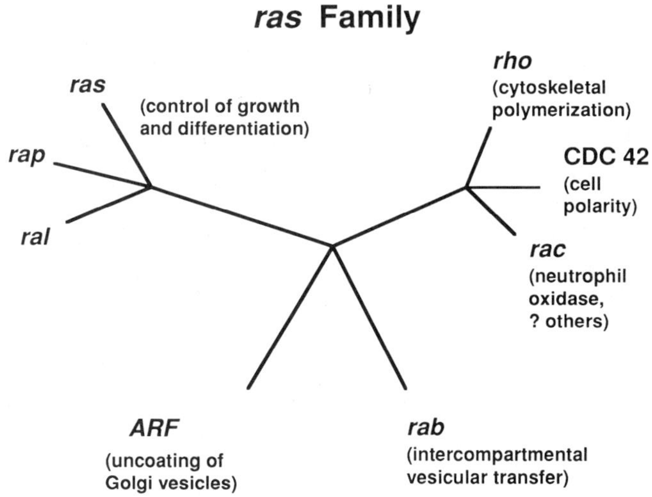

Figure 2 Ras family tree. Subfamilies are grouped according to general homology, conserved sequences around the GTP binding and hydrolyzing regions, and functional relatedness. No precise quantitative relations are implied by the lengths of connecting lines.

amino acid identity, respectively, with Ras proteins, play somewhat less well defined roles in cellular growth and differentiation (Section III). The mammalian Rab proteins and their yeast homologs, which currently number over 30, all appear to play similar regulatory roles at distinct steps of intracellular vesicular transport (Section IV). Rho proteins share approximately 55% amino acid identity with Rac proteins, and while there is substantial evidence implicating the former in regulation of cytoskeletal polymerization and the latter in regulation of the neutrophil NADPH oxidase (Section V), details of their molecular mechanisms are lacking at the time of this writing. Members of the last subfamily, the ARFs (ADP-ribosylation factors), take their peculiar name from the fact that they were known as cofactors for a bacterial toxin-catalyzed enzymatic reaction long before their physiologic cellular function was identified. Recent evidence implicates ARFs in regulation of the coating and uncoating of Golgi transport vesicles (Section VI).

II. GTP Binding and Hydrolysis: Structure and Function of a Ubiquitous Molecular Switch

All GTPase switches are thought to shuttle between two functional and conformational states, which are determined by whether the bound guanine nucleotide is GDP or GTP (Bourne et al., 1991; Wittinghofer and Pai, 1991; Milburn et al., 1990). The switching function results from the differing ability of the GTPase in each conformational state to interact with specific proteins. The interacting proteins include those which drive the cycle of nucleotide interchange, as well as the downstream effector of the GTPase. Proteins driving the cycle of nucleotide interchange are GEPs, for GDP/GTP exchange proteins, which may either accelerate or retard the release of bound GDP, and GAP, for GTPase accelerating protein, which accelerates the hydrolysis of bound GTP (Fig. 1). It is not yet known whether all GTPases are regulated by a full complement of individual GEPs and GAPs, nor is it know whether the downstream effector may in some cases be the GAP itself (McCormick, 1989, 1990; Wigler, 1990).

At present, it appears that all GTPase switches may be divided into two broad categories—those which function as "kinetic proofreaders," and those which function in information transfer. The best-studied example of the former is the bacterial protein EF-Tu, which binds to the mRNA-programed ribosome in a complex with aminoacyl-tRNA (Thompson, 1985). The kinetic proofreading function takes place when the dissociation rate of the tRNA codon-anticodon pair competes with the rates of GTP hydrolysis and GDP dissociation from EF-Tu. In the case of a correct codon match, the affinity is sufficiently high and the rate of dissociation sufficiently slow for the complex to stay assembled long enough for the full GTPase cycle, resulting in addition of aminoacyl-tRNA to the growing polypeptide chain. In the case of an incorrect codon match, the anticodon will dissociate due to its low affinity before completion of the GTPase cycle. This system now serves as a paradigm for GTPases such as those of the Rab and ARF families, which also appear to function in controlling the directionality and accuracy of macromolecular or vesicular transfer processes. These GTPases might not always be thought of as functioning in signal transduction, since they function in constitutive processes such as protein transport from ER to Golgi as well as functioning in regulated processes such as exocytosis and oxidase activation. As important regulatory switches, though, they and their GAPs and GEPs appear to be the targets both of physiologic phosphorylation events (Lapetina et al., 1989; Kaplan et al., 1990; Bailly et al., 1991; Kawata et al., 1989; Itoh et al., 1991; Hata et al., 1991) and of pathophysiologic ADP-ribosylation by bacterial toxins (Moss and Vaughan, 1990).

The other broad category of GTPases, typified by trimeric G proteins and Ras proteins, uses its timing function in a looser way to control the duration of generation (and hence the amplitude) of downstream signals. In addition to representing a basic difference in physiologic function, this broad division into proofreaders and transducers represents a basic difference in the molecular function of the GTPase switch. For the kinetic proofreaders, the GTPase must cycle in a precisely timed conditional fashion to function, and altering the cycle at any point by nonhydrolyzable or nonphosphorylable guanine nucleotide analogs (Thompson, 1985), toxin-catalyzed ADP-ribosylation (Collier, 1990), or mutations affecting GTP hydrolysis or nucleotide exchange, disrupts the regulated process. For the information transducers, in contrast, nucleotide analogs, toxins, and mutations may lock either the "off" or the "on" conformation, leading either to blockade or to constitutive activation of transduction pathways (Birnbaumer et al., 1990; Barbacid, 1987). For example, multiple natural and in vitro activating mutations of ras and of trimeric G proteins are transforming (Lyons et al., 1990; Landis et al., 1989; Barbacid, 1987). We have used a strategy of developing GTPase-specific synthetic guanine nucleotide analogs to control the activity state of individual GTPase-regulated processes (Fig. 3) (Shannon et al., 1992).

The Ras proteins are by far the best studied of the small GTPases, and through a variety of mutational, biochemical, and physical techniques, distinct domains have been mapped. Figure 4 shows these domains arrayed along a

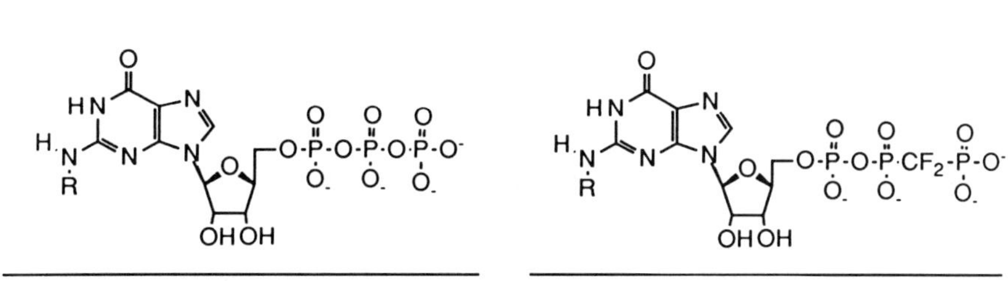

R = H, GTP

R = *p*-n-butylphenyl, BuPGTP

R = H, $GMPPCF_2P$

R = *p*-n-butylphenyl, $BuPGMPPCF_2P$

Figure 3 Structure of GTP, with N^2-butylphenyl and β,γ-methylene-substituted analogs. A series of guanine nucleotide analogs with substitutions at multiple sites were screened for differential binding to four small GTP-binding proteins of 23, 24, 25, and 26 kDa, partially purified from bovine lung lamellar bodies. One of these, $BuPGMPPCF_2P$, bound to the 25-kDa protein with an approximately 50-fold higher affinity than to the other proteins or to Ras. This empirically identified discriminant nucleotide was then systematically modified to further identify the atomic substitution that conferred the specificity, which was found to be the carbon for oxygen substitution at the β,γ-phosphoanhydride bond. These results also suggest that the Ras-cognate Asp^{119} of the highly conserved GTPase NKXD motif is able to effectively hydrogen bond from the 25-kDa protein to the butylphenyl-derivitized exocyclic amino group of the guanine base (see Fig. 5), and that there is a significant difference in the 25-kDa protein from Ras in the generally conserved β,γ-phosphate binding region (Fig. 5). (From Shannon et al., 1992.)

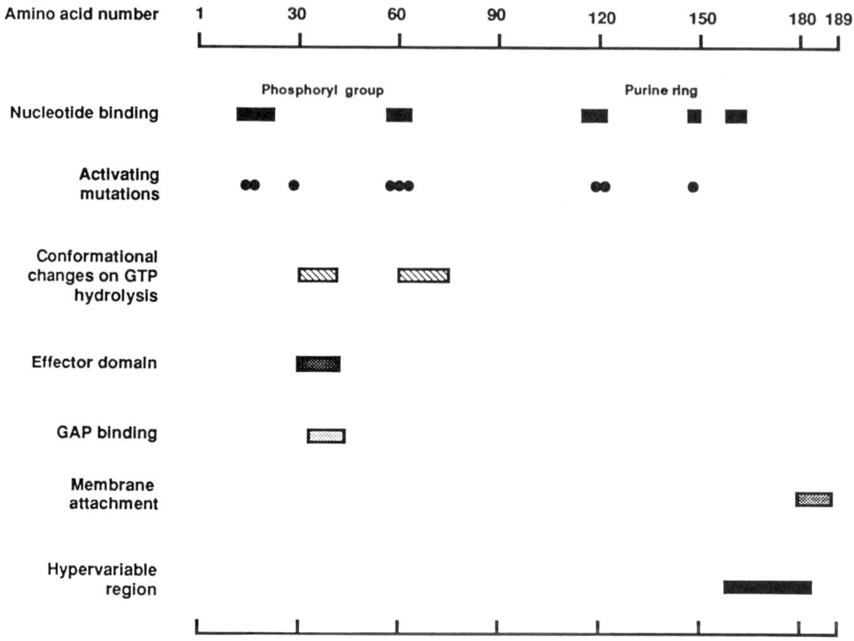

Figure 4 Linear map of Ras domains. One-dimensional representation of ras p21, illustrating the principal regions whose functions are understood from mutagenesis or crystallographic studies. (Adapted from Downward, 1990.)

one-dimensional representation of H-Ras, which serves as a general model for small GTPases. The nucleotide-binding regions are shared to some degree among all GTPases, including trimeric G proteins and protein synthetic factors. The degree of homology in these regions is even more extensive within the ras-related family, and amino acids adjacent to the nucleotide-binding sites are often conserved within ras-related subfamilies such as rho or rab whose members appear to be closely related functionally (Valencia et al., 1991). Activating mutations in ras genes are associated either with decreases in GAP activity (corresponding to the phosphoryl interactive regions) or with acceleration of nucleotide exchange (corresponding to the purine base interactive region) (Section III). H-Ras has been crystallized both with GDP and with nonhydrolyzable GTP analogs bound, and two areas of the protein undergo extensive conformational changes with changes in the bound nucleotide (Milburn et al., 1990; Wittinghofer and Pai, 1991). One of these flexible regions coincides in part with the so-called effector domain, defined by mutuational analysis as necessary for the transforming capability of Ras, and overlapping with one of the regions which interacts with GAP; the other flexible region abuts the phosphoryl-binding domain of Ras and is a possible candidate for a GEP-interactive region. As a result of the high-resolution crystallographic analysis of Ras (Pai et al., 1990; Milburn et al., 1990), NMR spectroscopic studies (Redfield and Papastavros, 1990), and time-resolved studies using caged GTP and fluorescent mant-GTP (Schlichting et al., 1990; Neal et al., 1990), a detailed understanding of nucleotide-protein interactions (Fig. 5) and of the mechanism of the GTPase switch is available (Wittinghofer and Pai, 1991). Continuing along the one-dimensional map (Fig. 4), the membrane attachment of all small GTPases except ARFs is mediated by some combination of carboxyl-terminal palmitoylation, isoprenylation, and carboxymethylation (Glomset et al., 1990). A hypervariable region adjacent to the lipidated tail shows the least sequence homology among small GTPases, and is a good candidate domain for mediating specific recognition processes.

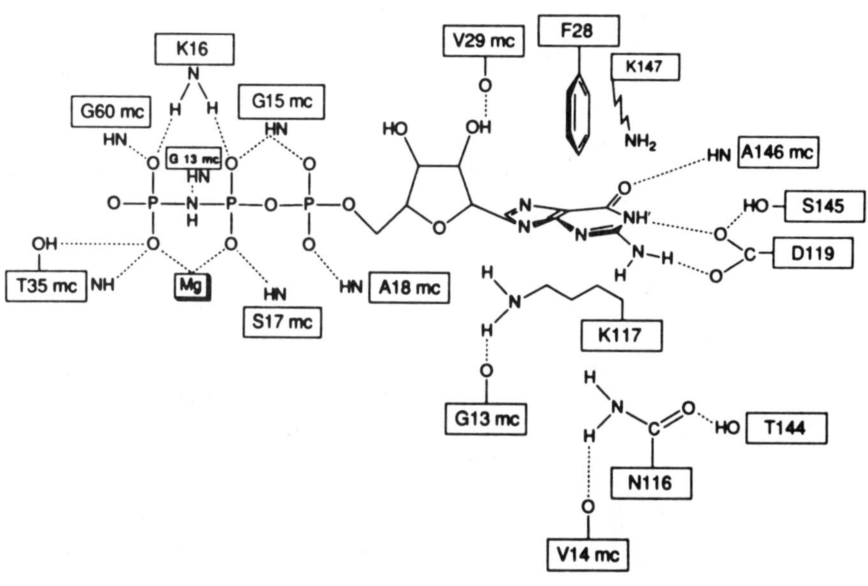

Figure 5 Schematic drawing of the most important nucleotide-protein interactions for H-Ras. All hydrogen bonds (dotted lines) up to a length of 3.1 Å have been included. (From Wittinghofer and Pai, 1990.)

III. Small GBP in the Control of Growth and Differentiation: Ras, Ral, Rap

A. Ras Proteins: Biochemical and Structural Properties

The three Ras proteins (products of the H-ras, Ki-ras, and N-ras genes, and known generically as ras p21) are ubiquitous mammalian small GTPases that have a high degree of evolutionary conservation with yeast RAS proteins, suggesting that these proteins serve essential cellular functions in eukaryotes (Barbacid, 1987; Lowy et al., 1991). Ras (for rat sarcoma) genes were initially identified in the 1960s as the transforming principle of two rat sarcoma viruses (Harvey, 1964; Kirsten and Mayer, 1967). Ras was subsequently "rediscovered" in 1982 when transforming alleles were identified as dominant oncogenes in human tumors. As the first oncogene products for which crystal structure was determined, the structural and biochemical properties of Ras proteins have been extensively characterized. However, their functional role in cellular signaling pathways controlling proliferation and differentiation remains largely unknown.

The biochemical properties of Ras proteins indicate that they are regulatory GTP-binding proteins, with slow rates of spontaneous GDP/GTP exchange and with low intrinsic GTPase activities (Hall, 1990). Recently, GDP/GTP exchange proteins (GEPs) (Wolfman and Macara, 1990; Huang et al., 1990; Downward et al., 1990b) and GTPase activating proteins (GAPs) (McCormick, 1990) have been identified that greatly accelerate these processes (see below). Oncogenic ras mutations that have been identified in a significant percentage (approximately 20%) of human cancers, including bladder, pancreas, colon, and lung, both reduce the intrinsic GTPase activity of ras proteins and render them unresponsive to GAP (Barbacid, 1987; Bourne et al., 1990; Adari et al., 1988). As a result, they remain in the active GTP-bound state and give an uncontrolled output signal, which can result in abnormal cell growth (Trahey and McCormick, 1987). Naturally occurring ras oncogenes have single point mutations at codon 12, 13, 59, or 61 (Barbacid, 1987), and crystallographic studies have demonstrated the importance of the two most frequently mutated colons, Gly^{12} and Gln^{61}, in GTP hydrolysis (Bourne et al., 1990; Wittinghofer and Pai, 1991). In contrast, in vitro ras oncogenic mutations at codons 28, 116, and 119 appear to increase GDP/GTP exchange rather than decrease GTPase activity (Barbacid, 1987; Walter et al., 1986; Sigal et al., 1986).

Mutational analysis has localized one site for interaction between the Ras proteins and their downstream effector molecules to amino acids 32 through 40, and this hydrophilic region closely coincides with one of three sites of interaction between GAP and Ras (Adari et al., 1988; Schaber et al., 1989;

McCormick, 1989). In addition, mutagenesis studies have identified a requirement for a cysteine in a conserved C-terminal CAAX motif (where A is any aliphatic amino acid, and X is any amino acid) for appropriate membrane localization and transforming activity of Ras proteins (Barbacid, 1987; Santos and Nebreda, 1989). This cycstein residue serves as a target for a series of posttranslational modifications that includes prenylation (farnesylation) of the C-terminal cysteine and removal of the three C-terminal amino acids (Gutierrez et al., 1989; Glomset et al., 1990). These modifications increase the hydrophobicity of Ras, although final localization to the plasma membrane requires an additional posttranslational palmitoylation of an upstream cysteine residue in the case of H-Ras or N-Ras (Glomset et al., 1990), or a specific hydrophobic region (residues 175–182) in the case of Ki-Ras (Hancock et al., 1990). These posttranslational lipidations may function in the dynamic regulation of Ras localization and function (Santos and Nebreda, 1989).

B. Ras Proteins: Biological Properties

Genetic and biochemical analysis of yeast RAS proteins that are highly homologous to mammalian ras p21 has provided some insight into the biological function of Ras. Disruption of both of the two *Saccharomyces cerevisiae* RAS genes results in nonviable cells that appear to be blocked at the G1 phase of the cell cycle, indicating an essential role for RAS proteins in normal yeast cellular proliferation (Barbacid, 1987). The normal growth and cycling of yeast cells is controlled by nutritional factors that regulate the cellular concentrations of cAMP. Although their exact function is yet unknown, yeast RAS proteins appear to be important components of a signal transduction pathway that culminates in the stimulation of adenylyl cyclase and production of cAMP (Broek et al., 1985). An upstream regulatory protein, the CDC25 gene product, appears to function as a GEP for yeast RAS (Broek et al., 1987), and proteins encoded by IRA1 and IRA2 genes, which have homology to mammalian GAP, may be involved in the down-regulation of RAS (Tanaka et al., 1990). The use of yeast RAS proteins as a model for mammalian ras p21 is supported by the partial functional interchangability of yeast and mammalian ras gene products (Santos and Nebreda, 1989). In addition to their similar biochemical properties, mammalian Ras genes can replace yeast genes for yeast cell growth, and the N-terminal portion of activated yeast genes can transform NIH3T3 fibroblasts.

Despite these insights from studies in *S. cerevisiae,* Ras does not appear to regulate adenylyl cyclase activity in any other eukaryotic cells, including the closely related fission yeast *Schizosaccharomyces pombe.* Disruption of the SPRAS gene in this fungus does not affect cAMP levels or cell proliferation, but rather interferes with sexual differentiation by blocking conjugation and inhibiting sporulation (Nadin-Davis et al., 1986), possibly by modulating protein phosphorylation (Nadin-Davis and Nasim, 1988). Studies of related Ras genes in organisms along the evolutionary scale such as slime molds, sponges, insects, amphibians, and fish suggest that Ras proteins are involved in the transduction of proliferative signals (Barbacid, 1987). In *Xenopus* oocytes, microinjection of activated Ras induces meiotic maturation, and studies with neutralizing ras p21 antibodies indicate that Ras plays a role in insulin-induced germinal vesicle breakdown (Barbacid, 1987). A role for ras p21 in mammalian cellular proliferation has been suggested by studies in fibroblasts. Microinjection of the neutralizing monoclonal antibody Y13-259 into fibroblasts blocks stimulation by a wide variety of growth factors. In addition, transfection with oncogenic Ras, overexpression of normal ras p21, or microinjection of recombinant oncogenic Ras proteins into quiescent cells all cause proliferation and transformation of NIH3T3 fibroblasts (Barbacid, 1987).

Microinjection studies using Y13-259 suggest that a variety of transduction pathways for proliferative signals may converge through ras p21 into a common pathway for activation of nuclear transcriptional factors. Microinjection of neutralizing antibodies causes transient reversion of cells transformed either by activating Ras mutants or by tyrosine kinase, membrane-associated oncogenes such as fms, src, or fes 1 (Santos and Nebreda, 1989). In contrast, transformation by cytoplasmic oncogenes with serine/threonine kinase activity, such as mos and raf, is not affected by anti-ras p21 antibodies (Barbacid, 1987), suggesting that these genes act either in different pathways or in the same pathway but downstream from Ras. The latter is supported by observations that transformation by Ras results in phosphorylation of Raf-1 and activation of its serine/threonine kinase activity (Kolch et al., 1991). Transformation by H-ras has been shown to result in activation of specific, jun-like transcriptional factors, which are similarly activated by phorbol esters, growth factors, and other transforming oncogenes (membrane and cytoplasmic) (Wasylyk et al., 1988).

In addition to a role in transducing growth factor signals to activation of transcription and cellular proliferation, ras p21 also appears to have a role in the control of differentiation processes in some cells. For example, whereas microinjection of activated Ras proteins causes proliferation and transformation of fibroblasts, microinjection of oncogenic Ras into rat pheochromocytoma PC12 cells (Barbacid, 1987) or into Schwann cells (Ridley et al., 1988) halts proliferation and results in terminal differentiation. Also, myogenic differentiation (Sternberg et al., 1989) and mammary epithelial cell differentiation (Andres et al., 1988) is blocked by constitutive expression of Ras oncogenes. Immunocytochemical studies in adult human tissues have shown ras p21 in actively proliferating tissues, but certain differentiated tissues, such as brain neurons, endocrine gland epithelium, and muscle cells, express very high levels of ras p21 as well (Furth et al., 1987). These patterns of expression of H-ras, Ki-ras, and N-ras suggest that these Ras proteins function differently at different points of cellular maturation in a tissue-specific manner.

C. Ras Proteins: Signal Transduction

The Role of GAP

The low intrinsic GTPase activity of normal ras p21 is stimulated 100-fold by interaction with GAP. Bovine and human ras p21 GAPs have been purified as approximately 120-kDa cytosolic proteins, whose corresponding cDNA clones have been isolated (McCormick, 1989; Gibbs et al., 1988; Vogel et al., 1988; Trahey et al., 1988). In addition, the IRA-1 and -2 proteins which show GAP activity for the yeast *S. cerevisiae* RAS protein have been identified (McCormick, 1990; Tanaka et al., 1990). Also, the protein encoded by the von Recklinghausen neurofibromatosis locus NF1 has homology to ras p21 GAP and the IRA proteins in the catalytic domain which stimulates GTPase activity (Carthon et al., 1990; Buchberg et al., 1990). Further, the NF1 protein has ras p21 GAP activity and can complement the IRA gene products (Xu et al., 1990).

Ras p21 GAP interacts with at least three regions of ras p21, one of which is the mutationally defined "effector domain." Oncogenic Ras proteins in their GTP-bound conformations also bind GAP, although their GTPase activities are not stimulated (McCormick, 1989). GAP activity has been shown to be inhibited by various membrane lipids whose metabolism is altered during mitogenic stimulation (Tsai et al., 1989). As a regulator of ras p21 GTPase activity, GAP may provide a link between growth-factor signaling pathways and Ras. In 3T3 fibroblasts, ras p21 GAP physically associates with the ligand-activated PDGF receptor and is phosphorylated by the PGDFβ tyrosine kinase (Kaplan et al., 1990). Of note, GAP fails to associate with the ligand-activated PDGF receptor in Ras-transformed cells (Kaplan et al., 1990). While the consequence of this phosphorylation is presently unknown, it is speculated that GAP phosphorylation may be a mechanism of regulation of ras p21 activity by receptor-coupled tyrosine kinases.

Several lines of evidence suggest that GAP may serve as a component of the downstream signaling effector complex as well as a regulator of ras p21 GTPase activity. First, a requirement for GAP interaction with oncogenic Ras proteins to achieve malignant transformation has been shown, even though this association does not activate the intrinsic GTPase (Barbacid, 1987; McCormick, 1989). Second, the highly homologous R-Ras protein (see below) has a fully conserved effector domain and efficiently binds GAP, but transfection with R-Ras containing an activating mutation does not transform cells (Lowe et al., 1988). Studies of chimeric proteins between H-ras and R-Ras genes indicate that the C-terminal portion of R-Ras prevents this molecule from having transforming ability (Lowe et al., 1988). Third, microinjection of a yeast RAS mutant that has a 100-fold higher affinity for GAP, but that also has a C-terminal deletion, blocks the effects of activated RAS on germinal vesicle breakdown in *Xenopus* oocytes (Gibbs et al., 1988). Similarly, a p21 Ras mutant with identical properties inhibited transformation of fibroblasts by oncogenic Ras (Johnston et al., 1991). In addition to this indirect evidence, GAP has been shown to be essential for ras p21 signaling in atrial membranes, where addition of exogenous GAP in conjunction with endogenous ras p21 stimulated the closure of muscarinically activated potassium channels (Johnston et al., 1991). Based on this evidence, it has been speculated that association of GAP with the "effector domain" might be involved in transmitting the GTP signal to the C-terminal domain, which is specific for each small GTPase and would provide for specific interactions with distinct cellular targets (Figure 4).

On the other hand, ras p21 with an altered "effector domain" that is temperature sensitive for transformation is fully sensitive to the catalytic activities of GAP and NF1 at both permissive and nonpermissive temperatures (Lowy et al., 1991), suggesting that the regulatory and downstream effector functions of GAP can at least be resolved by some criteria. Further, Ras activation in response to a cell surface ligand in T cells was caused by a decrease in GAP activity, suggesting that GAP acted upstream of Ras (Downward et al., 1990a). The loss of NF1 genes results in disordered growth, and NIH 3T3 transformation by Ras overexpression can be overcome by GAP overexpression, suggesting that Ras GAPs may act as negative regulators rather than effectors (Wigler, 1990). All of these interpretations, however, are currently confounded by the paradoxical effects of ras on growth and differentiation in different cell types.

The Role of Rap1A

Several mammalian proteins with relatively high amino acid homologies to ras p21 (Ral A and B, Rap1 A and B, Rap2, R-Ras, and TC21) have been isolated by biochemical and genetic techniques. Although the function of most of these proteins is not clear, overexpression of Rap1A (which is identical to Krev-1 and smg p21A) has been shown to suppress the Ki-ras-induced transformation of NIH3T3 fibroblasts, suggesting a role for Rap1A as a suppressor of ras p21 (Kitayama et al., 1990). Rap1A has an identical amino acid sequence to the "effector domain" of ras p21 and efficiently binds ras p21 GAP (Hata et al., 1990). Thus, Rap1A may compete for Ras GAP and block downstream signaling by ras p21 (Frech et al., 1990). Alternatively, Rap1A may participate in a different signaling pathway that may be antagonistic to ras p21 in some cells. A role for Rap1A in distinct signaling pathways is supported by the discovery of a Rap1A-specific GAP (Kikuchi et al., 1989; Garrett et al., 1989) and GEP (described as a GDP dissociation stimulator, or GDS) (Yamamoto et al., 1990; Kaibuchi et al., 1991). Further, Rap1A has been identified in many differentiated cell types, and in neutrophils has been shown to be associated with a plasma-membrane complex involved in superoxide generation (see Section V, below).

Rap1A has been shown to be phosphorylated by protein kinase A, which initiates GDS-stimulated GDP/GTP exchange and activates the protein (Lapetina et al., 1989; Kawata et al., 1989; Itoh et al., 1991; Hata et al., 1991). A-kinase (in addition to C-kinase) has also been shown to phosphorylate H-Ras and Ki-Ras proteins in some cell-free and intact cell systems (Ballester et al., 1987; Saikumar et al., 1988). Consequently, if Rap1A functions as a suppressor of ras p21 activity in vivo, cellular signals that raise intracellular cAMP may modulate ras p21 signaling either directly through A-kinase phosphorylation of ras p21, or through phosphorylation of Rap1A.

IV. Small GBP in the Control of Vesicular Traffic: Rab Proteins

A. Rab Proteins: Biochemical and Structural Properties

The Rab proteins also appear to be ubiquitous mammalian small GTPases. Studies of their yeast homologs YPT1 and SEC4 suggest that these proteins have a functional role in regulating intracellular vesicle transport. This protein family is distinguished from other small GTPases by two structural features. The first is the absence of a glycine residue at the equivalent position to codon 12 of ras p21 (Hall, 1990), which, as discussed previously, is critical for normal ras p21 GTPase activity and is the most common site for transforming point mutations in human Ras oncogenes. Second, the C-terminal CAAX domain conserved in the Ras proteins is replaced by either XXCC or CXC motifs. As with ras p21, these cysteine residues appear to be the site for posttranslational prenylation; however, they appear to be geranylgeranylated rather than farnesylated (Khosravi-Far et al., 1991; Kinsella and Maltese, 1991). While all of the details of the posttranslational processing of these residues are not yet known for the Rab proteins, these C-terminal motifs appear to be essential for membrane localization (Walworth et al., 1989). Rabs appear to have low GDP/GTP exchange rates and intrinsic GTPase activities that are similar to those for ras p21, despite the absence of Gly^{12}. In addition, regulatory GEP and GAP proteins have been identified for one Rab protein, Rab3A (identical to smg25A), and will likely be found for most, if not all, Rabs (Burstein et al., 1991) (discussed below).

B. Rab Proteins: Biological Properties

The strongest evidence for the role of small GTPases in intracellular vesicle transport processes derives from genetic and biochemical studies of the yeast small GTPases YPT1p and SEC4p (Burgoyne, 1988). Temperature-sensitive YPT1 mutants of *S. cerevisiae* show an early block in constitutive secretion, with the accumulation of vesicles and unprocessed secretory proteins within the Golgi (Segev et al., 1988). Immunocytochemical studies have localized the YPT1p to the Golgi (Segev et al., 1988), and blocking antibodies inhibit endoplasmic reticulum (ER) to Golgi transport in cell-free systems (Baker et al., 1990). Similar studies of SEC4 mutants reveal an accumulation of secretory vesicles between the Golgi complex and the plasma membrane (Salminen and Novick, 1987). Immunocytochemical studies have localized the SEC4 product predominantly to the yeast secretory vesicles and plasma membrane, with a small proportion (approximately 10%) found in the cytosol (Salminen and Novick, 1987; Goud et al., 1988).

Based on these findings, a model for the function of small GTPases in unidirectional intracellular vesicular transport has been proposed (Bourne, 1988; Walworth et al., 1989) (Fig. 6). Secretory vesicles leaving the donor membrane have an associated small GTPase in the GTP-bound form. Vesicles then translocate to the acceptor membrane, where the vesicle-small GTPase complex binds to a postulated "docking protein." Fusion of the secretory vesicle is permitted only after hydrolysis of GTP to GDP by the small GTPase protein, which subsequently dissociates into the cytosol and cycles back to the donor membrane. In this model, the small GTPase serves to ensure both the unidirectional transport of vesicles and, more important, the fusion of vesicles with the "correct" target membrane. By analogy with the role of the small GTPase EF-Tu in protein synthesis (Woolley and Clark, 1989), only high-affinity complexes of secretory vesicle-docking protein at the acceptor membrane can be maintained for the finite period of time required for GTP hydrolysis by the associated small GTPase; less correct matches will dissociate before GTP can be hydrolyzed. In addition, GEP and GAP proteins, as well as geranylgeranyltransferases, may be involved in regulating this cycle (discussed below).

Consistent with this model, transport from ER through the Golgi complex (Beckers and Balch, 1989; Melancon et al., 1987; Orci et al., 1989), exocytosis of adrenal chromaffin granules (Knight and Baker, 1985), and fusion of endocytic vesicles (Mayorga et al., 1989a, 1989b) have all been shown in mammalian cell-free systems to be blocked by poorly hydrolyzable GTP analogs. In addition, small GTPases have been shown to be associated with secretory vesicles in a number of mammalian cells, including primary and specific neutrophil granules (Dexter et al., 1990), secretory granules of adrenal

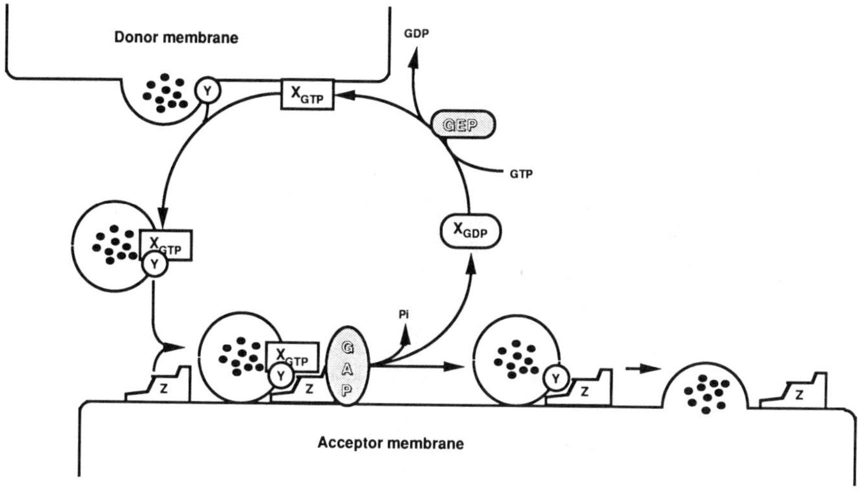

Figure 6 Model of Rab cycle. A soluble GDP-bound Rab (X) becomes associated with the donor membrane following exchange of GDP for GTP catalyzed by GDP/GTP exchange proteins (GEPs). [These may include a soluble GDP dissociation inhibitor (GDI) which could transport X_{GDP} through cytosol and a donor membrane-associated GDP dissociation stimulator (GDS).] X_{GTP} associates with a recognition protein Y in the nascent vesicle, which then docks with the target membrane when Y recognizes a docking protein Z. Z is also a Rab GTPase accelerating protein (GAP), and with assembly of the full XYZ complex, the timed switch begins its proofreading function in that an incorrectly assembled complex will dissociate before GTP hydrolysis occurs. GTP hydrolysis is postulated to be permissive for vesicle fusion, with GTPγS blocking fusion at the site shown. (Adapted from Bourne, 1988.)

chromaffin cells (Burgoyne and Morgan, 1989; Doucet et al., 1989), and synaptic vesicles (Matsuoka and Dolly, 1990; Mizoguchi et al., 1989; Bielinski et al., 1989). We have recently demonstrated multiple small GTPases associated with a preparation of purified secretory vesicles (lamellar bodies) from isolated rat type 2 alveolar epithelial cells (Rubins et al., 1991) (Fig. 7).

Rab proteins have been proposed as likely candidates for small GTPases regulating intracellular vesicle transport in mammalian cells largely based on their high degree of genetic homology with the yeast YPT1 and SEC4 proteins, especially in the region of the putative effector domain (Zahraoui et al., 1989). In particular, Rab1 shares approximately 75% sequence homology with YPT1p, and the rab1 can complement YPT1 deletions in *S. cerevisiae* (Segev et al., 1988; Haubruck et al., 1989). Moreover, the mammalian Golgi complex is recognized by antibody to the yeast YPT1p. The other 18 members of the

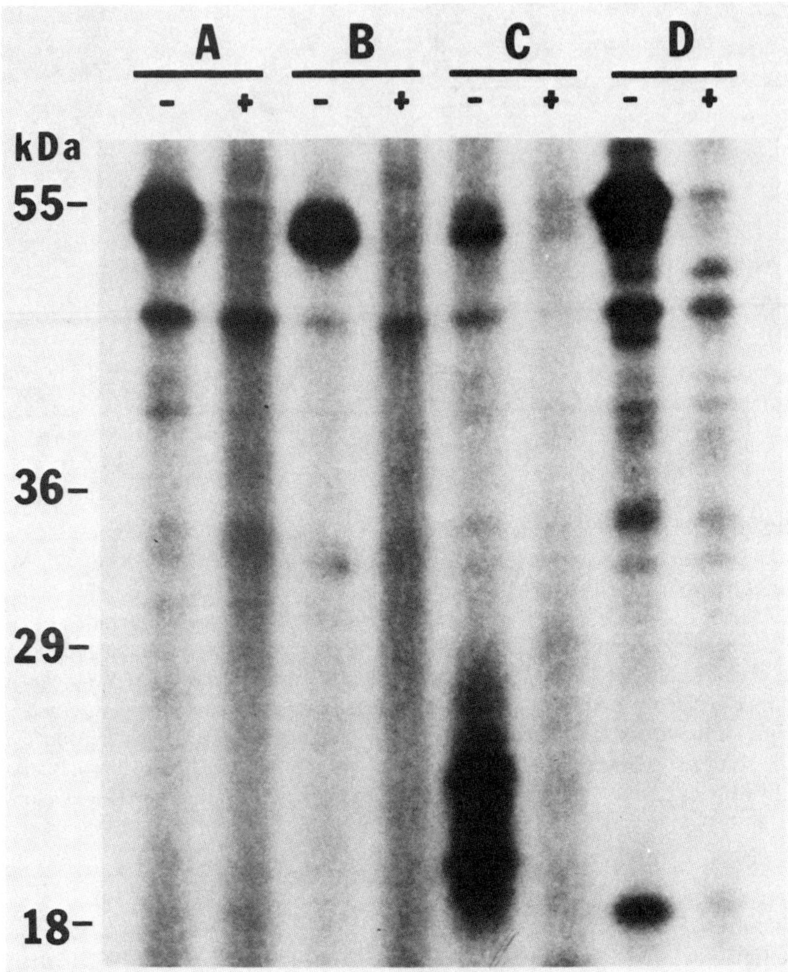

Figure 7 Small GTPases in lung lamellar bodies. Aliquots (25 μg protein) of rat type 2 pneumocyte homogenate (A), crude membranes (B), lamellar bodies (C), and cytosol (D) were photoaffinity labeled with 10 nM [α-^{32}P]GTP in the presence (+) or absence (−) of 10 μM unlabeled GTP, and separated by 14% SDS-PAGE. The autoradiogram was developed after 48 h using an intensifying screen. The figure shows enrichment of 20.5- and 21-kDa GTP-binding proteins in lamellar bodies by this technique; binding of [α-^{32}P]GTP to nitroblotted proteins reveals additional species with different distributions (not shown). (From Rubins et al., 1991.)

rab family that have been identified thus far share approximately 40% homology with each other, and lesser homology with SEC4, for which a true mammalian homolog has not yet been found.

Further support for the involvement of Rab proteins in vesicle transport in mammalian cells has been provided by immunocytochemical studies and experiments using synthetic peptides in permeabilized cells. Rab3A, the most extensively studied of the rab proteins, has been localized by immunocytochemical and immunochemical methods to synaptic vesicles in rat brain (Mizoguchi et al., 1990; Mollard et al., 1990). In addition, Rab3A has been identified in a number of secretory cells including bovine adrenal cells, rat endocrine and exocrine pancreatic cells, and rat salivary gland cells (Mizoguchi et al., 1989; Darchen et al., 1990). In these cells, Rab3A appears to be localized to the Golgi apparatus in exocrine cells, and to secretory granules in endocrine cells. More recently, other Rab proteins have been localized to early endosomes (Rab4, Rab5), late endosomes (Rab7), and the Golgi region (Rab2, Rab6) (Chavrier et al., 1990) (Fig. 7). Finally, studies in semi-intact cells have shown inhibition of ER to Golgi and intra-Golgi transport by synthetic peptides directed against the putative effector domain of the Rab protein family, whereas peptides against Ras, Ral, and Rho protein effector regions or against the Rab C-terminal domain had no inhibitory effect (Plutner et al., 1990).

A role for Rab-specific GAP and GEP proteins in regulating Rab-mediated intracellular vesicle transport has been postulated. As mentioned, Rab-specific GAP and GEP proteins have been identified for Rab3A (Burstein et al., 1991; Sasaki et al., 1990). The Rab3A GAP appears to have distinct properties from ras p21 GAP, Rapl GAP, and Rho GAP, but details of its regulation of rab3A are unknown (Burstein et al., 1991). The GEP identified at present, a GDP dissociation inhibitor (GDI), associates with the GDP-bound form of rab3A in a 1:1 molar ratio, thereby inhibiting dissociation of GDP from and binding of GTP to rab3A (Sasaki et al., 1990). Posttranslational prenylation of Rab3A appears to be required for interaction with GDI (Araki et al., 1991). Rab3A-specific GDI appears to regulate the reversible binding of the GDP-bound form of rab3A to synaptic vesicles. The GDI protein both inhibits binding of the GDP-bound form of rab3A to synaptic vesicles and also induces its dissociation from synaptic membranes and vesicles (Araki et al., 1990). It has been speculated that rab3A-GDI may be the target of intracellular signaling pathways, which induce the dissociation of the rab3A-GDI complex, resulting in binding of the GTP-bound form of rab3A to the synaptic vesicle. Activation of endo-exocytic processes by protein kinase C suggests that rab3A-GDI association may be regulated by phosphorylation. Certain Rab proteins are indeed phosphorylated by $p34^{CDC2}$ during mitosis (Bailly et al., 1991), consistent with the observed arrest of vesicular transport processes during mitosis; however, the modification of Rab proteins by other protein

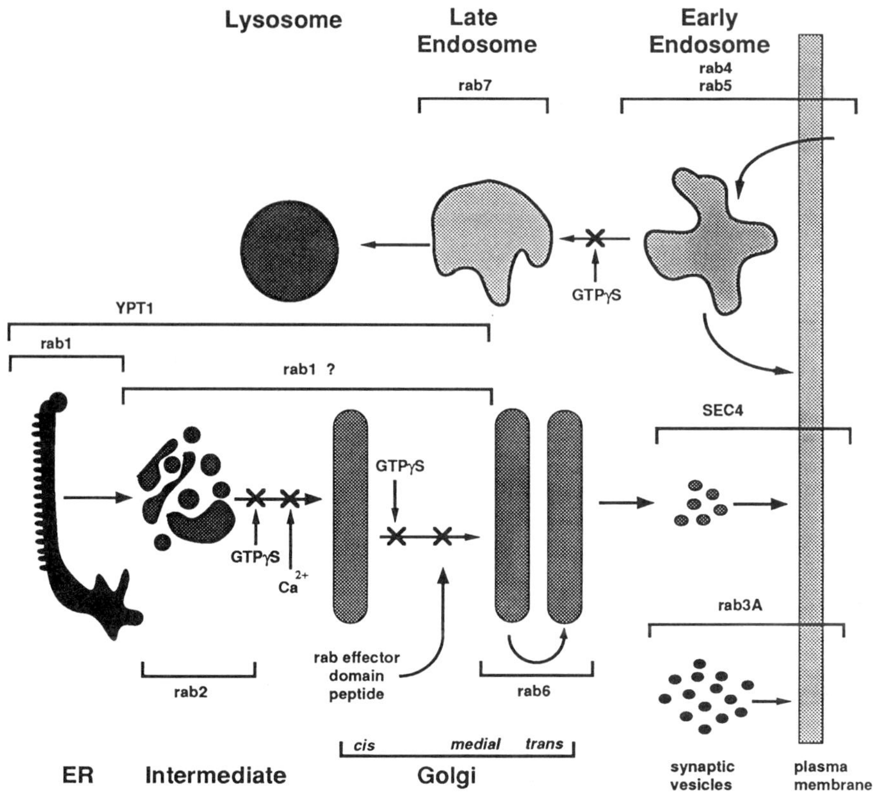

Figure 8 Subcellular localization of Rab proteins. Mammalian Rabs and the yeast proteins YPT1 and SEC4 have been localized to distinct compartments by a variety of functional and immunochemical techniques. Current evidence suggests the assignments shown. (Adapted from Balch, 1990.)

kinases and the biochemical consequences of phosphorylation have not yet been studied.

V. Rho, rac, Cytoskeletal Polymerization, and the Oxidative Burst

The rho gene was originally cloned from *Aplysia californica* (Madaule and Axel, 1985), but close homologs have subsequently been found in all eukaryotes examined, including yeast. Three mammalian genes have been cloned, and two of the isotypes are ubiquitously distributed in all tissues examined, while one is limited to neural tissue. A major clue to their cellular function was provided by the recent discovery that Rho proteins are the substrates of botulinum exoenzyme C3 ADP-ribosyltransferase (Aktories and Hall, 1989). Microinjection of exoenzyme C3 into mammalian cells causes disruption of actin filaments followed by cell rounding (Chardin et al., 1989), and this has been shown to be due to the inactivation of endogenous Rho by ADP-ribosylation (Paterson et al., 1989). Disruption of the yeast RHO gene is lethal (Hall, 1990). In contrast, a codon 14 mutation (cognate to an oncogenic Ras 12 mutation) acts as an activating mutation which prevents stimulation of the intrinsic GTPase by Rho-GAP (Garrett et al., 1989), and microinjection induces dramatic changes in cell morphology consistent with reorganization of the actin cytoskeleton (Paterson et al., 1990). Overexpression of one of the rho genes in fibroblasts reduces serum dependence for growth and is tumorigenic in nude mice. Recently, a Rho GDI (Ueda et al., 1991), a GDS (Isomura et al., 1990), and a Rho GAP with homology to Ras GAP (Garrett et al., 1991) have been purified, and it may be anticipated that the molecular details of the regulation of actin polymerization by Rho proteins will soon be elaborated. The fact that Rho and Rac proteins appear capable of constitutive activation (by nonhydrolyzable analogs in the case of Rac—see below—and by a GTPase-defective activating mutation or by overexpression in the case of Rho) suggests that this family will be shown to function more as information transducers (such as Ras or trimeric G proteins) than as kinetic proofreaders (such as EF-Tu or Rabs).

The myeloid-specific Rac proteins have also been shown to be weak substrates for exoenzyme C3 (Didsbury et al., 1989), but whether their ADP-ribosylation in vivo is physiologically significant is doubtful. Independent of this, however, they have recently been implicated in the regulation of the neutrophil NADPH oxidase. Nonhydrolyzable guanine nucleotide analogs and fluoride had previously been shown to stimulate oxidase activity in a cell-free system (Gabig et al., 1987; Seifert and Schultz, 1987; Ligeti et al., 1988), implicating a GTPase. Two components of the oxidase complex were recently identified by microsequencing as Rac1 and Rho GDI (Abo et al., 1991), and GTPγS-liganded Rac1 was found to reconstitute oxidase activation comparably to the previously identified neutrophil cytosolic fraction (Pick et al., 1989). These findings, however, remain to be resolved with the observation of the physical association of Rap1 with highly purified cytochrome b (Quinn et al., 1989), and the reconstitution of oxidase activation by Rap1 (Eklund et al., 1991).

Other members of this family include the Cdc42 protein (also known as Gp and G25k in mammalian cells), which appears to be involved in bud emergence in yeast (Bender and Pringle, 1989), and TC10 (Drivas et al., 1990), whose cellular function is as yet unknown.

VI. The Biology and Pathophysiology of ARF

A major step in elucidating the molecular mechanism of the diarrheal disease cholera was the finding in 1978 that cholera toxin enzymatically transfers the ADP-ribose moiety of NAD to the stimulatory G protein of adenylyl cyclase, Gs (Cassel and Pfeuffer, 1978; Gill and Meren, 1978). Covalent modification of Gs in this way results in constitutive activation of adenylyl cyclase, which in turns leads to chloride and water secretion by intestinal mucosal cells. During identification of this mechanism, it became apparent that a cytosolic cofactor was necessary for the efficient action of cholera toxin (Enomoto and Gill, 1980), and this cofactor has come to be generally known as ARF for ADP-ribosylation factor. ARF itself was then found to be a GTP-binding protein (Gill and Meren, 1983; Kahn and Gilman, 1986), which was purified in 1984 (Kahn and Gilman, 1984), and the first of several ARF cDNAs was cloned four years later (Price et al., 1988; Sewell and Kahn, 1988, Boback et al., 1989). These cDNAs code for closely related 21-kDa proteins which share homology with other GTPases in the GTP-binding and -hydrolyzing regions, but differ from most other small GTPases in containing an aspartic acid at the cognate glycine 12 position of H-Ras. Also unlike other small GTPases, which are lipidated by carboxy-terminal prenylation and/or palmitoylation, ARFs are myristoylated at the amino terminus, as are the α subunits of trimeric G proteins (Kahn et al., 1988).

While many of the details of the catalytic mechanism of ARF-supported cholera toxin action were also worked out during this time (Bobak et al., 1990), and cholera toxin-stimulated ADP-ribosylation of Gs was frequently employed as a valuable probe of G-protein function (see Chapter 3), the cellular physiology of ARF remained mysterious until recently. Clues to the cellular function of ARF proteins came from immunochemical and immunofluorescence data showing their localization in the Golgi apparatus (Stearns et al., 1990), and from the observation that the poorly hydrolyzed GTP analog, GTPγS, caused a functional inhibition of cis to medial Golgi transport in an in vitro system and the accumulation of non-clathrin-coated Golgi transport vesicles (Melancon et al., 1987). Most recently, ARF has been found to be a structural component of Golgi transport vesicles, that is, one of the coat proteins (Serafini et al., 1991). Based on the available evidence and by analogy with other GTPases whose functional cycles include interactions with nucleotide-exchange proteins (GEP) and GTPase-accelerating proteins, a model can be constructed of ARF's role in the regulation of Golgi transport vesicle coating and uncoating (Fig. 9). The purpose of employing a GTPase-regulated cycle of coating and uncoating appears to be, again by analogy with other GTPases, to control the accuracy and correct vectoral delivery of a piece of membrane from an appropriate donor organelle to an appropriate acceptor organelle within the same cytosolic compartment. The continuing discovery of

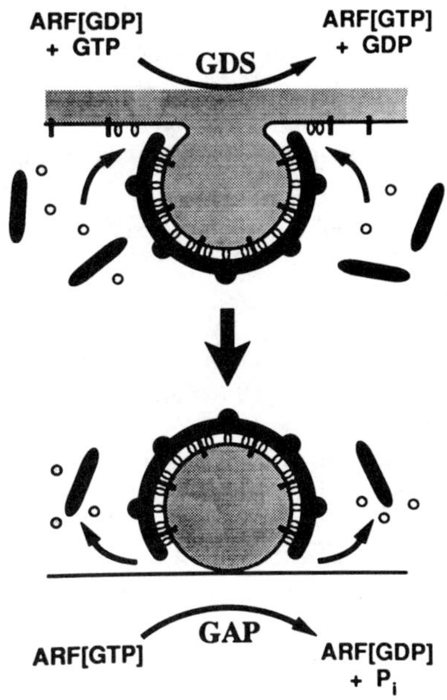

Figure 9 Hypothetical model of the mechanism of ARF action in vesicular transport. Soluble ARF (small circles) would bind to donor membranes (as small ovals, reflecting a conformational change) after exchange of GTP for GDP catalyzed by an exchange protein (GDS, for GDP dissociation stimulator) resident in the donor membrane. ARF bound to donor membranes, acting in concert with other membrane-associated proteins (small rectangles), would cause the assembly of the coat from coatamers (large ovals) during vesicle budding. After transfer of the coated vesicle to the target membrane, ARF would be induced to hydrolyze its bound GTP by a specific, localized GAP, causing release of soluble GDP-bound ARF and uncoating of the vesicle. ARF and coatamer molecules would then be available for another round of coat assembly. (Adapted from Serafini et al., 1991.)

increasing numbers of ARF family members would be consistent with their mediation of multiple distinct sorting events within the Golgi apparatus (Kahn et al., 1991). The relation between the roles of Rab1 and ARF in intra-Golgi sorting is not yet defined.

References

Abo A, Pick E, Hall A, Totty N, Teahan C, Segal A. Activation of the NADPH oxidase involves the small GTP-binding protein $p21^{rac1}$. Nature 1991; 353:668–70.

Adari H, Lowy DR, Willumsen BM, Der CJ, McCormick F. Guanosine triphosphatase activating protein (GAP) interacts with the p21 ras effector binding domain. Science 1988; 240:518–21.

Aktories K, Hall A. Botulinum ADP-ribosyltransferase C3; a new tool to study low molecular weight GTP-binding proteins. Trends Pharm Sci 1989; 10:415–8.

Andres AC, Vandervalk MA, Schonenberg CAS, Fluckiger F, Lemeur M, Groner B. Ha-ras and c-myc oncogene expression interferes with morphological and functional differentiation of mammary epithelial cells in single and double transgenic mice. Genes Dev 1988; 2:1486–95.

Araki S, Kaibuchi K, Sasaki T, Hata Y, Takai Y. Role of the C-terminal region of smg p25A in its interaction with membranes and the GDP/GTP exchange protein. Mol Cell Biol 1991; 11:1438–40.

Araki S, Kikuchi A, Hata Y, Isomura M, Takai Y. Regulation of reversible binding of smg25A, a ras p21-like GTP-binding protein, to synaptic plasma membranes and vesicles by its specific regulatory protein, GDP dissociation inhibitor. J Biol Chem 1990; 265:13007–15.

Bailly E, McCaffrey M, Touchot N, Zahraoui A, Goud B, Bornens M. Phosphorylation of two small GTP-binding proteins of the rab family by p34-CDC2. Nature 1991; 350:715–8.

Baker D, Wuestehube L, Schekman R, Botstein D, Segev N. GTP-binding Ypt1 protein and Ca2+ function independently in a cell-free protein transport reaction. Proc Natl Acad Sci USA 1990; 87:355–9.

Balch WE. Small GTP binding proteins in vesicular transport. Trends Biochem Sci 1990; 15:473–477.

Ballester R, Furth ME, Rosen OM. Phorbol ester- and protein kinase C-mediated phosphorylation of the cellular Kirsten ras gene product. J Biol Chem 1987; 262:2688–95.

Barbacid M. Ras genes. Ann Rev Biochem 1987; 56:779–827.

Beckers CJM, Balch WE. Calcium and GTP: essential components in vesicular trafficking between the endoplasmic reticulum and golgi apparatus. J Cell Biol 1989; 108:1245–56.

Bender A, Pringle JR. Multicopy suppression of the cdc24 budding defect in yeast by CDC42 and three newly identified genes including the ras-related gene RSR1. Proc Natl Acad Sci USA 1989; 86:9976–80.

Bielinski DF, Morin PJ, Dickey BF, Fine RE. Low molecular weight GTP-binding proteins are associated with neuronal organelles involved in rapid axonal transport and exocytosis. J Biol Chem 1989; 264:18363–7.

Birnbaumer L, Abramowitz J, Brown AM. Receptor-effector coupling by G-proteins. Biochim Biophys Acta 1990; 1031:163–224.

Bobak DA, Nightingale MS, Murtagh JJ, Price SR, Moss J, Vaughn M. Molecular cloning, characterization, and expression of human ADP ribosylation factors: two guanine nucleotide dependent activators of cholera toxin. Proc Natl Acad Sci USA 1989; 86:6101–5.

Bobak DA, Tsai S-C, Moss J, Vaughan M. Enhancement of cholera toxin ADP-ribosyltransferase activity by guanine nucleotide-dependent ADP ribosylation factors. In: Moss J, Vaughan M, eds. ADP-ribosylating toxins and G proteins. Washington, DC: American Society for Microbiology, 1990; 439–52.

Bourne HR. Do GTPases direct membrane traffic in secretion? Cell 1988; 53:669–71.

Bourne HR, Sanders DA, McCormick F. The GTPase superfamily: a conserved switch for diverse cell functions. Nature 1990; 348:125–32.

Bourne HR, Sanders DA, McCormick F. The GTPase superfamily: conserved structure and molecular mechanism. Nature 1991; 349:117–127.

Broek D, Samily N, Fasano O, Fujiyama A, Tamanoi F, Northup J, Wigler M. Differential activation of yeast adenylate cyclase by wild-type and mutant RAS proteins. Cell 1985; 41:763–9.

Broek D, Toda T, Michaeli T, Levin L, Birchmeier C, Zoller M, Powers S, Wigler M. The *S. cerevisiae* CDC25 gene product regulates the RAS/adenylate cyclase pathway. Cell 1987; 48:789–99.

Buchberg AM, Cleveland LS, Jenkins NA, Copeland NG. Sequence homology shared by neurofibromatosis type-1 gene and IRA-1 and IRA-2 negative regulators of the RAS cyclic AMP pathway. Nature 1990; 347:291–4.

Burgoyne RD. Yeast mutants illuminate the secretory pathway. Trends Biochem Sci 1988; 13:241–2.

Burgoyne RD, Morgan A. Low molecular mass GTP-binding proteins of adrenal chromaffin cells are present on the secretory granule. FEBS Lett 1989; 245:122–6.

Burstein ES, Linko-Stentz K, Lu Z, Macara IG. Regulation of the GTPase activity of the ras-like protein p25-rab3A. Evidence for a rab3A-specific GAP. J Biol Chem 1991; 266:2689–92.

Cassel D, Pfeuffer T. Mechanism of cholera toxin action: covalent modification of the guanyl nucleotide-binding protein of the adenylate cyclase system. Proc Natl Acad Sci USA 1978; 75:2669–73.

Cawthon R, Weiss R, Xu G, Viskochill D, Culver M, Stevens J, Robertson M, Dum D, Gesteland R, O'Connell P, White R. A major segment of the neurofibromatosis type 1 gene: cDNA sequence, genomic structure, and point mutations. Cell 1990; 62:193–201.

Chardin P, Boquet P, Madaule P, Popoff MR, Rubin EJ, Gill DM. The mammalian G protein rhoC is ADP-ribosylated by *Clostridium botulinum* exoenzyme C3 and affects actin microfilaments in Vero cells. EMBO J 1989; 8:1087–92.

Chavrier P, Parton RG, Hauri HP, Simons K, Zerial M. Localization of low molecular weight GTP binding proteins to exocytic and endocytic compartments. Cell 1990; 62:317–29.

Collier JR. Diphtheria toxin: structure and function of a cytocidal protein. In Moss J, Vaughan M, eds. ADP-ribosylating toxins and G proteins. Washington, DC: American Society for Microbiology, 1990; 3–19.

Darchen F, Zahraoui A, Hammel F, Monteils MP, Tavitian A, Scherman D. Association of the GTP-binding protein Rab 3A with bovine adrenal chromaffin granules. Proc Natl Acad Sci USA 1990; 87:5692–6.

Dexter D, Rubins JB, Manning EC, Khachatrian L, Dickey BF. Compartmentalization of low molecular mass GTP-binding proteins among neutrophil secretory granules. J Immunol 1990; 145:1845–50.

Didsbury J, Weber RF, Bokoch GM, Evans T, Snyderman R. rac, a novel ras-related family of proteins that are botulinum toxin substrates. J Biol Chem 1989; 264:16378–82.

Doucet JP, Fournier S, Parulekar M, Trifaro J-M. Detection of low molecular mass GTP-binding proteins in chromaffin granules and other subcellular fractions of chromaffin cells. FEBS Lett 1989; 247:127-31.

Downward J. The ras superfamily of small GTP-binding proteins. Trends Biochem Sci 1990; 15:469–72.

Downward J, Graves JD, Warne PH, Rayter S, Cantrell DA. Stimulation of p21 ras upon T-cell activation. Nature 1990a; 346:719–23.

Downward J, Riehl R, Wu L, Weinberg RA. Identification of a nucleotide exchange-promoting activity for $p21^{ras}$. Proc Natl Acad Sci USA 1990b; 87:5998–6002.

Drivas GT, Shih A, Coutavas E, Rush MG, D'Eustachio P. Characterization of four novel ras-like genes expressed in a human teratocarcinoma cell line. Mol Cell Biol 1990; 10:1793–8.

Eklund E, Marshall M, Gibbs J, Crean C, Gabig T. Resolution of a low molecular weight G protein in neutrophil cytosol required for NADPH oxidase activation and reconstitution of recombinant Krev-1 protein. J Biol Chem 1991; 266:13964–70.

Enomoto K, Gill DM. Cholera toxin activation of adenylate cyclase. Roles of nucleoside triphosphates and a macromolecular factorin ADP-ribocylation of the GTP-dependent regulatory component. J Biol Chem 1990; 255:1242–58.

Frech M, John J, Pizon V, Chardin P, Tavitian A, Clark R, McCormick F, Wittinghofer A. Inhibition of GTPase activating protein stimulation of ras-p21 GTPase by the Krev-1 gene product. Science 1990; 249:169–71.

Furth ME, Aldrich TH, Cordon-Cardo C. Expression of ras proto-oncogene proteins in normal human tissues. Oncogene 1987; 1:47–58.

Gabig TC, English D, Akard LP, Schell MJ. Regulation of neutrophil NADPH oxidase activation in a cell-free system by guanine nucleotides and fluoride. J Biol Chem 1987; 262:1685–90.

Garrett MD, Self AJ, Oers C, Hall A. Identification of distinct cytoplasmic targets for ras/R-ras and rho regulatory proteins. J Biol Chem 1989; 264:10–3.

Garrett MD, Major GN, Totty N, Hall A. Purification and N-terminal sequencing of the $p21^{rho}$ GTPase activating protein, rho-GAP Biochem J 1991; 276:833–836.

Gibbs JB, Schaber MD, Allard WJ, Signal IS, Scolnick EM. Purification of ras GTPase activating protein from bovine brain. Proc Natl Acad Sci USA 1988; 85:5026–30.

Gill DM, Meren R. ADP-ribosylation of membrane proteins catalyzed by cholera toxin: basis of the activation of adenylate cyclase. Proc Natl Acad Sci USA 1978; 75:3050–4.

Gill DM, Meren R. A second guanyl nucleotide-binding site associated with adenylate cyclase: distinct nucleotides activate adenylate cyclase and permit ADP-ribosylation by cholera toxin. J Biol Chem 1983; 258:11908–14.

Glomset JA, Gelb MH, Farnsworth CC. Prenyl proteins in eukaryotic cells: a new type of membrane anchor. Trends Biochem Sci 1990; 15:139–42.

Goud B, Salminen A, Walworth NC, Novick PJ. A GTP-binding protein required for secretion rapidly associates with secretory vesicles and the plasma membrane in yeast. Cell 1988; 53:753–68.

Gutierrez L, Magee AI, Marshall CJ, Hancock JF. Post-translational processing of p21 ras is two-step and involves carboxyl-methylation and carboxy-terminal proteolysis. EMBO J 1989; 8:1093–8.

Hall A. The cellular function of small GTP-binding proteins. Science 1990; 249:635–40.

Hancock JF, Paterson H, Marshall CJ. A polybasic domain or palmitoylation is required in addition to the CAAX motif to localize p21ras to the plasma membrane. Cell 1990; 63:133–9.

Harvey JJ. An unidentified virus which causes the rapid production of tumors in mice. Nature 1964; 204:1104–5.

Hata Y, Kaibuchi K, Kawamura S, Hiroyoshi M, Shirataki H, Takai Y. Enhancement of the actions of smg p21 GDP/GTP exchange protein by the protein kinase A-catalyzed phosphorylation of smg p21. J Biol Chem 1991; 266:6571–7.

Hata Y, Kikuchi A, Sasaki T, Schaber MD, Gibbs JB, Takai Y. Inhibition of the ras p21 GTPase-activating protein-stimulated GTPase activity of c-Ha-ras p21 by smg p21 having the same putative effector domain as ras p21s. J Biol Chem 1990; 265:7014–7.

Haubruck H, Plange R, Vorgias C, Gallwitz D. The ras-related mouse ypt1 protein can functionally replace the YPT1 gene product in yeast. EMBO J 1989; 8:1427–32.

Huang YK, Kung HF, Kamata T. Purification of a factor capable of stimulating the guanine nucleotide exchange reaction of ras proteins and its effect of ras-related small molecular mass proteins. Proc Natl Acad Sci USA 1990; 87:8008–12.

Isomura M, Kaibuchi K, Yamamoto T, Kawamura S, Katayama M, Takai Y. Partial purification and characterization of GDP dissociation stimulator (GDS) for the rho proteins from bovine brain cytosol. Biochem Biophys Res Commun 1990; 169:652–9.

Itoh T, Kaibuchi K, Sasaki T, Takai Y. The smg GDS-induced activation of smg p21 is initiated by cyclic AMP-dependent protein kinase-catalyzed phosphorylation of smg p21. Biochem Biophys Res Commun 1991; 177:1319–24.

Johnston C, Morris J, Hall A. The role of Ras in signal transduction. Biochem Soc Trans 1991; 19:296–9.

Kahn RA, Gilman AG. Purification of a protein cofactor required for ADP-ribosylation of the stimulatory regulatory component of adenylate4 cyclase by cholera toxin. J Biol Chem 1984; 259:6228–34.

Kahn RA, Gilman AG. The protein cofactor necessary for ADP-ribosylation of G_s by cholera toxin is itself a GTP binding protein. J Biol Chem 1986; 261:7906–11.

Kahn RA, Goddard C, Newkirk M. Chemical and immunological characterization of the 21-kDa ADP-ribosylation factor of adenylate cyclase. J Biol Chem 1988; 263:8282–7.

Kahn RA, Kern FG, Clark J, Gelmann EP, Rulka CU. Human ADP-ribosylation factors. J Biol Chem 1991; 266:2606–14.

Kaibuchi K, Mizuno T, Fujioka H, Yamamoto T, Kishi K, Fukumoto Y, Hori Y, Takai Y. Molecular cloning of the cDNA for stimulatory GDP/GTP exchange protein for smg p21s (ras p21-like small GTP-binding proteins) and characterization of stimulatory GDP/GTP exchange protein. Mol Cell Biol 1991; 11:2873–80.

Kaplan DR, Morrison DK, Wong G, McCormick F, Williams LT. PDGF β-receptor stimulates tyrosine phosphorylation of GAP and association of GAP with a signaling complex. Cell 1990; 61:125–33.

Kawata M, Kikuchi A, Hoshijima M, Yamamoto K, Hashimoto E, Yamamura H, Takai Y. Phosphorylation of smg p21, a ras p21-like GTP-binding protein, by cyclic AMP-dependent protein kinase in a cell-free system and in response to prostaglandin E1 in intact human platelets. J Biol Chem 1989; 264:15688–95.

Khosravi-Far R, Lutz RJ, Cox AD, Conroy L, Bourne JR, Sinensky M, Balch WE, Buss JE, Der CJ. Isoprenoid modification of rab proteins terminating in CC or CXC motifs. Proc Natl Acad Sci USA 1991; 88:6264–8.

Kikuchi A, Sasaki T, Araki S, Hata Y, Takai Y. Purification and characterization from bovine brain cytosol of two GTPase-activating proteins specific for smg p21, a GTP-binding protein having the same effector domain as c-ras p21s. J Biol Chem 1989; 264:9133–6.

Kinsella BT, Maltese WA. Rab GTP-binding proteins implicated in vesicular transport are isoprenylated in vitro at cysteines within a novel carboxy-terminal motif. J Biol Chem 1991; 266:8540–4.

Kirsten WH, Mayer LA. Morphologic responses to a murine erythroblastosis virus. J Natl Cancer Inst 1967; 39:311–35.

Kitayama H, Matsuzaki T, Ikawa Y, Noda M. Genetic analysis of the Kirsten-ras-revertant 1 gene: potentiation of its tumor suppressor activity by specific point mutations. Proc Natl Acad Sci USA 1990; 87:4284–8.

Knight DE, Baker PF. Guanine nucleotides and Ca-dependent exocytosis. Studies on two adrenal cell preparations. FEBS Lett 1985; 189:345–9.

Kolch W, Heidecker G, Lloyd P, and Rapp UR. Raf-1 protein kinase is required for growth of induced N1H/3T3 cells. Nature 1991; 349:426–8.

Landis C, Masters S, Spada A, Pace A, Bourne H, Vallar L. GTPase inhibiting mutations activate the α chain of G_s and stimulate adenylyl cyclase in human pituitary tumours. Nature 1989; 340:692–6.

Lapetina EG, Lacal JC, Reep BR, Molina y Vedia L. A ras-related protein is phosphorylated and translocated by agonists that increase cAMP levels in human platelets. Proc Natl Acad Sci USA 1989; 86:3131–4.

Ligeti E, Doussiere J, Vignais PV. Activation of the O_2^--generating oxidase in plasma membrane from bovine polymorphonuclear neutrophils by arachidonic acid, a cytosolic factor of protein nature, and nonhydrolyzable analogues of GTP. Biochemistry 1988; 27:193–200.

Lowe DG, Ricketts M, Levinson AD, Goeddel DV. Chimeric proteins define variable and essential regions of Ha-ras-encoded proteins. Proc Natl Acad Sci USA 1988; 85:1015–9.

Lowy DR, Zhang KH, DeClue JE, Willumsen BM. Regulation of $p21^{ras}$ activity. Trends Genet 1991; 7:346–9.

Lyons J, Landis C, Harsh G, Vallar L, Grunewald K, et al. Two G protein oncogenes in human endocrine tumors. Science 1990; 249:655–9.

Madaule P, Axel R. A novel ras-related gene family. Cell 1985; 41:31–40.

Matsuoka I, Dolly JO. Identification and localization of low-molecular-mass GTP-binding proteins associated with synaptic vesicles and other membranes. Biochim Biophys Acta 1990; 1026:99–104.

Mayorga LS, Diaz R, Colombo MI, Stahl PD. GTPγS stimulation of endosome fusion suggests a role for a GTP-binding protein in the priming of vesicles before fusion. Cell Regulation 1989a; 1:113–24.

Mayorga LS, Diaz R, Stahl PD. Regulatory role of GTP-binding proteins in endocytosis. Science 1989b; 244:1475–7.

McCormick F. Ras GTPase activating protein: signal transmitter and signal terminator. Cell 1989; 56:5–8.

McCormick F. The world according to GAP. Oncogene 1990; 5:1281–3.

Melancon P, Glick BS, Malhotra V, Weidman PJ, Serafini T, Gleason ML, Orci L, Rothman JE. Involvement of GTP-binding "G" proteins in transport through the golgi stack. Cell 1987; 51:1053–62.

Milburn MV, Tong L, DeVos AM, Brunger A, Yamaizumi Z, Nishimura S, Kim S-H. Molecular switch for signal transduction: structural differences between active and inactive forms of protooncogenic ras proteins. Science 1990; 247:939–45.

Mizoguchi A, Kim S, Ueda T, Takai Y. Tissue distribution of smg p25A, a ras p21-like GTP-binding protein, studied by use of a specific monoclonal antibody. Biochem Biophys Res Commun 1989; 162:1438–45.

Mizoguchi A, Kim S, Ueda T, Kikuchi A, Yorifuji H, Hirokawa N, Takai Y. Localization and subcellular distribution of smg p25A, a ras p21-like GTP-binding protein, in rat brain. J Biol Chem 1990; 265:11872–9.

Mollard GFV, Mignery GA, Baumert M, Perin MS, Hanson TJ, Burger PM, Jahn R, Sudhof TC. rab3 is a small GTP-binding protein exclusively localized to synaptic vesicles. Proc Natl Acad Sci USA 1990; 87:1988–92.

Moss J, Vaughan M, eds. ADP-ribosylating toxins and G proteins. Washington, DC: American Society for Microbiology, 1990; 1–567.

Nadin-Davis SA, Nasim A. A gene which encodes a predicted protein kinase can restore some functions of the ras gene in fission yeast. EMBO J 1988; 7:985–93.

Nadin-Davis SA, Yang RC, Narang SA, Nasim A. The cloning and characterization of a RAS gene from Schizosaccharomyces pombe. J Mol Evol 1986; 23:41–51.

Neal SE, Eccleston JF, Webb MR. Hydrolysis of GTP by p21 Nras, the NRAS protooncogene product, is accomplished by a conformational change in the wild-type protein: use of a single fluorescent probe at the catalytic site. Proc Natl Acad Sci USA 1990; 87:3562–5.

Orci L, Malhotra V, Amherdt M, Serafini T, Rothman JE. Dissection of a single round of vesicular transport: sequential intermediates for intercisternal movement in the golgi stack. Cell 1989; 56:357–68.

Pai EF, Krengel U, Petsko GA, Goody RS, Kabsch W, Wittinghofer A. Refined crystal structure of the triphosphate conformation of H-ras p21 at 1.35 A resolution: implications for the mechanism of GTP hydrolysis. EMBO J 1990; 9:2351–9.

Paterson HF, Self AJ, Garrett MD, Just I, Aktories K, Hall A. Microinjection of recombinant p21 rho induces rapid changes in cell morphology. J Cell Biol 1990; 111:1001–7.

Pick E, Kroizman T, Abo A. Activation of the superoxide-forming NADPH oxidase of macrophages requires two cytosolic components-one of them is also present in certain nonphagocytic cells. J Immunol 1989; 143:4180–7.

Plutner H, Schwaniger R, Pind S, Balch WE. Synthetic peptides of the Rab effector domain inhibit vesicular transport through the secretory pathway. EMBO J 1990; 9:2375–83.

Price SR, Nightingale M, Tsai S-C, Williamson KC, Adamik R, Chen H-C, Moss J, Vaughn M. Guanine nucleotide-binding proteins that enhance choleragen ADP-ribosyltransferase activity: nucleotide and deduced amino acid sequence of an ADP-ribosylation factor cDNA. Proc Natl Acad Sci USA 1988; 85:5488–91.

Redfield AG, Papastavrous MZ. NMR study of the phosphoryl binding loop in purine nucleotide proteins: evidence for strong hydrogen bonding in human N-ras p21. Biochemistry 1990; 29:3509–14.

Quinn MT, Parkos CA, Walker L, Orkin S, Dinauer MC, Jesaitis AJ. Association of a Ras-related protein with cytochrome b of human neutrophils. Nature 1989; 342:198–200.

Ridley AJ, Paterson HF, Noble M, Land H. Ras-mediated cell cycle arrest is altered by nuclear oncogenes to induce Schwann cell transformation. EMBO J 1988; 7:1635–45.

Rubins JB, Panchenko MP, Shannon T, Dickey BF. Identification of ras and ras-related low molecular mass GTP-binding proteins associated with rat lung lamellar bodies. 1992; 6:253–259.

Saikumar P, Ulsh LS, Clanton DJ, Huang KP, Shih TY. Novel phosphorylation of c-ras-p21 by proten kinases. Oncogene Res 1988; 3:213–22.

Salminen A, Novick PJ. A ras-like protein is required for a post-Golgi event in yeast secretion. Cell 1987; 49:527–38.

Santos E, Nebreda AR. Structural and functional properties of ras proteins. FASEB J 1989; 3:2151–63.

Sasaki T, Kikuchi A, Araki S, Hata Y, Isomura M, Kuroda S, Takai Y. Purification and characterization from bovine brain cytosol of a protein that inhibits the dissociation of GDP from and the subsequent binding of GTP to smg25A, a ras p21-like GTP-binding protein. J Biol Chem 1990; 265:2333–7.

Schaber MD, Garsky VM, Boylan D, Hill WS, Scolnick EM, Marshall MS, Sigal IS, Gibbs JB. Ras-interaction with the GTPase-activating protein (GAP). Proteins 1989; 6:306–15.

Segev N, Mulholland J, Botstein D. The yeast GTP-binding YPT1 protein and a mammalian counterpart are associated with the secretion machinery. Cell 1988; 52:915–24.

Seifert R, Schultz G. Fatty-acid-induced activation of NADPH oxidase in plasma membranes of human neutrophils depends on neutrophil cytosol and is potentiated by stable guanine nucleotides. Eur J Biochem 1987; 162:563–9.

Serafini T, Orci L, Amherdt M, Brunner M, Kahn RA, Rothman JE. ADP-ribosylation factor is a subunit of the coat of Golgi-derived COP-coated vesicles: a novel role for a GTP-binding protein. Cell 1991; 67:239–53.

Sewell JL, Kahn RA. Sequences of the bovine and yeast ADP-ribosylation factor and comparison to other GTP-binding proteins. Proc Natl Acad Sci USA 1988; 85:4620–4.

Schlichting I, Almo SC, Rapp G, Wilson K, Petratos K, Lentfer A, Wittinghofer A, Kabsch W, Pai EF, Petsko GA, Goody RS. Time-resolved X-ray crystallographic study of the conformational change in Ha-Ras p21 protein on GTP hydrolysis. Nature 1990; 345:309–15.

Shannon TM, Noonan TP, Hemant KM, Fishman JB, Rubins JB, Wright GE, Dickey BF. Differential binding of guanine nucleotide analogues to small GTP-binding proteins. Submitted 1991.

Sigal IS, Gibbs JB, D'Alonzo JS, Temeles GL, Wolanski BS, Socher SH, Scolnick EM. Mutant ras-encoded proteins with altered nucleotide binding exert dominant biological effects. Proc Natl Acad Sci USA 1986; 83:952–6.

Sternberg EA, Spizz G, Perry ME, Olson EN. A ras-dependent pathway abolishes activity of a muscle specific enhancer upstream from the muscle creatine kinase gene. Mol Cell Biol 1989; 9:594–601.

Stearns T, Willingham MC, Botstein D, Kahn RA. ADP-ribosylation factor is functionally and physically associated with the Golgi complex. Proc Natl Acad Sci USA 1990; 87:1238–42.

Tanaka K, Nakafuku M, Satoh T, Marshall MS, Gibbs JB, Matsumoto K, Kaziro Y, Toh-e A. *S. cerevisiae* genes IRA1 and IRA2 encode proteins that may be functionally equivalent to mammalian ras GTPase activating protein. Cell 1990; 60:803–7.

Thompson RC. EFTu provides an internal kinetic standard for translational accuracy. Trends Biochem Sci 1985; 13:91–4.

Trahey M, McCormick F. A cytoplasmic protein stimulates normal N-ras p21 GTPase, but does not affect oncogenic mutants. Science 1987; 238;542–5.

Trahey M, Wong G, Halenbeck R, Rubinfeld B, Martin GA, Ledner M, Long CM, Crosier WJ, Watt K, Koths K, McCormick F. Molecular cloning of two types of GAP complementary DNA from human placenta. Science 1988; 242:1697–700.

Tsai M-H, Yu C-L, Wei F-S, Stacey DW. The effect of GTPase activating protein upon ras is inhibited by mitogenically responsive lipids. Science 1989; 243:522–6.

Ueda T, Takeyama Y, Ohmori T, Ohyanagi H, Saitoh Y, Takai Y. Purification and characterization from rat liver cytosol of a GDP dissociation inhibitor (GDI) for liver 24kG, a ras p21-like GTP-binding protein, with proper ties similar to those of smg p25A GDI. Biochemistry 1991; 30:909–17.

Valencia A, Chardin P, Wittinghofer A, Sander C. The ras protein family: evolutionary tree and the role of conserved amino acids. Biochemistry 1991; 30(19):4637–48.

Vogel US, Dixon RAF, Schaber MD, Diehl RE, Marshall MS, Scolnick EM, Signal IS, Gibbs JB. Cloning of bovine GAP and its interaction with oncogenic ras p21. Nature 1988; 335:90–3.

Walter M, Clark SG, Levinson AD. The oncogenic activation of human p21ras by a novel mechanism. Science 1986; 233:649–52.

Walworth NC, Goud B, Kabcenell AK, Novick PJ. Mutational analysis of SEC4 suggests a cyclical mechanism for the regulation of vesicular traffic. EMBO J 1989; 8:1685.

Wasylyk C, Imler JL, Wasylyk B. Transforming but not immortalizing oncogenes activate the transcription factor PEA1. EMBO J 1988; 7:2475–83.

Wigler MH. GAPs in understanding Ras. Nature 1990; 346:696–7.

Wittinghofer A, Pai EF. The structure of Ras protein: a model for a universal molecular switch. Trends Biochem Sci 1991; 16:382–6.

Wolfman A, Macara IG. A cytosolic protein catalyzes the release of GDP from p21 ras. Science 1990; 248:67–9.

Woolley P, Clark BFC. Homologies in the structures of G-binding proteins: an analysis based on elongation factor EF-Tu. Biotechnology 1989; 7:913–20.

Xu G, Lin B, Tanaka K, Dunn D, Wood D, Gestland R, White R, Weiss R, Tamanoi F. The catalytic domain of the neurofibromatosis type 1 gene product stimulates ras GTPase and complements Ira mutants of S. cerevisiae. Cell 1990; 63:835–41.

Yamamoto T, Kaibuchi K, Mizuno T, Yiroyoshi M, Shirataki H, Takai Y. Purification and characterization from bovine brain cytosol of proteins that regulate the GDP/GTP exchange reaction of smg p21s, ras p21-like GTP-binding proteins. J Biol Chem 1990; 265:16626–34.

Zahraoui A, Touchot N, Chardin P, Tavitian A. The human rab genes encode a family of GTP-binding proteins related to yeast YPT1 and SEC4 products involved in secretion. J Biol Chem 1989; 264:3–8.

5

Phosphoinositides and Intracellular Calcium

FRANK S. MENNITI

Pfizer Central Research
Groton, Connecticut

KUTSUMI NOGIMORI

Kaken Pharmaceutical Co., Ltd.
Kyoto, Japan

MICHEL F. ROSSIER

University Hospital
Geneva, Switzerland

GARY ST. JOHN BIRD, MAJOR CLAY GLENNON, JOHNNY F. OBIE, KERRY G. OLIVER, and JAMES W. PUTNEY, JR.

National Institute of Environmental Health Sciences
Research Triangle Park, North Carolina

I. Introduction

The regulation of intracellular Ca^{2+} concentration ($[Ca^{2+}]_i$) is fundamental to the homeostasis of virtually all cells. Calcium, as an allosteric modulator, plays a role in the regulation of a wide variety of intracellular processes. In this regard, one of the most significant aspects of Ca^{2+} homeostasis is its regulation by signals external to the cell. The role for Ca^{2+} in the response of cells to external signals was first recognized by Ringer in 1883 in his studies on the Ca^{2+} requirement for cardiac contraction. It is now appreciated that Ca^{2+} signaling is involved across the phylogenetic spectrum and in such diverse processes as secretion, muscle contraction, cell growth and differentiation, and chemotaxis. Extracellular signals are transduced into changes in $[Ca^{2+}]_i$ by a number of different signaling systems. The most widespread of these is the phosphoinositide/Ca^{2+} cascade, which is the focus of this review. A schematic diagram of the components of this system is presented in Fig. 1.

Our current understanding of the phosphoinositide/Ca^{2+} cascade germinates from a review by Michell (Michell, 1975), who recognized the link between the "phosphoinositide effect" first described by the Hokins in 1955 (Hokin and Hokin, 1955) and Ca^{2+} mobilization in a variety of cells [see

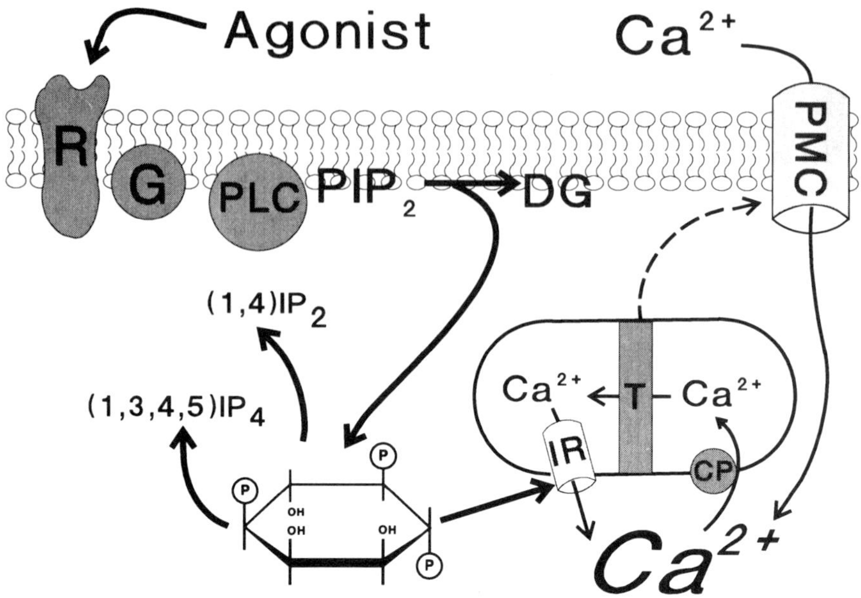

Figure 1 Schematic diagram of the phosphoinositide/Ca^{2+} signaling cascade. See text for detailed descriptions. R, plasma membrane receptor; G, heterotrimeric G protein; PLC, phospholipase C; IR, $(1,4,5)IP_3$ receptor and integral Ca^{2+} channel; CP, thapsigargin-sensitive Ca^{2+}-pumping ATPase; T, hypothetical translocation step between Ca^{2+} uptake and release sites in the $(1,4,5)IP_3$-regulated Ca^{2+} pool; PMC, plasma membrane Ca^{2+} entry mechanism, shown here as a Ca^{2+} channel, the dashed arrow indicating the pathway for communication between the $(1,4,5)IP_3$-regulated Ca^{2+} pool and this Ca^{2+} entry mechanism; PIP_2, phosphatidylinositol (4,5)*bis*phosphate; DG, diacylglycerol. The chemical structure is that of $(1,4,5)IP_3$. Note that for the tyrosine kinase family of receptors, it does not appear that an intermediary G protein is involved in the coupling of receptor and phospholipase C.

Michell (1986) for a historical review]. However, the molecular basis for this link was not clearly established until the discovery that the initial reaction in receptor-stimulated phosphoinositide turnover involved not the major isositol lipid but rather a minor phosphorylated derivative, phosphatidylinositol 4,5-*bis*phosphate (Abdel-Latif et al., 1977; Kirk et al., 1981) (Fig. 1). Thus, receptor activation, through stimulation of phospholipase C, causes the breakdown of this membrane lipid, and Berridge (Berridge, 1983) proposed that the water-soluble product of this reaction, inositol (1,4,5)*tris*phosphate [$(1,4,5)IP_3$], is the intracellular messenger which activates the Ca^{2+} mobilization process. He, with collaborators in England and Germany, demonstrated that $(1,4,5)IP_3$ in micromolar concentrations rapidly released Ca^{2+} from a nonmitochondrial store in permeabilized pancreatic acinar cells (Streb et al., 1983). This result was quickly confirmed by a number of different laboratories working with preparations from a variety of cells and established $(1,4,5)IP_3$ as the intracellular messenger which mediates receptor-stimulated Ca^{2+} mobilization.

With the identification of the role for $(1,4,5)IP_3$, it became apparent that receptor activation of phospholipase C yields a truly multifaceted intracellular signaling cascade. Nishizuka and colleagues had previously demonstrated that the other product of phosphatidylinositol 4,5-*bis*phosphate breakdown, diacylglycerol, is an activator of protein kinase C (Nishizuka, 1984). Consequently, diacylglycerol formation during receptor stimulation results in cell-specific protein phosphorylations. Furthermore, the 2-acyl position of phosphatidylinositol 4,5-*bis*phosphate is characteristically occupied by an arachidonyl moiety which is released from diacylglycerol through the activity of a diacylglycerol lipase. Arachidonic acid may itself have an intracellular signaling function and is the precursor to another family of signaling molecules, the eicosanoids. Very recently, a family of novel polyphosphoinositides has been discovered. These species arise from the action of a phosphatidylinositol 3-kinase which associates with, and appears to be activated by, certain protein tyrosine kinases such as ligand-activated PDGF receptor (Auger and Cantley, 1991). It has been suggested that these 3-phosphorylated inositol lipids mediate some of the effects of these protein tyrosine kinases.

In the years since the discovery of $(1,4,5)IP_3$ and its action, much about the phosphoinositide/Ca^{2+} signaling system has been learned. Here is presented a brief overview of these many developments. The reader will also be directed to selected references and more detailed reviews on specific aspects of this body of research.

II. The Ca^{2+} Response

In virtually all systems examined thus far, receptor-activated Ca^{2+} mobilization via the phosphoinositide/Ca^{2+} signaling system involves two phases: (1) Ca^{2+} release from an intracellular store, and (2) a more prolonged phase of extracellular Ca^{2+} entry (Putney, 1987). These two phases of the Ca^{2+} response can be demonstrated by stimulating cells with a phospholipase C-linked agonist in the presence or absence of extracellular Ca^{2+}. In the absence of extracellular Ca^{2+}, such stimulation results in only a transient increase in $[Ca^{2+}]_i$, originating from the release of a finite intracellular Ca^{2+} pool. In the presence of physiological (mM) extracellular Ca^{2+}, the elevation in $[Ca^{2+}]_i$ is sustained due to the additional component of extracellular Ca^{2+} influx. Extracellular Ca^{2+} influx is also the source for refilling of the intracellular Ca^{2+}

pool upon termination of receptor activation. The available evidence now strongly indicates that the initial internal Ca^{2+} release is signaled by (1,4,5)IP_3 (Berridge and Irvine, 1984). The mechanism by which (1,4,5)IP_3 releases intracellular Ca^{2+} involves its interaction with a specific intracellular membrane receptor and opening of a Ca^{2+} channel on the responsive organelle. This organelle was originally believed to be the endoplasmic reticulum (ER), but more recently it has been suggested to be either a highly specialized ER component or, in fact, a novel organelle termed "calciosome" (Volpe et al., 1988; see below). Much is now known about the mechanisms within the cell which mediate this initial, (1,4,5)IP_3-induced phase of Ca^{2+} mobilization; this is discussed in detail in subsequent sections of this review. In contrast, the regulation of the second, Ca^{2+} entry phase of Ca^{2+} mobilization is poorly understood. This aspect of Ca^{2+} mobilization is described below.

Injection of (1,4,5)IP_3 directly into a variety of cells produces a response pattern suggestive of an activation of both intracellular Ca^{2+} release and entry of Ca^{2+} from the extracellular space. However, this does not appear to be a direct effect of (1,4,5)IP_3 since, in the majority of studies, the direct application of (1,4,5)IP_3 to plasma membranes has no effect on Ca^{2+} permeability (Delfert et al., 1986; Dargemont et al., 1988). Thus, whereas (1,4,5)IP_3 may be responsible for the activation of Ca^{2+} entry in cells, it does not do so by acting directly. Instead, it appears that Ca^{2+} entry is activated as a result of the emptying of a (1,4,5)IP_3-sensitive pool (Putney, 1986). Specifically, when (1,4,5)IP_3 is being continuously maintained at an elevated level during agonist activation of phospholipase C, the Ca^{2+} pool is held empty and the pathway for Ca^{2+} entry from the extracellular space is open. This scenario, which has been termed capacitative Ca^{2+} entry (Putney, 1986), is indicated by two lines of experimental evidence: (1) an observed transient elevation in plasma membrane Ca^{2+} permeability during Ca^{2+} pool refilling under certain conditions, and (2) the activation of Ca^{2+} influx by the agent thapsigargin, which depletes the intracellular Ca^{2+} pool by a mechanism different from that of (1,4,5)IP_3. The model system in which these observations were made in our laboratory was the rat parotid acinar cell, as described below.

When parotid cells are incubated in a medium containing no added Ca^{2+}, the $[Ca^{2+}]_i$ decreases from 100 nM to about 50 nM, as measured with the intracellular fluorescent Ca^{2+} indicator, fura-2. Likewise, when extracellular Ca^{2+} is returned to physiological (mM) concentrations, a corresponding restoration of $[Ca^{2+}]_i$ occurs. These changes in $[Ca^{2+}]_i$ reflect the equilibration of $[Ca^{2+}]_i$ in response to the changed plasma membrane Ca^{2+} gradient and, therefore, a function of the permeability of the plasma membrane to Ca^{2+}. Such changes in $[Ca^{2+}]_i$ were used by Takemura and Putney (1989) to assay for changes in plasma membrane Ca^{2+} permeability as a function of the status of Ca^{2+} in the intracellular pool. The intracellular Ca^{2+} pool was depleted by stimulating parotid cells in the absence of extracellular Ca^{2+} with the muscarinic agonist, methacholine, which activates phospholipase C and $(1,4,5)IP_3$ production. The parotid cells were then treated with the muscarinic antagonist, atropine, after which elevated levels of inositol phosphates rapidly return to basal levels. However, since extracellular Ca^{2+} is the source for intracellular pool refilling, the intracellular pool remained depleted. Thus, conditions were established in which the effect on plasma membrane Ca^{2+} permeability of depleting the $(1,4,5)IP_3$-sensitive Ca^{2+} pool could be ascertained in the absence of receptor activation or elevation in second messengers. When plasma membrane Ca^{2+} permeability was assayed by addition of extracellular Ca^{2+}, it was observed that the $[Ca^{2+}]_i$ increased transiently to a higher level than that observed in cells in which the Ca^{2+} pool was not previously depleted. Thus, the plasma membrane Ca^{2+} permeability was elevated under conditions where the $(1,4,5)IP_3$-sensitive Ca^{2+} pool was depleted.

The above data provided strong evidence that depletion of the $(1,4,5)IP_3$-sensitive Ca^{2+} pool in the parotid cell increases plasma membrane permeability to Ca^{2+}. An independent approach which confirmed this finding came through the use of the non-phorbol ester tumor promoter, thapsigargin (Takemura et al., 1989). This agent releases intracellular Ca^{2+} by a mechanism which does not involve elevation of cellular inositol phosphates (Thastrup et al., 1990). Specifically, thapsigargin inhibits a certain fraction of microsomal Ca^{2+}-ATPases which includes the Ca^{2+} pump responsible for sequestration of Ca^{2+} into the $(1,4,5)IP_3$-sensitive pool (see below). Figure 2A shows that, in the absence of intracellular Ca^{2+}, thapsigargin depletes an intracellular Ca^{2+} pool corresponding to that released by agonist (in this case the muscarinic agonist, methacholine). In the presence of extracellular Ca^{2+}, this agent induces a sustained increase in $[Ca^{2+}]_i$ (Fig. 2B). Following the initial intracellular Ca^{2+} release, this sustained response becomes dependent on extracellular Ca^{2+}; i.e., it results from Ca^{2+} entry. Furthermore, methacholine, added to cells during thapsigargin-induced Ca^{2+} entry causes only a brief increase in intracellular $[Ca^{2+}]$, which then returns to the level achieved by thapsigargin alone (Fig. 2C), indicating that the Ca^{2+} entry induced by thapsigargin is not additive with that induced by agonist. Thus, it appears that thapsigargin and methacholine activate the same mechanism for Ca^{2+} entry.

The activation of Ca^{2+} entry by thapsigargin has now been observed in a wide variety of cell types (Putney, 1990). Taken with the previous observation discussed above, these actions of thapsigargin provide strong support for the capacitative Ca^{2+} entry hypothesis. However, the results of the above experiments do not reveal the mechanism(s) underlying this phenomenon. It is possible that the emptying of the intracellular pool could influence the permeability of the plasma membrane at some distance, in which case it must be considered

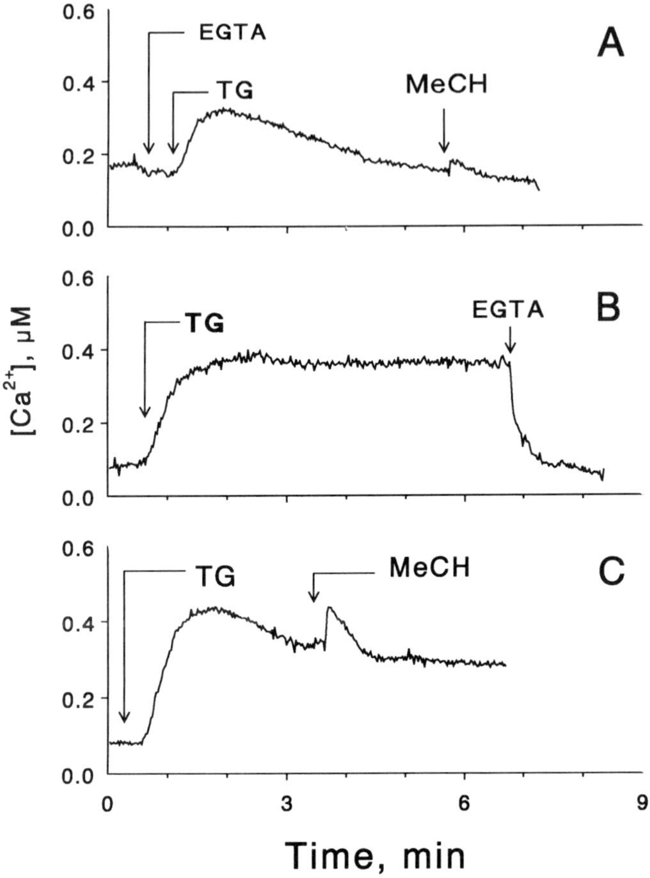

Figure 2 The effect of thapsigargin on $[Ca^{2+}]_i$ in the presence or absence of extracellular Ca^{2+} or the muscarinic agonist, methacholine, in rat parotid acinar cells. The $[Ca^{2+}]_i$ of populations of rat parotid acinar cells was determined using the Ca^{2+}-sensitive fluorescent dye, fura-2, as described by Takemura et al. (1989). (A) The extracellular Ca^{2+} concentration (1 mM) was first reduced to >100 nM by addition of 2 mM EGTA. Thapsigargin (TG, 2 μM) and methacholine (MeCH, 100 μM) were then added sequentially. (B) Thapsigarin and EGTA were added sequentially. (C) Thapsigargin and methacholine were added sequentially.

that this communication could be accomplished through some kind of intracellular chemical signaling. It has been suggested that, together with $(1,4,5)IP_3$, the phosphorylated $(1,4,5)IP_3$ metabolite $(1,3,4,5)IP_4$ (see below), modulates Ca^{2+} mobilization and/or may play a role in the regulation of Ca^{2+} influx (Changya et al., 1989; Morris et al., 1987). Data in support of this role for $(1,3,4,5)IP_4$ has been obtained from experiments using sea urchin eggs and mouse lacrimal cells (Irvine and Moor, 1987; Morris et al., 1987). Thus, in this latter system, it was found that $(1,4,5)IP_3$, when dialyzed into cells through patch pipettes, was capable of inducing a transient mobilization of Ca^{2+}, but the response could be sustained only if $(1,3,4,5)IP_4$ was also added. On the other hand, in this same cell type, thapsigargin induces Ca^{2+} influx (Kwan et al., 1990). Furthermore, Bird et al. (1991a) reported that direct injection into these cells of the poorly metabolized $(1,4,5)IP_3$ analog, $(2,4,5)IP_3$, was capable of fully activating both Ca^{2+} entry as well as intracellular Ca^{2+} release. Thus, an obligatory role for $(1,3,4,5)IP_4$ in Ca^{2+} signaling must be questioned until an explanation for these apparent paradoxes is obtained.

The debate over the role for $(1,3,4,5)IP_4$ notwithstanding, there is little experimental evidence which has provided a clue as to the mechanism for activation of Ca^{2+} influx. The study of this process has been hampered by the fact that the structural basis for the increase in Ca^{2+} permeability which underlies Ca^{2+} entry is not known. It is possible that it results from the opening of a Ca^{2+} channel in the plasma membrane; however, intensive study has yet to reveal such an entity, at least at the single channel level. Thus, the possibility must be entertained that Ca^{2+} entry is mediated by a nonchannel mechanism such as the activation of a Ca^{2+} exchange system (for example, a H^+/Ca^{2+} exchange). Recently, two compounds, SK&F 96365 and L-651,582, have been reported to block Ca^{2+} entry in a variety of different cell types by unknown mechanisms (Merritt et al., 1990; Felder et al., 1991; Hupe et al., 1991). Unfortunately, these compounds are somewhat nonspecific in action; both block voltage activated Ca^{2+} channels as well. Nonetheless, through further structure/activity studies, these compounds may spawn much-needed pharmacological tools for the identification of components critical to the mediation of Ca^{2+} entry.

Another level of complexity to the interplay between intracellular Ca^{2+} release and Ca^{2+} entry was revealed in 1986, when Woods et al. (1986) observed oscillations in $[Ca^{2+}]_i$ in single rat hepatocytes stimulated with low levels of agonists. This striking finding has now been confirmed for a wide variety of cells, and it seems likely to be a universal aspect of intracellular Ca^{2+} regulation. In fact, it now appears that oscillations in $[Ca^{2+}]_i$ are organized spatially as well as temporally in that Ca^{2+} oscillations are propagated throughout the cell in discrete Ca^{2+} waves (Rooney et al., 1990). Since the discovery of Woods et al., considerable advances have been made in the characterization of this phenomenon, as discussed below.

There are two types of Ca^{2+} oscillations, as shown in Fig. 3. The first type is a sinusoidal fluctuation in $[Ca^{2+}]_i$ which occurs on top of a sustained elevation in $[Ca^{2+}]_i$ (Fig. 3A). This type of oscillation has to date been observed only following muscarinic receptor stimulation. Although they occur at low muscarinic agonist concentrations, no oscillations are observed at levels of stimulation below a certain threshold. Beyond this threshold, increasing agonist concentration causes an increase in the amplitude of the oscillations, with little or no change in period. With further increases in stimulation, the sustained rise in $[Ca^{2+}]_i$ approaches the peak level of $[Ca^{2+}]_i$ attained during oscillation such that oscillations are no longer observed.

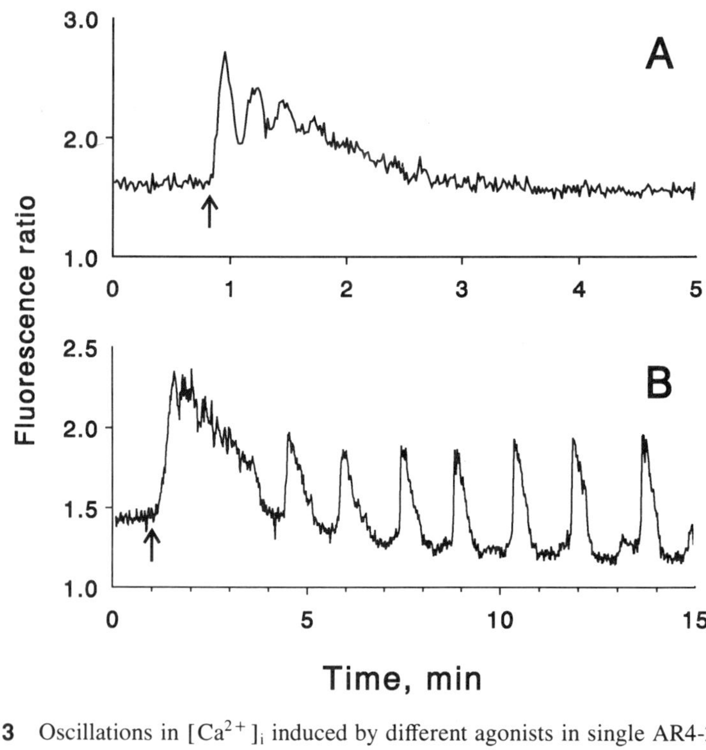

Figure 3 Oscillations in $[Ca^{2+}]_i$ induced by different agonists in single AR4-2J cells. Single AR4-2J cells (a rat pancreatoma cell line) were loaded with the fluorescent Ca^{2+} indicator fura-2 and $[Ca^{2+}]_i$ was determined as described by Bird et al. (1992). (A) A series of three sinusoidal Ca^{2+} oscillations induced by 0.5 μM methacholine. (B) Discrete Ca^{2+} transients induced by 10 pM angiotensin II.

The second type of Ca^{2+} oscillation is characterized by discrete spikes or transients in $[Ca^{2+}]_i$ (Fig. 3B). These occur in response to a wide variety of phospholipase C-linked agonists. $[Ca^{2+}]_i$ in the intervening periods between the Ca^{2+} transients remains at or close to basal level; in some, but not all, cases a small "pacemaker" elevation in $[Ca^{2+}]_i$ which precedes the Ca^{2+} transients has been observed. As with sinusoidal oscillations, Ca^{2+} transients generally occur only at low agonist concentrations; however, in this case the effect of agonist concentration is on the frequency of oscillation but not on the amplitude. Above a given threshold of stimulation, only a single spike occurs, which may or may not be followed by a sustained, nonoscillatory elevation in $[Ca^{2+}]_i$.

Both types of Ca^{2+} oscillations originate from the release of an intracellular Ca^{2+} pool. This is indicated by the fact that oscillations typically continue for several to many cycles before cessation if extracellular Ca^{2+} is removed. This observation also indicates that the oscillating intracellular Ca^{2+} pool is ultimately recharged by Ca^{2+} derived from extracellular Ca^{2+} influx. A role for extracellular Ca^{2+} influx in the regulation of Ca^{2+} oscillations is further indicated by the finding that increasing concentrations of extracellular Ca^{2+} typically intensify oscillatory patterns; i.e., this manipulation increases the amplitude of sinusoidal oscillations and the frequency of Ca^{2+} transients. In several cells types, such as hepatocytes (Rooney et al., 1990), it has also been observed that Ca^{2+} oscillations emanate from a discrete locus within a given cell via a self-propagating wave; i.e., the amplitude does not diminish with distance from the origin. Whether this reflects functionally distinct Ca^{2+} pools involved in wave initiation and propagation or the unique proximity of a portion of the Ca^{2+} pool to the machinery which generates the initiating signal is important from a mechanistic standpoint and remains to be determined.

As can be seen in Fig. 3, both transient and sinusoidal oscillations can occur in the same cell, albeit in response to activation of different receptors. Furthermore, the characteristics of transient oscillations vary with receptor type in a given cell. A case in point is the hepatocyte, where the kinetics and amplitude of the increase in $[Ca^{2+}]_i$ are independent of agonist type, but where the decrease in $[Ca^{2+}]_i$ has a time course which is characteristic of the activating agonist (Woods et al., 1987). These data indicate that the activated receptor contributes a modulatory input to the oscillatory pattern.

The mechanism(s) of oscillation in $[Ca^{2+}]_i$, including propagation, is not yet understood. The minimal requirements for the generation of sinusoidal Ca^{2+} oscillations are a stimulus to drive the process and an oscillating feedback-inhibitory mechanism. The minimal requirements for the generation of repetitive transients are more complex. Again, a stimulus to drive the process is required; however, in this case the generation of discrete spikes also requires a feed-forward mechanism activated at a certain stimulatory threshold. The feed-forward is then superseded by a mechanism to terminate the transient and return Ca^{2+} to basal levels. Several models have been proposed which can account for some, but not all, of the observations pertaining to these phenomena. These models focus on the much more widely observed transient Ca^{2+} oscillations. While it is beyond the scope of this review to give a detailed presentation and analysis of each of these various proposals, basic tenets are presented below. The reader is also directed to several more detailed reviews (Meyer and Stryer, 1991; Harootunian et al., 1991; Cobbold et al., 1991; Berridge, 1991).

Perhaps the most important question to be resolved is the nature of the oscillating component(s). Ca^{2+} oscillations could be generated by an oscillation in the level of $(1,4,5)IP_3$; thus, the repetitive changes in $[Ca^{2+}]_i$ would result from release of Ca^{2+} when $(1,4,5)IP_3$ level is high, followed by pool refilling when $(1,4,5)IP_3$ levels decline. As proposed by Cobbold and coworkers (Cobbold et al., 1991), oscillations in $(1,4,5)IP_3$ levels could be brought about by an effect at the level of the receptor, coupling guanine nucleotide-binding protein (see below), and/or phospholipase C. An additional component could be derived through the metabolism of $(1,4,5)IP_3$. Evidence for an increase in $(1,4,5)IP_3$ level associated with a discrete Ca^{2+} transient has been obtained in a population of synchronized REF52 cells (Harootunian et al., 1991). On the other hand, Ca^{2+} oscillations can be generated by a number of manipulations which do not apparently cause an increase in $(1,4,5)IP_3$ formation. These include infusion of a constant level of exogenous $(1,4,5)IP_3$ or poorly metabolized $(1,4,5)IP_3$ analogs (Wakui et al., 1989) or Ca^{2+} (Wakui et al., 1990); this latter effect was not blocked by infusion of the $(1,4,5)IP_3$ receptor antagonist heparin. Further evaluation of a possible role for oscillations in $(1,4,5)IP_3$ levels would be significantly advanced by a technique for measuring changes in $(1,4,5)IP_3$ in a single cell.

Proposals which do not require an oscillatory change in $(1,4,5)IP_3$ levels suggest instead that Ca^{2+} itself produces positive and/or negative feedback on the Ca^{2+} release process. Positive Ca^{2+} feedback encompasses the concept of Ca^{2+}-induced Ca^{2+} release (CICR), first described as a mechanism for Ca^{2+} release from sarcoplasmic reticulum (SR) of muscle (Endo, 1985). According to this proposal, during agonist activation the release of Ca^{2+} from the $(1,4,5)IP_3$-sensitive Ca^{2+} pool, coupled with extracellular Ca^{2+} influx, charges discrete CICR pools. As these pools reach capacity, a small, pacemaker change in extraorganellar and/or change in intraorganellar Ca^{2+} concentration triggers Ca^{2+} release. Dissipation of the charged pool, along with other negative feedback mechanisms, terminate the spike and the system undergoes further rounds of Ca^{2+} charging and release.

In muscle, CICR is mediated by the 400-kDa foot protein (ryanodine receptor) of the SR and has a complex pharmacology which includes the effects of ryanodine, ruthenium red, and caffeine. In some, but by no means all, nonmuscle cell types, these agents have been found to affect the pattern of Ca^{2+} oscillation (see, for example, Cheek et al., 1991; Malgaroli et al., 1990; Dehlinger-Kremer et al., 1991). In fact, evidence for the expression of ryanodine receptors in nonmuscle cells has recently been obtained (Nakai et al., 1990; Shoshan-Barmatz et al., 1990). These data raise the possibility that a musclelike CICR system in nonmuscle cells is principally responsible for the oscillatory changes in $[Ca^{2+}]_i$. In this scenario, the (1,4,5)IP_3-sensitive pool plays a role as pacemaker and/or in the regulation of Ca^{2+} influx to drive the process. Alternatively, it is possible that the (1,4,5)IP_3 receptor and associated Ca^{2+} channel could itself mediate CICR through effects of intraluminal (and/or extraluminal) Ca^{2+} on the regulation of Ca^{2+} release through this channel (Missiaen et al., 1991; see below). CICR is a particularly attractive mechanism to account for Ca^{2+} transients, since it can also easily account for the propagation of Ca^{2+} waves. However, CICR alone cannot fully account for these phenomena, since such a mechanism predicts the characteristics of the Ca^{2+} transients to be independent of the type of activating receptor.

Ca^{2+} may also play a feedback-inhibitory role on (1,4,5)IP_3-induced Ca^{2+} release. It has been found that the cytoplasmic $[Ca^{2+}]$ can affect the binding of (1,4,5)IP_3 to its receptor, as well as the efficacy of Ca^{2+} release in a wide variety of cell types (see, for example, Payne et al., 1990; Willems et al., 1990; Iino, 1990; Jean and Klee, 1986). Effects of the intraorganellar $[Ca^{2+}]$ have been proposed as well (Irvine, 1990; Cheek et al., 1991; Missiaen et al., 1991). Inhibitory effects of Ca^{2+} on (1,4,5)IP_3-induced Ca^{2+} release are insufficient in and of themselves to generate Ca^{2+} transients but may comprise a component of the mechanism for the termination of such events. Such inhibitory effects may also play an important role in the development of sinusoidal oscillations.

The above brief discussion delineates some of the functional characteristics of Ca^{2+} response(s) to phospholipase C-linked receptor activation. The next two sections describe the identification and characterization of some of the many cellular components which are involved in the regulation of these responses. The first section focuses on phospholipase C and the receptors to which it is coupled. The second section then describes the characteristics of the (1,4,5)IP_3-sensitive Ca^{2+} pool.

III. Phospholipase C and Receptor Activation

In mammalian cells, at least three distinct phospholipase C isozymes have been identified, phospholipase C-β, -δ, and -γ (see Meldrum et al., 1991, for a detailed review). The inclusion of the putative phospholipase C α as a member of this enzyme family is as yet unresolved. There is now also convincing evidence that each isozyme is represented by more than one subtype. The phospholipase C isozymes can be distinguished based on molecular weight and immunological criteria. Furthermore, the primary protein sequences for at least some of the subtypes of each isozyme have been deduced through molecular cloning techniques. Immunological and sequence data indicate that the isozymes share only modest sequence homology. The notable exception is two domains, termed X (150 aa) and Y (240 aa). It is an attractive hypothesis that these two domains may represent common substrate-binding and/or catalytic sites (Meldrum et al., 1991).

The phospholipase C subtypes can be further differentiated based on catalytic activity. These phospholipases are specific for inositol lipids and can attack, in vitro, each of the naturally occurring phosphorylated derivatives (with the exception of the newly discovered phosphatidylinositol 3-phosphates). The subtypes differ, however, with respect to catalytic preference for these species (Meldrum et al., 1991). On the other hand, it appears that exclusively phosphatidylinositol 4,5-*bis*phosphate is the physiological substrate for these enzymes; i.e., the accumulation of inositol *bis*- and *mono*phosphates during agonist activation are accounted for by the breakdown of the phosphatidylinositol 4,5-*bis*phosphate product, (1,4,5)IP_3 (Hughes and Putney, 1989). Thus, the significance of the differences in catalytic activity and differential preferences for the inositol lipids remains to be determined. The question of which of these various phospholipase C isozymes is activated by receptors is discussed below.

The role of $[Ca^{2+}]_i$ in the regulation of phospholipase C remains controversial. It has been proposed that Ca^{2+} activation of phospholipase C in intact cells may provide a component of the feed-forward mechanism to induce Ca^{2+} transients (Meyer and Stryer, 1988). However, it would appear that changes in $[Ca^{2+}]_i$ in the physiological range have little effect on phospholipase C activity in the majority of cell types. For example, ionomycin-induced increases in $[Ca^{2+}]_i$ were found to have little or no effect on phospholipase activity in pancreatic acinar cells or hepatocytes, respectively (Taylor et al., 1986; Renard et al., 1987). More broadly, it has been widely observed that thapsigargin, which fully activates the agonist-regulated Ca^{2+} mobilization pathway (see above), fails to increase inositol phosphate formation (Thastrup, 1990). Nonetheless, given the importance to the overall regulation of $[Ca^{2+}]_i$ in all cells, the regulation of phospholipase C activity by $[Ca^{2+}]_i$ requires further evaluation.

Receptors for a wide variety of ligands have been found to couple to phospholipase C and induce inositol phosphate formation and Ca^{2+} mobilization. These include receptors for neurotransmitters and neuromodulators, chemoattractants, and hormones and growth factors occurring in a vast number of eukaryotic cell types. Virtually every mammalian cell examined has been found to express phospholipase C-linked receptors. These receptors fall into two broad categories: those which couple via an intermediary guanine nucleotide-binding protein (G-protein), and those of the protein tyrosine kinase family of receptors.

The phospholipase C/G-protein coupling motif is shared with receptors linked to other signaling systems including adenylate cyclase, phospholipase A_2, cGMP phosphodiesterase, and G-protein regulated ion channels (Birnbaumer, 1991). The first intimation that receptor coupling to phospholipase C occurs via an intermediary G protein came from the demonstrations of Cockcroft and Gomperts (1985), who showed that the nonhydrolyzable GTP analog, GTPγS, activated phospholipase C in mast cells, and of Litosch et al. (1985), who demonstrated that guanine nucleotides were required for agonist to activate phospholipase C in blowfly salivary gland. Similar results have subsequently been observed with a variety of phospholipase C-linked receptors. By analogy with other receptor/G-protein-coupled systems, these data suggested that receptor activation of phospholipase C is mediated by a member of the family of heterotrimeric G proteins (Birnbaumer, 1991).

The isolation and characterization of the phospholipase C-linked G protein(s) has lagged well behind that of the much more intensively studied group of G proteins involved in the regulation of adenylate cyclase. In some few instances, it has been found that receptor activation of phospholipase C is inhibited by pertussis toxin, a hallmark of receptor-effector coupling mediated by a Gα_i (Birnbaumer, 1991). However, in the majority of cases it would appear that the putative G-protein(s) coupling to phospholipase C is distinct from those involved in the regulation of other signaling systems. Recently, Taylor et al. (1991) and Smrcka et al. (1991) have described the isolation of Gα subunits capable of activating phospholipase C in reconstitution systems. These α subunits correspond to a new class, termed Gα_q, which was also identified by Strathmann and Simon (1990) using molecular cloning techniques. The Gα_q protein(s) is of similar molecular weight ($\sim$42 kDa) to that of other members of the Gα family but appears to bear little sequence homology outside of several domains which are highly conserved across this entire Gα family (Strathmann and Simon, 1990). Gα_q has low intrinsic GTPase activity, binds guanine nucleotides poorly, and is not significantly activated by GTPγS. This may reflect a very high affinity for GDP with a low rate of dissociation. As noted by Smrcka et al. (1991), these properties of Gα_q, coupled with the fact that this subunit has a relatively low apparent K_m for phospholipase C activation, have probably contributed to the difficulty in isolating this species. Northern analysis of various mouse tissues indicate that Gα_q (and the closely related Gα_{11}) is ubiquitously expressed, as would be expected for a molecule involved with the phosphoinositide/Ca^{2+} cascade, and suggests that there may be multiple forms of the protein (Strathmann and Simon, 1990). However, it should also be noted that in many broken cell preparations, phospholipase C activity is readily stimulated by GTPγS addition, presumably through activation of a G protein (Martin, 1989). Whether this reflects a complex regulation for the G_q family or indicates the activity of other G proteins coupling to phospholipase C remains to be determined.

Receptors which couple to phospholipase C via G proteins appear to be members of the receptor "superfamily" whose effects are also mediated through intermediary G proteins. A number of the receptors in this family, including several phospholipase C-linked receptors, have been cloned, and primary protein sequences have been deduced. Many of the structural details of this receptor type have been illuminated through the analysis of the rhodopsin, β-adrenergic, and muscarinic receptors (see O'Dowd et al., 1989, for a review). This receptor family is characterized by an N-terminal extracellular domain, an intracellular C terminus, and seven transmembrane spanning domains linked by various intra- and extracellular loops. Some regions of the seven transmembrane segments are conserved across the family, suggesting that they may play a structural role and/or comprise common sites for interaction with effectors such as G proteins. Additional regions of the membrane spanning domains appear to form the ligand-binding domain (Tota et al., 1991). Of the other domains, it appears that the third cytoplasmic loop, as well as that portion of the C terminus which emerges from the membrane, are involved in G-protein coupling (O'Dowd et al., 1989). The third cytoplasmic loop, and in some cases the C terminus, also contain sites for phosphorylation which may be involved in receptor regulation.

Evidence is accumulating that the β isozyme of phospholipase C is activated by receptors through G protein(s). Taylor et al. (1991) reported that the $\beta 1$, but not the $\delta 1$ or $\gamma 1$, phospholipase C isozyme was activated by $G\alpha_q$. Although it was not examined directly, Smrcka et al. (1991) reported that the phospholipase C preparation used in their reconstitution studies contained detectable β isozyme but not δ or γ isozyme. However, these data do not exclude the possibility that other phospholipase C isozymes are also coupled to receptors via an intermediary G protein(s) of the $G\alpha_q$ subtype or other subtype.

The second category of receptors which activate phospholipase C are those of the protein tyrosine kinase family. Ligand binding to several members of this family of receptors has been demonstrated to result in the activation of phospholipase C (see Rhee, 1991, for a review). This includes receptors for EGF, PDGF, bFGF, and the *trk* oncogene product. In many cells, the characteristics of phospholipase C activation through tyrosine kinase receptors differ significantly from those associated with putative G-protein-linked receptors (see, for example, Hepler et al., 1990). In fact, evidence is accumulating that coupling between the tyrosine kinase class of receptors and phospholipase C can occur through a direct interaction of the ligand-bound receptor and phospholipase C. Furthermore, it appears to be the phospholipase C γ isozyme that is involved in this interaction.

Phospholipase C γ contains two "*src* homology" domains (SH-2 and SH-3), unlike any of the other phospholipase C isozymes characterized to date (Rhee, 1991). These sequences are shared with a diverse group of proteins which associate with ligand-bound tyrosine kinase receptors. This association, in turn, is thought to be critical for the interaction and activation of this group of proteins by these receptors. Experimentally, phospholipase C γ can be coprecipitated with antibody directed at the EGF and PDGF receptor in a ligand-dependent fashion, indicating that a physical association between the receptor and the phospholipase C occurs upon ligand binding (Meisenhelder et al., 1989). Phospholipase C γ is also tyrosine phosphorylated by these activated receptors. In most cases, however, it has been observed that tyrosine phosphorylation fails to result in a corresponding activation of phospholipase C γ in vitro. This has led to the suggestion that perhaps such phosphorylation alters the interaction of phospholipase C γ with an inhibitory element in vivo. Recently, it has been reported that profilin, an abundant cytoplasmic actin-binding protein, binds specifically and with high affinity to phosphatidylinositol 4,5-*bis*phosphate, inhibiting its breakdown by phospholipase C (Goldschmidt-Clermont et al., 1991). Interestingly, this inhibition is relieved for phospholipase C γ following tyrosine phosphorylation by the EGF receptor in vitro. Thus, the role of profilin and other similar elements in the regulation of phospholipase C γ, and other phospholipase C isozymes, may prove to be interesting.

The above discussion notwithstanding, there are several reports which suggest that for the EGF receptor, the activation of signaling pathways including phospholipase C may also be accomplished through an intermediary G protein (see Liang and Garrison, 1991, and references therein). This is surprising, since the EGF receptor bears no obvious structural homology to receptors of the above-mentioned superfamily of receptors known to couple to effectors via G proteins. Nonetheless, the data which suggest an EGF receptor/G protein interaction are compelling, including sensitivity to pertussis toxin of EGF effects in intact cells, an apparent GTP requirement for EGF action in membrane preparations, and effects of EGF on GTP/GDP exchange in membranes. Although such effects have been observed in only a limited number of cell types, these data raise a number of intriguing questions. These include the structure and mechanism of coupling of the EGF receptor, the nature of the G protein and the phospholipase C with which it interacts, and whether G-protein interactions occur with other members of the protein tyrosine kinase class of receptors.

The end result of receptor activation of phospholipase C through the various machinery described above is the liberation of the water-soluble messenger, $(1,4,5)IP_3$. The next section describes the cellular components which, upon binding of this molecule, participate in the mobilization of intracellular Ca^{2+}.

IV. The $(1,4,5)IP_3$-Sensitive Intracellular Ca^{2+} Pool

As described above, the initiating event in agonist-stimulated Ca^{2+} mobilization is the release of Ca^{2+} from a finite intracellular Ca^{2+} pool through the action of $(1,4,5)IP_3$. This agonist-sensitive-, $(1,4,5)IP_3$-sensitive-Ca^{2+} pool also appears to be involved in both the regulation of extracellular Ca^{2+} influx as well as in the generation of oscillations in $[Ca^{2+}]_i$. Given this pivotal role in the regulation of Ca^{2+} signaling, there has been a great deal of interest in the structural and functional characterization of this entity. In a number of important ways, the $(1,4,5)IP_3$-sensitive Ca^{2+} pool of nonmuscle cells is analogous to the SR of muscle. Thus, as with the SR, this organelle is thought to be comprised of two systems: (1) a system to accumulate and concentrate Ca^{2+} in a specialized, membrane-delimited intracellular organelle; and (2) a system to release this Ca^{2+}. Much progress has been made in defining these systems. This work is outlined briefly below.

Ca^{2+} is accumulated into the $(1,4,5)IP_3$-sensitive Ca^{2+} pool in an ATP-dependent fashion by a Ca^{2+}-ATPase(s) located in the membrane of the Ca^{2+}-storing organelle. This Ca^{2+}-ATPase is likely to be one of a family of structurally and functionally similar intracellular Ca^{2+} pumps which includes the Ca^{2+} pumps of skeletal and smooth muscle SR (Green and MacLennan, 1989; Burk et al., 1989). A distinguishing characteristic of the Ca^{2+}-ATPase of the $(1,4,5)IP_3$-sensitive Ca^{2+} pool is its unique sensitivity to the agent thapsigargin (Thastrup, 1990). Molecular cloning of Ca^{2+} pumps in this family from different tissues and species has been accomplished and indicates that different subtypes of these pumps exist; such structural variation may underlie functional specialization (Burk et al., 1989). With regard to nonmuscle tissues, Burgoyne et al. (1989) identified two putative Ca^{2+} pumps in bovine adrenal chromaffin cells of MW 100 kDa and 140 kDa based on differential recognition by monoclonal antibodies raised against skeletal muscle Ca^{2+}-ATPase. The latter appeared to be associated with the $(1,4,5)IP_3$-sensitive Ca^{2+} pool, whereas the former was associated with a distinct Ca^{2+} pool which may be involved in CICR.

A second component is also involved in the accumulation of Ca^{2+} in the (1,4,5)IP_3-sensitive Ca^{2+} pool, namely, an intraorganellar Ca^{2+}-binding protein(s). The expectation of such an entity was deduced by analogy with the SR, where such proteins are thought to act as important Ca^{2+} buffers (i.e., these proteins allow a large quantity of Ca^{2+} to be accumulated within the organelle while maintaining a free Ca^{2+} concentration which avoids the formation of Ca^{2+}-anion precipitates). In fact, the first intimation for the existence of such a Ca^{2+}-buffering protein associated with the (1,4,5)IP_3-sensitive Ca^{2+} pool came from the immunocytochemical studies of Volpe et al. (1988) using an antibody raised against the SR protein, calsequestrin. Subsequent studies indicate that the major Ca^{2+}-binding protein associated with the (1,4,5)IP_3-sensitive Ca^{2+} pool in nonmuscle tissue is more likely to be the functionally similar protein, calreticulin (Treves et al., 1990). Calreticulin is one of a group of five ubiquitous proteins found to be localized to ER (Koch, 1987); this protein is also localized in skeletal muscle SR, where it had been identified as the high-affinity Ca^{2+}-binding protein (HACBP) (Fliegel et al., 1989). Calreticulin has recently been purified, and primary sequence data has been derived by molecular cloning techniques (Fliegel et al., 1989). Calreticulin contains a concentration of acidic amino acids (net negative charge of -57) which are thought to comprise a low-affinity Ca^{2+}-binding domain; 25 mol Ca^{2+} can be bound per mole of protein with $K_d \geqslant 1$ mM. The protein also contains a single high-affinity Ca^{2+}-binding site ($K_d = 3$ μM). These properties, and its widespread distribution, suggest that calreticulin is the major Ca^{2+}-binding protein of ER and the (1,4,5)IP_3-sensitive Ca^{2+} pool (see below for a discussion of the relationship between these latter two entities). It is interesting to draw an analogy between calreticulin and some of the functional properties of the more thoroughly studied calsequestrin. Calsequestrin undergoes dynamic changes upon binding and release of Ca^{2+} which, in turn, affect its interaction with other proteins in the lumen of the SR. In particular, Ikemoto et al. (1989) have found that calsequestrin interacts in a Ca^{2+}-dependent fashion with the 400-kDa foot protein (ryanodine receptor) of the junctional face membrane thought to be involved in Ca^{2+} release. In fact, the Ca^{2+}-dependent association of calsequestrin with junctional face membrane proteins correlated with changes in the kinetics of Ca^{2+} release. These data raise the possibility that the Ca^{2+}-dependent protein-protein interactions of calsequestrin participate in the regulation of Ca^{2+} flux in the SR. As mentioned above, it has been suggested that the intraluminal Ca^{2+} concentration of the (1,4,5)IP_3-sensitive Ca^{2+} pool plays a role in the regulation of the sensitivity of the (1,4,5)IP_3 receptor to (1,4,5)IP_3 as well as the permeability to Ca^{2+} at the plasma membrane. It is interesting to speculate that Ca^{2+}-dependent protein-protein interactions for calreticulin might be a mechanism for the transduction of this information. Thus, the role of the intraluminal

Ca^{2+} binding proteins may go beyond that of Ca^{2+} buffer to integral component of the Ca^{2+} signaling mechanism.

The release of sequestered Ca^{2+} is initiated in nonmuscle tissue by $(1,4,5)IP_3$. As noted above, the $(1,4,5)IP_3$-sensitive element was first identified based on function in 1983 (Streb et al., 1983). The next significant advance in its characterization came in 1988 with the purification of an $(1,4,5)IP_3$-binding site from rat cerebellum (Supattapone et al., 1988b). Subsequently, this $(1,4,5)IP_3$-binding protein was reconstituted into lipid vesicles and demonstrated to contain both the $(1,4,5)IP_3$-binding site and an integral Ca^{2+} channel (Ferris et al., 1989). Primary sequence data has now been obtained through molecular cloning techniques (Furuichi et al., 1989; Mignery et al., 1990), and the cloned $(1,4,5)IP_3$ receptor has been subjected to mutational analyses (Mignery et al., 1990).

The $(1,4,5)IP_3$ receptor and integral Ca^{2+} channel is a large molecule (320 kDa) which appears to contain three domains. The large, cytoplasmic N-terminal domain contains a single, high-affinity $(1,4,5)IP_3$-binding site (Mignery et al., 1990; Miyawaki et al., 1991). The C terminus contains a series of hydrophobic regions which comprise as many as eight membrane spanning domains (Furuichi et al., 1989). This region of the $(1,4,5)IP_3$-receptor protein shares a significant homology with the transmembrane regions of the ryanodine receptor of skeletal muscle. In fact, it appears that these transmembrane regions form the Ca^{2+} channels associated with these two molecules (Mignery et al., 1990; Nakade et al., 1991). The $(1,4,5)IP_3$-binding and transmembrane channel regions of the $(1,4,5)IP_3$ receptor are linked by a long (>1400 aa) segment which, nonetheless, is capable of communicating $(1,4,5)IP_3$ binding to the channel (Mignery and Sudhof, 1990). In fact, in this latter study, gel filtration experiments indicated that $(1,4,5)IP_3$ binding had profound effects on the conformation of the protein. The functional $(1,4,5)IP_3$-receptor/Ca^{2+} channel is comprised of four identical subunits organized to form a central pore (Mignery and Sudhof, 1990; Maeda et al., 1991). This organization is strikingly similar to that for the ryanodine receptor. It is also interesting in light of the findings of Meyer et al. (1988), who showed that, functionally, the receptor contained four independent $(1,4,5)IP_3$-binding sites but that the opening of the Ca^{2+} channel was highly cooperative and required at least three bound molecules of $(1,4,5)IP_3$. A number of potential mechanisms for the regulation of $(1,4,5)IP_3$ receptor function have been identified at the level of the protein structure. The protein is a substrate for phosphorylation by cAMP-dependent protein kinase (Supattapone et al., 1988a; Mignery et al., 1990), Ca^{2+}/calmodulin-dependent protein kinase II, and protein kinase C (Ferris et al., 1991). It also contains an ATP-binding site (Ferris et al., 1990; Maeda et al., 1991). In addition, alternative forms of the $(1,4,5)IP_3$ receptor have been identified which may have altered regulatory properties. Thus, a distinct neuronal form of the receptor as well as a form which is expressed differentially during development have been detected (Danoff et al., 1991; Nakagawa et al., 1991). Sequence analysis of cDNA encoding the region linking the receptor and channel domains reveals that several of the potential regulatory sites occur in this region. This is also the region in which the sequence of different forms of the receptor have been found to diverge as a result of alternative splicing. An alternative splice site has also been identified in the N-terminal domain of the protein which could result in variants with altered $(1,4,5)IP_3$ binding (Nakagawa et al., 1991).

The regulation of the $(1,4,5)IP_3$ receptor has also been intensively studied at a functional level. One potentially significant regulator is Ca^{2+} itself. As noted above, studies of $(1,4,5)IP_3$-induced Ca^{2+} release in permeabilized cells indicate that in certain tissues Ca^{2+} can inhibit the release of Ca^{2+} from the $(1,4,5)IP_3$-sensitive store in a concentration-dependent fashion for Ca^{2+} concentrations in the physiological range (Payne et al., 1990; Willems et al., 1990; Iino, 1990; Jean and Klee, 1986). Ca^{2+} has also been found in some cases to have a biphasic effect on Ca^{2+} release (Iino, 1990; Bezprozvanny et al., 1991). An intriguing possibility has also been raised that the intraluminal Ca^{2+} concentration plays a role in the regulation of the $(1,4,5)IP_3$ receptor (Irvine, 1990). Such regulation has been demonstrated recently by Missiaen et al. (1991), who observed in permeabilized rat hepatocytes that Ca^{2+} uptake ("overload") into the $(1,4,5)IP_3$-sensitive Ca^{2+} pool could cause a sensitization of the $(1,4,5)IP_3$ receptor such that Ca^{2+} was spontaneously released from the store at ambient $(1,4,5)IP_3$ levels. This group has suggested that such an effect of intraluminal Ca^{2+} could provide a mechanism whereby the $(1,4,5)IP_3$ receptor can behave as a CICR channel. The mechanisms for the effects of Ca^{2+} on $(1,4,5)IP_3$-induced Ca^{2+} release are largely unknown. The functional effects of $(1,4,5)IP_3$-receptor phosphorylation by Ca^{2+}/calmodulin-dependent protein kinase II and protein kinase C have not yet been reported. In neuronal tissue, a membrane-bound protein, calmedin, appears to mediate effects of Ca^{2+} on $(1,4,5)IP_3$ binding (Danoff et al., 1988); however, this protein does not appear to be expressed outside of the central nervous system and/or does not interact with the $(1,4,5)IP_3$ receptor of nonneuronal tissues. Whether a protein(s) corresponding to calmedin exists in nonneuronal tissues remains to be determined.

Whereas effects of Ca^{2+} on $(1,4,5)IP_3$-induced Ca^{2+} release may be a component in the dynamic regulation of intracellular Ca^{2+} release, it must be noted that these effects may be tissue specific. Thus, in both rat parotid acinar cells (Menniti et al., 1991) and lacrimal cells (Bird and Putney, unpublished observations) the Ca^{2+} concentration to which permeabilized cells were exposed had no effect on the efficacy of $(1,4,5)IP_3$-induced Ca^{2+} release. Importantly, the manipulation of the extraluminal Ca^{2+} concentration in these studies also altered the intraluminal Ca^{2+} content. Data such as these suggest that, although Ca^{2+} modulation of $(1,4,5)IP_3$ receptor function may be important in many tissues, it may not be one of the factors which is fundamental to the dynamic regulation of Ca^{2+} release.

Several other potential regulatory factors for the $(1,4,5)IP_3$ receptor have been identified. In the study of Missiaen et al. (1991) discussed above, it was also observed that the sensitization of the $(1,4,5)IP_3$ receptor was enhanced by oxidized glutathione and other oxidized thiol reagents. Rooney et al. (1991) have similarly concluded that, in intact rat hepatocytes, sensitization of the $(1,4,5)IP_3$ receptor by the agent *tert*-butyl hydroperoxide occurs through elevation of oxidized glutathione levels. This, in turn, induces Ca^{2+} oscillations in these cells, presumably triggered at resting $(1,4,5)IP_3$ levels. The mechanism(s) whereby oxidized thiol reagents decrease the threshold for $(1,4,5)IP_3$-induced Ca^{2+} release, and whether this occurs physiologically, is at present not known. Nonetheless, it is tempting to speculate that the redox cycling of thiol groups on the $(1,4,5)IP_3$ receptor or an associated protein might function as part of the regulation of $(1,4,5)IP_3$-induced Ca^{2+} release. The $(1,4,5)IP_3$ receptor is also a substrate for cAMP-dependent protein kinase. In cerebellar microsomes, phosphorylation of the $(1,4,5)IP_3$ receptor by this kinase results in a decrease in the efficacy of Ca^{2+} release by $(1,4,5)IP_3$ (Supattapone et al., 1988a; Miyawaki et al., 1991). In contrast, in hepatocytes, activation of cAMP-dependent protein kinase apparently increases both the potency and efficacy of $(1,4,5)IP_3$ for Ca^{2+} release (Burgess et al., 1991). Although it was not demonstrated directly in this latter study that the effect of the kinase was due to phosphorylation of the $(1,4,5)IP_3$-receptor, these data suggest that the net effect of activation of this kinase is tissue specific. Another potential regulator, ATP, has been reported to increase $(1,4,5)IP_3$-induced Ca^{2+} flux via the purified cerebellar $(1,4,5)IP_3$ receptor (Ferris et al., 1990; Maeda et al., 1991). However, these effects occur at ATP concentrations well below those which occur physiologically. Nonetheless, it has been suggested that localized changes in intracellular ATP concentration, if such occur, could have an impact on the regulation of this system.

Paralleling the isolation and characterization of the individual components of the $(1,4,5)IP_3$-induced Ca^{2+} release mechanism has been the intensive study of the organization of these various components within the cell. The earliest studies on $(1,4,5)IP_3$-induced Ca^{2+} release broadly defined the $(1,4,5)IP_3$-sensitive Ca^{2+} pool as ER, since $(1,4,5)IP_3$-induced Ca^{2+} release occurred in permeabilized cell preparations in the presence of various mitochondrial poisons. Subsequent investigations have attempted to refine this assignation, and it has been proposed that this Ca^{2+} pool resides either in a highly specialized component of the ER or, in fact, in a unique intracellular organelle termed "calciosome" (Volpe et al., 1988). From a teleological standpoint, some separation of the Ca^{2+} signaling function from the other specialized functions of the ER might be expected to have evolved (however, see Koch, 1990). Similarly, cell-specific specialization of Ca^{2+} signaling components might be expected to have evolved to meet the needs of different cell types. Evidence is accumulating to support both of these ideas (see Rossier and Putney, 1991; Krause, 1991, for reviews). This evidence comes from two experimental approaches: (1) the analysis of the functional organization of the $(1,4,5)IP_3$-sensitive Ca^{2+} pool; and (2) identification of its localization within the cell and relation to other cellular components. These are discussed in turn below.

It has generally been observed that upon permeabilization of a wide variety of cell types, only a fraction of the total, nonmitochondrial Ca^{2+} pool can be released by $(1,4,5)IP_3$. This observation minimally indicates some degree of segregation of the $(1,4,5)IP_3$-regulated Ca^{2+} release sites from the entirety of the Ca^{2+} uptake and storage capacity available within the cell. It is likely that part of the nonmitochondrial, $(1,4,5)IP_3$-insensitive Ca^{2+} pools are also involved in Ca^{2+} signaling as, for example, in CICR. In fact, investigation of this phenomenon in the laboratories of Dawson (Dawson and Comerford, 1989) and Gill (Gill et al., 1990) has lead to the realization this segregation is, at least in part, an "artifact" of permeabilization; i.e., in intact cells the apparent $(1,4,5)IP_3$-sensitive and -insensitive Ca^{2+} pools may be in communication via a mechanism which is disrupted by the permeabilization process. This conclusion derives from the observation that GTP can specifically cause a link to be formed between $(1,4,5)IP_3$-sensitive and -insensitive Ca^{2+} pools in a variety of permeabilized cell systems. Thus, GTP appears to cause the recruitment and/or fusion of vesicles containing functional $(1,4,5)IP_3$-gated release channels with vesicles capable of accumulating Ca^{2+} but which lack this Ca^{2+} channel. An apparent segregation of Ca^{2+} release and uptake sites within rat parotid acinar cells has also observed under somewhat different circumstances (Menniti et al., 1991). Following permeabilization of these cells, the release of Ca^{2+} was induced either by $(1,4,5)IP_3$, under conditions in which its metabolism was minimized, or by poorly metabolized $(1,4,5)IP_3$ analogs. Surprisingly, released Ca^{2+} was rapidly resequestered into the $(1,4,5)IP_3$-sensitive pool. This reuptake was not secondary to a desensitization of the $(1,4,5)IP_3$ response. Instead, it appeared to result from Ca^{2+} uptake into a segment of the pool structurally separated from that segment involved in $(1,4,5)IP_3$-induced Ca^{2+} release. Since Ca^{2+} reuptake into the $(1,4,5)IP_3$-sensitive pool in the presence of $(1,4,5)IP_3$ did not occur in intact parotid cells, it was concluded that a kinetic delay in translocation of Ca^{2+} from uptake to release sites was caused or exaggerated by cell permeabilization. This was due to a disruption of the translocation of Ca^{2+} between the uptake and release sites.

The experiments described above indicate that there is at least some separation of (1,4,5)IP_3-sensitive Ca^{2+} release and uptake sites. Such a separation is reminiscent of that which has evolved in skeletal muscle SR and, based on the observed effects of GTP in a wide variety of permeabilized cells, it is likely to be a general phenomenon in nonmuscle cells as well. At present, the physiological significance of this separation is not known. It may reflect a requirement for localized Ca^{2+} release within the cell together with a more uniform requirement for Ca^{2+} reuptake and scavenging throughout the cytoplasm. Evidence for such a localized Ca^{2+} release in parotid acinar cells has been presented by Foskett et al. (1989), who observed that the activation of Ca^{2+}-regulated plasma membrane K^+ channels occurs before the observation of an increase in $[Ca^{2+}]_i$ as measured with the cytoplasmic Ca^{2+} dye, fura-2. Furthermore, Rossier et al. (1991) have found that, in liver, (1,4,5)IP_3 receptor binding, but not ATP-driven Ca^{2+} uptake, co-purified with plasma membrane markers during subcellular fractionation, again indicating a separation of Ca^{2+} uptake and release sites. Interestingly, the association of the plasma membrane and (1,4,5)IP_3 receptor-binding [and presumably the (1,4,5)IP_3-sensitive Ca^{2+}-storing organelle] appeared to be through actin microfilaments. Thus, it was suggested that such cytoskeletal elements may mediate communication between these two structures. It is also tempting to speculate that communication between the segregated Ca^{2+} pools may be regulated in the intact cell. The formation of the link between (1,4,5)IP_3-sensitive and -insensitive Ca^{2+} pools induced by GTP requires GTP hydrolysis, since nonhydrolyzable GTP analogs (such as GTPγS) are ineffective and block the effect of GTP. This has led to the speculation that this link is mediated by a member of the family of small G proteins known to be involved in membrane trafficking in a variety of systems (Dawson and Comerford, 1989; Gill et al., 1990). On the other hand, GTP has no apparent effect on the kinetics of translocation of Ca^{2+} between uptake and release sites in permeabilized rat parotid cells (Menniti et al., 1991), suggesting that different mechanisms may be involved in this system.

Another functional characteristic of the (1,4,5)IP_3-sensitive Ca^{2+} pool is that (1,4,5)IP_3-induced Ca^{2+} release occurs in an all-or-none or "quantal" fashion. This is most clearly demonstrated in experiments with permeabilized cells, where it has been shown that the fraction of the (1,4,5)IP_3-sensitive Ca^{2+} pool released, and not the rate of release of the total pool, is dependent on the concentration of (1,4,5)IP_3 (Taylor and Potter, 1990; Meyer and Stryer, 1990). Quantal Ca^{2+} release could arise from a highly segmented Ca^{2+} pool with different sensitivities to (1,4,5)IP_3 or, alternatively, from a complex regulation of the (1,4,5)IP_3 receptor/Ca^{2+} channel by the intra- and/or extraorganellar Ca^{2+} concentration. Understanding this process is an essential step in the full characterization of this complex structure. A question intimately related to the functional organization of the (1,4,5)IP_3-sensitive Ca^{2+} pool is its localization within the cell with respect to other cellular structures. Subcellular distribution studies have supported the notion that the (1,4,5)IP_3-sensitive Ca^{2+} pool is a distinct intracellular entity. Thus, the subcellular distribution of components of the (1,4,5)IP_3-sensitive Ca^{2+} pool differ in the majority of cell types from plasma membrane markers, mitochondria, and, significantly, classical ER markers (see Rossier and Putney, 1991; Krause, 1991). In fact, there is little correlation between the content of classical ER markers and markers for an (1,4,5)IP_3-sensitive Ca^{2+} pool across cell types or under different experimental conditions within a given cell type. The localization of the (1,4,5)IP_3-sensitive Ca^{2+} pool has also been examined using immunocytochemical techniques. Indeed, the concept of the calciosome arose from the visualization in several nonmuscle cell types of intracellular vacuoles containing a calsequestrinlike protein which were distinct from other recognized cellular structures (Volpe et al., 1988; Hashimoto et al., 1988). A tissue that has been intensively studied using immunocytochemical techniques is the cerebellum, in which the Purkinje cells contain an unusually high content of the (1,4,5)IP_3 receptor (see Burgoyne and Cheek, 1991). These studies to date confirm that the various components of the (1,4,5)IP_3-sensitive Ca^{2+} pool are not localized to mitochondria or plasma membrane. Furthermore, high concentrations of (1,4,5)IP_3 receptor have been localized to dendritic spines (among other areas), where they may mediate Ca^{2+} release in response to glutamatergic stimulation. However, as can be gathered from the preceding discussion, the identification of an (1,4,5)IP_3-sensitive Ca^{2+} pool based solely on immunocytochemical data may be incomplete, since it is likely that elements of this pool may have distinct subcellular localizations. As has been suggested (Rossier and Putney, 1991; Krause, 1991), it also appears likely that the organization of these elements will be found to be tissue specific. Thus, our full understanding of the range of entities encompassed by the concept of calciosome awaits combined functional and immunological analyses across a number of model systems.

V. Formation and Metabolism of the Inositol Phosphates

With the acceptance of the role of (1,4,5)IP_3 in Ca^{2+} signaling, the investigation of mechanisms for the inactivation of (1,4,5)IP_3 became of obvious interest. Such investigation focused on the metabolism of (1,4,5)IP_3 and has provided a detailed, although by no means complete, description of the pathways involved. Surprisingly, these studies also revealed that (1,4,5)IP_3 is the first intermediate in a pathway for the formation of a family of highly phosphorylated inositol phosphates, some of which may also have distinct intracellular functions. These two aspects of $(1,4,5)IP_3$ metabolism are briefly reviewed below. Readers are also directed to a more detailed review (Shears, 1989).

There are two routes for the metabolism of $(1,4,5)IP_3$ in mammalian cells: dephosphorylation via a 5-phosphatase and phosphorylation via a 3-kinase. The 5-phosphatase pathway clearly serves to inactivate $(1,4,5)IP_3$, since the product, $(1,4)IP_2$, does not release Ca^{2+}. $(1,4)IP_2$ is further metabolized by a series of phosphatases to salvage inositol (Fig. 4). $(1,4,5)IP_3$ is also phosphorylated by a 3-kinase to form $(1,3,4,5)IP_4$. While $(1,3,4,5)IP_4$ does not release Ca^{2+} directly from the $(1,4,5)IP_3$-sensitive Ca^{2+} pool, it has been suggested that this molecule may participate in the Ca^{2+}-mobilizing process, as discussed above. Data from various systems suggest that the 5-phosphatase is of lower affinity but of greater capacity than the 3-kinase. Thus, these parallel metabolic pathways may be expected to tightly control the level of $(1,4,5)IP_3$ over a wide range of intracellular concentrations. In fact, it has been observed that the level of $(1,4,5)IP_3$ decreases very rapidly following the cessation of phospholipase C activation; for example, the $t_{1/2}$ for the metabolism of $(1,4,5)IP_3$ is approximately 7 s in rat parotid cells (Hughes et al., 1988).

The regulation of $(1,4,5)IP_3$ metabolism in intact cells may be important for the regulation of $[Ca^{2+}]_i$ during agonist activation, as mentioned above. Of the two metabolic pathways, more potential mechanisms have been described for the 3-kinase, including regulation through protein kinase C, Ca^{2+}/calmodulin, and through the calpain system. Note that the activity of each of these regulatory systems may be affected during $(1,4,5)IP_3$ formation, i.e., by either the $(1,4,5)IP_3$-induced rise in $[Ca^{2+}]_i$ or the concomitant formation of diacylglycerol. Thus, the dynamic regulation of 3-kinase activity, and thereby the level of $(1,4,5)IP_3$, may contribute to the temporal (i.e., oscillatory) patterns of changes in $[Ca^{2+}]_i$ during agonist activation observed in a variety of cells. Unfortunately, by its very nature, an oscillatory pattern of change in 3-kinase activity will be difficult to measure, and a role for these regulatory mechanisms in the regulation of $[Ca^{2+}]_i$ will be difficult to discern.

As mentioned above, one of the outcomes of the investigation into the metabolism and inactivation of $(1,4,5)IP_3$ has been the realization that this molecule is also the first intermediate in a pathway for the formation of a family of highly phosphorylated inositol phosphates. $(1,3,4,5)IP_4$, the product of the $(1,4,5)IP_3$ 3-kinase reaction, is but one of several inositol *tetrakis*phosphates which may be increased during phospholipase C activation in mammalian cells. It has also been found that mammalian cells contain very high levels of IP_5 [primarily $(1,3,4,5,6)IP_5$] and IP_6. The discovery of this family of molecules immediately spurred interest in their metabolic interrelationships

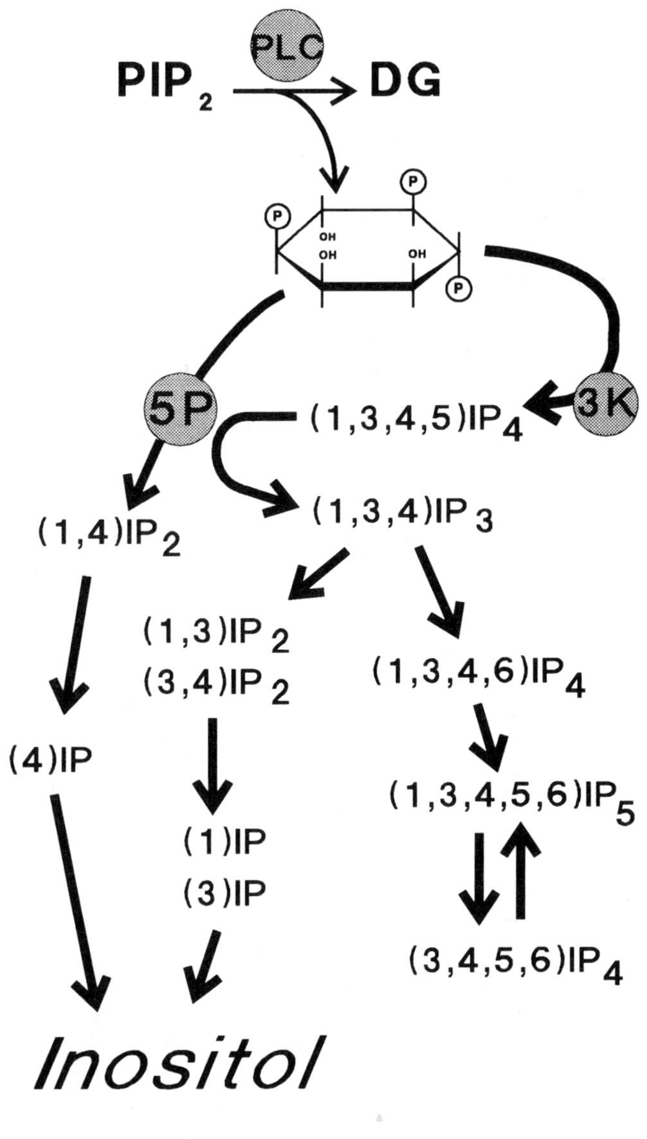

Figure 4 Putative pathway for the metabolism of $(1,4,5)IP_3$ in mammalian cells. The chemical structure is $(1,4,5)IP_3$. $(1,4,5)IP_3$ is formed by the action of phospholipase C (PLC) on phosphatidylinositol (4,5)*bis*phosphate (PIP_2). The second product of this reaction is diacylglycerol (DG). 5P, $(1,4,5)IP_3$/$(1,3,4,5)IP_4$ 5-phosphatase; 3K, $(1,4,5)IP_3$ 3-kinase.

and raised the intriguing question of possible function(s). Considerable progress has been made in the area of the metabolism of these molecules, as is outlined briefly below. Although this discussion will focus on mammalian cells, it is noted that our current knowledge of the mammalian system owes much to the detailed studies of inositol phosphate metabolism carried out in avian erythrocytes and dictyostelium by Stephens and co-workers (see Shears, 1989, for references). An outstanding question is the comparability between the pathways described for these latter systems and for mammalian cells; resolution of this question may ultimately depend on discovering the function of this family of molecules.

It would appear that in mammalian cells, the first step in the pathway for the formation of the highly phosphorylated inositol phosphates is the phosphorylation of $(1,4,5)IP_3$ to $(1,3,4,5)IP_4$. The primary route for the metabolism of $(1,3,4,5)IP_4$ is then via dephosphorylation by a 5-phosphatase to form $(1,3,4)IP_3$ (Fig. 4). It is thought that the phosphatase involved in this reaction is that which also metabolizes $(1,4,5)IP_3$. $(1,3,4,5)IP_4$ is also attacked in vitro by a 3-phosphatase; however, the physiological significance of this activity has recently been called into question with the purification of the 3-phosphatase from rat liver and determination of its substrate specificities (Nogimori et al., 1991). In this latter study, it was found that the affinities of the 3-phosphatase for both $(1,3,4,5,6)IP_5$ and IP_6 were very much greater than that toward $(1,3,4,5)IP_4$ and, given the relatively high concentrations of $(1,3,4,5,6)IP_5$ and IP_6 in mammalian cells, it was concluded that these latter molecules are the likely physiological substrates for the enzyme.

The product of $(1,3,4,5)IP_4$ dephosphorylation, $(1,3,4)IP_3$, is further metabolized by both dephosphorylation and phosphorylation. Dephosphorylation appears to be the major route and is accomplished by two phosphatases, a 4-phosphatase and a 1-phosphatase; both of the predicted products of these enzymes [$(1,3)IP_2$ and $(3,4)IP_2$, respectively] accumulate in intact cells upon phospholipase C activation. As with $(1,4)IP_2$, the dephosphorylation product of $(1,4,5)IP_3$, these inositol *bis*phosphates are further dephosphorylated to salvage inositol. The minor, but perhaps equally important, route of $(1,3,4)IP_3$ metabolism is its phosphorylation by a 6-kinase to form $(1,3,4,6)IP_4$. Based on the purification and characterization of this kinase, it would appear that its activity may be highly constrained in vivo (Hughes et al., 1989). Nonetheless, $(1,3,4,6)IP_4$ is generally found to accumulate upon phospholipase C activation. The constraints to the formation of $(1,3,4,6)IP_4$ take on significance, however, in that this molecule is in turn a precursor for $(1,3,4,5,6)IP_5$ (Shears, 1989).

$(1,3,4,5,6)IP_5$ (and IP_6, about which little is known) is the major inositol phosphate found in mammalian cells. In the rat pancreatoma cell line, AR4-2J, it is estimated that the intracellular $(1,3,4,5,6)IP_5$ concentration is 50 to 100 μM (unpublished observation). Data from various broken cell preparations indicate that mammalian cells express a 5-kinase capable of phosphorylating $(1,3,4,6)IP_4$ to $(1,3,4,5,6)IP_5$. In addition, it has also been found that this IP_5 can be derived via the phosphorylation of $(3,4,5,6)IP_4$. Until recently, the life cycle of this latter inositol *tetrakis*phosphate had been something of a puzzle. To gain insight into the metabolic interrelationships among these molecules in an intact cell, Menniti et al. (1990) developed a technique for preferentially labeling with ^{3}H-*myo*inositol rapidly and slowly turning over inositol polyphosphates in AR4-2J cells. Then, through comparison of relative specific radioactivities of the ^{3}H-inositol lipids and ^{3}H-inositol polyphosphates across labeling conditions in these cells, metabolic relationships among the inositol polyphosphates were deduced, including the relationship between $(1,3,4,6)IP_4$, $(3,4,5,6)IP_4$, and $(1,3,4,5,6)IP_5$.

Two groups of inositol polyphosphates were identified in the AR4-2J cells (Menniti et al., 1990). The first group consisted of $(1,4,5)IP_3$, $(1,3,4,5)IP_4$, $(1,3,4)IP_3$, and $(1,3,4,6)IP_4$; these shared a similar relative specific radioactivity across labeling conditions, consistent with a metabolic pathway for their interconversion as outlined above. The second group, consisting of $(3,4,5,6)IP_4$ and $(1,3,4,5,6)IP_5$, were of similar relative specific radioactivity to one another but incorporated radioactivity from ^{3}H-*myo*inositol much more slowly than did the first group. Since differential effects of the labeling protocol on the incorporation of radioactivity into $(1,3,4,5,6)IP_5$ were observed, it was deduced that only $(1,3,4,6)IP_4$ could serve as a precursor for de-novo $(1,3,4,5,6)IP_5$ synthesis. Furthermore, it was concluded that $(3,4,5,6)IP_4$ is the product of agonist-activated $(1,3,4,5,6)IP_5$ breakdown. This conclusion stemmed from the observation that no ^{3}H-inositol phosphate other than ^{3}H-$(1,3,4,5,6)IP_5$ was of an equivalent specific radioactivity and of sufficient mass to account for this increase in ^{3}H-$(3,4,5,6)IP_4$ during agonist activation.

The above-outlined results gave the first clue as to the metabolic origin of $(3,4,5,6)IP_4$ in a mammalian cell and suggested the existence of an as-yet-unidentified $(1,3,4,5,6)IP_5$ 1-phosphatase which may be under hormonal control. Since the level of $(1,3,4,5,6)IP_5$ is very high and decreases slightly during phospholipase C-stimulation in the AR4-2J cells, the activation of $(3,4,5,6)IP_4$ formation from $(1,3,4,5,6)IP_5$ must result from a regulated change in one or more of the intrinsic kinetic properties for the enzyme(s) involved in this pathway. At present the functions of the $(3,4,5,6)IP_4$ and $(1,3,4,5,6)IP_5$ are uncertain. Vallejo et al. (1987) have suggested that $(1,3,4,5,6)IP_5$ (and IP_6) may act as extracellular signaling molecules, at least in neurons. However, in AR4-2J cells, as in other cell types where it has been detected, although the mass of IP_5 is relatively high, its level during agonist activation is affected to a much lesser degree than that of the less highly phosphorylated inositol phosphates, including (3,4,5,6)IP_4. On the other hand, it is more generally found that agonists stimulate severalfold increases in (3,4,5,6)IP_4. Thus, (3,4,5,6)IP_4 has been proposed as a candidate for a modifier of intracellular regulatory processes in response to receptor activation (Menniti et al., 1990).

VI. Conclusions

The discussion above presents a general outline of the phosphoinositide/Ca^{2+} signaling cascade based on the wealth of information about this system which has accumulated since the discovery in 1983 of the role of (1,4,5)IP_3 in this process. It is now clear that this system is involved in the transduction of extracellular signals into cell responses in an enormously wide variety of eukaryotic cell types. With this in mind, one should view with some circumspection such generalizations as are made here, since they may not reflect specialization of this system which has evolved with the needs of diverse cell functions. The illumination of these specializations will undoubtably contribute substantially to our basic understanding of this system in the future.

References

Abdel-Latif AA, Akhtar R, Hawthorne JN. Acetylcholine increases the breakdown of triphosphoinositide of rabbit iris muscle prelabelled with [^{32}P] phosphate. Biochem J 1977; 162:61–73.

Auger KR, Cantley LC. Novel polyphosphoinositides in cell growth and activation. Cancer Cells 1991; 3:263–70.

Berridge, MJ. Rapid accumulation of inositol trisphosphate reveals that agonists hydrolyse polyphosphoinositides instead of phosphatidylinositol. Biochem J 1983; 212:849–58.

Berridge MJ. Cytoplasmic calcium oscillations: a two pool model. Cell Calcium 1991; 12:63–72.

Berridge MJ, Irvine RF. Inositol trisphosphate, a novel second messenger in cellular signal transduction. Nature 1984; 312:315–21.

Bezprozvanny I, Watras J, Ehrlich BE. Bell-shaped calcium-response curves of Ins(1,4,5)P_3^- and calcium-gated channels from endoplasmic reticulum of cerebellum. Nature 1991; 351:751–4.

Bird GSJ, Rossier MF, Hughes AR, Shears SB, Armstrong DL, Putney JW Jr. Activation of Ca^{2+} entry into acinar cells by a non-phosphorylatable inositol trisphosphate. Nature 1991; 352:162–5.

Bird GStJ, Takemura H, Thastrup O, Putney JW Jr, Menniti FS. Mechanism of activated Ca^{2+} entry in the rat pancreatoma cell line, AR4-2J. Cell Calcium 1992; 13:49–58.

Birnbaumer L. G proteins in signal transduction. In: George R, Cho AK, Blaschke TF, eds. Annual review of pharmacology and toxicology. Palo Alto, CA: Annual Reviews, Inc. 1991; 675–705.

Burgess GM, Bird GSJ, Obie JF, Putney JW, Jr. The mechanism for synergism between phospholipase C- and adenylylcyclase-linked hormones in liver. Cyclic AMP-dependent kinase augments inositol trisphosphate-mediated Ca^{2+} mobilization without increasing the cellular levels of inositol polyphosphates. J Biol Chem 1991; 266:4772–81.

Burgoyne RD, Cheek TR. Locating intracellular calcium stores. TIBS 1991; 16:319–20.

Burgoyne RD, Cheek TR, Morgan A, O'Sullivan AJ, Moreton RB, Berridge MJ, Mata AM, Colyer J, Lee AG, East JM. Distribution of two distinct Ca^{2+}-ATPase-like proteins and their relationships to the agonist-sensitive calcium store in adrenal chromaffin cells. Nature 1989; 342:72–4.

Burk SE, Lytton J, MacLennan DH, Shull GE. cDNA cloning, functional expression, and mRNA tissue distribution of a third organellar Ca^{2+} pump. J Biol Chem 1991; 264:18561–8.

Changya L, Gallacher DV, Irvine RF, Potter BVL, Peterson OH. Inositol 1,3,4,5-trisphosphate is essential for sustained activation of the Ca^{2+}-dependent K+ current in single internally perfused lacrimal cells. J Membrane Biol 1989; 109:85–93.

Cheek TR, Barry VA, Berridge MJ, Missiaen L. Bovine adrenal chromaffin cells contain an inositol 1,4,5- trisphosphate-insensitive but caffeine-sensitive Ca^{2+} store that can be regulated by intraluminal free Ca^{2+}. Biochem J 1991; 275:697–701.

Cobbold PH, Sanchez-Bueno A, Dixon CJ. The hepatocyte calcium oscillator. Cell Calcium 1991; 12:87–95.

Cockcroft S, Gomperts B. Role of guanine nucleotide binding protein in the activation of polyphosphoinositide phosphodiesterase. Nature 1985; 314:534–6.

Danoff SK, Ferris CD, Donath C, Fischer GA, Munemitsu S, Ullrich A, Snyder SH, Ross CA. Inositol 1,4,5-trisphosphate receptors: distinct neuronal and nonneuronal forms derived by alternative splicing differ in phosphorylation. Proc Nat Acad Sci USA 1991; 88:2951–5.

Danoff SK, Supattone S, Snyder SH. Characterization of a membrane protein from brain mediating the inhibition of inositol 1,4,5-trisphosphate binding by calcium. Biochem J 1988; 254:701–5.

Dargemont C, Hilly M, Claret M, Mauger J-P. Characterization of Ca^{2+} fluxes in rat liver plasma-membrane vesicles. Biochem J 1988; 256:117–24.

Dawson AP, Comerford JG. Effects of GTP on Ca^{2+} movements across endoplasmic reticulum membranes. Cell Calcium 1989; 10:343–50.

Dehlinger-Kremer M, Zeuzem S, Schulz I. Interaction of caffeine-, IP-3 and vanadate-sensitive Ca^{2+} pools in acinar cells of the exocrine pancreas. J Membr Biol 1991; 119:85–100.

Delfert DM, Hill S, Pershadsingh HA, Sherman WR. *myo*-Inositol 1,4,5-trisphosphate mobilizes Ca^{2+} from isolated adipocyte endoplasmic reticulum but not from plasma membranes. Biochem J 1986; 236:37–44.

Endo M. In: Bonner F, ed. Current topics in membranes and transport. New York: Academic Press, 1985; 181–230.

Felder CC, Ma AL, Liotta LA, Kohn EC. The antiproliferative and antimetastatic compound L651582 inhibits muscarinic acetylcholine receptor-stimulated calcium influx and arachidonic acid release. J Pharmacol Exp Ther 1991; 257:967–71.

Ferris CD, Huganir RL, Bredt DS, Cameron AM, Snyder SH. Inositol trisphosphate receptor: phosphorylation by protein kinase C and calcium calmodulin-dependent protein kinases in reconstituted lipid vesicles. Proc Natl Acad Sci USA 1991; 88:2232–5.

Ferris CD, Huganir RL, Snyder SH. Calcium flux mediated by purified inositol 1,4,5-trisphosphate receptor in reconstituted lipid vesicles is allosterically regulated by adenine nucleotides. Proc Natl Acad Sci USA 1990; 87:2147–51.

Ferris CD, Huganir RL, Supattapone S, Snyder SH. Purifies inositol 1,4,5-trisphosphate receptor mediates calcium influx in reconstituted lipid vesicles. Nature 1989; 342:87–9.

Fliegel L, Burns K, MacLennan DH, Reithmeier RAF, Michalak M. Molecular cloning of the high affinity calcium-binding protein (calreticulin) of skeletal muscle sarcoplasmic reticulum. J Biol Chem 1989; 264:21522–8.

Foskett JK, Gunter-Smith PJ, Melvin JE, Turner RJ. Physiological localization of an agonist-sensitive pool of Ca^{2+} in parotid acinar cells. Proc Natl Acad Sci USA 1989; 86:167–71.

Furuichi T, Yoshikawa S, Miyawaki A, Wada K, Maeda N, Mikoshiba K. Primary structure and functional expression of the inositol 1,4,5-trisphosphate-binding protein P_{400}. Nature 1989; 342:32–8.

Gill DL, Mullaney JM, Ghosh TK, Tarazi FI. Inositol phosphate- and guanine nucleotide-activated calcium translocation within cells. In: Naccache PH, ed. G-proteins and calcium mobilization. Boca Raton, FL: CRC Press, 1990; 95–119.

Goldschmidt-Clermont PJ, Kim JW, Machesky LM, Rhee SG, Pollard TD. Regulation of phospholipase C-γ1 by profilin and tyrosine phosphorylation. Science 251:1231–3.

Green NM, MacLennan DH. ATP driven ion pumps: an evolutionary mosaic. Biochem Soc Trans 1989; 17:819–22.

Harootunian AT, Kao JPY, Paranjape S, Tsien RY. Generation of calcium oscillations in fibroblasts by positive feedback between calcium and IP_3. Science 1991; 251:75–8.

Hashimoto S, Bruno B, Lew DP, Pozzan T, Volpe P, Meldolesi J. Immunocytochemistry of calciosomes in liver and pancreas. J Cell Biol 1988; 107:2523–31.

Hepler JR, Jeffs RA, Huckle WR, Outlaw HE, Rhee SG, Earp HS, Harden TK. Evidence that the epidermal growth factor receptor and non-tyrosine kinase hormone receptors stimulate phosphoinositol hydrolysis by independent pathways. Biochem J 1990; 270:337–44.

Hokin LE, Hokin MR. Effects of acetylcholine on the turnover of phosphoryl units in individual phospholipids of pancreas slices and brain cortex slices. Biochem Biophys Acta 1955; 18:102–10.

Hughes AR, Putney JW Jr. Source of [^{3}H]-labelled inositol bis- and monophosphates in agonist-activated rat parotid acinar cells. J Biol Chem 1989; 264:9400–7.

Hughes, PJ, Hughes AR, Putney JW Jr, Shears SB. The regulation of the phosphorylation of inositol 1,3,4-trisphosphate in cell free preparations and its relevance to the formation of inositol 1,3,4,6-tetrakisphosphate in agonist-stimulated rat parotid acinar cells. J Biol Chem 1989; 264:19871–8.

Hughes AR, Takemura H, Putney JW Jr. Kinetics of inositol 1,4,5-trisphosphate and inositol cyclic 1:2,4,5-trisphosphate metabolism in intact rat parotid acinar cells: relationship to calcium signalling. J Biol Chem 1988; 263:10314–9.

Hupe DJ, Boltz R, Cohen CJ, Felix J, Ham E, Miller D, Soderman D, Van Skiver D. The inhibition of receptor-mediated and voltage-dependent calcium entry by the antiproliferative L-651,582. J Biol Chem 1991; 266:10136–42.

Iino M. Biphasic Ca^{2+} dependence of inositol 1,4,5-trisphosphate-induced Ca^{2+} release in smooth muscle cells of the guinea pig taenia caeci. J Gen Physiol 1990; 95:1103–22.

Ikemoto N, Ronjat M, M'száros LG, Koshita M. Postulated role of calsequestrin in the regulation of calcium release from the sarcoplasmic reticulum. Biochemistry 1989; 28:6764–71.

Irvine RF. 'Quantal' Ca^{2+} release and the control of Ca^{2+} entry by inositol phosphates—a possible mechanism. FEBS Lett 1990; 263:5–9.

Irvine RF, Moor RM. Inositol(1,3,4,5)tetrakisphosphate-induced activation of sea urchin eggs requires the presence of inositol trisphosphate. Biochem Biophys Res Commun 1987; 146:284–90.

Jean T, Klee CB. Calcium modulation of inositol 1,4,5-trisphosphate-induced calcium release from neuroblastoma x glioma hybrid (NG108-15) microsomes. J Biol Chem 1986; 261:16414–20.

Kirk CJ, Creba JA, Downes CP, Michell RH. Hormone-stimulated metabolism of inositol lipids and its relationship to hepatic receptor function. Biochem Soc Trans 1981; 9:377–9.

Koch GLE. Reticuloplasmins: a novel group of proteins in the endoplasmic reticulum. J Cell Sci 1987; 87:491–2.

Koch GLE. The endoplasmic reticulum and calcium storage. BioEssays 1990; 12:527–31.

Krause K-H. Ca^{2+}-storage organelles. FEBS Lett 1991; 285:225–9.

Kwan CY, Takemura H, Obie JF, Thastrup O, Putney JW Jr. Effects of methacholine, thapsigargin and La^{3+} on plasmalemmal and intracellular Ca^{2+} transport in lacrimal acinar cells. Am J Physiol 1990; 258:C1006–15.

Liang M, Garrison JC. The epidermal growth factor receptor is coupled to a pertussis toxin-sensitive guanine nucleotide regulatory protein in rat hepatocytes. J Biol Chem 1991; 266:13342–9.

Litosch I, Wallis C, Fain JN. 5-Hydroxytryptamine stimulates inositol phosphate production in a cell-free system from blowfly salivary glands. Evidence for a role of GTP in coupling receptor activation to phosphoinositide breakdown. J Biol Chem 1985; 260:5464–71.

Maeda N, Kawasaki T, Nakade S, Tokota N, Taguchi T, Kasai M, Mikoshiba K. Structural and functional characterization of inositol 1,4,5- trisphosphate receptor from mouse cerebellum. J Biol Chem 1991; 266:1109–16.

Malgaroli A, Fesce R, Meldolesi J. Spontaneous $[Ca^{2+}]$i fluctuations in rat chromaffin cells do not require inositol 1,4,5-trisphosphate elevations but are generated by a caffeine- and ryanodine-sensitive intracellular Ca^{2+} store. J Biol Chem 1990; 265:3005–8.

Martin TFJ. Lipid hydrolysis by phosphoinositidase C: enzymology and regulation by receptors and guanine nucleotides. In Michell RH, Drummong AH, Downes CP, eds. Inositol lipids in cell signalling. London: Academic Press, 1989; 81–112.

Meisenhelder J, Suh P-G, Rhee SG, Hunter T. Phospholipase C-γ is a substrate for the PDGF and EGF receptor protein-tyrosine kinases in vivo and in vitro. Cell 1989; 57:1109–22.

Meldrum E, Parker PJ, Carozzi A. The PtdIns-PLC superfamily and signal transduction. Biochim Biophys Acta 1991; 1092:49–71.

Menniti FS, Bird GStJ, Takemura H, Thastrup O, Potter BVL, Putney JW Jr. Mobilization of calcium by inositol trisphosphates from permeabilized rat parotid acinar cells. Evidence for translocation of calcium from uptake to release sites within the inositol 1,4,5-trisphosphate- and thapsigargin-sensitive calcium pool. J Biol Chem 1991; 266:13646–53.

Menniti FS, Oliver KG, Nogimori K, Obie JF, Shears SB, Putney JW Jr. Origins of *myo*inositoltetrakisphosphates in agonist-stimulated rat pancreatoma cells: stimulation by bombesin of *myo*inositol (1,3,4,5,6)pentakisphosphate to *myo*inositol (3,4,5,6)tetrakisphosphate. J Biol Chem 1990; 265:11167–76.

Merritt JE, Armstrong WP, Benham CD, Hallam TJ, Jacob R, Jaxa-Chamiec A, Leigh BK, McCarthy SA, Moores KE, Rink TJ. SK&F 96365, a novel inhibitor of receptor-mediated calcium entry. Biochem J 1990; 271:515–22.

Meyer T, Stryer L. Molecular model for receptor-stimulated calcium spiking. Proc Natl Acad Sci USA 1988; 85:5051–5.

Meyer T, Stryer L. Transient calcium release induced by successive increments of inositol 1,4,5-trisphosphate. Proc Natl Acad Sci USA 1990; 87:3841–5.

Meyer T, Stryer L. Calcium spiking. Ann Rev Biophys Biophys Chem 1991; 20:153–74.

Meyer T, Holowka D, Stryer L. Highly cooperative opening of calcium channels by inositol 1,4,5-trisphosphate. Science 1988; 240:653–6.

Michell RH. Inisotol phospholipids and cell surface receptor function. Biochim Biophys Acta 1975; 415:81–147.

Michell RH. Inositol lipids and their role in receptor function: history and general principles. In: Putney JW Jr, ed. Phosphoinositides and receptor mechanisms. New York: Alan R. Liss, Inc., 1986; 1–24.

Mignery GA, Sudhof TC. The ligand binding site and transduction mechanism in the inositol-1,4,5-trisphosphate receptor. EMBO J 1990; 9:3893–8.

Mignery GA, Newton CL, Archer BT III, Sudhof TC. Structure and expression of the rat inositol 1,4,5-trisphosphate receptor. J Biol Chem 1990; 265:12679–85.

Missiaen L, Taylor CW, Berridge MJ. Spontaneous calcium release from inositol trisphosphate-sensitive calcium stores. Nature 1991; 352:241–4.

Miyawaki A, Furuichi T, Ryou Y, Yoshikawa S, Nakagawa T, Saitoh T, Mikoshiba K. Structure-function relationships of the mouse inositol 1,4,5-trisphosphate receptor. Proc Natl Acad Sci USA 1991; 88:4911–5.

Morris AP, Gallacher DV, Irvine RF, Petersen OH. Synergism of inositol trisphosphate and tetrakisphosphate in activating Ca^{2+}-dependent K^+ channels. Nature 1987; 330:653–5.

Nakade S, Maeda N, Mikoshiba K. Involvement in the C-terminus of the inositol 1,4,5-trisphosphate receptor in Ca^{2+} release analysed using region-specific monoclonal antibodies. Biochem J 1991; 277:125–31.

Nakagawa T, Okano H, Furuichi T, Aruga J, Mikoshiba K. The subtypes of the mouse inositol 1,4,5-trisphosphate receptor are expressed in a tissue-specific and developmentally specific manner. Proc Natl Acad Sci USA 1991; 88:6244–8.

Nakai J, Imagawa T, Hakamata Y, Shigekawa M, Takeshima H, Numa S. Primary structure and functional expression from cDNA of the cardiac ryanodine receptor/calcium release channel. FEBS Lett 1990; 271:169–77.

Nishizuka Y. The role of protein kinase C in cell surface signal transduction and tumour promotion. Nature 1984; 308:693–8.

Nogimori K, Hughes PJ, Hodgson ME, Putney JW Jr, Shears SB. Purification of an inositol (1,3,4,5) tetrakisphosphate 3-phosphatase activity from rat liver, and determination that its physiologically relevant substrate is inositol hexakisphosphate. J Biol Chem 1991; 266:16499–506.

O'Dowd BF, Lefkowitz RJ, Caron MG. Structure of the adrenergic and related receptors. Ann Rev Neurosci 1989; 12:67–83.

Payne R, Flores TM, Fein A. Feedback inhibition by calcium limits the release of calcium by inositol trisphosphate in Limulus ventral photoreceptors. Neuron 1990; 4:547–55.

Putney JW Jr. A model for receptor-regulated calcium entry. Cell Calcium 1986; 7:1–12.

Putney JW Jr. Calcium-mobilizing receptors. Trends Pharmacol Sci 1987; 8:481–6.

Putney JW Jr. Capacitative calcium entry revisited. Cell Calcium 1990; 11:611–24.

Renard D, Poggioli J, Berthon B, Claret M. How far does phospholipase C activity depend on the cell calcium concentration? A study in intact cells. Biochem J 1987; 243:391–8.

Rhee SG. Inositol phospholipid-specific phospholipase C: interaction of the γ 1 isoform with tyrosine kinase. TIBS 1991; 16:297–301.

Rooney TA, Renard DC, Sass EJ, Thomas SP. Oscillatory cytosolic calcium waves independent of stimulated inositol 1,4,5-trisphosphate formation in hepatocytes. J Biol Chem 1991; 266:12272–82.

Rooney TA, Sass EJ, Thomas AP. Agonist-induced cytosolic calcium oscillations originate from a specific locus in single hepatocytes. J Biol Chem 1990; 265:10792–6.

Rossier MF, Putney JW Jr. The identity of the calcium storing inositol 1,4,5-trisphosphate-sensitive organelle in non-muscle cells: calciosome, endoplasmic reticulum, . . . or both? Trends Neurosi 1991; 14:310–4.

Rossier MF, Bird GStJ, Putney JW Jr. Subcellular distribution of the calcium-storing inositol 1,4,5-trisphosphate-sensitive organelle in rat liver. Possible linkage to the plasma membrane through the actin microfilaments. Biochem J 1991; 274:643–50.

Shears SB. Metabolism of the inositol phosphates produced by receptor activation. Biochem J 1989; 260:313–24.

Shoshan-Barmatz V, Zhang GH, Garretson L, Kraus-Friedmann N. Distinct ryanodine- and inositol 1,4,5-trisphosphate-binding sites in hepatic microsomes. Biochem J 1990; 268:699–705.

Smrcka AV, Hepler JR, Brown KO, Sternweis PC. Regulation of polyphosphoinositide-specific phospholipase C activity by purified G_q. Science 1991; 251:804–7.

Strathmann M, Simon ML. G protein diversity: a distinct class of α subunits is present in vertabrates and invertabrates. Proc Natl Acad Sci USA 1990; 87:9113–7.

Streb H, Irvine RF, Berridge MJ, Schulz I. Release of Ca^{2+} from a nonmitochondrial store in pancreatic cells by inositol-1,4,5-trisphosphate. Nature 1983; 306:67–8.

Supattapone S, Danoff SK, Theibert A, Joseph SK, Steiner J, Snyder SH. Cyclic AMP-dependent phosphorylation of a brain inositol trisphosphate receptor decreases its release of calcium. Proc Natl Acad Sci USA 1988a; 85:8747–50.

Supattapone S, Worley PF, Baraban JM, Snyder SH. Solubilization, purification, and characterization of an inositol trisphosphate receptor. J Biol Chem 1988b; 263:1530–4.

Takemura H, Putney JW Jr. Capacitative calcium entry in parotid acinar cells. Biochem J 1989; 258:409–12.

Taylor CW, Potter BVL. The size of inositol 1,4,5-trisphosphate-sensitive Ca^{2+} stores depends on inositol 1,4,5-trisphosphate concentration. Biochem J 1990; 266:189–94.

Takemura H, Hughes AR, Thastrup O and Putney JW Jr. Activation of calcium entry by the tumor promoter, thaspigargin, in parotid acinar cells. Evidence that an intracellular calcium pool and not an inositol phosphate, regulates calcium fluxes at the plasma membrane. J Biol Chem 1989; 264:12266–71.

Taylor CW, Merritt JE, Putney JW Jr, Rubin RP. Effects of Ca^{2+} on phosphoinositide breakdown in exocrine pancreas. Biochem J 1986; 238:765–72.

Taylor SJ, Chae HZ, Rhee SG, Exton JH. Activation of the $\beta 1$ isozyme of phospholipase C by α subunits of the Gq class of G proteins. Nature 1991; 350:516–8.

Thastrup O. Role of Ca^{2+}-ATPase in regulation of cellular Ca^{2+} signalling, as studied with the selective microsomal Ca^{2+}-ATPase inhibitor, thapsigargin. Agents and Actions 1990; 28:8–15.

Thastrup O, Cullen PJ, Drobak BK, Hanley MR, Dawson AP. Thapsigargin, a tumor promoter, discharges intracellular Ca^{2+} stores by specific inhibition of the endoplasmic reticulum Ca^{2+}-ATPase. Proc Natl Acad Sci USA 1990; 87:2466–70.

Tota MR, Candelore MR, Dixon RA, Strader CD. Biophysical and genetic analysis of the ligand-binding site of the beta-adrenoceptor. Trends Pharmacol Sci 1991; 12:4–6.

Treves S, De Mattei M, Landfredi M, Villa A, Green NM, MacLennan DH, Meldolesi J, Pozzan T. Calreticulin is a candidate for a calsequestrin-like function in Ca^{2+}-storage compartments (calciosomes) of liver and brain. Biochem J 1990; 271:473–80.

Vallejo M, Jackson T, Lightman S, Hanley MR. Occurrence and extracellular actions of inositol pentakis- and hexakisphosphate in mammalian brain. Nature 1987; 330:656–8.

Volpe P, Krause K-H, Hashimoto S, Zorzato F, Pozzan T, Meldolesi J, Lew DP. "Calcioisome," a cytoplasmic organelle: the inositol 1,4,5-trisphosphate-sensitive Ca^{2+} store of nonmuscle cells? Proc Natl Acad Sci USA 1988; 85:1091–5.

Wakui M, Osipchuk YV, Petersen OH. Receptor-activated cytoplasmic Ca^{2+} spiking mediated by inositol trisphosphate is due to Ca^{2+}-induced Ca^{2+} release. Cell 1990; 63:1025–32.

Wakui M, Potter BVL, Petersen OH. Pulsatile intracellular calcium release does not depend on fluctuations in inositol trisphosphate concentration. Nature 1989; 339:317–20.

Willems PHGM, De Jong MD, de Pont JJHHM, Van Os CH. Ca^{2+}-sensitivity of inositol 1,4,5-trisphosphate-mediated Ca^{2+} release in permeabilized pancreatic acinar cells. Biochem J 1990; 265:681–7.

Woods NM, Cuthbertson KS, Cobbold PH. Repetitive transient rises in cytoplasmic free calcium in hormone-stimulated hepatocytes. Nature 1986; 319:600–2.

Woods NM, Cuthbertson KS, Cobbold PH. Agonist-induced oscillations in cytoplasmic free calcium concentration in single rat hepatocytes. Cell Calcium 1987; 8:79–100.

6

Protein-Tyrosine Kinases

TIMOTHY O'HARE, JOHNGSOON LEE, and PAUL F. PILCH

Boston University School of Medicine
Boston, Massachusetts

I. Introduction

The protein kinase family (i.e. those enzymes that catalyze the phosphorylation of proteins) comprises a large number of enzymes displaying considerable heterogeneity with regard to structure, mechanism of activation, subcellular localization, interaction with other cellular proteins, and substrate specificity. It is this last feature, that of specificity, that constitutes the major division of protein kinases. One very large class phosphorylates serine and threonine residues and includes the well-studied cAMP-dependent protein kinase and protein kinase C. The other class of protein kinases phosphorylates proteins on tyrosine residues and includes viral oncogene products and various growth factor receptors. Despite these differences, the entire family of protein kinases contains a highly conserved catalytic domain which links the members evolutionarily (Hanks et al., 1988).

The group of enzymes that catalyzes serine and threonine phosphorylation of proteins was first described in the late 1950s. Since 1979, evidence has been mounting for the existence of protein kinase activity which includes tyrosyl residues as well as the usual serine and threonine modifications (reviewed by Srivastava, 1990). First, tyrosine phosphorylation was documented in preparations containing the Rous sarcoma virus oncogene product $pp60^{v\text{-}src}$, and the viral transforming protein from the Yamaguchi avian sarcoma virus, $p90^{gag\text{-}yes}$ (reviewed by Brickell, 1991). Subsequently, a number of growth factor receptors were shown to co-purify with protein-tyrosine kinase (PTK) activity, the first being the receptor for epidermal growth factor (EGF) (reviewed by Yarden and Ullrich, 1988). From 1980 to 1991 the rate of discovery of new protein kinases accelerated rapidly, due largely to the discovery of numberous protein-tyrosine kinases. Initially, these new protein-tyrosine kinases were identified by enzymatic assay and phosphoamino acid analysis of purified proteins, but subsequently, a larger proportion of them were identified as putative protein-tyrosine kinase (PTKs) based on sequence homology. The method used most often employs low-stringency screening of cDNA libraries using probes derived from the conserved catalytic domain which all protein kinases share. There are two well-recognized regions of the catalytic domain, in which the sequence differs between protein kinases with differing specificity. The protein kinases with serine/threonine specificity share a consensus sequence of Asp-Leu-Lys-Pro-Glu-Asn, whereas the PTKs have the sequence Asp-Leu-Arg-Ala-Ala-Asn or Asp-Leu-Ala-Ala-Arg-Asn. The use of different oligonucleotide probes, based on these regions, can result in selection of PTKs or serine/threonine protein kinases. Most of the *src* subfamily of PTKs were discovered with probes based on the latter sequence. In addition, screening of expression libraries using antiphosphotyrosine antibodies as well as the application of the polymerase chain reaction have been used to clone PTKs (Raz et al., 1991, and references within). Because of this new methodology, a number of PTKs were identified by sequence only, and not biochemically. Over the years, it has been demonstrated that all of these putative PTKs, when tested biochemically, contain kinase activity. Therefore it is generally accepted that the presence of a catalytic domain within an appropriate protein sequence indicates a true protein-tyrosine kinase.

The members of the protein-tyrosine kinase seem to have pivotal roles in the regulation of growth, metabolism, and differentiation of cells. Cells may contain a number of different tyrosine kinases, including growth factor receptors and cytoplasmic PTKs. Many PTKs of both classes are protooncogenes, and the latter are called cytoplasmic because most express little or no sequence outside the cell. However, almost all of the cytoplasmic PTKs are membrane bound, via myristoylation at a conserved amino-terminal glycine residue (Resh, 1990). Several cytoplasmic PTKs were partially purified from normal mammalian tissues in the early 1980s based on the high level of tyrosine kinase activity observed in certain tissues (e.g., spleen and brain) and have been characterized (reviewed by Srivastava, 1990). However, normally the level of tyrosine phosphorylation in most cells is extremely low and accounts for a small percentage of cellular phosphorylation ($<0.1\%$). An important point is that PTKs are involved in normal cellular function. Paradoxically, although the tyrosine kinase activity of the PTKs has been associated with the transformed state (i.e., cancer) and dysregulated growth, the highest levels of tyrosine kinase activity are found in the adult brain and other nonproliferative tissue. Therefore, it would seem that the regulation of normal cellular activities by the PTKs is complex, and is not solely related to the level of tyrosine kinase activity within a cell. For obvious reasons, there is considerable interest in the PTKs and the type of biochemical signaling pathways they utilize.

The classification of the PTKs is becoming increasingly complex as new PTKs are discovered and reported with a high frequency. For this reason, a careful search of the literature was made to include as many of these as possible, though the speed at which new PTKs are discovered will likely lead to omissions. The rest of this chapter summarizes the current understanding of the signaling pathways of the PTKs and their role in transformation and oncogenesis.

II. Classification

The ability to compare amino acid sequences and build phylogenetic trees based on these sequences has proved useful in separating the different protein kinases and in predicting potential function. The PTKs can be separated into two classes based on the presence or absence of a transmembrane segment; however, it is also useful to separate the PTKs based on a phylogenetic comparison of the catalytic domain (Hanks et al., 1988). This has more general usefulness, as more tyrosine kinases are discovered which are not easily placed into either the receptor or nonreceptor PTKs. The classification used is based on sequence analysis of both catalytic and noncatalytic sequences. The nonreceptor PTKs are divided into three subfamilies, whereas the receptor PTKs are currently divided into four subfamilies. There are several PTKs which do not belong to one of these seven subfamilies and may represent members of new subfamilies.

There is an interesting permutation to classifying and quantifying the number of PTKs which has to do with heterogeneity. All of the nonreceptor protein tyrosine kinases discovered so far are expressed as at least two different forms generated by alternative splicing. Neuronal cells express an altered form of $p60^{c\text{-}src}$ which contains a six-amino acid insert. This suggests a tissue-specific role for this protein, designated $p60^{c\text{-}src(+)}$ (Brickell, 1991). Both the *arg* (for *abl*-related gene) and *c-abl* (for Abelson murine leukemia virus) gene products are expressed as two forms consisting of different amino termini, only one of which contains the myristoylation site. There is a testis-specific gene product of *fer* (for *fps/fes*-related, see below), presumably via alternative splicing (Hao et al., 1991), as most of the PTKs demonstrate alternative splicing. For example, the insulin receptor and FGF (fibroblast growth factor) receptors. In the case of the FGF receptor, five forms of mRNA have been discovered, one of which appears to be expressed only in neural tissue (Eisemann et al., 1991). Similarly, the *trk*B (see below) gene gives rise to a number of different mRNA forms, and some of these code for truncated receptors with highly specific tissue expression (Middlemas et al., 1991). In addition to different genes and alternative splicing, there is also evidence that (1) insulin/insulinlike growth factor hybrid receptors are formed in vivo, (2) functional platelet-derived growth factor receptors can be comprised of products from two different genes, and (3) there can be postreceptor interaction between two different types of receptor PTKs (Ullrich and Schlessinger, 1990).

A. Src

Nine members of the Src subfamily have currently been described (see Table 1). The viral oncogene of the Rous avian sarcoma virus pp60$^{v\text{-}src}$ (or p60$^{v\text{-}src}$ was the first PTK discovered and has been extensively described. The cellular homolog or protooncogene p60$^{c\text{-}src}$ has also been extensively characterized, but the normal biological role(s) that this protein performs are unclear. Next identified were the viral oncogene product of the Yamaguchi 73 avian sarcoma virus p90$^{gag\text{-}yes}$, the Esh sarcoma virus-type *yes* oncogene protein product p80$^{gag\text{-}yes}$, and p72$^{gag\text{-}fgr}$, the oncogene product produced by the Garder-Rasheed feline sarcoma virus. It is interesting that the first three members of the Src subfamily were the viral transduced versions of *src, yes*, and *fgr*, whereas the remaining members do not yet have identified viral oncogene counterparts. The *lck* PTK (for lymphoid cell kinase) was first identified from the particulate fraction of a lymphoma cell line LSTRA (reviewed by Sefton, 1991), where the expression of p56^{lck} is abundant. It is not clear if overexpression of the *lck* gene, by itself, is responsible for the phenotypic change of this cell line. However, overexpression of the wild-type p56^{lck} in mice reproducibly leads to thymoma production (Abraham et al., 1991). The *hck* gene (hematopoietic cell kinase) was identified using a *lck* cDNA probe (Ziegler et al., 1991), and codes for p55^{hck}. Similar probes, based on *lck* catalytic sequences, led to identification of additional members of the *src* subfamily of PTKs, and their respective protein products p59^{fyn}, p56^{lyn}, and p55^{blk} (Dymecki et al., 1990). The name designation is usually based on relatedness to known oncogenes (*lyn* for *lck/yes*-related novel gene) or for selective tissue/cell expression (*blk* for B lymphoid kinase). The protooncogene *syn* (for *src/yes*-related novel gene) was identified by screening a human placenta genomic library with a *v-yes* DNA segment (see references in Brickell, 1991). The genes *lck, hck, fyn, lyn, syn,* and *blk* are considered protooncogenes by most

Table 1 The Protein-Tyrosine Kinase Family

Receptor			
INS receptor	EGF receptor	PDGF receptor	EPH
INS R	EGF R/TGF-α R	PDGFR	EPH
IGF-1 R	HER2/NEU	CSF-1 R/FMS	ECK
IR R	HER3	FGF R/FLG	ELK
ROS		KIT	
NGF R/TRK/TRKB		FLT	
HGF R/MET		RET	
LTK/TYK1			
SEA			

Nonreceptor/other			
SCR	ABL	FPS/FES	OTHER
SRC	ABL	FPS	TYK2
YES	ARG	FES	JAK
FGR		FER	TEC
LCK		FLK	KLG
HCK			YPK1
FYN			
LYN			
BLK			
SYN			

investigators, despite the fact that no naturally occurring virally transformed counterparts have been identified. Evidence to support this belief is based on experiments which show that expression of mutants which either lack the carboxy-terminal tyrosine, or are mutated at one of the conserved tyrosines within the kinase domain, will transform various cell lines or cause tumors in vivo.

B. Abl

There are two members of the Abl subfamily. Abl was first identified as the transforming gene of the Abelson murine leukemia and the Hardy-Zuckerman 2 feline sarcoma viruses. One of the gene products of *v-abl* (there are several) is $p120^{gag\text{-}abl}$, and the cellular protooncogene *c-abl* codes for $p150^{c\text{-}abl}$. In addition, the *c-abl* locus on chromosome 9 is the site of a translocation with chromosome 22, resulting in the Philadelphia chromosome, which is found in almost all patients with chronic myelogenous leukemia and in a minority of patients with acute lymphoblastic leukemia (Shore et al., 1990). The second member of this subfamily is the gene *arg* for (Abelson related gene). The most striking feature of this gene is its high degree of homology with *c-abl*, with regard to amino acid sequence and gene product expression of two forms, each with a different amino-terminus (Kruh et al., 1990).

C. Fes/Fps

The Fes/Fps of PTKs is comprised of three members. The *c-fes/fps* protooncogene is the common cellular homolog to numerous viral transduced versions. The human protooncogene codes for $p92^{c\text{-}fes}$, and the avian locus codes for the related protein $p98^{c\text{-}fps}$. Both the Gardner-Arstein and Synder-Thilen feline sarcoma viruses contain transduced versions of *fes* which code for various gag-*fes* fusion oncogene products. Likewise, a number of avian sarcoma viruses (e.g., Fujinami) have transduced versions of *fps* which therefore code for gag-*fps* fusion transforming proteins with differing amounts of gag and *fps* sequences. The avian form of the transforming gene *v-fps* and the feline form of *v-fes* represent the same oncogene. The *fer* PTK was discovered by low-stringency hybridization screening with both *v-abl* and *c-fms* kinase domain probes, which accounts for the two names (i.e., *fer* and *tyk3*) (Krolewski et al., 1990). The *fer* gene codes for a 94-kDa cellular phosphoprotein originally designated NCP94, which had been identified with antiserum directed against *v-fps* specific peptide sequences (Hao et al., 1991). The rat homologous gene is *flt*. The gene *flk* (*fps/fes*-like tyrosine kinase) has been only partially sequenced (Letwin et al., 1988).

D. Epidermal Growth Factor Receptor

The cognate receptor for epidermal growth factor (EGF) defines another subfamily of PTKs. The connection between tyrosine kinase activity, growth factor receptors, and viral oncogenes became intertwined when it was discovered that portions of the amino acid sequence of affinity-purified epidermal growth factor receptor (EGFR) were nearly identical to the transforming protein encoded by the *erb*B, the oncogene of the avian erythroblastosis virus (Yarden and Ullrich, 1988). These data, and the fact that the transforming protein of the avian leukosis virus also derives from the EGF receptor, demonstrates a common mechanism by which viruses use PTKs to induce transformation. The epidermal growth factor receptor is therefore designated *c-erbB-1*, since related PTKs have been discovered. The human and chicken forms are designated by HER (or HER1) and CER, respectively. The human EGF receptor can be designated as HER, HER1, *c-erbB*, or *c-erbB-1* and can also bind

TGF-α with high affinity. Whether or not TGF-α has its own receptor remains unknown. The HER2 or *erbB-2* gene encodes a receptor-type PTK with structural homology to the protein products of both HER and the chicken or avian virus oncogene *v-erbB* (Ullrich and Schlessinger, 1990). Note that this is not designated *c-erbB-2*, because there is no known virally transduced homolog. The ligand for this putative receptor, designated $p185^{HER2}$, has not been identified. The murine homolog is the *neu* protooncogene, which encodes for $p185^{neu}$. This gene becomes oncogenic due to a single base substitution when rats are treated with the chemical carcinogen ethylnitrosourea, and causes the formation of neuroectodermal tumors (Nishimura and Sekiya, 1987). A related human gene with extensive sequence homology to HER and HER2, and which maps to a different chromosome, has been designated *erbB-3* or HER3 (Kraus et al., 1989). Interestingly, *Drosophila* also has an EGF receptor of as yet unknown physiological role.

E. Platelet-Derived Growth Factor Receptor

The platelet-derived growth factor (PDGF) receptor defines a subfamily of PTKs that includes the receptors for macrophage colony-stimulating factor (CSF-1) and one or more of the fibroblast growth factors, *kit, ret,* and *flt* (Williams, 1989). This subfamily is characterized by an extracellular domain with consensus sequences that predict an extracellular immunoglobulinlike structure, and a catalytic domain that is split in two by a kinase insert of variable length. Similar to the receptors for EGF, insulin, and insulinlike growth factor-one (IGF-1), the entire protein sequence of the PDGF receptor was predicted from the cDNA clones following affinity purification, protein sequencing, and cDNA library screening using oligonucleotides based on the amino acid sequences. Two distinct genes for the PDGF receptor exist, and the protein products have been labeled the α and β subunits (Williams, 1989). The dimeric ligand PDGF is found in three isoforms, representing the three possible combinations of the A and B chains of PDGF. Thus there are nine possible permutations of receptor-ligand interaction in this system, since both the ligand and the receptor exist as dimers. However, the $\beta\beta$ receptor binds only PDGF-BB with high affinity, whereas the $\alpha\alpha$ receptor binds all three PDGF isoforms. The $\alpha\beta$ PDGF receptor binds PDGF-BB and PDGF-AB with high affinity (Williams, 1989).

The receptor for mononuclear phagocyte colony-stimulating factor-one (CSF-1) or *c-fms*, is the protooncogene which the McDonough strain of feline sarcoma virus uses to give rise to the *v-fms* oncogene product $gp140^{v\text{-}fms}$ (Yarden and Ullrich, 1988). The protooncogene *c-kit* encodes a related receptor $p145^{c\text{-}kit}$. The *v-kit* oncogene is found in the genome of the Harvey-Zuckerman-4 strain of feline sarcoma virus, which encodes a carboxy- and amino-terminally truncated form (Yarden et al., 1987). The *c-kit* product is thought to bind a 30-kDa stem cell factor derived from various mesenchymal-derived cells (Williams et al., 1990). The *c-kit* gene has been found to be allelic with the dominant *white spotting* locus (W) of the mouse and the w^{42} mutation, which affects hematopoiesis, and is associated with a *c-kit* product with defective tyrosine kinase activity (Tan et al., 1990).

Both basic and acidic fibroblast growth factor (bFGF and aFGF, respectively) elicit tyrosine phosphorylation when added to cells which express the high-affinity receptor (Mr = 110 to 150 kDa). The bFGF receptor was isolated from chicken embryos and partially sequenced and cloned based on sequence homology to the human gene *flg* (*fms*-like gene) (Ruta et al., 1989). Three related PTKs sequences were cloned from chicken embryos termed *cek*1-3 (for chicken embryo kinase). *Cek*1 codes for the chicken bFGF receptor and the *flg* gene is the human homolog. *Cek*2 and *cek*3 code for receptor PTKs related to the bFGF receptor, and *cek*3 is probably the chicken homolog to the mouse gene *bek* (for bacterially expressed kinase) (Pasquale, 1990). As stated previously, there is evidence that like the *c-src* gene, the mouse mRNA for the bFGF receptor, is expressed as alternatively spliced products (see above). Also related to this subfamily of PTKs are *ret* (Takahashi et al., 1991), a protooncogene which becomes transformed by recombination leading to a chimeric gene, and *flt* (for *fms*-like tyrosine kinase), a newly discovered human gene (Shibuya et al., 1990).

F. Insulin Receptor

The receptors for insulin and insulinlike growth factor-one (IGF-1) define yet another subfamily of PTKs and display a distinctive structure (Yarden and Ullrich, 1988). These two receptors and the *Drosophila* homolog are comprised of tetramers, generated by proteolytic cleavage and disulfide linkage of two primary translation products. The other members of this subfamily are *ros, trk, met, sea,* and *ltk*, which are closely related, based on sequence comparison. Another human gene has been isolated and sequenced which shows extensive homology to both the insulin and IGF-1 receptors. This gene is called IRR, for insulin receptor related, and it maps to a different chromosome than insulin or IGF-1 receptors (Shier and Watt, 1989). However, it is minimally, if at all expressed in normal tissues, and therefore its physiological role remains unknown. The protooncogene *c-ros* is the cellular counterpart of the transforming gene *v-ros* of avian sarcoma virus UR2 (Chen et al., 1991). The transforming protein is $p68^{gag\text{-}ros}$, in which all but eight amino acids of the extracellular domain have been deleted, whereas the normal cellular counterpart is a much larger 280-kDa glycoprotein. In addition, the *c-ros* gene is highly homologous to the *Drosophila sevenless* (*sev*) gene, which is essential for the development of R7 photoreceptor cells. The human *c-ros* and the *Drosphila sev* genes both code for receptor PTKs with very large extracellular domains (Chen et al., 1991).

The *trk* gene (for tropomyosin receptor kinase) was first discovered as a somatic chromosomal rearrangement in human colon carcinoma cells, where the transmembrane and cytoplasmic PTK sequences were fused with sequences from nonmuscle tropomyosin (Middlemas et al., 1991, and references within). The human *trk* protooncogene codes for gp140^{trk} and gp110^{trk}, and the related *trk*B protooncogene codes for gp145^{trk}B and gp95^{trk}B. The *trk*B protooncogene lacks a catalytic domain and also is expressed in a different pattern of tissues than *trk* (Middlemas et al., 1991). *Drosophila* has a homologous gene designated as D*trk*. Recently it has been shown that the *trk* protooncogene product is the nerve growth factor (NGF) receptor (Kaplan et al., 1991). NGF induces tyrosine phosphorylation and activates the tyrosine kinase activity of gp140^{trk}, binds with high affinity to gp140^{trk}, and can be chemically cross-linked to the *trk* protooncogene product.

The *met* oncogene was first identified in a human osteogenic sarcoma cell line which had been treated in vitro with the chemical carcinogen N-methyl-N′-nitro-N-nitrosoguanidine (Giordano et al., 1989). The somatic recombination involves separate chromosomes, and results in replacement of the extracellular and transmembrane domains with unrelated sequences. Interestingly, the *met* protooncogene contains a consensus sequence for proteolytic cleavage, similar to the insulin and IGF-1 receptors (Giordano et al., 1989). The structure of the expressed *met* protooncogene product is a heterodimer of a 50-kDa α and a 145-kDa β subunit which are disulfide-linked together. Recently it has been shown that the *met* protooncogene may be the receptor for hepatocyte growth factor (HGF) (Bottaro et al., 1991).

The S13 strain of avian erythroblastosis virus contains the oncogene *v-sea*, which encodes for a fusion protein of viral *env* sequences and PTK sequences. Although the fusion protein is anchored to the membrane by a membrane spanning region encoded by the *env* gene, the catalytic domain shares the greatest homology with the insulin receptor subfamily, in particular the human *met* protooncogene (Smith et al., 1989). The *ltk* gene was identified in the mouse genome using *v-ros* as a probe, and was designated *ltk* (leukocyte tyrosine kinase) based on the limited expression of this gene (Krolewski et al., 1990). The human homolog h*ltk* (or *tyk1*) and the rat *ltk* gene encode a PTK with an extracellular domain of only eight amino acids (Maru et al., 1990).

G. Eph

The first member of the Eph subfamily to be discovered was the *eph* gene by low-strigency screening of ETL-1, a erythropoietin-producing human hepatoma cell line, with a *v-fps* probe. There are two related genes, *elk* (*eph*-like kinase) isolated from a rat brain cDNA library (Letwin et al., 1988), and *eck* (epithelial cell kinase) isolated from a human keratinocyte cDNA library (Lindberg and Hunter, 1990). These three genes are placed into a separate subfamily based on the structure of the protein products and the phylogenetic analysis of the catalytic domain. Interestingly, this receptor PTK subfamily is phylogenetically more closely related to the nonreceptor PTKs (Hanks et al., 1988).

H. Other

The *tyk2* gene was identified by screening a cDNA library from human lymphoid cells with a *c-fms* probe (Krolewski et al., 1990). Overall it does not resemble the receptor or nonreceptor PTKs. A reanalysis of the entire sequence suggests that it is comprised of two separate catalytic domains preceded by sequences that have homology to the nonreceptor PTKs, and seems to be more related to the feline *c-fes* (Bernards, 1991). Two recently described PTKs, JAK1 and JAK2, contain one classical catalytic domain, and a catalytic-like domain (Wilks et al., 1991). The *tec* gene was identified in a murine liver cDNA library with a *v-fps* probe and codes for a nonreceptor PTK (Mano et al., 1990). However, except for the catalytic domain, there is minimal sequence homology with the other nonreceptor PTKs. The *tec* (tyrosine kinase expressed in hepatocellular carcinoma) gene is expressed preferentially in hepatocytes and in two of the four human hepatocullular carcinoma cell lines that were examined. The *klg* gene (kinaselike gene) was isolated from the genome of chicken embryos and encodes a novel receptor PTK, $p140^{klg}$, which contains seven immunoglobulinlike loops in the extracellular domain (Chou and Hayman, 1991). The PTKs in the PDGF receptor subfamily also have immunoglobulin loops in the extracellular domain. Within the kinase domain, the amino acid aspartate, within the conserved triplet of Asp-Phe-Gly, is replaced by the triplet of Ala-Leu-Ser. The aspartate is also missing from the EGF receptor-related gene *erbB-3* (Kraus et al., 1989). This single amino acid change results in the absence of kinase activity. The *c-kit* PTK becomes altered at the same critical aspartate residue in the w^{42} mouse, and is kinase inactive (see above and Tan et al., 1990).

Recently, the PTK responsible for the phosphorylation of tyrosine 527 of the $p60^{c\text{-}src}$ has been cloned (Nada et al., 1991). This PTK, termed CSK for c-SRC kinase, is unusual because, although it contains a tyrosine kinase domain and other sequences homologous to the nonreceptor PTKs, it lacks the conserved tyrosines within the kinase domain.

III. Structure

A. General Structure

The division between receptor and nonreceptor PTKs is useful for a number of reasons. Except for the catalytic domain, there are striking differences in structure which become manifested in functional terms (e.g., ability to bind a ligand). The receptor PTKs reside in the plasma membrane as integral membrane proteins and function as cellular sensors of the extracellular environment. Only a few of the receptor PTKs have unidentified natural ligands. Therefore, the normal biological role of the receptor PTK is known, at least in part, yet this is not true of the nonreceptor PTKs, and knowledge of their role in normal cellular function is not as clear.

Figure 1 shows a schematic of the receptor PTKs, arranged into the four subfamilies. The EGF receptor subfamily is characterized by a large extracellular domain which contains two cysteine-rich regions, whereas the PDGF receptor subfamily is comprised of scattered cysteines and sequences that predict an immunoglobulinlike structure. The insulin receptor subfamily is the most diverse. The receptor for both insulin and IGF-1 are comprised of two α

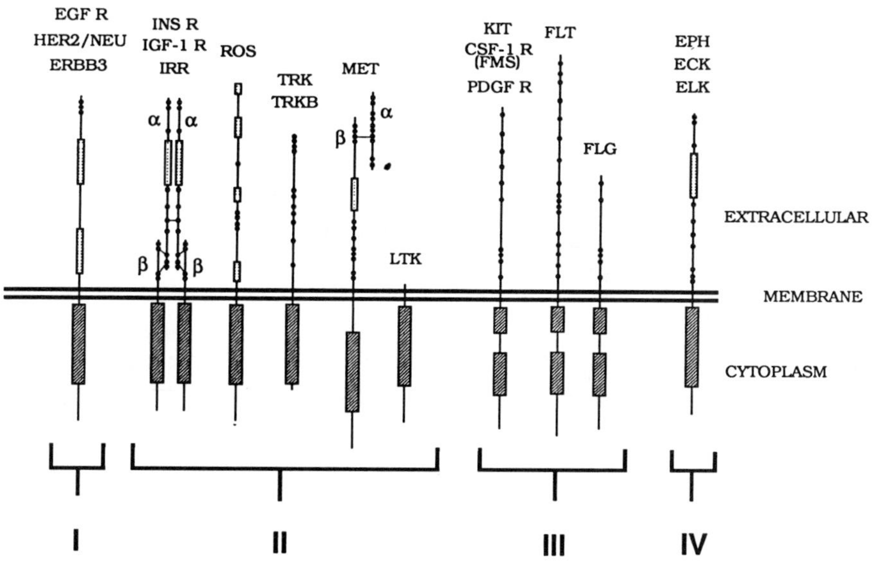

Figure 1 Recptor protein-tyrosine kinases. Symbols: 1. Dark circles = cysteines and thin boxes are cysteine rich regions; 2. Large dark boxes = conserved tyrosine kinase domains. 3. Lines between α and β subunits indicate disulfide bonds.

subunits which are entirely extracellular and are disulfide-linked to two β subunits. The β subunits contain the transmembrane and cytoplasmic entities. However, each $\alpha\beta$ half-receptor is synthesized as a single polypeptide translation product prior to processing. Lastly, the *eph, elk,* and *eck* gene products are shown which resemble a fusion of portions of an α and β subunit of the insulin receptor. Figure 2 indicates the domain structure of the three nonreceptor PTK subfamilies. The nine current members of the *src* subfamily are structurally very comparable, a function of the similar exon organization, indicating a common evolutionary precursor (Brickell, 1991). There is high sequence homology in several regions in addition to the catalytic domain, and there is considerable divergence at the amino and carboxy termimi. The members of the *fps/fes* subfamily are characterized by lengthy amino-terminal regions preceding the conserved sequences, whereas members of the *abl* subfamily have extensive carboxy-terminal sequences. Presumably, the divergent areas in the primary structure are involved in allowing the common kinase domain to exert influence in a tissue-, species-, or cellular-specific manner. There is accumulating evidence that this is due to interaction with other cellular proteins.

B. Catalytic Domain

As stated previously, the catalytic domain is the most highly conserved domain of the PTKs, with 11 defined subdomains (Hanks et al., 1988). It is comprised

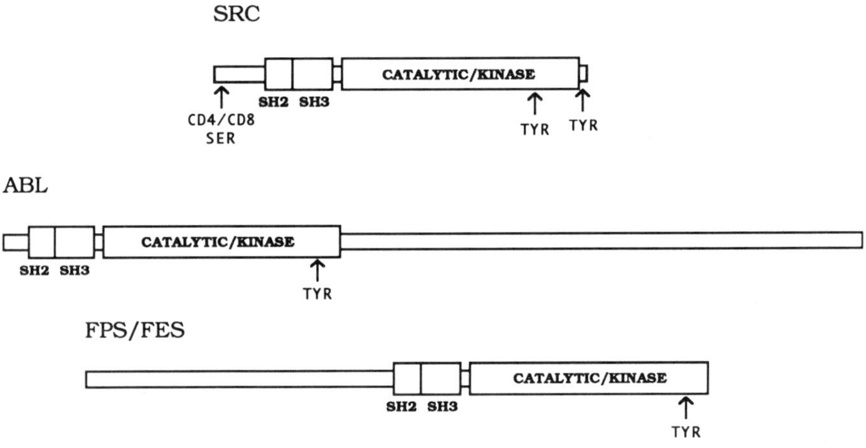

Figure 2 Nonrecptro protein-tyrosine kinases. Features noted (see text) are SH2 and SH3 domains, modified amino acids and interaction sites with cell surface molecules (*lck* with CD4/CD8).

of 250 to 300 amino acids and can be isolated as a functional entity either by protease digestion or by expression of a truncated mRNA. The kinase domain is generally carboxy-terminal and contains certain short sequences that are essential for enzymatic activity. Both the consensus Gly-X-Gly-X-X-Gly-X-X-Val (the valine is usually present) and the lysine 15 to 20 amino acids downstream from this are involved in ATP binding. In addition, an Asp-Asn and Asp-Phe-Gly in subdomains 6 and 7, respectively, are also thought to be involved in binding nucleotide and are necessary for kinase activity (Tan et al., 1990). Subdomains 6 and 8 contain short sequences that are specifically conserved in either serine/threonine protein kinases or the PTKs (Hanks et al., 1988). The catalytic domain also contains known tyrosine autophosphorylation site(s) for all members of the $p60^{src}$, the EGF receptor, and the insulin receptor families as well as for most other PTKs (see below).

C. Extracellular Domain

The receptor PTKs contain, by definition, a domain placed on the cell exterior for ligand recognition. Thus, the extracellular domain has two functions: to bind ligand and to convert that binding event into an intracellular activation signal for cellular response. In the case of the receptors for EGF and PDGF, the ligand-binding domain and the kinase domain are located on the same polypeptide, whereas the receptors for insulin and IGF-1 express these domains on separate subunits. For the *met* gene product it is not yet known to what subunit(s) the presumed ligand (hepatocyte growth factor) binds. The extracellular domain is missing in the PTK derived from the *ltk* gene, and for this reason it is postulated that this "truncated" receptor PTK interacts with additional membrane proteins that recognize and bind a ligand (Maru et al., 1990). Alternatively, this type of receptor PTK might bind to either a membrane protein on another cell or to extracellular matrix proteins, and in this way be involved in cell-cell and cell-matrix interactions.

The extracellular domains are typically heavily glycosylated using some or all of the potential glycosylation sites. This type of posttranslational modification does not seem to influence either binding of ligand or kinase activity. It is in this domain that the receptor PTKs are most divergent, which reflects the unique role that each plays in cellular function. In fact, the ability to manipulate the genes for the various receptor PTKs has led to the production of a large number of chimeric receptors, in which the extracellular domain of one receptor is linked to the cytoplasmic domain of another (Yarden and Ullrich, 1988). The data from these studies indicate that each domain can function independently of the other. For example, when chimeras containing the extracellular domain of the EGF receptor and the cytoplasmic domain of HER2 are expressed, the addition of EGF will lead to kinase activation of the HER2 catalytic domain. Another particular feature of these domains is the high number of cysteines and, more important, the conservation of the spacing of cysteine residues. These cysteines are responsible for covalent stabilization of a highly folded structure which forms the pocket to which the specific ligand binds.

D. Transmembrane Domain

The transmembrane domain is responsible for anchoring the receptor in the plasma membrane and there is no sequence homology in this domain that is shared between any of the receptor PTKs. However, the transmembrane domain of every receptor PTK is (1) hydrophobic and (2) of sufficient length (20 to 26 amino acids) to span a lipid bilayer, and (3) has a group of basic amino acids immediately distal to the stretch of hydrophobic residues. Based on this and a number of studies with chimeric receptors and either partial additions or deletions of this region, it is thought that this domain plays a relatively passive role in transmembrane signaling (Ullrich and Schlessinger, 1990). The notable exception to this thinking is the fact that the activation of the *neu* protooncogene is due to a point mutation which leads to a single amino acid change (valine to glutamic acid) within the transmembrane domain.

E. Juxtamembrane Domain

One of the working postulates about the PTKs is that regions without a high degree of sequence homology are involved in receptor-specific function within a cell. There is in fact evidence to suggest that the juxtamembrane region, comprised of 40 to 50 amino acids between the transmembrane and catalytic domains, does have regulatory roles. In the case of the EGF receptor, a threonin (position 654) in this region is phosphorylated by protein kinase C, which leads to loss of high-affinity binding and decreased receptor autophosphorylation and kinase activity (Ullrich and Schlessinger, 1990). The insulin receptor contains a tyrosine (Tyr 960) in this region that was originally thought to be an autophosphorylation site. When site-directed mutants were made and expressed in which the tyrosine was changed to a phenylalanine, there was no effect on the level of autophosphorylation or kinase activity; however, a presumptive insulin receptor substrate p185 does not get phosphorylated (Backer et al., 1990, and references therein). In addition, a short consensus sequence of Asn-Pro-X-Tyr (NPXY) in the juxtamembrane domain of the other non-PTK receptors is thought to play a role in endocytosis via coated pits. The insulin receptor contains this motif at Tyr 960, and its deletion led to decreased intracellular accumulation of the insulin receptor (Backer et al., 1990). Therefore, in the case of the insulin receptor, this region is probably involved in receptor-substrate interaction and endocytosis. The role that this region plays in the overall functioning of each receptor PTK may be different and will be an area of future research.

F. Kinase Insert Domain

The kinase insert domain is comprised of 75 to 110 amino acids which occur between subdomains V and VI and is unique to the PDGF receptor subfamily, although the other subfamilies of the receptor PTKs have an "insert" in the corresponding position of only 10 to 20 amino acids (Hanks et al., 1988). There is limited information on the role of this region, but as previously noted, the divergent domains will likely be involved in receptor-specific function, and it is postulated that the role of this region is to interact with receptor substrates and/or modulators. The data from deletions of all or part of this domain and mutation of a tyrosine (postion 751 in the PDGF receptor) suggest that this region can influence kinase activation and interaction with cytoplasmic substrates (Williams, 1989).

G. Carboxy-Terminal Domain

The carboxy-terminal domain is defined as the region between the end of the catalytic domain and the extreme carboxy terminus. The length and the sequence is quite variable among each of the subfamilies. The *src, fps/fes,* and *abl* subfamilies have 10 to 20, none, and 650 amino acids, respectively. For *abl*, this domain appears not to be necessary for kinase activity; however, the specific role that this region is involved in is currently unknown. The *src* subfamily members contain a tyrosine phosphorylation site here (527 in *src*, 505 in *lck,* 501 in *hck,* 522 in *fgr,* etc.), and it has been shown for *src* and *lck* that phosphorylation of this tyrosine reduces kinase activity (Brickell, 1991, and references therein). Also, this is not an autophosphorylation site, because the phosphorylation of this residue is performed by another cellular enzyme (Nada et al., 1991; Thomas et al., 1991). In the case of the receptor PTKs, this region varies from 75 to 200 amino acids, and is divergent even among the same subfamily. Tyrosines 1068, 1148, and 1173 in the EGF receptor and 1316 and 1322 in the insulin receptor, and 969 in the CSF-1 receptor, are known autophosphorylation sites, and in general are conserved in the other members of these two subfamilies (Ullrich and Schlessinger, 1990). In the case of the insulin receptor, these sites do *not* correlate with the acquisition of kinase activity, and deletion of this region does not affect kinase activity nor is there any major effect on insulin action (reviewed by O'Hare and Pilch, 1990). Similar findings have been observed in deletion mutants and tyrosine to phenylalanine mutants, in that the mutant receptors displayed no change in enzymatic or biological function.

H. Nonreceptor, Noncatalytic Domain

The nonreceptor PTKs lack a transmembrane region but still contain amino acids that are amino-terminal to the catalytic domain. In the case of *src* and *abl*, within this domain are two homologous regions termed SH2 and SH3, for *Src* homology region 2 (as the kinase domain is homology region 1) and *Src* homology region 3 (Brickell, 1991). The members of the *fps/fes* subfamily contain an SH2, but lack an SH3 region. The SH2 region is not required for subcellular localization or kinase activity, though it may be necessary for expression of the fully transformed state induced by the oncogenic counterparts of *c-src* and *c-fps*. More recently, it has been discovered that SH2 (and SH3) regions can be found in other proteins such as phospholipase C-gamma 1 (PLC-gamma 1), which catalyzes the hydrolysis of phosphatidylinositol 4,5-bisphosphate to the second messengers, diacylglycerol and inositol triphosphate. The GTPase activator protein (GAP), which can mediate signals generated by *ras*, also contains an SH2 domain, as does α-spectrin, and the product from the avian oncogene v-*crk* (Koch et al., 1991). There is accumulating evidence that these SH2 regions are involved in protein-protein association with cellular proteins that are tyrosine phosphorylated. Deletions and mutations in the SH3 region can influence morphology in transformed chicken embryo fibroblasts. However, kinase activity and ability to transform cells is unaffected.

It has been demonstrated that p60c^{src} is predominantly membrane bound, and that localization to the cell membrane plays a role in transformation (reviewed by Resh, 1990). Localization to the plasma membrane is mediated by myristate, which is covalently linked to the amino-terminal glycine of p60c^{src}. All of the nonreceptor PTKs contain a myristoylated amino-terminal glycine. In the case of *abl* and *arg*, alternative splicing results in two transcripts which lead to two protein products, only one of which contains the glycine residue. It is not yet known if this has functional significance.

IV. Mechanism of Activation

There is considerable interest in elucidating the mechanism by which the PTKs are activated and transduce a signal. The receptor PTKs utilize their intrinsic tyrosine kinase activity as part of the transduction apparatus. This enzymatic activity and therefore signaling potential is functional in all cell types that possess these receptor PTKs. To what extent the kinase domain is responsible or necessary for all the pleiotropic effects of ligand is currently not clear. However, based on the data so far, there is a close association between a functioning kinase domain and the cellular responses to the appropriate ligand. In addition, there is considerable evidence that both classes of PTKs are involved in the pathogenesis of certain cancers, and detailed knowledge of activation mechanisms is a prerequisite for potential therapeutic intervention.

A. Receptor Protein-Tyrosine Kinases

Ligand Binding

The initial event in the process which leads to kinase activation and biologic function is the specific interaction of receptor and ligand (O'Hare and Pilch, 1990, and references therein). This association is typically specific, reversible, and of high affinity, with a K_D measured in the nanomolar to subnomolar range. Information pertaining to the binding-site location on the receptor PTK is limited to the receptors for EGF, insulin, and IGF-1. When the EGF receptor is digested with CNBr after labeling with ^{125}I-EGF, a large fragment of the extracellular domain is obtained corresponding to domain III, the region between the two cysteine-rich regions (domains II and IV). Moreover, through the use of human and chicken EGF receptor chimeras, it was determined that the ability to bind EGF with high affinity seems to reside in domain III (Lax et al., 1989). However, domain III may not bind EGF directly, but may simply function as a scaffolding, allowing domains II and/or IV to interact with EGF. It is important to keep in mind that a domain based on primary structure (i.e., amino acid sequence or intron/exon boundaries) may not have a functional correlate. Protease digestion of insulin receptor cross-linked to ^{125}I-insulin generates stable fragments of differing sizes depending on the experimental conditions. Based on sequence analysis of the fragments, the use of antipeptide antisera, and exhaustive proteolysis of labeled insulin receptor, both the amino-terminus and cysteine-rich regions are involved in binding of insulin (reviewed by Pilch et al., 1989). Interestingly, the predicted secondary structure of both the EGF and insulin receptors indicate that there are two hypervariable regions, one at the amino terminus and one in the cysteine-rich region. In the case of immunoglobulins, it is the hypervariable regions of the heavy and light chains which determine specificity. Chimeras produced from insulin and IGF-1 receptors indicate that exon 1, which codes for the amino terminus, determines insulin specificity, whereas exon 3, which codes for the bulk of the cysteine-rich region, determines IGF-1 specificity (Zhang and Roth, 1991). More detailed knowledge of ligand-receptor interaction will have to await three-dimensional analysis of crystals.

Another level of complexity exists for the receptors for insulin and IGF-1 as the native receptor contains two insulin-binding sites per tetrameric receptor. However, Scatchard analysis of binding data shows a curvilinear plot indicating negative cooperativity or negative homotropic allosteric regulation between the two binding sites. This raises a number of interesting questions with regard to the mechanism of transmembrane signaling in light of the critical role of receptor dimerization (see below).

Dimerization

Although overall there is a paucity of information of the details of transmembrane signaling utilized by the receptor PTKs, one aspect is well documented. Receptor-receptor interaction (i.e., dimerization) seems to be a shared feature of all the receptor PTKs and plays a critical role in signal transduction (Yarden and Ullrich, 1988; Ullrich and Schlessinger, 1990; Pilch et al., 1989). The insulin and EGF receptors have been the focus of many studies on this point. In fact, the insulin receptor has served as a paradigm for the model of dimerization, as this receptor exists as a disulfide-linked dimer of two $\alpha\beta$ subunits. Like the *met* protooncogene, the primary transcript encodes a long polypeptide which contains both the α and β subunits. Only after posttranslational modification by proteolysis do they become distinct subunits. The receptors for insulin and IGF-1 are unique in that the disulfide-linked $\alpha\beta$ subunits normally occur as a disulfide-linked tetrameric complex, $\alpha_2\beta_2$. Dimerization for the other receptors is induced by ligand, brings about conformational changes, and can be correlated with functional consequences including activation of the catalytic domain. So far, dimerization has been demonstrated for each type of receptor PTK except for the *eph* subclass (Ullrich and Schlessinger, 1990). It is not known why the insulin and IGF-1 receptors exist natively as dimers, but it is postulated that this may facilitate the acute (within seconds) metabolic effects of this hormone.

The $\alpha_2\beta_2$ tetrameric insulin receptor can be reduced to form functional $\alpha\beta$ subunits which show low-affinity noncooperative binding, and that insulin can potentiate the reassociation of two $\alpha\beta$ subunits with reacquisition of high-affinity binding. In addition, (1) insulin can stimulate autophosphorylation of the $\alpha\beta$ species when in solution but not when immobilized to a substratum, and (2) receptor autophosphorylation is receptor concentration independent for tetrameric receptors but concentration dependent for the $\alpha\beta$ form (Pilch et al., 1989). These data have been extended to the IGF-1 receptor and, moreover, hybrid receptors consisting of an $\alpha\beta$ from the insulin receptor and an $\alpha\beta$ from the IGF-1 receptor have been prepared in vitro and have been demonstrated to exist in vivo (Ullrich and Schlessinger, 1990, and references therein). EGF has been shown to induce the formation of EGF receptor dimers by sucrose gradient density centrifugation and by chemical cross-linking and gel electrophoresis. More recently, it has been shown that a lysine-to-alanine mutant of the EGF receptor, which results in an inactive kinase domain, can become phosphorylated in the presence of EGF when mixed with wild-type EGF receptor (Honegger et al., 1989). In the case of the PDGF receptor, dimerization of the two forms (α and β) leads to three isoforms ($\alpha\alpha$, $\beta\beta$, and $\alpha\beta$) and this dimerization is (1) necessary for high-affinity binding, (2) determines PDGF specificity, and (3) is associated with kinase activation via autophosphorylation (Kelly et al., 1991). The mechanism of dimerization is also utilized by the CSF-1 receptor (Lee and Nienhuis, 1990). There have been a large number of chimeric receptor constructs, including chimeras of the EGF receptor and HER2, the insulin receptor and (1) the IGF-1 receptor, (2) the EGF receptor, (3) the *ros* oncogene product, and (4) the *Escherichia coli* aspartate receptor. It is clear from all of these studies that the PTK receptors utilize similar signaling mechanisms, however, dimerization was not always examined directly. Regardless, it is widely accepted that dimerization is involved in the signaling mechanism of the receptor PTKs, and it is tempting to speculate that all receptor PTKs utilize receptor dimerization, although other mechanisms may exist. The model of receptor-receptor interaction may also be applicable to the cytoplasmic PTKs based on experiments with the *lck* protooncogene and the T-cell receptor complex (see below).

Lastly, there have been some interesting observations when various mutated and kinase-inactive receptors have been expressed in cells which express their own native and kinase-active receptors. In general, there is a diminution in the ability of a ligand to cause the typical cellular responses, and this has been attributed to the dimerization of a native functional receptor with one of the mutated receptors. As stated previously, the protooncogene *c-kit* is allelic with the murine *white spotting* locus. Mice homozygous for this allele die prematurely in the perinatal period, whereas the heterozygous mouse lives with a severe macrocytic anemia, loss of coat pigment, and gonadal atrophy (Tan et al., 1990). The only change in the *c-kit* protein product is the diminished tyrosine kinase activity, as would be expected from the missense mutation which replaces a conserved aspartic acid with an alanine at position 790. The heterozygote is much more severely affected than would be predicted if there were only a 50% reduction in $p145^{c\text{-kit}}$ receptor kinase levels. Therefore this result can be taken as indirect evidence that, in vivo, the function of the *c-kit* protooncogene product requires receptor-receptor interaction, and is negatively affected in a transdominant fashion by mutation of one allele. Similar data exist for certain patients with severe insulin resistance who express one normal and one mutant allele (Kadowaki et al., 1990). Since insulin receptors must be expressed as covalent dimers, and since both halves must be normal for the overall receptor function (O'Hare and Pilch, 1990), the 25% of normal receptors expressed under these circumstances are eventually overwhelmed by the kinase-inactive receptors overrepresented at the cell surface (see above), and severe insulin resistant diabetes mellitus results.

Autophosphorylation

The conservation of amino acid sequence within the kinase domain of all protein kinases predicts similar functional characteristics, and this extends to the ability of most protein kinases to phosphorylate themselves. Insulin induces autophosphorylation of several tyrosyl residues within seconds of binding. These sites are contained within three separate regions of the β-subunit cytoplasmic region and are 1146, 1150, 1151, 1316, 1322, and at least one additional site (O'Hare and Pilch, 1990). There are three tyrosines located within the catalytic domain (Tyr 1146, 1150, and 1151) which become rapidly phosphorylated. However, any given receptor may display different levels of phosphorylation at these three sites. For instance, one receptor may have all three sites phosphorylated (the triply phosphorylated form) or only two of the sites phosphorylated (one of the three possible doubly phosphorylated forms). Since each receptor contains two β subunits, the degree of complexity can be considerable; however, what is clear is that the triply phosphorylated form (i.e., sites 1146, 1150, and 1151 are phosphorylated) correlates best with activation of the catalytic domain toward substrates. The carboxy-terminal tyrosine phosphorylation sites (1316 and 1322) do not correlate with activation of the kinase domain. Mutants in which these tyrosines are deleted or changed to phenyelalanine display normal catalytic activites. The role of autophosphorylation in kinase activation has also been examined for IGF-1, PDGF, NGF (*trk*), HGF (*met*), and CSF-1 (*fms*) receptors. In each case, the mechanism of kinase activation by receptor autophosphorylation is due to an increase in the V_{max} of substrate phosphorylation (Ullrich and Schlessinger, 1990). It is expected that phosphorylation of the tyrosine(s) at the conserved region within the catalytic domain will correlate with kinase activation.

The receptor of EGF behaves somewhat differently, and it has not been easy to account for activation of the kinase domain solely upon autophosphorylation (Ullrich and Schlessinger, 1990). The major autophosphorylation site in the EGF receptor is at the carboxy terminus (Tyr 1173), whereas the conserved tyrosine in the kinase domain (Tyr 845) is not significantly phosphorylated. Although there have been conflicting data on the effect of EGF receptor autophosphorylation on kinase activity, the more recent data suggests that carboxy-terminal autophosphorylation does have a positive effect on kinase activity. The EGF receptor may utilize conformational changes induced by EGF binding, in addition to conformational changes attributable to autophosphorylation. The insulin receptor kinase domain, once activated by autophosphorylation, remains active even in the absence of insulin. The same does not seem to be true for the EGF receptor, and that EGF-induced dimerization is critical to initiate and sustain the catalytic activity.

Protein-Tyrosine Kinase Activity

The role of the enzymatic activity of the receptor PTKs is well established for some but not all of the cellular processes in which these receptors are involved. There have been a large number of studies involving most of the receptor PTKs and the functional effect of site-directed and/or deletion mutants. One category of functional effects is biochemical, such as transport of glucose and amino acids, activation of cellular enzymes, and DNA synthesis. A second category involves cell biological functioning such as endocytosis and cellular transport of the endocytosed receptors. A third category has been the connection between protein-tyrosine kinase activity and transformation/oncogenesis.

The enzymatic activity of all of the receptor PTKs that have been examined show a uniform conclusion that this activity is necessary for stimulation of the biochemical effects. The kinase domain is not sufficient by itself to carry out the receptor's functions, since mutations elsewhere can alter receptor-mediated cellular effects. For instance, the mutated insulin receptor in which the tyrosine at position 960 is replaced by a phenylalanine displays normal kinase function but fails to phosphorylate an endogenous substrate, and also does not induce DNA synthesis (Backer et al., 1990). In addition, there are a number of insulin receptor species which are mutated in the extracellular domain, some with only a single amino acid change, that have no kinase activity or biological function.

Ligand- or antibody-induced endocytosis may not depend on a functional kinase domain, but the targeting of both the EGF and insulin receptors is dependent on normal kinase activity. That is, kinase defective receptor mutants endocytose normally, but instead of being targeted to the lysosome, these receptors are prematurely sorted back to the plasma membrane. However, there are conflicting data on the role of the kinase domain in the process of endocytosis. The ability of the receptor PTKs to tranform cells is often related to structural alterations (Yarden and Ullrich, 1988). The activation of the receptors for EGF and CSF-1 and the *c-kit* protooncogene product involve carboxy-terminal truncation. The *v-erbB* and *v-kit* oncogenes also have amino-terminal deletions, whereas the activation of the *neu* protooncogene involves a point mutation in which the valine at position 664 (a site within the transmembrane region) is substituted by a glutamine. In each case the structural change is associated with constitutive kinase activity. There are other mechanisms by which the receptor PTKs may be involved in transformation, such as participation in an autocrine or paracrine loop, or overexpression of the PTK. In addition, a defect in the turning-off mechanism, such as a deletion in one of the tyrosine-specific phosphatases, could also potentially result in transformation.

B. Nonreceptor Protein-Tyrosine Kinases

Relative to the receptor PTKs, there is a lack of information concerning the mechanism of activation of the nonreceptor PTKs, although recent progress has been made for $p60^{c\text{-}src}$. The comparison of retroviral oncogenes and their cellular counterpoints has proven most useful in understanding the structural basis for activation. Comparing the structure of $p60^{c\text{-}src}$ to $p60^{v\text{-}src}$ shows that the carboxy-terminal 19 amino acids of $p60^{c\text{-}src}$ are replaced by an unrelated sequence of 12 amino acids in $p60^{v\text{-}src}$ (Piwnica-Worms et al., 1987). Contained within this terminal sequence is tyrosine 527, whereas tyrosine 416, which resides in the kinase domain, is conserved among almost all of the PTKs. The pattern of phosphorylation of $p60^{c\text{-}src}$ and $p60^{v\text{-}src}$ is different. In vitro, $p60^{c\text{-}src}$ is phoshoratd on tyrosine 416, whereas in vivo, it is phosphorylated predominantly at tyrosine 527. There is an increased tyrosine kinase activity and an increased ability to transform cells by (1) $p60^{v\text{-}src}$, which shows in vivo phosphorylation on tyrosine 416 but not tyrosine 527; (2) $p60^{c\text{-}src}$ from polyoma-transformed cells in which the polyoma middle T antigen binds to $p60^{c\text{-}src}$ and influences phosphorylation such that tyrosine 416, but not tyrosine 527 is modified; and (3) $p60^{c\text{-}src}$ in which tyrosine 527 is mutated to phenlalanine. Therefore in the case of *src*, autophosphorylation at tyrosine 416 will activate the kinase domain (Piwnica-Worms et al., 1987). Phosphorylation of tyrosine 527, performed by another PTK, has a negative influence on kinase activity (Nada et al., 1991; Thomas et al., 1991). Thus, it appears that *c-src* is maintained in an inactive state by a novel PTK that phosphorylated *c-src* at Tyr 527. Removal of this phosphorylation by a tyrosine phosphatase would then allow autophosphorylation of Tyr 416 and activation of the PTK. However, the regulatory orchestration of these events remains unknown.

The conserved tyrosine within the kinase domain of $p130^{gag\text{-}fps}$ is tyrosine 1073. Based on site-directed mutagenesis, differences between the transforming and nontransforming versions of *fps*, and biochemical analysis, autophosphorylation at this conserved site activates the kinase domain. The *lck* protooncogene encodes $p56^{lck}$, which contains tyrosine residues at position 394 and 505, corresponding to tyrosine 416 and 527 of $p60^{c\text{-}src}$, respectively. Site-directed mutagenesis of tyrosine 505 to phenylalanine results in the ability to transform fibroblasts and leads to an increase in tyrosine phosphorylation of cellular proteins (reviewed by Sefton, 1991). Recently, $p56^{lck}$ has been implicated in signaling through the T-cell receptor (TCR), and when anti-CD4 antibodies are used to activate the TCR, there is increased autophosphorylation of $p56^{lck}$ at tyrosine 394, and a concomitant increase in kinase activity (Klausner and Samelson, 1991).

The *v-abl* oncogene, when compared to *c-abl*, shows a loss of amino acids at the amino terminus, resulting in the loss of the SH3 domain. In addition, at least two strains of the Abelson murine leukemia virus have a G-to-A point mutation in codon 832. This second mutation is synergistic with the amino-terminal truncation and may account for those patients with stable Philadelphia chromosome-positive CML that acutely turns into a more malignant course (Shore et al., 1990).

Purified p60$^{c\text{-}src}$ and p56^{lck} behave like most other purified PTKs except that p60$^{c\text{-}src}$ can utilize GTP as well as ATP as the phosphate donor. The kinetic analysis of catalytic function yield typical values for turnover number of K_m for ATP. In addition, there is confirmatory data that autophosphorylation activates the intrinsic kinase activity. In the case of p56^{lck}, this involves self-phosphorylation at tyrosine 394 and another uncharacterized site. This is similar to data derived from experiments utilizing both a cytoplasmic bovine spleen PTK and the expressed cytoplasmic domain of the insulin receptor.

V. Signaling Mechanisms

There have been significant advances in our understanding of the signaling pathways utilized by receptor and nonreceptor PTKs. What has been surprising is how similar these pathways are for both receptor and nonreceptor PTKs. The activation of *lck* seems to occur by dimerization of the cell surface T-cell receptor complex. There seem to be common substrates among the different classes of PTKs. This is probably a consequence of the newly discovered function of the SH2 domains, that of binding to tyrosine-phosphorylated proteins. This is one of the areas of PTK research which is most active, as new potential roles for PTKs are being rapidly discovered. The functional aspects of SH2 and SH3 domains in linking the signal transduction pathways for the receptor PTKs and the nonreceptor PTKs is the subject of an excellent recent review (Koch et al., 1991).

A. Nonreceptor Protein-Tyrosine Kinases

It had been surmised that the cytoplasmic PTKs probably were involved in transmembrane signaling events based on their location within the cell (at the cytoplasmic face of the plasma membrane via myristoylation). One of the receptors that probably utilizes the cytoplasmic PTKs for signal transduction is the T-cell receptor complex. The T-cell receptor is comprised of a number of different polypeptides, including an $\alpha\beta$ heterodimer and several polypeptides which form CD3 (Klausner and Samelson, 1991). The T-cell surface molecules which recognize the major histocompatibility antigens are the T-cell receptor (TCR) accessory proteins, CD4 and CD8. Activation of the T-cell receptor with antigen, anti-CD4/CD8 antibodies (Sefton, 1991), the human immunodeficiency virus gp 120 glycoprotein (Juszczak et al., 1991), or anti-CD3 antibodies (Park et al., 1991) results in an increase in tyrosine phosphorylation within the cell (including the zeta subunit of the TCR), cell proliferation, lymphokine secretion, Ca^{2+} accumulation, and the breakdown of phosphoinositides. A search for the responsible PTK(s) led to the discovery of a noncovalent association of $p56^{lck}$ with both CD4 and CD8, and the inducement of autophosphorylation and activation of $p56^{lck}$ by antibodies to CD4 and CD8. Based on experiments with chimeric proteins, the $p56^{lck}$-binding region of CD4 and CD8 resides in their carboxy terminus, whereas binding to CD4/CD8 can be eliminated by altering the amino-terminal 32 amino acids of $p56^{lck}$ (Sefton, 1991). The physiological role of CD4/CD8-lck association is obscured by the fact that antigen stimulation of T cells does not activate $p56^{lck}$ and leads to a different pattern of phosphorylated proteins. Also, only a small percent of the total $p56^{lck}$ can be co-immunoprecipitated with anti-CD4/CD8 antibodies. Despite these reservations, there is accumulating evidence that *lck* plays a role in T-cell signaling, including activation of the phospholipase C-gamma 1 (PLC-gamma 1). The treatment of the human T-cell line Jurkat with anti-CD3 antibody leads to the phosphorylation of PLC-gamma 1 on serine and tyrosine residues (Park et al., 1991). Moreover, these are the same sites that are phosphorylated in response to treatment of the cells with EGF and PDGF. Therefore, both the receptor and nonreceptor PTK classes may utilize, at least in part, the same or similar signaling pathways. The activation of $p56^{lck}$ involves phosphorylation at tyrosine 394 and an absence of tyrosine phosphorylation at position 505. The dephosphorylation of tyrosine 505 may be catalyzed by CD45, a hematopoietic cell surface antigen and protein-tyrosine phosphatase. If the Jurkat T-cell line is made CD45 deficient, the TCR cannot mediate activation as assessed by tyrosine phosphorylation, phosphatidylinositol turnover, and interleukin 2 production (Koretzky et al., 1991). Therefore, autophosphorylation at Tyr 394 and dephosphorylation at Tyr 505 of $p56^{lck}$ may both be required for T-cell activation.

Since the discovery of the association of *lck* with the TCR, additional associations have been examined. In the case of the TCR, it has been demonstrated that $p59^{fyn}$, but not $p56^{lck}$, can be co-immunoprecipitated with anti-TCR antibody, and that $p59^{fyn}$ can phosphorylate the TCR zeta chain (Samelson et al., 1990). The *lck* gene product may also play a role in interleukin-2 signaling of lymphocytes (Hatakeyama et al., 1991). The in vitro kinase assay after co-immunoprecipitation with anti-interleukin-2 receptor antibody leads to tyrosine phosphorylation of the β chain of the interleukin receptor. Also, a B-cell-specific PTK *lyn* is speculated to be involved in antigen-induced signal transduction events via a B-cell surface receptor (Yamanashi et al., 1990).

B. Receptor Protein-Tyrosine Kinases

The various growth factors and insulin induce pleiotropic biochemical changes in cells, including altered transport of amino acids and glucose, the generation of metabolites from the phosphatidylinositol pathway, and changes in the phosphorylation state and activity of a number of cellular proteins. A major approach to discover endogenous substrates used antiphosphotyrosine antibodies, and a number of cellular proteins were identified, initally only by mobility on sizing electrophoresis. These included p36, which belongs to the lipocortin family, p42 (a protein-serine/threonine kinase), and p185 (a novel cytoplasmic protein), which undergoes tyrosine phosphorylation in response to insulin and IGF-1. In trying to map the signal pathway from PDGF to increases in inositol 1,4,5-triphosphate generation, it was discovered that the PDGF receptor can associate and phosphorylate PLC-gamma 1 in a ligand- and kinase-dependent manner (Ullrich and Schlessinger, 1990). Later, it was shown that tyrosine phosphorylation leads to activation of PLC-gamma 1 (Nishibe et al., 1990). These results have been extended to the receptor for EGF, the *c-kit* gene product (Lev et al., 1991), and the NGF receptor (Vetter et al., 1991).

Another enzyme, phosphatidylinositol 3-kinase, has been shown to be associated with the receptors for PDGF, EGF, CSF-1, and insulin (Escobedo et al., 1991, and references therein). The enzymatic activity of phosphatidylinositol 3-kinase is also increased upon addition of each of these ligands. However, the role that this enzyme plays in growth factor signaling events is in question. A point-mutated CSF-1 receptor cannot stimulate mitogenesis, yet it can still associate with phosphatidylinositol 3-kinase (Roussel et al., 1990). Also, a kinase insert-mutated PDGF receptor, which cannot bind to phosphatidylinositol 3-kinase, is still able to effect mitogenic and chemotactic signal transduction (Yu et al., 1991). The guanosine triphosphatase activator protein (GAP), which enhances the GTPase activity of *ras*, is another substrate of activated growth factor receptors (Ullrich and Schlessinger, 1990).

Another method for elucidating the signaling mechanism of the receptor PTKs has been to start at a known cellular effect of mitogen and work backward, toward the receptor. In this way, the mitogen-activated protein-serine/threonine kinases have been identified as intermediates in a phosphorylation cascade. S6, a 40S ribosomal protein, undergoes multiple serine phosphorylation in response to both mitogenic substances and oncogene products (reviewed by Erikson, 1991). There are two classes of S6 kinases based on size. One class of S6 kinases, with Mr = 90 and 92 kDa, is called the RSK family (ribosomal S6 kinase). The other class has one member and it is called the 70-kDa S6 kinase. There is convincing data which show that the S6 kinases function in vivo. For example, in vitro S6 kinase activity from mitogen-activated cells (in vitro phosphorylation of S6), correlates with in vivo levels of S6 phosphorylation. Initially, it was hypothesized that the S6 kinases were phosphorylated and activated by a receptor PTK. However, it was disappointing when it was discovered that the S6 kinases were activated by serine, not tyrosine phosphorylation. This meant that the S6 kinases are not direct substrates of the PTKs, and there was at least one, if not several intermediates between the receptor PTKs and S6 kinase. One of those intermediates is MAP kinase (for mitogen-activated protein kinase), although it used to stand for the exogenous substrate microtubule-associated protein-2. Two closely related 42-kDa and 44-kDa MAP kinases exist (or $p42^{mapk}$ and $p44^{mapk}$), though they are also referred to as S6 kinase kinases. It has been shown that $p42^{mapk}$ can phosphorylate and activate S6 kinase in vitro, and that this leads to phosphorylation of the same sites on S6 kinase that become phosphorylated in vitro. Tyrosine phosphorylation of the MAP kinases is increased in cells treated with mitogens, or transformed by a number of the PTK oncogenes. However, there has been no direct demonstration of tyrosine phosphorylation by any of the receptor PTKs. Another mitogen-stimulated protein-serine/threonine kinase, Raf-1, the homolog to *v-raf* (a murine sarcoma virus oncogene), is also serine phosphorylated in vivo, in response to mitogens. Raf-1 can be found in noncovalent association with some of the receptor PTKs, and the tyrosine kinase activity of the receptor seems essential for Raf-1 activation, though it is clear that Raf-1 is not a direct substrate of the receptor PTKs. Similar to the S6 kinase situation, Raf-1 kinase (Raf-1 protein kinase *kinase*) activity can be partially purified from insulin- and mitogen-stimulated cells using Raf-1 as a substrate. In vitro, the Raf-1 kinase kinase activity leads to phosphorylation of the same sites on Raf-1 that occur during in vivo stimulation with mitogens (Lee et al., 1991).

VI. Role in Oncogenesis

In the early 1900s, Peyton Rous first demonstrated that a nonfilterable agent, when injected into mice, would cause tumors, namely, sarcomas. The discovery of the Rous sarcoma virus, and the presence within its genome of *v-src*, were responsible in part for the discovery of additional oncogenes and their cellular counterparts. A large number of tumor viruses have been discovered which cause tumors in vivo and the in vitro correlate, cell transformation. In addition, there are a number of chemically induced and presumably spontaneous activations of protooncogenes. The transformation of cells is characterized by a number of biochemical and morphological changes which are related to alterations in growth control, such as decreased growth factor requirement, anchorage-independent growth, and immortalization. Protein-tyrosine kinases constitute the majority of characterized protooncogenes. Cellular proto-oncogenes can be grouped into several functional classes (Table 2), and it is apparent that they all are involved in the process of intracellular signaling. The decreased growth factor requirement conferred by the oncogene has a multitude of potential mechanisms, including (1) increased growth factor production, (2) growth factor receptor alterations, and (3) dysregulation of cytoplasmic and nuclear proteins which serve as transducers of growth factor signals.

The role of oncogenes, in particular protein-tyrosine kinases, in human malignancies is currently an area of extreme interest because of the potential for prognostic information and therapeutic intervention. Some cellular oncogenes are expressed in a wide variety of tumor types, and these include *myc, fos,* and *ras*. In addition, it is very common for two or more cellular oncogenes to be transcribed in any given tumor (Nishimura and Sekiya, 1987). This is consistent with the view that carcinogenesis is a slow process of initiation and promotion, in that over time there is an acquisition of cellular and nuclear insult which leads to sequential activation of multiple proto-oncogenic loci and loss of tumor-suppressor gene function. A logical interpretation of this is that the simple overexpression of a single oncogene does not lead to tumor formation or transformation. There must be other genetic changes, for example, a mutation in a growth factor receptor or other proteins which allows the receptor-generated signal to be active continually. Studies which show transformation of NIH 3T3 fibroblasts by simple overexpression should be viewed cautiously, as this cell line is already partially transformed.

Table 2 Known Oncogenes Separated by Function

Growth factors	*sis*		
Protein-tyrosine kinases			
	src	*yes*	*fgr*
	abl	*fes/fps*	*erb*B-1
	neu	*fms*	*kit*
	ros	*trk/trk*B	*met*
Protein-serine/ threonine kinases	*raf*	*mos*	
GTP-binding proteins	H-*ras*	K-*ras*	N-*ras*
Nuclear proteins	*myc*	*fos*	*myb*
	p53	*ski*	
Steroid receptors	*erbA*		

An autocrine mechanism is thought to play a role in oncogenesis in some human malignancies including those involving the lung. Both small cell and carcinoid lung cancer cell lines express bombesinlike peptides which act as mitogens and which activate tyrosine-phosphorylation of p115, a bombesin receptor-associated protein (Moody et al., 1990). The non-small cell lung cancer cell lines express the EGF receptor at high levels and also secrete TGF-α. One has to be cautious about the interpretation of data on a small number of cell lines, but it was shown in a recent article (Kern et al., 1990) that when 131 primary lung adenocarcinomas were examined, the majority of them expressed TGF-α, EGF, or the EGF receptor at high levels. In the EGF receptor-positive cases, the survival at 5 years was significantly reduced if the primary tumor also expressed TGF-α (36% versus 85%). This suggests that an autocrine pathway is operational in a subset of lung cancers and that in this subset, prognosis is worse. However, the stage of lung cancer was also more advanced in this subset, and because of this, it is not clear if the expression of both TGF-α and EGF receptor was responsible for or a consequence of tumor progression. Regardless, this additional information, so-called biochemical staging, may prove useful in the selection of patients who might benefit from either earlier intervention with standard chemotherapeutics or treatment with molecules which affect the growth factor receptor pathway. In another immunohistological analysis of primary non-small cell lung cancer, the expression of the HER2/*neu* protooncogene p183^{neu} in adenocarcinomas was significantly associated with shortened survival, independent of stage (Tateishi et al., 1990). The "receptor" status of breast cancer has been used to determine treatment options, and it is hoped that a similar situation will exist for other tumor sites.

There are structural changes in some of the receptor PTKs which are associated with transformation, though this has been outlined in more specific reviews and will not be covered here (Yarden and Ullrich, 1988). There are numerous studies which examine the relative expression of the mRNA of a particular protooncogene in human malignancies. For example, it was shown that the human counterpart to the murine *ltk* proto-oncogene is expressed in 10 of 18 leukemia cell lines but none of 18 nonleukemia cell lines (Maru et al., 1990). In this particular report, the level of expression and the size of the transcripts were not altered from expression in normal cells. Although there is ever-increasing knowledge with regard to expression of oncogenes in human cancers, there is a need to test the importance of this expression directly. That is, to what extent does the expression of PTKs contribute to the transformed phenotype? It is interesting that there are relatively few reports which show structural abnormalities of the PTKs in human malignancy. The best-understood example of this is the alteration of the *c-abl* protooncogene product p145$^{c\text{-}abl}$. Philadelphia chromosome-positive chronic myelogenous leukemia (CML) cells express both the normal product and p210$^{bcr\text{-}abl}$, a fusion protein resulting from a chromosomal translocation. This product expresses higher levels of tyrosine kinase activity, and is constitutively activated. Recently it has been shown that there is an additional fusion product, p185$^{\text{bcr-}abl}$, which consistently associated with a more malignant form of CML, and this fusion protein displays an even greater degree of intrinsic kinase activity (Lugo et al., 1990).

References

Abraham KM, Levin SD, Marth JD, Forbush KA, Perlmutter RM. Thymic tumorigenesis induced by overexpression of p56^{lck}. Proc Natl Acad Sci USA 1991; 88:3977–81.

Backer JM, Kahn CR, Cahill DA, Ullrich A, White MF. Receptor-mediated internalization of insulin receptor requires a 12-amino acid sequence in the juxtamembrane region of the insulin receptor β-subunit. J Biol Chem 1990; 265:16450–4.

Bernards A. Predicted *tyk*2 protein contains two tandem protein kinase domains. Oncogene 1991; 6: 1185–7.

Bottaro DP, Rubin JS, Faletto DL, Chan AML, Kmiecik TE, Vande Woude GF, Aaronson SA. Identification of the hepatocyte growth factor receptor as the c-*met* proto-oncogene product. Science 1991; 251:802–4.

Brickell PM. Current status review the c-*src* family of protein-tyrosine kinases. Int J Exp Pathol 1991; 72:97–108.

Chen J, Heller D, Poon B, Kang L, Wang LH. The proto-oncogene c-*ros* codes for a transmembrane tyrosine protein kinase sharing sequence and structural homology with sevenless protein of *Drosophila melanogaster*. Oncogene 1991; 6:257–64.

Chou YH, Hayman MJ. Characterization of a member of the immunoglobulin gene superfamily that possibly represents an additional class of growth factor receptor. Proc Natl Acad Sci USA 1991; 88:4897–901.

Daily D, Schieven GL, Lim MY, Marquardt H, Gilmore T, Thorner J, Martin GS. Novel yeast protein kinase (*YPK1* gene product) is a 40-kilodalton phosphotyrosyl protein associated with protein-tyrosine kinase activity. Mol Cell Biol 1990; 10:6244–56.

Dymecki SM, Niederhuber JE, Desiderio SV. Specific expression of a tyrosine kinase gene, *blk*, in B lymphoid cells. Science 1990; 247:332–6.

Eisemann A, Ahn JA, Graziani G, Tronick SR, Ron D. Alternative splicing generates at least five different isoforms of the human basic-FGF receptor. Oncogene 1991; 6:1195–02.

Erikson RL. Structure, expression, and regulation of protein kinases involved in the phosphorylation of ribosomal protein S6. J Biol Chem 1991; 266:6007–10.

Escobedo JA, Kaplan DR, Kavanaugh WM, Turck CW, Williams LT. A phosphatidylinositol-3 kinase binds to platelet-derived growth factor receptors through a specific receptor sequence containing phosphotyrosine. Mol Cell Biol 1991; 11:1125–32.

Giordano S, Di Renzo MF, Narsimhan RP, Cooper CS, Rosa C, Comoglio PM.

Biosynthesis of the protein encoded by the c-*met* proto-oncogene. Oncogene 1989; 4:1383–8.

Hanks SK, Quinn AM, Hunter T. The protein kinase family: Conserved features and deduced phylogeny of the catalytic domains. Science 1988; 241:42–52.

Hao Q-L, Ferris DK, White G, Heisterkamp N, Groffen J. Nuclear and cytoplasmic location of the FER tyrosine kinase. Mol Cell Biol 1991; 11:1180–3.

Hatakeyama M, Kono T, Kobayashi N, Kawahara A, Levin SD, Perlmutter RM, Taniguchi T. Interaction of the IL-2 receptor with the *src*-family kinase $p56^{lck}$: Identification of novel intermolecular association. Science 1991; 252:1523–8.

Honegger AM, Kris RM, Ullrich A, Schlessinger J. Evidence that autophosphorylation of solubilzed receptors for epidermal growth factor is mediated by intermolecular cross-phosphorylation. Proc Natl Acad Sci USA 1989; 86:925–9.

Juszczak RJ, Turchin H, Truneh A, Culp J, Kassis S. Effect of human immunodeficiency virus gp 120 glycoprotein on the association of the protein tyrosine kinase $p56^{lck}$ with CD4 in human T lymphocytes. J Biol Chem 1991; 266:11176–83.

Kadowaki T, Kadowaki H, Accili D, Taylor SI. Substitution of lysine for asparagine at position 15 in the α-subunit of the human insulin receptor. J Biol Chem 1990; 265:19143–50.

Kaplan DR, Hempstead BL, Martin-Zanca D, Chao MV, Parada LF. The *trk* proto-oncogene product: a signal transducing receptor for nerve growth factor. Science 1991; 252:554–63.

Kelly JD, Haldeman BA, Grant FJ, Murray MJ, Seifert RA, Bowen-Pope DF, Cooper JA, Kazlauskas A. Platelet-derived growth factor (PDGF) stimulated PDGF receptor subunit dimerization and intersubunit *trans*-phosphorylation. J Biol Chem 1991; 266:8987–2.

Kern JA, Schwartz DA, Nordberg JE, Weiner DB, Greene MI, Torney L, Robinson RA. $p185^{neu}$ Expression in human lung adenocarcinomas predicts shortened survival. Cancer Res 1990; 50:5184–91.

Klausner RD, Samelson LE. T cell antigen receptor activation pathways: the tyrosine kinase connection. Cell 1991; 64:875–8.

Koch CA, Anderson D, Moran MF, Ellis C, Pawson T. SH2 and SH3 domains: elements that control interactions of cytoplasmic signaling proteins. Science 1991; 252:668–74.

Koretzky GA, Picus J, Schultz T, Weiss A. Tyrosine phosphatase CD45 is required for T-cell antigen receptor and CD2-mediated activation of a protein tyrosine kinase and interleukin 2 production. Proc Natl Acad Sci USA 1991; 88:2037–41.

Kraus MH, Issing W, Miki T, Popescu NC, Aaronson SA. Isolation and characterization of ERBB3, a third member of the ERBB/ epidermal growth factor receptor family: evidence for overexpression in a subset of human mammary tumors. Proc Natl Acad Sci USA 1989; 86:9193–7.

Krolewski JJ, Lee R, Eddy R, Shows TB, Dalla-Favera R. Identification and chromosomal mapping of new human tyrosine kinase genes. Oncogene 1990; 5:277–82.

Kruh GD, Perego R, Miki T, Aaronson SA. The complete coding sequence of *arg* defines the Abelson subfamily of cytoplasmic tyrosine kinases. Proc Natl Acad Sci USA 1990; 87:5802–6.

Lax I, Bellot F, Howk R, Ullrich A, Givol D, Schlessinger J. Functional analysis of the ligand binding site of EGF-receptor utilizing chimeric chicken/human receptor molecules. EMBO J 1989; 8:421–7.

Lee AWM, Nienhuis AW. Mechanism of kinase activation in the receptor for colony-stimulating factor 1. Proc Natl Acad Sci USA 1990; 87:7270–4.

Lee RM, Rapp UR, Blackshear PJ. Evidence for one or more Raf-1 kinase kinase(s) activated by insulin and polypeptide growth factors. J Biol Chem 1991; 266:10351–7.

Letwin K, Yee SP, Pawson T. Novel protein-tyrosine kinase cDNAs related to *fps/fes* and *eph* cloned using anti-phosphotyrosine antibody. Oncogene 1988; 3:621–7.

Lev S, Givol D, Yarden Y. A specific combination of substrates is involved in signal transduction by the *kit*-encoded receptor. EMBO J 1991; 10:647–54.

Lindberg RA, Hunter T. cDNA cloning and characterization of *eck*, an epithelial cell receptor protein-tyrosine kinase in the *eph/elk* family of protein kinases. Mol Cell Biol 1990; 10:6316–24.

Lugo TG, Pendergast AM, Muller AJ, Witte ON. Tyrosine kinase activity and transformation potency of *bcr-abl* oncogene products. Science 1990; 247:1079–82.

Mano H, Ishikawa F, Nishida J, Hirai H, Fumumaro T. A novel protein-tyrosine kinase, *tec*, is preferentially expressed in liver. Oncogene 1990; 5:1781–6.

Maru Y, Hirai H, Takaku F. Human *ltk*: gene structure and preferential expression in human leukemic cells. Oncogene Res 1990; 5:199–204.

Middlemas DS, Lindberg RA, Hunter T. *trk*B, a neural receptor protein-tyrosine kinase: evidence for a full-length and two truncated receptors. Mol Cell Biol 1991; 11:143–53.

Moody TW, Lee M, Kris RM, Bellot F, Bepler G, Oie H, Gazdar AF. Lung carcinoid cell lines have bombesin-like peptides and EGF receptors. J Cell Biochem. 1990; 43:139–47.

Nada S, Okada M, MacAuley A, Cooper JA, Nakagawa H. Cloning of a complementary DNA for a protein-tyrosine kinase that specifically phosphorylates a negative regulatory site of $p60^{c\text{-}src}$. Nature 1991; 351:69–72.

Nishibe S, Wahl MI, Hernandez-Sotomayor SMT, Tonks NK, Rhee SG, Carpenter G. Increase of the catalytic activity of phospholipase C-gamma 1 by tyrosine phosphorylation. Science 1990; 250,1253–56.

Nishimura S, Sekiya T. Human cancer and cellular oncogenes. Biochem J 1987; 243:313–27.

O'Hare T, Pilch PF. Intrinsic kinase activity of the insulin receptor. Int J Biochem 1990; 22: 315–24.

Park DJ, Rho HW, Rhee SG. CD3 stimulation causes phosphorylation of phospholipase C-gamma 1 on serine and tyrosine residues in a human T-cell line. Proc Natl Acad Sci USA 1991; 88:5453–6.

Pasquale EB. A distinctive family of embryonic protein-tyrosine kinase receptors. Proc Natl Acad Sci USA 1990; 87:5812–6.

Pilch PF, Waugh SM, O'Hare T. The structural basis for insulin receptor function. In: Mond JJ, Cambier JC, Weiss W, eds. Advances in regulation of cell growth, volume 1: regulation of cell growth and activation. New York: Raven Press, 1989; 203–32.

Piwnica-Worms H, Saunders KB, Cheng SH. Tyrosine phosphorylation regulates the biochemical and biological properties of pp60$^{c\text{-}src}$. Cell 1987; 49:75–82.

Raz V, Kelman Z, Avivi A, Neufeld G, Givol D, Yarden Y. PCR-based identification of new receptors: molecular cloning of a receptor for fibroblast growth factors. Oncogene 1991; 6:753–60.

Resh MD. Membrane interactions of pp60$^{v\text{-}src}$: a model for myristoylated tyrosine protein kinases. Oncogene 1990; 5:1437–44.

Roussel MF, SHurtleff SA, Downing JR, Sherr CJ. A point mutation at tryosine-809 in the human colony-stimulating factor 1 receptor impairs mitogenesis without abrogating tyrosine kinase activity, association with phosphatidylinositol 3-kinase, or induction of c-*fos* and *junB* genes. Proc Natl Acad Sci USA 1990; 87:6738–42.

Ruta M, Burgess W, Givol D, Epstein J, Neiger N, Kaplow J, Crumley G, Dionne C, Jaye M, Schlessinger J. Receptor for acidic fibroblast growth factor is related to the tyrosine kinase encoded by the *fms*-like gene (FLG). Proc Natl Acad Sci USA 1989; 86:8722–6.

Samelson LE, Phillips AF, Luong ET, Klausner RD. Association of the *fyn* protein-tyrosine kinase with the T-cell antigen receptor. Proc Natl Acad Sci USA 1990; 87:4358–62.

Sefton BM. The *lck* tyrosine protein kinase. Oncogene 1991; 6:683–6.

Shibuya M, Yamaguchi S, Yamane A, Ikeda T, Tojo A, Matsushime H, Sato M. Nucleotide sequence and expression of a novel human receptor-type tyrosine kinase gene (*flt*) closely related to the *fms* family. Oncogene 1990; 5:519–24.

Shier P, Watt, VM. Primary structure of a putative receptor for a ligand of the insulin family. J Biol Chem 1989; 264:14605–8.

Shore SK, Bogart SL, Reddy EP. Activation of murine c-*abl* protoonocogene: effect of a point mutation on oncogenic activation. Proc Natl Acad Sci USA 1990; 87:6502–6.

Smith DR, Vogt PK, Hayman MJ. The v-*sea* oncogene of avian erythroblastosis retrovirus S13: another member of the protein-tyrosine kinase gene family. Proc Natl Acad Sci USA 1989; 86:5291–5.

Srivastava AK. Non-receptor protein tyrosine kinases of normal tissues. Int J Biochem 1990; 22:1229–34.

Takahashi M, Buma Y, Taniguchi M. Identification of the *ret* proto-oncogene products in neuroblastoma and leukemia cells. Oncogene 1991; 6:297–301.

Tan JC, Nocka K, Ray P, Traktman P, Besmer P. The dominant W^{42} spotting phenotype results from a missense mutation in the c-kit receptor kinase. Science 1990; 247:209–12.

Tateishi M, Ishida T, Mitsudomi T, Kaneko S, Sugimachi K. Immunohistochemical evidence of autocrine growth factors in adenocarcinoma of the human lung. Cancer Res 1990; 50:7077–80.

Thomas JE, Soriano P, Brugg JS. Phosphorylation of c-*src* on tyrosine 527 by another protein tyrosine kinase. Science 1991; 254:568–71.

Ullrich A, Schlessinger J. Signal transduction by receptors with tyrosine kinase activity. Cell 1990; 61:203–12.

Vetter ML, Martin-Zanca D, Parada LF, Bishop JM, Kaplan DR. Nerve growth factor rapidly stimulates tyrosine phosphorylation of phospholipase C-gamma 1 by a kinase activity associated with the product of the *trk* protooncogene. Proc Natl Acad Sci USA 1991; 88:5650–54.

Wilks AF, Harpur AG, Kurban RR, Ralph SJ, Zurcher G, Ziemiecki A. Two novel protein-tyrosine kinases, each with a second phosphotransferase-related catalytic domain, define a new class of protein kinase. Mol Cell Biol 1991; 11:2057–65.

Williams DE, Eisenman J, Baird A, Ruch C, Van Ness K, March CJ. Identification of a ligand for the *c-kit* proto-oncogene. Cell 1990; 63:167. Williams LT. Signal transduction by the platelet-derived growth factor receptor. Science 1989; 243:1564–70.

Wittbrodt J, Adam D, Malitschek B, Maueler W, Raulf F, Telling A, Robertson SM, Schartl. Novel putative receptor tyrosine kinase encoded by the melanoma-inducing *Tu* locus in *Xiphophorus*. Nature 1989; 341:415–21.

Yamanashi Y, Kakiuchi T, Mizuguchi J, Yamamoto T, Toyoshima K. Association of B cell antigen receptor with protein tyrosine kinase lyn. Science 1990; 251:192–4.

Yarden Y, Ullrich A. Growth factor receptor tyrosine kinases. Ann Rev Biochem 1988; 57:443–78.

Yarden T, Kuang WJ, Yang-Feng T, Coussens L, Munemitsu S, Dull TJ, Chen E, Schlessinger J, Francke U, Ullrich A. Human proto-oncogene *c-kit*: a new cell surface receptor tyrosine kinase for an unidentified ligand. EMBO J 1987; 6:3341–51.

Yu JC, Heidaran MA, Pierce JH, Gutkind JS, Lombardi D, Ruggiero M, Aaronson SA. Tyrosine mutations within the A platelet-derived growth factor receptor kinase insert domain abrogate receptor-associated phosphatidylinositol-3 kinase activity without affecting mitogenic or chemotactic signal transduction. Mol Cell Biol 1991; 11:3780–85.

Zhang B, Roth RA. Binding properties of chimeric insulin receptors containing the cysteine-rich domain of either the insulin-like growth factor I receptor or the insulin receptor related receptor. Biochemistry 1991; 30:5113–7.

Ziegler SF, Pleiman CM, Perlmutter RM. Structure and expression of the murine *hck* gene. Oncogene 1991; 283–8.

7

Protein Tyrosine Phosphatases: A Multigene Family of Diverse Structure and Function

DEBORAH E. COOL

University of Washington
Seattle, Washington

I. Introduction

Signal transduction pathways involving tyrosine phosphorylation are commonly stimulated by growth factors such as EGF or PDGF. Ligands bind to their receptors and trigger their intracellular tyrosine kinase activity, which ultimately leads to enhanced protein synthesis, cell growth, and cell proliferation (see Bishop, 1991). The family of tyrosine kinases, reviewed elsewhere in this volume, includes transmembrane receptor forms and the intracellular $pp60^{c\text{-}src}$-related kinases. Although numerous tyrosine kinases are found in cells, the level of tyrosine phosphorylation in vivo is very low, representing less than 0.1% of the total phosphate incorporated into proteins (Hunter and Sefton, 1980). Cells harboring constitutively active tyrosine kinases, such as $pp60^{v\text{-}src}$ found in the transforming Rous sarcoma virus (RSV) (Collett and Erikson, 1978), are no longer dependent on mitogen stimulation for growth. The level of tyrosine phosphorylation in vivo remains elevated, and transformation ensues.

This phenomenon has generated considerable interest in the processes that normally reverse tyrosine kinase activity. Since the first detection of tyrosine phosphorylation (Hunter and Sefton, 1980), numerous groups have provided evidence for tyrosine dephosphorylation of proteins as well. Early evidence was shown in A431 cells expressing high levels of EGF receptor (Carpenter et al., 1979; Ushiro and Cohen, 1980), in which isolated membrane fractions stimulated with EGF became phosphorylated on tyrosine residues but underwent rapid dephosphorylation when MgATP was removed. Similar data were obtained with avian cells infected with a RSV-pp60$^{v\text{-}src}$ ts (temperature-sensitive) mutant (Sefton et al., 1980); the level of tyrosine phosphorylation diminished rapidly when these cells were shifted to nonpermissive temperatures. These preliminary studies suggest that tyrosine dephosphorylation is physiological and probably requires the action of specific enzymes.

Over the past decade, protein tyrosine phosphatases have been partially purified and characterized from various tissues and cell lines (Shriner and Brautigan, 1984; Okada et al., 1986; Horlein et al., 1982; Nelson and Branton, 1984). Rapid expansion of the field followed with the purification and sequencing of placenta PTP1B (Tonks et al., 1988a,b; Charbonneau et al., 1989). PTP1B is not related to the Ser/Thr protein phosphatases or to the acid or alkaline phosphatases, but sequence identity is found within the two homologous cytoplasmic domains of the leukocyte common antigen, CD45 (Charbonneau et al., 1988). Over 40 PTP sequences have been identified using PCR and low-stringency cloning techniques. PTPs now comprise a family of enzymes including low-molecular-weight intracellular forms and transmembrane, receptorlike proteins. In view of their considerable structural diversity, it is anticipated that these enzymes will have multiple functions and will be involved in all aspects of cell physiology and development (Cool et al., 1989; Brown-Shimer et al., 1990; Guan et al., 1990; Chernoff et al., 1990; Sap et al., 1990; Kaplan et al., 1990; Matthews et al., 1990; Krueger et al., 1990; Streuli et al., 1989).

II. Structure and Enzymatic Properties of PTP1B and TC.PTP

PTP1B is isolated from human placenta as a 37-kDa molecule that distributes evenly between the soluble and particulate fractions. The enzyme is inhibited by vanadate, molybdate, Zn^{2+}, and by nanomolar concentrations of polyanionic compounds such as heparin and random copolymers of Glu:Tyr (Tonks et al., 1988a,b).

PTP1B is absolutely specific for phosphotyrosyl residues. It displays a high specific activity (20,000–40,000 nmol P_i/min/mg) toward phosphorylated reduced, carboxamidomethylated, and maleylated lysozyme (RCML) as the substrate (Tonks et al., 1988). In comparison, the specific activity of a potent tyrosine kinase, v-abl, is about 3,000 nmol P_i/min/mg toward angiotensin II as the substrate (Foulkes et al., 1985). Considering the large disparity in specific activities, one might expect the PTPs to be tightly regulated. A natural problem is to determine which structural features of the PTPs are involved in regulation.

All PTPs are related through a core of approximately 240 residues containing a highly conserved 11-residue segment with the consensus sequence [I/V]HCXAGXXR[S/T]G, a hallmark of the PTPs (Fischer et al., 1991). The conserved cysteinyl residue is essential for catalysis; hence, in vivo, all PTPs require reducing agents for activity (Streuli et al., 1990). Among the nonreceptor PTPs, the major differences arise from extensions outside the catalytic core (Fig. 1), which may be important to their differential regulation, localization, and function.

A human T-cell cDNA (TC.PTP) was isolated that encodes a 48-kDa protein displaying 85% identity within the catalytic core and 65% identity overall with human placenta PTP1B. The main difference is that the T-cell enzyme has an 11-kDa extension at the carboxy terminus that is not found in the purified 37-kDa PTP1B molecule (Cool et al., 1989), suggesting that the latter form may have been proteolytically truncated. It was subsequently reported that the cDNA for PTP1B encodes a 50-kDa species, confirming that the enzyme purified is a fragment of a larger portion (Brown-Shimer et al.,

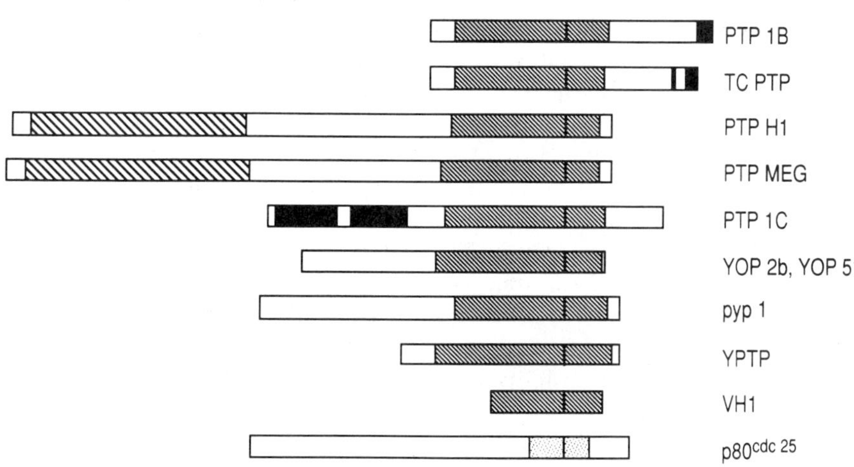

Figure 1 Structural organization of intracellular PTPs. The conserved catalytic domains are shown as narrow cross bars except for p80cdc25, whose catalytic domain is indicated by a dotted pattern. The wide solid bars in PTP1B and TC.PTP represent the hydrophobic segment at their carboxy termini; the polybasic region in TC.PTP is marked by a narrow bar. The cytoskeletal homologies are shown by a wide cross bar in PTPH1 and PTPMEG; the SH2 domains are represented by solid regions in PTP1C.

1990; Guan et al., 1990; Chernoff et al., 1990). It is not clear if this cleavage represents a physiological process. Although the greatest sequence variability between PTP1B and TC.PTP lies within the carboxyl extension, they are structurally related (Cool et al., 1989; Brown-Shimer et al., 1990; Guan et al., 1990; Chernoff et al., 1990). Most notable are the last 19 residues, which form a very hydrophobic segment in both enzymes. Preceding this segment is a hydrophilic region which, in the T-cell enzyme, contains the polybasic sequence RKRKR (Cool et al., 1989; Brown-Shimer et al., 1990; Guan et al., 1990; Chernoff et al., 1990). Although this sequence resembles a nuclear localization signal (Van Etten et al., 1989), there is currently no evidence that these enzymes localize to the nucleus.

In baby hamster kidney (BHK) cells selected to examine the effects of overexpression, the 48-kDa enzyme localizes predominantly to the particulate fraction, requiring detergent for extraction. Following gel filtration of a Triton X-100 extract, the full-length enzyme is recovered in a complex with a Mr > 600 k. When RCML is used as substrate, activity is detected only after limited trypsinolysis; this treatment cleaves the carboxy-terminal segment, generating a 33-kDa fragment with properties similar to PTP1B as isolated from placenta (Cool et al., 1990).

A carboxy-terminal, truncated 37-kDa form of TC.PTP (ΔC11.PTP) was constructed by introducing a premature stop codon into the cDNA and then transfecting into BHK cells. In contrast to the 48-kDa form, this enzyme is fully active when assayed with RCML as substrate, distributes evenly between the soluble and particulate fractions, and elutes at its expected Mr following gel filtration (Cool et al., 1990). In Sf9 cells, distribution and gel filtration properties of TC.PTP and ΔC11.PTP were the same as those observed in the BHK cell system. However, the full-length enzyme eluted from the gel filtration column in a fraction with an Mr of 220 k (Zander et al., 1991).

The 48- and 37-kDa forms of TC.PTP were purified from Sf9 cells and shown to exhibit distinct substrate specificities. The 37-kDa form was five times more active toward RCML as a substrate, while the full-length enzyme was 12 times more active toward myelin basic protein (MBP). After limited trypsinolysis, the 48-kDa form displays the substrate preference of the 37-kDa form (Zander et al., 1991).

III. Yersinia/Vaccinia

PTPs have been implicated in bacterial and viral pathogenicity. The genus *Yersinia* consists of three bacterial species, all known to cause severe infectious diseases in humans, including bubonic plague (*Y. pestis*). These bacteria carry a 70-kb virulence plasmid that codes for several secreted proteins called yersinia outer-membrane proteins (YOPs), whose function is to overcome host defence mechanisms. YOP 2b and YOP 51, expressed by *Y. pseudotuberculosis* and *Y. enterocolitica,* respectively, share 99% sequence identity, and possess catalytic domains that are homologous to that of PTP1B. In contrast to PTP1B and TC.PTP, YOP 2b and YOP 51 have extensive amino-terminal segments that probably function in membrane anchoring and secretion. In vitro, YOP 2b specifically dephosphorylates phosphotyrosine, and its expression in a macrophage model system correlates with tyrosine dephosphorylation of 120-kDa and 60-kDa proteins (Guan and Dixon, 1990; Bliska et al., 1991).

A role for PTPs in viral pathogenesis was demonstrated when VH1, a protein (Mr of ca. 20 k) encoded by the H1 gene of vaccinia virus, was found to be 20% identical with the catalytic core region of the low Mr PTPs. This phosphatase retains the ability to dephosphorylate both phosphoseryl and phosphotyrosyl residues, with vanadate completely inhibiting both activities (Guan et al., 1991a). Given the recent discovery of several kinases with specificity toward S/T and Y residues, perhaps there exists a corresponding family of dual-specificity phosphatases, represented by VH1.

IV. Involvement of PTPS in Cell Cycle

The action of the protein kinase $p34^{cdc2}$ is central to cell cycle progression (Nurse, 1990). In *S. pombe*, $p34^{cdc2}$ is essential for transition across both the G1/S and the G2/M boundaries (Nurse and Bissett, 1981; Nurse, 1975; Nurse and Thuroaux, 1980) whereas in higher aukaryotes distinct $p34^{cdc2}$-like molecules have been proposed for each of these control points (Fang and Newport, 1991). The protein kinase activity of $p34^{cdc2}$ is subject to exquisite control that, at the G2/M transition, involves both phosphorylation/dephosphorylation and complex formation with a mitotic cyclin (Gould and Nurse, 1989; Booher et al., 1989). In *S. pombe*, mik 1^+, wee 1^+, and cdc 25^+, whose products affect the phosphorylation state of $p34^{cdc2}$, are required for normal cell cycle progression (Russell and Nurse, 1986, 1987; Lundgren et al., 1991).

In late G2, $p34^{cdc2}$ is phosphorylated on tyrosine 15, located in the ATP-binding site. Activation of kinase activity and concomitant entry into mitosis is accompanied by association with a cyclin and dephosphorylation of the tyrosyl residue (Gould and Nurse, 1989). However, in higher eukaryotes, Thr 14 in the ATP binding site of the $p34^{cdc2}$ must also be dephosphorylated for kinase activation. Wee 1^+ and mik 1^+ are functionally redundant genes whose simultaneous deletion results in rapid dephosphorylation of tyrosine 15, a phenotype that is hypermitotically lethal. Expression studies of $p107^{wee1}$ suggest that wee 1^+ may play a direct role in phosphorylating $p34^{cdc2}$ on both Y and S/T residues (Lundgren et al., 1991; Featherstone and Russell, 1991; Parker et al., 1991). The wee 1^+ kinase, which belongs to a new family of dual-specificity kinases, may be important in higher eukaryotes, such as frogs and humans, where $p34^{cdc2}$ homologs are phosphorylated on Thr 14 and Tyr 15 at the G2/M transition (Krek and Nigg, 1991).

S. pombe $p80^{cdc25}$ appears to oppose the inhibitory action of the wee 1^+ and mik 1^+ gene products (Russell and Nurse, 1986; Lundgren et al., 1991). At restrictive temperatures, cells containing a ts $p80^{cdc25}$ mutation are arrested in late G2, where $p34^{cdc2}$ remains phosphorylated on Tyr 15. Upon returning to a permissive temperature, the level of $p34^{cdc2}$ tyrosine phosphorylation decreases, and cells progress through the cycle. A tyr 15-to-Phe mutation in $p34^{cdc2}$ allows the cdc25 ts mutant cells to enter M phase even at restrictive temperatures (Gould and Nurse, 1989). Furthermore, expression of the 48-kDa form of TC.PTP rescues the ts cdc 25 mutant (Gould et al., 1990). While $p80^{cdc25}$ displays some sequence identity with the PTP family, it is most closely related to the vaccinia virus VH1 enzyme, showing 21% identity over a segment of approximately 90 residues (Moreno and Nurse, 1991). Indeed, in higher eukaryotics, the evidence suggests that the cdc25 protein is a phosphatase with dual specificity capable of removing the inhibitory phosphatases from Thr 14 and Tyr 15 (Dunphy and Kumagai, 1991; StrausFeld et al., 1991; Gautier et al., 1991).

The identification of two additional yeast tyrosine phosphatases suggests a more extensive role for tyrosine phosphorylation in the physiology of this organism than was previously suspected. The *S. pombe* $pypl^+$ and the *S. cerevisiae* YPTP1 genes are more closely related to mammalian PTPs than to $cdc25^+$. However, the sequence of these enzymes diverges greatly outside of the catalytic domain, with $pypl^+$ possessing an extensive amino terminal segment. In vitro, $p39^{YPTP1}$ is 100-fold more active toward phosphotyrosine- than phosphoserine-containing substrates. The in vivo role for these phosphatases is unknown, as disruption of either gene results in viable cells with no readily distinguishable phenotype. This may indicate that another gene product can functionally complement the YPTP1 or $pypl^+$ protein (Ottilie et al., 1991; Guan et al., 1991b).

V. Receptor PTPs

The complete amino acid sequence of seven transmembrane PTPs has been derived from cDNA isolated from human, mouse, and *Drosophila* libraries (Fig. 2) (reviewed in Fischer et al., 1991; Saito and Streuli, 1991). Except for RPTPβ, the intracellular segments of all receptor PTPs identified so far have a

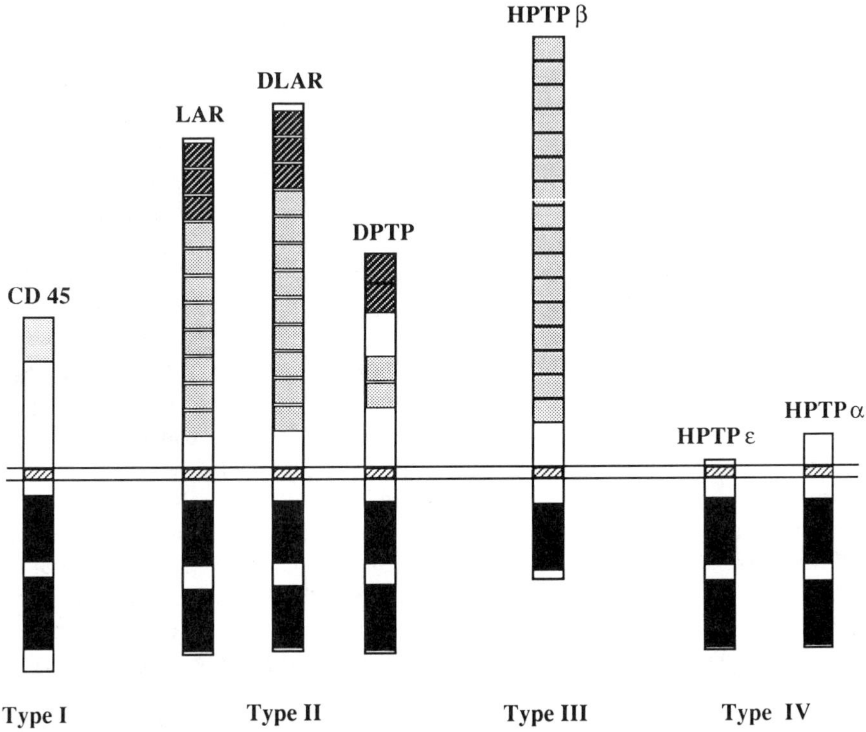

Figure 2 Structural organization of transmembrane PTPs. Conserved catalytic domains are shown as solid segments. The region of CD45 encoded by differentially expressed exons is indicated by a segment possessing horizontal bars. Shaded boxes indicate the position of type III, whereas crosshatched boxes designate immunoglobulin-like domains. Reprinted from Fischer et al. (1991).

tandem duplication of the PTP catalytic domain located 78 to 95 residues from the transmembrane segment. Each domain exhibits 30% to 40% sequence identity with that of PTP1B. All the RPTPs are widely distributed except for the leukocyte common antigen (LCA), or CD45, which is restricted to cells of hematopoietic lineages.

The structures of the extracellular domains provide a basis for classification into at least four types (Fig. 2). Type I, such as CD45 (Thomas et al., 1985, 1987; Ralph et al., 1987), bears no sequence homology to other proteins. It is highly O- and N-glycosylated and contains cysteine-rich segments that could be involved in ligand binding. Type II, which includes leukocyte common antigen-related (LAR) PTPs (Streuli et al., 1988, 1989), has immunoglobulinlike and fibronectin (FN) type III homology domains as found in the neutral cell adhesion molecules (NCAMs), fasciclin II, and L1. These homologies are based on sequence identity and positioning of conserved cysteinyl residues required for tertiary structure. FN type III repeats are found in other transmembrane molecules such as the receptors for human growth hormone, interleukins, and erythropoietin (Patthy, 1990). Type III, represented by PTPβ (Kreuger et al., 1990), which contains a single PTP domain, has 16 FN type III repeats in its extracellular segment. Type IV, represented by HPTPα (Kaplan et al., 1990; Krueger et al., 1990) and HPTPϵ (Kreuger et al., 1990), have very small extracellular regions of unrelated structure to other known proteins.

What might be the nature of the ligands for receptor PTPs? Recently the B-cell specific antigen CD22 was described as a ligand for CD45 (Stamenkovic et al., 1991). However, an effect of CD22 on the enzyme was not reported. Ligands have not been identified for the other receptor PTPs, but the extracellular segment of the type II form is structurally related to the N-CAMs. These molecules engage in homophilic cell-cell interaction and play a role in neural cell development (Edelman, 1985, 1988; Cunningham, 1987; Edelman, 1988). By analogy, the extracellular segments of the type II RPTPs could also interact with one another during cell contact, affecting their activity or localization. This might, in turn, block the mitogenic pathways and prevent cellular growth.

A few of the receptors, such as CD45 (Tonks et al., 1988a), LAR (Cho et al., 1991; Pot et al., 1991) and R-PTPα (Daum et al., 1991) have been purified and their enzymatic properties examined. In contrast to low-Mr PTPs, CD45 and R-PTPα exhibit a very high specific activity toward MBP and are 30 times less active toward RCML. The substrate preference of R-PTPα is dependent on cationic effectors such as Mn^{2+} and spermine, which increase the activity toward RCML by an order of magnitude. This suggests that substrate specificity of the enzyme may be subject to regulation inside the cell.

Like the low-Mr PTPs, the receptor forms also contain a conserved cysteinyl residue in their catalytic domain and require sulfhydryl compounds for activity in vitro. Mutation of this cysteinyl residue in either catalytic domain of CD45 or LAR showed that only the first PTP site (domain 1), closest to the membrane, is catalytically active (Streuli et al., 1990). It is possible that the second catalytic domain (domain 2), which is nearest to the carboxy terminus, recognizes different substrates or requires regulatory molecules to manifest its activity. However, expression in *E. coli* of the intracellular portion of RPTPα or either of its two catalytic domains separately revealed that the holoenzyme as well as each domain by itself are active (Wang and Pallen, 1991). Domain 2 alone is least active but within the holoenzyme, domain 1 enhances its activity and may modify its substrate specificity. The need for two PTP catalytic domains is still perplexing; determination of their physiological function will be required to understand the role of receptor PTPs in signal transduction mechanisms.

VI. Role of CD45 in Lymphocyte Signal Transduction

CD45 represents a family of transmembrane glycoproteins, ranging in Mr from 180 to 220 kDa depending on the cell lineage, which contributes to 10% of the lymphocyte surface area (reviewed in Thomas, 1989). Antibodies raised against the extracellular segment of CD45 and various other surface antigens can be used to stimulate signaling pathways. Cross-linking biotinylated antibodies with avidin shows that formation of heteroconjugates between CD4 and CD45 augments the release of intracellular calcium stores (Ledbetter et al., 1989). Interestingly, formation of CD45 homoconjugates inhibits interleukin-2-mediated pathways (Gilliland et al., 1990) yet enhances cytokine release from monocytes (Webb et al., 1990), indicating a diversity of CD45 function in lymphocyte action.

Direct evidence that CD45 participates in signal transduction has been obtained in a murine T-cell clone lacking surface expression of CD45 (Pingel and Thomas, 1989). These cells are unresponsive to stimulation by cross-linking CD3 antibodies or directly through the T-cell receptor (TCR). Normal TCR function is restored in a spontaneous $CD45^{+}$ revertant. Similarly, stimulation through the TCR is lost in a human leukemic $CD45^{-}$ HPB-ALL cell line (Koretsky et al., 1990). Release of intracellular Ca^{2+} and production of IP3 are blocked, although the phosphatidyl inositol turnover pathway is functional since a normal response is elicited when these cells are stimulated with phorbol esters. Finally, a $CD45^{-}$ Jurkat cell line shows no increase in the tyrosine phosphorylation normally observed following either TCR- or CD2-mediated stimulation (Koretsky et al., 1991). These mutants resume their normal phenotype when transfected with CD45 cDNA. These studies suggest that CD45 is necessary for coupling the TCR to the phosphatidyl inositol pathway and for tyrosine kinase activation.

Upon stimulation of T cells, the zeta chain of the TCR-CD3 complex and many other proteins are rapidly phosphorylated on tyrosine (see reviews by Bolen and Veillette, 1989; Mustelin and Altman, 1989; Klausner and Samelson, 1991). The kinases involved include $p56^{lck}$, known to associate with CD4 and CD8 (Veillette et al., 1989), and a closely related tyrosine kinase, $p59^{fyn}$, which forms a complex directly with the TCR (Samelson et al., 1990). These two lymphoid-specific, src-family tyrosine kinases contain a carboxy-terminal tyrosyl residue which, in $p56^{lck}$, is inhibitory when phosphorylated; mutation of this residue to phenylalanine in $p56^{lck}$ produces a constitutively active enzyme that is transforming in NIH 3T3 cells (Marth et al., 1988) and tumorigenic in transgenic mice (Abraham et al., 1991). In normal cells, CD45 may be responsible for dephosphorylation at this site and for activation of the enzyme. This role for CD45 has been shown in a number of CD45-deficient T-cell lines. For example, the activity of $p56^{lck}$, determined by autophosphorylation in isolated membranes from a lymphoma line, increases twofold over the $CD45^{-}$ mutant cells (Mustelin et al., 1989). Also, in other $CD45^{-}$ lymphoma cells, levels of tyrosine phosphorylation in $p56^{lck}$ are elevated in vivo when compared to $CD45^{+}$ cells (Ostergaard et al., 1989). Presumably, $p56^{lck}$ activity increases as a result of tyrosine 505 dephosphorylation by CD45 (Mustelin and Altman, 1990; Ostergaard and Trowbridge, 1990).

VII. SH2 Domains in PTPs

Src-homology 2 (SH2) domains were originally identified as regions of approximately 100 residues near the amino termini of v-Fps and v-Src whose mutation altered the kinase's catalytic and transforming properties (reviewed in Koch et al., 1991). Various other intracellular proteins contain SH2 domains. These include PLCγ-1, GAP, and PI-3kinase, which are substrates for activated EGF or PDGF receptor kinases (see Cantley et al., 1991). After stimulation with growth factors, the receptor kinases undergo autophosphorylation and the SH2 domains of the kinase substrates bind to these sites with high affinity. Therefore, one function of these domains is to mediate protein-protein interactions, allowing stable complex formation to occur. How the PTPs gain access to the phosphorylated components of the complex remains unknown. Recently, an intracellular, 66-kDa PTP was identified (PTPIC, Fig. 1) (Shen et al., 1991) that contains two SH2 domains, both sharing 45% sequence identity with homologous regions in other proteins. In vitro, these regions of PTPIC bind to the activated EGF receptor, suggesting that this PTP may participate in the EGF-mediated signal transduction pathway.

VIII. PTPs in Cytoskeleton

The processes that link signal transduction to cytoskeletal reorganization remain obscure. Cells undergo considerable changes in morphology in response to growth factors or transformation (see Ben-Ze'ev, 1986; Burridge et al., 1988; Kellie, 1988). Infection by RSV causes avian cells to round up and grow uncontrollably. Localization of $pp60^{v\text{-}src}$ to adhesion plaques correlates with the enhanced phosphorylation of several proteins found within these structures (Kellie et al., 1991) such as the beta chain of the integrin receptor, vinculin, paxillin, and possibly tensin (Davis et al., 1991). The physiological significance of these phosphorylation events has not been established, although phosphorylation of the beta subunit of the integrin receptor reduces its affinity for fibronectin (Tapley et al., 1989). Additionally, phosphorylation of the cytoskeleton components in the adhesion plaque may cause their dissociation, resulting in detachment of the cell from the matrix (Horvath et al., 1990; Otey et al., 1991).

Since tyrosine phosphorylation is intimately implicated in cytoskeletal organization, could specific PTPs be involved? Preliminary evidence indicates that overexpression of the 48-kDa form of TC.PTP in BHK cells renders the actin fibers resistant to cytochalasin B, a compound that normally induces actin fiber disassembly (Andreassen P, Cool DE, Margolis B, Fischer EH, unpublished data). More direct evidence supporting the involvement of PTPs in cytoskeletal organization comes from recent reports (Gu et al., 1991; Yang and Tonks, 1991) of two 95-kDa intracellular enzymes that contain regions of homology to an erythrocyte cytoskeletal protein (band 4.1), ezrin, and talin. This region of approximately 325 residues in the amino-terminal segment of PTPMEG and PTPH1 (Fig. 1) displays 45% and 33% sequence identity, respectively, with band 4.1. Also, they have a segment with a number of putative phosphorylation sites for CKII and $pp34^{cdc2}$, suggesting that the serine/threonine kinases may influence PTP involvement in cytosekeleton organization. The putative cytoskeletal properties of PTPMEG and PTPH1 could allow their co-localization with the oncogenic kinase $pp60^{v\text{-}src}$ in adhesion plaques of transformed cells, possibly enabling the PTPs to reverse cell transformation.

IX. PTPs in Cell Transformation

An early hypothesis proposed that elevating tyrosine phosphatase activity in transformed cell lines might have an antiproliferative effect and perhaps reverse the transformed phenotype. The diversity in structure and potential function of PTPs suggests that this theory is probably too simplistic. For example, overexpression of the full-length, 48-kDa TC.PTP in BHK cells did not hinder growth nor alter their transformed state (Cool DE, Fischer EH, Krebs EG, unpublished data), nor did TC.PTP have any notable effect on v-fms-transformed Rat 2 cells (Zander N, Cool DE, Fischer EH, unpublished data). This is in contrast to other studies, in which NIH 3T3 cells overexpressing the 50-kDa form of PTP1B were resistant to transformation when challenged with a retrovirus containing the neu oncogene (Brown-Shimer et al., 1991). Obviously, defining the mechanisms which regulate PTP activity during cell growth will be crucial to their application as tumor suppressor proteins.

Correlation of PTP gene deletions with known cancers may provide the best evidence for their potential role as antioncogenes. To date, the only known example is R-PTPγ, which has been mapped to chromosome region 3p21, which is associated with renal and lung carcinomas (LaForgia et al., 1991). However, it should be kept in mind that PTPs may not only reduce the level of tyrosine phosphorylation in vivo and return the cell to a nonstimulated or detransformed state: they can also act positively on cell proliferation, such as has been shown for CD45 in T-cell activation and for cdc25 in cell cycle progression.

Acknowledgments

I would like to thank Ed Fischer and Ed Krebs for their kind and ceaseless support for the work done by the author in their labs. Also, I would like to thank James Lorenzen for his contribution to this chapter, and Carmen Westwater for preparation of the manuscript.

References

Abraham KM, Levin SD, Marth JD, Forbush KA, Perlmutter RM. Thymic tumorigenesis induced by overexpression of p56lck. Proc Natl Acad Sci USA 1991; 88:3977–81.

Ben-Ze'ev A. The relationship between cytoplasmic organization, gene expression and morphogenesis. TIBS 1986; 11:478–81.

Bishop JM. Molecular themes in oncogenesis. Cell 1991; 64:235–48.

Bliska JB, Guan K, Dixon JE, Falkow S. Tyrosine phosphate hydrolysis of host proteins by an essential Yersinia virulence determinant. Proc Natl Acad Sci USA 1991; 88:1187–91.

Bolen JB, Veillette A. A function for the lck proto oncogene. TIBS 1989; 14:404–7.

Booher RN, Alfa CE, Hyams JS, Beach DH. The fission yeast cdc/1/cdc13/suc1 protein kinase: regulation of catalytic activity and nuclear localization. Cell 1989; 58:485–97.

Brown-Shimer S, Johnson KA, Hill DE, Bruskin A. Effect of protein tyrosine phosphatase 1B expression on transformation by the human neu oncogene. Cancer Res 1992; 52:1235–1236.

Brown-Shimer S, Johnson KA, Lawrence JB, Johnson C, Bruskin A, Green NR, Hill DE. Molecular cloning and chromosome mapping of the human gene encoding protein phosphotyrosyl phosphatase 1B. Proc Natl Acad Sci USA 1990; 87:5148–52.

Burridge K, Fath K, Kelly T, Nuckolls G, Turner C. Focal adhesions: transmembrane junctions between the extracellular matrix and the cytoskeleton. Ann Rev Cell Biol 1988; 4:487–525.

Cantley LC, Auger KR, Carpenter C, Duckworth B, Graziani A, Kapeller R, Soltoff S. Oncogenes and signal transduction. Cell 1991; 64:281–302.

Carpenter G, King L, Cohen S. Rapid enhancement of protein phosphorylation in A-

431 cell membrane preparations by epidermal growth factor. J Biol Chem 1979; 254:4884–91.

Charbonneau H, Tonks NK, Kumar S, Diltz CD, Harrylock M, Cool DE, Krebs EG, Fischer EH, Walsh KA. Human placenta protein-tyrosine-phosphatase: amino acid sequence and relationship to a family of receptor-like proteins. Proc Natl Acad Sci USA 1989; 86:5252–6.

Charbonneau H, Tonks NK, Walsh KA, Fischer EH. The leukocyte common antigen (CD45): a putative receptor-linked protein tyrosine phosphatase. Proc Natl Acad Sci USA 1988; 85:7182–6.

Chernoff J, Schievella AR, Jost CA, Erickson RL, Neal BG. Cloning of a cDNA for a major human protein-tyrosine phosphatase. Proc Natl Acad Sci USA 1990; 87:2735–9.

Cho H, Ramer SE, Itoh M, Winkler DG, Kitas E, Bannwarth W, Burn P, Saito H, Walsh CT. Purification and characterization of a soluble catalytic fragment of the human transmembrane leukocyte antigen related (LAR) protein tyrosine phosphatase from an *Escherichia coli* expression system. Biochemistry 1991; 30:6210–6.

Collett MS, Erikson RL. Protein kinase activity associated with the avian sarcoma virus src gene product. Proc Natl Acad Sci USA 1978; 75:2021–4.

Cool DE, Tonks NK, Charbonneau H, Fischer EH, Krebs EG. Expression of a human T-cell protein-tyrosine-phosphatase in baby hamster kidney cells. Proc Natl Acad Sci USA 1990; 87:7280–4.

Cool DE, Tonks NK, Charbonneau H, Walsh KA, Fischer EH, Krebs EG. cDNA isolated from a human T-cell library encodes a member of the protein-tyrosine-phosphatase family. Proc Natl Acad Sci USA 1989; 86:5257–61.

Cunningham BA, Hemperley BA, Murray BA, Prediger EA, Brackenberg R, Edelman GM. Neural cell adhesion molecule: structure, immunoglobulin-like domains, cell surface modulation, and alternative RNA splicing. Science 1987; 236:799–806.

Daum G, Zander NF, Morse B, Hurwitz D, Schlessinger J, Fischer EH. Characterization of a human recombinant receptor-linked protein tyrosine phosphatase. J Biol Chem 1991; 266:1211–1215.

Davis S, Lu MI, Ko SH, Lin S, Butler JA, Druker BJ, Roberts TH, An Q, Chen LB. Presence of an SH2 domain in the actin binding protein tensin. Science 1991; 252:712–5.

Dunphy WG, Kumagai A. The cdc25 protein contains an intrinsic phosphatase activity. Cell 1991; 67:189–96.

Edelman GM. Cell adhesion and the molecular processes of morphogenesis. Ann Rev Biochem 1985; 54:135–69.

Edelman GM. Morphoregulatory molecules. Biochemistry 1988; 27:3533–43.

Fang F, Newport JW. Evidence that the G1-S and G2-M transitions are controlled by different cdc2 proteins in higher eukaryotes. Cell 1991; 66:731–42.

Featherstone C, Russell P. Fission yeast p107weel mitotic inhibitor in a tyrosine/service kinase. Nature 1991; 349:808–11.

Fischer EH, Charbonneau H, Tonks NK. Protein tyrosine phosphatases: A diverse family of intracellular and transmembrane enzymes. Science 1991; 253:401–6.

Foulkes JG, Chow M, Gorka C, Frackelton AR, Baltimore D. Purification and characterization of a protein-tyrosine kinase encoded by the Abelson murine leukemia virus. J Biol Chem 1985; 260:8070–7.

Gautier J, Solomon MJ, Booher RN, Bazan JF, Kirschner MW. cdc25 is a specific tyrosine phosphatase that directly activates p34cdc2. Cell 1991; 67:197–211.

Gilliland LK, Schieven GL, Grosmaire LS, Damle NK, Ledbetter JA. CD45 ligation in T-cells regulated signal transduction through both the interleukin-2 receptor and the CD3/Ti T-cell receptor complex. Tissue Antigens 1990; 35:128–35.

Gould KL, Nurse P. Tyrosine phosphorylation in the fission yeast cdc2$^+$ protein kinase regulates entry into mitosis. Nature 1989; 342:39–45.

Gould KL, Moreno S, Tonks NK, Nurse P. Complementation of the mitotic activator, p80cdc25, by a human protein-tyrosine phosphatase. Science 1990; 250:1573–6.

Gu M, York JD, Warshawsky I, Majerus PW. Identification cloning and expression of a cytosolic megakaryocyte protein tyrosine phosphatase with sequence homology to cytoskeletal protein 4.1. Proc Natl Acad Sci USA 1991; 88:5867–71.

Guan K, Dixon JE. Protein tyrosine phosphatase activity of an essential virulence determinant in Yersinia. Science 1990; 249:553–6.

Guan K, Broyles SS, Dixon JE. A Tyr/Ser protein phosphatase encoded by Vaccinia virus. Nature 1991a; 350:359–62.

Guan K, Deschenes RJ, Qiu H, Dixon JE. Cloning and expression of a yeast protein tyrosine phosphatase. J Biol Chem 1991b; 266:12964–70.

Guan K, Haun RS, Watson SJ, Geahlen RL, Dixon JE. Cloning and expression of a protein-tyrosine phosphatase. Proc Natl Acad Sci USA 1990; 87:1501–5.

Horlein D, Gallis B, Brautigan DL, Bornstein P. Partial purification and characterization of phosphotyrosyl-protein phosphatase from Ehrlich ascites tumor cells. Biochemistry 1982; 21:5577–84.

Horvath AR, Elmore MA, Kellie S. Differential tyrosine-specific phosphorylation of integrin in Rous sarcoma virus transformed cells with differing transformed phenotype. Oncogene 1991; 5:1349–57.

Hunter T, Sefton BM. Transforming gene product of Rous sarcoma virus phosphorylates tyrosine. Proc Natl Acad Sci USA 1980; 77:1311–5.

Kaplan R, Morse B, Huebner K, Croce C, Howk R, Ravera M, Ricca G, Jaye M, Schelssinger J. Cloning of three human tyrosine phosphatases reveals a multigene family of receptor-linked protein-tyrosine-phosphatases expressed in brain. Proc Natl Acad Sci USA 1990; 87:7000–4.

Kellie S. Cellular transformation, tyrosine kinase oncogenes, and the cellular adhesion plague. Bio Essays 1988; 8:25–30.

Kellie S, Horvath AR, Elmore MA. Cytoskeleton targets for oncogene tyrosine kinases. J Cell Sci 1991; 99:207–11.

Klausner RD, Samelson LE. T-cell antigen receptor activation pathways: the tyrosin kinase connection. Cell 1991; 64:875–8.

Koch CA, Anderson D, Moran MF, Ellis C, Pawson T. SH2 and SH3 domains: elements that control interactions of cytoplasmic signaling proteins. Science 1991; 252:668–74.

Korestsky GA, Picus J, Schultz T, Weiss A. Tyrosine phosphatase CD45 is essential for coupling T-cell antigen receptor to the phosphatidyl inositol pathway. Nature 1990; 346:66–8.

Koretsky GA, Picus J, Schultz T, Weiss A. Tyrosine phosphatase CD45 is required for T-cell antigen receptor and CD2-mediated activation of a protein tyrosine kinase and interleukin 2 production. Proc Natl Acad Sci USA 1991; 88:2037–41.

Krek W, Nigg EA. Differential phosphorylation of vertebrate p34cdc2 kinase at the G1/S and G2/M transitions of cell cycle: identification of major phosphorylation sites. EMBO J 1991; 10:305–16.

Krueger NK, Streuli M, Saito H. Structural diversity and evolution of human receptor-like protein tyrosine phosphatases. EMBO J 1990; 9:3241–52.

LaForgia S, Morse B, Levy J, Barnea G, Cannizzaro LA, Li F, Nowell PC, Boghosian-Sell L, Glick J, Weston A, Harris CC, Drabkin H, Patterson D, Croce CM, Schlessinger J, Huebner K. Receptor protein-tyrosine phosphatase gamma is a candidate tumor suppressor gene at human chromosome region 3p21. Proc Natl Acad Sci USA 1991; 88:5036–40.

Ledbetter JA, Tonks NK, Fischer EH, Clark EA. CD45 regulates signal transduction and lymphocyte activation by specific association with receptor molecules on T or B cells. Proc Natl Acad Sci USA 1989; 85:8628–32.

Lundgren K, Walworth N, Booher R, Dembsk M, Kirschner M, Beach D. mik1 and wee1 cooperate in the inhibitory tyrosine phosphorylation of cdc2. Cell 1991; 64:1111–22.

Marth JD, Cooper JA, King CS, Ziegler SF, Tinker DA, Overell RW, Krebs EG, Perlmutter RM. Neoplastic transformation induced by an activated lymphocyte-specific protein tyrosine kinase ($pp56^{lck}$). Mol Cell Biol 1988; 8:540–50.

Matthews RJ, Cahir ED, Thomas ML. Identification of an additional member of the protein-tyrosine-phosphatase family: evidence for alternative splicing in the tyrosine phosphatase domain. Proc Natl Acad Sci USA 1990; 87:4444–8.

Moreno S, Nurse P. Clues to action of cdc25 protein. Nature 1991; 351:194.

Mustelin T, Altman A. Do CD4 and CD8 control T-cell activation via a specific tyrosine protein kinase? Immunol. Today 1989; 10:189–92.

Mustelin T, Altman A. Dephosphorylation and activation of the T-cell tyrosine kinase pp56lck by the leukocyte common antigen (CD45). Oncogene 1990; 5:809–14.

Mustelin T, Coggeshall KM, Altman A. Rapid activation of the T-cell tyrosine protein kinase $pp56^{lck}$ by the CD45 phosphotyrosine phosphatase. Proc Natl Acad Sci USA 1989; 86:6302–6.

Nelson RL, Branton PE. Identification, purification, and characterization of phosphotyrosine-specific protein phosphatases from cultured chicken embryo fibroblasts. Mol Cell Biol 1984; 4:1003–12.

Nurse P. Genetic control of cell size at cell division in yeast. Nature 1975; 256:547–51.

Nurse P. Universal control mechanism regulating onset of M-phase. Nature 1990; 344:503–8.

Nurse P, Bissett Y. Gene required in G1 for commitment to cell cycle and in G2 for control of mitosis in fission yeast. Nature 1981; 292:558–60.

Nurse P, Thuroaux P. Regulatory genes controlling mitosis in the fission yeast *S. pombe*. Genetics 1980; 96:627–37.

Okada M, Owada K, Nakagawa H. [Phosphotyrosine] protein phosphatase in rat brain. A major [phosphotyrosine] protein phosphatase is a 23 kDa protein distinct from acid phosphatase. Biochem J 1986; 239:155–62.

Ostergaard HL, Trowbridge IS. Coclustering CD45 and CD4 or CD8 alters the phosphorylation and kinase activity of p56lck. J Exp Med 1990; 172:347–50.

Ostergaard HL, Shackelford DA, Hurley TR, Johnson P, Hyman R, Sefton BM, Trowbridge IS. Expression of CD45 alters phosphorylation of the lck tyrosine protein kinase in murine lymphoma T-cell lines. Proc Natl Acad Sci USA 1989; 86:8959–63.

Otey C, Pavalko TM, Burrdige K. An interaction betweeen A-actinin and the β1 integrin subunit in vitro. J Cell Biol 1990; 111:721–9.

Ottilie S, Chernoff J, Hannig G, Hoffman CS, Erikson RL. A fission yeast gene encoding a protein with features of protein tyrosine phosphatases. Proc Natl Acad Sci USA 1991; 88:3455–9.

Parker LL, Atherton-Fessler S, Lee MS, Ogg S, Falk JL, Swenson KI, Piwnica-Worms H. Cyclin promotes the tyrosine phosphorylation of p34cdc2 in a wee1+ dependent manner. EMBO J 1991; 10:1255–63.

Patthy L. Homology of a domain of the growth hormone/prolactin receptor family with type III modules of fibronectin. Cell 1990; 61:13–4.

Pingel JT, Thomas ML. Evidence that the leukocyte common antigen is required for antigen induced T lymphocyte proliferation. Cell 1989; 58:1055–65.

Pot DA, Woodford TA, Remboutsika E, Haun RS, Dixon JE. Cloning, bacterial expression, purification and characterization of the cytoplasmic domain of rat LAR, a receptor-like protein tyrosine phosphatase. J Biol Chem 1991; 266:19688–96.

Ralph SJ, Thomas ML, Morton CC, Trowbridge IS. Structural variants of human T200 glycoprotein (leukocyte-common antigen). EMBO J 1987; 6:1251–7.

Russell P, Nurse P. cdc25+ functions as an inducer in the mitotic control of fission yeast. Cell 1986; 45:145–53.

Russell P, Nurse P. The mitotic inducer nim1+ functions in a regulatory network of protein kinase homologs controlling the initiation of mitosis. Cell 1987; 49:569–76.

Saito H, Streuli M. Molecular characterization of protein tyrosine phosphatases. Cell Growth and Differ. 1991; 2:59–65.

Samelson LE, Phillips AF, Luong ET, Klausner RD. Association of fyn protein tyrosine kinase with the T-cell antigen receptor. Proc Natl Acad Sci USA 1990; 87:4358–62.

Sap J, D'Eustachio P, Givol D, Schlessinger J. Cloning and expression of a widely expressed receptor tyrosine phosphatase. Proc Natl Acad Sci USA 1990; 87:6112–6.

Sefton BM, Hunter T, Beemon K, Eckhart W. Evidence that the phosphorylation of tyrosine is essential for cellular transformation by Rous sarcoma virus. Cell 1980; 20:807–17.

Shen S-H, Bastien L, Posner BI, Chretien P. A protein-tyrosine phosphatase with sequence similarity to the SH2 domain of the protein-tyrosine kinases. Nature 1991; 352:736–9.

Shriner CL, Brautigan DL. Cytosolic protein phosphotyrosine phosphatases from rabbit kidney: purification of two distinct enzymes that bind to Zn^{2+}-iminodiacetate agarose. J Biol Chem 1984; 259:11383–90.

Stamenkovic I, Sgroi D, Aruffo A, Sy MS, Anderson T. The B lymphocyte adhesion molecule CD22 interacts with leukocyte common antigen CD45RO on T cells and A2-6 sialyl transferase, CD75, on B cells. Cell 1991; 66:1133–44.

StrausFeld U, Labbe JC, Fesquet D, Cavadore JC, Picard A, Sadhu D, Russell P, Doree M. Dephosphorylation and activation of a p34cdc2/B cyclin complex in vitro by human cdc25 protein. Nature 1991; 351:242–5.

Streuli M, Krueger NX, Hall LR, Schlossman SF, Saito H. A new member of the immunoglobulin superfamily that has a cytoplasmic region homologous to the leukocyte common antigen. J Exp Med 1988; 168:1523–30.

Streuli M, Krueger NX, Thai T, Tang M, Saito H. Distinct functional roles of the two intracellular phosphatase like domains of the receptor-linked protein tyrosine phosphatases LCA and LAR. EMBO J 1990; 9:2399–407.

Streuli M, Kreuger NX, Tsai AYM, Saito H. A family of receptor-linked protein tyrosine phosphatases in humans and *Drosophila*. Proc Natl Acad Sci USA 1989; 86:8698–702.

Tapley P, Horwitz A, Buck C, Duggan K, Rohrschneider L. Integrins isolated from Rous sarcoma virus transformed chicken embryo fibroblasts. Oncogene 1989; 4:325–33.

Thomas ML. The leukocyte common antigen family. Ann Rev Immunol 1989; 7:339–69.

Thomas ML, Barclay AN, Gagnon J, Williams AF. Evidence from cDNA clones that the rat leukocyte-common antigen (T200) spans the lipid bilayer and contains a cytoplasmic domain of 80,000 Mr. Cell 1985; 41:83–93.

Thomas ML, Reynold PJ, Chain A, Ben-Neriah Y, Trowbridge IS. B-cell variant of mouse T200 (Ly-5): evidence for alternative mRNA splicing. Proc Natl Acad Sci USA 1987; 84:5360–3.

Tonks NK, Charbonneau H, Diltz CD, Fischer EH, Walsh KA. Demonstration that the leukocyte common antigen CD45 is a protein tyrosine phosphatase. Biochemistry 1988a; 27:8695–701.

Tonks NK, Diltz CD, Fischer EH. Characterization of the major protein-tyrosine-phosphatases of human placenta. J Biol Chem 1988b; 263:6731–7.

Ushiro H, Cohen S. Identification of phosphotyrosine as a product of epidermal growth factor-activated protein kinase in A-431 cell membranes. J Biol Chem 1980; 255:8363–5.

Van Etten RA, Jackson P, Baltimore D. The mouse type IV c-abl gene product is a nuclear protein, and activation of transforming ability is associated with cytoplasmic localization. Cell 1989; 58:669–78.

Veillette A, Bookman MA, Horak EM, Samelson LE, Bolen JB. Signal transduction through the CD4 receptor involves the activation of the internal membrane tyrosine-protein kinase p56lck. Nature 1989; 338:257–9.

Wang Y, Pallen CJ. The receptor-like protein tyrosine phosphatase HPTPÁ has two active catalytic domains with distinct substrate specificity. EMBO J. 1991; 10:3231–7.

Webb DSA, Shimizu Y, Gijs A, Sventor V, Shaw S, Gerrard TL. LFA-3, CD44 and CD45: physiological triggers of human monocytes TNF and IL-1 release. Science 1990; 249:1295–7.

Yang O, Tonks NK. Isolation of a cDNA clone encoding a human protein tyrosine phosphatase with homology to the cytoskeletal associated proteins band 4.1, ezrin and talin. Proc Natl Acad Sci USA 1991; 88:5949–53.

Zander NF, Lorenzen JA, Cool DE, Tonks NK, Daum G, Krebs EG, Fischer EH. Purification and characterization of a human recombinant T-cell protein tyrosine phosphatase from a baculovirus expression system. Biochemistry 1991; 30:6964–70.

8

Purinergic Signal Transduction: Extracellular ATP and Adenosine as Agonists for Receptor-Mediated Signaling

GEORGE R. DUBYAK

Case Western Reserve University
Cleveland, Ohio

I. Background

All biologists recognize the fundamental roles of intracellular nucleotides and nucleosides in energy metabolism, nucleic acid synthesis, and enzyme regulation. However, extracellular nucleotides and nucleosides also exert significant biological actions in many tissues and cells. Shortly after the initial identification and isolation of ATP, Drury and Szent-Gyorgyi (1929) observed that extracellular ATP and other adenine-containing compounds could affect cardiovascular performance and other physiological functions. The following 50 years witnessed the accumulation of a large pharmacological literature describing effects of extracellular adenine nucleotides on functional responses in various intact organ systems, isolated tissues, and purified cell preparations. In certain tissues, the responses were selectively induced by low concentrations of ATP per se. However, the responses of other tissues to ATP were shown to be absolutely dependent on catabolism of the nucleotide to adenosine. Given the results of extensive studies on neurotransmission, which was resistant to conventional adrenergic and cholinergic antagonists, Burnstock (1971) has proposed that ATP/adenosine, released at synaptic junctions, might mediate this "nonadrenergic, noncholinergic" signaling. He also suggested the possible existence of so-called purinergic nerves, which release ATP/adenosine as putative neurotransmitters. Finally, he proposed that the target cells of these purinergic nerves expressed cell surface "purinergic" receptors which were activated either by ATP or by the breakdown product, adenosine. The ATP-selective receptors were functionally classified as P_2-purinergic, while the receptors for adenosine were termed P_1-purinergic. While adenosine receptors are now routinely identified by the A_1-type and A_2-type nomenclature, the term P_2-purinergic has persisted as a useful "umbrella" classification for putative receptors for ATP.

II. Overview of Purinergic Signal Transduction

An overall perspective (Fig. 1) on "purinergic signal transduction" should consider: (1) characterization of the sources and metabolism of extracellular nucleotides and nucleosides; (2) delineation of key biological functions which can be stimulated/modulated in various cells and tissues exposed to extracellular nucleotides and nucleosides; (3) classification of the putative receptor subtypes which mediate the diverse biological responses to extracellular nucleotides and nucleosides; and (4) definition, at the cellular and subcellular levels, of the various signal transduction pathways which can be regulated by extracellular nucleotides and nucleosides.

A. Sources and Metabolism of Extracellular ATP and Adenosine

The diversity which characterizes ATP- and adenosine-induced changes in biological function has prompted both investigation and speculation as to the possible physiological sources of extracellular nucleotides and nucleosides. While ATP is present in millimolar concentrations in the cytosol of all eukaryotic cell types, extracellular levels of the nucleotide will normally be maintained at extremely low levels by several mechanisms. First, there is minimal permeation of either ATP^{4-} or $MgATP^{2-}$ (the predominant cytosolic form) across lipid bilayers. Second, ubiquitous ecto-ATPases and ecto-phosphatases rapidly and efficiently hydrolyze extracellular nucleotides. Thus, as is the case for any putative signaling agent, appreciable levels of extracellular ATP should occur only transiently and in response to specific physiological and/or pathological conditions. Four major sources of extracellular nucleotides/nucleosides may be considered. First, ATP can be rapidly released by specific exocytotic processes in a variety of neuronal and secretory cell types. This is consistent with the fact ATP (and/or ADP) is co-packaged with "conventional" neurotransmitters/local mediators within the specific exocytotic granules of many cell types (Unsworth and Johnson, 1990). Thus, the ATP packaged within both adrener-

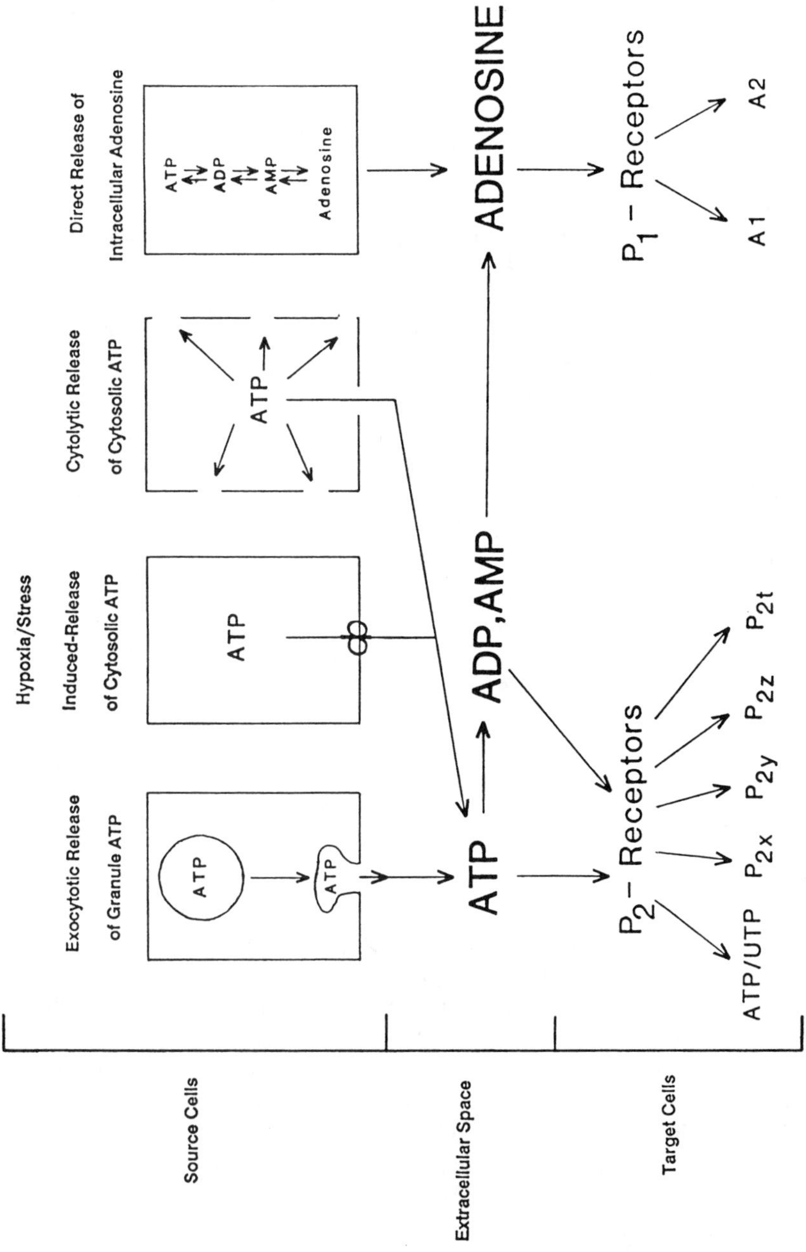

Figure 1 Overview of purinergic signal transduction.

gic and cholinergic neurotransmitter granules is released during neurotransmission into synaptic spaces. Chromaffin granules from adrenal medulla also contain substantial amounts of sequestered ATP (~125 mM or 560 nmol/mg protein ATP); these stores are rapidly released to the circulation upon cholinergic stimulation of the adrenal chromaffin cells. Likewise, ATP is co-packaged with serotonin in platelet dense granules and is released in very significant amounts during platelet activation and thrombus formation. A second (and obvious) source of extracellular ATP is cytosolic ATP (~3 to 5 mM in most cells), which can be released by sudden breakage of intact cells such as might occur during rupture of blood vessels and other types of tissue trauma. As noted below, ATP receptors appear to be expressed by both phagocytic cell types (neutrophils, monocytes, macrophages) and vascular endothelial cells, the vessel elements with which phagocytes interact at vascular lesions. Third, Forrester (1990) has demonstrated that cytosolic ATP can be released from intact cells, such as cardiac myocytes and erthyrocytes, which are metabolically compromised by hypoxia or ischemia. This hypoxia-induced release may involve transport of cytosolic ATP through specific transport proteins given the attenuation of this release by inhibitors of anion transporters (e.g., stilbene derivatives) and nucleoside transporters such as nitrothiobenzylinosine or dipyridamole. These various sources (exocytotic release from neurosecretory cells, release from lysed cells, release from metabolically compromised cells) suggest that significant amounts of extracellular ATP may accumulate locally within: (1) vascular sites of thrombus formation; (2) vascular and nonvascular sites of tissue trauma and infection/inflammation; and (3) confined extracellular spaces in close proximity to neuronal synapses.

It should be stressed that extracellular ATP is rapidly catabolized by a variety of extracellular ATPases and nucleotidases. These "ecto-ATPase" and "ecto-nucleotidase" activities are variably expressed on the cell surfaces of different tissues and cells. For example, Barankiewicz et al. (1988) have noted that human B lymphocytes, but not T lymphocytes, express high levels of ecto-ATPase and ecto-5′-AMPase (also known as ecto-nucleotidase). Not surprisingly, vascular cell types express very active ecto-enzymes which catalyze rapid catabolism of ATP and ADP. Using perifused, cultured cells, Slakey et al. (1990) have performed elegant and extensive kinetic studies of extracellular nucleotide metabolism by vascular endothelial and smooth muscle cells. Lin (1990) has isolated a cDNA clone encoding an ecto-ATPase which is preferentially expressed in canalicular plasma membranes isolated from liver. These various ecto-enzymes serve two major roles. The first is termination of ATP/ADP-induced signal transduction. The rapid breakdown of ATP (and ADP) will act to reduce the extracellular concentrations of these nucleotides within the vicinity of P_2-type (i.e., nucleotide-selective) purinergic receptors which mediate the biological actions of these nucleotides. Second, these ecto-enzymes catalyze the generation of adenosine (and adenosine metabolites) which, following reuptake into the cells, serve to replete intracellular purine nucleoside and nucleotide pools.

The adenosine which is generated extracellularly, as a product of ecto-ATPase/nucleotidase reactions, can also serve as an agonist for P_1-type (i.e., nucleoside-selective) purinergic receptors. For this reason, caution should be exercised when assuming that a particular cellular response to extracellular ATP involves the activation of P_2-type purinergic receptors per se. While the direct generation of extracellular adenosine, via hydrolysis of released ATP, may play a role in stimulation of adenosine receptors in certain tissues, there is very considerable evidence which indicates that cytosolic adenosine (generated via hydrolysis of intracellular ATP/ADP) can be specifically released from intact cells under a variety of physiological or pathological situations (Bruns, 1990). Particular attention has been focused on the release of intracellular adenosine which rapidly accumulates in cells during conditions of decreased energy supply (e.g., hypoxia, ischemia) or increased energy demand (e.g., exercise). As is the case for ATP, extracellular adenosine has a short half-life (subminute). This is due to existence of: (1) efficient nucleoside transport mechanisms for reuptake into the cells; and/or (2) utilization of adenosine by enzymes such as adenosine deaminase.

B. Major Biological Actions of Extracellular Adenine Nucleotides and Nucleosides

Biological responses to extracellular ATP and/or extracellular adenosine have been documented in virtually every major organ/tissue system which has been studied. This is a massive literature, which can be only cursorily addressed in a short review. Several recent books, symposia proceedings, and review articles should be consulted for more comprehensive treatments of these topics (Burnstock and Kennedy, 1986; Gordon, 1986; Dubyak and Fedan, 1991; Jacobson et al., 1990; Linden, 1991; O'Connor et al., 1991; Stiles, 1991). However, several biological responses to purinergic stimulation are particularly worth noting.

Biological Responses Mediated by P_2-Purinergic Receptors for ATP

As noted in previous reviews/symposium proceedings (Burnstock and Kennedy, 1986; Gordon, 1986; Dubyak and Fedan, 1990), ATP can affect many organ/tissue systems by stimulating a variety of P_2-type purinergic receptors which, in turn, are functionally coupled to: (1) stimulation or inhibition of contractility in most vascular and visceral smooth muscle types; (2) stimulation or modulation of the release of neurotransmitters, hormones, or other factors from platelets and multiple endocrine/neuronal/exocrine tissue types; (3)

stimulation of the release of prostacyclin, endothelium-derived relaxing factor (EDRF), and other vasoactive agents from vascular endothelial cells; (4) stimulation or "priming" of the functional responses of neutrophils, monocytes, and macrophages to inflammatory stimuli; and (5) modulation of voltage-operated ion channels and membrane potential in several excitable cells, including cardiac myocytes, developing skeletal muscle, and certain neuronal cell types. Recently, Mason et al. (1991) have demonstrated that ATP receptors can activate transepithelial ion transport in human respiratory epithelia. Given the wide diversity of organ systems and tissues which exhibit altered function in response to extracellular ATP, it appears likely that P_2-type purinergic receptors are expressed in a wide variety of phenotypically distinct cell types. Recent studies utilizing highly homogeneous cultured cell lines or isolated single-cell preparations have confirmed this supposition. Extracellular ATP (and other nucleotides), at nanomolar to micromolar concentrations, can induce responses in isolated fibroblasts, epithelial cells, endothelial cells, phagocytic leukocytes, smooth muscle cells, cardiac myocytes, glial cells, and neurons.

Biological Responses Mediated by P_1-Purinergic Receptors for Adenosine

Specific adenosine-induced responses have also been noted in a wide variety of organs and tissues (reviewed in Bruns, 1990; Linden, 1991; Stiles, 1991). Particularly salient responses include: (1) inhibition of excitatory neurotransmitter release in the brain; (2) inhibition of inotropic and chronotropic responses by the heart; (3) inhibition of the inflammatory response by activated phagocytes; and (4) decreased lipolysis by adipose tissue. As is the case with ATP, adenosine has been shown to alter the function of many discrete cell types which have been studied as isolated and highly homogeneous preparations. Moreover, given the availability of selective radiolabeled agonists, antagonists, and photoaffinity probes with high affinity for P_1 receptor, these functional observations have been complemented by many direct measurements of ligand binding to these receptors.

C. Classification and Subtyping of Purinergic Receptors for Extracellular Nucleotides and Nucleosides

Pharmacological classification of receptor subtype (e.g., of the multiple adrenergic receptors) has traditionally involved establishment of the "rank order of potency," which characterizes the ability of naturally occurring or synthetic ligands (either agonists or antagonists) to interact with particular receptor subtypes. These agents can be used either to characterize the biological response presumably initiated by a given receptor subtype, or, in combination with selective high-affinity radiolabeled probes, to characterize the ligand-binding properties of the receptor directly.

Classification of P_1-Type Purinergic Receptors

A wide variety of adenosine analogs have been synthesized for the characterization of P_1-purinergic receptor subtypes (Linden, 1991; Stiles, 1991). These include the R and S isomers of phenylisopropyladenosine (R-PIA and S-PIA), cyclohexyladenosine (CHA), 5′-N-ethylcarboxamide adenosine (NECA), and N^6-cyclopentyladenosine (CPA). The use of these agonists in both functional studies and in ligand-binding studies has permitted the unequivocal differentiation of two major P_1-purinergic subtypes, the A_1- and A_2-adenosine receptors. Table 1 details information concerning the rank orders of potency and the signal transduction pathways which are activated by these receptors. Recent developments have included the synthesis of several antagonists which exhibit high selectivity for either the A_1- or A_2-type receptor; these antagonists, or variations thereof, may prove useful for further characterization of yet additional P_1-purinergic subtypes. The possible existence of multiple A_1-receptor "sub-subtypes" has been proposed.

Classification of P_2-Type Purinergic Receptors

Functional studies in a wide variety of tissues and cells have provided strong support for the existence of multiple ATP receptor subtypes. While bona fide antagonists of these receptors have yet to be identified or synthesized, various other nucleotides (both natural and synthetic) have been shown to mimic, with different potencies, the ability of ATP to elicit diverse biological responses. Nucleotides can be structurally modified in three major domains: the base moiety, the ribose (or deoxyribose) group, and the polyphosphate chain. Particularly useful ATP analogs are those containing substitutions at 2-′OH of the ribose or substitutions of methylene bridges for the normal oxygen bridges of the polyphosphate chain. As summarized in Table 1 and in previous reviews by Gordon (1986) and by O'Connor et al. (1991), at least four classes of ATP receptors can be grouped on the basis of their selectivities for various nucleotides; three of these subtypes have been (semiofficially) termed P_{2x}- P_{2y}-, and P_{2z}-type purinergic receptors. Platelets express an ADP-selective receptor which has been denoted as P_{2t} (for "thrombocyte"). As noted in Table 1, these pharmacologically defined classes or subtypes can also be correlated with the functionally defined subtypes described above. A fourth class of nucleotide receptor can be activated by both ATP and certain nonadenine nucleotides;

Table 1 Classification of Purinergic Receptors

Subtype	Agonist selectivity	Signal transduction mechanism: Primary	Signal transduction mechanism: Secondary
P_1-purinergic:	Adenosine >> AMP > ADP, ATP		
A_1-adenosine	CPA ⩾ R-PIA ⩾ 2-CADO > NECA > S-PIA	G protein activation (G_s)	Inhibition of adenylate cyclase Activation of K^+ channels Inhibition of Ca^{2+} channels
A_2-adenosine	NECA >> CPA > R-PIA > S-PIA	G protein activation (G_i)	Activation of adenylate cyclase
P_2-purinergic:	ATP, ADP >> AMP > adenosine		
P_{2t}-purinergic	2-MeSADP > ADP (ATP antag.)	? (Channel protein ?)	Elevation of $[Ca^{2+}]$ Inhibition of adenylate cyclase
P_{2x}-purinergic	L-AMPP-C-P ⩾ AMP-C-PP = ATP = ADP > AMP	Intrinsic ion channel	Elevation of $[Ca^{2+}]$ Membrane depolarization
P_{2y}-purinergic	2-MeSATP = ADP S > ATP = ADP >> AMP-C-PP = AMPP-C-P = UTP	G protein activation	Activation of PI-PLC Elevation of $[Ca^{2+}]$ Activation of protein kinase C Activation of other phospholipases
P_{2z}-purinergic	ATP = ATP S >> ADP, AMP	Intrinsic channel/pore	Elevation of $[Ca^{2+}]$ Loss of endogenous metabolites
$P_{2?}$/nucleotide	UTP ⩾ ATP = ATP S > ADP > 2-MeSATP > AMPP-C-P = AMP-C-PP	G protein activation	Activation of PI-PLC Elevation of $[Ca^{2+}]$ Activation of protein kinase C Activation of other phospholipases

UTP is a particularly potent and efficacious agonist for these receptors. Functional studies suggest that this yet-to-be-named receptor may be one of the most widely expressed of the P_2-type purinergic (or more appropriately, purinergic/pyrimidinergic) receptors.

III. General Overview of Receptor-Mediated Signal Transduction

Research during the past several years has greatly increased our appreciation of the number and diversity of cell surface receptors which activate or attenuate a variety of signal transduction pathways. In particular, the application of molecular biological methods has facilitated the identification of either cDNA species or genomic DNA species which encode many of these receptors. This structural data, in conjunction with the results of functional (both physiological and biochemical) studies, has indicated (Hollenberg, 1991) that signal-transducing receptors can be broadly categorized into several major classes: (1) the receptor ion channels; (2) the G-protein-coupled receptors with 7-transmembrane spanning domains; (3) the receptor tyrosine kinases; and (4) receptors (e.g., for certain cytokines and growth factors) which exhibit structural features (or lack thereof) which do not permit their inclusion in the above categories. In general, these latter receptor types (Waldmann, 1989), as well as the receptor tyrosine kinases (Yarden and Ulrich, 1988), recognize ligands which are large polypeptide growth factors, hormones, or cytokines. In contrast, the receptor ion channels (Numa, 1989) bind ligands which are small organic ligands which function as neurotransmitters; these ligands include acetylcholine, gamma-aminobutyric acid (GABA), glutamate, and glycine. As recently reviewed by Dohlman et al. (1991), G-protein-coupled receptors constitute a large "superfamily" of receptors for many structurally diverse ligands, including small organic molecules (e.g., catecholamines and prostaglandins), small (3 to 10 amino acids) peptides (e.g., vasopressin); and larger (20 to 40 amino acids) polypeptide hormones (e.g., glucagon). As will be discussed below, the recent isolation of cDNA clones encoding both A_1- and A_2-adenosine receptors permits unequivocal inclusion of these latter purinergic receptors within the superfamily of seven-transmembrane-spanning domain, G-protein-coupled receptors. Given that ATP is a relatively small organic ligand, it is reasonable to postulate that ATP receptors may also be likely to belong to the superfamily of G-protein-coupled receptors or to the superfamily of receptor ion channels. As will be discussed below, this hypothesis is indeed consistent with the results of functional studies. Discussion of these functional data may be aided by a brief review of the "common" structural features of both the G-protein-coupled receptors with seven transmembrane spanning domains and the receptor ion channels.

A. G-Protein-Coupled Receptors with Seven Transmembrane Spanning Domains

Guanine nucleotide-binding regulatory proteins (G proteins) play an obligatory role in mediating signal transduction by receptors for a broad range of neurotransmitters, hormones, and local mediators. Biochemical and/or molecular biological analyses indicate that many of these G-protein-coupled receptors may constitute a superfamily of structurally and functionally related membrane proteins (Dohlman et al., 1991). The shared structural characteristics of these single-polypeptide receptors include: (1) an amino-terminal extracellular domain; (2) seven putative transmembrane domains (each consisting of 20 to 25 predominantly hydrophobic amino acids; (3) three intracellular and three extracellular loops which connect the various transmembrane segments; and (4) an intracellular carboxy terminus. Two segments within the third intracellular loop appear critical for interaction of the receptors with G proteins. G proteins also constitute a superfamily of related proteins (Gilman, 1987). As with G-protein-coupled receptors, the number and diversity of G proteins and G-protein subunits continues to increase. This diversity underlies the multiplicity of the signal transducing effector systems which are known (or have been hypothesized) to be regulated by G-protein-coupled receptors. These effector systems include adenylate cyclase, a variety of ion channels, and an increasing number of phospholipases.

B. Multisubunit Receptor Ion Channels

Functional receptor ion channels (Numa, 1989) exist as a complex of several polypeptide subunits such that the functional domain is formed by the interaction of several discrete subunits to form a channel through which cations or anions can pass with varying degrees of ion selectivity. Such receptors may exist as either homopolymeric or heteropolymeric complexes. The component polypeptide subunits are generally intrinsic membrane proteins which contain multiple transmembrane spanning domains. Each oligomeric complex forms a channel which stochastically flickers between open and closed states. Ligand binding to one or more subunits alters the kinetics of channel opening. The altered ion conductances then result in changes in the membrane potential of the target cell. The physiological role of such receptors is the rapid (millisecond time scale) transmission of signals from nerve cell to nerve cell or from nerve cell to muscle cell.

IV. Mechanisms Underlying Signal Transduction by P_1-Purinergic Receptors

A. A_2-Adenosine Receptors

Agonist occupation of A_2-adenosine receptors is associated with stimulation of adenylate cyclase activity in isolated plasma membranes and with accumulation of cyclic AMP in intact cells (reviewed in Stiles, 1991). The ability of A_2 agonists to activate the adenylate cyclase effector enzyme in membranes is absolutely dependent on the presence of GTP or nonhydrolyzable GTP analogs. These observations have led to the hypotheses: (1) that A_2-receptors are functionally, and preferentially, coupled to the G_s type of GTP-binding regulatory protein; and (2) that the various biological responses to A_2-receptor agonists are primarily, if not exclusively, mediated by increased cyclic AMP and the stimulation of cyclic AMP-dependent protein kinase(s).

B. A_1-Adenosine Receptors

Based on early functional studies, signal transduction via A_1 receptors has been associated with inhibition of adenylate cyclase activity in isolated membranes and with depressed cyclic AMP levels in intact cells. These findings indicated that these receptors were functionally, and preferentially, coupled to the G_i-type G proteins (of which there are three major isoforms, G_{i-1}, G_{i-2}, G_{i-3}). Consistent with this hypothesis, most biological responses to A_1 agonists are inhibited in tissues/cells which have been pretreated with pertussis toxin. However, the pertussis toxin-sensitive G_i-type G proteins are now known to additionally regulate signal-transducing effector proteins other than adenylate cyclase (reviewed in Linden, 1991; Stiles, 1991). These include K^+ channels, Ca^{2+} channels, and a variety of phospholipases. Likewise, recent studies have demonstrated that agonist occupation of A_1-adenosine receptors can induce activation of these effector enzymes/channels in various cell types. The activation of G_i-regulated K^+ channels, which results in hyperpolarization of target cells, is likely to represent a major mechanism underlying adenosine action on excitability in heart and brain.

V. Mechanisms Underlying Signal Transduction by P_2-Purinergic Receptors

As noted previously, extracelluar ATP elicits diverse responses in many tissues and cell types. Additional research has indicated that many of the functional responses to extracellular ATP can be ascribed to, or correlated with, alterations in cellular Ca^{2+} homeostasis. Utilization of fluorescent Ca^{2+} indicator dyes and appropriate spectrofluorimetric methods has been particularly useful in defining ATP-induced signal transduction in numerous cell types. Many studies during the last few years have documented that extracellular ATP rapidly (within seconds) increases cytosolic $[Ca^{2+}]$ in the vast majority of cell/tissues which exhibit ATP-induced changes in function. In turn, the ATP-induced increase in cytosolic $[Ca^{2+}]$ can activate a host of secondary signal transduction processes, including: (1) stimulation of phospholipase A_2 with consequent production of various eicosanoids; (2) stimulation of diverse Ca^{2+}-dependent regulatory enzymes including calmodulin-dependent kinases; and (3) activation or inhibition of various plasma membrane ion channels (e.g., Ca^{2+}-activated K^+ channels) with consequent changes in membrane potential and excitability. However, comparison of the ATP-induced increases in cytosolic $[Ca^{2+}]$ observed in diverse cell types has indicated that these functional responses appear to involve three very different mechanisms of action. This strongly suggests the existence of at least three distinct ATP receptor subtypes. The diverse mechanisms underlying ATP-induced increases in cytosolic $[Ca^{2+}]$ are outlined in Fig. 2 and are discussed in detail below.

A. ATP Receptors Coupled to G Proteins and Phospholipase Effector Enzymes

In many cell types, the functional effects elicited by exposure to extracellular ATP can be correlated with an ATP-induced mobilization of intracellular Ca^{2+} stores (reviewed in O'Connor et al., 1991). Consistent with this Ca^{2+}-mobilizing action, extracellular ATP and other nucleotides elicit rapid activation of inositol phospholipid hydrolysis in such cells. Thus, certain ATP receptor subtypes are likely to belong to the superfamily of cell surface receptors which are functionally coupled to inositol phospholipid-specific phospholipase C (PI-PLC) effector enzymes (Rhee et al., 1989). Similar to the other receptor types which belong to this superfamily, P_2-purinergic receptors appear to activate PI-PLC effector enzymes indirectly via the mediation of guanine nucleotide-binding regulatory proteins (G proteins). In some, but not all, ATP-responsive cell types, the ability of P_2-purinergic agonists to elicit inositol polyphosphate accumulation is inhibited or attenuated upon pretreatment of the cells with pertussis toxin (Fine et al., 1989; Cowen et al., 1990). Moreover, several cell-free systems have been described for the characterization of ATP receptor/G-protein/PI-PLC signaling cascades under defined in-vitro conditions. These preparations include membrane ghosts from turkey erythrocytes (Boyer et al., 1989) and washed membranes isolated from HL-60 promyelocytic leukemia cells (Cowen et al., 1990). In such membrane preparations, the ability of P_2-purinergic agonists to activate PI-PLC is absolutely dependent on

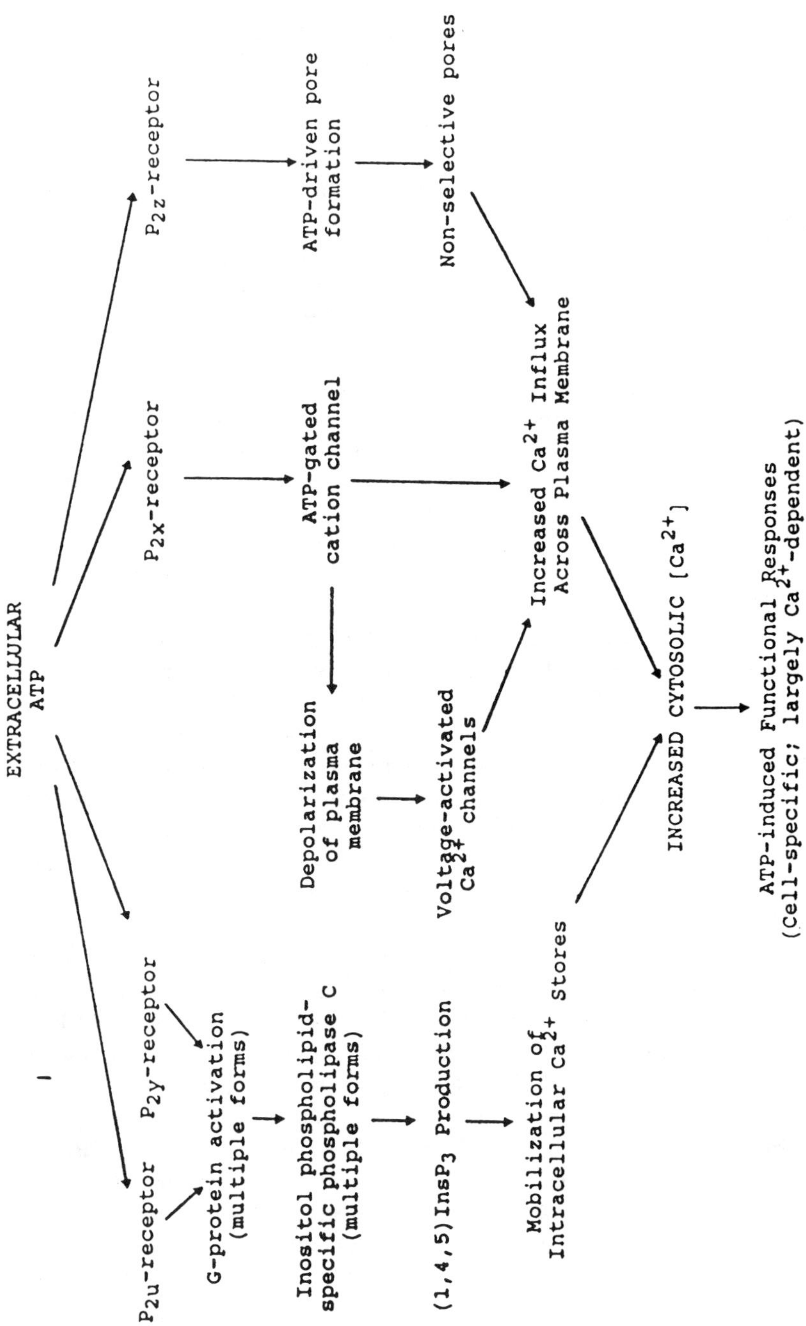

Figure 2 Mechanisms by which extracellular ATP can increase cytosolic $[Ca^{2+}]$.

the additional presence of guanine nucleotides. Moreover, Cooper et al., (1989) have successfully used the ADP analog, ^{35}S-ADPβS, as a radioligand probe of the G-protein-coupled P_2-purinergic receptors expressed in turkey erythrocytes. High-affinity binding of this nucleotide to the erythrocyte membranes is inhibited in the presence of guanine nucleotides. The ability of guanine nucleotides to modulate the high-affinity binding of receptor agonists is a hallmark of G-protein-coupled receptors (Dohlman et al., 1991).

Fairly extensive agonist selectivity studies have been performed to define the putative ATP receptor subtype(s) involved in the activation of guanine nucleotide-dependent PI-PLC effector enzymes. Such studies indicate that P_2-purinergic activation of PI-PLC/Ca^{2+} mobilization in various cell types can be further subcategorized into either of two groups with very distinctive nucleotide selectivities (reviewed in O'Connor et al., 1991). In certain cells, including hepatocytes, endothelial cells, and avian erythrocytes, ADP is equipotent to ATP. In these cells, nonadenine (both purine and pyrimidine) nucleotides are generally inactive or much less potent than ATP/ADP. This nucleotide selectivity is characteristic of the P_{2y}-purinergic receptor subtype. Conversely, in many other cell types (including phagocytic leukocytes, macrophages, fibroblasts, and airway epithelial cells), ADP is several orders of magnitude less potent than ATP in stimulating PI-PLC and Ca^{2+} mobilization. However, these latter cell types can be activated by relatively low concentrations of certain nonadenine nucleotide triphosphates. In particular, UTP has been shown to be either equipotent to, or more potent than, ATP is in all such cells.

Four salient hypotheses may be formed from the preceding observations: First, certain P_2-purinergic receptor subtypes are likely to be structurally related to the superfamily of G-protein-coupled receptors. Second, there appear to be at least two distinct (the ATP/ADP-selective versus the ATP/UTP-selective) types of P_2-purinergic receptors which activate PI-PLC signaling pathways via G proteins. This is characteristic of many small organic ligands (e.g., acetylcholine, serotonin, dopamine, epinephrine) known to activate G-protein-coupled receptors (Dohlman et al., 1991). Third, it is likely that the ATP receptors which activate guanine nucleotide-dependent PI-PLC enzymes may also activate or inhibit other G-protein-regulated signal transduction events. For example, ATP has also been shown to inhibit, via a guanine nucleotide-dependent mechanism, the so-called voltage-dependent M current (carried by a specific K^+ channel), characteristic of certain neurons (Brown, 1988). It is likely that many of the diverse biological responses to ATP observed in particular tissues and cell types will eventually be associated with the activation of these and other G-protein-dependent signal transduction pathways. Fourth, these PI-PLC-coupled ATP receptors should also activate a variety of other phospholipase-based signal transduction reactions. Extracellular nucleotides have been reported to stimulate phospholipase A_2 (Gupta et al.,

1990) and phospholipases (both C and D), which selectively hydrolyze phosphatidylcholine (Irving and Exton, 1987; Martin and Michaelis, 1989). Increased activity of these latter enzymes, in concert with the stimulation of PI-PLC, may underlie the large ATP-induced increases in diacylglycerol, with consequent activation of protein kinase C, observed in certain cells.

B. ATP Receptors That Activate Ion Channels Directly

ATP-activated cation conductance pathway(s) appear to be present in the plasma membranes of a variety of cells, including cardiac myocytes (Friel and Bean, 1988), smooth muscle cells (Benham and Tsien, 1987), sensory neurons (Krishtal et al. 1988), and developing myotubes (Thomas and Hume, 1990). The ability of nanomolar to micromolar concentrations of extracellular ATP to activate these conductance pathways has been documented using state-of-the-art patch-clamp recording of both whole-cell currents and single-channel currents. While there are some differences in the ATP-activated conductance pathways observed in these various cell types, there are several common features. These include: (1) cation specificity with relatively little selectivity for different monovalent cations; (2) substantial activity even at hyperpolarized membrane potentials; (3) subsecond activation and rapid inactivation; (4) an apparent lack of involvement of either G proteins or diffusible second messengers (Ca^{2+}, cAMP) in generating in the increased ionic conductances; and (5) a substantial Ca^{2+} permeability in some cell types. In certain (but not all) cells, the nucleotide selectivity which characterizes these ATP-activated conductances is characteristic of the so-called P_{2x}-type receptors.

Activation of these ATP-sensitive channels can lead to increased cytosolic $[Ca^{2+}]$ in two ways. In certain cells [e.g., rabbit ear artery (Benham and Tsien, 1987)], there is sufficient influx of Ca^{2+} directly via the ATP-activated cation channel to elevate cytosolic $[Ca^{2+}]$. In other cell types [e.g., cardiomyocytes (DeYoung and Scarpa, 1989)], the opening of ATP-activated cation channels results primarily in depolarization of the plasma membrane with secondary activation of voltage-dependent Ca^{2+} channels. Flux through these latter channels then leads to increased cytosolic $[Ca^{2+}]$, which, in turn, can activate mobilization of intracellular, ryanodine-sensitive Ca^{2+} stores. These data strongly suggest that ATP-activated cation channels may be common features of many excitable cells and may be likely candidates for explaining at least some aspects of "nonadrenergic, noncholinergic" neurotransmission. Moreover, the extremely rapid activation of these ionic conductances by ATP suggests that the presumed receptor and the regulated ion channel are parts of a preexisting functional complex. As noted above, this is characteristic of the receptors for a variety of excitatory and inhibitory neurotransmitters (Numa, 1989). Thus, it is likely that certain ATP receptor subtypes may be structurally related to the superfamily of the multisubunit receptor ion channels.

C. ATP Receptors That Induce Formation of Nonselective Pores

There is a longstanding observation that extracellular ATP can effect a "permeabilization" of some cell types to molecules up to 1000 Da in size, including Ca^{2+}, other ions, and endogenous metabolites. Cells in which this response has been extensively characterized include mast cells (Cockcroft and Gomperts, 1980), macrophages (Greenberg et al. 1988), and most recently, certain T-lymphocyte subtypes (DiVirgilio et al., 1989). Prolonged opening of these ATP-induced "pores" will lead not only to equilibration of transmembrane ionic gradients (hence, the increase in the intracellular Ca^{2+}) but also to loss of the critical intracellular metabolites (e.g., nucleotides) and eventual cell death. Greenberg et al., (1988) have utilized this cytotoxic response to generate mutant sublines of the J774 mouse macrophage line which are resistant to the cytotoxic effects of extracellular ATP. Recent studies using these mutant cell lines have indicated that this response can be dissociated from the ATP-induced mobilization of intracellular Ca^{2+} stores discussed above. Thus, the presumed ATP receptors which mediate pore formation appear to be distinct from the G-protein-coupled ATP receptors which activate PI-PLC. Recent patch-clamp electrophysiological studies of the ion conductance properties of these ATP-induced pores in mast cells indicate that the pores can be functionally distinguished from ATP-activated cation channels also described above (Tatham and Lindau, 1990). The nucleotide selectivity which characterizes ATP-induced pore formation is characteristic of the P_{2z}-type purinergic receptors.

Induction of these pores requires the presence of near-millimolar concentrations of extracellular ATP when cells are incubated in saline containing physiological concentrations of divalent cations. This contrasts with the activation of G-protein-coupled ATP receptors or ion channel-coupled ATP receptors by nanomolar to micromolar [ATP]. For this reason, ATP-induced permeabilization or pore formation has been considered something of a phenomenonological curiosity with minimal physiological significance. However, it should be noted that the size permeability of these pores is similar to that of the gap junction channels which mediate cell-cell coupling. Significantly, the mutant J774 macrophages, which are deficient in ATP-induced pore formation, are also characterized by greatly reduced expression (both mRNA and protein) of the connexin-43 gap junction protein (Beyer and Steinberg, 1991).

It is also interesting to note that this ATP-induced pore formation is expressed in cell types (T lymphocytes, mast cells, macrophages) involved in cell-mediated inflammatory or immune responses. Moreover, the ATP-induced permeabilization response is similar to the permeabilization or lytic responses triggered by various insect toxins as well as that initiated by cytotoxic T-cells. Recent studies have raised the extremely intriguing possibility that these ATP-induced pores may be involved in cell-mediated cytotoxicity. DiVirgilio et al., (1989) have reported that cytotoxic T cells and killer cells are themselves highly resistant to the pore-forming and cytotoxic effects in extracellular ATP. Moreover, Fillipini et al., (1990) have demonstrated that these same cytotoxic T cells release ATP when incubated with activating ligands such as concanavalin A or monoclonal antibodies against the T-cell antigen receptor. These findings have prompted speculation that ATP may play a physiological role in cell-mediated immune responses and cytotoxicity. Such models are predicted on the known formation of tight cell-cell junctions between the cytotoxic cells and target cells. ATP may be released into the restricted spaces formed by the intercellular junction. This ATP can then induce pore formation in the target cells with resulting loss of intracellular ions and nucleotides. While it is speculative, this model is consistent with many disparate findings and should provoke interesting experimental approaches in future studies.

VI. Structural Characterization of Receptors for ATP and Adenosine

A. Structure of P_1-Type Purinergic Receptors

A_1-Adenosine Receptors

Through the use of specific photoaffinity probes and ligand-affinity chromatography, A_1-adenosine receptors from the brain have been purified to homogeneity by two research groups (Barrington et al., 1989; Olah et al., 1990; Nakata, 1989). The purified receptor, which is a 36-kDa glycoprotein, binds adenosine analogs with a rank order of potency characteristic of the in-situ receptor. Upon deglycosylation, the receptor migrates as a 32-kDa protein during SDS-gel electrophoresis (Stiles, 1991). A full-length cDNA clone which encodes an A_1-adenosine receptor has very recently been isolated from a dog thyroid cDNA library by Libert et al., (1989, 1991). COS cells transfected with this cDNA express A_1-adenosine receptors which exhibit signal transduction and ligand-binding properties characteristic of normal brain A_1-type receptors. The amino acid sequence indicated from the cDNA yields a 36-kDa protein with seven predicted transmembrane spanning domains and N-linked glycosylation sites on the second extracellular loop. Northern analysis indicates that the receptor is encoded by a predominant 2.4-kb mRNA transcript. Particularly high levels of these RNA transcripts were observed in brain and thyroid gland; lower levels were noted in heart, kidney, and testis.

A_2-Adenosine Receptors

A_2-adenosine receptors have not yet been purified to homogeneity. However, Barrington et al., (1990) have utilized A_2-selective photoaffinity probes to covalently label a 45-kDa glycoprotein from brain striatal membranes. Significantly, Libert et al., (1989, 1991) have used a dog thyroid cDNA library to clone cDNA which, when transfected into COS cells, induces expression of an adenosine receptor with ligand-binding properties very similar to those characteristic of brain striatal A_2-adenosine receptors. Analysis of the amino acid coding sequence indicated from this cDNA predicts the synthesis of a 45-kDa protein with seven transmembrane-spanning domains. Northern analysis indicates that abundant expression of mRNA transcripts encoding this A_2-adenosine receptor is limited to the brain and thyroid gland. Within the central nervous system, particularly high transcript levels were observed in the striatum, nucleus accumbens, and olfactory tubercle (Maenhaut et al., 1990). Thus, both the A_1- and A_2-adenosine receptors belong to the superfamily of G-protein-coupled receptors with seven transmembrane spanning regions.

B. Structure of P_2-Type Purinergic Receptors

Much less is known concerning the structures of P_2-purinergic receptors for ATP. Indeed, most of our knowledge concerning the various subtypes of this receptor family is based on functional responses in either intact tissue or isolated cells. Detailed classification of P_2-purinergic subtypes upon such functional responses has been limited by the absence of high-affinity selective antagonists. Moreover, with several exceptions, there is little ligand-binding data of the type usually employed to define subtypes for other well-characterized receptor families (e.g., muscarinic or adrenergic). Notably, Cooper et al., (1989) have successfully utilized ^{35}S-ADPβS as a reasonably selective radioligand for the labeling of the P_{2y} receptors expressed in turkey erythrocytes. ^{3}H-α,β-methylene ATP has been reported to be useful for the labeling of P_{2x} receptors in smooth muscle (Ruggieri et al., 1990). However, given the extremely large number of membrane proteins which recognize ATP as either a substrate or an allosteric ligand, it has generally proven difficult to develop nucleotide analogs which will specifically (or at least primarily) recognize only ATP receptor protein.

Nucleotide photoaffinity probes have been used, with some success, to covalently label presumed P_{2x}-purinergic receptors in vas deferens muscle (Fedan et al. 1985), P_{2z}-purinergic receptors in 3T6 fibroblasts (Erb et al. 1990), and the G-protein-coupled P_{2y}-purinergic receptors in turkey erythrocytes (Boyer et al., 1990). Treatment of turkey erythrocyte membranes with ^{32}-P-labeled benzoylbenzoic-ATP (BzATP) yields a light-catalyzed, preferential incorporation of ^{32}P into a 53-kDa membrane protein (Boyer et al., 1990).

Jeffs et al., (1991) have also been successful in solublizing affinity-labeled P_{2y}-type ATP receptors from turkey erythrocytes. These solubilized receptors should prove to be useful starting points for the eventual purification (by conventional protein chemistry) of P_{2y}-type ATP receptors. Partial amino acid sequence derived from the purified P_{2y} receptor can then be used to construct oligonucleotide probes to screen cDNA libraries. In principle, this strategy might also be used for the identification, isolation, and cloning of other P_2-type purinergic receptors. However, most receptor proteins that are expressed in very low amounts are also intrinsic membrane proteins which, de facto, are much more difficult to isolate and purify than abundant soluble or structural proteins. In the case of the P_2-purinergic receptors, this general problem is compounded by the paucity of highly selective ligands. Alternatively, expression cloning strategies have been used successfully for the isolation of cDNA species encoding signal transducing receptors in the absence of prior purification of the receptor protein (reviewed in Snutch, 1988). This approach has been particularly useful for the isolation and cloning of many receptors known to belong to the superfamily of G-protein-coupled receptors (Dohlman et al., 1991). Murphy and Tiffany (1990) have very recently reported the expression of PI-PLC coupled ATP/UTP receptors in *Xenopus* oocytes injected with mRNA isolated from differentiated HL-60 granulocytes. Nuttle et al., (1991) have also expressed functional ATP/UTP receptors in *Xenopus* oocytes injected with mRNA isolated from either undifferentiated HL-60 cells, a human leukocyte cell line, or DDT_1-MF2 cells, a hamster smooth muscle cell line. The application of expression and homology cloning methods should prove invaluable for determining the sequence and structure of the various P_2-purinergic receptor subtypes. In turn, such structural information will facilitate ongoing characterization of the physiological/pathological roles of such receptors in regulating signal transduction and in diverse cells and tissues.

VII. Purinergic Signal Transduction in the Lung

Table 2 summarizes the role of purinergic signal transduction in the lung. The major lung cell types which exhibit altered function in response to extracellular ATP or adenosine are type II alveolar pneumocytes, airway epithelial cells, airway and parenchymal smooth muscle cells, and resident alveolar macrophages. Under pathological conditions, invading neutrophils will also constitute a major population of ATP- and adenosine-responsive cells. Recent studies indicate that ATP receptors, in particular, can modify key functions in the lung, including surfactant secretion by type II pneumocytes (Rice, 1990) and chloride secretion by airway epithelial cells (Mason et al., 1991; Knowles et

Table 2 Purinergic Signal Transduction in the Lung

Cell type	Biological action	Purinergic receptor	References
Type II alveolar pneumocytes	Activation of surfactant secretion	P_2 (P_{2y}/ATP/UTP)	Rice, 1990
	Activation of PI-PLC		
	Elevation of [Ca^{2+}]		
Tracheal smooth muscle	Relaxation	P_1 (A_1-adenosine?)	Christie and Satchell, 1980
Lung parenchymal	Contraction	P_2 (P_{2y}?)	Satchell, 1990
Airway epithelial	Activation of SCC	P_2 (ATP/UTP, other?)	Mason et al., 1991
	Activation of Cl^- conductance		Knowles et al., 1991
	Elevation of [Ca^{2+}]		Clarke and Boucher, 1990
Lung phagocytes:	Activation/priming of superoxide release	P_2 (ATP/UTP)	Ward et al., 1990
Neutrophils	Activation of PI-PLC		Nakanishi et al., 1991
Macrophages	Elevation of [Ca^{2+}]		Cowen et al., 1989
	Activation of PC-PLD		Xie and Dubyak, 1991
Neutrophils	Inhibition of phagocyte activation	P_1 (A_1-adenosine)	Cronstein et al., 1990

al., 1991). Both of these responses, as well as the activation of neutrophils, monocytes, and macrophages (Cowen et al., 1989; Ward et al., 1990) appear to involve the G-protein-coupled ATP/UTP receptors which are functionally coupled to inositol phospholipid hydrolysis and/or Ca^{2+} mobilization. The importance of these receptors in both normal and pathological respiratory function is underscored by the recent in-vivo human studies of Knowles et al., (1991). These investigators noted that inclusion of ATP or UTP (in the presence of amiloride to block Na^+ reabsorption) during nasal perfusion of normal or cystic fibrosis subjects significantly increased chloride secretion. Given the critical role of chloride secretion in limiting net fluid secretion in cystic fibrosis airways, it may be possible to use ATP/UTP receptor agonists as part of an inhalation therapy for such patients.

Acknowledgments

This review was written during support of the author as an Established Investigator of the American Heart Association.

References

Barankiewicz J, Dosch H-M, Cohen A. Extracellular nucleotide catabolism in human B and T lymphocytes: the source of adenosine production. J Biol Chem 1988; 263:7094–8.

Barrington WW, Jacobson KA, Stiles GL. Demonstration of distinct agonist and antagonist coformations of the A_1-adenosine receptor. J Biol Chem 1989; 264:13157–64.

Barrington WW, Jacobson KA, Stiles GL. The glycoprotein of the A_2-adenosine receptor binding subunit, Mol Pharmacol 1990; 38:177–83.

Benham CD, RW Tsien. A novel, receptor-operated Ca^{2+}-permeable channel activated by ATP in smooth muscle. Nature 1987; 328:275–8.

Beyer EC, Steinberg TH. Evidence that the gap junction protein connexin-43 is the ATP-induced pore of mouse macrophages. J Biol Chem 1991; 266:7971–4.

Boyer JL, Cooper CL, Harden TK. [^{32}P]3′-0-(4-benzoyl)benzoyl adenosine 5′-triphosphate as a photoaffinity label for a phospholipase C-coupled P_{2Y}-purinergic receptor. J Biol Chem 1990; 265:13515–20.

Boyer JL, Downes CP, Harden TK. Kinetics of activation of phospholipase C by P_{2y}-purinergic receptor agonists and guanine nucleotides. J Biol Chem 1989; 264:884–90.

Brown DA. M-currents. In: Narahashi T, ed. Ion channels, Vol. 1. New York: Plenum, 1988; 55–94.

Bruns RF. Adenosine receptors: roles and pharmacology. Ann NY Acad Sci 1990; 603:211–26.

Burnstock G. Neural nomenclature. Nature 1971; 229:282–3.

Burnstock G, Kennedy C. A dual function for adenosine 5′-triphosphate in the regulation of vascular tone. Excitatory cotransmitter with noradrenaline from perivascular nerves and locally released inhibitory intravascular agent. Circ Res 1986; 58:319–30.

Christie J, Satchell DG. Purine receptors in the trachea: is there a receptor for ATP? Br J Pharmacol 1980; 70:512–4.

Clarke LL, Boucher RC. Activation of anapical Cl conductance by exogenous ATP in cystic fibrosis airway epithelium. Pediatr Pulmonol 1990; Suppl. 5:216 (abstr).

Cockcroft S, Gomperts BD. The ATP^{4-} receptor of rat mast cells. Biochem J 1980; 188:789–98.

Cooper CL, Morris AJ, Harden TK. Guanine nucleotide-sensitive interaction of a radiolabeled agonist with a phospholipase C-linked P_{2Y}-purinergic receptor. J Biol Chem 1989; 264:6202–6.

Cowen DS, Baker B, Dubyak GR. Pertussis toxin produces differential inhibitory effects on basal, P_2-purinergic-, and chemotactic peptide-stimulated inositol phospholipid breakdown in HL-60 cells and HL-60 cell membranes. J Biol Chem 1990; 265:16181–9.

Cowen DS, Lazarus HM, Shurin SB, Stoll S, Dubyak GR. Extracellular adenosine triphosphate activates calcium mobilization in human phagocytic leukocytes and neutrophil/monocyte progenitor cells. J Clin Invest 1989; 83:1651–60.

Cronstein BN, Daguma L, Nichols D, Hutchison AJ, Williams M. The adenosine/neutrophil paradox resolved: human neutrophils possess both A_1 and A_2 receptors that promote chemotaxis and inhibit O_2 generation, respectively. J Clin Invest 1990; 85:1150–7.

DeYoung MB, Scarpa A. ATP receptor-induced C+ transients in cardiac myocytes: sources of mobilized C+. Am J Physiol 1989; C750–8.

Drury AN, Szent-Gyorgyi A. The physiological activity of adenine compounds with special reference to their action upon the mammalian heart. J Physiol 1929; 68:213–37.

DiVirgilio F, Bronte V, Callavo D, Zanovello P. Responses of mouse lymphocytes to extracellular adenosine 5′-triphosphate [ATP]. Lymphocytes with cytotoxic activity are resistant to the permeabilizing effects of ATP. J Immunol 1989; 143:1955–60.

Dohlman HG, Thorner J, Caron MG, Lefkowitz RJ. Model systems for the study of seven-transmembrane segement receptors. Ann Rev Biochem 1991; 60:653–88.

Dubyak GR, Fedan JS, eds. Biological actions of extracellular ATP. Ann NY Acad Sci Vol. 603. 1991.

Erb L, Lustig KD, Ahmed AH, Gonzalez FA, Weisman GA. Covalent incorporation of 3′-0-(4-benzyl)benzoyl-ATP into a P_2-purinoreceptor in transformed mouse fibroblasts. J Biol Chem 1990; 7424–31.

Fedan JS, Hogaboom GK, O'Donnell JP, Jeng SJ, Guillory RJ. Interaction of $[^3H]$arylazido aminopropionyl ATP ($[^3H]ANAPP_3$) with P_2-purinergic receptors in the smooth muscle of the isolated guinea-pig vas deferens. Eur J Pharmacol 1985; 108:49–61.

Fillipini A, Taffs RE, Sitkovsky MV. Extracellular ATP in T-lymphocyte activation: possible role in effector functions. Proc Natl Acad Sci USA 1990; 87:8267–71.

Fine, J, Cole P, Davidson JS. Extracellular nucleotides stimulate receptor-mediated calcium mobilization and inositol phosphate production in human fibroblasts. Biochem J 1989; 263:371–6.

Forrester T. Release of ATP from heart: presentation of a release model using human erythrocyte. Ann NY Acad Sci 1990; 603:335–52.

Friel DD, Bean BP. Two ATP-activated conductances in bullfrog atrial cells. J Gen Physiol 1988; 91:1–27.

Gilman AG. G-proteins: transducers of receptor-generated signals. Ann Rev Biochem 1987; 56:615–24.

Gordon JL. Extracellular ATP: effects, sources, and fate. Biochem J 1986; 233:309–19.

Greenberg, S, DiVirgilio F, Steinberg TH, Silverstein SC. Extracellular nucleotides mediate C+ fluxes in J774 macrophages by two distinct mechanisms. J Biol Chem 1988; 263:10337–43.

Gupta SK, Diaz E, Heasley LE, Osawa S, Johnson GL. A G-protein mutant that inhibits thrombin and P_2-purinergic receptor activation of phospholipase A_2. Science 1990; 249:662–6.

Hollenberg MD. Structure-activity relationships for transmembrane signaling: the receptor's turn. FASEB J 1991; 5:178–86.

Irving HR, Exton JH. Phosphatidylcholine breakdown in rat liver plasma membranes: roles of guanine nucleotides and P_2-purinergic agonists. J Biol Chem 1987; 262:3440–3.

Knowles MR, Clarke LL, Boucher RC. Activation by extracellular nucleotides of chloride secretion in the airway epithelia of patients with cystic fibrosis. N Engl J Med 1991; 325:533–8.

Krishtal OA, Marchenko SM, Obukov AG, Volkova TM. Receptors for ATP in rat sensory neurons: the structure-function relationship for ligands. Br J Pharmacol 1988; 95:1057–62.

Jacobson KA, Daley JW, Mangiello V, eds. Purines in cellular signaling: targets for new drugs. Berlin: Springer-Verlag, 1990.

Jeffs RA, Cooper CL, Harden TK. Solubilization of a guanine nucleotide-sensitive form of the P_{2y}-purinergic receptor. Mol Pharmacol 1991; 40:85–92.

Libert F, Parmentier M, Lefort A, Dinsart C, Van Sande J, Maenad C, Simon MJ, Dormant JD, Vassal G. Selective amplification and cloning of four new members of the G-protein-coupled receptor family. Science 1989; 244:569–72.

Libert F, Schiffmann SN, Lefort A, Parmentier M, Gerard C, Dumont JE, Vanderhaeghen J-J, Vassart G. The orphan receptor cDNA RDC7 encodes an A_1-adenosine receptor. EMBO J 1991; 10:1677–82.

Lin S-H. Liver plasma membrane ecto-ATPase: purification, localization, cloning, and functions. Ann NY Acad Sci 1990; 603:394–400.

Linden J. Structure and function of A_1-adenosine receptors. FASEB J 1991; 5:2668–76.

Maenhaut C, VanSande J, Libert F, Abramowicz M, Parmentier M, Vanderhaeghen J-J, Dumont JE, Vassart G, Schiffmann SN. RDC8 codes for an adenosine A_2 receptor with physiological constitutive activity. Biochem Biophys Res Commun 1990; 173:1169–78.

Martin TW, Michaelis K. P_2-purinergic agonists stimulate phosphodiesteric cleaveage of phosphatidylcholine in endothelial cells. Evidence for activation of phospholipase D. J Biol Chem 1989; 264:8847–56.

Mason SJ, Paradiso AM, Boucher RC. Regulation of transepithelial ion transport and intracellular calcium by extracellular adenosine triphosphate in human normal and cystic fibrosis airway epithelium. Br J Pharmacol 1991; 103:1649–56.

Murphy PM, Tiffany HL. Characterization of phagocyte P_2 nucleotide receptors expressed in *Xenopus* oocytes. J Biol Chem 1990; 265:11615–21.

Nakanishi M, Takihara H, Minoru Y, Yagawa K. Extracellular ATP itself elicits superoxide generation in guinea pig peritineal macrophages. FEBS Lett 1991; 282:91–4.

Nakata H. Purification of A_1-adenosine receptor from rat brain membranes. J Biol Chem 1989; 264:16545–51.

Numa S. A molecular view of neurotransmitter receptors and ion channels. Harvey Lect 1989; 83:121–65.

Nuttle LC, Schomisch S, Dubyak GR. Expression of P_2-purinergic receptors in *Xenopus* oocytes injected with mRNA from HL-60 and DDT_1-MF2 cell lines. FASEB J 1991; 5:A1574 (abstr).

O'Connor SE, Dainty IA, Leff P. Further classification of ATP receptors based on agonist studies. TIPS 1991; 12:137–41.

Olah ME, Jacobson KA, Stiles GL. Purification and characterization of bovine cerebral cortex A_1-adenosine receptors. Arch Biochem Biophys 1990; 283:440–6.

Rhee SG, Suh PG, Ryu S-H, Lee SY. The family of inositol phospholipid-specific phospholipase C enzymes. Science 1989; 244:546–50.

Rice WR. Effects of extracellular ATP on surfactant secretion. Ann NY Acad Sci 1990; 603:64–75.

Ruggieri MR, Whitmore KE, Levin RM. Bladder purinergic receptors. Ann NY Acad Sci 1990; 603:458–60.

Satchell D. The effects of ATP and related nucleotides on visceral smooth muscle. Ann NY Acad Sci 1990; 603:53–63.

Slakey LL, Gordon EM, Pearson JD. A comparison of ectonucleotidase activities on vascular endothelial and smooth muscle cells. Ann NY Acad Sci 1990; 603:366–79.

Snutch TP. The use of *Xenopus* oocytes to probe synaptic communication. Trends Neurosci 1988; 11:250–6.

Stiles GL. Adenosine receptors: physiological regulation and biochemical mechanisms. NIPS 1991; 6:161–5.

Tatham PER, Lindau M. ATP-induced pore formation in the plasma membrane of rat peritoneal mast cells. J Gen Physiol 1990; 95:459–476.

Thomas SA, Hume RI. Permeation of both cations and anions through a single class of ATP-activated ion channels in developing chick skeletal muscle. J Gen Physiol 1990; 95:569–90.

Unsworth CD, Johnson RG. ATP compartmentation in neuroendocrine secretory vesicles. Ann NY Acad Sci 1990; 603:353–65.

Waldman TA. The multi-subunit interleukin 2 receptor. Ann Rev Biochem 1989; 58:875–911.

Ward PA, Walker BAM, Hagenlocker BE. Functional consequences of interactions between human neutrophils and ATP, ATPγS, and adenosine. Ann NY Acad Sci 1990; 603:108–19.

Xie M, Jacobs LS, Dubyak GR. Regulation of phospholipase D and primary granule secretion by P_2-purinergic- and chemotactic peptide receptor agonists is induced during granulocytic differentiation of HL-60 cells. J Clin Invest 1991; 88:45–54.

Yarden Y, Ulrich A. Growth factor receptor tyrosine kinases. Ann Rev Biochem 1988; 443–78.

Part Two

EPITHELIAL CELLS

9

Epithelial Cell Na^+ and Cl^- Channels

SADIS MATALON, YOUNGSUK OH, and DALE J. BENOS

University of Alabama at Birmingham
Birmingham, Alabama

I. Introduction

The primacy of Na^+ and Cl^- channels in the generation of both absorptive and secretory fluids of epithelial tissues is impressive. Yet, despite knowledge of the functional existence of such biological entities, woefully little information is available concerning the biochemistry and molecular biology of these epithelial ion channels. It is true that efforts along these lines have been made difficult by the lack of suitable molecular probes (e.g., inhibitors) and by the absence of appropriately rich sources of biological material for isolation purposes. However, in recent years, several laboratories have succeeded in isolating, purifying, and reconstituting epithelial Na^+ and Cl^- channels.

This chapter first briefly summarizes the current state of knowledge of epithelial Na^+ and Cl^- channels, focusing primarily on what is known about their molecular architecture. Second, we discuss the involvement of amiloride-sensitive Na^+ transport and its regulation in the mammalian alveolar epithelium. Third, we summarize our attempts at the identification of amiloride-binding proteins in extracts of rat alveolar type II (ATII) cells. Throughout, we highlight areas in need of experimental work. We conclude with a brief perspectus on future experimentation.

II. Biochemistry of Epithelial Na^+ and Cl^- Channels

Table 1 represents the various experimental approaches that have been used to isolate ion channels from a variety of sources.

The application of many of these approaches to epithelial channels is just beginning. We will discuss the application of these different protocols to, first, epithelial Na^+ channels and then, epithelial Cl^- channels. Although a great deal of electrophysiological information is available concerning epithelial K^+ channels (see Benos et al., 1987, for a review) and a K^+ channel from rat kidney has been identified by expression cloning (Takumi, 1988), no epithelial K^+ channel protein has yet been purified. Hence, we will not discuss K^+ channels in this chapter.

A. General Properties of Epithelial Na^+ Channels

Amiloride-sensitive sodium channels have been identified in the apical membranes of a variety of epithelial tissues and are an important class of proteins that regulate Na^+ transport and fluid homeostasis (Benos, 1988; Benos et al., 1991, 1992; Garty and Benos, 1988; Smith and Benos, 1991). Currently, two distinct categories of Na^+ channels have been identified (Smith and Benos, 1991). Channels in the first category, present in the apical membranes of frog skin, toad urinary bladder, A6 cells, bovine renal papillary collecting ducts, and other epithelial tissues (Garty and Benos, 1988), display high affinity to benzamil and amiloride binding [inhibitory constants (K_i) in the presence of physiological external Na^+ concentrations: 15 nM and 115 nM, respectively] but much lower affinity for EIPA [5-(N-ethyl-N-isopropyl)-2′-4′-amiloride; $K_i = 10\ \mu M$; (Bridges et al., 1988, 1989)]. Channels in the second category, present in LLC-PK_1, MDCK cell lines, and in rat and porcine brain endothelia (Moran et al., 1988; Smith and Benos, 1991), have much lower affinity for

Table 1 Approaches to the Isolation of Ion Channels

1. Specific high-affinity ligand binding
   a. Photoaffinity labeling
   b. Equilibrium binding
2. Antibody recognition/immunoaffinity chromatography
3. Genetic approaches
   a. Transposon mutation analysis
   b. Expression cloning
4. Biochemical purification

amiloride, and show a different panel of pharmacological sensitivity to its analogs than do the high-affinity channels ($K_{i(phenamil)} = 66\,\mu M$; $K_{i(amil)} = 30\,\mu M$; $K_{i(EIPA)} = 5\,\mu M$). Evidence to be presented below indicates the possible existence of these low-amiloride-affinity Na^+ channels in the plasma membranes of ATII cells.

Benos and co-workers (Benos et al., 1987; Sariban-Sohraby et al., 1988) have purified a high-amiloride-affinity epithelial Na^+ channel from amphibian renal A6 cells and bovine kidney papilla using a combination of the first and fourth approaches, i.e., specific high-affinity ligand binding coupled with classical biochemical purification. The general protocol for the purification of this channel consisted of preparing apical membrane vesicles, and then solubilizing the proteins of these vesicles with the zwitterionic detergent CHAPS. The solubilized material was then passed over a lectin (i.e., wheat germ agglutinin) chromatography column, the protein that bound to the column eluted, and then subjected to gel filtration using size-exclusion high-performance liquid chromatography (HPLC). At each stage of purification, the equilibrium binding of an analog of amiloride (^{3}H-methylbromoamiloride) was measured, and functional activity assessed in reconstitution studies. A 5000-fold enrichment at greater than 90% homogeneity of the channel protein was obtained using this protocol.

When the purified material was radiolabeled with ^{125}I, separated on a nonreducing SDS gel, and an autoradiograph made of the gel, the native protein was found to consist of a single broad band averaging 730 kDa. SDS-PAGE of the native channel protein under reducing conditions revealed that this single, large-molecular-mass protein consisted of at least six major polypeptides with average molecular masses of 315, 150, 95, 70, 55, and 40 kDa. This finding was true for the channel protein isolated from either A6 cells or the bovine kidney.

In subsequent experiments, it has been shown that the 300-kDa subunit of this channel can be specifically phosphorylated both in vitro in the presence of the catalytic subunit of protein kinase A and in vivo in response to vasopressin (antidiuretic hormone) (Sariban-Sohraby et al., 1988). In preliminary experiments, addition of the catalytic subunit of protein kinase A to the cytoplasmic surface of an excised A6 cell apical membrane patch in the presence of ATP results in the activation of amiloride-sensitive Na^+ channels by increasing their open probability. These results are similar to those obtained earlier by Frings et al. (1988) in toad urinary bladder. These observations suggest that phosphorylation of the 300-kDa subunit may in fact increase the transport of Na^+ through these ion channels, thus accounting for the increase in transepithelial Na^+ transport produced by vasopressin without involving recruitment of channels from cytoplasmic pools.

In photoaffinity-labeling studies, it was found that ^{3}H-methylbromo-amiloride bound to only the 150-kDa subunit. In order to verify this result, we used anti-amiloride antibodies generated by Kleyman and his colleagues (Kleyman et al., 1986). These antibodies recognize the pyrazine ring moiety of amiloride. Purified Na^+ channel protein was irradiated in the presence of 5 nM 2′-methoxy-5′-nitrobenzamil (NMBA), another photolabile amiloride analog (Kleyman et al., 1989), separated on an SDS gel, transferred to nitrocellulose, and probed with these anti-amiloride antibodies. The antibodies recognized only the 150-kDa polypeptide band. Further, photoaffinity studies with ^{125}I-labeled NMBA and antiidiotypic antibodies (Kleyman et al., 1991) revealed that binding occurred only to the 150-kDa subunit. On the basis of these three independent experiments, we conclude that it is the 150-kDa subunit to which amiloride binds. An amiloride-binding protein that presumably represents an amiloride-sensitive Na^+ channel has been purified from pig and human kidney cortex vesicles using another amiloride analog, namely, ^{3}H-phenamil, by Barbry and co-workers (1987). SDS-PAGE of the purified phenamil binding protein, under nonreduced conditions, revealed a single polypeptide of molecular mass 180 kDa. Under reducing conditions, only a single 90- to 100-kDa band was seen. Interestingly, this protein from pig kidney, when reconstituted into lipid vesicles, displayed conductive $^{22}Na^+$ uptake that was inhibited by phenamil and/or amiloride with a K_i of ~ 10 μM. These investigators have recently reported the cloning and sequencing of this channel protein from a human kidney cDNA library (Barbry et al., 1986). Surprisingly, while there are glycosylation sites, there are no consensus phosphorylation sites nor any hydrophobic spanning regions based on hydropathy plots. If the mRNA coding for this protein is transfected into mammalian cells or *Xenopus* oocytes, amiloride-binding activity can be recovered, but no amiloride-sensitive Na^+ transport measured. The authors suggest that their phenamil-binding protein may be a component of a larger-molecular-mass complex. Cloning of high-amiloride-affinity Na^+ channels is currently in progress in several laboratories (Cunningham et al., 1991; Staub, 1990).

Little is known concerning the functional capacities or purposes of the other subunits. It is known that the 40-kDa subunit is a G protein, specifically $\alpha_{i\text{-}3}$ (Ausiello et al., 1992). G proteins themselves are known to regulate Na^+ channel activity in epithelia, and it may be that the association of a G protein with the isolated complex may indicate a unique mechanism of signal transduction and regulation in these cells. The 70-kDa subunit may be an aldosterone-induced protein (Cox, 1991), but little else is known about this polypeptide. It has been shown recently that the native Na^+ channel complex is attached to the underlying cortical cytoskeleton via ankyrin attachment to actin filaments (Smith et al., 1991). Ankyrin appears to bind directly to the 150-kDa subunit of the channel protein. A labile interaction between this Na^+ channel and the cytoskeleton may also provide another mechanism for regulation of physiological activity (Cantiello et al., 1991).

Another important question involves the minimal number of subunits required for ion conduction. In an abstract published in 1990, Sariban-Sohraby and Fisher (1990) isolated the 150-kDa subunit from A6 cells and reconstituted it into giant liposomes. They were able to reveal in these liposomes, via patch clamp, the presence of amiloride-sensitive Na^+ channels with a conductance of 9pS. If this result is verified, it would indicate that this subunit may be all that is required for ion channel activity. In another abstract, Cunningham et al. (1991) found that when the cRNA coding for the 55-kDa subunit of this channel was injected into oocytes, Na^+-specific currents were produced. In addition, and not surprisingly, these currents were not sensitive to amiloride. Further work on this important question is currently underway in these laboratories.

B. General Properties of Epithelial Cl^- Channels

In contrast to amiloride-sensitive Na^+ channels, very little is known concerning the biochemistry of epithelial anion channels. A major problem in this regard has been the lack of suitable Cl^- channel ligands with sufficiently high affinity to be used in purification protocols in spite of a large effort in the development of such molecules (Cabantchik and Greger, 1992). One approach to channel purification has been to isolate Cl^- channel-containing cell membrane fractions, and then to assay for channel activity in reconstitution studies (approach 4 of Table 1). Membrane vesicle fractions enriched for Cl^- transport activity have been prepared from bovine trachea (Dubinsky and Monti, 1986; Langridge-Smith et al., 1984). The apical membrane fraction was solubilized with octylglucoside, and the solubilized protein reconstituted into vesicles. Reconstitution of apical membrane vesicles into planar lipid bilayers revealed the presence of three Cl^- channels that were inhibited by stilbene disulfonate molecules, such as DIDS.

Using a similar approach, Ran and Benos (1991) extracted membrane proteins from the apical membrane of bovine trachea with Triton X-100. The solubilized protein was subjected to ion-exchange chromatography and separation on SDS gels. Channel activity was assessed by the measurement of conductive DIDS-sensitive transport into liposomes. An enriched protein fraction with molecular mass of 38 kDa was discovered. When this protein was electroeluted from a gel and run a second time over another gel, an additional minor doublet of 62 to 64 kDa was observed (as well as the original 38-kDa protein). Antibodies were prepared against the 38-kDa protein, and were used for immunoaffinity purification. When protein that specifically bound to the immobilized antibody column was run under nonreducing conditions, a major band of 140 kDa was eluted. Upon subsequent sulfhydryl group reduction, bands at 62 kDa, 64 kDa, and finally 38 kDa were revealed with increasing amounts of DTT added. Furthermore, immunoaffinity-purified protein reconstituted into planar lipid bilayers revealed a major 40-pS anion-selective channel. These authors suggest that this native channel is composed of four polypeptides of 38–36 kDa in size yielding a functional channel of ~140 kDa. This channel may correspond to the so-called ICOR (intermediate-conductance, outwardly rectifying) chloride channel of Greger (Tilman et al., 1991).

Sodium cholate-solubilized protein isolated from membrane vesicles of the thick ascending loop of Henle in the pig kidney has yielded similar results (Breuer, 1990). In this study, membrane vesicles enriched for Cl^- transport were labeled by incubation with ^{3}H-DIDS. Two bands of molecular mass 65 and 31 kDa were labeled. Preston and co-workers (1992) have reported that three protein bands of 40, 67, and 200 kDa are associated with Cl^- channel activity as assessed by bilayer studies in bovine trachea, renal cortex, intestine, and pancreas. However, in these studies it is likely that this channel arises from intracellular sources, as the vesicles are insensitive to ouabain and do not possess alkaline phosphatase activity.

In another study, another group of Cl^- channel blockers, namely, the indanyloxy-acetic acids (IAA compounds), have been used by Landry and co-workers (1989) to isolate four membrane-associated proteins from bovine trachea and cortex microsomes. These four proteins were of molecular mass 97, 64, 40, and 27 kDa. When all of these proteins were precipitated and reconstituted into planar bilayers, three different Cl^- channels were observed.

Another approach to isolate Cl^- channels is to raise antibodies against short synthetic peptides whose amino acid sequences correspond to a defined region of the protein (Table 1, approach 2). Obviously, this technique is applicable only if the amino acid sequence of the protein is already known. However, this technique can be used if there is good reason to suppose that an unknown protein shares a common region of homology with another protein whose primary structure is known. This strategy has been adopted by Sorscher et al. (1992), who have raised a polyclonal antibody against a synthetic peptide corresponding to a region in the murine erythrocyte band 3, and thought to be involved in DIDS binding. This antibody recognized an approximately 60- to 64-kDa protein in membrane preparations of a model Cl^- secretory epithelia, the human colonic T-84 cell line. Interestingly, immunocytochemical studies reveal that the antibody labeled small vesicles inside the cells that were transported to the membrane surface following Cl^- secretion activation by forskolin. The time course of this translocation process was identical to that observed for the forskolin-induced stimulation of Cl^- transport in these cells. These results suggest that cAMP-induced increases in Cl^- activity may be due to the recruitment of new channels from the cell interior.

Finn and collaborators (1989) have identified two candidate Cl$^-$ channel proteins in the gall bladder epithelium of the mud puppy *Necturus*. Monoclonal antibodies were raised against whole-cell homogenates and then screened for their ability to inhibit Cl$^-$ conductance as assessed from microelectrode measurements. Five of these antibodies were found to satisfy this requirement. In Western blots, one antibody recognized two bands at 219 and 69 kDa. It is interesting to point out that several of the aforementioned studies have identified proteins in the 60- to 70-kDa range as being potential Cl$^-$ channel proteins. As a first step, it will be important to compare the antibody-recognition patterns of all of these isolated proteins to see whether any of these putative Cl$^-$ channels are identical. Ultimately, of course, a direct comparison of primary amino acid sequences will establish homology.

It is important to point out that a voltage-gated anion channel from Torpedo electric organ has been recently cloned by Jentsch and colleagues (1990) using an expression cloning system. The cDNA codes for a protein of 805 amino acids with a predicted molecular mass of 89 kDa. Whether this channel will prove to be unique or to be related to a family of anion channels including those of epithelial cells remains to be elucidated.

III. Amiloride-sensitive Na$^+$ Transport and Its Regulation in Mammalian Alveolar Epithelia

The most important function of the mammalian lung is gas exchange. Oxygen and carbon dioxide move from the alveolar gas into the mixed venous blood and vice versa by diffusion, because of the existing partial pressure differences. To this end, the morphometric characteristics of the blood-gas barrier of the adult lung are optimized for gas exchange in that there is a large surface area and a thin (0.5-μm) alveolar capillary membrane (Weibel, 1963). The alveolar space is free from fluid, except for a very thin layer that forms a subphase between the surfactant layer and the alveolar epithelium (Clements, 1977). This last property is vital, because the presence of fluid will increase the mean diffusion distance and result in compromised gas exchange.

The mechanisms that account for the maintenance of this "dry" alveolar state have been the subject of intense study. Until recently, it was thought that the reabsorption of the fluid that had entered the alveolar space was dictated entirely by passive oncotic and hydrostatic forces. Thus, the absence of intraalveolar fluid was attributed to a combination of the low permeability of the alveolar epithelium to plasma proteins (Matalon and Egan, 1981; Taylor et al., 1965) and the existence of a subatmospheric hydrostatic interstitial pressure (Meyer et al., 1968). Both of these factors favor the reabsorption of fluid from the alveolus into the interstitial space. In addition, the presence of pulmonary surfactant decreases the surface tension of the blood-gas interface, which tends to draw fluid into the alveolar space (Clements, 1961).

Recent evidence, however, indicates that the adult alveolar epithelium is capable of actively transporting solute, with water following passively. Evidence for the existence of active sodium transport both in vivo and ex vivo was provided by several groups who demonstrated the reabsorption of isotonic fluid from the alveolar space into the pulmonary circulation against an oncotic pressure gradient, and the inhibition of this process by amiloride or ouabain (Basset et al., 1987; Berthiaume et al., 1987; Effros et al., 1987; Goodman et al., 1987; Mathay et al., 1982). This active transepithelial Na^+ reabsorption process may play an important role in the maintenance of alveolar fluid balance, especially in pathological situations in which passive forces have been compromised. Active chloride secretion, a prominent feature of the fetal lung (Olver and Strang, 1974) and of the adult canine mammalian trachea (Al-Bazzaz and Al-Awqati, 1979), does not occur across the adult alveolar epithelium (Cott et al., 1986; Goodman et al., 1983).

Because of the complex nature of the mammalian lung and limitations in experimental design, the above cited in-vivo and ex-vivo studies, although clearly demonstrating the existence of active solute transport across the alveolar epithelium, failed to provide information as to the nature and regulation of specific transport pathways. Presently, these questions have been partially answered by performing bioelectric, immunofluorescence, tracer uptake, and patch-clamp measurements in freshly isolated or cultured alveolar type II (ATII) pneumocytes, the most abundant mammalian alveolar epithelial cell.

A. Evidence for Active Solute Transport Across ATII cells

Dome Formation

The first indirect evidence that ATII cells are capable of active solute transport was provided by the observation that when these cells are cultured to confluence on nonporous surfaces, they form fluid-filled domes. These structures are thought to result because of active solute transport from the medium above the cell monolayer to the substratum, with water following passively (Goodman and Crandall, 1982; Lever, 1979). Demonstration of dome formation by ATII cells was provided by Goodman and co-workers (Goodman and Crandall, 1982; Goodman et al., 1983) and by Mason et al. (1982). Goodman and Crandall (1982) demonstrated that addition of ouabain or amiloride to the culture medium (1 mM final concentration) resulted in disappearance of these domes. The effects of amiloride and ouabain were reversible, indicating that they were not due to irreparable injury to cell function. These authors concluded that dome formation in their monolayers was due to the net active transport of Na^+ from the cell medium to the substratum. Additional evidence for this hypothesis was provided by the fact that substitution of extracellular Na^+

with choline led to a virtual disappearance of domes 6 h later. On the other hand, substitution of extracellular Cl^- with gluconate, or addition of Cl^- transport inhibitors (furosemide: 0.5–1 mM; DIDS: 0.5 mM; MK-196: 0.2–0.5 mM) had no effect on dome formation. It should be noted that at the concentration used (1 mM), amiloride may have a number of nonspecific effects in addition to blocking Na^+ entry through apical pathways (Benos, 1988).

Mason et al. (1982) reported that both the density and the size of these domes increased significantly when ATII cells were cultured on an extracellular matrix formed by corneal endothelial cells. These findings suggested that the culture substratum may affect the bioelectric properties of the ATII cells. In contrast to the findings of Goodman and co-workers (Goodman and Crandall, 1982; Goodman et al., 1984) discussed above, addition of 1 mM ouabain or 0.1 mM amiloride had no immediate effect on dome formation. Long-term incubation with these agents (24 h) resulted in irreversible inhibition of dome formation and cell destruction. Another major difference between these two studies was the variable response of these domes to agents that enhanced intracellular cAMP levels (see below).

The reasons for these experimental discrepancies are not entirely clear. Dome formation is a multifocal phenomenon, and the observed changes in dome size may be the result of differences in cell-substratum adhesiveness (Rabito et al., 1980). In a later study from Mason's group, Sugahara et al. (1984) documented the appearance of amiloride- and ouabain-blockable Na^+ currents flowing out of the ATII cell domes. These authors concluded that domes are not specialized areas of active sodium transport but rather areas of weak attachment of the monolayer to the substratum. In spite of these shortcomings, these studies provided the first qualitative evidence for the existence of active Na^+ transport across ATII monolayers cultured on impermeable supports. However, the aforementioned limitations of this model, and the lack of access to the basolateral membrane, prevented the investigators from identifying the precise mechanisms responsible for active Na^+ transport across alveolar ATII cells.

Measurements of Short-Circuit Current and Ion Fluxes Across ATII Cell Monolayers

More detailed information on the nature of ion transport across ATII cells was derived by culturing these cells on porous supports, mounting them in Ussing chambers, and measuring short-circuit current (Isc) and ion fluxes under voltage-clamp conditions. ATII cells, isolated as previously described, were cultured either on Millipore filters previously coated with rat tail collagen and fixed with glutaraldehyde (Mason et al., 1982), human amniotic basement membrane (Cott, 1989), or tissue culture-treated Nuclepore filters (Cheek et al., 1989). Confluent monolayers were formed 3 to 4 days after seeding. The filters were mounted as flat sheets in Ussing chambers, bathed on both sides with identical physiological solutions maintained at 37°C and pH 7.4, and bubbled continuously with 95% air/5% CO_2. Short-circuit current (Isc, a measure of net Na^+ reabsorption) was measured by clamping the transepithelial potential difference (V_t) across the monolayers to zero and measuring the applied current. Isc decreased between 1 and 6 days in culture, reaching a steady state between 6 and 8 days (Cott, 1989). It should be noted, however, that during this same time period, ATII cells undergo extensive phenotypic changes (Diglio and Kikkawa, 1977; Dobbs et al., 1985; Liley et al., 1988; Panus et al., 1989), including decreased synthesis of pulmonary surfactant. Thus, it is unclear to what extent their properties reflect those of ATII cells in vivo.

After 6 days in culture, control Isc values ranged from 2 to 9 $\mu A/cm^2$, V_t from 0.2 to 10 mV (apical side negative), and R_t from 200 to more than $2000\,\Omega \cdot cm^2$. It is interesting to note that the spontaneous V_t values across the ATII monolayers are similar to those observed in vivo between the alveolar lumen and the pleural surface by micropuncture techniques [3.5–10 mV; alveolar lumen negative (Nielson, 1986)]. The higher R_t values obtained by Cheek et al. (1989) were attributed to the use of tissue-treated Nuclepore filters; plating ATII cells under identical conditions on other supports resulted in much lower R_t values. These findings and those of Cott (1989) indicate that the nature of the substratum on which ATII cells are cultured, the culture time, and the initial seeding density affect their bioelectric properties. Thus, one may surmise that alterations in the nature of the lung interstitial matrix by a number of agents may also alter the transport properties of the alveolar epithelium.

Further insight as to the nature of the ion flux responsible for the Isc was obtained by ion-substitution experiments, direct measurements of radiolabeled sodium fluxes, and by noting the effects of specific inhibitors of ion transport to ATII cell monolayer bioelectric properties. The value of Isc corresponded closely to the net apical to basolateral Na^+ flux as measured by the movement of radiolabeled Na^+ across the monolayer (Cheek et al., 1989). Substitution of sodium by equimolar concentrations of choline in the apical side of the monolayer quickly decreased Isc to zero. On the other hand, substitution of gluconate for Cl^- did not alter the value of Isc. Addition of $10\,\mu M$ amiloride to the apical membrane decreased the value of Isc to 20% of each control value within 2 to 3 min. It should be noted that the time course of this inhibition was considerably slower than what has been observed in other epithelial tissues (Benos, 1988). Furosemide (0.1 mM), which is known to inhibit Cl^- transport by its action on the membrane-bound Na^+-$2Cl^-$-K^+ co-transporter, had no effect on the value of Isc. Lastly, 1 mM ouabain added to the basolateral, but not the apical side, decreased Isc to zero (Cheek et al., 1989).

B. Pathways for Na^+ Entry into ATII Cells

Recent studies indicate that Na^+ may enter ATII cells through a variety of different pathways, including a Na^+/H^+ antiporter (Nord et al., 1987), a

Na^+-D-glucose symport (Kerr et al., 1981, 1982), Na^+-amino acid cotransporters (Brown et al., 1985; Lin and Schanker, 1981), and by electrodiffusion through cation channels. Below we summarize evidence for the existence of conductive pathways in ATII cells capable of transporting Na^+.

Measurements of Na^+ Influx Across ATII Cell Membrane Vesicles and Its Inhibition by Amiloride and Its Structural Analogs

Experiments as to the nature of the Na^+ pathways were performed in plasma membrane vesicles of freshly isolated rabbit ATII cells (Matalon et al., 1991). $^{22}Na^+$ uptake into these vesicles was measured in the presence of a negative inside membrane potential, produced by the addition of the K^+ ionophore valinomycin ($10\,\mu M$) after all external K^+ was removed. Valinomycin-sensitive electrogenic Na^+ uptake (V_{Na}) was defined as the difference in uptake in the presence and absence of valinomycin. V_{Na}, normalized per milligram of protein, was twice as high across ATII cells as across alveolar macrophage membrane vesicles, was inhibited by amiloride in a dose-response fashion, and was decreased in the presence of an outwardly directed proton gradient, suggesting that it was not mediated by a Na^+/H^+ antiporter. Furthermore, in the range of 0.1 to $10\,\mu M$, V_{Na} was equally inhibited by amiloride, benzamil and EIPA but became more sensitive to EIPA at higher concentrations of this amiloride analog. The electrogenic nature of this transporter, the lack of glucose or amino acids in the medium, and the fact that it was equally inhibited by amiloride and benzamil argue against this transport pathway being a glucose or amino acid co-transporter or Na^+/Ca^{2+} exchanger. Instead, the pharmacological profile of this transporter is similar to the aforementioned low-affinity amiloride channel that has been identified in a variety of epithelial tissues.

Immunofluorescence Studies in ATII Cell

The spatial distribution of binding of a polyclonal antibody raised against Na^+ channel protein purified from beef kidney papilla [NaAb (Sorscher et al., 1988)] to the plasma membranes of freshly isolated rat ATII cells was quantified using computer-assisted, image-intensified video microscopy (Matalon et al., 1992). Rat ATII cells, isolated by elastase digestion of lung tissue and purified by density-gradient centrifugation, were cytospun on glass slides, fixed and permeabilized, and incubated with NaAb, followed by a secondary antibody (goat antirabbit IgG conjugated to FITC). ATII cells exhibited specific polarized staining, with NaAb localized to the level of the plasma membrane. The intensity of binding diminished threefold when nonspecific rabbit IgG was substituted for the NaAb. In contrast, alveolar macrophages (AM), an abundant nonepithelial-origin alveolar cell, exhibited equal amounts of staining with either IgG or NaAb. In a number of experiments, ATII cells were labeled both with the aforementioned antibodies and the apical surface marker *Maclura pomifera* agglutinin (Weller and Karnovsky, 1989). The colocalization of the two fluoroprobes in the plasma membrane region of ATII cells suggest that the NaAb binds to apical regions of ATII cell membranes. Tousson et al. (1989) has previously shown that NaAb binds exclusively to the apical membranes of Na^+-transporting epithelial cells, known to express Na^+ channels with high affinity to amiloride.

Inhibition of ATII Cell Na^+ Transport by Amiloride and Its Analogs

The extent of inhibition of Na^+ transport across ATII cells by amiloride remains in dispute. Basset et al. (1987) found that instillation of 100 μM amiloride in isolated perfused adult lungs decreased net Na^+ flux across the alveolar epithelium from 3.3 to 2.1 pmol/(cm^2 · s). Clemens et al. (1991) reported that 500 μM amiloride, added to the apical side of an Ussing chamber containing ATII cell monolayers, caused a 61% decrease in Isc. The amiloride K_i (concentration of amiloride needed to block 50% of this maximum decrease) was 1.14 μM. Similar results were obtained by Matalon et al. (1991) in ATII cell membrane vesicles. On the other hand, Cheek et al. (1989) reported that addition of 10 μM amiloride to the apical side of an Ussing chamber containing a monolayer of ATII cells, cultured under identical conditions as those of Clemens et al. (1991), diminished Isc by more than 80%. Bubien et al. (1990) and Matalon et al. (1992), have identified the existence of outwardly rectified Na^+ currents in freshly isolated rabbit ATII cells when patch-clamped in the whole-cell mode. These currents were blocked by 10 μM amiloride at a very slow rate (8% per min). Culture-induced phenotypic changes to these cells may account for differences between freshly isolated and cultured cells. Alternatively, freshly isolated ATII cells may be injured by the isolation procedure. Recently, Hu et al. (1991) reported differences in the pattern of binding of the NaAb between freshly isolated and cultured ATII pneumocytes: After 48 h in culture, mainly intracellular binding was observed.

Critical evaluation of these data reveals many unanswered questions. For example, the selectivity of this conductance to various cations has not been determined. It is possible that this conductance is either a potassium (K^+) or a cation channel. The presence of two different populations of voltage-activated K^+ channels (low- and high-threshold types) in freshly isolated and cultured rat ATII cells has been identified from whole-cell patch-clamp recordings (DeCoursey, 1990; Peers et al., 1990). Low threshold currents were activated when the potential difference across the cell was between -40 and -20 mV and their amplitude was reduced by 20 mM tetraethylammonium (TEA) and 2 mM 4-aminopyridine. The second population of currents (high-threshold type) were activated when the cell potential was greater than -20 mV and were more sensitive to TEA. In addition, Bland and Boyd (1986) have identified the presence of an ouabain- and furosemide-insensitive component of $^{86}Rb^+$

uptake in freshly isolated adult ATII cells and concluded that these cells contain cation channels. Presently, single-channel measurements, have been performed only in frog lungs but not in ATII cells. However, recent findings from our laboratories (Matalon, 1992) indicate that freshly isolated ATII cells, patch-clamped in the whole-cell mode, have conductive pathways that are three times more selective for Na^+ than K^+ ($P_{Na+}/P_{K+} = 2.8 \pm 0.3$ mean $\pm$ SEM; $n = 5$). Furthermore, whole-cell Na^+ currents were not inhibited by 10 mM tetraethylammonium, a well known inhibitor of K^+ channels (Yue et al., 1992). Thus, it is unlikely that Na^+ ions crossed the cell membranes through K^+ channels. Rather these results indicate the presence of multiple types of conductive pathways in the plasma membranes of ATII cells.

In the amphibian lung, Fisher et al. (1989), using the technique of noise analysis, reported a single-channel sodium current around 0.3 pA with an amiloride-inhibitory constant of 1.23 μM. It should be noted that there are important differences between the frog and the mammalian alveolar epithelium. For example, the frog lung secretes chloride, while the alveolar mammalian epithelium reabsorbs Na^+. Furthermore, cAMP does not increase Na^+ current in the frog lung, while it does in the mammalian lung. Additional patch-clamp studies in both freshly and cultured ATII cells, at both the whole-cell and single-channel levels, are needed to characterize better the biophysical properties of this channel.

C. Hormonal Regulation of Na^+ Transport in Pneumocytes

cAMP and Aldosterone

The studies of Goodman et al. (1984) and Cott et al. (1986) have shown that substances that increase intracellular cAMP, either via the adenylate cyclase system (β-adrenergic agonists, cholera toxin), or directly (8-bromo-cAMP), also increase dome formation and Isc across ATII cell monolayers. Similar results were obtained with phosphodiesterase inhibitors, compounds which prevent the hydrolysis of cAMP. A threefold increase in Isc was observed within 30 min of the addition of 1 mM of 8-bromo-cAMP in the Ussing chamber containing ATII cell monolayers. It is interesting to note that the increase in Isc was transient, its value returning toward the baseline after this time. A similar behavior is seen for amphibian epithelial Isc and may be due to the interplay of different factors. Alpha-agonists decreased dome formation, perhaps by decreasing intracellular cAMP; dibutyryl guanosine 3′-5′-cyclic monophosphate (dbcGMP) did not affect Na^+ transport. An increase in intracellular cGMP also failed to alter Na^+ transport across the toad urinary bladder (Sahib et al., 1978). Surprisingly, antidiuretic hormone (ADH) significantly increased dome formation (Goodman et al., 1984), but had no effect on Isc across ATII cell monolayers (Cott et al., 1986). The reasons for this discrepancy are not clear.

These findings are at variance with the observations of Mason et al. (1982), who reported that cholera toxin or terbutaline failed to increase dome formation in ATII cell monolayers. However, careful examination of their data reveals a high degree of variability. In one experiment, for example, 10 μM terbutaline increased dome formation by over 165%. Furthermore, 3 mM of sodium butyrate, an agent known to increase intracellular cAMP, significantly enhanced dome formation. As mentioned above, dome formation is a complex process and may depend on a number of factors in addition to active ion transport. It has been suggested that dbcAMP may increase or decrease the attachment of epithelial cells to the substratum, which may affect dome formation (Johnson and Pastan, 1972; Rabito et al., 1980). In another study, discussed above (Cott et al., 1986), investigators from the same laboratory reconfirmed that an increase in cAMP augments Isc across ATII cells.

Presently, the effects of aldosterone on transcellular Na^+ transport across ATII cells have not been adequately investigated. In the only study on this subject, Cott et al. (1986) reported that overnight incubation of ATII cell monolayers with aldosterone concentrations ranging between 10^{-7} and 10^{-4}M did not increase Isc. More work needs to be done to explore potential biochemical differences between Na^+ channels in ATII cells and those in other epithelia with respect to aldosterone.

Glucocorticoids

ATII cells are known to have glucocorticoid receptors (Ballard, 1986), and dexamethasone has been shown to accelerate both fetal and postnatal surfactant maturation (Massaro et al., 1985; Mendelson et al., 1986). The results of Young and Silbajoris (1986), Floros et al. (1989), and Eik-Nes et al. (1987) have shown that intravenous dexamethasone increases the amount of lipid stored in lamellar bodies, the synthesis of SPA (a specific surfactant apoprotein) and the secretion of surfactant by ATII cells. Dexamethasone is also known to induce high-amiloride affinity Na^+ channels in rat colon (Bridges et al., 1988), a tissue of the same embryonic origin as the lung epithelium. Presently, the effects of this class of hormones on sodium transport across ATII cells have not been investigated.

IV. Amiloride-Binding Proteins in ATII Cells

Polyclonal antibodies against the purified Na^+ channel protein from bovine kidney have been used on Western blots of apical membrane vesicular proteins obtained from rat alveolar type II cells (Oh et al., 1991). These antibodies have also been used to immunolocalize Na^+ channels in rat lung alveolar type II cells (Matalon, 1992; Hu et al., 1991). On Western blots, these antibodies cross-reacted with a single 135-kDa peptide from the alveolar type II cell membrane proteins. Antiidotypic antibodies generated against antiamiloride antibodies and photolabeling studies using the photolabile amiloride analog NMBA consistently recognize a protein of the same molecular mass (i.e., 135 kDa), suggesting that this polypeptide could be an amiloride-binding subunit of a low-amiloride-affinity type. Because these antibodies react with at least four different subunits of the Na^+ channel protein obtained from bovine kidney and two polypeptides from the Na^+ channel protein isolated from A6 cells, it is unlikely that there exists a single type of Na^+ channel with two different amiloride-binding sites. These results are consistent with the hypothesis that there exist different amiloride-sensitive Na^+ channels with distinct biochemical structures. This proposal would also be consistent with the experimental observations of variable kinetic and pharmacological characteristics of amiloride-sensitive Na^+ channels. But because the size of the actual amiloride-binding protein is very close between the different channel types, it is not yet possible to exclude the possibility that a substantial homology between these amiloride-binding proteins exists within these different channel types.

V. Summary and Perspectus

The examination of Na^+ transport properties of mammalian ATII cells is a field of study just beginning. With the recent success in identifying the proteins mediating Na^+ and Cl^- conductive transport in other epithelia and with the eventual molecular cloning of their respective genes, many new and powerful probes are becoming available for application to the mammalian lung. Questions such as the involvement of Na^+ and Cl^- channels during the fetal-adult transition, the influence of hormones on the expression and regulation of these channels at different stages of maturation, and the influence of various pathophysiological conditions on these transport systems can be addressed at the molecular level. It is apparent that much exciting and important new information concerning the physiology, biochemistry, and molecular biology of mammalian lung function will be obtained in the coming years.

Acknowledgments

This work was support by grants from the National Institute of Health (HL31197 and DK37206), a Career Investigator Award from the American Lung Association to SM, and a grant from the Office of Naval Research (SM and DJB). We thank Cathy Guy for typing the manuscript.

Because this project is sponsored in part by the Department of the Navy, Office of the Chief of Naval Research, the content of the information does not necessarily reflect the position or the policy of the government, and no offered endorsement should be inferred.

References

Al-Bazzaz FJ, Al-Awqati Q. Interaction between sodium and chloride transport in canine tracheal mucosa. J Appl Physiol 1979; 46:111–9.

Ausiello DA, Stow JL, Cantiello HF, Benos DJ. Purified epithelial Na^+ channel complex contains the pertussis toxin-sensitive $G\alpha_{i\text{-}3}$ protein. J Biol Chem 1992; in press.

Ballard PL. Hormones and lung maturation. In: Monographs on endocrinology, Vol. 28. New York: Springer-Verlag, 1986; 1–354.

Barbry P, Chassande O, Vigne P, Frelin C, Ellory C, Cragoe E Jr, Lazdunski M. Purification and subunit structure of [^{3}H]-phenamil receptor associated with the renal apical Na^+ channel. Proc Natl Acad Sci USA 1987; 84:4836–40.

Barbry P, Frelin C, Vigne P, Cragoe EJ Jr, Lazdunski M. [^{3}H]-phenamil, a radiolabeled diuretic for the analysis of the amiloride-sensitive Na^+ channels in kidney membranes. Biochem Biophys Res Commun 1986; 135:25–32.

Basset G, Crone C, Saumon G. Fluid absorption by rat lung in situ. Pathways for sodium entry in the luminal membrane of alveolar epithelium. J Physiol 1987; 384:325–45.

Benos DJ. Amiloride: chemistry, kinetics, and structure-activity relationships. In: Grinstein S, ed. Na^+/H^+ exchange. Boca Raton, Florida: CRC Uniscience, 1988:121–36.

Benos DJ, Cunningham S, Baker RR, Beason KB, Oh Y, Smith PR. Molecular characteristics of amiloride-sensitive sodium channels. Rev Physiol Biochem Pharmacol 1992; Vol. 120 Springer-Verlag: 1–113.

Benos DJ, Saccomani G, Sariban-Sohraby S. The epithelial sodium channel. Subunit number and location of the amiloride binding site. J Biol Chem 1987; 262:10613–8.

Benos DJ, Sorscher EJ. Transport proteins: ion channels. In: Seldin DW and Giebisch G, eds. The kidney: physiology and pathophysiology, 2d ed. New York: Raven Press, 1992:587–623.

Berthiaume Y, Staub NC, Matthay MA. Beta-adrenergic agonists increase lung liquid clearance in anesthetized sheep. J Clin Invest 1987; 79:335–43.

Bland RD, Boyd CAR. Cation transport in lung epithelial cells derived from fetal, newborn, and adult rabbits. J Appl Physiol 1986; 61(2):507–15.

Bridges RJ, Cragoe EJ Jr, Frizzell RA, Benos DJ. Inhibition of colonic Na^+ transport by amiloride analogues. Am J Physiol 1989; 256:C67–74.

Bridges RJ, Garty H, Benos DJ, Rummel W. Na^+ uptake into colonic enterocyte membrane vesicles. Am J Physiol 1988; 254:C484–90.

Breuer W. Reconstitution of a kidney chloride channel and its identification by covalent labelling. Biochim Biophys Acta 1990; 1022:229–36.

Brown SES, Kim KJ, Goodman BE, Wells JR, Crandall ED. Sodium-amino acid cotransport by type II alveolar epithelial cells. J Appl Physiol 1985; 59(5):1616–22.

Bubien JD, Benos DJ, Matalon S. Identification of amiloride-inhibitable sodium currents across freshly-isolated alveolar type II pneumocytes. Am Rev Respir Dis 1990; 141:A842 (abstr).

Cabantchik ZI, Greger R. Chemical probes for anion transporters of mammalian cell membranes. Am J Physiol: Cell Physiol 1992; in press.

Cantiello HF, Stow JL, Prat AG, Ausiello DA. Actin filaments regulate epithelial Na$^+$ channel activity. Am J Physiol: Cell Physiol 1991; 261:C882–8.

Cheek JM, Kim K-J, Crandall ED. Tight monolayers of rat alveolar epithelial cells: bioelectric properties and active sodium transport. Am J Physiol 1989; 256(25):C688–93.

Clemens JW, Weaver EJ, Goodman BE. Dose-response relationship for inhibition of short-circuit current by amiloride in cultured alveolar epithelium. Am Rev Respir Dis 1991; 143:A208 (abstr).

Clements JA. Pulmonary edema and permeability of alveolar membranes. Arch Environ Health 1961; 2:280–3.

Clements JA. Functions of the alveolar lining. Am Rev Respir Dis 1977; 115:67–71.

Cott GR. Modulation of bioelectric properties across alveolar type II cells by substratum. Am J Physiol 1989; 257(26):C678–88.

Cott GR, Sugahara K, Mason RJ. Stimulation of net active ion transport across alveolar type II cell monolayers. Am J Physiol 1986; 250(19):C222–7.

Cox M. Relationship of the aldosterone-induced protein GP-70 to the renal epithelial conductive Na$^+$ channel. In: Bonvalet JP, Farman N, Lombes M, Rafestin-Oblin ME, eds. Aldosterone: fundamental aspects. Paris, France: John Libbey Eurotext, 1991; 215:249–57.

Cunningham SA, Worrell RT, Benos DJ, Frizzell RA. camp-stimulated ion currents in Xenopus oocytes expressing CFTR cRNA. Am J Physiol (Cell Physiol) 1992; 262:C783–8.

DeCoursey TE. State-dependent inactivation of K$^+$ currents in rat type II alveolar epithelial cells. J Gen Physiol 1990; 95:617–46.

Diglio CA, Kikkawa Y. The type II epithelial cells of the lung. IV. Adaptation and behavior of isolated type II cells in culture. Lab Invest 1977; 37:622–31.

Dobbs LG, Williams MC, Brandt AE. Changes in biochemical characteristics and pattern of lectin binding of alveolar type II cells with time in culture. Biochim Biophys Acta 1985; 846:155–66.

Dubinsky WP, Monti LB, Solubilization and reconstitution of a chloride transporter from tracheal apical membrane. Am J Physiol 1986; 251:C713–20.

Effros RM, Mason GR, Sietsema K, Silverman P, Hukkanen J. Fluid reabsorption and glucose consumption in edematous rat lungs. Circ Res 1987; 60:708–19.

Eik-Nes KB, Marthinsen ABL, Lofthus R, Haug E. Coricosteroids and lung surfactant levels in adult male rates. J Steroid Biochem 1987; 27:1035–41.

Finn AL, Tsai L-M, Falk RJ. Monoclonal antibodies to the apical chloride channel in *Necturus* gallbladder inhibit the chloride conductance. Proc Natl Acad Sci USA 1989; 86:7649–52.

Fischer H, Van Driessche W, Clauss W. Evidence for apical sodium channels in frog lung epithelial cells. Am J Physiol 1989; 256:C769–71.

Floros J, Phelps DS, Harding HP, Church S, Ware J. Postnatal stimlation of rat surfactant protein A synthesis by dexamethasone. Am J Physiol 1989; 257(1):L137–43.

Frings S, Purves RD, Macknight ADC. Single channel recordings from the apical membrane of the toad urinary bladder epithelial cell. J Membr Biol 1988; 106:157–72.

Garty H, Benos DJ. Characterization and regulatory mechanisms of the amiloride-blockable Na^+ channel. Physiol Rev 1988; 68:309–73.

Goodman BE, Crandall ED. Dome formation in primary cultured monolayers of alveolar epithelial cells. Am J Physiol 1982; 243(12):C96–100.

Goodman BE, Brown SES, Crandall ED. Regulation of transport across pulmonary alveolar epithelial cell monolayers. J Appl Physiol 1984; 57(3):703–10.

Goodman BE, Fleischer RS, Crandall ED. Evidence for active Na^+ transport by cultured monolayers of pulmonary alveolar epithelial cells. Am J Physiol 1983; 245(14):C78–83.

Goodman BE, Kim KJ, Crandall ED. Evidence for active sodium transport across alveolar epithelium of isolated rat lung. J Appl Physiol 1987; 62:2460–6.

Hu P, Oh Y, Jilling T, Benos DJ, Matalon S. Immunofluorescent localization of sodium conductance in cultured rat alveolar type II pneumocytes (ATII). Am Rev Respir Dis 1991; 143:A208 (abstr).

Jentsch TJ, Steinmeyer K, Schwartz G. Primary structure of a *Torpedo marmorata* chloride channel isolated by expression cloning in *Xenopus* oocytes. Nature 1990; 348:510–4.

Johnson GS, Pastan DI. cAMP increases adhesion of fibroblasts to substratum. Nature Lond New Biol 1972; 236:247–9.

Kerr JS, Fisher AB, Kleinzeller A. Transport of glucose analogues in rat lung. Am J Physiol (Endocrinol Metabol 4) 1981; 241:E191–5.

Kerr JS, Reicherter J, Fisher AB. 2-Deoxy-D-glucose uptake by rat granular pneumocytes in primary culture. Am J Physiol (Cell Physiol 12) 1982; 243:C14–9.

Kleyman TR, Cragoe EJ, Kraehenbuhl J-P. The cellular pool of Na^+ channels in the amphibian cell line A6 is not altered by mineralocorticoids. J Biol Chem 1989; 264:11995–2000.

Kleyman TR, Kraehenbuhl J-P, Ernst SA. Characterization and cellular localization of the epithelial Na^+ channel. Studies using an anti-Na^+ channel antibody raised by an anti-idiotypic route. J Biol Chem 1991; 266:3907–15.

Kleyman TR, Ragagopalan R, Cragoe EJ Jr, Erlanger BF, Al-Awaqati Q. New amiloride analogue as hapten to raise anti-amiloride antibodies. Am J Physiol 1986; 250:C165–70.

Landry DW, Akabas MH, Redhead C, Edelman A, Cragoe EJ, Al-Awqati Q. Purification of reconstitution of chloride channels from kidney and trachea. Science 1989; 244:1469–72.

Langridge-Smith JE, Field M, Dubinsky WP. Cl^- transport in apical plasma membrane vesicles isolated from bovine tracheal epithelium. Biochim Biophys Acta 1984; 777:84–92.

Lever JE. Regulation of dome formation in differential epithelial cell cultures. J Supra Struct 1979; 12:259–72.

Liley HG, Ertsey R, Gonzales LW, Odom MW, Hawgood S, Dobbs LG, Ballard PL.

Synthesis of surfactant components by culture type II cells from human lung. Biochim Biophys Acta 1988; 961:86–95.

Lin YJ, Schanker LS. Pulmonary absorption of amino acids in the rat: evidence of carrier transport. Am J Physiol 1981; 240:C215–21.

Mason RJ, Williams MC, Widdicombe JH, Sanders MJ, Misfeldt BS, Berry LD Jr. Transepithelial transport by pulmonary alveolar type II cells in primary culture. Proc Natl Acad Sci USA 1982; 79:6033–7.

Massaro D, Teich N, Maxwell S, Massaro GD, Whitney P. Postnatal development of alveoli, regulation and evidence for a critical period in rats. J Clin Invest 1985; 76:1297–305.

Matalon S, Egan EA. Effects of 100% O_2 breathing on permeability of alveolar epithelium to solute. J Appl Physiol 1981; 50:859–63.

Matalon S, Bridges RJ, Benos DJ. Amiloride-inhibitable Na^+ conductive pathways in alveolar type II pneumocytes. Am J Physiol 1991; 260(4):L90–6.

Matalon S, Kirk KL, Bubien J, Oh Y, Hu P, Yue G, Shoemaker RL, Cragoe EF, and Benos DJ. Immunocytochemical and patch clamp characterization of Na^+ conductance in alveolar pneumocytes. Am J Physiol (Cell Physiol 31) 1992; 262:C1228–38.

Matthay MA, Landolt CC, Staub NC. Differential liquid and protein clearance from the alveoli of anesthetized sheep. J Appl Physiol 1982; 53:96–104.

Mendelson CR, Chen C, Boggaram V, Zacharias C, Snyder JM. Regulation of the synthesis of the major surfactant apoprotein in fetal rabbit lung tissue. J Biol Chem 1986; 261:9938–43.

Meyer BJ, Meyer A, Guyton AC. Interstitial fluid pressure V: Negative pressure in the lungs. Circ Res 1968; 22:263–71.

Moran A, Asher C, Cragoe EJ Jr, Garty H. Conductive sodium pathway with low affinity to amiloride in LLC-PK_1 cells and other epithelial cells. J Biol Chem 1988; 263:19586–91.

Nielson DW. Electrolyte composition of pulmonary alveolar subphase in anesthetized rabbits. J Appl Physiol 1986; 60(3):972–9.

Nord EP, Brown SES, Crandall ED. Characterization of Na^+-K^+ antiport in type II alveolar epithelial cells. Am J Physiol: Cell Physiol 1987; 252:C490–8.

Oh Y, Matalon S, Kleyman TR, and Benos DJ. Biochemical evidence for the presence of amiloride binding protein in adult alveolar type II pneumocytes. J Biol Chem, 1992; in press.

Olver RE, Strang LB. Ion fluxes across the pulmonary epithelium and the secretion of lung liquid in the fetal lamb. J Physiol 1974; 241:327–57.

Panus PC, Matalon S, Freeman BA. Responses of type II pneumocyte antioxidant enzymes to normoxic and hyperoxic culture. In Vitro Cell & Dev Biol 1989; 25:821–9.

Peers C, Kemp PJ, Boyd CAR, Nye PCG. Whole-cell K^+ currents in type II pneumocytes freshly isolated from rat lung: pharmacological evidence for two subpopulations of cells. Biochim Biophys Acta 1990; 1052:113–8.

Preston CL, Calenzo MA, Dubinsky WP. Isolation of a chloride channel enriched membrane fraction from bovine tracheal epithelia and renal cortex. Am J Physiol:Cell Physiol 1992; in press.

Rabito CA, Tchao R, Valentich J, and Leighton J. Effect of cell-substratum interaction on hemicyst formation by MDCK cells. In Vitro 1980; 16:461–8.

Ran S, Benos DJ. Isolation and functional reconstitution of a 38-kDa chloride channel protein from bovine tracheal membranes. J Biol Chem 1991; 266:4782–8.

Sahib MK, Schwartz JH, Handler JS. Inhibition of toad urinary bladder sodium transport by carbamycholine: possible role of cyclic GMP. Am J Physiol 1978; 235(4):F586–91.

Sariban-Sohraby S, Fisher RS. Single channel activity by the amiloride binding subunit of the epithelial Na^+ channel. Biophys J 1990; 57:87a (abstr).

Sariban-Sohraby S, Sorscher EJ, Brenner BM, Benos DJ. Phosphorylation of a single subunit of the epithelial Na^+ channel following vasopressin treatment of A6 cells. J Biol Chem 1988; 263:13875–9.

Smith PR, Benos DJ. Epithelial Na^+ channels. Ann Rev Physiol 1991; 53:509–30.

Smith PR, Saccomani G, Joe E-H, Angelides KJ, Benos DJ. Amiloride-sensitive sodium channel is linked to the cytoskeleton in renal epithelial cells. Proc Natl Acad Sci USA 1991; 88:6971–5.

Sorscher EJ, Accavitti MA, Keeton D, Steadman E, Frizzell RA, Benos DJ. Antibodies against purified epithelial channel proteins from bovine renal papilla. Am J Physiol 1988; 255:C835–43.

Sorscher EJ, Fuller CM, Bridges RJ, Tousson A, Marchase RB, Brinkley BR, Frizzell RA, Benos DJ. Isolation of a 60 kDa membrane protein from T-84 cells using antibodies made against a synthetic DIDS-binding peptide. Am J Physiol:Cell Physiol 1992; 262:C136–47.

Staub O, Verrey F, Rossier BC, Kraehenbuhl J-P. Gene expression of a kidney apical protein (AP) in *Xenopus laevis* oocyte and during early development. J Cell Biol 1990; 111:310a.

Sugahara K, Caldwell JH, Mason RJ. Electrical currents flow out of domes formed by cultured epithelial cells. J Cell Biol 1984; 99:1541–6.

Takumi T, Ohkubo H, Nakanishi S. Cloning of a membrane protein that induces a slow voltage-gated potassium current. Science 1988; 242:1042–5.

Taylor AE, Guyton AC, Bishop VC. Permeability of the alveolar membrane to solutes. Circ Res 1965; 16:353–62.

Tilman M, Kunzelmann K, Frobe U, Cabantchik I, Lang HJ, Englert HC, Greger R. Different types of blockers of the intermediate-conductance outwardly rectifying chloride channel in epithelia. Pflugers Arch 1991; 418:556–63.

Tousson A, Alley CD, Sorscher EJ, Brinkley BR, Benos DJ. Immunocytochemical localization of amiloride-sensitive sodium channels in sodium transporting epithelia. J Cell Sci 1989; 93:349–62.

Weibel ER. Morphometry of the human lung. New York: Academic Press, 1963.

Weller KN, Karnovsky MJ. Identification of a 185 kd *Maclura pomifera* agglutinin binding glycoprotein as a candidate for a differentiation marker for alveolar type II cells in adult rat lung. Am J Pathol 1989; 134:277–85.

Young SL, Silbajoris R. Dexamethasone increases adult rat lung surfactant lipids. J Appl Physiol 1986; 60:1665–72.

Yue G, Shoemaker RL, Benos DJ, Matalon S. Patch clamp characterization of Na^+ currents (I_{Na}) on cultured rat alveolar type II (ATII) cells. Am Rev Respir Dis 1992; 145(4):A365.

10

Protein Phosphorylation and Dephosphorylation Systems in Alveolar Epithelial Cells

DAVID WARBURTON, NORBERT BERNDT and FREDERICK L. HALL

University of Southern California School of Medicine
and Childrens Hospital Los Angeles
Los Angeles, California

I. Introduction

All cells must communicate with their environment and respond to extracellular signals. It is now clear that the reversible phosphorylation of proteins is a ubiquitous mechanism which regulates many if not most biological processes. The enzymes which catalyze the transfer of the gamma phosphate of ATP to protein substrates are termed protein kinases and comprise an evolutionarily related family of enzymes which has become highly elaborated throughout evolution (Hunter, 1987). A homologous catalytic core amino acid sequence is conserved among protein kinases, while the molecular size, substrate specificity, mechanisims of activation, subunit composition and subcellular localization, and substrate specificity of protein kinases are "customized" to the regulation of specific substrates (Hanks et al., 1988). The steady-state level of protein phosphorylation reflects the relative activities of protein kinases and protein phosphatases, which often are themselves regulated by phosphorylation and dephosphorylation mechanisms. Phosphorylation of serine, threonine, and/or tyrosine residues results in small conformational changes in protein kinase substrate proteins which may cause profound alterations in their biological properties (Cohen, 1989).

Strong presumptive evidence indicates that protein phosphorylation and dephosphorylation reactions play a key role in the regulation of several essential aspects of alveolar epithelial cell function. Surfactant production by the fetal lung is known to be regulated by hormonal and metabolic factors including hyperinsulinemia, hyperglycemia, glucocorticoids, thyroid hormones, and beta-adrenergic agonists (Warburton 1983; Warburton et al., 1988, 1989). The proliferation of alveolar epithelial cells, both in normal development and in response to lung injury, is regulated by polypeptide growth factors and cytokines (King et al., 1989, Warburton, 1991). Thus, not only alveolar cell growth but also differentiated functions, such as alveolar solute and fluid homeostasis and surfactant secretion, are regulated by growth factor- and second messenger-activated pathways that are mediated largely by protein phosphorylation and dephosphorylation. In this chapter we concisely and selectively review the structure and function of protein kinases and protein phosphatases, and relate this information to the regulation of alveolar epithelial cell functions, including the proliferation of alveolar epithelial cells during lung development and/or lung repair and the secretion of pulmonary surfactant by differentiated type II alveolar epithelial cells.

II. Protein Kinase Structure and Function

Protein kinases comprise a large family of enzymes which have been elaborated and diversified during evolution to perform a multitude of highly specialized functions (Krebs, 1985; Hunter, 1987). Historically, the first protein kinase to be characterized in any detail was phosphorylase kinase, which activates glycogen phosphorylase by the phosphorylation of a single serine residue (Krebs et al., 1959). The discovery of phosphorylase kinase, later named cyclic AMP-dependent protein kinase (PKA) (Walsh et al., 1968), can be considered as the take-off point of protein phosphorylation research. Ever since, the list of both protein kinases and phosphoprotein substrates has been growing.

In the absence of cAMP, PKA exists as a holoenzyme complex consisting of two regulatory (R) subunits and two catalytic subunits (C). Activation of cAMP binding to PKA involves the dissociation of an R_2 subunit dimer from two free and enzymatically active C subunits. The crystal structure of the catalytic subunit of PKA has recently been elucidated, providing a template for viewing the conserved catalytic core of all eukaryotic protein kinases (Knighton et al., 1991). The PKA C-subunit molecule is bilobed in shape, with a deep cleft between the two lobes. The core of the small lobe is NH_2-terminal, while the core of the large lobe is COOH-terminal. The cleft is the site of binding both of ATP and of the PKA-inhibitory peptide PKI. The floor of the cleft is the site of catalysis. Two highly conserved loops, as well as a triad of invariant charged residues, appear to be important for nucleotide binding and catalysis, while the peptide substrate recognition sequences are variable. These features of the catalytic core are highly conserved among more than 100 protein kinases, including several growth factor receptors and viral oncoproteins (Knighton et al., 1991).

III. Protein Kinase Substrate Recognition Motifs

Sequence analysis of phosphoproteins suggested early on that the amino acid sequence surrounding the phosphorylation sites of substrates may be essential in enabling kinases to recognize specific substrates. Following the study of several PKA substrates, including studies with synthetic peptides such as IRRASLG (Kemptide), it became clear that the recognition sequence of the PKA C subunit is Arg-Arg-X-Ser[Thr]-Y, where X is any small residue and Y is a large hydrophobic group. Analysis of substrate recognition sequences for numerous protein kinases has revealed several unifying themes of regulation (Kennely and Krebs, 1991; Pearson and Kemp, 1991). In sum, the recognition of protein substrates by a number of protein kinases is determined by the pattern of basic or acidic amino acids surrounding the phosphorylation site. PKA and protein kinase C (PKC) provide a paradigm for site-specific substrate recognition by flanking basic regions, while casein kinase II (CKII) specificity is directed by flanking acidic regions. Multiple or hierarchical protein phosphorylation may either activate or inhibit the activity of individual substrates (Roach, 1991). The first example of hierarchical regulation to be worked out at the level of site-specific phosphorylation was the phosphorylation of glycogen synthase by casein kinase as a prerequisite for subsequent phosphorylation and inactivation of glycogen synthase activity by glycogen synthase kinase 3 (Roach, 1990). The proline-directed protein kinase (PDPK) is another recently described growth factor-sensitive serine/threonine kinase which recognizes XS/TPX as its minimal consensus site specificity (Vuillet et al., 1899, Hall et al., 1990). However, it is becoming increasingly clear that the XS/TPX beta-turn motif is the fundamental determinant for both the cytosolic PDPK and the nuclear histone H1 kinase (Pines and Hunter, 1989; Moreno and Nurse, 1990). This motif is widely found in gene-regulatory proteins adjacent to or excluded from the protein/DNA- or protein/protein-binding regions (Suzuki, 1989). This motif also occurs in the cytoplasmic domain of several growth factor receptors, steroid hormone receptors, and plasma membrane-associated tyrosine kinases. As an example, the progesterone receptor is phosphorylated at three XS/PX sites which affect the rate of receptor-mediated transcription control (Denner et al., 1990). The challenge now before us is to determine which, where, and when individual regulatory elements act in signal transduction pathways, so that the networks of regulation by protein phosphorylation and dephosphorylation can be understood in specific physiological systems.

There is also increasing evidence of crosstalk between tyrosine and serine/threonine kinases. Several serine/threonine kinases are directly activated by phosphorylation on tyrosine residues (Hall and Vuillet, 1991). For example, the microtubule-associated or mitogen-activated protein kinase (MAP kinase) is phosphorylated and activated in mammalian cells stimulated with epidermal growth factor (EGF), platelet-derived growth factor (PDGF), insulinlike growth factor (IGF)-II, or phorbol esters (Rossamondo et al., 1989). MAP kinase is a cytosolic serine/threonine kinase which is activated only when phosphorylated on both tyrosine and threonine residues (Anderson et al., 1990). MAP kinase in turn phosphorylates and activates the "growth-associated" (40S ribosomal protein) S6 kinases (Sturgill et al., 1990). MAP kinase has recently been cloned, and the deduced amino acid sequence, known as extracellular signal-regulated kinase (ERK-1), has high homology to the yeast kinases KSS1 and FUS3 that mediate growth responses to polypeptide mating factors (Boulton et al., 1990). MAP kinase is present in type II alveolar epithelial cells, and may play a role in transducing growth factor signals following the activation of growth factor receptor tyrosine kinases (Warburton and Hall, unpublished results).

Thus, protein kinase families provide complex and interacting networks of regulatory enzymes which modify the properties of substrate proteins by the addition of negative charges in the form of phosphate groups at critical sites. Despite the increasing complexity of possible modes of regulation by protein phosphorylation, biochemical tools are now at hand, including specific antibodies, specific pharmacological probes, analysis of site-specific phosphorylation, and site-directed mutagenesis of specific protein kinases and substrates, for investigating the molecular basis of cell regulation by protein phosphorylation.

IV. Protein Phosphatase Structure and Function

Until very recently, protein phosphatases were considered to be housekeeping enzymes providing the necessary means of removing phosphate groups from proteins that were previously phosphorylated by protein kinases. Within the last two years, however, a different picture has emerged, suggesting that phosphatases may be involved in the control of cellular proliferation. In this section, the discussion is limited to serine/threonine phosphatases (for discussion of tyrosine phosphatases, see Chapter 7).

All of the major protein phosphatase activity in mammalian cells can be attributed to four types termed, PP1, PP2A, PP2B, and PP2C (Cohen, 1989).

The contribution of the latter two is usually minor. Not only are the protein phosphatases themselves related to each other (except 2C), but they are also the most highly conserved enzyme family described to date. Although molecular cloning studies have revealed more phosphatases all showing a relatively high homology to the PP1, 2A, and 2B prototypes (Cohen, 1991), it is clear that the number of protein phosphatases is far outstripped by the kinases. Conversely, the number of substrates handled by a protein phosphate must be far greater than that of a kinase. This is reflected by the fact that, in vitro, the substrate specificity of most phosphatases is broad and even overlapping. On a biochemical level, the best-understood enzymes are PP1 and to a lesser extent PP2A. They consist of relatively small (37 and 36 kDa, respectively) catalytic subunits that are normally complexed to targeting subunits directing the activity toward the correct location and substrates in the cell. These targeting subunits may do for the phosphatases what consensus phosphorylation sites do for the kinases. Genetic studies in yeast (Okhura et al., 1989, Kinoshita et al., 1990) and Drosophila (Axton et al., 1990) have revealed that both PP1 and PP2A control mitosis, although the biochemical mechanism of their action is not yet clear. Two additional recent findings provide further support for the view that protein phosphatases are involved in the regulation of cellular proliferation. First, okadaic acid, a marine toxin isolated from sponges, is not only a powerful tumor promoter (Bialojan and Takai, 1988), but also specifically inhibits PP1 and 2A (Suganuma et al., 1988). The question is whether these two effects (tumor promotion and phosphatase inhibition) are causally related. Second, the products of a recently described gene class, the tumor suppressors, appear to be inactivated by dephosphorylation, such as the Rb tumor-susceptibility gene product (see below). Since protein kinases that phosphorylate tumor-suppressor proteins have been termed oncogenic, it is fair to say that whichever protein phosphatase dephosphorylates these substrates must be antioncogenic.

Increases in protein phosphatase 1 and 2A activities coincide with perinatal surfactant ontogeny in the fetal rat lung (Warburton and Cohen, 1988), suggesting that protein phosphatases may be important in regulating type II alveolar epithelial cell function. Okadaic acid, the marine sponge toxin, which has specific inhibitory activity against protein phosphatases 1 and 2A, produces marked increases in the phosphorylation states of several proteins in type II pneumocytes (Fig. 1) (Warburton et al., 1991). Similar effects of okadaic acid have also been reported in alveolar macrophages, adipocytes, and hepatocytes. Okadiac acid stimulates myosin phosphorylation and shape change in alveolar macrophages, glycogenolysis and gluconeogenesis in hepatocytes, and lipolysis and the conversion of glucose to glyceride-glycerol in adipocytes (Haystead et al., 1989). Shape change has also been described in type II alveolar cells in response to okadaic acid (Warburton et al., 1991).

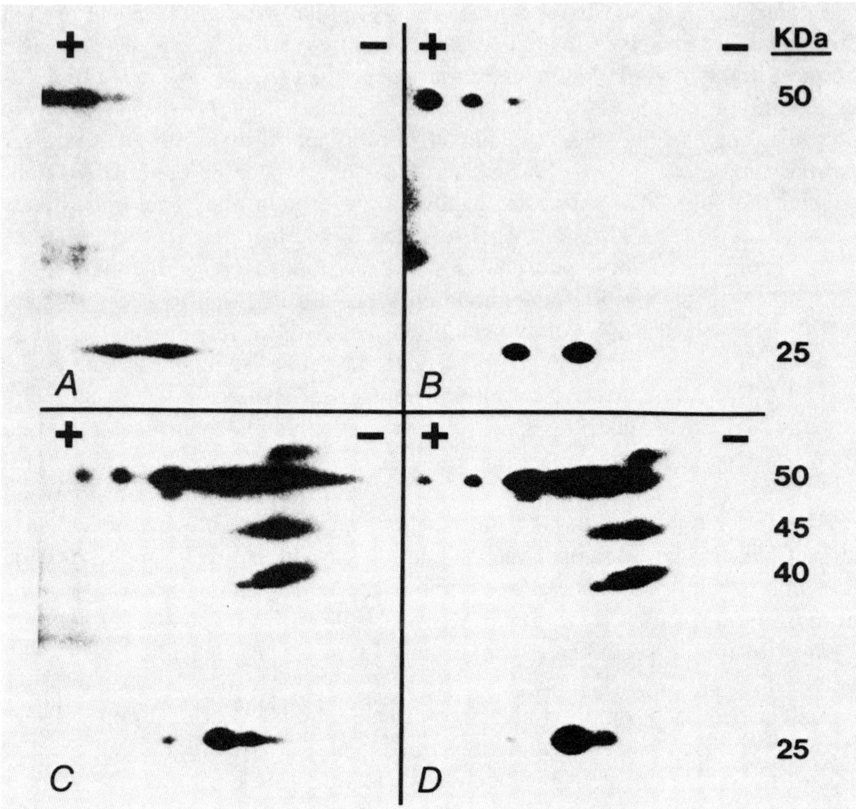

Figure 1 Protein kinase C-dependent phosphorylation and protein phosphatase-dependent dephosphorylation of specific phosphoproteins in rat type II alveolar epithelial cells in culture (Warburton et al., 1991). Culture type II alveolar epithelial cells were labeled with ^{32}Pi for 4 h. Cells were then treated with: A, control; B, 12-O-tetradecanoyl phorbol 13-acetate (TPA 10^{-4} M) for 15 min; C, 1 μM okadaic acid (OA) for 2.0 h; D, 1 μM OA for 2.5 h, followed by TPA (10^{-4} M) for 15 min. Protein extracts were separated by isoelectric point in an electrofocusing tube gel (pH gradient 5–7.7) and then 5–15% SDS-PAGE. Gels were then dried and autoradiographed. Significant increases in protein phosphorylation of specific substrates were found in response to PKC activation with TPA (B) and in response to protein phosphatase inhibition with OA (C). In D, additional phosphorylation by TPA was found in the presence of protein phosphatase blockade with OA.

The recent advent of new and specific pharmacological approaches to this study of protein phosphatase structure and function including okadaic acid, together with the recent cloning and expression of protein phosphatases (Berndt et al., 1987, 1990), provide exquisite molecular probes with which the molecular structure and biological function of protein phosphatases can be examined in detail.

V. Protein Phosphorylation and Dephosphorylation Cascades

Regulation of intracellular processes by extracellular signals is accomplished by the activation of protein kinase cascades that can be divided conceptually into two major regulatory themes, the first of which we have termed the second messenger-activated protein kinase cascade and the second of which we have referred to as the direct activation protein kinase cascade. These concepts are illustrated in Fig. 2.

Processes such as neuronal secretion, or almost all intermediary metabolism pathways, often require rapid regulation. The extracellular signals are generally transduced by receptors which activate the synthesis or release of second messengers, including cAMP, phosphatidyl inositols, diacylglycerol, and intracellular calcium (C_i^{2+}). The second messengers activate the respective serine/threonine kinases, which in turn phosphorylate and regulate the activity of specific intracellular protein substrates. Some of these substrates may be other kinases, which may themselves be either positively or negatively regulated by serine/threonine phosphorylation. To prevent futile cycles, protein phosphatases are often rendered inactive when kinases are active, and vice versa. Protein phosphatases may themselves be protein kinase substrates, providing another level of coordinated regulation. Fluctuations in the levels of second messengers in response either to the initial or to subsequent stimuli modulate the activity of the initial protein kinases in the cascade, enhancing the sensitivity of this mode of regulation.

VI. Second Messenger-Activated Serine/Threonine Phosphorylation and Alveolar Epithelial Cell Function

Both beta-adrenergic and P1-purinergic (A2-adenosinergic) signals are transduced by receptor-mediated, GTP-binding protein-coupled activation of cAMP production by adenylate cyclase in the type II alveolar epithelial cell (Dobbs et al., 1979; Gilfillian and Rooney, 1987; Warburton et al., 1989). PKA is activated in type II pneumocytes by cAMP (Whitsett et al., 1983, 1985). In-vitro phosphorylation of several families of phosphoproteins with purified

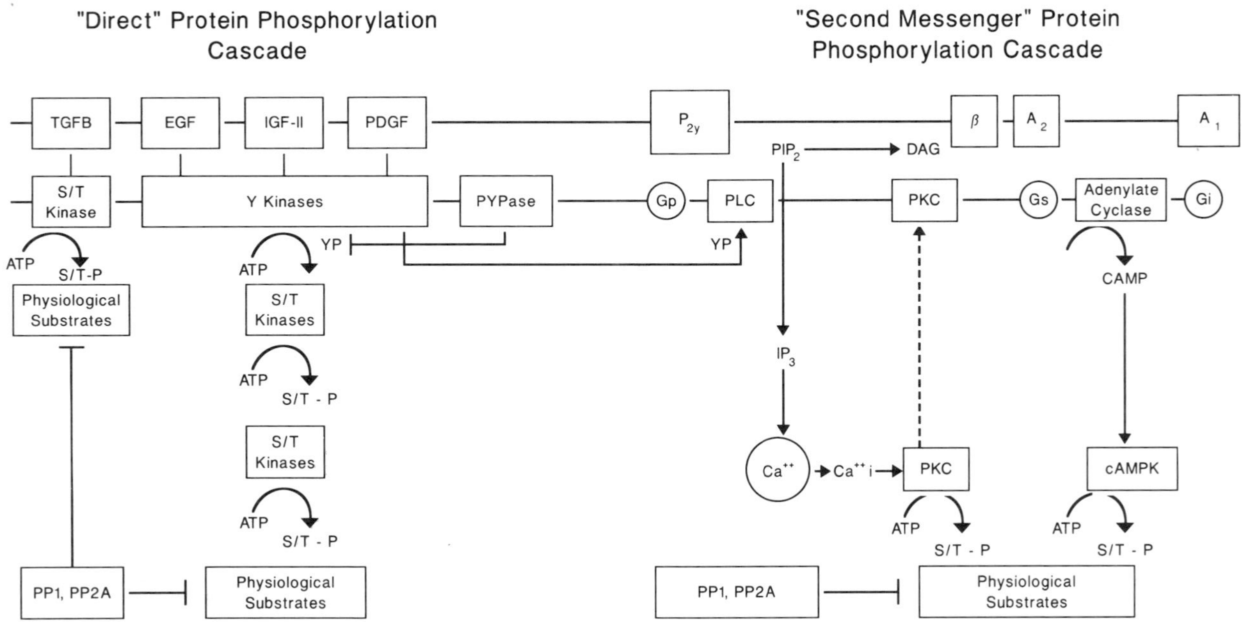

Figure 2 Regulation of intracellular processes by extracellular signals is accomplished by the activation of protein kinase cascades that can be divided conceptually into two major regulatory themes, the first of which we have termed the second messenger-activated protein kinase cascade and the second of which we have referred to as the direct-activation protein kinase cascade. These concepts are illustrated in the figure. Abbreviations: TGFB, transforming growth factor beta receptor; EGF, epidermal growth factor receptor; IGF-II, insulinlike growth factor-II receptor; PDGF, platelet-derived growth factor receptor; P2y, class P2y purinergic receptor; B, beta-adrenergic receptor; A2, A1, respective classes of adenosinergic receptor; S/T kinase, serine/threonine kinase; PYPase, tyrosine phosphatase; Gp, Gs, Gi, respective classes of GTP-dependent binding proteins; PLC, phospholipase C; PKC, protein kinase C; S/T-P, phosphorylation of serine/threonine groups; YP, phosphorylation of tyrosine groups; cAMPK, cAMP-dependent protein kinase (PKA in the text); PP1, PP2A, respective classes of protein serine/threonine phosphatases.

PKA catalytic subunit has been demonstrated in rat lung cytosol (Whitsett et al., 1985). Moreover, cAMP-dependent phosphorylation of actin in the developing rat lung is developmentally regulated (Whitsett et al., 1983).

Phorbol esters are also potent agonists of surfactant secretion from type II alveolar epithelial cells in culture, and at the same time rapidly induce the activation and translocation of protein kinase C from cytosol to membrane bound forms (Sano et al., 1985). Several putative phosphoprotein substrates of protein kinase C have been mapped in type II alveolar epithelial cells in culture by two-dimensional SDS-PAGE and autoradiography (Warburton et al., 1991) (Fig. 1).

Extracellular ATP is another potent receptor stimulant or surfactant secretion which stimulates cAMP production via a P1-purinergic receptor-mediated pathway and inositol polyphosphate and diacylglycerol formation by a P2y-purinergic receptor-mediated pathway (Warburton et al., 1989). Inositol 1,4,5-triphosphate is formed rapidly in response to extracellular ATP. A rapid rise in intracellular calcium also occurs (Tio et al., 1991). Diacylglycerol and calcium are necessary for activation of protein kinase C. The potency of ATP as a surfactant secretagogue could be explained by activation of PKA and PKC, since secretion of surfactant appears to be positively coupled between these two serine/threonine kinases. Several isoforms of PKC with slightly different amino acid sequences and substrate specificities have now been isolated from several tissues. It remains to be established which PKC isoforms are expressed in type II alveolar epithelial cells.

VII. "Direct" Phosphorylation and Alveolar Epithelial Cell Function

Biological processes which require a commitment over longer periods of time, such as commitment to cell growth and proliferation, are generally mediated by growth factor receptors containing tyrosine kinase domains. Activated receptor tyrosine kinases phosphorylate and activate serine/threonine kinases, which may be in turn phosphorylate and regulate additional serine/threonine protein kinases in a direct kinase-to-kinase-to-kinase cascade. Tyrosine phosphorylation of the receptor kinase substrates (including autophosphorylation) in the direct cascade is counteracted by tyrosine phosphatases, while subsequent serine/threonine phosphorylation is counteracted by serine/threonine phosphatases.

Type II alveolar epithelial cells respond to several factors, including EGF, TGF-alpha, and IGF by increasing the rate of thymidine incorporation into DNA (Leslie et al., 1985, 1989). These signals are known to be transduced by growth factor receptors containing tyrosine kinase domains which are intrinsic to the growth factor-transducing activity of the receptor. EGF receptors, which transduce both the EGF and TGF-alpha signal both by tyrosine phosphorylation and activation of phosphatidylinositol hydrolysis, have been demonstrated in type II alveolar epithelial cells in culture (Raaberg et al., 1991).

The role of tyrosine phosphatases has not yet been extensively studied in alveolar epithelial cells. However, these enzymes have been extensively characterized in other systems. It is likely that tyrosine phosphatases, which exhibit much higher turnover numbers than tyrosine kinases, exert a tonic negative regulation of cellular tyrosine phosphorylation.

VIII. New Directions in Alveolar Epithelial Cell Protein Phosphorylation Research

The alveolar epithelial cell is rather unique in that it exists in the adult lung as a highly differentiated phenotype with numerous specialized functions including surfactant synthesis, secretion, re-uptake and processing, fluid and electrolyte transport, and production of growth factors and cytokines. Yet, following the denudation of the alveolar epithelium which occurs during lung injury, the type II alveolar epithelial cell proliferates and repopulates the damaged alveolar epithelium. Recent studies have shown that both fetal and adult alveolar epithelial cells in culture respond to several growth factors, including EGF, IGF-II, PDGF, and as yet uncharacterized factors in broncoalveolar alveolar fluid and macrophage supernatants, by increasing that rate of thymidine incorporation into DNA and cell numbers. Thus, mechanisms of regulation of the cell cycle in alveolar epithelial cells are of great fundamental significance not only for normal lung morphogenesis but also for alveolar repair as well as pulmonary alveolar cell carcinogenesis.

A major control point in the mammalian cell cycle is located at the G1-S phase transition. This G1 control is analogous to the Start (i.e., late G1) of the yeast cell cycle, at which external mating factors regulate cell cycle progression. It is during G1 that growth factors such as EGF, IGF-I and PDGF act to stimulate the transition to S-phase (Pardee, 1989), and after which growth factors are no longer required to complete the mammalian cell cycle. Recent studies have identified a number of key regulatory proteins that appear to govern this critical cell cycle transition. These key regulatory proteins are $p34^{cdc2}$, cyclins, and the retinoblastoma tumor suppressor protein. From genetic studies in yeast, $p34^{cdc2}$ is predicted to serve a vital function at Start, in addition to its role as a promoter of M-phase (Norbury and Nurse, 1989). Recent studies using immunodepletion (Blow and Nurse, 1990) and antisense oligodeoxynucleotide strategies (Furukawa et al., 1990) have confirmed that p34^{cdc2} related enzymes play a central role in the initiation of DNA replication in higher eukaryotes.

Recently, excitement in this field has centered on the observation that the prototype tumor-suppressor protein, the retinoblastoma-susceptibility (Rb) gene product, is phosphorylated in a cell-cycle-dependent manner (Cooper, 1990). The Rb gene has been found to be mutated or deleted in numerous types of cancer, including lung cancer. In some cancers the Rb protein, although apparently intact is hyperphosphorylated. In normal cells the Rb protein is underphosphorylated in G0 and early G1, becomes highly phosphorylated at the G1-S boundary, and remains phosphorylated during S and G2 (Mihara et al., 1989; Ludlow et al., 1990). This finding has led to the hypothesis that phosphorylation of the Rb protein inactivates its growth-suppressing function and may be required for the transition from G1 to S phase. Further studies have shown that Rb phosphorylation is first detectable in G1 and can be inhibited by the action of transforming growth factor-beta (TGF-beta) (Laiho et al., 1990), or by other treatments which lead to cytodifferentiation (Chen et al., 1989; Stein et al., 1990). The TGF-beta receptor has recently been identified as homologous to the activin receptor, which possesses serine/threonine kinase activity. Activation of the TGF-beta receptor both in mink lung epithelial cells and in the A549 human alveolar carcinoma cell line inhibits the phosphorylation of the Rb gene product as well as cell cycle progression (Laiho et al., 1990). It is tempting to speculate that site-specific phosphorylation of serine/threonine kinase substrates by the activated TGF-beta receptor may play a role in negative regulation of the cell cycle at the locus of the Rb gene product.

The identity of the physiological Rb kinases and phosphatases is therefore a compelling but largely unresolved question. Active p34^{cdc2} complexes phosphorylate Rb protein at multiple sites in vitro (Chen et al., 1989; Stein et al., 1990). In a recent study of human T cells, both induction of p34^{cdc2} and phosphorylation of Rb as well as the transition into S phase were blocked by antisense oligodeoxynucleotides (Furukawa et al., 1990), whereas blockade of cyclin B by the same strategy had no effect on DNA synthesis.

The role of G1-S phase-specific cyclins in cell cycle control is also a dynamic area of research at this time. Genetic, molecular, and biochemical evidence has suggested that an operationally defined S-phase-promoting factor (SPF) may exist in mammalian cells, analogous to the well-defined maturation promoting factor of oocytes (MPF) (Murray and Kirschner, 1989a, 1989b; Cross et al., 1989; Lewin, 1990; Pines and Hunter, 1990). It has been postulated that the active enzymatic component of SPF would be p34^{cdc2} complexed to a G1-specific cyclin, (Murray and Kirschner, 1989a), a cyclin-linker protein (Lewin, 1990; Pines and Hunter, 1990), or possibly cyclin A itself (Murray and Kirschner, 1989b). Cyclin A has only recently been characterized in mammals. In *Drosophila* embryos, cyclin A accumulates exclusively in the embryonic cytoplasm during interphase, and translocates to the nuclei during prophase, where it is thought to participate in chromosome condensation and nuclear envelope breakdown. Thus, after several embryonic cell divisions, cyclin A but not cyclin B becomes essential for continued cell division (Lehner and O'Farrell, 1990). Addition of cyclin A to G1 cell extracts has also recently been shown to initiate DNA replication in in-vitro assays (Durso et al., 1990). Moreover, the growth factor-sensitive, proline-directed protein kinase (PDPK) that is active in the cytosol of proliferative cells has recently been characterized as a heterodimer comprised of $p34^{cdc2}/p58^{cyclin\ A}$ (Hall et al., 1991). These and other recent findings have confirmed that cyclin A is active much earlier in the cell cycle than had been previously thought (Hall and Vuillet, 1991), and that cyclin A must therefore be a prime candidate as a regulatory component of SPF. The demonstration of physical association between the $p34^{cdc2}/p58^{cyclin\ A}$ PDPK and the Rb protein (Williams et al., 1991) provides additional support for this concept.

Unlike the regulation of protein phosphorylation carried out by the classical serine/threonine protein kinases (PKA, PKC, calcium-calmodulin-dependent protein kinase), which respond to rapid fluctuations in intracellular second messenger concentrations, the $p34^{cdc2}$ protein kinase complexes which are proposed to control both S phase ($p34^{cdc2}/p58^{cyclin\ A}$ [PDPK]) and M phase ($p34^{cdc2}/p62^{cyclin\ B}$) are designed to remain active for prolonged periods of time. The activity of $p34^{cdc2}$ catalytic subunit is itself regulated by protein phosphorylation and dephosphorylation, and appears to be relatively independent of ionic fluctuations and second messengers. Each of the $p34^{cdc2}$-containing protein kinases also contains an inducible regulatory subunit, which is itself regulated by covalent modification. Once triggered, the $p34^{cdc2}$ kinase complexes may remain active for prolonged periods of time, until the activating cyclin regulatory subunits are proteolytically degraded. The persistent activity of these enzymes correspond to biochemical "drive states" as opposed to reversible steady states, the strategic modulation of which remains a challenge for both scientists and clinicians.

IX. Summary

Alveolar epithelial cell function is regulated by two cascades of reversible protein phosphorylation. Differentiated functions such as surfactant secretion are regulated positively by activation of serine/threonine kinases such as PKA and PKC in response to release of second messengers including cAMP, phosphatidyl inositols, and calcium. Protein synthesis and cell proliferation, on the other hand, are regulated principally by activation of growth factor receptor tyrosine kinases, which in turn activate serine/threonine kinases such as ERK-1 and PDPK. Cross-talk between these two broad cascades of protein phosphorylation occurs at multiple levels. The major challenges of modern regulatory biology are to identify the executive kinases which initiate cascades of protein phosphorylation and dephosphorylation that affect cellular physiology. New tools are now at hand, including new and specific pharmacological reagents, antibodies, cloned and expressed protein kinases and protein phosphatases, site-directed mutagenesis, and site-specific phosphorylation analysis. These tools provide exquisite molecular probes with which the molecular structure and function of protein phosphorylation and dephosphorylation systems can be investigated at the level of molecular detail.

Acknowledgments

Work in David Warburton's laboratory is supported by grants HL44977 and 44060 from the National Institutes of Health. Norbert Berndt's laboratory is supported by the Margaret E. Early Medical Research Trust, by the Faculty Research and Innovation Fund of the University of Southern California, and by NIH Biomedical Research Grant support to Children's Hospital Los Angeles. Frederick L. Hall's work is supported by the National Science Foundation, the John C. Wilson Endowment, and Vernon Tolo.

References

Anderson N, Maller J, Tonks N, Sturgill T. Requirement for integration of signals from two distinct phosphorylation pathways for activation of MAP kinase. Nature 1990; 343:651-3.

Axton JM, Dombradi V, Cohen PTW, Glover DM. One of the protein phosphatase 1 isoenzymes in Drosophila is essential for mitosis. Cell 1990; 63:33–46.

Berndt N, Campbell DG, Caudwell FB, Cohen P, da Cruz e Silva EF, da Cruz e Silva OB, Cohen PTW. Isolation and sequence analysis of a cDNA clone encoding a type-1 protein phosphatase catalytic subunit; homology with protein phosphatase 2A. FEBS Lett 1987; 223:340–6.

Berndt N, Cohen PTW. Renaturation of protein phosphatase 1 expressed at high levels in intact cells using a baculovirus vector. Eur J Biochem 1990; 190:291–7.

Bialojan C, Takai A. Inhibitory effect of a marine sponge toxin, okadaic acid, on protein phosphatases. Biochem J 1988; 256:283–90.

Blow J, Nurse P. A cdc2-like protein is involved in the initiation of DNA replication in Xenopus egg extracts. Cell 1990; 62:855–62.

Boulton T, Yancopoulos G, Gregory J, Slaughter C, Moomaw C, Hsu J, Cobb M. An insulin-stimulated protein kinase similar to yeast kinases involved in cell cycle control. Science 1990; 249:64–7.

Chen P, Scully Shew J, Wang J, Lee W. Phosphorylation of the retinoblastoma gene product is modulated during the cell cycle and cellular differentiation. Cell 1989; 58:1193–8.

Cohen P. The role of protein phosphorylation in the hormonal control of enzyme activity. Eur J Biochem 1985; 151:439–48.

Cohen P. The structure and regulation of protein phosphatases. Ann Rev Biochem 1989; 58:453–508.

Cohen PTW. Molecular cloning of protein-serine/threonine phosphatases. Methods Enzymol 1991; 201:398–401.

Cooper J. Oncogenes and anti-oncogenes. Curr Opin Cell Biol 1990; 2:285–95.

Cross F, Roberts J, Weintraub H. Simple and complex cell cycles. Ann Rev Cell Biol 1989; 5:341–95.

Denner I, Wiegel N, Maxwell B, Schrader W, O'Malley B. Regulation of progesterone receptor-mediated transcription by phosphorylation. Science 1990; 250:1740–3.

Dobbs LG, Mason RG. Pulmonary alveolar type II cells isolated from rats. Release of phosphatidylcholine in response to beta-adrenergic stimulation. J Clin Invest 1979; 63:378–87.

Durso G, Marraccino R, Marshak D, Roberts J. Cell cycle control of DNA replication by a homologue from human cells of the $p34^{cdc2}$ protein kinase. Science 1990; 250:786–91.

Edelman AM, Blumenthal DK, Krebs EG. Protein serine/threonine kinases. Ann Rev Biochem 1987; 56:567–613.

Furukawa Y, Piwinica-Worms H, Ernst T, Kanakura Y, Groffin JD. cdc2 gene expression at the G1 to S transition in human T lymphocytes. Science 1990; 250:805–8.

Gilfillian AM, Rooney SA. Purinoceptor agonists stimulate phosphatidylcholine secretion in primary cultures of rat type II pneumocytes. Biochim Biophys Acta 1987; 917:18–23.

Hall FH, Vuillet PR. Proline-directed protein phosphorylation and cell cycle regulation. Curr Opinion Cell Biol 1991; 3:176–84.

Hall FL, Braun RK, Mihara K, Fung YKT, Berndt N, Carbonaro-Hall DA, Vuillet PR. Characterization of the cytoplasmic proline-directed kinase in proliferative cells and tissues as a heterodimer comprised of $p34^{cdc2}$ and $p58^{cyclin\ A}$. J Biol Chem 1991; 266:17430–40.

Hall FL, Mitchell JP, Vuillet PR. Phosphorylation of synapsin I at a novel site by proline-directed protein kinase. J Biol Chem 1990; 265:6944–8.

Hanks SK, Quinn AM, Hunter T. The protein kinase family: conserved features and deducted phylogeny of the catalytic domains. Science 1988; 241:42.

Haystead TAJ, Sim ATR, Carling D, Honor RC, Tsukitani Y, Cohen P, Hardie DG. Effects of the tumor promoter okadaic acid on intracellular protein phosphorylation and metabolism. Nature 1989; 337:78–81.

Hunter T. A thousand and one protein kinases. Cell 1987; 50:823–9.

Kennely PJ, Krebs EG. Consensus sequences as substrate specificity determinants. J Biol Chem 1991; 266:15555–8.

Kinoshita N, Including H, Yanagida M. Distinct, essential roles of type 1 and 2A protein phosphatases in the control of fission yeast cell division cycle. Cell 1990; 63:405–15.

Krebs EG. The phosphorylation of proteins: a major mechanism for biological regulation. Fourteenth Sir Frederick Gowland Hopkins Memorial Lecture. Biochem Soc Trans 1985; 13:813.

Knighton DR, Zheng J, Ten Eyck LF, Ashford VA, Xuong NH, Taylor SS, Sowadski JM. Crystal structure of the catalytic subunit of cyclic adenosine monophosphate-dependent protein kinase. Science 1991; 253:407–14.

Laiho M, DeCaprio, Ludlow J, Livingston D, Massague J. Inhibition by TGF-beta linked to suppression of retinoblastoma protein phosphorylation. Cell 1990; 62:175–85.

Lehner C, O'Farrell P. The roles of Drosophila cyclins A and B in mitotic control. Cell 1990; 61:535–47.

Leslie CC, McCormick-Shannon K, Cook JL, Mason RJ. Macrophages stimulate DNA synthesis in rat alveolar type II cells. Am Rev Respir Dis 1985; 132:1246–52.

Leslie CC, McCormick-Shannon K, Mason RJ. Bronchoalveolar lavage fluid from normal rats stimulates DNA synthesis in rat alveolar type II cells. Am Rev Respir Dis 1989; 139:360–6.

Lewin B. Driving the cell cycle: M phase kinase, its partners and substrates. Cell 1990; 61:743–52.

Ludlow J, Shon J, Pipias J, Livingston D, DeCaprio J. The retinoblastoma susceptibility gene product undergoes cell cycle-dependent dephosphorylation and binding to and release from SV 40 large T. Cell 1990; 60:387–96.

Mihara K, Cao X, Yen A, Chandler S, Driscoll B, Murphree A, T'Ang A, Fung Y. Cell cycle-dependent regulation of phosphorylation of the human retinoblastoma gene product. Science 1989; 246:1300–3.

Moreno S, Nurse P. Substrates for $p34^{cdc2}$: in vivo veritas? Cell 1990; 61:549–51.

Murray A, Kirschner M. Cyclin synthesis drives the early embryonic cell cycle. Nature 1989a; 339:275–80.

Murray A, Kirschner M. Dominoes and clocks: the union of two views of the cell cycle. Science 1989b; 246:614–21.

Norbury C, Nurse P. Control of the higher eukaryote cell cycle by $p34^{cdc2}$ homologues. Biochim Biophys Acta 1989; 989:85–95.

Okhura H, Kinoshita N, Miyatani S, Toda T, Yanagida M. The fission yeast dis^{2+} gene required for chromosome disjoining encodes one of two putative type 1 protein phosphatases. Cell 1989; 57:997–1007.

Pardee A. G1 events and regulation of cell proliferation. Science 1989; 246:603–8.

Pearson RB, Kemp BE. Protein kinase phosphorylation site sequence and consensus specificity motifs. Methods Enzymol 1991; 200:63–81.

Pines J, Hunter T. $p34^{cdc2}$: the S and M kinase? New Biol 1990; 2:389–401.

Raaberg L, Nexo E, Buckley S, Luo W, Snead ML, Warburton D. Epidermal growth factor transcription, translation and signal transduction by rat type II pneumocytes in culture. Am J Respir Cell Mol Biol 1991; in press.

Rice WR, Hull WM, Dion CA, Holinger BA, Whitsett JA. Activation of cAMP-dependent protein kinase during surfactant release from type II pneumocytes. Exp Lung Res 1985; 9:135–49.

Roach PJ. Control of glycogen synthase by hierarchical protein phosphorylation. FASEB J 1990; 4:2961–8.

Roach PJ. Multisite and hierarchical protein phosphorylation. J Biol Chem 1991; 266:14139–42.

Rossamondo A, Payne D, Weber M, Sturgill T. Evidence that pp42, a major tyrosine kinase target protein is a mitogen-activated serine/threonine kinase. Proc Natl Acad Sci USA 1989; 86:6940–3.

Sano K, Voelker DR, Mason RJ. Involvement of protein kinase C in pulmonary surfactant secretion from alveolar type II cells. J Biol Chem 1985; 260:12725–9.

Stein G, Beeson M, Gordon I. Failure to phosphorylate the retinoblastoma gene product in senescent human fibroblasts. Science 1990; 249:666–9.

Sturgill T, Ray L, Erikson E, Maller J. Insulin-stimulated MAP-2 kinase phosphorylates and activates ribosomal protein S6 kinase II. Nature 1988; 334:715–8.

Suganuma M, Fujiki H, Suguri H, Yoshikawa S, Hirota M, Nakayasu M, Ojika M, Wakamatsu K, Tamada K, Sugimura T. Okadaic acid; an additional nonphorbol-12-O-tetradecanoate-13-acetate-type tumor promoter. Proc Natl Acad Sci USA 1988; 85:1768–71.

Suzuki M. SPXX, a frequent sequence motif in gene regulatory proteins. J Mol Biol 1989; 207:61–84.

Tio D, Tayag A, Rehn M, Warburton D. Role of inositol triphosphate isomer formation in type II pneumocyte signal transduction. Am J Respir Cell Mol Biol 1991; 4:514–8.

Vuillet PR, Hall FL, Mitchell JP, Hardie DG. Identification of a novel proline-directed serine/threonine-specific protein kinase in rat pheochromocytoma. J Biol Chem 1989; 264:16293–8.

Walsh DA, Perkins JP, Krebs EG. An adenosine 3′,5′-monophosphate-dependent protein kinase from rabbit skeletal muscle. J Biol Chem 1968; 243:3763.

Warburton D. Chronic hyperglycemia with secondary hyperinsulinemia inhibits the maturational response of fetal lamb lung to cortisol. J Clin Invest 1983; 72:433–40.

Warburton D. Epigenetic autocrine and paracrine factors regulating lung morphogenesis: a paradigm for lung repair. Chest 1991; 99:15S–8S.

Warburton D, Cohen P. Ontogeny of protein phosphatase 1 and 2 A in developing rat lung. Pediatr Res 1988; 24:25–7.

Warburton D, Buckley S, Cosico L. P1 and P2 purinergic receptor signal transduction in rat type II pneumocytes. J Appl Physiol 1989; 66:901–5.

Warburton D, Parton L, Buckley S, Cosico L, Enns G, Saluna T. Combined effects of corticosteroid, thyroid hormone and beta agonist on surfactant, lung mechanics and beat receptor binding in fetal lamb lung. Pediatr Res 1988; 24:164–70.

Warburton D, Tayag A, Buckley S, Cosico L, Seth R. Protein phosphorylation and dephosphorylation in type II pneumocytes. Am J Physiol 1991; 260:L548–54.

Whitsett JA, Hull DW, Dion C, Lessard J. cAMP dependent actin phosphorylation in developing rat lung and type II epithelial cells. Exp Lung Res 1985; 9:191–209.

Whitsett JA, Matz S, Darovec-Beckerman C. cAMP dependent protein kinase and protein phosphorylation in developing rat lung. Pediatr Res 1983; 17:959–96.

Williams RT, Carbonaro-Hall DA, Hall FL. Copurification of $p34^{cdc2}/p58^{cyclin\ A}$ proline-directed protein kinase and the retinoblastoma tumor susceptibility gene product: interaction of an oncogenic serine/threonine protein kinase with a tumor suppressor protein. Oncogene 1991; in press.

11

Growth Factors and Pulmonary Epithelia

RITA M. RYAN and JACOB N. FINKELSTEIN

University of Rochester School of Medicine and Dentistry
Rochester, New York

I. Pulmonary Epithelium

The epithelium of the lung is a continuous lining which interfaces with air from the trachea to the alveoli; for reviews, see Rennard et al. (1991) and Mason and Williams (1991). Throughout the lung, the epithelium consists of both lining cells and secretory cells. The epithelial cells are polarized, with the apical cell membrane associated with an extracellular surface lining layer and the basal cell membrane associated with a basement membrane which separates it from the underlying interstitium. In the larger conducting airways, the bronchial epithelium has ciliated cells responsible for movement of mucus, mucin-secreting goblet cells, and seromucous glands. The surface-lining layer is biphasic, with an aqueous phase containing proteins, lipids, and ions, and a gel phase containing mucus. As one moves down the conducting airways, the epithelium becomes thinner, the density of the cilia is decreased, goblet cells gradually disappear, and, in bronchioles, the secretory cell is the Clara cell. The surface-lining layer of bronchioles is thinner and is composed of an aqueous layer and a surfactant film.

The epithelium changes radically at the transition from bronchioles to the gas-exchanging alveoli. The lining cell of the alveolar epithelium is the greatly attenuated type I pneumocyte that covers 96% of the alveolar surface. The secretory cells are the cuboidal type II cells which, even though somewhat more numerous than type I cells, cover only 4% of the alveolar surface. Type II cells are the cells which produce and secrete surfactant.

During development and during repair after lung injury, the pulmonary epithelium undergoes both proliferation and differentiation. In the tracheobronchial epithelium, the source of stem cells is thought to be the basal cell; however, after an injury, it appears that the mucous cell is also capable of contributing to repair of the epithelium. In the alveolus, it once was thought that type II cells were precursors of type I cells during development; this is presently an area of intense study. However, after an injury to the alveolar epithelium, the type I cells are damaged and the often-spared type II cells proliferate and differentiate to repopulate the injured epithelium. During these processes there are important interactions between the epithelium and its neighboring interstitium, including the extracellular matrix (ECM). These interactions are likely to be mediated in part by polypeptide growth factors. Additionally, cells of the immunological or inflammatory systems, e.g., the alveolar macrophage, are important producers of growth factors; these cells can also be in close proximity to the pulmonary epithelium.

II. Growth Factors and Epithelia

A. Polypeptide Growth Factors

In the past 10 years, major progress has been made in the identification and isolation of a large number of polypeptide growth factors and their respective receptors; for review, see James and Bradshaw (1984). Genomic and mRNA sequences have been determined, and certain important observations such as the serine/threonine or tyrosine kinase activity of many growth-factor receptors have added to the general understanding of growth-factor effects. However, the detailed mechanisms of action of any of the known growth factors are only beginning to be understood.

In general, most growth factors consist of primarily polypeptide material. Although the term "growth factor" is used because in many cases the result of ligand-receptor interaction is known to cause a mitogenic response, "growth factors" are now known also to be growth inhibitors or to have the ability to alter differentiated characteristics. The initial step in the interaction of these growth factors with target cells is the rapid and high-affinity binding to receptors present on the plasma membrane. These receptors are usually intrinsic membrane glycoproteins that show both high affinity and specificity for their ligands. In many cases, the ligand-receptor complex is internalized by receptor-mediated endocytosis. The majority of growth factors act in a paracrine mode; however, there are important examples of endocrine and autocrine substances as well.

All polypeptide growth factors do not elicit the same responses. For example, mitogenic growth factors have been subdivided into progression factors and competence factors. Progression factors stimulate cells to pass into S (DNA synthesis) phase of the cell cycle in the absence of other factors. In contrast, competence factors are able to prime the cells but require additional factors or different conditions to induce DNA synthesis. As another example, some growth factors are able to induce mitogenesis rapidly, whereas other growth factors require long exposures. Also, even the same growth factor may cause different responses in two different cell types. Thus, the nature of the signal induced by the ligand-receptor complex is different for different growth factors, and possibly for the same growth factor acting on different target cells.

Another important determinant of the target cell response is the cell surface clustering and internalization of the ligand-receptor complex and its course in the cell, whether it be lysosomal degradation, recycling, or a balance between these two. This balance may regulate the sensitivity of the cell to the growth factor by altering cell surface receptor number. In addition to the signaling which occurs after simple ligand-receptor coupling at the plasma membrane, internalization of the ligand and/or receptor itself seems to be important. Nuclear receptors have been described for certain growth factors. Translocation of the cell surface receptors to the nucleus or vice versa would add another means of regulation of cell surface receptor number as well as a potential site of action for an undegraded ligand or as yet unsecreted ligand.

The phosphorylation of a cascade of proteins is often a crucial consequence of many receptor-ligand interactions. The long-term effects of these growth factors are at the molecular level, with alteration of gene transcription and/or mRNA translation. Many growth factors induce various oncogenes as well as other proteins. Often, phosphorylation of a specific protein is one of the steps required for induction of a given gene.

B. Epidermal Growth Factor and Transforming Growth Factor α

Growth factors that are known to affect epithelia include epidermal growth factor (EGF), transforming growth factor α (TGFα), transforming growth factor β (TGFβ), and insulinlike growth factors I and II (IGF-I and II). EGF, Mr = 6000, is a polypeptide growth factor which affects growth and differentiated functions of a variety of cells and tissues; for review, see Stoscheck and King (1986). EGF is known to stimulate the proliferation of numerous cell types in vitro and epithelial cells, in particular, in vivo. Urogastrone is the human form of EGF. There is some sequence homology across species (70% between human and mouse), but EGF is less conserved than some of the other growth factors or receptors. EGF is derived from a very large precursor of 1217-aa, Mr = 127,000, based on mRNA sequencing. The amino end of the precursor has a hydrophobic signal sequence. The mRNA sequence also predicts a membrane spanning region. In vivo, EGF can be associated with arginine esterase, which cleaves the precursor at both ends, resulting in mature EGF. Interestingly, much of the pro-EGF molecule is composed of nine sequences, each of which has significant homology to EGF itself.

The EGF receptor is detectable on a large variety of cell types or tissues. This receptor mediates the biological signals of not only EGF, but also of at least two other EGF-like growth factors, TGFα and vaccinia virus growth factor (VGF). The three ligands are only 22% identical in amino acid sequence, but bind to the EGF receptor with nearly identical affinities.

The amino acid sequences of mouse, rat, and human TGFα, Mr = 5400, show approximately 90% similarity [for reviews, see Lyons and Moses (1990); Massague (1990); Saloman et al., (1990)] but only 30–40% similarity with either mouse or human EGF. However, alignment of the six cysteine residues suggests that TGFα can fold into a tertiary structure identical to that of EGF. TGFα allows normal cells to grow in soft agar, a property usually associated with malignancy. Cloning of the human and rat TGFα cDNAs indicated that TGFα is synthesized as a 160-aa (human) or 159-aa (rat) precursor. Pro-TGFα contains a 23-aa hydrophobic signal sequence followed by a domain which includes the 50-aa mature growth-factor sequence as well as potential glycosylation sites, a second hydrophobic domain which spans the plasma membrane, and, finally, a cysteine-rich domain at the carboxyl terminus. Mature TGF-α is released from the extracellular domain of pro-TGFα through proteolytic cleavage. The mature TGFα peptide is flanked one either end by Ala-Val dipeptides, which probably serve as cleavage sites for an elastaselike protease. The hydrophobic sequence of 25-aa that functions as a transmembrane domain is flanked on either side by a pair of dibasic amino acids, which may be cleavage sites for a serine protease. Various larger forms of TGFα have been reported and appear to be due to incomplete cleavage or various glycosylated forms. The TGFα precursor also has biological activity. Membrane-imbedded pro-TGFα is able to induce tyrosine autophosphorylation of EGF receptors on indicator cells. TGFα is somewhat unique among growth factors because it can induce transmembrane signaling as either a soluble or membrane-anchored form. With this property, TGFα can support cell-cell adhesion, mediate stimulation through cell-cell contact, and yield, by cleavage, soluble growth factors. There is a potential mode of intercellular communication established by the binding of a membrane-anchored growth factor on one cell to its receptor on the adjacent cell that has been termed "juxtacrine." The role of the TGFα precursor in vivo remains uncertain, although it has been suggested that biologically active pro-TGFα may provide a highly localized and/or prolonged autocrine or "juxtacrine" growth signal.

Originally, TGFα was found produced by transformed cells or in embryonic tissues. The discovery that normal adult cells both produce and respond to TGFα provided strong evidence that normal cell proliferation also may be regulated in an autocrine manner. There is auto-induction of TGFα message which has been proposed as a mechanism of signal amplification which may be necessary to propagate proliferative signals throughout an organ or tissue. Although TGFα was originally associated with transformed cells, it has also been found in nonmalignant tissues undergoing proliferation. For example, TGFα is increased in the mammary gland during pregnancy and in the liver after partial hepatectomy. The autocrine hypothesis is no longer regarded solely as leading to uncontrolled growth, but is believed to be important in the regulation of normal cell proliferation.

EGF and TGFα are members of a much larger family of EGF-related proteins of diverse functions. Other proteins in the EGF family include VGF, urokinase, tissue plasminogen activator, clotting factors, and low-density lipoprotein (LDL) receptor. One cardinal feature of the proteins in this family is a stretch of six cysteine residues spaced at defined intervals within a sequence of appoximately 40-aa. EGF and TGFα have many biological activities, including functioning as autocrine and paracrine growth factors, stimulating chemotaxis, regulating ECM production, stimulating wound healing, and stimulating precocious eyelid opening and tooth eruption in rodent neonates. Sometimes TGFα is more potent than EGF, for example, in stimulating angiogenesis, regulating cell shape by tropomyosin production, stimulating bone resorption, and inhibiting smooth muscle contraction. The different biological activities of TGFα and EGF may be due to their ability to bind to different regions of the EGF receptor and to differences in receptor internalization and degradation. For example, TGFα binds with 110-fold higher affinity than EGF to the chicken EGF receptor. There is some evidence that the two growth factors may interact with different sites in the extracellular domain of the receptor, or cause different conformational changes in the receptor, thereby leading to variable tyrosine kinase activity. Finally, TGFα, unlike EGF, decreases EGF receptor down-regulation and degradation in pancreatic tumor cells; lower concentrations of TGFα may be more effective than EGF in eliciting a biological response of comparable magnitude.

Since EGF and TGFα both bind to the EGF receptor, it is logical to assume that characteristics common to the two proteins are important for binding. Striking is the conservation of the cysteine residues, implying that disulfide bonds are important for the conformation of the protein to bind to the EGF receptor. The receptor binds to EGF with an affinity of 10^{-11} to 10^{-8} M. Addition of EGF results in the rapid activation of protein kinase activity, specific for tyrosine residues.

The receptor then functions to transmit the signal from the cytoplasmic face of the receptor to other proteins in the cell. This is carried out by the tyrosine-specific protein kinase activity of the receptor. The receptor phosphorylates itself as well as a wide variety of proteins in vitro, including histone, tubulin, insulin receptor, troponin I, and human growth hormone, and a smaller number of proteins in vivo. For EGF, as well as platelet-derived growth factor (PDGF) and insulin, receptor-ligand interaction induces stimulation of these tyrosine kinase receptors, which leads to phosphorylation of the raf-1 protooncogene product and activation of its associated serine/threonine kinase. Raf-1 is present in most tissues and cell lines.

The EGF receptor from several tissues and species has been shown to have a molecular mass of 170,000 Da. The receptor is a transmembrane protein with an extracellular EGF binding domain and an intracellular protein kinase domain. It is glycosylated extracellularly and phosphorylated intracellularly. After cDNA cloning, the deduced amino acid sequence predicts a protein of ~1200-aa with a molecular mass of ~132,000 for the mature receptor molecule. The remaining molecular weight is due to attached carbohydrate residues. There is a hydrophobic sequence representing the putative membrane-spanning region. The mRNA sequence also predicts a small stretch of hydrophobic amino acids resembling a signal peptide. The cytoplasmic domain includes the ATP-binding site and the tyrosine kinase activity, considered to be the primary effector system in the transmembrane signaling process. Also included in the cytoplasmic domain are three autophosphorylation sites.

One mole of EGF receptor binds one mole of EGF. Most cells display average receptor numbers within the range of 20,000 to 200,000 per cell except for some epithelial carcinoma cell lines that have approximately 2,000,000. Maximal activation of EGF receptor tyrosine kinase activity in vitro requires saturating concentrations of EGF, and most concentration curves suggest a linear relationship between the ligand and kinase activation. The EGF receptor protein kinase utilizes ATP or GTP as substrates and a metal cofactor (Mn^{2+}, preferably, or Mg^{2+}) is required for activity. It appears that ATP binding is preceded by substrate binding and followed by the release of phosphorylated substrate and then ADP.

C. Transforming Growth Factor β

Transforming growth factor β (TGFβ) was a term originally used to describe the ability of medium conditioned by a murine sarcoma virus-transformed cell line to induce the soft agar growth of normally anchorage-dependent cells; for reviews, see Roberts and Sporn (1989), Segarini (1991), and Sporn et al. (1987). The original isolation of TGFβ from human platelets resulted in the identification of a single form of the peptide, a homodimer with Mr = 25,000, now called TGFβ_1; a second form, TGFβ_2, has now been purified, and a small amount of TGF$\beta_{1,2}$ heterodimer can be purified from porcine platelets. The 25-kDa TGFβ dimer is the biologically active molecule, and reducing agents result in a total loss of activity. Both these peptides have been cloned, revealing that TGFβ_1 is encoded as a 390-aa precursor and TGFβ_2 as a 412-aa precursor. These precursors contain a typical hydrophobic signal sequence which is cleaved. The processed 112-aa chains of the two peptides are 72% identical, including conservation of all nine cysteine residues. The binding characteristics and biological activities of TGFβ_1 and TGFβ_2 are very similar. Three additional forms of TGFβ have been identified by screening of cDNA libraries. None of these putative peptides has been isolated from natural sources, although corresponding mRNAs have been demonstrated. TGFβ_3 is the most abundant mRNA expressed in developing chick embryos. TGFβ_4 was cloned from a chicken chondrocyte library; it is the only TGFβ whose precursor lacks the characteristic signal peptide sequence. This form of TGFβ may have a unique biological activity, perhaps functioning within the cell. TGFβ_5 was cloned from a frog oocyte library. All of the five TGFβs have 64–82% homology and share essential structural features such as synthesis from a long precursor and conservation of all nine cysteine residues in the processed peptide. Each of these different TGFβs is 98% to 100% conserved among species.

The TGFβ genes are part of a larger TGFβ supergene family. Other members include inhibins, activins, Mullerian inhibiting substance, predicted product of the *Drosophila* decapentaplegic gene complex, DPP-C, and amphibian gene in frog oocytes, Vg1. All peptides belonging to the family are encoded as larger precursors. There is 30–40% homology, including conservation of seven of the nine cysteine residues. A unifying feature of the biology of these peptides is their ability to regulate developmental processes. It appears that each of these peptides has its own unique receptor.

TGFβ is released from degranulating platelets and secreted from nearly all cells in a biologically inactive or latent form which is unable to bind to cellular receptors and is not recognized by antibodies to TGFβ. The peptide can be activated by acidification, alkalinization, or action of chaotropic agents in vitro. The structure of the latent TGFβ_1 complex includes mature, processed dimeric TGFβ_1 in a noncovalent association with the amino-terminal (minus the signal peptide) glycopeptide portion of the TGFβ precursor (latency protein) resulting in a 110-kDa latent complex. The latency protein is highly conserved; TGFβ_1 and TGFβ_3 are 91% and 85% conserved between human and chicken. In some instances, the latent complex includes a 125- to 160-kDa protein which is disulfide-linked to latency protein. In serum, a latenet form of TGFβ_1 exists consisting of a complex of mature TGFβ_1 and alpha$_2$ macroglobulin with no latency protein. Latent TGFβ can be activated by proteases, including plasmin, a protease produced by alveolar macrophages. Activation of

TGFβ appears to be the result of any event which leads to an unstable conformation of the latent TGFβ complex, enabling the release of mature TGFβ.

The action of TGFβ is mediated through binding to specific cell membrane receptors. Essentially all cells bind TGFβ, but the parameters of binding vary with each cell type. Dissociation constants range from 1 to 60 pM, and there may be as few as 200 to as many as 100,000 receptors on a cell. TGFβ appears to mediate metabolic changes at minimal receptor occupancy. The number of receptors available for TGFβ binding is modulated by a variety of conditions that include cell density, differentiation, treatment with retinoic acid, or addition of various hormones.

Cross-linking of TGFβ to membrane receptors with disuccinimidyl suberate (DSS) has revealed three distinct receptor types. The types I and II receptor proteins are glycoproteins of Mr = 50,000-65,000 and Mr = 80,000-110,000 respectively, and contain N-linked sugars. Type III receptors are generally considered to be dimeric and to be composed of proteoglycan subunits of Mr = 250,000-350,000. Type I and II receptors, like most growth-factor receptors, are glycoproteins. In contrast to all other known polypeptide receptors, the type III receptor, also known as betaglycan, is a proteoglycan consisting predominantly of heparin sulfate glycosaminoglycan chains with a smaller amount of chondroitin sulfate or dermatan sulfate attached to a core protein of approximately 100–140 kDa. The binding site for TGFβ resides in this core protein. The type III receptor/betaglycan has also been detected in a soluble form. Because of its unusual structure, the type III receptor may be important in ECM interactions. Studies indicate that the type I receptor is required for TGFβ sensitivity. No reports of kinase activity, intracellular calcium fluxes, or inositol phosphate involvement have been associated with the TGFβ receptors.

The types I and II receptors are found on all cell types, while the type III receptor/betaglycan is not detected on primary cells of endothelial or epithelial origin. The type III protein is the major TGFβ-binding component found on primary cells of mesenchymal origin (fibroblasts, chondroblasts, and osteoblasts) but may also be found on cultured and transformed cells of epithelial origin. The type III receptor is not a true receptor because, although it does bind TGFβ, it does not transmit a signal for metabolic change. The type III proteoglycan is not involved in transduction of the signal mediating changes in cell morphology, or in the synthesis of DNA, collagen, and fibronectin, leaving control of these activities to the type I receptors. The type II receptor may also be involved, perhaps as a component of a receptor complex that includes the type I receptor.

The biochemical mechanism of action of TGFβ is not yet fully elucidated. The ability of TGFβ to alter expression of receptors for other growth factors and hormones is related potentially to its mode of action on these cells.

TGFβ interferes with the action of many different mitogens. TGFβ induces c-sis mRNA and secretion of PDGF-like peptides. Although TGFβ blocks the mitogenic effects of certain growth factors, it does not block the various early signals induced by these mitogens, suggesting that TGFβ has its effect after some initial processes have occurred. Several lines of evidence suggest that a GTP-binding protein is involved in TGFβ signal transduction, including blockage of the mitogenic action of TGFβ by pertussis toxin, stimulation of membrane GTPase activity by TGFβ, and reduction of the receptor-binding affinity of TGFβ by Gpp(NH)p, known to dissociate G proteins. TGFβ results in arrest of cells in the middle to late G_1 phase, delaying entry into S phase. TGFβ delays entry into S phase, but cells that have passed the G_1/S border, or are in G_2, are no longer responsive to inhibition.

D. Insulinlike Growth Factors

Insulinlike growth factor-I (IGF-I), or somatomedin-C, is structurally related to proinsulin and, in human plasma, is bound to a carrier protein; for reviews, see De Pablo et al. (1990) and Phillips et al. (1990). Its receptor is highly homologous to the insulin receptor and has both tyrosine kinase and ATPase activity. IGF-II is homologous to multiplication-stimulating activity in the rat and has been implicated as a fetal growth factor. Its mitogenic activity appears to be mediated through the IGF-I receptor. Both IGF-I and II consist of A and B chains of amino acids with intrachain disulfide bridges. The A and B chains exhibit approximately 50% sequence homology with insulin. The tertiary structure is thought to be similar to that of insulin as well. Two forms of IGF receptors have been identified: a heterotetrameric type I receptor which resembles the insulin receptor and a monomeric type II receptor which has a structure completely different from that of the insulin receptor.

Both IGFs have broad, anabolic, insulinlike actions, including stimulation of multiplication of a wide variety of cells, promotion of processes related to energy storage, including amino acid transport and protein synthesis in muscle, glucose transport and fatty acid synthesis in liver, and stimulation of processes related to skeletal elongation, including synthesis of RNA, DNA, protein, and proteoglycan in chondrocytes and cartilage explants. Actions of the IGFs include negative feedback on growth hormone release as well. Levels of IGF-II appear to be high in fetal life; adult animals generally have low circulating and tissue levels except in the central nervous system. Levels of IGF-I are low in fetal life and peak in humans during adolescence.

The type IIGF receptor resembles the insulin receptor and has particular affinity for IGF-I. By analogy to insulin, binding of IGF-I to the type I IGF-I receptor may activate a cascade of cellular events including generation of phosphoinositide-derived messenger proteins, which may mediate slower processes such as activation or inactivation of different genes. It is likely that IGF-I promotes the action of phospholipase C which causes formation of diacylglycerol (which activates protein kinase C) and inositol triphosphate (which promotes mobilization of Ca^{2+} from intracellular stores). As with other polypeptide growth factors, IGF-I induced substrate phosphorylation is thought to be the prime effector of ligand-receptor interaction. There is direct negative feedback of IGF-I, which decreases the release of growth hormone from the pituitary. This negative feedback appears to be mediated by a decrease in the rate of transcription of the gene encoding growth-hormone mRNA. The bioavailability of IGFs is also regulated by the binding of IGFs to several IGF-binding proteins.

Growth facilitation by IGF-I involves not only release by the liver and action in an endocrine mode, but also local production by a wide variety of tissues and action in an autocrine-paracrine mode. Cultured fibroblasts released immunoreactive IGF-I. This release is facilitated by the addition of growth hormone. Addition of growth hormone to cultured fibroblasts increases cell number; this is blocked by addition of a monoclonal antibody to IGF-I. Thus, local formation of IGF-I by fibroblasts is a necessary part of the mechanism by which growth hormone stimulates cell multiplication in this system. This is an example of IGF-I action as a local autocrine-paracrine growth factor which amplifies tissue response to growth hormone.

There are many similarities between insulin and IGF-I receptor binding. Both activate an endogenous tyrosine-specific protein kinase. Receptor autophosphorylation enhances kinase activity with increased phosphorylation of cellular proteins presumed to be essential physiological intermediates. The endogenous substrates that can be phosphorylated by each of the receptors are similar. It is a matter of intense study as to whether and how the insulin receptor and the IGF-I receptor differ in their intrinsic ability to mediate metabolic and growth responses. In addition to binding to its own receptor with high affinity, each of the ligands bind to the receptor of its homolog with a reduced affinity but full intrinsic activity.

E. Summary

In summary, much is known about some of the polypeptide growth factors that affect epithelia. Protein structure and sequence, as well as gene structure and sequence for both growth factors and their receptors, is often available. However, the mechanisms of signal transduction are known less precisely. In addition, there are less-well-characterized growth factors known to affect lung epithelia that have been described and only partially characterized. These include lung epithelial cell growth factors produced by macrophages and fibroblasts.

III. Effects of Growth Factors on Lung Epithelia

A. Growth

Various injuries to the lung can damage both the airway and alveolar epithelium. Both epithelia have the ability to repopulate themselves. During repair after injury to the airway epithelium, reepithelialization occurs through the recruitment and proliferation of epithelial cells. After damage to the alveolar epithelium, type II cells are required to both proliferate, and then differentiate, to replace the lining type I pneumocytes. It is likely that growth factors play a role in mediating these processes. Isolated bronchial or tracheal epithelial cells and isolated type II alveolar epithelial cells have been studied in culture for the ability to proliferate under various conditions. Certain growth factors have been found to be mitogenic for these cells in vitro. In addition to the growth factors classically associated with responses by epithelial cells, there are many additional growth factors that have been examined in these proliferation studies, including growth factors more traditionally associated with fibroblast proliferation. Also commonly studied are the media conditioned by pulmonary cell types, e.g., fibroblasts and macrophages.

Tracheobronchial Epithelium

For an overall review on growth factor effects on tracheobronchial epithelium, see Jetten (1991). Using bovine bronchial epithelial cells (BEC), Takizawa et al. (1990) found that pulmonary macrophage-conditioned medium (CM) caused an increase in cell number and also stimulated ^{3}H-thymidine (^{3}H-TdR) uptake by nuclei. Co-culture of pulmonary macrophages with BECs also led to an increase in cell number. Bovine BECs were also stimulated to proliferate by media conditioned by human fetal lung fibroblasts, as demonstrated by Shoji et al. (1990). The CM caused an increase in cell number, ^{3}H-TdR uptake labeling index, and metaphase arrest. The growth-promoting activities found in the macrophage CM or fibroblast CM have not been completely purified. These studies suggest that other lung cells produce factors that may act to stimulate growth of BECs in a paracrine fashion.

Others, also studying primary isolated airway epithelial cells, have examined the effects of known growth factors on these cells in culture. Retsch-Bogart et al. (1990) examined in detail the effects of insulin and IGF-I on canine tracheal epithelial cells. These cells did not exhibit baseline proliferation but increased in both cell number and ^{3}H-TdR incorporated specifically into DNA. IGF-I was much more potent than insulin, suggesting that the effect was mediated via the IGF-I receptor. Additionally, there was no additive effect of IGF-I and insulin, implying that their effects were via the same mechanism. An antibody, αIR3, which inhibits binding of insulin or

IGF-I to the IGF-I receptor, inhibited both insulin and IGF-I-induced cell proliferation. IGF-I receptors were demonstrated by both DSS cross-linking and Northern analysis using a human IGF-I receptor riboprobe. Siegfried (1987) studied normal human BECs and used colony-forming efficiency (CFE) to assess proliferation. TGFα and EGF increased bronchial cell CFE by 13- to 14-fold over control. IGF-I and insulin increased CFE to a lesser degree; 1.6-fold and 2.3-fold, respectively. TGFβ decreased CFE 3-fold. All of these growth factors are known to be produced by cells in the lung, and any are likely to play a role in growth of the tracheobronchial epithelium during development or after injury. Much work is still required to determine which cells specifically are affected by these growth factors, and whether these effects are consistent in vivo.

Alveolar Epithelium

Similar studies have been carried out on type II alveolar epithelial cells isolated from several species at different ages. Again, investigators have examined the growth-stimulating effects of media conditioned by lung cells as well as the ability of known growth factors to alter type II cell proliferation.

The alveolar macrophage is recruited in large number to the lung after lung injury, and this cell is also known to have the ability to produce multiple growth factors. Thus, media conditioned by alveolar macrophages (AM) has been examined by several groups for the ability to affect type II cell proliferation. Leslie et al. (1985a) isolated and cultured rat and human alveolar macrophages and demonstrated that media conditioned by these AMs increased ^{3}H-TdR incorporation into isolated rat type II cells twofold over 10% fetal bovine serum (FBS). Co-culture of AMs with type II cells also increased ^{3}H-TdR incorporation. These type II cell cultures were stained with tannic acid to identify lamellar bodies of type II cells; this was followed by autoradiography to identify ^{3}H-TdR-labeled nuclei, and a labeling index was determined. Co-culture with macrophages also increased the labeling index as compared to control. Brandes and Finkelstein (1989) also studied the effect of AM CM on type II cell proliferation but used rabbit cells instead of rat. Type II cell proliferation was assessed by ^{3}H-TdR incorporation, cell number, and labeling index. All three were increased by media conditioned by AMs that were stimulated by serum-treated zymosan (STZ) in a dose-dependent fashion; there was no effect from unstimulated AM CM. AM/type II cell co-cultures gave similar results. It appears that there is a novel type II cell growth factor present in AM CM that has a molecular mass of approximately 32,000 Da by gel filtration, is heat and trypsin sensitive, but is stable to reduction by β-mercaptoethanol and acid (pH 2.5).

Related to AM CM promoting type II cell growth is a report by Leslie et al. (1989) that rat bronchoalveolar lavage fluid (BALF) stimulates DNA synthesis in rat type II cells. It is thought that the growth factors found in BALF are AM derived. This group showed that rat BALF increased type II cell ^{3}H-TdR incorporation in a dose-dependent fashion. Again, this activity was heat and trypsin sensitive. However, the activity was also sensitive to reduction by dithiothreitol and acidification to pH 2.5. Further characterization of this activity revealed a molecular mass of approximately 100,000 Da by gel filtration.

Alveolar macrophages are a likely source for type II mitogens after injury, but fibroblasts are also in close proximity to epithelial cells, and it is felt that mesenchymal epithelial interactions are important in both development and injury. Everett et al. (1990) demonstrated that media conditioned by hamster lung fibroblasts isolated after in-vivo exposure to 100% oxygen contained growth factors for rat type II cells. Type II cell proliferation was assessed by cell number and by ^{3}H-TdR incorporation. This group also showed that the effect of the hyperoxic fibroblast CM was greater the lower the density of the plated type II cells. Unlike the previous work discussed, where the presence of serum enhanced the proliferative effect of AM CM or BALF, this hyperoxic fibroblast CM had less stimulatory effect in 1% of 10% FBS (although lesser amounts of FBS were required). Media conditioned by fibroblasts isolated from normal, nonhyperoxia-exposed animals did not increase type II cell proliferation. This work suggests that hyperoxia, through unknown mechanisms, results in the production by fibroblasts of type II cell growth factors. These factors could participate in the type II cell proliferation that is known to occur after hyperoxic lung injury.

As it was logical to investigate whether lung cells produced type II cell growth factors, it was also logical to determine if type II cells responded to known growth factors. Many growth factors are known to be produced by lung cells [for review, see King et al., (1989)], including those produced by platelets, macrophages, and fibroblasts. Messmer et al. (1982) reported that primary isolated adult rat type II cells that have been subcultured respond to EGF with an increase in ^{3}H-TdR incorporation and labeling index. Fibroblast growth factor (FGF) had no effect. Leslie et al. (1985b) studied the effect of various growth factors and conditions on the proliferation of isolated adult rat type II cells. After 24 h in culture, ^{3}H-TdR incorporation was increased by EGF, insulin, and cholera toxin. Endothelial cell growth supplement (extract of bovine hypothalamus), but not PDGF or interleukin-1 (IL-1), also increased ^{3}H-TdR incorporation by type II cells. DNA content was measured and increased under different serum and matrix conditions; there were no changes in cell number. Since endothelial cell growth supplement (ECGS) had been stimulatory, this group (Leslie et al., 1990) investigated heparin-binding growth factors associated with neural tissue. Alpha- and beta-endothelial cell growth factor (ECGF) both stimulated DNA synthesis as measured by ^{3}H-TdR incorporation. Acidic FGF (aFGF) was quite stimulatory, increasing ^{3}H-TdR incorporation 16-fold; type II cells were much less responsive to basic FGF (bFGF). Serum was required for aFGF to have its effect. Interestingly, it appears that high-density lipoprotein (HDL) is one of the factors in serum that works synergistically with aFGF. Acidic FGF and combinations of other stimulatory growth factors also caused an increase in cell number (measured by hemacytometer) and DNA content.

Tanswell et al. (1991) also studied the proliferative response of adult rat type II cells, examining the effects of the IGFs. These cells were cultured on laminin and, after initial plating, were maintained in a serum-free defined medium. When cells were plated at higher density, there was an increase in cell number in response to IGF-I. However, cells plated at lower density showed no increase in cell number with IGF-I over control. IGF-II did not increase cell number.

Brandes and Finkelstein (1988,1989) have also studied the response of isolated adult type II epithelial cells to various known growth factors, but used rabbit cells instead of rat cells. EGF and TGFα both caused proliferation of type II cells as exhibited by increased ^{3}H-TdR incorporation, cell number, and labeling index. Cell number was counted using a gridded eyepiece while cells remained attached. PDGF, IL-1, IGF-I, and tumor necrosis factor alpha (TNFα) did not cause increased ^{3}H-TdR incorporation. TGFβ caused inhibition of baseline proliferation of isolated adult rabbit type II cells and inhibited the mitogenic effects of TGFα or EGF. In contrast to the work discussed previously using rat cells, further studies by this group (Brandes and Finkelstein 1988) demonstrate no effect of aFGF or bFGF, or ECGFα or ECGFβ, on rabbit type II cell proliferation.

Recently, Mason et al. (1990) have studied human alveolar type II cells isolated from surgical specimens. When growth factors were tested on cells in 10% FBS, insulin and ECGS were stimulatory, but cholera toxin and aFGF were inhibitory. In 0.4% serum, stimulation of ^{3}H-TdR incorporation was seen with insulin, EGF, ECGS, aFGF, and bFGF. Cholera toxin had no effect. Labeling index as measured by bromodeoxyuridine incorporation was also increased, and mitotic figures could be seen by electron microscopy.

In contrast to studies using adult cells, Stiles et al. (1986) isolated 19-day-gestation fetal rat type II cells and demonstrated increased ^{3}H-TdR incorporation with 10% FBS, and with PDGF, but not with EGF or IGF-I. Since fetal cells respond to PDGF but not EGF, and adult cells respond to EGF but not PDGF, it is not clear when the transition from the fetal phenotype to the adult occurs. Ryan et al. (1991) have demonstrated that type II cells isolated from newborn rabbits (day 1–2 of age) respond similarly to cells isolated from adult animals. Specifically, newborn cells increase in cell number and ^{3}H-TdR incorporation after exposure to EGF or TGFα; TGFα is more potent than EGF. In contrast, TGFβ inhibits cell growth compared to control cells.

Finally, several investigators have studied the effects of growth factors on lung epithelial-derived cell lines. Siegfried (1987) demonstrated that media conditioned by A549 cells, a pulmonary adenocarcinoma cell line, was stimulatory for A549 cell growth as measured by CFE. Imanishi et al. (1989) studied the same cell line and found that the addition of anti-TGFα antibody inhibited A549 cell growth as determined by the 3-(4,5-dimethyl-2-thiazolyl)-2,5-diphenyltetrazolium bromide (MTT) dye assay, which reflects changes in cell number. Their results imply that endogenous TGFα is stimulatory for A549 cell growth.

Two groups have studied EGF receptors on lung-derived cell lines. Ferriola et al. (1991) used immortalized rat tracheal epithelial (RTE) cell lines. They found that these cell lines expressed the mature and immature forms of the EGF receptor and that the amount was less or equal to that of primary rat tracheal cells. There was phosphorylation of these receptors in response to EGF. Tyrphostin, which inhibits EGF receptor tyrosine kinase, was inhibitory in a dose-dependent fashion to RTE cell proliferation as assessed by cell number (measured by hemacytometer) and ^{3}H-TdR incorporation. Lastly, these cells were inhibited by an anti-TGFα antibody, again indicating that endogenous TGFα was stimulatory to these transformed cells. Oberg and Carpenter (1989) used a fetal rat lung epithelial cell line to study EGF receptors. Specifically, they examined the effect of dexamethasone on EGF binding. Dexamethasone caused a 40–50% decrease in ^{125}I-EGF binding capacity in a dose-dependent fashion without changing affinity. Total numbers of receptors dropped from 29,000 per cell to 16,000. Retinoic acid increased EGF receptor synthesis. Thus, dexamethasone and retinoic acid were found to alter EGF-binding capacity by regulating EGF receptor synthesis.

Summary

Overall, there are a moderate number of studies that demonstrate that proliferation of isolated primary airway and alveolar epithelial cells is altered by exogenous growth factors, media conditioned by other lung cells, or, in fact, endogenously produced growth factors. Further work is required to determine how important these effects are in vivo. Also, studies that have been initiated on cell lines need to be performed on the primary, nontransformed cells to begin to determine which signal transduction pathways are important in these cells. Lastly, direct comparison of these studies should be done cautiously. For the type II cell proliferation studies, for example, it appears that not only obvious differences such as species (rat, rabbit, human) or the age of the animal (fetal, newborn, adult), but also more subtle differences such as cell plating density or the amount of serum present, can affect greatly which growth factors elicit a type II cell response and how large that response may be. Another critical factor not discussed is the matrix on which the cells are cultured. For studying any of the epithelial cell responses including tracheobronchial or transformed cells, the conditions under which the experiments are performed appear to be quite important.

B. Maturation/Differentiation

Just as it is likely that cell-cell communication via growth factors is important in the regulation of lung epithelial cell growth, it also plays a role in epithelial cell maturation. Many growth factors are known to be regulators of both cell proliferation and cell differentiation. Also, alterations in differentiated cell functions often are associated with a cell's phase of the cell cycle. Thus, both bronchial epithelial cells and alveolar type II cells have been studied, both in vitro and in vivo, to begin to determine what controls maturation of these cell types.

Tracheobronchial Epithelium

Masui et al. (1986) demonstrated that TGFβ was a potent growth inhibitor and induced normal human bronchial epithelial cells (BEC) to undergo squamous differentiation. Growth was assessed by clonal growth and ^{3}H-TdR incorporation. In addition to inducing the squamous phenotype, TGFβ also increased several markers of terminal squamous differentiation, including Ca^{2+} ionophore-induced formation of cross-linked envelopes, extracellular activity of plasminogen activator, and cell surface area. ^{125}I-TGFβ binding demonstrated a K_d of 13 pM with 10,000 binding sites per cell. Attempting to study a potential second messenger, this group demonstrated that TGFβ did not alter cAMP levels.

Also studying human BECs, Ke et al. (1990) showed that cell density altered the ability of isolated BECs or a BEC transformed cell line (S.6) to respond to TGFβ. Colony-forming efficiency (CFE) was used as an indicator of squamous differentiation. At lower density (2000 cells/cm^2), the CFE was 0%; that is, all of the cells underwent squamous differentiation. Cells plated at higher density (>5000 cells/cm^2) had significantly greater CFEs. ^{3}H-TdR incorporation studies were consistent with the CFE data.

In addition to the effects of growth factors on BEC terminal squamous differentiation, others in this group (Gerwin et al., 1990) have studied the effect of TGFβ on specific gene expression by BECs. TGFβ increased the mRNAs for plasminogen activator inhibitor type 1 (PAI-1) and urokinase 50-fold and 5-fold, respectively, within 4 h. This effect was dose dependent. PAI-1 protein was also increased in media conditioned by TGFβ-treated BECs. Interestingly, an immortalized BEC cell line that was resistant to TGFβ causing squamous differentiation, R.1, did not demonstrate a TGFβ-induced increase in PAI-1 or urokinase mRNA. The block in the TGFβ signal transduction pathway appears to be after receptor binding, since TGFβ binding was the same in this nonresponsive cell line as in the S.6 TGFβ-responsive cell line.

In beginning to examine the signal transduction mechanisms of growth factors on BECs, still other members of this laboratory (Miyashita et al., 1989) have studied fluctuations in calcium concentration after addition of serum, TGFβ or 12-O-tetradecanoyl-phorbol-13-acetate (TPA). They measured ionized cytosolic calcium concentration ($[Ca^{2+}]_i$) using the intracellularly trapped fluorescent probe, fura 2. Eight percent calcium-free FBS induced an increase in $[Ca^{2+}]_i$ in both normal human BECs and the adenovirus transformed BEC cell line; these conditions also induced squamous differentiation of these cells. Interestingly, S.6, the serum-sensitive (i.e., responds to serum with squamous differentiation) subclone of the BEC cell line, responded with a greater increase in $[Ca^{2+}]_i$ than did the serum-resistant R.1 subclone. TGFβ is felt to be an important serum component in the serum-induced squamous transformation, and itself induces BEC squamous differentiation; however, $[Ca^{2+}]_i$ is not increased by TGFβ. TPA also induces squamous transformation but did not cause an increase in $[Ca^{2+}]_i$. The different response of $[Ca^{2+}]_i$ in the S.6 and R.1 cell lines suggest that an increase in $[Ca^{2+}]_i$ is associated with serum-induced squamous transformation; however, this alteration in cytosolic calcium is not necessary for TGFβ or TPA to cause squamous differentiation.

Several investigators have examined the in-vivo effects of exogenous administration of growth factors on tracheobronchial epithelial maturation, focusing on development; for review, see Warburton (1991). Stahlman et al. (1989) observed hyperplasia of the airway epithelium and squamous metaplasia after treatment with EGF. St. George et al. (1991), using Rhesus monkeys at a different time in gestation, demonstrated that EGF induced airway maturation. Experimental fetuses at 75% of gestation who were EGF-treated had a taller tracheal epithelium and had significantly more secretory cells and less intermediate cells, although no increase in cell number was observed. In addition, EGF-treated animals had greater amounts of respiratory secretory product measured within the epithelial and submucosal glands, as well as that detected in lung lavage and amniotic fluid. The appearance of the tracheal epithelium and the increase in secretory product were more characteristic of adult monkey airway.

Alveolar Epithelium

The effects of growth factors on type II epithelial cell maturation have been studied extensively. These studies have examined primarily one type II cell function: the production of pulmonary surfactant. Studies have involved whole animals, lung explants, and isolated type II cells, and have been performed in many different species. With explant studies as a bridge, the whole-animal studies have examined the effect of certain factors on the type II cell while in its native milieu, whereas the studies utilizing isolated cells have attempted to focus on the mechanisms related directly to the type II cell itself. For an excellent review on the overall regulation of fetal lung maturation, see Gross (1990).

Although many hormones and factors can influence lung maturation, the effects of only several specific growth factors have been studied. In the whole animal, early studies by Catterton et al. (1979) and Sundell et al. (1980) demonstrated that EGF enhanced fetal pulmonary maturation by morphological and physiological measurements. Several groups have studied specifically type II maturation using fetal lung explants. Gross et al. (1986) demonstrated an increase in the rate of choline incorporation and phosphatidylcholine (PC) and disaturated PC (DSPC) by EGF. Thyroid hormone had an additive effect to EGF. Whitsett et al. (1987) reported that EGF stimulated the fetal type II cell in explant culture, but not when isolated, to produce surfactant-associated protein A (SP-A). This could also be manipulated by TGFβ in explant culture: TGFβ decreased basal SP-A production and inhibited the ability of EGF to increase SP-A production.

The effect of insulin on maturation of the surfactant system has also been studied. Smith et al. (1975) used monolayer cultures prepared from fetal rabbits at 28 days of gestation to examine the effects of insulin in the presence or absence of cortisol on ^{3}H-choline incorporation into lecithin. Insulin (10–100 μU/mL) increased choline incorporation. Normally, cortisol (5.5×10^{-6}) increases choline incorporation into lecithin. However, when cortisol was present, all concentrations of insulin (1–100 μU/mL) reduced lecithin synthesis. Interestingly, insulin did not block the normal inhibitory effect of cortisol on cell growth as measured by DNA, suggesting that the effect on lecithin synthesis is somewhat selective. In contrast to this work, but using a different model system, Snyder and Mendelson (1987) demonstrated that, in human fetal explant cultures, treatment with a combination of insulin and cortisol increased PC synthesis. SP-A was also examined. Normally, SP-A is not present until after 2 days of explant culture and increases thereafter. After 6 days of incubation with insulin (2.5–250 ng/mL), there was a dose-dependent decrease in SP-A in the explant homogenates.

In addition to studies on the direct effects of growth factors on type II cell maturation, a number of investigators have proposed a mechanism by which such growth factors could influence intercellular communication in the developing lung. Fetal lung fibroblasts produce a growth factor, which has been partially purified, termed fibroblast pneumocyte factor (FPF) (Smith and Post, 1989) which induces maturation of the type II epithelial cell manifested by increased surfactant synthesis. FPF is produced only during a specific time during gestation. Neilsen (1989) hypothesized that EGF could alter type II cell maturation by influencing the time period during which FPF is produced. Isolated fetal (day 19 gestation) rat type II cells were used after 48 h in culture and surfactant synthesis was measured as the incorporation of ^{3}H-choline into DSPC. FPF activity was determined by measuring DSPC synthesis induced by fibroblast CM. FPF activity is present in fetal day 18 fibroblasts but not day 16 or day 17 fibroblasts. EGF (10 ng/mL) increased FPF activity by day 17 and day 18 fibroblasts. Cortisol increases FPF production by day 17 fibroblasts isolated from female fetuses, and in both day 18 male and female fibroblasts. EGF advanced the appearance of this cortisol stimulated FPF activity: Now day 16 female fibroblasts and both male and female day 17 fibroblast CM increased DSPC synthesis. EGF (0.1–50 ng/mL) alone had no significant effect on type II cell DSPC synthesis. Also, EGF added to cortisol-stimulated fibroblast CM did not enhance DSPC synthesis. These studies confirm the direct effect of FPF on type II cell maturation as well as demonstrate one possible way for EGF to affect type II cell maturation indirectly.

Again, studying FPF activity, Torday and Kourembanas (1990) reported the effects of fetal fibroblast CM on type II cells. Day 21 rat fetal fibroblasts produce FPF which caused an increase in ^{3}H-saturated PC (SPC) synthesis by day 19 rat fetal type II cells; CM from day 17 fetal fibroblasts blocked this activity. If antibody to TGF-β is present, the day 17 fibroblast CM no longer blocked the ability of day 21 fibroblast CM to cause an increase in ^{3}H-SPC synthesis by the day 19 type II cells; this was a dose-response effect. TGF-β itself had no effect on baseline synthesis of ^{3}H-SPC, indicating its mechanism of action is as an antagonist. Once stimulated by FPF, TGFβ had no effect.

In addition to studies on the effects of growth factors on the maturation of the surfactant system by alveolar epithelial cells, a number of investigators have examined some aspects of the production and degradation of extracellular matrix components. The two types of epithelial cells that have been studied are fetal rat lung epithelial cells and A549 cells, a type II cell-like adenocarcinoma cell line. Studies by Federspiel et al. (1990, 1991), using a fetal rat lung epithelial cell line (FRLE), focused on the effects of EGF and retinoic acid on the production of different collagen types. In both acute (Federspiel et al., 1990) and chronic (Federspiel et al., 1991) exposure experiments, the authors found significant alteration in specific collagen subtype production. Under either culture condition, EGF tended to decrease collagen production unless retinoic acid was also present. Chronic exposure to EGF also enhanced cell growth and inhibited the appearance of lamellar body-like structures. Additional studies using this cell line were carried out by DiMari et al. (1991), examining acute and chronic effects of TGFβ on growth, differentiation, and collagen synthesis. Initially cells were subcultured and placed in the presence or absence of TGFβ (3 ng/mL). After 8 days in culture, some of the control cells were exposed acutely to TGFβ for 24 h. In contrast to EGF, TGFβ did not alter growth or morphology (including lamellar body-like structures) of these cells. However, exposure to TGFβ did increase collagen synthesis. Acute TGFβ exposure increased collagen synthesis by 50%; chronic TGFβ exposure by 300%; types I and III were increased more than types IV and V. This group of studies demonstrates that growth factors can affect the growth, maturation, and collagen synthesis of a fetal type II pneumocyte-derived cell line. Since the type II cell's environment, including the surrounding matrix, can alter type II cell differentiated functions, it is possible that some of the growth factor effects on type II cells in vivo are a result of growth-factor effects on type II cell ECM production.

Keski-Oja et al. (1988a, 1988b) used A549 cells to examine the effects of TGFβ on production of urokinase plasminogen activator (u-PA), type I plasminogen activator inhibitor (PAI-1), fibronectin, and type I procollagen. This group of investigators (Keski-Oja et al., 1988a) compared the responses of A549 cells and WI-38 cells (human lung fibroblasts). Cells were used at confluence after 6 h in serum-free media with TGFβ added at 1–20 ng/mL and proteolytic activity was determined by caseinolysis. TGFβ stimulated proteolytic activity with a submaximal effect at 5 ng/mL. By 48 h of treatment there was a six-fold increase in activity accompanied by increased u-PA mRNA levels. There was a concomitant stimulation of PAI-1. In contrast, in WI-38 fibroblasts, TGFβ decreased μ-PA. IN addition to effects of TGFβ on proteolytic components, TGFβ is also known to influence the production of ECM proteins. Other studies (Keski-Oja, 1988b) demonstrated that procollagen α(i) mRNA was increased by TGFβ only in the fibroblast cell line. However, TGFβ increased fibronectin mRNA production by both A549 cells and WI-38 fibroblasts. PAI-1 mRNA was also induced in both cell lines, but the induction in A549 cells was of shorter duration. These studies provide important information on the growth factor regulation of ECM production and destruction by a lung epithelial cell line. Control of the ECM by growth-factor communication between lung cells in close proximity is likely to be important for repair of the lung after an injury as well as for normal development and normal cell turnover.

Summary

In summary, it is clear that growth factors affect both airway and alveolar epithelial cell growth and various cellular functions including morphologic differentiation, maturation, and extracellular matrix production. This implies that growth factors produced not only by these cells themselves, but by cells in close proximity such as inflammatory cells or mesenchymal cells in the underlying interstitium, as well as those growth factors produced in cells farther away but that can be carried to the epithelium, all can affect lung epithelial cells. The study of the specifics of signal transduction in lung epithelial cells remains in its infancy. The response by epithelial cells to these factors under various conditions, at different times in development and at different stages of repair after a lung injury, will be an exciting area of intense research in years to come. It is possible that different responses at different times will determine whether a lung injury resolves or develops into fibrosis, or whether a cell can go on to become invasive and ultimately malignant.

IV. Production of Growth Factors by Lung Epithelia

A. Epithelial Cells and Cell Lines

Although a significant amount of data exists using in-vivo models to detect the production of growth factors by specific lung cells, somewhat less data exists regarding production by cells in culture. The major experimental concern of most investigators previously has been to characterize the normal growth of these cells in culture focusing on the response of cells to a specific factor rather than endogenous growth-factor production. However, the one area where investigators have examined the ability of cultured lung cells to produce polypeptide factors has been in the area of aberrant growth control. In particular, lung carcinoma cell lines have been examined for the production of autocrine growth factors such as TGFα. The work of Imanishi et al. (1989) and Damstrup et al. (1989) suggests that a number of human lung carcinoma cell lines produce high levels of TGFα. Among these lines is the A549 cell line, which has also been shown by Danielpour et al. (1991) to produce TGFβ isoforms. In an early comprehensive study, Derynck et al. (1987) demonstrated that messenger RNA for TGFα, the EGF receptor, and TGFβ all were produced by A549 cells; TGFα measured by ELISA was present in A549 cells and their CM. In so far as the A549 is thought to be derived from the alveolar type II cell and retain some characteristics of this cell, it is interesting to speculate about the ability of this cell to synthesize these factors that may regulate ultimately its own growth and differentiation.

In addition to the information about human cell lines, some information has also been gathered from studies of rodent cells in culture. Rabbit tracheal epithelial cells immortalized by SV40-T antigen and transformed by v-Ha-ras were found to grow independent of exogenous EGF (Jetten, 1991); cells that were immortalized but not transformed still required EGF for growth. This was thought to be due to increased constitutive expression of TGFα by the transformed cells. Similarly, rat tracheobronchial epithelial cells transformed through the action of chemical carcinogens have also been shown to express increased amounts of TGFα protein and mRNA; these cells also express the EGF receptor (Ferriola et al., 1991).

TGFβ has also been shown to be produced by a number of pulmonary epithelial cells. As discussed earlier, the A549 adenocarcinoma cell line is known to produce large quantities of TGFβ (Danielpour et al., 1991; Derynck et al., 1987; Wakefield et al., 1987). In addition, normal human bronchiolar epithelial cells have been shown to secrete TGFβ (Wakefield et al., 1987) although the amounts produced appear to be significantly less. There have been scattered preliminary reports of TGFβ production by normal type II cells but as yet no definitive studies have been published.

B. In Vivo

Development

Several groups have used immunohistochemistry to localize specific growth factors in developing lungs. Stahlman et al. (1989) examined human fetuses and newborns throughout gestation for immunoreactive EGF staining in the trachea and lung. In normal early-gestation trachea, there were many nonciliated lining epithelial cells which were positively stained for EGF. After the appearance of ciliated cells in the lining epithelium, the staining of nonciliated cells became more scattered. By 24 weeks gestation, there was little lining epithelial staining. As soon as submucosal glands appeared, staining was present there; these glands remained positive for the remainder of gestation. EGF staining never was identified in the bronchiolar epithelium of fetuses or newborns without lung pathology; staining here could be seen in infants with chronic lung disease. This group continued its work (Johnson et al., 1989) by examining the ontogeny of the EGF receptor in the ovine lung. At 41 days gestation, immunostaining was faint. By 55 days, EGF receptor immunoreactivity appeared in the bronchial lining epithelium; this staining was extensive by 75 days, still present through 129 days, but absent by near term ($\geqslant$ 141 days). As soon as bronchial submucosal glands were seen at 90 days, staining was present there and remained throughout gestation. Bronchiolar epithelial staining was present from 75 days through term. The deep folds of the bronchial and bronchiolar epithelium were the most intensely stained. The junction of bronchiole and alveolus was identified at 90 days, but immunoreactivity was not seen until near term (144–145 days). This is consistent with the studies of Raaberg et al., (1991), who demonstrated that EGF immunoreactivity was localized to type II cells in late fetal rats.

Heine et al. (1990) utilized mouse embryos (day 11 to day 18 of gestation) to study the appearance of TGFβ_1 during lung morphogenesis. Polyclonal antibodies that recognize intracellular TGFβ, extracellular TGFβ, collagen I, collagen III, or fibronectin were used. As early as day 11 of gestation, intracellular TGFβ was seen in the cytoplasm of both major lung cell types present at that time in development, stromal cells and epithelial cells of primordial ducts. Extracellular TGFβ was found in low concentration in the stroma. By day 13, extracellular TGFβ staining was much more intense; the entire stroma was deeply stained. By day 15, there was differentiation of the ducts into the proximal bronchiolar and distal alveolar components. Overall mesenchymal staining was decreased. Intracellular TGFβ was detected in the cytoplasm of both bronchial and alveolar lining epithelial cells and in smooth muscle cells surrounding bronchial epithelium. Extracellular TGFβ was localized to the cytoplasm of the bronchiolar duct and adjacent smooth muscle. Now the alveolar areas were negative. For the most part, staining for extracellular matrix proteins (collagen types I and III, fibronectin, and proteoglycan) co-localized with extracellular TGFβ, particularly during the period of differentiation into bronchiolar and alveolar ducts from day 13 to day 15. These studies suggest that TGFβ plays an important role in lung morphogenesis, both stimulating or inhibiting growth or differentiation of various cell types during different periods of development.

Injury

A second important area where expression of growth factors by epithelial cells can be critical is that of response and recovery from injury. From studies of wound healing (Rappolee and Werb, 1990), it is clear that activated inflammatory cells produce large amounts of growth factors. Less data are available about the production of such factors by parenchymal cells in the lung. However, based on our knowledge of the critical role of the pulmonary epithelium in the process of lung repair (Witschi, 1991), it would seem logical that one aspect of the epithelial response to injury would be production of factors that would influence growth, differentiation, and matrix protein production.

In bleomycin-induced pulmonary fibrosis, an increase in total lung TGFβ mRNA (Hoyt and Lazo, 1988) as well as an increase in TGFβ production (Khalil et al., 1989) has been observed. Using immunohistochemical techniques, TGFβ was found to be elevated in bronchial epithelium soon after exposure, eventually returning to baseline levels while macrophage expression increased. Recently, this group has continued to examine the expression of TGFβ mRNA and protein, utilizing tissue obtained from patients suffering from idiopathic pulmonary fibrosis. Khalil et al., (1991) found a marked and consistent increase in TGFβ production in epithelial cells. As with the bleomycin injury studies, prominent immunostaining was observed in the bronchiolar epithelium. In addition, hyperplastic type II cells were found to be intensely stained. They concluded, from the spatial arrangement of extracellular and intracellular forms of TGFβ, that the majority of this material found in advanced fibrosis was secreted by the epithelial cells.

In a recent series of preliminary studies in a model of radiation pneumonitis, Finkelstein et al. (1991) have detected increased secretion of TGFβ and TGFα by type II cells isolated from in-vivo irradiated lungs as well as from control type II cells irradiated in vitro.

Taken together, these studies suggest that the epithelium may play a significant role in the processes of repair of the lung after injury through increased production of polypeptide factors. It is clear that future studies in this area will be extremely illuminating.

V. Summary

This chapter attempted to provide an overview of the area of pulmonary epithelial cells and growth factors. As the field of study of these cells is still relatively young and in a phase of rapid growth, little detailed information about specific signal transduction mechanisms is currently available. However, insofar as we have attempted to show that the majority of the responses of the pulmonary epithelium are similar to cellular responses from other organ systems, one would expect that similar transduction pathways will be operating. What we have tried to do is provide evidence for the responsiveness of the pulmonary epithelium to specific polypeptide growth factors and used this as evidence for the existence of appropriate receptors for these factors. The fact that these growth factors exist in the lung, either during development of the pulmonary parenchyma or during the process of repair after injury, is not a matter of speculation. What remains to be clarified is the precise site of synthesis, and the role of the epithelium in regulating growth factor mediated processes.

Acknowledgments

The authors gratefully acknowledge the support from the National Heart Lung and Blood Institute (HL36543, HL37388 to J.N.F., and Clinical Investigator Award K08-HL02630-01 to R.M.R.), from the National Cancer Institute (CA27791 to J.N.F.), and from the American Lung Association (R.M.R.).

References

Brandes ME, Finkelstein JN. Production of growth modulators for type II epithelial cells by activated alveolar macrophages. FASEB J 1988; 2:A1010.

Brandes ME, Finkelstein JN. Stimulated rabbit alveolar macrophages secrete a growth factor for type II pneumocytes. Am J Respir Cell Mol Biol 1989; 1:101–9.

Catterton WZ, Escobedo MB, Sexson WR, Gray ME, Sundell HW, Stahlman MT. Effect of epidermal growth factor on lung maturation in fetal rabbits. Pediatr Res 1979; 13:104–8.

Damstrup L, Rorth M, Poulson HS. Growth factors and growth factor receptors in human malignancies, with special reference to human lung cancer: a review. Lung Cancer 1989; 5:49–68.

Danielpour D, Kim KY, Winokur TS, Sporn MB. Differential regulation of the expression of transforming growth factor-βs 1 and 2 by retinoic acid, epidermal growth factor, and dexamethasone in NRK-49F and A549 cells. J Cell Physiol 1991; 148:235–44.

De Pablo F, Scott LA, Roth J. Insulin and insulin-like growth factor I in early development: peptides, receptors and biological events. Endocr Rev 1990; 11(4):558–77.

Derynck R, Goeddel DV, Ullrich A, Gutterman JU, Williams RD, Bringman TS, Berger WH. Synthesis of messenger RNAs for transforming growth factors α and β and the epidermal growth factor receptor by human tumors. Cancer Res 1987; 47:707–12.

DiMari SJ, Howe AM, Haralson MA. Effects of transforming growth factor-β on collagen synthesis by fetal rat lung epithelial cells. Am J Respir Cell Mol Biol 1991; 4:455–62.

Everett MM, King RJ, Jones MB, Martin HM. Lung fibroblasts from animals breathing 100% oxygen produce growth factors for alveolar type II cells. Am J Physiol 1990; 259 (Lung cell mol physiol 3):L247–54.

Federspiel S, DiMari SJ, Guerry-Force ML, Haralson MA. Extracellular matrix biosynthesis by cultured fetal rat lung epithelial cells; II. effects of acute exposure to epidermal growth factor and retinoic acid on collagen biosynthesis. Lab Invest 1990; 63(4):455–66.

Federspiel SJ, DiMari SJ, Howe AM, Guerry-Force ML, Haralson MA. Extracellular matrix biosynthesis by cultured fetal rat lung epithelial cells; III. effects of chronic exposure to epidermal growth factor on growth, differentiation and collagen biosynthesis. Lab Invest 1991; 64(4):463–73.

Ferriola PC, Earp HS, Di Augustine R, Nettesheim P. Role of TGFα and its receptor in proliferation of immortalized rat tracheal epithelial cells: studies with tyrphostin and TGFα antisera. J Cell Physiol 1991; 147:166–75.

Finkelstein JN, Rubin P, Kramer CM, Paulhamous L. Production of growth factors by isolated lung cells after irradiation in vivo and in vitro. Am Rev Respir Dis 1991; 143(4;2):A546.

Gerwin BI, Keski-Oja J, Seddon M, Lechner JF, Harris CC. TGF-β_1 modulation of urokinase and PAI-1 expression in human bronchial epithelial cells. Am J Physiol 1990; 259 (Lung cell mol physiol 3):L262–9.

Gross I. Regulation of fetal lung maturation. Am J Physiol 1990; 259 (Lung cell mol physiol 3):L337–44.

Gross I, Dynia DW, Rooney SA, Smart DA, Warshaw JB, Sissom JF, Hoath SB. Influence of epidermal growth factor on fetal rat lung development in vitro. Pediatr Res 1986; 20(5):473–7.

Heine UI, Munoz EF, Flanders KC, Roberts AB, Sporn MB. Colocalization of TGF-beta 1 and collagen I and III, fibronectin and glycosaminoglycans during lung branching morphogenesis. Development 1990; 109:29–36.

Hoyt DG, Lazo JS. Alterations inpulmonary mRNA encoding procollagens, fibronectin, and transforming growth factor-β precede bleomycin-induced pulmonary fibrosis in mice. J Pharmacol Exp Ther 1988; 246:765–71.

Imanishi K-I, Yamaguchi K, Kuranami M, Kyo E, Hozumi T, Abe K. Inhibition of growth of human lung adenocarcinoma cell lines by anti-transforming growth factor-α monoclonal antibody. J Natl Cancer Inst 1989; 81:220–3.

James R, Bradshaw RA. Polypeptide growth factors. Ann Rev Biochem 1984; 53:259–92.

Jetten AM. Growth and differentiation factors in tracheobronchial epithelium. Am J Physiol 1991; 260 (Lung Cell Mol Physiol 4):L361–73.

Johnson MD, Gray ME, Carpenter G, Pepinsky RB, Sundell H, Stahlman MT. Ontogeny of epidermal growth factor receptor/kinase and of lipocortin-1 in the ovine lung. Ped Res 1989; 25(5):535–41.

Ke Y, Gerwin BI, Ruskie SE, Pfeifer AMA, Harris CC, Lechner JF. Cell density governs the ability of human bronchial epithelial cells to recognize serum and transforming growth factor beta-1 as squamous differentiation-inducing agents. Am J Pathol 1990; 137(4):833–43.

Keski-Oja J, Blasi F, Leof EB, Moses HL. Regulation of the synthesis and activity of urokinase plasminogen activator in A549 human lung carcinoma cells by transforming growth factor-β. J Cell Biol 1988a; 106:451–9.

Keski-Oja J, Raghow R, Sawdey M, Loskutoff DJ, Postlethwaite AE, Kang AH, Moses HL. Regulation of mRNAs for type-1 plasminogen activator inhibitor, fibronectin, and type I procollagen by transforming growth factor-β; divergent responses in lung fibroblasts and carcinoma cells. J Biol Chem 1988b; 263(7):3111–5.

Khalil N, Bereznay O, Sporn M, Greenberg AH. Macrophage production of transforming growth factor β and fibroblast collagen synthesis in chronic pulmonary inflammation. J Exp Med 1989; 170:727–37.

Khalil N, O'Connor RN, Unruh HW, Warren PW, Flanders KC, Kemp A, Bereznay OH, Greenberg AH. Increased production and immunohistochemical localization of transforming growth factor-β in idiopathic pulmonary fibrosis. Am J Respir Cell Mol Biol 1991; 5:155–62.

King R, Jones MB, Minoo P. Regulation of lung cell proliferation by polypeptide growth factors. Am J Physiol 1989; 257 (Lung Cell Mol Physiol 1):L23–38.

Leslie CC, McCormick-Shannon K, Cook JL, Mason RJ. Macrophages stimulate DNA synthesis in rat alveolar type II cells. Am Rev Respir Dis 1985a; 132:1246–52.

Leslie CC, McCormick-Shannon K, Mason RJ. Bronchoalveolar lavage fluid from normal rats stimulated DNA synthesis in rat alveolar type II cells. Am Rev Respir Dis 1989; 139:360–6.

Leslie CC, McCormick-Shannon K, Mason RJ. Heparin-binding growth factors stimulate DNA synthesis in rat alveolar type II cells. Am J Respir Cell Mol Biol 1990; 2:99–106.

Leslie CC, McCormick-Shannon K, Robinson PC, Mason RJ. Stimulation of DNA synthesis in cultured rat alveolar type II cells. Exp Lung Res 1985b; 8:53–66.

Lyons RM, Moses HL. Transforming growth factors and the regulation of cell proliferation. Eur J Biochem 1990; 187:467–73.

Mason RJ, Williams MC. Alveolar type II cells. In: Crystal RG, West JB, eds. The lung, scientific foundations. New York: Raven Press, 1991; 235–46.

Mason RJ, Apostolou S, Power J, Robinson P. Human alveolar type II cells: stimulation of DNA synthesis by insulin and endothelial cell growth supplement. Am J Respir Cell Mol Biol 1990; 3:571–7.

Massague J. Transforming growth factor-α. J Biol Chem 1990; 265(35):21393–6.

Masui T, Wakefield LM, Lechner JF, LaVeck MA, Sporn MB, Harris CC. Type β transforming growth factor is the primary differentiation-inducing serum factor for normal human bronchial epithelial cells. Proc Natl Acad Sci USA (Cell Biology) 1986; 83:2438–42.

Messmer TO, Armour R, Holley RW. Factors influencing the growth of alveolar type II epithelial cells isolated from rat lungs. Exp Cell Res 1982; 142:417–26.

Miyashita M, Smith MW, Willey JC, Lechner JF, Trump BF, Harris CC. Effects of serum, transforming growth factor type β, or 12-O-tetradecanoyl-phorbol-13-acetate on ionized cytosolic calcium concentration in normal and transformed human bronchial epithelial cells. Cancer Res 1989; 49:63–7.

Nielsen HC. Epidermal growth factor influences the developmental clock regulating maturation of the fetal lung fibroblast. Biochim Biophys Acta 1989; 1012:201–6.

Oberg KC, Carpenter G. Dexamethasone acts as a negative regulator of epidermal growth factor receptor synthesis in fetal rat lung cells. Mol Endocrinol 1989; 3(6):915–22.

Phillips LS, Harp JB, Goldstein S, Klein J, Pao CI. Regulation and action of insulin-like growth factors at the cellular level. Proc Nutr Soc 1990; 49:451–8.

Raaberg L, Poulsen SS, Nexo E. Epidermal growth factor in the rat lung. Histochemistry 1991; 95:471–5.

Rappolee DA, Werb Z. mRNA phenotyping for studying gene expression in small numbers of cells: platelet-derived growth factor and other growth factors in wound-derived macrophages. Am J Cell Mol Biol 1990; 2:3–10.

Rennard SI, Beckman JD, Robbins RA. Biology of airway epithelial cells. In: Crystal RG, West JB, eds. The lung, scientific foundations. New York: Raven Press 1991; 157–67.

Retsch-Bogart GZ, Stiles AD, Moats-Staats BM, Van Scott MR, Boucher RC, D'Ercole AJ. Canine tracheal epithelial cells express the type I insulin-like growth factor receptor and proliferate in response to insulin-like growth factor I. Am J Respir Cell Mol Biol 1990; 3:227–34.

Roberts AB, Sporn MB. The transforming growth factors-betas. In: Sporn MB, Roberts AB, eds. Peptide growth factors and their receptors. Handbook of experimental pharmacology, Vol. 95. Heidelberg: Springer-Verlag, 1989.

Ryan RM, Mineo-Kuhn MM, Finkelstein JN. Newborn rabbit type II cells proliferate in culture and respond to growth factors. Am Rev Respir Dis 1991; 143(4;2):A301.

Saloman DS, Kim N, Saeki T, Ciardiello F. Transforming growth factor-α: an oncodevelopmental growth factor. Cancer Cells 1990; 2(12):389–97.

Segarini PR. A system of transforming growth factor-β receptors. Am J Respir Cell Mol Biol 1991; 4:395–6.

Shoji S, Rickard KA, Takizawa H, Ertl RF, Linder J, Rennard SI. Lung fibroblasts produce growth stimulatory activity for bronchial epithelial cells. Am Rev Respir Dis 1990; 141:433–9.

Siegfried JM. Detection of human lung epithelial cell growth factors produced by a lung carcinoma cell line: use in culture of primary solid lung tumors. Cancer Res 1987; 47:2903–10.

Smith BT, Post M. Fibroblast-pneumonocyte factor. Am J Physiol 1989 257 (Lung Cell Mol Physiol 1):L174–8.

Smith BT, Giroud CJP, Robert M, Avery ME. Insulin antagonism of cortisol action on lecithin synthesis by cultured fetal lung cells. J Pediatr 1975; 87(6):953–5.

Snyder JM, Mendelson CR. Insulin inhibits the accumulation of the major lung surfactant apoprotein in human fetal lung explants maintained in vitro. Endocrinology 1987; 120:1250–7.

Sporn MB, Roberts AB, Wakefield LM, de Crombrugghe B. Some recent advances in the chemistry and biology of transforming growth factor-beta. J Cell Biol 1987; 105:1039–45.

Stahlman MT, Orth DN, Gray ME. Immunocytochemical localization of epidermal growth factor in the developing human respiratory system and in acute and chronic lung disease in the neonate. Lab Invest 1989; 60(4):539–47.

St. George JA, Read LC, Cranz DL, Tarantal AF, George-Nascimento C, Plopper CG. Effect of epidermal growth factor on the fetal development of the tracheobronchial secretory apparatus in Rhesus monkey. Am J Respir Cell Mol Biol 1991; 4:95–101.

Stiles AD, Smith BT, Post M. Reciprocal autocrine and paracrine regulation of growth of mesenchymal and alveolar epithelial cells from fetal lung. Exp Lung Res 1986; 11:165–77.

Stoscheck CM, King LE Jr. Functional and structural characteristics of EGF and its receptor and their relationshiop to transforming proteins. J Cell Biochem 1986; 31:135–52.

Sundell HW, Gray ME, Fredrik SS, Escobedo MB, Stahlman MT. Effects of epidermal growth factor on lung maturation in fetal lambs. Am J Pathol 1980; 100:707–26.

Takizawa H, Beckmann JD, Shoji S, Claasen LR, Ertl RF, Linder J, Rennard SI. Pulmonary macrophages can stimulate cell growth of bovine bronchial epithelial cells. Am J Respir Cell Mol Biol 1990; 2:245–55.

Tanswell AK, Byrne PJ, Han RN, Edelson JD, Han VK. Limited divisions of low-density adult rat type II pneumocytes in serum-free culture. Am J Physiol 1991; 260 (6 Pt 1):L395–402.

Torday JS, Kourembanas S. Fetal rat lung fibroblasts produce a TGFβ homolog that blocks alveolar type II cell maturation. Dev Biol 1990; 139:35–41.

Wakefield LM, Smith DM, Masui T, Harris CC, Sporn MB. Distribution and modulation of the cellular receptor for transforming growth factor-beta. J Cell Biol 1987; 105:965–75.

Warburton D. Epigenetic autocrine and paracrine factors regulating lung morphogenesis; a paradigm for lung repair. Chest 1991; 99(3, suppl.):15S-8S.

Whitsett JA, Weaver TE, Lieberman MA, Clark JC, Daugherty C. Differential effects of epidermal growth factor and transforming growth factor-β on synthesis of Mr = 35,000 surfactant-associated protein in fetal lung. J Biol Chem 1987; 262(16):7908–13.

Witschi H. Role of epithelium in lung repair. Chest 1991; 99(3):22S-5S.

12

Expression and Regulation of Chemotactic Cytokine Genes by Nonimmune Pulmonary Cells

STEVEN L. KUNKEL

University of Michigan Medical School
Ann Arbor, Michigan

THEODORE J. STANDIFORD, MARK W. ROLFE, and ROBERT M. STRIETER

University of Michigan Medical Center
Ann Arbor, Michigan

I. Introduction

The recruitment of leukocytes from the lumen of a venule into injured or inflamed tissue is a general hallmark of a variety of inflammatory reactions. The initiation and maintenance of a successful chemotactic reaction is critically dependent on a preset cascade of events, which must occur with complete fidelity in order for white blood cells to arrive at a site of inflammation. Although many of these steps have been clearly identified, a complete understanding of this process still remains to be illuminated. For example, it is now obvious that early events which occur within the first minutes of an inflammatory reaction are important in dictating many distal events. One key proximal event appears to be altered blood flow to a localized area of tissue. This may reduce blood pH, lower oxygenation, and promote both leukocyte-leukocyte and leukocyte-endothelial cell interactions.

While many of these alterations are physical changes, a number of biochemical events are rapidly being engaged. Initial phases of the inflammatory response are known to greatly increase protein synthesis in a number of cells, resulting in the expression of a variety of important polypeptides. Two early-response cytokine gene products, interleukin-1 (IL-1) and tumor necrosis factor-alpha (TNF), are known to play a leading role in the continued progression of leukocyte elicitation (Matsushima et al., 1988; Larsen et al., 1989; Strieter 1989a, 1989b; Thornton et al., 1990; Baggiolini et al., 1989). Both IL-1 and TNF have received particularly close attention, as numerous studies have identified their ability to initiate the adherence of inflammatory cells to the vascular endothelium. Under normal stimulation, this interaction "commits" the leukocyte to an area of inflammation and may be an important regulatory mechanism in restricting the white blood cells to a designated area of reactivity. These adherence proteins, including endothelial cell-derived ELAM, ICAM, and VLAM, as well as leukocyte-derived CD11/CD18, have all been shown to be "up-regulated" by IL-1 and TNF.

Interestingly, the adherence of white blood cells to the vascular endothelium is transient. The binding of leukocytes to the endothelium must be a temporary affair, as this reversible adherence allows the inflammatory cells to congregate, accumulate, and activate as they pass through the endothelial cell wall. Next, the leukocytes must migrate through a matrix barrier of basement membrane and begin to recognize chemical signals which serve as a homing mechanism, directing the cells to the precise tissue location. Although this process appears rather uncomplicated when assessed histologically through the lens of a microscope, many of the steps leading to a chemotactic response remain a fundamental enigma of immunologic reactions. Many questions remain unanswered regarding inflammatory cell recruitment including: the source of the proximal mediators IL-1 and TNF; the mechanism(s) whereby inflammatory cells bind reversibly to the endothelium, the mechanism(s) of diapedesis; the ability to pass the basement membrane barrier; the generation of chemotactic signals; and the ability of leukocytes to follow chemical gradients.

While this introduction has touched briefly on a variety of important steps which are needed for successful chemotaxis, the remainder of this chapter will be devoted to a discussion of a novel class of chemotatic factors, chemotactic cytokines. An area which will be especially analyzed is the expression of two key chemotaxins, interleukin-8 (IL-8) and monocyte chemotactic protein-1 (MCP), by pulmonary type II-like epithelial cells.

II. Chemotactic Factors

A. A Redundant System of Host Response

Upon reviewing the chemotactic factor literature, one is struck with the diversity of chemotactic agents that represent one of the most redundant systems involved in host defense (Fernandez et al., 1978; Showell et al., 1976; Samuelsson, 1983; Thelen et al., 1988). Peptides, polypeptides, fragments of polypeptides, and lipids have all been shown to possess an activity associated with the movement of inflammatory cells (Table 1). This heterogeneous array of chemotactic factors underscores the importance of leukocyte recruitment to a successful inflammatory response. It is likely that specific responses may depend more heavily on particular chemotactic factors, with additional collateral mediators as backup mechanisms. Although many of these factors demonstrate little in the way of cell specificity, many inflammatory reactions are still characterized by set waves of relatively similar inflammatory cells. This fact suggests that either regulatory mechanisms are acting to limit the induction of only a few specific chemotactic factors, or that elicited cells respond only to certain "window" concentrations of chemotactic agents. The end result of a successful chemotactic response is the elicitation and localization of leukocytes to an inflamed/injured tissue site.

An additional observation regarding the activation of inflammatory cells via chemotactic factors is the general phenomenon of receptor specificity. The occupancy of a given receptor by one chemotactic agent does not preclude the ability of an additional chemotactic signal to bind to its specific receptor, resulting in further inflammatory cell activation. In other words, inflammatory cells demonstrate little in the way of cross-desensitization when chemotactic factors bind to their receptors. This observation again stresses the importance of a successful chemotactic response during inflammation. Not only are there multiple chemotactic mediators, there are multiple specific receptors. When coupled appropriately to its ligand, these receptor-ligand interactions can induce a classical signal transduction response, resulting in inflammatory cell activation. Most granulocyte chemotactic factors bind to their receptors and activate via a calcium and protein kinase C-dependent GTP-binding protein (Thelen et al., 1988). Therefore, inhibition of chemotactic factor-induced leukocyte activation events can be observed using agents such as staurosporine, an inhibitor of protein kinase C. While a variety of chemotactic agents can bind to different cell receptors, these diverse receptor-ligand interactions appear to activate a common intercellular pathway leading to directed cell movement.

Table 1 A Partial List of Known Chemoattractants, Comprised of a Variety of Biologically Active Agents

C5a	Leukotriene-B4
Fibronectin products	Interleukin-8
Transforming growth factor-beta	Monocyte chemotactic protein-1
Bacterial peptides	RANTES
GRO/MGSA	NAP-2

Recent studies have demonstrated that the recruitment of cells into an area of inflammation is dependent not only on complement, leukotrienes, and bacterial-derived peptides, but also on a novel group of small proteins with relatively specific chemotactic activity (Matsushima et al., 1988; Larsen et al. 1989; Strieter et al. 1989a, 1989b; Thornton et al., 1990; Baggiolini et al. 1989; Lindley et al., 1988; Oppenheim et al., 1991; Matsushima and Oppenheim, 1989; Kunkel et al. 1991). As shown in Table 2, these small-molecular-weight polypeptides are synthesized by a variety of cells and belong to a novel supergene family of inflammatory mediators. This family of related cytokines can be broadly divided into two groups dependent on the location of four cysteine amino acids. One subgroup is characterized by two of the four cysteines in juxtaposition to one another (C-C family) and the other subgroup is characterized by two of the four cysteine residues being separated by one amino acid (C-X-C family). Recently it has been suggested that the C-C family, which can all be located on chromosome 4 in humans, be placed in a subfamily labeled intercrine-alpha (Oppenheim et al., 1991). The C-X-C subfamily, located on human chromosome 17, would in turn be labeled intercrine-beta (Oppenheim et al., 1991). One of the prototypes of the C-C family is monocyte chemotactic protein-1 (MCP), and a characteristic prototype of the C-X-C family is interleukin-8 (IL-8).

Interleukin-8

Interleukin-8 was originally defined as a monocyte-derived neutrophil chemotactic factor (MDNCF). However, it is now known that this polypeptide is derived from many cellular sources. Monocytes (Matsushima et al., 1988), alveolar macrophages (Strieter et al., 1990a), endothelial cells (Strieter et al., 1989a), hepatocytes (Thornton et al., 1990), fibroblasts (Strieter et al., 1988), and epithelial cells (Standiford et al., 1990) have all been identified as cellular sources of IL-8. While the first three cells in the above list are susceptible to a number of stimuli, the latter three possess an interesting degree of stimulus

Table 2 Selected Members of Two Subdivisions of a Supergene Family of Small-Molecular-Weight Polypeptides Comprising the C-C Family and the C-X-C Family

C-C/intercrine-alpha group	C-X-C/intercrine-beta group
Interleukin-8	Monocyte chemotactic protein-1
GRO/MGSA	RANTES
NAP-2	ACT-2
IP-10	LD-78

specificity. Endothelial cells and mononuclear phagocytes can readily respond to lipopolysaccharide (LPS), IL-1, and TNF with an increase in the synthesis of IL-8. However, fibroblasts and epithelial cells are not susceptible to LPS stimulation, but can respond to IL-1 and TNF. It appears that fibroblasts and epithelial cells must wait for the host immune system to recognize and respond to an inflammatory insult before these cells can secrete effector mediators, such as IL-8. This phenomenon will be expanded upon in the context of the lung below.

The physical and chemical aspects of IL-8 place this cytokine in a unique position among inflammatory mediators. Interleukin-8 maintains a large degree of bioactivity after being subjected to a number of denaturing conditions (Thelen et al., 1988). For example, IL-8 is relatively resistant to changes in pH, temperature, denaturing chemicals, and modest enzymatic digestion. Little biological activity of IL-8 is lost when this molecule is exposed to either pH 2 or pH 10, temperatures in excess of 56°C, or high concentrations of guanidine. In addition, exposure of IL-8 to limited trypsinization did not abolish the activity of this inflammatory mediator. The stability of IL-8 is likely conferred by its low molecular weight, amino acid composition, and folded structure. The importance of the four cysteine amino acids is ultimately responsible for the stability of IL-8, as reducing agents, such as beta-mercaptoethanol, can reduce the activity of this chemotaxin. Reduction and alkylation of IL-8 can significantly alter the activity of cytokine, suggesting that full activity of this molecule is dependent on an oxidized state. The stability of IL-8 suggests that this factor may be a long-lived molecule at a site of inflammation. The potential longevity of IL-8 in vivo may be buffered by the fact that neutrophils possess up to 20,000 receptors/cell for IL-8 and bind this chemotactic factor with pico-nanomolar affinities. The IL-8 receptor is also known to be recycled on these cells. Thus, certain inflammatory cells may prove to be a sink for IL-8 and provide for a convenient in-vivo mechanism for the clearance of the polypeptide.

Interestingly, the neutrophil appears to be a key cell in that it not only binds IL-8 with a high affinity and responds to IL-8 effectively, but also can synthesize IL-8 (Strieter et al., 1990b). Purified human polymorphonuclear cells can generate IL-8 in response to IL-1, TNF, and LPS. Picogram-per-milliliter concentrations of IL-1 and TNF are effective in increasing the expression of neutrophil-derived IL-8 as determined by a sensitive ELISA. Further studies demonstrated that steady-state levels of IL-8 mRNA from either LPS-, IL-1-, or TNF-stimulated neutrophils were detected by 0.5 h post-challenge. These findings demonstrate that peripheral blood neutrophils can also actively participate as effector cells during inflammation via the production of de-novo polypeptide mediators.

Monocyte Chemotactic Protein-1

An additional chemotactic cytokine, which is a member of the above C-X-C supergene family, is monocyte chemotactic protein-1 (MCP) (Furutani et al., 1989; Toshimura et al., 1989; Zachariae et al., 1990). This chemoattractant has also been shown to activate monocytes and has been alternatively named monocyte chemotactic and activating factor (MCAF). Monocyte chemotactic protein was originally isolated from a number of cell sources, including a human glioma cell line, a human myelomonocytic cell line (THP-1), and PHA-stimulated human blood mononuclear cells. Monocyte chemotactic protein can be detected as two proteins, MCP-alpha and MCP-beta. However, the two MCP polypeptides possess the exact same amino acid composition and are likely the same gene product. The differences may be due to posttranslational modification. The molecular weight of MCP is approximately 8700 Da, but has been reported to be higher due to glycosylation.

Although MCP and IL-8 broadly belong to the same supergene family, there are a number of differences between these polypeptides. For example, they share only 28% homology at the nucleotide level, only MCP is glycosylated, and only the 3′-untranslated region of IL-8 mRNA contains the UUAUUUAU repeat. The differences in the 3′-untranslated region of IL-8 and MCP mRNA are likely responsible for some of the disparate regulatory mechanisms observed between these two chemotactic cytokines. Finally, the biological activity of the polypeptides is quite different. Monocytes are the primary peripheral blood leukocytes which respond to MCP with directed movement, while IL-8 is a chemoattractant for either polymorphonuclear leukocytes or lymphocytes. Scientific information directed at understanding IL-8 has been advanced on a number of fronts, since this cytokine is a product of a number of cell types and this polypeptide is expressed in abundance upon stimulation. However, the discovery that MCP is identical to the cell competence factor JE has provided a body of information regarding the expression and regulation of the MCP gene product (Rollins et al., 1988). As with IL-8, MCP is a product of a variety of immune and nonimmune cells. In the context of the lung, the expression of IL-8 by pulmonary endothelial cells, epithelial cells, and fibroblasts likely represents an important mechanism for the rapid recruitment of leukocytes into the alveolar and interstitial space.

III. Expression of IL-8 and MCP by Pulmonary Type-II-Like Epithelial Cells

A. IL-1 and TNF Are Key Cytokines for the Expression of Pulmonary Epithelial Cell (A549)-Derived IL-8

The recruitment of leukocytes to the lung is dependent on the generation of local chemotactic factors that can establish a gradient from the vascular compartment to the alveolar space. Although a variety of chemoattractants have been isolated from the lung, the cellular or tissue source of these factors have not been clearly identified. The alveolar macrophage was once presumed to be the sole source of chemotactic factors, but recent evidence has identified other cells of the alveolar capillary unit as important contributors to the chemotactic response (Standiford et al., 1990; Kunkel and Streiter, 1990). A human type II-like pneumocyte (A549) has been identified as a cell type which can respond to a specific cytokine challenge (IL-1 or TNF) with the subsequent generation of IL-8. As shown in Fig. 1, IL-1 (20 ng/mL)- and TNF (20 ng/mL)-treated epithelial cells rapidly expressed steady-state levels of IL-8 mRNA. Detectable levels of IL-8 mRNA could be observed by 30 min. The expression of IL-8 mRNA peaked by 4 h and remained elevated for the next 24 to 48 h. A similar scenario was observed with regard to the generation of biological activity. While unstimulated A549 epithelial cells did not constitutively release a chemotactically active mediator, the addition of IL-1 or TNF to the epithelial cells caused a time-dependent increase in the production of chemotactic activity (Fig. 2). Kinetic analysis of bioactivity demonstrated significant levels by 4 h, a peak in activity by 8 h, and continued expression for the remaining 24 to 48 h of the study period.

Neutralization studies, using high-titer polyclonal antibodies directed against human IL-8, demonstrated that approximately 50% and 20%, respec-

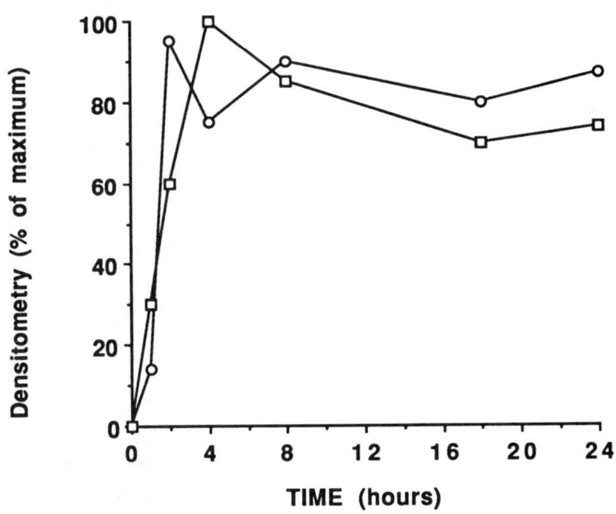

Figure 1 Laser densitometry taken of autoradiographs of Northern blots demonstrating the kinetics of steady-state levels of IL-8 mRNA by IL-1- or TNF-treated A549 pulmonary epithelial cells. As shown in this figure, the induction of IL-8 mRNA by these epithelial cells was rapid and sustained.

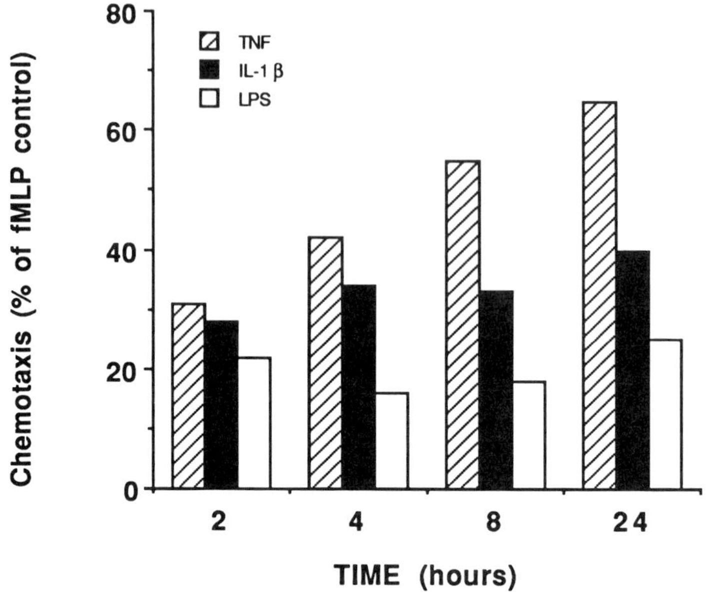

Figure 2 Kinetic analysis of chemotactic activity generated by either IL-1- or TNF-stimulated A549 epithelial cells. Significant levels of neutrophil chemotactic activity were observed in the supernatant of the A549 epithelial cells by 4 h (compared to non-IL-8 associated migration in response to LPS-stimulated cell supernatants).

tively, of the IL-1- or TNF-induced activity was due to IL-8. To further assess the production of epithelial cell-derived IL-8, an additional kinetic analysis was executed to identify the synthesis of antigenic IL-8 in response to either IL-1 or TNF. Using a sensitive IL-8 ELISA (sensitivity 10 pg/mL), statistically significant levels of IL-8 were not observed until 8 h post-addition of IL-1 or TNF (Fig. 3). After 24 h of stimulation, the levels of epithelial cell-derived antigenic IL-8 reached 25 and 45 ng/mL in response to TNF and IL-1, respectively. Interleukin-1 has always proven to be more potent than TNF in the induction of IL-8 antigenic protein and other related members of the supergene family.

An interesting difference can be observed between the production of a biologically active chemotactic factor(s) (Fig. 2) and antigenic IL-8 (Fig. 3) by the pulmonary epithelial cells. Taken together, the data in these two figures suggests that other neutrophil chemotactic factors are being generated in response to either IL-1 or TNF. Recent studies from our laboratory have begun the initial characterization of an additional low-molecular-weight polypeptide derived from IL-1-treated human epithelial cells (A549) that has

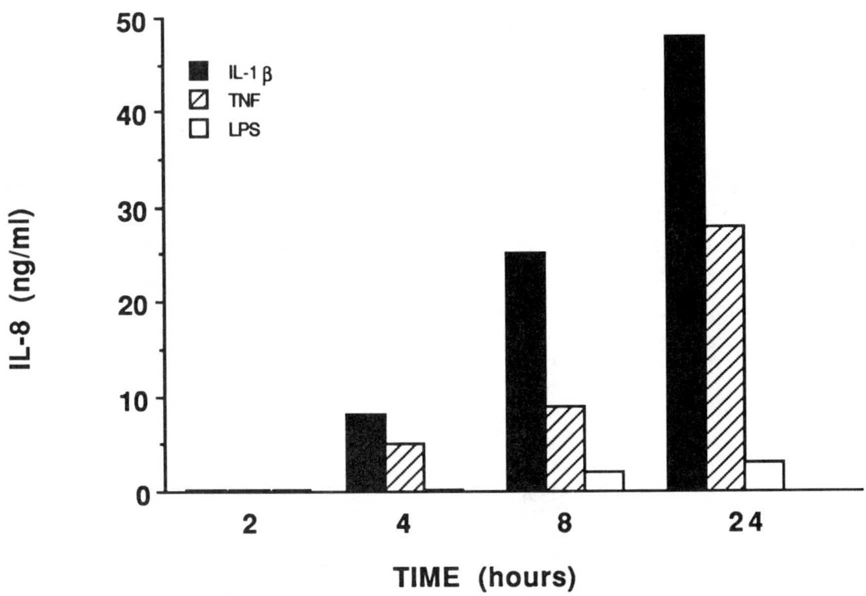

Figure 3 Kinetic analysis of the synthesis of antigenic IL-8 by a sensitive ELISA (sensitivity 10 pg/mL). The antigenic data demonstrated that significant levels of IL-8 did not occur until 8 h post-challenge with IL-1 or TNF. LPS was not an effective signal for the expression of antigenic IL-8.

chemotactic and activating properties for neutrophils. Initial amino acid analysis of this chemotaxin demonstrates that it also is a member of the IL-8 supergene family, as it belongs to the C-X-C family. The synthesis of IL-8 by type II-like epithelial cells appears to be dependent on stimulus specificity, as LPS is not an effective agonist for the expression of IL-8 (Fig. 3). Yet LPS can induce the production of low levels of chemotactic activity from the pulmonary epithelial cells (Fig. 2). This information further demonstrates the importance of a successful chemotactic response to a foreign challenge and supports the concept of continued redundancy of chemotactic factors.

B. The Expression of Pulmonary Epithelial Cell (A549)-Dervied IL-8 via Cytokine Networks

The above data suggest that the participation of lung epithelial cells in the recruitment of leukocytes is dependent on the prior generation of proximal mediators. In an in-vivo setting, the epithelial cells may exert maximal effector function for leukocyte elicitation only after the host has responded to the inciting insult. Thus, alveolar and interstitial lung macrophages are likely to respond rapidly to a Gram-negative bacterial challenge with the generation of IL-1 and/or TNF. The synthesis of these early response cytokines can then "network" with surrounding epithelial lung cells and induce the expression of other inflammatory mediators, such as IL-8. To assess this hypothesis, A549 cells were incubated with conditioned media from LPS-treated human alveolar macrophages. After 16 h of incubation, the type II-like epithelial cells were studied for the expression of steady-state levels of IL-8 mRNA. As shown by the laser densitometry scans in Fig. 4, soluble factors generated by LPS-triggered human alveolar macrophages could induce IL-8 gene expression from the lung epithelial cells. Interestingly, conditioned supernatants from alveolar macrophages cultured in the absence of LPS also stimulated the

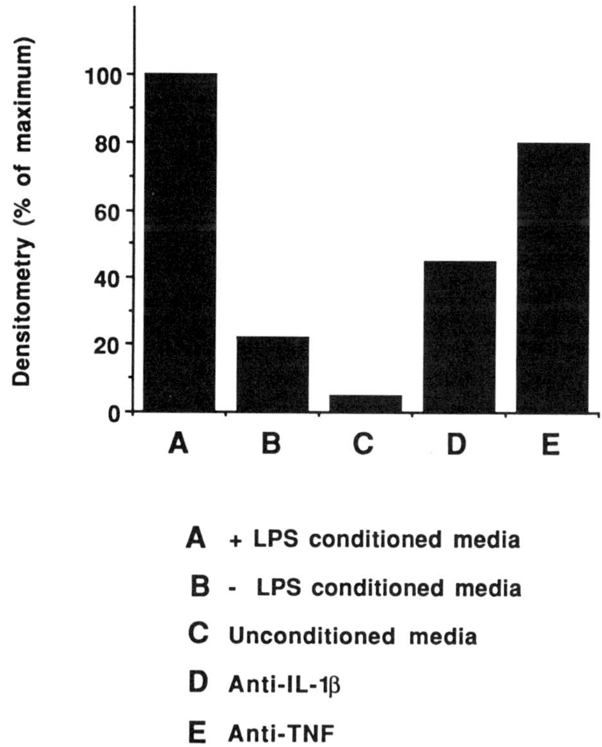

Figure 4 Cytokine networking between soluble factors generated by LPS-treated alveolar macrophages and A549 epithelial cells. Conditioned media from LPS- and non-LPS-treated macrophages was shown to induce steady-state levels of IL-8 mRNA. The laser densitometry of Northern blots from the conditioned media-treated A549 epithelial cells demonstrated significant expression of IL-8 mRNA, which was due in part to IL-1 and TNF.

expression of low but significant levels of steady-state IL-8 mRNA. No constitutive expression of IL-8 mRNA was found in the absence of any type of alveolar macrophage-conditioned media. Further experiments were conducted to assess the role of specific neutralizing antibody to either IL-1-beta or TNF-alpha. These studies were designed to determine the role of specific alveolar macrophage-derived cytokines on the expression of epithelial cell-derived IL-8 mRNA. The addition of either neutralizing IL-1 beta or TNF-alpha antibodies to the LPS-conditioned media resulted in a 55% or 20% reduction, respectively, in steady-state levels of IL-8 mRNA (Fig. 4). The above data support the hypothesis that cytokine networking is likely to occur between IL-1/TNF derived from resident tissue macrophages and the normal nonimmune cellular components of a given tissue. In the lung, both alveolar and interstitial macrophages are likely sources of proximal cytokines which can stimulate the production of IL-8 from surrounding fibroblasts, endothelial, and epithelial cells.

C. MCP Expression by Pulmonary Epithelial Cells (A549)

Since MCP belongs to the same supergene family as IL-8 and MCP has been shown to be a potent chemotaxin for monocytes, studies were designed to assess the expression of this factor from type II-like epithelial cells (Standiford et al., 1991). The addition of IL-1 or TNF to A549 epithelial cells resulted in a time-dependent increase in the expression of monocyte chemotactic activity. The addition of either cytokine to the epithelial cells caused a rapid increase (60 to 120 min) in the accumulation of monocyte chemotactic activity. The bioactivity synthesized by the type II-like cells appeared to peak by 8 h and persist for the next 24 h. The ability of neutralizing MCP antibody to suppress the chemotactic activity derived from the stimulated epithelial cells confirmed that the bioactivity was indeed MCP. In additional studies, a kinetic assessment of steady-state MCP mRNA demonstrated the appearance of MCP mRNA by 1 h and a plateau in MCP mRNA expression that occurred between 4 and 8 h. The induction of MCP mRNA from the pulmonary epithelial-like cells was not only time dependent, but was also dependent on the dose of IL-1 or TNF. Concentrations of IL-1 in the low picogram-per-milliliter range were found to be effective in stimulating steady-state levels of MCP mRNA from the cells. As with the expression of IL-8, MCP was not produced by LPS-challenged epithelial cells. This phenomenon suggests that the ability of pulmonary epithelial cells to synthesize MCP is dependent on the generation of early-response cytokines (IL-1/TNF) prior to the generation of chemotactic cytokines. In an in-vivo setting, the production of MCP from pulmonary epithelial cells is likely dependent on cytokine networking. To test this hypothesis, alveolar macrophages from nonsmoking/nondisease volunteers were stimulated with LPS and the conditioned media recovered from the cells was overlaid on A459 epithelial cells. After 4 h, the epithelial cells were assessed for the expression of MCP mRNA by Northern analysis. The conditioned media from the LPS-treated alveolar macrophages was efficient in stimulating the A549 cells to express significant steady-state levels of MCP mRNA, while supernatant recovered from unstimulated alveolar macrophages was not an effective stimulus. Neutralization studies were performed on the alveolar macrophage-conditioned media to assess the relative contribution of either IL-1 or TNF for epithelial cell-derived IL-8 expression. The addition of IL-1-beta-neutralizing antibodies to the LPS-conditioned media was efficacious in reducing steady-state levels of epithelial cell-derived MCP, while TNF-neutralizing antibodies were only partially effective. However, the combination of both IL-1-beta- plus TNF-neutralizing antibodies resulted in a synergistic reduction in altering the activity of the LPS-conditioned media for the expression of A549 epithelial cell-derived MCP. Using laser densitometry of the Northern blots in the neutralization studies, a 57%, 15%, and 72% suppression in steady-state levels of MCP mRNA was observed using antibodies directed against either IL-1-beta, TNF, or a combination of antibodies, respectively. This scenario is analogous to the production of IL-8 as described above.

IV. Signal Transduction Mechanisms and the Expression of Pulmonary Epithelial Cell-Derived IL-8 and MCP

The next set of studies was directed at understanding the potential signal transduction mechanisms that are active during the induction of epithelial cell-derived IL-8 and MCP. While a number of studies have elegantly described the signal transduction mechanisms that occur during inflammatory cell activation, little is known regarding signal transduction that dictates cytokine expression in nonimmune cells. In preliminary studies, A549 epithelial cells were treated with 2 ng/mL of IL-1 in the presence or absence of either cholera toxin, pertussis toxin, or staurosporine. These reagents have proven useful to assess the role of G-binding proteins and protein kinase C during cell activation. As shown in Fig. 5, both cholera toxin and staurosporine resulted in a dose-dependent reduction in the synthesis of IL-8 by IL-1-stimulated A549 epithelial cells, while pertussis toxin was not effective over the same dose range. Using 500 ng/mL of either cholera toxin or staurosporine, the synthesis of IL-8 by IL-1-challenged A549 epithelial cells was reduced by 45% and 50%, respectively. The expression of MCP by IL-1-treated A549 epithelial cells was also suppressed by cholera toxin or staurosporine (Fig. 5). Pertussis toxin over a wide concentration range did not alter the production of MCP by IL-1-treated A549 epithelial cells. A concentration of 500 ng/mL of either cholera toxin or staurosporine reduced the expression of MCP by approximately 50%. These preliminary investigations suggest that protein kinase C

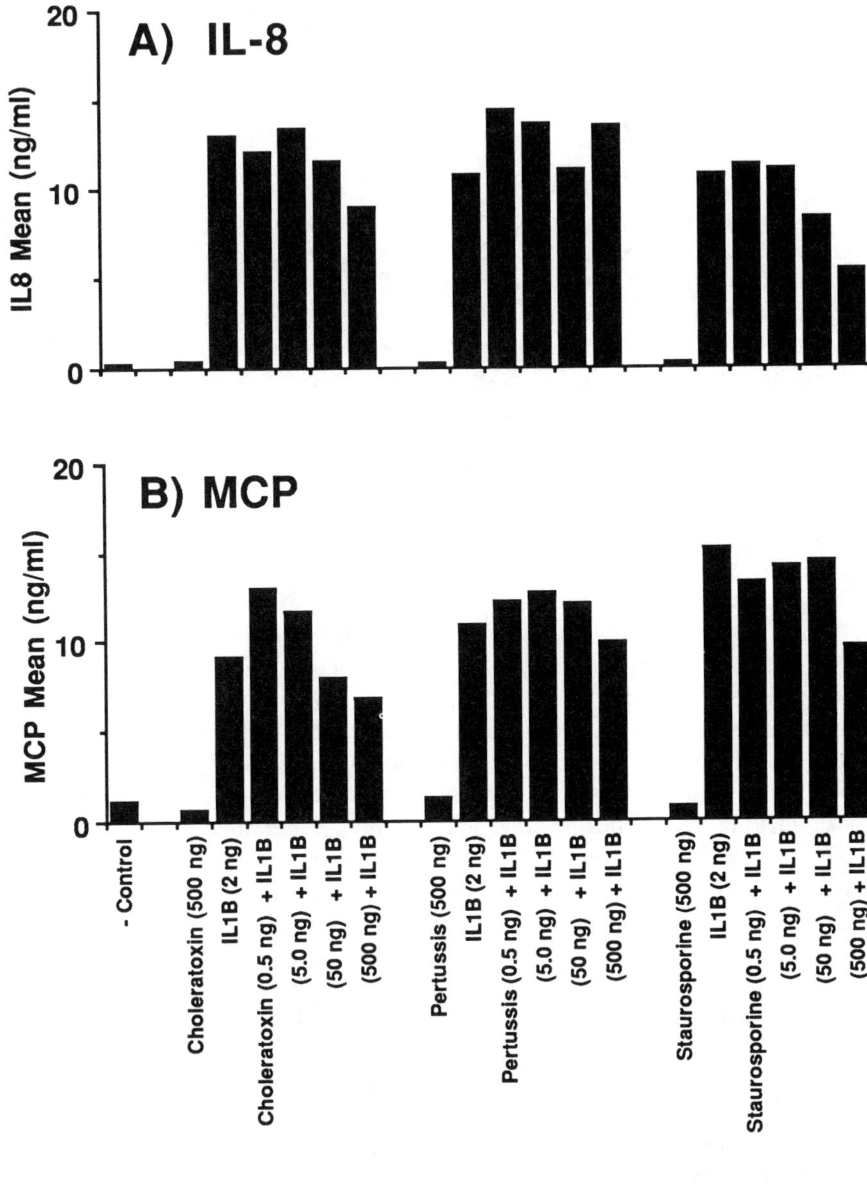

Figure 5 The concentration-dependent effects of pertussis toxin, cholera toxin, and staurosporine on the expression of IL-1-induced steady-state levels of IL-8 and MCP mRNA.

and certain G-binding proteins are important in IL-1-induced activation of epithelial cells.

V. Conclusion

The participation of cytokines in the general area of health and disease has generated much interested in different scientific disciplines. Not only are immunologists acutely interested in understanding the role of these polypeptides during certain diseases, but physiologists are also studying cytokines as they dictate normal homeostasis. This latter aspect is underscored by recent advances that have linked cytokine generation to sleep patterns, appetite regulation, and temperature rhythms. Yet the real explosion of research concerning cytokines has been derived from the immune/inflammation field. This area of study has truly flourished during the past decade, as mechanisms behind the activities of various cytokines have been uncovered. For example, the critical role that cytokine networks play during inflammation has generated renewed interest in the effector activity of many noninflammatory cells. A common theme that is evolving from these investigations is that nonimmune resident cells which comprise a given tissue can no longer be viewed as only targets of inflammation or passive "bystanders." In actuality, the resident cells are likely key players in the maintenance of an inflammatory reaction. The multifunctional role of these cells transcends their function as a normal epithelial cell, endothelial cell, fibroblast, or hepatocyte, since these cells can actively participate in an inflammatory response via cytokine networking. Once the host's sentinel cells, such as fixed macrophages and circulating monocytes, respond to an inciting agent and generate a repertoire of proximal mediators (IL-1/TNF), resident cells can quickly become activated and respond with the expression of distal cytokines (IL-8/MCP). Thus, the perpetuation of an inflammatory lesion becomes dependent on cytokine networking. This phenomenon is likely very important in the context of the lung, where a number of inflammatory disease states are characterized by the rapid elicitation of leukocytes into the pulmonary interstitium and alveolar space. In total, immune reactivity is a complex interaction of various cells and mediators that dictate the initiation, maintenance, and final resolution of the response. Future investigations into the mechanisms of these circuits in pulmonary inflammation will not only increase our understanding of lung diseases, but will likely result in novel therapeutic approaches.

Acknowledgments

The authors acknowledge the expert technical assistance of Marie Burdick and Holly Evanoff. These studies were supported in part by NIH grants HL 31693,

HL 35276, and HL 02401. Robert M. Strieter is an RJR Nabisco Research Scholar.

References

Baggiolini M, Walz A, Kunkel SL. Neutrophil activating peptide-1/interleukin-8, a novel cytokine that activated neutrophils. J Clin Invest 1989; 84:1045–9.

Fernandez HN, Henson PM, Otani A, Hugli TE. Chemotactic response to human C3a and C5a anaphylatoxins. J Immunol 1978; 120:109–13.

Furutani Y, Nomura H, Notake M, Oyamada Y, Fukui T, Yamada M, Larsen CG, Oppenheim JJ, Matsushima K. Cloning and sequencing of the cDNA for human monocyte chemotactic and activating factor (MCAF). Biochem Biophys Res Commun 1989; 159:249–55.

Kunkel SL, Strieter RM. Cytokine networks in lung inflammation. Hospital Practice 1990; 25:233–47.

Kunkel SL, Standiford T, Kasahara K, Strieter RM. Interleukin-8 (IL-8): the major chemotactic factor in the lung. Exp Lung Res 1991; 17:17–23.

Larsen C, Anderson AO, Oppenheim JJ, Matsushima K. Production of interleukin-8 by human dermal fibroblasts and keratinocytes in response to interleukin-1 or tumor necrosis factor. Immunology 1989; 68:31–6.

Lindley I, Aschauer H, Seifert JM, Lam C, Brunowsky W, Kownatski E, Thelen M, Peveri P, Dewald B, von Tscharner V, Walz A, Baggiolini M. Synthesis and expression in *E. coli* of the gene encoding monocyte-derived neutrophil chemotactic factor: biological equivalence between natural and recombinant neutrophil activating factor. Proc Natl Acad Sci USA 1988; 85:9199–203.

Matsushima K, Oppenheim JJ. Interleukin-8 and MCAFL. Novel inflammatory cytokines inducible by IL-1 and TNF. Cytokine 1989; 1:2–13.

Matsushima K, Morishita K, Yoshimura T, Lavu S, Kobayashi Y, Lew W, Apella E, Leonard EJ, Oppenheim JJ. Molecular cloning of a human monocyte-derived neutrophil chemotactic factor (MDNCF) and the induction of MDNCF mRNA by interleukin-1 and tumor necrosis factor. J Exp Biol 1988; 167:1883–93.

Oppenheim JJ, Zachariae CO, Mukaida N, Matsushima K. Properties of the novel proinflammatory supergene "intercrine" cytokine family. Ann Rev Immunol 1991; 9:617–48.

Rollins RJ, Morrison ED, Stiles SD. Cloning and expression of JE, a gene inducible by platelet-derived growth factor and whose product has cytokine-like activities. Proc Natl Acad Sci USA 1988; 85:3738–42.

Samuelsson B. Leukotrienes: mediators of immediate hypersenstivity reactions and inflammation. Science 1983; 220:568–71.

Showell HJ, Freer RJ, Zigmond SH, Schiffmann E, Aswanikumar S, Corcoran B, Becker EL. The structure-activity relations of synthetic peptides as chemotactic factors and inducers of lysosomal enzyme secretion for neutrophils. J Exp Med 1976; 143:1154–59.

Standiford TJ, Kunkel SL, Basha MA, Chensue SW, Lynch JP, Toews GB, Westwick J, Strieter RM. Interleukin-8 gene expression by a pulmonary epithelial cell line. A model for cytokine networks in the lung. J Clin Invest 1990; 86:1945–53.

Standiford TJ, Kunkel SL, Phan SH, Rollins BJ, Strieter RM. Alveolar macrophage-derived cytokines induce monocyte chemoattractant protein-1 expression from human pulmonary type II-like epithelial cells. J Biol Chem 1991; 266:9912–8.

Strieter RM, Chensue SW, Basha MA, Standiford TJ, Lynch JP, Kunkel SL. Human alveolar macrophages interleukin-8 gene expression by TNF-alpha, LPS, and IL-1 beta. Am Rev Respir Cell Mol Biol 1990a; 2:321–6.

Strieter RM, Kasahara K, Allen R, Chensue SW, Showell HJ, Standiford TJ, Kunkel SL. Human neutrophils exhibit disparate chemotactic factor gene expression. Biochem Biophys Res Commun 1990b; 173:725–30.

Strieter RM, Kunkel SL, Showell HJ, Marks RM. Monokine-induced gene expression of a human endothelial cell-derived neutrophil chemotactic factor. Biochem Biophys Res Commun 1988; 156:1340–5.

Strieter RM, Kunkel SL, Showell HJ, Remick DG, Phan SH, Ward PA, Marks R. Endothelial cell gene expression of a neutrophil chemotactic factor by TNF, LPS, and IL-1. Science 1989a; 243:1467–9.

Strieter RM, Phan SH, Showell HJ, Remick DG, Lynch JP, Genord M, Raiford C, Eskandari M, Marks R, Kunkel SL. Monokine-induced neutrophil chemotactic factor gene expression in human fibroblasts. J Biol Chem 1989b; 264:10621–6.

Thelen M, Peveri P, Kernen V, von Tscharner V, Walz A, Baggiolini M. Structure/function of NAP-1/IL-8. FASEB J 1988; 2:2702–10.

Thornton A, Strieter RM, Lindley I, Baggiolini M, Kunkel SL. Cytokine-induced gene expression of a neutrophil chemotactic factor/IL-8 by human hepatocytes. J Immunol 1990; 144:2609–13.

Toshimura T, Robinson EA, Tanaka S, Appella E, Kuratsu J, Leonard EJ. Purification and amino acid analysis of two human glioma-derived monocyte chemoattractants. J Exp Med 1989; 169:1449–159.

Zachariae CO, Anderson AO, Thompson HL, Appella E, Mantovani A, Oppenheim JJ, Matsushima K. Properties of monocyte chemotactic and activating factor (MCAF) purified from a human fibrosarcoma cell line. J Exp Med 1990; 171:2177–82.

13

Cellular and Biochemical Mechanisms of Airway Epithelial Mucin Release

K. CHUL KIM

University of Maryland School of Pharmacy
Baltimore, Maryland

I. Introduction

Mucus lining the airway luminal surface plays an important role in host defense against airborne particles and chemicals. The protective function of mucus is due mainly to the physicochemical properties of constituent mucous glycoproteins (mucins). Therefore, abnormalities in either the quality or the quantity of airway mucins not only affects removal of respired particles and bacteria, but may also impair host defenses, leading to further pathology. Airway mucins are multimillion-dalton glycoproteins containing extremely heterogeneous carbohydrate structures, and it is these sugar structures which contribute to their important biological functions in health and disease. As yet, their protein structures are not clearly understood. In humans, airway mucins are derived from two different cell types; goblet cells of surface epithelium and mucous cells of the submucosal gland. Although various agents influencing the secretion of submucosal mucins have been discovered (see reviews by Reid et al., 1983; Spicer and Martinez, 1984; Adler, 1986; Marin, 1986), the subcellular biochemical mechanisms of mucin secretion are not clearly understood due mainly to the unavailability of a characterized cell culture system. On the other hand, despite the recent availability of well-characterized cell culture systems (Kim and Brody, 1989), our understanding of the regulation of surface epithelial cell mucins is also very much limited, due presumably to the "intrinsic" nature of airway goblet cell mucin secretion; goblet cells are free of autonomic innervation and have been shown to be unresponsive to a majority of common secretagogues (Neutra et al., 1982; Kim, 1991). The biochemistry and pharmacology of the surface epithelial cell mucins have been recently reviewed (Kim, 1991). Although the main theme of this series deals with signal transduction of lung cells, no specific mechanism has yet been defined for the regulation of airway mucin secretion. Therefore, in this chapter, we describe recent developments which suggest the involvement of extracellular ATP in the regulation of mucin release via a signal transduction mechanism.

II. Mechanical Strain on Cells Can Stimulate Mucin Release

Production of mucins or mucinlike glycoproteins by cultured tracheal epithelial cells requires at least two conditions, growing on a thick collagen gel matrix and the presence of vitamin A in the culture medium (Kim, 1985; Wu et al., 1985; Rearick et al., 1987). Hamster tracheal surface epithelial (HTSE) cells cultured under this condition undergo dramatic morphological changes, from flat, proliferating cells to extremely compact, cuboidal to columnar confluent cells (Wasano et al., 1988). As cells become compact, secretory granules appear inside cells and mucins are released into the culture medium (Wasano et al., 1988). This observation has led us to suggest that the change of cell shape may be a necessary stimulus for mucous cell differentiation during cell growth and the constitutive mucin secretion following mucous cell differentiation. We tested the hypothesis by the following series of experiments.

When the thick collagen gel on which confluent cells were maintained was released and floated in the culture dish, the gel contracts concomitant with an increased release of mucins (100 ± 3% for control versus 248 ± 25% for treatment). The mucin increase during the gel floating was a result of the physical contraction of the gel but not a result of cell membrane damage during the gel contraction. After the gel floating, cell shape changed dramatically from cuboidal to columnar cells (Kim and Brody, 1987). Thus, the result of this experiment suggested that the mechanical strain on cells during gel contraction may be a stimulus for mucin release. This hypothesis was further tested using the following indirect methods.

EGTA (4 mM), which interrupts the tight junction and changes cells from flat to cuboidal shape (Pitelka and Taggart, 1983; Pitelka et al., 1983), enhanced mucin release by 51%. Collagenase (320 U/mL, Type IV, Worthington Biochemical Corp.), which interrupts the attachment of the basal cell surface to the collagen gel matrix, also resulted in a 114% increase in mucin release. Finally, cytochalasin B (0.1 μM), which dissembles actin filaments, also stimulated mucin release by 44%. Since these conditions dramatically change the cell shape, thus likely causing a strain on the plasma cell membrane, the results seem to support the hypothesis that mechanical strain on cells can stimulate mucin release. Finally, we examined if a direct application of mechanical strain can affect the mucin release.

A commercially available computerized vacuum-operated strain instrument (Flexcell Corp., McKeesport, PA) was used to mechanically strain confluent HTSE cells by maintaining cells on a rubber support and applying a constant vacuum (15 kPa) for 10 min (Banes et al., 1985). During this strain period, mucin release was greatly enhanced (100 $\pm$ 10% for control versus 264 $\pm$ 11% for treatment) without any noticeable membrane damage judging from LDH release. Therefore, it was concluded that mechanical strain on cells can stimulate the release of mucins.

III. Are GTP-Binding Proteins of Exocytosis Involved in Mucin Release?

It has been reported that there is a guanine nucleotide-binding protein(s) or G protein(s) in the plasma membrane the stimulation of which results in exocytosis without increasing intracellular Ca^{2+} level (Burgoyne, 1987). This plasma membrane protein was refered to as Ge (G protein of exocytosis) (Burgoyne, 1987). Since mucin release from HTSE cells is resistant to a Ca^{2+} ionophore, A23187, but responsive to mechanical strain, we suspected the involvement of "Ge" in the mechanical strain-induced mucin release. Thus, we examined whether or not such a mechanism is present for mucin release from HTSE cells. HTSE cells were radiolabeled with ^{3}H-glucosamine for 24 h and chased in the presence of 5 U/mL streptolysin O (Sigma) and various guanine nucleotides. Streptolysin O was used to generate pores on the membrane (Howell and Gomperts, 1987; Stuchfield and Cockcroft, 1988). As shown in Fig. 1, GTP$_\tau$S, a nonhydrolyzable GTP analog, stimulated mucin release in a dose-dependent fashion, suggesting the presence of the G protein. In order to confirm the involvement of the G protein, we further examined whether the GTP$_\tau$S effect could be blocked by GDP$_\beta$S. As seen in Fig. 2A, GDP$_\beta$S did not block the effect of GTP$_\tau$S effect but rather slightly increased the effect due to its own intrinsic stimulatory effect. Moreover, GTP showed the same potency as GTP$_\tau$S (Fig. 2B). Thus, the result did not seem to support the involvement of a G protein in mucin release. This was confirmed by using ATP$_\tau$S, which does not activate G proteins. ATP$_\tau$S (200 μM) stimulated mucin release with a magnitude even higher than that of 200 μM GTP$_\tau$S (224 $\pm$ 16% for ATP$_\tau$S

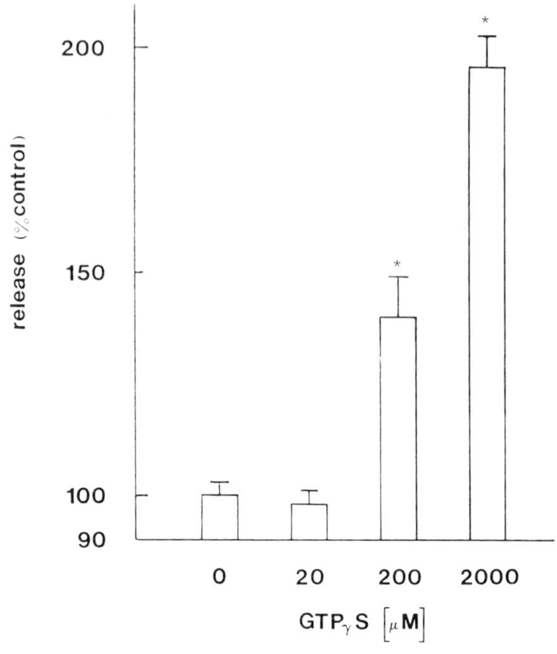

Figure 1 Effect of GTP$_\gamma$S on ^{3}H-mucin release. Confluent cultures (16 mm) which had been pretreated with streptolysin O were pulsed with ^{3}H-glucosamine for 24 h and chased for 30 min in the presence of varying concentrations of GTP$_\gamma$S. Ratios of the amount of ^{3}H-mucin released during the chase period to the amount of ^{3}H-mucin released during the pulse period was compared between control and treated groups using Student's t test for unpaired samples as previously described (Kim and Brody, 1989). Data represent means + S.E. of four culture dishes. *$P < 0.01$.

and 140 ± 5% for GTP$_\gamma$S). Therefore, the result indicated that the GTP$_\gamma$S effect was not mediated via G proteins but likely via cell surface receptors for purine analogs, so-called purinoceptors (Burnstock, 1978).

IV. ATP Stimulates Mucin Release by Activating Type II Purinoceptors Located on the Cell Surface

Purinergic receptors have been classified as type I and type II based on the potency order of adenine nucleotides, the dependence on intracellular cAMP levels, and its susceptibility to xanthines (Burnstock, 1978). As can be seen in Fig. 3, mucin release induced by adenine nucleotides appeared to involve type II or P2 purinoceptors. Type II purinoceptors are subclassified into P_{2x} and P_{2y}

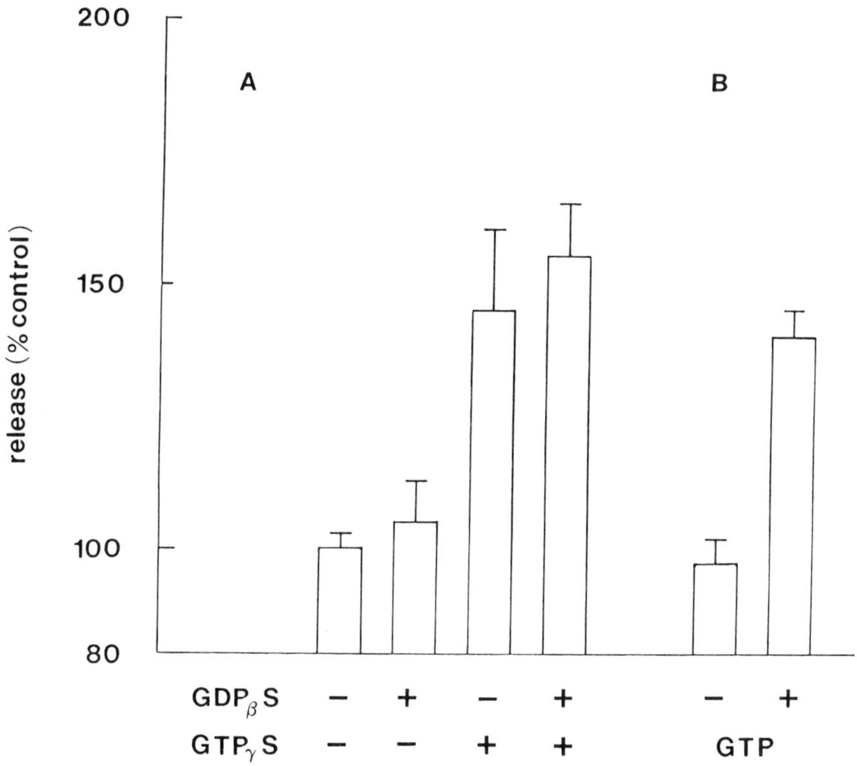

Figure 2 Effect of GDP$_\beta$S, GTP$_\gamma$S, and GTP on ^{3}H-mucin release. Confluent cultures (16 mm) which had been pretreated with streptolysin O were pulsed with ^{3}H-glucosamine from 24 h and chased for 30 min in the presence of guanine nucleotides (200 μM GTP$_\gamma$S, 200 μM GDP$_\beta$S, and 200 μM GTP). In A, GDP$_\beta$S was added immediately before GTP$_\gamma$S. Data represent means + S.E. of four culture dishes. Mucin release by GTP$_\gamma$S was not significantly ($p > 0.05$)[a] different in the presence or absence of GDP$_\beta$S.

[a] based on Student's t test for unpaired samples.

based on the pharmacological potency of ATP analogs (Burnstock and Kennedy, 1985) and into P_{2z} (Gordon, 1986) if pharmacological effects are mediated by ATP^{4-} (a tetravalent form) through plasma membrane permeabilization (Cockcroft and Gomperts, 1980). Based on the potency order of the ATP analogs (Fig. 4), the ATP-induced mucin release did not seem to involve either

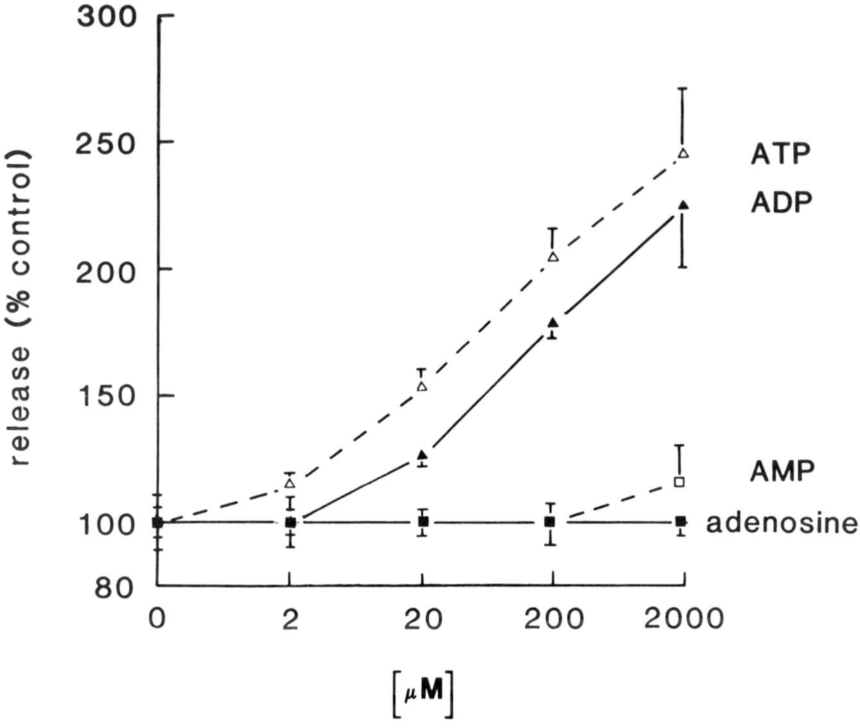

Figure 3 Effect of ATP, ADP, AMP, and adenosine on ^{3}H-mucin release. Confluent cultures (16 mm) were pulsed with ^{3}H-glucosamine for 24 h and chased for 30 min in the presence of the adenine analogs. Data represent means + S.E. of four culture dishes. (Data from Kim and Lee, 1991.)

P_{2x} or P_{2y} types. Moreover, reactive blue 2, a putative antagonist for P_{2y} receptor (Burnstock and Warland, 1987, Rice and Singleton, 1989), could not inhibit the effect of ATP (Kim and Lee, 1991). Finally, the presence of high concentrations of Mg^{2+} did not alter the effect of ATP, indicating that P_{2z} receptor is not responsible for ATP-induced mucin release (Kim and Lee, 1991). Therefore, it was concluded that the ATP-induced mucin release involves the P2 purinoceptor, which appears to be somewhat different from the already subclassified types. A similar P2 purinoceptor which seems to be involved in the regulation of transepithelial ion transport has been shown to be present on human nasal epithelial cells (Mason et al., 1991).

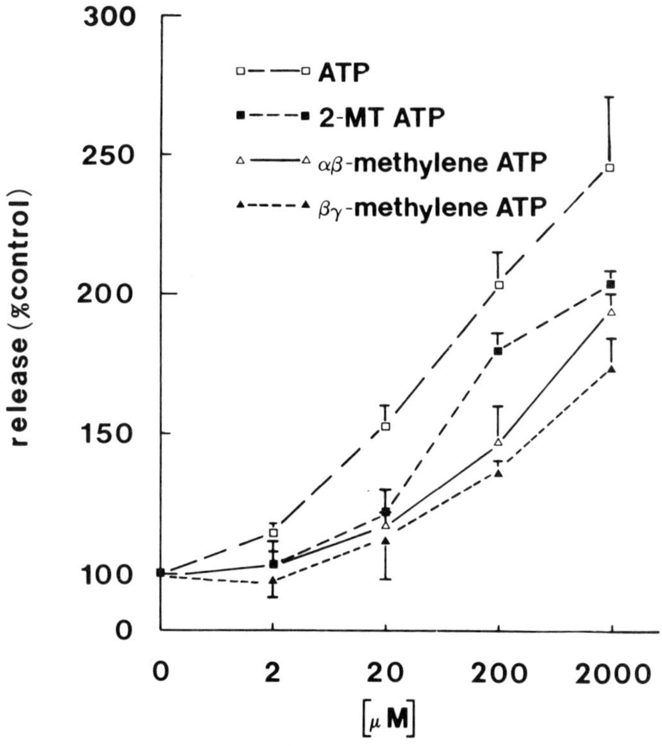

Figure 4 Effect of ATP, 2-methylthio ATP, α,β-methylene ATP, and β,τ-methylene ATP on ^{3}H-mucin release. Confluent cultures (16 mm) were pulsed with ^{3}H-glucosamine for 24 h and chased for 30 min in the presence of the ATP analogs. Data represent means + S.E. of four culture dishes. (Data from Kim and Lee, 1991.)

V. ATP-Induced Mucin Release Involves a Signal Transduction Mechanism via a Pertussis Toxin-Sensitive G Protein(s) Coupled to Phospholipase C

Stimulation of P_2 receptors has been shown to result in an increase in phosphatidylinositol (PI) turnover. In some cell types, the stimulated PI turnover is inhibited or attenuated by pretreatment with pertussis toxin (Pirotton et al., 1987; Tawada et al., 1987; Dubyak et al., 1988; Okajima et al., 1988). Therefore, in some cells, ATP stimulates P_2 receptors, which transduce a signal through pertussis toxin-sensitive G protein(s) into phospholipase C (PLC), which in turn generates inositol 1,4,5-triphosphate (IP3), a second messenger responsible for various cellular responses. We examined whether this signal transduction mechanism is involved in the ATP-induced mucin release. Confluent cells were labeled for 24 h with either ^{3}H-glucosamine (for mucin release) or ^{3}H-myoinositol (for PI turnover) in the presence or absence of pertussis toxin and chased in the presence or absence of 200 μM ATP. ATP increased both PI turnover (67 $\pm$ 2%) and mucin release (120 $\pm$ 10%). However, both ATP-induced PI turnover and mucin release were partially blocked in the presence of pertussis toxin, 46% and 42%, respectively. The results are consistent with other P_2 purinoreceptor systems (Dubyak et al., 1988; Okajima et al., 1988) and indicate that both PI turnover and mucin release take place distal to the site of action of pertussis toxin. Therefore, it is concluded that ATP-induced mucin release from HTSE cells involves a signal transduction mechanism in which pertussis toxin-sensitive G protein(s) and PLC serve as a transducer and an effector, respectively. The inositol phosphates thus produced will increase a cytosolic Ca^{2+} level which might be responsible for mucin release. The role of cytosolic Ca^{2+} in the release of mucins from HTSE cells is still unclear and will be discussed further below.

VI. Perspectives: Mechanisms of Airway Goblet Cell Mucin Release

Airway goblet cells are easily identified by their unique appearance, i.e., large, abundant PAS-positive secretory granules. Despite their distinct structure and location in the airway, we still do not understand how release of these granules is regulated under normal physiologic conditions. Considering that mucins are an essential component of lung defense and should be produced throughout the whole lifetime in order to compensate for their continuous loss as a result of constant mucociliary clearance of the airway fluid, it is conceivable that mucins must be produced continuously via some constitutive mechanisms under normal physiologic conditions. At present, little is known about the source of this "constitutive" mucin secretion or its regulation.

What stimulates this continuous synthesis and release of mucins from goblet cells? Several possible mechanisms can be thought of: (1) neuronal control, (2) humoral control, and (3) some nonneuronal, nonhumoral control. Since there is no nerve innervation into the airway surface epithelium, it is very unlikely that neural transmitters are involved in the regulation of epithelial mucin release. On the other hand, only limited humoral substances, such as proteases (Klinger et al., 1984; Adler et al., 1986; Niles et al., 1986; Kim et al., 1987), platelet-activating factor (Adler et al., 1987; Rieves et al., 1991), and purine nucleotides (Kim and Lee, 1991), have been shown to influence mucin release from surface epithelial cells. Since these products would be expected to be found mainly during airway inflammation, their role in the physiologic regulation of mucin release seems quite unlikely. Therefore, the fact that mechanical strain can stimulate mucin release is of great interest and may be of importance in understanding the physiologic regulation of airway epithelial mucin secretion.

How is the mechanical strain-induced mucin release related to the physiological regulation of airway epithelial mucins? The airway is heavily innervated by autonomic nerves, and the activity of smooth muscles in the airway is under the control of the autonomic nervous system. Since the surface epithelium is physically associated with underlying smooth muscles, contraction and relaxation of airway smooth muscles may mechanically strain cells in the epithelium. Thus, it is possible that under physiologic conditions, the airway smooth muscle tones may influence mucin release from surface epithelial cells. At present, biochemical mechanisms of mechanical strain-induced mucin release are not known.

Although we do not understand either the sources or concentrations of extracellular ATP under physiologic conditions, it is expected that, during airway inflammation, local extracellular ATP concentrations in the airway can reach high levels from lysed epithelial or inflammatory cells, since cytosolic ATP concentrations are greater than 5 mM (Gordon, 1986). Figure 5 describes a proposed model of the regulation of airway goblet cell mucin release by extracellular ATP. In this model, ATP binds to specific P2 purinoceptors on the cell surface. The ATP-bound P2 receptors then activate guanine nucleotide-binding proteins in the plasma membrane, which in turn activate

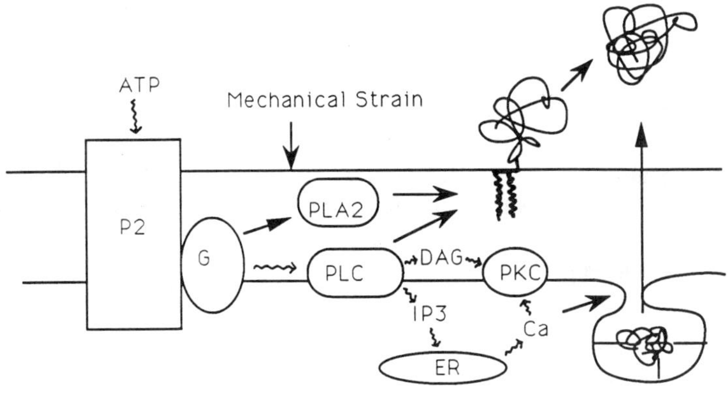

Figure 5 A proposed model of biochemical mechanisms of ATP-induced mucin release from airway surface epithelial cells.

phospholipase C. It is interesting that pertussis toxin blocked both the PI turnover and mucin release only partially, but both to the same extent. The result suggests (1) a cause-effect relationship between PI turnover and mucin release, and (2) the involvement of pertussis toxin-insensitive PLC as well as toxin-sensitive PLC in ATP-induced mucin release. As shown in Fig. 5, three pathways could be involved in PLC-mediated mucin release. First, the increased IP3 as a result of activation of PLC could release Ca^{2+} from intracellular Ca^{2+} pools, and the increased cytosolic Ca^{2+} could cause release of secretory granules as shown in various endocrine cells. The fact that mucin release from HTSE cells was not responsive to a Ca^{2+} ionophore, A23187, is intriguing at this moment. Since A23187 is a negatively charged hydrophobic molecule while mucins are also highly negatively charged through sialic acids and sulfates (Kim et al., 1985) as well as extremely hydrophobic (Kim and Singh, 1990), it might be possible that A23187 could not enter the plasma membrane due to its physicochemical interactions with mucins present on the cell surface (Kim et al., 1987). Thus, the question as to whether or not intracellular Ca^{2+} is necessary for mucin release remains to be answered. Second, diacylglycerol produced as a result of activation of PLC may activate protein kinase C along with cytosolic Ca^{2+} which becomes available through the first pathway. Although the role of PKC in mucin release is not clear in HTSE cells, activation of protein kinase C by phorbol esters and diacylglycerol have been shown to stimulate mucin secretion in rat submandibular gland cells (Fleming et al., 1986). Third, the activated PKC may be able to cleave mucins anchored on the cell surface via a specific catalytic action (Low and Saltiel, 1988). Some mucins are present on the apical cell surface of HTSE cells and releaseable by neutrophil elastase via its specific catalytic action (Kim et al., 1987). We have also found that highly purified mucins from HTSE cells contain palmitic acid as well as inositol, suggesting the presence of a glycosyl phosphatidylinositol moiety as an anchorage molecule (unpublished data). However, both the anchorage molecule and the specific PLC need to be purified.

To summarize, our present knowledge of the regulation of airway epithelial mucin secretion is primitive and far from explaining mucin release at the biochemical level. Continued research on these two prominent stimuli for mucin release, mechanical strain, and ATP will likely establish a basis for understanding the biochemical mechanism of the airway goblet cell mucin secretion in health and disease.

Acknowledgment

I thank Jeffrey Berman for his critical review of the manuscript. Preparation of this article was supported by a grant from the NIH, HL-47125.

References

Adler KB. Mucin secretion by explants of respiratory tissue in vitro. In: Schiff LJ, ed. In vitro models of respiratory epithelium. Boca Raton, FL: CRC Press, 1986; 27–50.

Adler KM, Hendley DD, David GS. Bacteria associated with obstructive pulmonary disease elaborate extracellular products that stimulate mucin secretion by explants of guinea pig airways. Am J Pathol 1986; 125:501–514.

Adler KB, Schwarz JE, Anderson WH, Welton AF. Platelet activating factor stimulates secretion of mucin by explants of rodent airways in organ culture. Exp Lung Res 1987; 13:25–43.

Banes AJ, Gilbert J, Taylor D, Monbureau O. A new vacuum-operated stress-providing instrument that applies static or variable duration cyclic tension or compression to cells in vitro. J Cell Sci 1985; 75:35–42.

Burgoyne RD. Control of exocytosis. Nature 1987; 328:112–3.

Burnstock G. A basis for distinguishing two types of purinergic receptor. In: Straub RW, Bolis L, eds. Cell membrane receptors for drugs and hormones. A multidisciplinary approach. New York, Raven Press, 1978; pp. 107–18.

Burnstock G, Kennedy C. Is there a basis for distinguishing two types of P2-purinoceptor. Gen Pharmacol 1985; 16:433–40.

Burnstock G, Warland JJI. P2-purinoceptors of two subtypes in the rabbit mesenteric artery: reactive blue 2 selectively inhibits responses mediated via the P2y-but not P2x-purinoceptor. Br J Pharmacol 1987; 90:383–91.

Cockcroft S, Gomperts BD. The ATP^{-4} receptor of rat mast cells. Biochem J 1980; 188:789–98.

Dubyak GR, Cowen DS, Meuller LM. Activation of inositol phospholipid breakdown in HL60 cells by P2-purinergic receptors for extracellular ATP. Evidence for mediation by both pertussis toxin-sensitive and pertussis toxin-insensitive mechanisms. J Biol Chem 1988; 263:18018–117.

Fleming N, Bilan PT, Sliwinski-Lis E. Effects of phorbol ester and diacylglycerols on secretion of mucin and arginine esterase by rat submandibular gland cells. Pfugers Arch 1986; 406:6–11.

Gordon JL. Extracellular ATP: effects, sources and fate. Biochem J 1986; 233:309–19.

Howell TW, Gomperts BD. Rat mast cells permeabilised with streptolysin-O secrete histamine in response to Ca^{2+} at concentrations buffered in the micromolar range. Biochim Biophys Acta 1987; 927:177–83.

Kim KC. Possible requirement of collagen gel substratum for production of mucin-like glycoproteins by primary rabbit tracheal epithelial cells in culture. In Vitro 1985; 21:617–21.

Kim KC. Biochemistry and pharmacology of mucin-like glycoproteins produced by cultured airway epithelial cells. Exp Lung Res 1991; 17:533–45.

Kim KC, Brody JS. Gel contraction causes mucin release in primary hamster tracheal epithelial cells growing on a collagen gel. J Cell Biol 1987; 105:158a.

Kim KC, Brody JS. Use of primary cell culture to study regulation of airway surface epithelial mucus secretion. In: Chantler EN, Ratcliffe NA, eds. Mucus and related topics. Cambridge U.K.: Company of Biologists Limited, 1989; 231–9.

Kim KC, Lee BC. P2 purinoceptor regulation of mucin release by airway goblet cells in primary culture. Br J Pharmacol 1991; 103:1053–6.

Kim KC, Singh BN. Hydrophobicity of mucin-like glycoproteins secreted by cultured tracheal epithelial cells: association with lipids. Exp Lung Res 1990; 16:279–92.

Kim KC, Rearick JI, Nettesheim P, Jetten AM. Biochemical characterization of mucous glycoproteins synthesized and secreted by hamster tracheal epithelial cells in primary culture. J Biol Chem 1985; 260:4021–7.

Kim KC, Wasano K, Niles RM, Schuster JE, Stone PJ, Brody JS. Human neutrophil elastase releases cell surface mucins from primary cultures of hamster tracheal epithelial cells. Proc Natl Acad Sci USA 1987; 84:9304–8.

Klinger JD, Tandler B, Liedke CM, Boat TF. Proteases of *Pseudomonas aeruginosa* evoke mucin release by tracheal epithelium. J Clin Invest 1984; 74:1669–78.

Low MG, Saltiel AR. Structural and functional roles of glycosylphosphatidylinositol in membranes. Science 1988; 239:268–75.

Marin MG. Pharmacology of airway secretion. Pharmacol Rev 1986; 38:273–89.

Mason SJ, Paradiso AM, Boucher RC. Regulation of transepithelial ion transport and intracellular calcium by extracellular ATP in human normal and cystic fibrosis airway epithelium. Br J Pharmacol 1991; 103:1649–56.

Neutra MR, O'Malley LJ, Specian RD. Regulation of intestinal goblet cell secretion. II. A survey of potential secretagogues. Am J Physiol 1982; 242:G380–7.

Niles RM, Christensen TG, Breuer R, Stone PJ, Snider GL. Serine proteases stimulate mucous glycoprotein release from hamster tracheal ring organ culture. J Lab Clin Med 1986; 108:489–97.

Okajima F, Sho K, Kondo Y. Inhibition of islet-activating protein, pertussis toxin, of P2-purinergic receptor-mediated iodide efflux and phosphoinositide turnover in FRTL-5 cells. Endocrinology 1988; 123:1035–43.

Pirotton S, Erneux C, Boeynaems JM. Dual role of GTP-binding proteins in the control of endothelial prostacyclin. Biochem Biophys Res Commun 1987; 147:1113–20.

Pitelka DR, Taggart BN. Mechanical tension induces lateral movement of intramembrane components of the tight junction: studies on mouse mammary cells in culture. J Cell Biol 1983; 96:606–12.

Pitelka DR, Taggart BN, Hamamoto ST. Effects of extracellular calcium depletion on membrane topography and occluding junctions of mammary epithelial cells in culture. J Cell Biol 1983; 96:613–24.

Rearick JI, Deas M, Jetten AM. Synthesis of mucous glycoproteins by rabbit tracheal cells in vitro. Modulation by substratum, retinoids and cyclic AMP. Biochem J 1987; 242:19–25.

Reid L, Bhaskar K, Coles S. Control and modulation of airway epithelial cells and their secretion. Exp Lung Res 1983; 4:157–70.

Rice WR, Singleton FM. Reactive blue 2 selectively inhibits P2y-purinoceptor-stimulated surfactant phospholipid secretion from rat isolated alveolar Type II cells. Br J Pharmacol 1989; 97:158–62.

Rieves D, Wu T, Logun C, Shelhamer J. Platelet activating factor stimulates mucin glycoprotein release from feline tracheal epithelial cells. Am Rev Respir Dis 1991; 143:A152.

Spicer SS, Martinez JR. Mucin biosynthesis and secretion in the respiratory tract. Environ Health Perspect 1984; 55:193–204.

Stuchfield J, Cockcroft S. Guanine nucleotides stimulate polyphosphoinositide phosphodiesterase and exocytotic secretion from HL60 cells permeabilized with streptolysin O. Biochem J 1988; 250:375–82.

Tawada Y, Furukawa KI, Shigekawa MJ. ATP-induced calcium transient in cultured rat aortic smooth muscle cells. J Biochem 1987; 102:1499–509.

Wasano K, Kim KC, Niles RM, Brody JS. Membrane differentiation markers of airway epithelial secretory cells. J Histochem Cytochem 1988; 36:167–78.

Wu R, Nolan E, Turner C. Expression of tracheal differentiated function in serum-free hormone-supplemented medium. J Cell Physiol 1985; 125:167–81.

14

Dysregulation of Cellular Genes by Latent Viral Genes

LOIS J. GEIST and GARY W. HUNNINGHAKE

University of Iowa Hospitals and Clinics
Iowa City, Iowa

I. Introduction

Throughout life, the average person is exposed to multiple viruses. Some of these viruses cause self-limited diseases and are eradicated by the immune system. However, there are a number of viruses that, once they gain entry, are somehow able to escape eradication by the immune system and exist in a latent state. The mechanism of latency is not well understood. It is clear that the viral genome is intact, as evidenced by reactivation during times of stress and immunosuppression. It is also clear that during the latent state the virus is not inactive but rather demonstrates a degree of restricted replication. The viral gene products generated during latent infection could induce disease by one of three mechanisms: If they are expressed on the cell surface, they could function as antigens to drive an "autoimmune" response; they may mimic other immune response genes through sequence homology; or, finally, these gene products may act as *trans*-acting factors to control the expression of genes involved in the immune response. In any one of these ways, the virus may trigger an inflammatory response without ongoing viral replication.

The purpose of this chapter is to review some of the information related to the *trans*-activation of cellular genes by viral gene products. We will focus on three separate virus families: the herpes viruses (HSV, CMV, and EBV), the retroviruses (HTLV-1 and HIV), and adenovirus. These viruses have several things in common: (1) They all cause clinical disease in the actively replicating state; (2) they all establish latent infections; and (3) they all express *trans*-acting factors now shown to interact with cellular genes.

II. Viral Replication

For an excellent overview of viral replication, in general, and, specifically, of the viruses discussed here, we refer the reader to Fields' *Virology* (1990).

In general, viral particles are relatively simple structures which carry all of the necessary genetic material to replicate after enlisting the help of the host cell. We will approach this topic by first giving a brief overview of viral replication and the function of the *trans*-acting factors. Then we will summmarize the data from our laboratory and others that demonstrate the interaction of cell-derived genes with the viral *trans*-acting factors. Finally, we will bring these concepts together with what is known about a few diseases of unknown etiology as well as diseases that are caused directly by these viruses.

Viral gene products are identified according to their appearance relative to the time of infection: immediate-early (IE), early (E), and late (L). The IE genes code for proteins which function in *trans* to up- or down-regulate other viral genes. The E genes contain the information for viral-specific polymerases. Coat proteins and other structural proteins are encoded by the late genes.

The concept of latency is widely accepted (Pagano, 1975; Klein, 1989). How latency is established and maintained remains a question. The site of latent infection is specific for each virus. For example, adenovirus sets up latency in the adenoidal tissue, primarily B cells; cytomegalovirus prefers macrophages and T lymphocytes; HIV is trophic for CD4+ T lymphocytes; herpes simplex occupies neuronal cells in the mucous membranes. Clearly, there is a degree of specificity, but the message which determines that specificity is not as clear.

Viral reactivation is often related to immunologic stress, either exogenous or endogenous. The classic example of exogenous stress is the development of CMV pneumonitis in the transplant recipient receiving immunosuppressive agents. The development of herpetic oral lesions in college students during final exams is an excellent example of endogenous stress causing reactivation.

But what role does the latent virus play in disease? Specifically, what is the role of viral *trans*-acting factors in disease? The latent virus is not inactive and it does have limited gene expression. Cytomegalovirus is probably the best studied. During latency, it expresses the gene products of its immediate early 1 and 2 regions. The adenovirus *trans*-activatng gene products are derived from the designated E1A region. HIV and HTLV *trans*-activators are designated *tat* and *tax,* respectively.

Trans-acting factors are proteins which either bind directly to DNA by specific consensus sequences or which interact with other DNA-binding proteins to modify activity (Ptashne, 1988; Ptashne and Gann, 1990). These factors may affect autologous as well as heterologous promoters. The modulation may occur in either a positive or negative fashion (Diamond et al., 1990). In the normal cascade of an immunologic response to the virus, these factors may serve to modulate the response to an appropriate manner. However, in the absence of an obvious need for inflammation, these *trans*-acting factors may result in aberrant production of inflammatory mediators which subsequently cause tissue damage and disease.

III. Viral *Trans*-Activation of Cellular Genes

A. Herpesviridae

The herpesviridae are a family of double-stranded DNA viruses that have the same general structure. The core DNA is surrounded by an icosadeltahedral capsid of approximately 100 to 110 nm in diameter. Surrounding the capsid is an amorphous tegement. The outer envelope is a trilaminar structure with glycoprotein spikes scattered on its surface.

Cytomegalovirus

Cytomegalovirus (CMV) is a β-herpes virus with a genome of approximately 240 kb. It demonstrates restricted cell tropism and upon infection produces a classic cell enlargement with intranuclear inclusions. After the infectious state, CMV may establish a latent state, primarily in lymphocytes and macrophages; there it exhibits a restricted expression of its genome (Rice et al., 1984). Reactivation in the setting of immunosuppression is a major clinical problem in both solid organ (Pollard, 1988) and bone marrow transplant recipients (Meyers, 1986).

Products of the CMV immediate early (IE) region are latency-associated proteins which also have *trans*-activating capabilities. The different CMV IE proteins are generated via differential splicing of the primary IE transcript. Viral replication and gene expression are regulated by the products of the IE genes. IE1 upregulates its own (IE) promoter, as well as the promoters of the early and late genes (Malone et al., 1990, Stenberg et al., 1989). A second, less abundant IE gene product, IE2, down-regulates the IE promoter (Stenberg and Stinski, 1985). Interestingly, it is the IE2 gene product which has been shown to up-regulate heterologous viral promoters (Pizzorno et al., 1988).

In a series of experiments from our laboratory, we demonstrated that the IE gene products up-regulate cell derived cytokine genes. The IE1 gene product markedly up-regulates the promoter of interleukin-1β (IL-1β) gene (Iwamoto et al., 1990). This effect of IE1 appears to be specific for the region 3′ of -131 in the IL-1β promoter, since progressive truncation of the promoter down to -131 base pairs had no effect on IE1 upregulation (Crump et al., 1992). Further mutational analysis of the -131 to $+14$ promoter region is underway to provide insight into the mechanism of IE1 upregulation of the IL-1β gene.

IE gene products also up-regulate expression of the interleukin-2 (IL-2) and IL-2 receptor (IL-2R) genes. Using transient transfections in the human T-cell line Jurkat, we were able to demonstrate that IE2, but not IE1, up-regulates both IL-2 and IL-2R (Geist et al., 1991). Studies using a variety of deletions within the IL-2R promoter demonstrated that an area between -317 and -281 was important for IE2 up-regulation (unpublished data). This area has internal sequence homology with a nine-base-pair region in the HCMV IE promoter which mediates the IE2 regulation of its own promoter (Lui et al., 1991; Pizzorno and Hayward, 1990; Cherrington et al., 1991). Of note, these latter studies were unable to demonstrate direct binding of IE2 to this motif. This observation suggests that IE2 may act as a "co-activator" by binding to cellular transcription factors. The IE gene products are not nonspecific in their actions. They do not effect housekeeping genes, such as actin.

Herpes Simplex Virus

Herpes simplex virus (HSV) was the first human herpes virus to be discovered. It is an alpha-herpes virus with a genome of approximately 150 kb. After infection, HSV establishes latency in host neuronal cells that supply the inoculated region.

There are five HSV alpha (α) genes which are considered immediate early genes. These α genes also function in a *trans*-acting fashion to control subsequent expression of other HSV genes. The protein product designated ICP4 has been shown to have DNA-binding properties, primarily recognizing sequences within the HSV genome (Michael et al., 1988; Kattar-Cooley and Wilcox, 1989). Blaho and Roizman (1991) have shown that ICP4 contains four regions which possess GTP-binding elements, suggesting that ICP4 may be one of a class of GTP-binding proteins involved in *trans*-activation. In addition, ICP4 has been found associated with the inner surface of both the nuclear and plasma membranes (Yao and Courtney, 1991). This membrane association may also allow the HSV to be involved in intracellular signaling.

Epstein-Barr Virus

Epstein-Barr virus (EBV) is a gamma-herpes virus with a genome of approximately 170 kb. After infection, the virus establishes latency in the host B lymphocytes. EBV is probably best recognized for its oncogenic role in Burkitt's lymphoma and some types of nasopharyngeal carcinoma.

The primary latency products of EBV are the Epstein-Barr nuclear antigens (EBNAs). EBNA-2, in particular, appears to be associated with the transforming properties of EBV (Cohen et al., 1989). There has also been noted increased levels of latent membrane protein (LMP), another latency product, in association with the epithelial-derived tumor nasopharyngeal carcinoma (Dawson et al., 1990) and with Hodgkins lymphoma (Herbst et al., 1991). These tumors occur in the absence of replicating virus and may arise as a consequence of deregulation caused by the latent viral gene products.

There is additional evidence that gene products from the productive phase of EBV expression may regulate cellular genes indirectly. The BZLF1 gene product appears to have sequence homology to the components of the AP-1 transcription factor, *c-fos* and *c-jun.* In addition, the BZLF1 gene product recognizes the AP1 consensus binding site (Packham et al., 1990). In this way the virus could modify cellular gene activity by functioning through the AP1 regulatory site known to be present in a variety of cellular genes.

Human Herpes Virus 6

Human herpes virus 6 (HHV-6) is the most recently recognized of the herpes viridae. The viral genome is approximately 160 kb. Little is known about the viral gene products. However, it is clear that it also has a latent phase similar to the other herpes viruses.

HHV-6 infects T lymphocytes. Infection with HHV-6 was shown to up-regulate IL-1β and tumor necrosis factor (TNF) from peripheral blood mononuclear cells, but had no effect on interleukin-6 (IL-6) (Flamand et al., 1991). This up-regulation of IL-1β and TNF protein was most likely due to an effect of the immediate early gene products, as neither UV irradiation nor inhibition of viral RNA polymerase interfered with cytokine induction. In addition, this interaction between viral and cellular genes was specific, as there was no effect on IL-6 production.

B. Retroviridae

The retroviridae are a heterogeneous group of viruses with a number of subfamilies. These are small RNA viruses with a genomic size of 7 to 10 kb. There are usually three structural genes; *gag, pol,* and *env.* The virion is enveloped and has a single protein on its surface.

Human T-Cell Leukemia Virus

Human T-cell leukemia virus (HTLV) belongs to the subfamily oncovirinae. As the name implies, it has been associated with human T-cell leukemia as a causative agent. The viral promoter or long terminal repeat (LTR) carries the information for the *tax trans*-acting factor.

The *tax* gene product has been shown to up-regulate both IL-2 and IL-2R transcription (Cross et al., 1987; Siekevitz et al., 1987). This interaction is felt to be the mechanism whereby HTLV-1 causes T-cell leukemias. The evidence suggests that the *tax* gene product interacts with these cellular genes by activating the NF-κB nuclear binding protein. Both IL-2 and its receptor have κB consensus sequences in their respective 5′ enhancer regions (Bohnlein et al., 1988; Hoyos et al., 1989). In addition, the IL-2R enhancer element also appears to have a recognition site for the *tax* gene product separate from NF-κB site (Ballard et al., 1989). This allows the viral gene product two separate and distinct ways to impact on IL-2R gene transcription.

The *tax* gene product has also been shown to up-regulate endogenous *c-fos* production. Using stable transfectants of the *tax* gene product, *c-fos* mRNA was increased, while there was no effect on other cellular protooncogenes (Nagata et al., 1989). Chronic stimulation of *c-fos* production triggered by interaction with the *tax* gene product could then result in abnormal cell proliferation.

Human Immunodeficiency Virus

Human Immunodeficiency Virus (HIV) is a member of the lentivirinae subfamily. This virus is now recognized as the causative agent in the acquired immunodeficiency syndrome (AIDS). Its primary cell tropism is for the CD4+ T cell, with the CD4 antigen serving as a receptor for the virus. Within its LTR is the message for the HIV *trans*-activator, *tat*.

The *tat* protein appears to function through the NF-κB element, similar to *tax*. Suzan et al. (1991) demonstrated that *tat* could replace other cellular stimuli to result in the induction of active NF-κB. This also provides the virus some level of self-stimulation, as the viral LTR also contains NF-κB-binding sites (Nabel and Baltimore, 1987).

The *tat* gene product appears to have effects on a variety of cellular genes. Whether these effects are direct or are mediated through NF-κB is not entirely clear. The *tat* protein was shown to induce TNF production from B cells (Sastry et al., 1990). In addition, mononuclear cells obtained from the lungs of AIDS patients appear to constitutively produce TNF (Agostini et al., 1991; Millar et al., 1991). This is an interesting finding in a disease characterized by severe wasting of the infected individual, a clinical finding which may be mediated by TNF. Other indirect evidence that HIV is involved in cellular gene regulation was demonstrated by Birx et al. (1990). These authors showed that monocytes infected with HIV produced significantly greater quantities of IL-6 than their uninfected counterparts.

HIV infection may also serve as a primer. In the resting state, infected monocytes do not produce increased levels of IL-1β. However, after stimulation with LPS, there is a more marked increase in IL-1β production in infected cells compared to noninfected cells (Yamato et al., 1990). The effect on cytokine gene production may also be a function of the type of stimulation. D'Addario et al. (1990) demonstrated that HIV-infected monocytes co-infected with Sendai virus produced more interferon (IFN), but there was no effect on either IL-1β or TNF. On the other hand, stimulation with LPS resulted in enhanced production of IL-1β and TNF but not IFN. In either case, chronic infection with HIV resulted in a greater response of the cytokine genes to exogenous stimuli. The precise role of the virus is not clear. Presumably the virus plays a *trans*-acting role and either up-regulates the cellular gene directly or interacts with another transcription factor to result in up-regulation in ways similar to those noted for other viruses. An additional role may lie in signal transduction, suggested by the variable effects of different stimuli on the cellular genes in the presence of the virus.

As mentioned above, the CD4 antigen on T cells appears to be the primary receptor by which HIV gains access to the cell. The *tat* gene product is able to up-regulate expression of the CD4 antigen. This upregulation occurred at the level of gene transcription (Koka et al., 1988). In addition, HIV infection also was shown to up-regulate $\alpha5\beta$ integrin expression, resulting in increased T-cell adherence to tissues and providing a mechanism whereby the virus can mediate tissue-specific injury (Weeks et al., 1991).

C. Adenovirus

The adenovirus is a DNA virus with a nonenveloped regular icosohedron shape of approximately 65 to 80 nm. In its actively replicating state, it is a frequent cause of acute respiratory illness. The primary control of viral replication and gene transcription is exerted by the adenovirus E1A gene product(s). Different E1A gene products are generated by differential splicing of the primary transcript of the E1A gene. Most of the *trans*-acting responses attributed to the adenovirus E1A gene product(s) appear to involve protein-protein interactions rather than direct DNA binding.

The products of the E1A gene also appear to interact, indirectly, with host cell genes via protein-protein interactions. One of the E1A gene products has been shown to interact with the eukaryotic transcription factor IID (TFIID), which binds to the TATA box (Horikoshi et al., 1991). Binding of TFIID, and other proteins associated with the transcription complex, to the

TATA box is necesssary for these genes to initiate transcription by RNA polymerase II (Fire et al., 1984). An E1A gene product has also been shown to interact with the *c-fos* promoter (Simon et al., 1990) and is a positive regulator of the *c-jun* gene (deGroot et al., 1991). These effects of E1A alter regulatory control of the cell-derived AP-1 transcriptional activator, which is composed of *fos-jun* or *jun-jun* dimers. AP-1 is an important regulator of many cellular genes and provides another mechanism whereby E1A can exert control over cellular gene expression.

In addition, E1A gene product(s) have also been shown to modulate transcription of the immunoglobulin κ-chain gene. This latter effect is cell specific, with repression of transcription in lymphoid cells and activation of transcription in fibroblasts (Bergman and Shavit, 1988). Other studies have demonstrated a role for adenovirus E1A gene products in dysregulation of cell growth which is mediated by its ability to bind to the retinoblastoma (RB) gene product (Whyte et al., 1988). RB is a known cellular antioncogene which regulates cellular proliferation (Huang et al., 1988). RB functions, in part, by acting as a "sink" for the cellular transcription factor, E2F (Bagchi et al., 1991). Following phosphorylation of RB, or binding of RB by E1A, E2F is released and translocates to the nucleus, where it initiates gene transcription and DNA synthesis (Chellappan et al., 1991; Raychaudhuri et al., 1990; Hiebert et al., 1991). Through these above mechanisms the E1A gene product can control cellular gene transcription at several levels.

IV. Summary

In this review we have attempted to point out the ways in which viral gene products may alter the expression of cellular genes. Specifically, we focused on the *trans*-activating properties of viral gene products with regard to their influence on cellular gene production. In some instances the mechanisms are necessarily vague, which is a reflection of current knowledge in the field. However, as can be noted from the references, much knowledge has been gained in the past few years. Clearly, viral gene products interact in some fashion with cellular genes. These interactions may provide etiologies for previously undefined or idiopathic illnesses. As investigation continues, more insight will be gained into the mechanism(s) of interaction and the ways to intervene at a cellular level to block these effects.

References

Agostini C, Zambello R, Trentin L, Garbisa S, DiCelle PF, Bulian P, Onisto M, Poletti V, Spiga L, Raise E, Foa R, Semenzato G. Alveolar macrophages from patients with AIDS and AIDS-related complex constitutively synthesize and release tumor necrosis factor alpha. Am Rev Respir Dis 1991; 144:195–201.

Bagchi S, Weinman R, Raychaudhuri P. The retinoblastoma protein copurifies with E2F-I, and E1A-regulated inhibitor of the transcription factor E2F. Cell 1991; 65:1063–72.

Ballard DW, Bohnlein E, Hoffman JA, Bogerd HP, Dixon EP, Franza BR, Greene WC. Activation of the interleukin-2 receptor α gene: regulatory role for DNA-protein interactions flanking the kB enhancer. The New Biologist 1989; 1:83–92.

Bergman Y, Shavit D. Regulation of the Ig κ-chain enhancer by the adenovirus E1A gene products. J Immunol 1988; 140:2073–80.

Birx DL, Redfield RR, Tenar K, Fowler A, Burke DS, Tosato G. Induction of interleukin-6 during human immunodeficiency virus-related lung disease. Am J Respir Cell Mol Biol 1990; 5:144–8.

Blaho JA, Roizman B. ICP4, the major regulatory protein of herpes simplex virus, shares features common to GTP-binding proteins and is adenylated and guanylated. J Virol 1991; 65:3759–69.

Bohnlein E, Lowenthal JW, Siekevitz M, Ballard DW, Franza BR, Greene WC. The same inducible nuclear proteins regulates mitogen activation of both interleukin-2 receptor-alpha gene and type I HIV. Cell 1988; 53:827–36.

Chellappan SP, Hiebert S, Mudryi M, Horowitz JM, Nevins JR. The E2F transcription factor is a cellular target for the RB protein. Cell 1991; 65:1053–61.

Cherrington JM, Khoury EL, Mocarski ES. Human cytomegalovirus IE2 negatively regulates alpha gene expression via a short target sequence near the transcription start site. J Virol 1991; 65:887–96.

Cohen JI, Wang F, Mannick J, Kieff E. Epstein-Barr virus nuclear protein 2 is a key determinant of lymphocyte transformation. Proc Natl Acad Sci USA 1989; 86:9558–62.

Cross SL, Feinberg MB, Wolf JB, Holbrook NJ, Wong-Staal F, Leonard WJ. Regulation of the human interleukin-2 receptor alpha chain promoter: activation of a nonfunctional promoter by the transactivator gene of HTLV-I. Cell 1987; 49:47–56.

Crump JW, Geist LJ, Auron PE, Webb AC, Stinski MF, Hunninghake GW. The immediate early genes of human cytomegalovirus require only proximal promoter elements to upregulate expression of interleukin-1 beta. Am J Respir Cell Mol Biol 1992; 6:674–7.

D'Addario M, Poulston A, Wainberg MA, Hiscott J. Coordinate enhancement of cytokine gene expression in human immunodeficiency virus type 1-infected promonocytic cells. J Virol 1990; 64:6080–9.

Dawson CW, Rickinson AB, Young LS. Epstein-Barr latent membrane protein inhibits human epithelial cell differentiation. Nature 1990; 344:777–80.

deGroot R, Foulkes N, Mulder M, Kruijer W, Sassone-Corsi P. Positive regulation of *jun*/AP-1 by E1A. Mol Cell Biol 1991; 11:192–201.

Diamond MI, Miner JN, Yoshinaga SK, Yamamoto KR. Transcription factor interactions: selectors of positive or negative regulation from a single DNA element. Science 1990; 249:1266–72.

Fields BN, ed. Virology, 2nd ed. New York: Raven Press, 1990.

Fire A, Samuels M, Sharp PA. Interactions between RNA polymerase II factors and template leading to accurate transcription. J Biol Chem 1984; 259:2509–16.

Flamand L, Gosselin J, E'Addario M, Hiscott J, Ablasti DV, Gallo RC, Menezes J. Human herpes virus 6 induces interleukin-1β and tumor necrosis factor alpha, but not interleukin-6, in peripheral blood mononuclear cultures. J Virol 1991; 65:5105–10.

Geist LJ, Monick MM, Stinski MF, Hunninghake GW. The immediate early genes of human cytomegalovirus upregulate expression of the interleukin-2 and interleukin-2 receptor genes. Am J Respir Cell Mol Biol 1991; 5:292–6.

Herbst H, Dallenbach F, Hummel M, Niedobitek G, Pileri S, Muller-Lantzsch N, Stein H. Epstein-Barr virus latent membrane protein expression in Hodgkin and Reed-Sternberg cells. Proc Natl Acad Sci USA 1991; 88:4766–70.

Hiebert SW, Blake M, Azizkahn J, Nevins JR. Role of E2F transcription factor in E1A-mediated *trans*activation of cellular genes. J Virol 1991; 65:3547–52.

Horikoshi N, Maguire K, Kralli A, Maldonado E, Reinberg D, Weinmann R. Direct interaction between adenovirus E1A protein and the TATA box binding transcription factor IID. Proc Natl Acad Sci USA 1991; 88:5124–8.

Hoyos B, Ballard DW, Bohnlein E, Siekevitz M, Greene WC. Kappa-B specific binding proteins: role in the regulation of human interleukin-2 gene expression. Science 1989; 244:456–60.

Huang H-JS, Yee J-K, Shew J-Y, Chen P-L, Bookstein R, Friedmann T, Lee EY-HP, Lee WH. Suppression of the phenotype by replacement of the RB gene in human cancer cells. Science 1988; 242:1563–6.

Iwamoto GK, Monick MM, Stinski MF, Hunninghake GW. Modulation of interleukin-1 beta gene expression by the immediate early genes of human cytomegalovirus. J Clin Invest 1990; 85:1853–7.

Kattar-Cooley P, Wilcox KW. Characterization of the DNA-binding properties of herpes simplex virus regulatory protein ICP4. J Virol 1989; 63:696–704.

Klein G. Viral latency and transformation: the strategy of Epstein-Barr virus. Cell 1989; 58:5–8.

Koka P, Yunis J, Passarelli AL, Dubey DP, Faller D, Yunis EJ. Increased expression of CD4 molecules on Jurkat cells mediated by human immunodeficiency virus *tat* protein. J Virol 1988; 62:4353–7.

Lui B, Hermiston TW, Stinski MF. A *cis*-acting element in the major immediate-early (IE) promoter of human cytomegalovirus is required for negative regulation by IE2. J Virol 1991; 65:897–903.

Malone CL, Vesole DL, Stinski MF. Transactivation of a human cytomegalovirus early promoter by gene products from the immediate-early gene IE2 and augmentation by IE1: mutational analysis of the viral proteins. J Virol 1990; 64:1498–506.

Meyers JD. Infection in bone marrow transplant recipients. Am J Med 1986; 81(1A):27–38.

Michael N, Spector D, Mavromara-Nazos P, Kristie TM, Roizman B. The DNA-binding properties of the major regulator protein α4 of herpes simplex viruses. Science 1988; 239:1531–4.

Millar AB, Miller RF, Foley NM, Meager A, Semple SJG, Rook GAW. Production of tumor necrosis factor-α by blood and lung lymphocytes from patients with human immunodeficiency virus-related lung disease. Am J Respir Cell Mol Biol 1991; 5:144–58.

Nabel G, Baltimore D. An inducible transcription factor activates expression of human immunodeficiency virus in T cells. Nature 1987; 326:711–3.

Nagata K, Ohtani K, Nakamura M, Sugamura K. Activation of endogenous *c-fos* proto-oncogene expression by human T-cell leukemia virus type 1-encoded $p40^{tax}$ protein in the human T-cell line, Jurkat. J Virol 1989; 63:3220–6.

Packham G, Economou A, Rooney CM, Rowe DT, Farrell PJ. Structure and function of the Epstein-Barr virus BZLF1 protein. J Virol 1990; 64:2110–6.

Pagano JS. Diseases and mechanisms of persistent DNA virus infection: latency and cellular transformation. J Inf Dis 1975; 132:209–23.

Pizzorno MC, Hayward GS. The IE2 gene products of human cytomegalovirus specifically down-regulate expression from the major immediate-early promoter through a target sequence located near the cap site. J Virol 1990; 64:6154–65.

Pizzorno MC, O'Hare PC, Sha L, LaFemina RL, Hayward GS. *Trans*-activation and autoregulation of gene expression by the immediate-early region 2 gene products of human cytomegalovirus. J Virol 1988; 62:1167–79.

Pollard RB. Cytomegalovirus infections in renal, heart, heart-lung and liver transplantations. Pediatr Infect Dis J 1988; 7:S97–102.

Ptashne M. How eukaryotic transcriptional activators work. Nature 1988; 335:683–9.

Ptashne M, Gann AAF. Activators and targets. Nature 1990; 356:329–31.

Raychaudhuri P, Bageni S, Neill SD, Nevins JR. Activation of the E2F transcription factor in adenovirus-infected cells involves E1A-dependent stimulation of DNA-binding activity and induction of cooperative binding mediated by an E4 gene product. J Virol 1990; 64:2702–10.

Rice GPA, Schrier RD, Oldstone MBA. Cytomegalovirus infects human lymphocytes and monocytes: virus expression is restricted to immediate early gene products. Proc Natl Acad Sci USA 1984; 81:6134–8.

Sastry KJ, Reddy RHR, Pandita R, Totpal K, Aggarwal BB. HIV-1 *tat* gene induces tumor necrosis factor-β (lymphotoxin) in a human β-lymphoblastoid cell line. J Biol Chem 1990; 265:20091–3.

Siekevitz M, Feinberg MB, Holbrook N, Wong-Staal F, Greene WC. Activation of interleukin 2 and interleukin 2 recept (tac) promoter expression by the transactivator (tat) gene product of human T-cell leukemia virus, type I. Proc Natl Acad Sci USA 1987; 84:5389–93.

Simon MC, Rooney RJ, Fisch TM, Heintz N, Nevins JR. E1A-dependent transactivation of the *c-fos* promoter requires the TATAA sequence. Proc Natl Acad Sci USA 1990; 87:513–7.

Stenberg RM, Stinski MF. Autoregulation of the human cytomegalovirus major immediate early gene. J Virol 1985; 56:676–82.

Stenberg RM, Depto AS, Fortney J, Nelson JA. Regulated expression of early and late RNAs and proteins from the human cytomegalovirus immediate early gene region. J Virol 1989; 63:2699–2708.

Suzan M, Salaun D, Neuveut C, Spire B, Hirsch I, LeBouteiller P, Querat G, Sire J.

Induction of NF-κB during monocyte differentiation by HIV type I infection. J Immunol 1991; 146:377–83.
Weeks BS, Klotman ME, Dhawan S, Kibbey M, Rappaport J, Kleinman HK, Yamada KM, Klotman PE. HIV-1 infection of human T lymphocytes results in enhances $\alpha 5 \beta$ integrin expression. J Cell Biol 1991; 114:847–53.
Whyte P, Buchovich JJ, Horowitz JM, Friend SH, Raybuck M, Weinberg RA, Harlow E. Association between an oncogene and an anti-oncogene: the adenovirus E1A proteins bind to the retinoblastoma gene product. Nature 1988; 334:124–9.
Yamato K, El-Hajjaoui Z, Simon K, Koeffler HP. Modulation of interleukin-1β RNA in monocytic cells infected with human immunodeficiency virus-1. J Clin Invest 1990; 86:1109–14.
Yao F, Courtney RJ. Association of a major transcriptional regulator protein, ICP4, of herpes simplex virus type 1 with the plasma membrane of virus-infected cells. J Virol 1991; 65:1516–24.

Part Three

SMOOTH MUSCLE CELLS

15

Role of Calcium in the Regulation of Smooth Muscle Tone

KATHLEEN G. MORGAN

Harvard Medical School
Boston, Massachusetts

I. $[Ca^{2+}]_i$ Profiles During Tonic Vascular Contraction

A. Methodology

Three major types of techniques have been used to measure intracellular ionized calcium ($[Ca^{2+}]_i$) levels. These are (1) calcium indicators, (2) calcium-sensitive microelectrodes, and (3) nuclear magnetic resonance. Of the calcium indicators there are, again, three basic types: metallochromic, luminescent, and fluorescent. The metallochromic indicators undergo a change in absorbance in response to a change in $[Ca^{2+}]_i$ and have the disadvantage of displaying pronounced artifactual signals in contractile cells because of the changes in path length which occur during cell contraction. For this reason, the metallochromic indicators have not been used extensively in smooth muscle preparations.

Although Ca-sensitive microelectrodes have been used fairly extensively in larger cell types, particularly before the advent of luminescent and fluorescent calcium indicators, so far only one laboratory has been able to apply this technology to smooth muscle cells. In 1986, Yamaguchi used Ca-sensitive microelectrodes to measure $[Ca^{2+}]_i$ in smooth muscle cells. The average resting $[Ca^{2+}]_i$ as determined by this technique was 163 nM. This technique has not been pursued extensively in smooth muscle cells, primarily because of the technical concern that membrane damage from impaling small (2- to 7-μM-diameter) smooth muscle cells with the microelectrodes might artifactually increase $[Ca^{2+}]_i$ through a leak pathway.

It has become possible in recent years to measure $[Ca^{2+}]_i$ with the use of symmetrically substituted difluro-derivatives of BAPTA [1,2-bis (0-aminophenoxy-ethane-N,N,N′, N′-tetraacetic acid) as NMR probes (Smith et al., 1983); however, thus far, relatively high concentrations of this probe have been required in order to obtain adequate signals in reasonable sampling times. This results in the concern that buffering of Ca may result in altered levels of $[Ca^{2+}]_i$ as well as a change in the kinetics of Ca transients. With methodological improvements, this technique may become more popular in the future, but current efforts appear to be focused on perfecting the technique in cardiac and skeletal muscle.

Thus, at the present time, the only technique in general use for measuring $[Ca^{2+}]_i$ levels in smooth muscle are the luminescent Ca indicator aequorin and the fluorescent Ca indicators.

B. $[Ca^{2+}]_i$ Measurements in Smooth Muscle

Aequorin

Aequorin is a luminescent calcium indicator, i.e., photons are emitted on the addition of Ca ions without the need for an excitation light. The relative advantages and disadvantages of luminescent and fluorescent calcium indicators have been presented elsewhere (Blinks et al., 1982; Morgan and Brozovich, 1989). Aequorin was used first in smooth muscle primarily because its development preceded that of the currently popular fura-2 and indo-1. The first measurements of smooth muscle agonist-induced $[Ca^{2+}]_i$ profiles were made in aequorin-loaded aortas of *Amphiuma tridactylum* (Morgan and Morgan, 1982). These initial aequorin signals (Fig. 1) were surprising, since they indicated that, on addition of a stimulant, $[Ca^{2+}]_i$ levels rose to relatively high values during the development of tonic contractions, but during the plateau phase of "tone maintenance," $[Ca^{2+}]_i$ levels appeared to return to baseline levels. This qualitative finding was subsequently confirmed in mammalian smooth muscle from ferret portal vein (Morgan and Morgan, 1984), ferret aorta (DeFeo and Morgan, 1985), hog carotid artery (Rembold and Murphy, 1986), hog coronary artery (Bradley and Morgan, 1987), and bovine (Takuwa et al., 1987) and canine (Gerthoffer et al., 1989) tracheal muscle. However, because of the nonlinear properties of aequorin, these results remained controversial until confirmed independently with fluorescent calcium indicators.

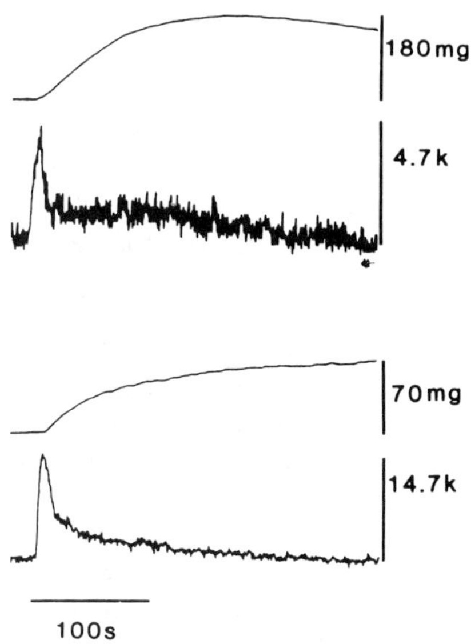

Figure 1 Simultaneous force (upper trace, in mg) and $[Ca^{2+}]_i$-dependent luminescence (lower trace, in kcounts from photon counter) from aequorin-loaded *Amphiuma* aortas, Top pair of traces: 2×10^{-7} M angiotensin added at start of trace. Lower pair of traces: 10^{-4} M phenylephrine added at start of traces.

Fura-2 Measurements

In 1985, a new generation of fluorescent calcium indicators was introduced which allowed the use of fluorescence ratio measurements (Grynkiewicz et al., 1985). The use of ratio measurements largely eliminated artifacts due to changes in cell thickness in contractile cells, which had complicated the use of previous fluorescent dyes such as quin2. The use of fura-2 was developed in phasic gastric smooth muscle cells of the toad stomach (Williams et al., 1985), but, shortly thereafter, was extended to cultured vascular (Griendling et al., 1986; Hassid, 1986; Reynolds and Dubyak, 1986) and tracheal (Taylor et al., 1989) smooth muscle cells and, later, to enzymatically isolated vascular smooth muscle cells (DeFeo et al., 1987). Initial results from these preparations were complicated by the lack of a parallel measurement of contractile force as well as questions concerning modification of smooth muscle phenotype in tissue culture and damage to cells by enzymatic isolation procedures. Between 1986 and 1990, however, techniques were developed to allow simultaneous measurement of contractile force and $[Ca^{2+}]_i$ from smooth muscle strips which had been loaded with fura-2 (Bruschi et al., 1986; Himpens and

Somlyo, 1988; Ozaki et al., 1990). The fura-2-loaded strips made simultaneous force recordings feasible but also presented additional technical problems, including: (1) unknown amounts of the indicator loaded into various cells of the strips; (2) the necessity for high dye concentration and long loading times which resulted in the loading of dye into intracellular organelles; (3) generation of bleached Ca-insensitive fluorescent intermediates to variable amounts in various cells of the strip; as well as (4) the generation of partially hydrolyzed calcium-insensitive fluorescent intermediates of the acetoxymethylester form of fura-2 to varying amounts throughout the strip (for a review, see Roe et al., 1990). In general, however, the qualitative $[Ca^{2+}]_i$ profiles were surprisingly similar to those previously reported with the use of aequorin. Again, $[Ca^{2+}]_i$ signals in response to the addition of agonists were seen to peak in less than a minute, while contractions were maintained for up to an hour. The aequorin and fura-2 techniques are vastly different, and the techniques are expected to produce different types of artifacts. For example, aequorin signals are generally expected to overestimate $[Ca^{2+}]_i$, while fura-2 signals are expected to underestimate $[Ca^{2+}]_i$. Thus, the similarity of the findings in smooth muscle strongly indicated that the observed dissociation between $[Ca^{2+}]_i$ and tension was real.

As both the aequorin and fura-2 methodologies were refined, quantitative values were found to be somewhat different. The differences appear to be a result of: (1) different tissues being compared; (2) different agonists being compared; (3) inaccurate calibration methods; or (4) the intrinsic differences in the way in which $[Ca^{2+}]_i$ is reported by the two types of indicators. The luminescent calcium indicator aequorin produces light that is roughly proportional to the 2.5th power of the change in $[Ca^{2+}]_i$ and, thus, aequorin will magnify small changes in $[Ca^{2+}]_i$ and give light signals containing very large, sharp peaks (Blinks et al., 1982). In contrast, fura-2 is a somewhat more linear Ca indicator with a K_d near the resting levels of $[Ca^{2+}]_i$. Since fura-2 saturates at around micromolar $[Ca^{2+}]_i$, large spikelike increases in $[Ca^{2+}]_i$ will be blunted, whereas small, plateaulike elevations slightly above resting levels will appear relatively larger compared to the same $[Ca^{2+}]_i$ signal recorded with aequorin (Grynkiewicz et al., 1985).

II. Agonist-Specific Nature of $[Ca^{2+}]_i$ Profiles

We have found that not all agonists produce the same $[Ca^{2+}]_i$ profile. Qualitatively different types of $[Ca^{2+}]_i$ profiles are seen with different agonists, and these can be grouped into three basic types of patterns which are simplistically illustrated in Fig. 2. The simplest $[Ca^{2+}]_i$-force relationship occurs during potassium depolarization, which would be expected to increase $[Ca^{2+}]_i$ by

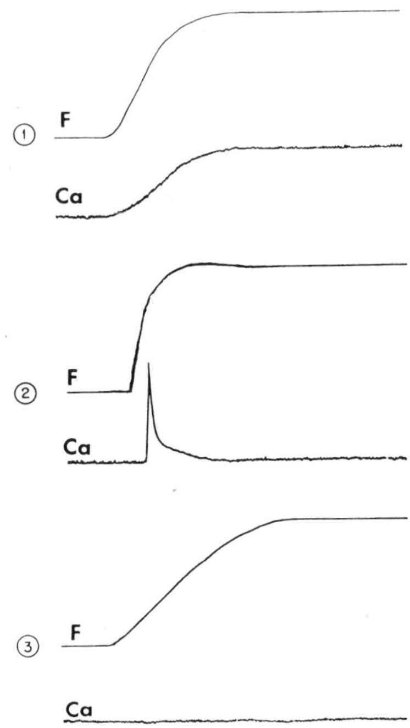

Figure 2 Three types of $[Ca^{2+}]_i$ profiles associated with tonic contractions of smooth muscle. In each pair of traces, F denotes force and Ca is the $[Ca^{2+}]_i$ as indicated by aequorin or fura-2. Modified with permission from Morgan (1990).

opening voltage-dependent Ca channels without the generation of any additional second messengers. During low levels of potassium depolarization (i.e., up to 50 mM K, where there are minimal effects on tonicity, sodium depletion, or transmitter release), the changes in $[Ca^{2+}]_i$ parallel the change in force (Morgan and Morgan, 1984) (Fig. 2, pattern 1). $[Ca^{2+}]_i$ and force rise together and stay elevated as long as the cell is depolarized. Both responses return to baseline in parallel when the muscle is returned to a normal physiological saline solution. The only other agonist observed so far which may produce such a simple $[Ca^{2+}]_i$-force relationship is histamine in the hog coronary artery (Bradley and Morgan, 1987), but this needs to be confirmed.

In many other cases, such as during the action of angiotensin, carbachol, vasopressin, or the alpha agonist phenylephrine, in most tissues a typical large spikelike increase in $[Ca^{2+}]_i$ is seen during the development of force, which then falls rapidly back toward basal levels during force maintenance (Fig. 2, pattern 2).

Finally, it has been recently found that during the action of certain agonists in specific tissues, such as phorbol esters in ferret aorta (Jiang and Morgan, 1987), prostaglandin $F_{2\alpha}$ in hog coronary artery (Bradley and Morgan, 1987), and phenylephrine in rat aorta (Papageorgiou and Morgan, 1991), a third type of profile is seen where there is no detectable change in $[Ca^{2+}]_i$ even though there is a robust contraction of the tissue (Fig. 2, pattern 3). Thus, three qualitatively and dramatically different $[Ca^{2+}]_i$ profiles are observed even though the tonic contractions are often indistinguishable. The only difference in the force profiles that has been observed is that in some instances where there is no change in intracellular [Ca], the rate of force development is slowed, such as is shown in Fig. 2, pattern 3. Conversely, a large initial $[Ca^{2+}]_i$ spike (Fig. 2, pattern 2) has been associated with an increased rate of force development (Papageorgiou and Morgan, 1991).

The apparent dissociation between steady-state $[Ca^{2+}]_i$ and force levels suggests that the $[Ca^{2+}]_i$ requirement for contraction is a variable with respect to the time point during the contraction and the choice of agonist. In general, there are two ways in which a decrease in the $[Ca^{2+}]_i$ requirement for contraction could occur; either (1) an increase in the $[Ca^{2+}]_i$ affinity of a specific $[Ca^{2+}]_i$ binding site occurs; or (2) a pathway is recruited that is entirely different and possesses a second Ca binding site that has an increased Ca affinity, or perhaps no Ca requirement at all. The question then becomes where is/are the $[Ca^{2+}]_i$ binding site(s) in smooth muscle cells?

III. Role of Myosin Light-Chain Phosphorylation

A. Importance of Myosin Light-Chain Phosphorylation

In 1977, Sobieszek demonstrated the importance of Ca-dependent myosin light-chain (MLC) phosphorylation in the regulation of actin-activated myosin ATPase activity in smooth muscle. This observation has been followed by reports that Ca-dependent phosphorylation of the 20,000-Da MLC precedes and appears to be necessary for contraction of smooth muscle preparations in a variety of situations (Hoar et al., 1979; Adelstein and Eisenberg, 1980; and Kamm and Stull, 1985). The kinase was orginally reported to be activated in a simple manner by Ca and calmodulin (CaM), and dephosphorylation was reported to occur by the action of an unregulated phosphatase (Fig. 3). In the last 10 years, however, it has become apparent that major dissociations can occur between $[Ca^{2+}]_i$ and MLC phosphorylation levels as well as between phosphorylation and tension levels.

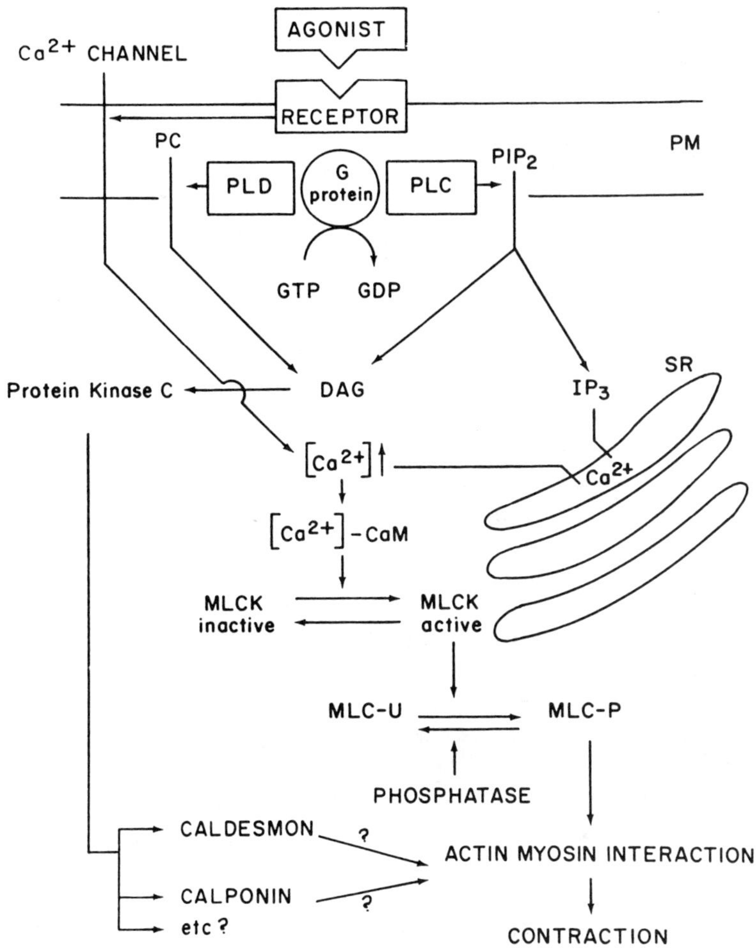

Figure 3 Putative mechanism of agonist-induced activation of contraction. CaM = calmodulin; DAG = diacylglycerol; IP_3 = inositol 1,4,5-trisphosphate; MLCK = myosin light-chain kinase; MLC-U = myosin light chain, unphosphorylated; MLC-P = myosin light chain, phosphorylated; PC = phosphatidyl choline; PIP = phosphatidyl inositol 1,4,5-bisphosphate; PLC = phospholipase C; PLD = phospholipase D; PM = plasma membrane; SR = sarcoplasmic reticulum. From Morgan and Suematsu (1990), with permission.

B. Variability of Ca Requirement for MLC Phosphorylation

We have recently reported that during prostaglandin $F_{2\alpha}$ contractions in ferret aorta, under conditions where there is no detectable change in $[Ca^{2+}]_i$, a contraction can still occur and, in parallel, there is a proportionate increase in MLC phosphorlyation (Suematsu et al., 1991a). This suggests that either the response of MLC kinase to a constant resting $[Ca^{2+}]_i$ level is increased, or, alternatively, somehow the activity of the phosphatase has been decreased. We have also seen an uncoupling between $[Ca^{2+}]_i$ and MLC phosphorlyation in the presence of forskolin (Suematsu et al., 1991b), and similar shifts in the $[Ca^{2+}]_i$-phosphorylation relationship have been previously reported by Rembold (1990) and Stull et al. (1990). Biochemical studies have established that cyclic AMP-dependent protein kinase, cyclic GMP-dependent protein kinase, protein kinase C, and Ca^{2+}-CaM-dependent protein kinase II (CaM kinase II) can cause phosphorylation of MLC kinase in vitro (Conti and Adelstein, 1981; Nishikawa et al., 1984, 1985; Ikebe et al., 1985; Ikebe and Reardon, 1990). Phosphorylation at a specific site called "site A" appears to decrease the affinity of MLC kinase for Ca^{2+}-CaM. Stull et al. (1990) have reported that during agonist-induced stimulation of tracheal muscle, only CaM kinase II appears to be capable of significant phosphorylation of site A. Thus, a mechanism is proposed whereby when $[Ca^{2+}]_i$ is elevated to high levels; this will result in the activation of CaM kinase II, which will phosphorylate MLC kinase, decreasing its Ca sensitivity, and thus decreasing the amount of MLC phosphorylation. This negative feedback mechanism could possibly explain the relatively low ratio between MLC phosphorylation and $[Ca^{2+}]_i$ during steady-state potassium depolarization-mediated contractions. This could also explain the relatively high ratio of MLC phosphorylation to $[Ca^{2+}]_i$ levels during the action of prostaglandin $F_{2\alpha}$ in ferret aorta. However, during the action of agonists such as α agonists, where a transient, spikelike increase in Ca is observed (Fig. 2, pattern 2), the MLC phosphorylation profile mirrors that of the Ca profile; i.e., myosin phosphorylation increases to high levels transiently during force development, but then falls toward basal levels during tone maintenance (Jiang and Morgan, 1989). Thus, in these cases, there is no dissociation between $[Ca^{2+}]_i$ and phosphorylation levels, but rather the challenge is in explaining the dissociation between MLC phosphorylation levels and contractile force.

C. Latch Hypothesis

One explanation that has been proposed to explain the dissociation between phosphorylation and tension is the "latch hypothesis" of Hai and Murphy (1988). This hypothesis is based on the original observations of Dillon et al. (1981), who reported that the temporal profile of MLC phosphorylation did not parallel the force profile in hog carotid artery and that force maintenance often occurred in the presence of near-baseline or falling levels of MLC phosphorylation. Murphy and co-workers have suggested that dephosphorylation of the light chains in response to decreasing $[Ca^{2+}]_i$ levels can lead to the formation of an attached, noncycling, or slowly detaching crossbridge termed the "latch" bridge. A kinetic model has been developed by Hai and Murphy (1988) and also by Drika et al. (1989) in which there are seven rate constants describing the phosphorylation/dephosphorylation and attachment/detachment of myosin crossbridges. With appropriate choices for the seven rate constants in the model, many examples of phosphorylation-tension dissociations can be explained; however, clear-cut exceptions do exist, and the model remains difficult to test because of the lack of an isolated biochemical analog of the "latchbridge."

IV. Thin Filament-Linked Regulation of Vascular Smooth Muscle

A. Evidence for Thin Filament Regulation

The calcium-dependent regulation of actomyosin ATPase activity through phosphorylation of the 20,000-Da myosin light chain is a unique aspect of smooth muscle excitation-contraction coupling. Both skeletal and cardiac muscle are primarily regulated by a thin filament-linked mechanism in which Ca-dependent conformational changes in the troponin-tropomyosin complex on the actin filament represents the main mode of regulation of contractile force by $[Ca^{2+}]_i$. The possibility of the existence of thin filament-linked regulation in smooth muscle was suggested by the work of Driska and Hartshorne (1975) as well as by the work in the laboratory of Ebashi and co-workers (1977). Some of this earlier work was questioned because of possible contamination of the native tropomyosin fractions with myosin light-chain kinase and CaM, but more clear-cut experiments utilizing a myosin competition test (Lehman and Szent-Györgyi, 1975) clearly showed that invertebrate smooth muscle utilizes double regulation through both myosin and the thin filaments. Marston et al. (1980), using a method they developed to minimize degradation while isolating thin filaments from blood vessels, have shown Ca regulation of thin filaments from sheep aorta in combination with rabbit skeletal muscle myosin. Half maximal activation in this reconstituted system occurred at 3 μM $[Ca^{2+}]$.

If we are to consider the possibility of thin filament-linked regulation of vascular smooth muscle excitation-contraction coupling, the question then becomes the nature of the regulatory proteins involved. Currently, three major hypotheses exist. The first two involve the two leading candidates for troponin-like substances in smooth muscle (i.e., caldesmon and calponin). The third possibility is a more complicated hypothesis suggesting that force can be maintained by connections between actin and the intermediate filament lattice.

B. Caldesmon

Caldesmon is a CaM, tropomyosin, and actin binding protein that exists in two forms. In contractile smooth muscle, the larger form (molecular weight 120–150 kDa) is predominant. In nonmuscle cells such as platelets and in cultured smooth muscle cells, a lower-molecular-weight form (70–80 kDa) is found. The biochemical properties of caldesmon have been recently reviewed (Sobue and Sellers, 1991). In the absence of Ca, caldesmon is known to interfere with the interaction of myosin and actin, thereby inhibiting actin-activated myosin ATPase activity. In the presence of elevated levels of Ca/CaM, the inhibition is removed. It was initially thought that in high-Ca solutions, CaM was physically removed from the actin filaments, but it is now thought such high Ca concentrations are never reached in the intact smooth muscle cell. It has now been suggested that a simple conformational change removes the inhibition of actomyosin ATPase activity. There is still considerable debate, however, as to whether the Ca/Cam concentration is ever high enough in the intact cell to cause even this conformational change. The K_d of caldesmon for CaM is in the micromolar range, and this would be the upper limit of the current estimates of free CaM concentrations in the smooth muscle cell. However, it should be emphasized that we do not have an accurate measurement of the intracellular free [CaM].

An alternative proposal is that caldesmon activity is regulated not by Ca, but rather by phosphorylation. Caldesmon is phosphorylated by at least four kinases, PKC, CaM kinase II, casine kinase II, and cdc2 protein kinase. Phosphorylation of caldesmon, at least by PKC and CaM kinase II, has been associated with a reversal of caldesmon's inhibition of actomyosin ATPase activity. Caldesmon phosphorylation has been measured in intact tissues during agonist-induced contraction, but, thus far, the sites of phosphorylation in this particular tissue did not appear to correlate with the known phosphorylation sites of an identifiable kinase (see Adam et al., 1989, and Sobue and Sellers, 1991, for review).

It has recently also been demonstrated in vitro that whereas the C-terminal end of the caldesmon molecule is responsible for inhibiting actomyosin ATPase activity, the N-terminal fragment possesses a myosin binding site, probably on the S2 subfragment. It has been suggested that the intact caldesmon molecule can tether actin and myosin together, perhaps providing a basis for latchbridges and explaining the Ca-independent maintenance of vascular tone (Ikebe and Reardon, 1988; Lash et al., 1986). This hypothesis remains untested by physiological experiments at the present time.

C. Calponin

Calponin has only recently been characterized by Takahashi and colleagues (1988a, 1988b, 1991). This is a 34- to 35-kDa molecule with Ca, CaM, tropomyosin, and actin binding sites. Calponin also shares sequence homology with an abundant protein termed "SM 22," which, as yet, has no known function and for which the binding affinities are undefined. The Ca and CaM binding properties of calponin appear to be even weaker than those for caldesmon, making it unlikely that this is a Ca regulatory protein in vivo. However, the binding to actin is high-affinity, resulting in actomyosin ATPase activity inhibition. Calponin, like caldesmon, can also be a substrate for several kinases (Winder and Walsh, 1990). The main kinases involved appear to be protein kinase C and CaM kinase II. Phosphorylation of calponin by either of these kinases has been shown to be capable in vitro of reversing the inhibition of actomyosin ATPase activity. However, there is no evidence as of yet for this process in vivo. Barany et al. (1991) have measured calponin phosphorylation levels in vivo during agonist-induced contractions of porcine carotid arteries, and no calponin phosphorlyation was detectable during the time course of the contraction. In abstract form, however, positive results were obtained in tracheal tissue, where a correlation was reported between the time course of calponin phosphorylation and contraction (Pohl et al., 1991). More experiments along these lines are needed.

D. Force Maintenance by Actin Intermediate Filament Interactions

Studies from Small's laboratory (Fürst et al., 1986) using immunofluorescence microscopy combined with electron microscopic immunocytochemistry have resulted in the description of two spacially distinct domains of microfibrils. Small has described an inner zone containing actin, myosin, and tropomyosin as well as caldesmon, and an outer zone containing actin and intermediate filament proteins, such as desmin, vimentin, and synemin. The outer intermediate filament zone contains a number of actin binding proteins, such as filamin. In the outer zone, the filaments are anchored to the cell surface through structures referred to as "dense plaques," which contain proteins such as vinculin, plectin, and filamen (Small, 1985). Through an intermediate filament network, these outer structures are linked to dense bodies intracellularly. The dense bodies are rich in alpha actinin and probably represent the smooth muscle analog of the Z line in striated muscle. It has been suggested that actin binding proteins may be able to "glue" together the interdigitating arrays of actin filaments and intermediate filaments, allowing the maintenance of tension (Small et al., 1986; Rasmussen et al., 1987). Another actin binding protein, which has been shown to be Ca-dependent, is gelsolin, which could be capable during elevations of Ca of loosening interactions between actin and intermediate filaments, but then allowing the reestablishment of these interactions when Ca returns to new, lower levels such as is often seen in pattern 2-type agonists (Small et al., 1986; Rasmussen et al., 1987).

V. Summary

Ca-dependent MLC phosphorylation represents a major pathway of excitation-contraction coupling in vascular smooth muscle. The existence of other Ca-dependent pathways remains a possibility, but measurements of $[Ca^{2+}]_i$ during agonist-induced contractions also raises the possibility of Ca-independent regulatory pathways.

References

Adam LP, Haeberle JR, Hathaway DR. Phosphorylation of caldesmon in arterial smooth muscle. J Biol Chem 1989; 264:7698–703.

Adelstein R, Eisenberg E. Regulation and kinetics of the actin-myosin-ATP interaction. Ann Rev Biochemm 1980; 49:921–56.

Barany M, Rokolya A, Barany K. Absence of calponin phosphorylation in contracting or resting arterial smooth muscle. FEBS Lett 1991; 279:65–8.

Blinks JR, Wier WG, Hess P, Prendergast FG. Measurement of Ca^{2+} concentrations in living cells. Prog Biophys Mol Biol 1982; 40:1–114.

Bradley AB, Morgan KG. Alteration in cytoplasmic calcium sensitivity during porcine coronary arterial contractions as detected by aequorin. J Physiol (London) 1987; 385:437–48.

Bruschi G, Bruschi ME, Regotishi G, Cavatora A, Borghetti A. Relationship between myoplasmic Ca^{2+} and contraction in stimulated arterial smooth muscle. J Hypertension 1986; 4:S328–30.

Conti MA, Adelstein RS. The relationship between calmodulin binding and phosphorylation of smooth muscle myosin kinase by the catalytic subunit of 3′:5′ cAMP-dependent protein kinase. J Biol Chem 1981; 256:3178–81.

DeFeo TT, Morgan KG. Calcium-force relationships as detected with aequorin in two different vascular smooth muscles of the ferret. J Physiol (London) 1985; 369:269–82.

DeFeo TT, Briggs GM, Morgan KG. Ca^{++} signals obtained with multiple indicators in mammalian vascular smooth muscle cells. Am J Physiol 1987; 253:H1456–61.

Dillon PF, Aksoy MO, Driska SP, Murphy RA. Myosin phosphorylation and the cross-bridge cycle in arterial smooth muscle. Science 1981; 211:495–7.

Driska SP, Hartshorne DJ. The contractile proteins of smooth muscle. Properties and components of Ca^{2+}-sensitive actomyosin from chicken gizzard. Arch Biochem Biophys 1975; 167:203–12.

Driska SP, Stein PG, Porter R. Myosin dephosphorylation during rapid relaxation of hog carotid artery smooth muscle. Am J Physiol 1989; 256:C315–21.

Ebashi S, Mikawa T, Hirata M, Toyo-Oka T, Nonomura Y. Regulatory proteins of smooth muscle. In: Excitation-contraction coupling in smooth muscles. Casteels R, Godfraind T, Ruegg JC, eds. Amsterdam: Elsevier 1977; 325–34.

Fürst DO, Cross RA, Demey J, Small JV. Caldesmon is an elongated flexible molecular localized in the actomyosin domain of smooth muscle. EMBO J 1986; 5:251–7.

Gerthoffer WT, Murphey KA, Gunst SJ. Aequorin luminescence, myosin phosphorylation and active stress in tracheal smooth muscle. Am J Physiol 1989; 257:C1062–8.

Griendling KK, Rittenhouse SE, Brock TA, Ekstein LS, Gimbrone MA, Alexander RW. Sustained diacylglycerol formation from inositol phospholipids in angiotensin II-stimulated vascular smooth muscle cells. J Biol Chem 1986; 261:5901–6.

Grynkiewicz G, Poenie M, Tsien RY. A new generation of Ca^{2+} indicators with greatly improved fluorescence properties. J Biol Chem 1985; 260:3440–50.

Hai C-M, Murphy RA. Cross-bridge phosphorylation and regulation of the latch state in smooth muscle. Am J Physiol 1988; 254:C99–106.

Hassid A. Atriopeptin II decreases cytosolic free Ca in cultured vascular smooth muscle cells. Am J Physiol 1986; 251:C681–6.

Hemric ME, Chalovich JM. Characterization of caldesmon binding to myosin. J Biol Chem 1990; 265:19672–8.

Himpens B, Somlyo AP. Free-calcium and force transients during depolarization and pharmacomechanical coupling in guinea pig smooth muscle. J Physiol 1988; 395:507–30.

Hoar PE, Kerrick WGL, Cassidy PS. Chicken gizzard: relation between myosin phosphorylation and contraction. Science 1979; 204:503–6.

Ikebe M, Reardon S. Binding of caldesmon to smooth muscle myosin. J Biol Chem 1988; 263:3055–8.

Ikebe M, Reardon S. Phosphorylation of smooth muscle myosin light chain kinase by smooth muscle Ca^{2+}/calmodulin-dependent multifunctional protein kinase. J Biol Chem 1990; 265:8975–8.

Ikebe M, Inagaki M, Kanamaru K, Hidaka H. Phosphorylation of smooth muscle myosin light chain kinase by Ca^{2+}-activated, phospholipid-dependent protein kinase. J Biol Chem 1985; 260:4547–50.

Jiang MJ, Morgan KG. Intracellular calcium levels in phorbol ester-induced contractions of vascular muscle. Am J Physiol 1987; 253:H1365–71.

Jiang MJ, Morgan KG. Agonist-specific myosin phosphorylation and intracellular calcium during isometric contraction of arterial smooth muscle. Pfluegers Archiv 1989; 413:637–43.

Kamm KE, Stull JT. The function of myosin and myosin light chain kinase phosphorylation in smooth muscle. Ann Rev Pharmacol Toxicol 1985; 25:593–620.

Lash JA, Sellers JR, Hathaway DR. The effects of caldesmon on smooth muscle heavy actomeromyosin ATPase activity and binding of heavy meromyosin to actin. J Biol Chem 1986; 261:16155–60.

Lehman W, Szent-Györgyi AG. Regulation of muscular contraction. Distribution of actin control and myosin control in the animal kingdom. J Gen Physiol 1975; 66:1-30.

Marston SB, Trevett RM, Walters M. Calcium ion-regulated thin filaments from vascular smooth muscle. Biochem J 1980; 185:355-65.

Morgan KG. The role of calcium in the control of vascular tone as assessed by the Ca^{2+} indicator aequorin. Cardiovasc Drugs and Therap 1990; 4:1355-62.

Morgan KG, Brozovich FV. Intracellular calcium levels and arterial smooth muscle function. In: Aoki K, Frohlich ED, eds. Essential hypertension and calcium: calcium ions, calcium channels, calcium agonists and antagonists. New York: Academic Press, 1989.

Morgan KG, Suematsu E. Effect of calcium on vascular smooth muscle tone. Am J Hypertension 1990; 3:2915-85.

Morgan JP, Morgan KG. Vascular smooth muscle: the first recorded Ca^{2+} transients. Pfluegers Archiv 1982; 395:75-7.

Morgan JP, Morgan KG. Stimulus-specific patterns of intracellular calcium levels in ferret portal vein smooth muscle. J Physiol (London) 1984; 351:155-67.

Nishikawa MP, De Lanerolle P, Lincoln TM, Adelstein RS. Phosphorylation of mammalian myosin light chain kinases by the catalytic subunit of cyclic AMP-dependent protein kinase and by cyclic GMP-dependent protein kinase. J Biol Chem 1984; 259:8429-36.

Nishikawa MP, Shirakawa S, Adelstein RS. Phosphorylation of smooth muscle myosin light chain kinase by protein kinase C. J Biol Chem 1985; 260:8978-83.

Ozaka H, Kwon S-C, Tajimi M, Karaki H. Changes in cytosolic Ca^{2+} and contraction induced by various stimulants and relaxants in canine tracheal smooth muscle. Pfluegers Archiv 1990; 416:351-9.

Papageorgiou P, Morgan KG. Increased Ca^{2+} signaling following alpha adrenoceptor activation in vascular hypertrophy. Circ Res 1991; 68:1080-4.

Pohl J, Walsh MP, Gerthoffer WT. Calponin and caldesmon phosphorylation in canine tracheal smooth muscle. Biophys J 1991; 59:58a.

Rasmussen H, Takuwa Y, Park S. Protein kinase C in the regulation of smooth muscle contraction. FASEB J 1987; 1-177-98.

Rembold CM. Modulation of the $[Ca^{2+}]$ sensitivity of myosin phosphorylation in intact swine arterial smooth muscle. J Physiol Lond 1990; 429:77-94.

Rembold CM, Murphy RA. Myoplasmic calcium, myosin phosphorylation and regulation of the crossbridge cycling in swine arterial smooth muscle. Circ Res 1986; 58:803-15.

Reynolds EE, Dubyak GR. Agonist-induced calcium transients in cultured smooth muscle cells: measurements with fura-2 loaded monolayers. BBRC 1986; 136:927-34.

Roe MW, Lemasters JJ, Herman B. Assessment of Fura-2 for measurements of cytosolic free calcium. Cell Calcium 1990; 11:63-73.

Small JV. Geometry of actin-membrane attachments in the smooth muscle cell: the localizations of vinculin and alpha-actinin. EMBO J 1985; 4:45-9.

Small JV, Fürst DO, Demey J. Localization of filamin in smooth muscle. J Cell Biol 1986; 102:210-20.

Smith GA, Hesketh RT, Metcalfe JC, Feeney J, Morris PG. Intracellular calcium measurements by ^{19}FNMR of fluorine labelled chelators. Proc Natl Acad Sci USA 1983; 80:7178–82.

Sobieszek A. Ca-linked phosphorylation of a light chain of vertebrate smooth muscle myosin. Eur J Biochem 1977; 73:477–83.

Sobue K, Sellers JR. Caldesmon, a novel regulatory protein in smooth muscle and nonmuscle actomyosin systems. J Biol Chem 1991; 266:12115–8.

Stull JT, Hsu LC, Tansey MG, Kamm KE. Myosin light chain kinase phosphorylation in tracheal smooth muscle. J Biol Chem 1990; 265:16683–90.

Suematsu E, Resnick M, Morgan KG. Change of Ca^{2+} requirement for myosin phosphorylation by protaglandin $F_{2\alpha}$. Am J Physiol 1991a; 261:C253–8.

Suematsu E, Resnick M, Morgan KG. Ca^{2+}-independent change in phosphorylation of the myosin light chain during relaxation of ferret aorta by vasodilators. J Physiol (London) 1991b; 440:85–93.

Takahashi K, Nadal-Ginard B. Molecular cloning and sequence analysis of smooth muscle calponin. J Biol Chem 1991; 266:13284–88.

Takahashi K, Abe M, Hiwada K, Kokubu T. A novel troponin T-like protein (calponin) in vascular smooth muscle: Interaction with tropomyosin paracrystal. J Hypertension 1988a; 6:S40–3.

Takahashi K, Hiwada K, Kokubu T. Vascular smooth muscle calponin: a novel troponin T-like protein. Hypertension 1988b; 11:620–6.

Takuwa Y, Noriko T, Rasmussen H. Measurement of cytoplasmic free Ca^{2+} concentration in bovine tracheal smooth muscle using aequorin. Am J Physiol 1987; 253:C817–27.

Taylor DA, Bowman BF, Stull JT. Cytoplasmic Ca^{2+} is a primary determinant for myosin phosphorylation in smooth muscle cells. J Biol Chem 1989; 264:6207–13.

Williams DA, Fogarty KE, Tsien RY, Fay FS. Calcium gradients in single smooth muscle cells revealed by the digital imaging microscope using Fura-2. Nature 1985; 318:558–61.

Winder SJ, Walsh MP. Smooth muscle calponin: Inhibition of actomyosin MgATPase and regulation by phosphorylation. J Biol Chem 1990; 265:10148–55.

Yamaguchi H. Recording of intracellular Ca^{2+} from smooth muscle cells by submicron tip, double-barrelled Ca^{2+}-sensitive microelectrodes. Cell Calcium 1986; 7:203–19.

16

Role of Myosin Phosphorylation in Smooth Muscle and Macrophage Signaling

PRIMAL de LANEROLLE, GAYLE GORGAS, BALAZS DEZSO, KIMBERLY SCHLUNS, and ALLISON K. WILSON

University of Illinois at Chicago
Chicago, Illinois

I. Introduction

Signal transduction in cells that perform work requires a mechanism for converting chemical energy into mechanical work. In most mammalian cells this function is performed by the cyclic interaction of actin and myosin. The energy released when ATP is hydrolyzed following the interaction of actin and myosin is used to move thin filaments, composed primarily of actin, past myosin-containing thick filaments, as described by the sliding filament model. To be physiologically useful, the actin-myosin interaction must be regulated, and this is the function of calcium. Two calcium-dependent mechanisms for regulating the actin-myosin interaction have been described. One mechanism involves the binding of calcium to troponin, a thin-filament protein found in skeletal and cardiac muscle cells. The binding of calcium to troponin causes a conformational change in the thin filament that permits the acto-myosin interaction, which leads to the hydrolysis of ATP, the movement of filaments past each other, and muscle contraction. Relaxation results from dissociation of calcium from troponin when intracellular calcium decreases following the cessation of the contractile stimulus. This appears to be a specialized mechanism that has been described only in striated muscles. The second regulatory mechanism involves the activation of myosin by phosphorylation of one of the myosin subunits by the calcium/calmodulin-dependent enzyme myosin light-chain kinase. This is the more general of the two mechanisms because it is thought to regulate smooth muscle contraction, secretion, and cell motility, and it has been suggested to be involved in cell division. This chapter reviews what is known about the role of myosin phosphorylation in regulating the physiological responses of airway smooth muscle and alveolar macrophages.

II. The Biochemistry of Myosin Phosphorylation

The current interest in myosin phosphorylation can be traced back to a seminal publication in which Adelstein and Conti reported that phosphorylation of the 20-kDa light-chain subunit of myosin (MLC_{20}) resulted in the stimulation of the actin-activated ATPase activity of myosin (AM ATPase) purified from human platelets (Adelstein and Conti, 1975). This observation was of great importance because it had been known that smooth muscle and nonmuscle myosins, unlike myosins purified from striated muscles, have a low rate of ATP hydrolysis that did not appear to be simulated by the addition of actin. The implication of the data available prior to 1975 was that nonstriated muscle myosins require an activation step, but the mechanism for this activation was not known. The data published by Adelstein and Conti resolved this issue.

Two additional papers have also proven to be quite important. The first of these was a publication by Sobieszek demonstrating that MLC_{20} phosphorylation stimulated the AM ATPase activity of smooth muscle myosin (Sobieszek, 1977). Despite the importance of the studies on platelet myosin phosphorylation, little progress had been made in characterizing the biochemical properties of this reaction because of the expense and difficulty of purifying myosin from human platelets. In contrast, it is quite cheap and relatively easy to obtain large amounts of actin, myosin, and the other contractile proteins from smooth muscles. This made it possible to study the kinetic and regulatory properties of MLC_{20} phosphorylation in detail. Moreover, Sobieszek's data demonstrated that nonmuscle cells and smooth muscles have a common regulatory mechanism and that studies on smooth muscles could serve as a paradigm for nonmuscle cells.

Biochemical studies using smooth muscle proteins resulted in the third important observation on myosin phosphorylation when Hartshorne and colleagues demonstrated that myosin light-chain kinase (MLCK), the enzyme that phosphorylates MLC_{20}, requires calcium and calmodulin for activity (Dabrowska et al., 1978). Calmodulin is a small, acidic, calcium-binding protein that is similar to troponin C, the troponin subunit that binds calcium

(Ebashi and Endo, 1968). Furthermore, while calmodulin is abundant in smooth muscles and nonmuscle cells, no convincing evidence has been presented for the presence of troponin in these tissues. Thus, the observation by Hartshorne and colleagues suggested that activation of MLCK by the calcium/calmodulin complex was the Ca^{2+}-dependent regulatory mechanism for activating the contractile element in smooth muscle and nonmuscle cells.

Based on these and many other biochemical experiments (see Hartshorne, 1987, for review), the mechanism depicted schematically in Fig. 1

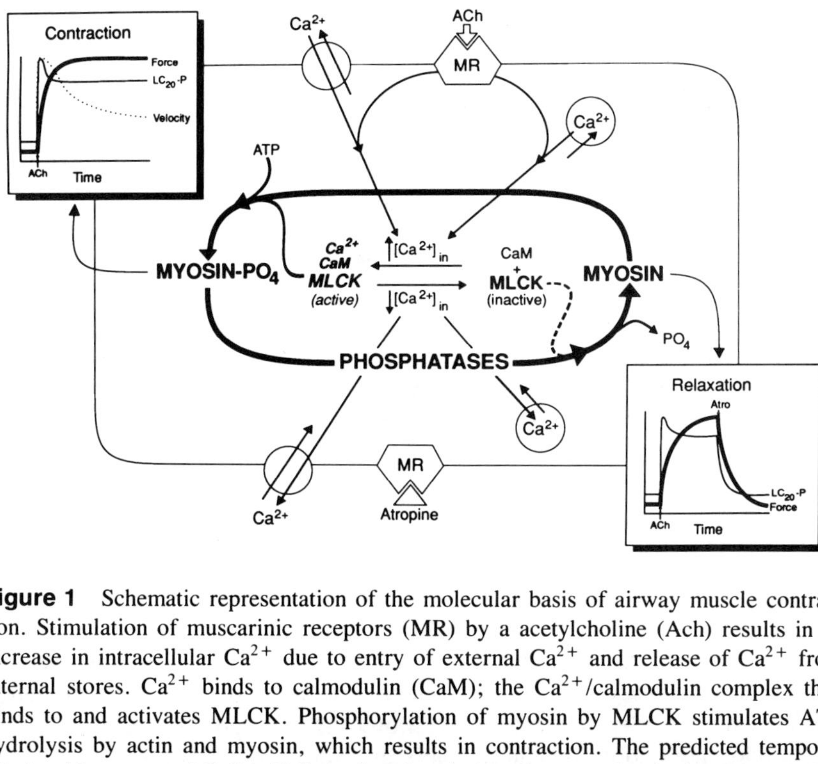

Figure 1 Schematic representation of the molecular basis of airway muscle contraction. Stimulation of muscarinic receptors (MR) by a acetylcholine (Ach) results in an increase in intracellular Ca^{2+} due to entry of external Ca^{2+} and release of Ca^{2+} from internal stores. Ca^{2+} binds to calmodulin (CaM); the Ca^{2+}/calmodulin complex then binds to and activates MLCK. Phosphorylation of myosin by MLCK stimulates ATP hydrolysis by actin and myosin, which results in contraction. The predicted temporal relationships among MLC_{20} (LC_{20}) phosphorylation, force, and velocity (V_{us}) during contraction are shown at top left. Relaxation begins with cessation of agonist stimulation, in this case due to atropine (Atro) binding to muscarinic receptors. This results in a decrease in intracellular Ca^{2+}, dissociation of Ca^{2+} from calmodulin, inactivation of MLCK due to dissociation of calmodulin from the enzyme, dephosphorylation of myosin by phosphoprotein phosphatases, and relaxation. The broken line from inactive MLCK signifies that MLCK is passively involved in dephosphorylation. The temporal relationship between force and myosin dephosphorylation during relaxation is shown at bottom right. (Adapted from de Lanerolle and Paul, 1991.)

is thought to regulate the contractile properties of smooth muscles and contractile events (e.g., cell motility, secretion, shape changes) in nonmuscle cells. This mechanism suggests that the activation of stimulatory receptors, in this case muscarinic receptors on an airway muscle cell, leads to an increase in intracellular calcium, which results in occupancy of the calcium-binding sites on calmodulin and the binding of the calcium/calmodulin complex to MLCK. The binding of the calcium/calmodulin complex activates MLCK and results in the phosphorylation of MLC_{20}. This, in turn, stimulates the rate of ATP hydrolysis by actin and myosin and the movement of filaments past each other. The end result of this process is smooth muscle contraction. Relaxation is initiated by the withdrawal of agonist stimulation. The subsequent decreases in intracellular calcium and calcium/calmodulin complexes leads to the dissociation of calcium/calmodulin complexes from MLCK, the inactivation of MLCK, and the dephosphorylation of MLC_{20} by phosphoprotein phosphatases.

Such a mechanism predicts specific relationships between myosin phosphorylation and smooth muscle contraction and between myosin dephosphorylation and smooth muscle relaxation. As shown in the insets in Fig. 1, (1) myosin phosphorylation should precede force generation (Fig. 1, top left), (2) there should be a relationship between the steady-state levels of MLC_{20} phosphorylation and maintained force, and (3) MLC_{20} dephosphorylation should precede or be simultaneous with relaxation (Fig. 1, bottom right). Many physiological experiments have been performed on intact and membrane-permeabilized smooth muscles to test these predictions, and this is the subject of the next section.

III. Myosin Phosphorylation and Smooth Muscle Contraction

The basic approach in these studies has been to correlate the kinetics and stoichiometry of MLC_{20} phosphorylation with the contractile properties of smooth muscles. Many of these studies have been performed on membrane-permeabilized (skinned) smooth muscles because skinned fibers permit precise control of intracellular calcium levels. Moreover, it is possible to manipulate MLC_{20} phosphorylation in skinned fibers by using antibodies (de Lanerolle et al., 1991), protein kinases (Pfitzer et al., 1986; Ruegg and Paul, 1982) and phosphoprotein phosphatases (Haeberle et al., 1985). These studies, as well as those on intact muscle, have provided important tests of the myosin phosphorylation hypothesis. For the most part, such experiments have demonstrated that myosin phosphorylation precedes force generation and that it peaks and then decays to a stable level prior to the generation of maximal force (Fig. 2). They have also suggested that the level of myosin phosphorylation determines the force generated by a smooth muscle, although this point is more controversial.

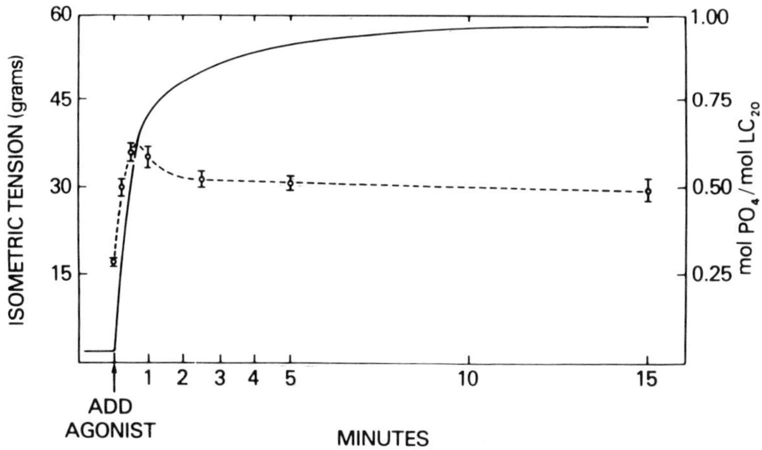

Figure 2 Myosin phosphorylation and tension generation by tracheal smooth muscle contracted with 10^{-4} M methacholine: The solid line represents tension generated by a representative muscle contracted for 15 min; the broken line represents MLC_{20} phosphorylation in muscles frozen at the times indicated. Error bars indicate the S.E. for MLC_{20} phosphate determinations on at least five different pieces of tissue at each time point. (From de Lanerolle et al., 1982.)

The relationship between myosin dephosphorylation and relaxation has been studied, and these experiments demonstrated that smooth muscle relaxation correlates with dephosphorylation of MLC_{20} (Fig. 3).

Although there is unanimity of opinion that myosin phosphorylation is essential for initiating contraction and that activation of MLCK is the calcium-dependent mechanism for turning on the contractile element, there are data that suggest the presence of other mechanisms that modulate and/or regulate smooth muscle contraction. For instance, one of the basic tenets of the mechanism depicted in Fig. 1 is that only phosphorylated myosin has significant AM ATPase activity. However, there are reports that unphosphorylated myosin purified from aortic and gizzard smooth muscles have significant AM ATPase activities when assayed under precise conditions (Wagner and Vu, 1986, 1987). In addition, many studies have demonstrated that the unloaded velocity of shortening (V_{us}) peaks and then decreases to a low level while most smooth muscles, including tracheal muscle, are still in the process of generating force (Fig. 1, inset top left). V_{us} is an indirect measure of the cycling rate of actomyosin cross-bridges (Edman, 1979). Since cycling rate is an index of the activation of myosin molecules, V_{us} should also reflect the level of MLC_{20} phosphorylation. Thus, a decay in V_{us} suggests a change in the

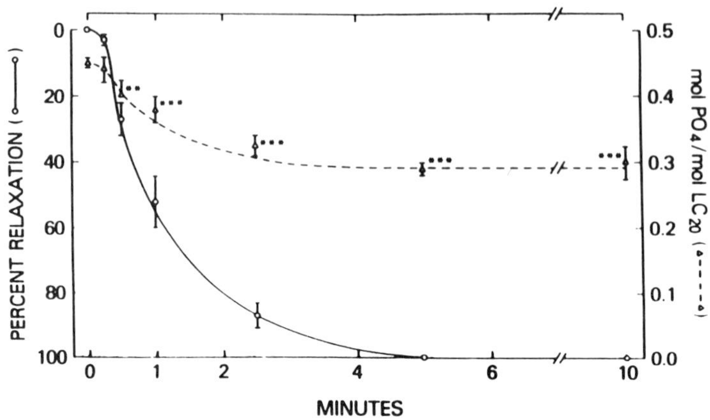

Figure 3 MLC_{20} (LC_{20}) phosphorylation (broken line) and relaxation (solid line) of tracheal smooth muscles. Canine tracheal muscles were stimulated with 10^{-6} M methacholine for 15 min and relaxed by adding 4×10^{-7} M atropine at time 0. Muscles were frozen at defined times and assayed for myosin phosphate content. Present relaxation was calculated from the expression ($\Delta T/\Delta M$) 100, where ΔT = decrease in tension at each time point after addition of atropine and ΔM = increase in tension at 15 min after addition of methacholine and just before adding atropine. The error bars represent the S.E. for measurements performed on at least five pieces of tissue assayed at each time point. Statistical significance of the myosin phosphorylation data was evaluated by performing a Student's t test (*$P \leqslant 0.05$, **$P \leqslant 0.01$, ***$P \leqslant 0.001$) and an analysis of variance ($P < 0.001$). (Modified from de Lanerolle, 1988.)

nature of the cross-bridge that may be related to the presence of a second regulatory mechanism. Finally, caldesmon (Sobue et al., 1981) and calponin (Winder and Walsh, 1990), two proteins whose functions are affected by the presence of calcium, have been purified from smooth muscles. Since these proteins are abundant in smooth muscles, it has been suggested that they may play a role in regulating smooth muscle contraction.

In summary, the activation of MLCK appears to be the major mechanism for calcium regulation of smooth muscle contraction and the phosphorylation of MLC_{20} appears to be the dominant mechanism for activating the contractile element in smooth muscles. It is now generally accepted that MLC_{20} phosphorylation is essential for generating force, and there is a growing consensus that MLC_{20} phosphorylation determines that force output of a smooth muscle (see de Lanerolle and Paul, 1991, for a more complete review of the role of myosin phosphorylation in regulating smooth muscle contraction). However, MLC_{20} phosphorylation is clearly dissociated from V_{us}, and data have been presented in support of the existence of other regulatory mechanisms in smooth muscles (see Gerthoffer, 1991, for a more complete review of these putative regulatory mechanisms), although the precise manner in which they affect smooth muscle contractility is unclear at this time.

IV. Myosin Phosphorylation and Macrophage Motility

Macrophage motility, like muscle contraction, also requires a transduction mechanism for converting the energy released when ATP is hydrolyzed into mechanical work. Actin and myosin have been postulated to perform this function in motile cells for many years (Korn, 1978). However, the mechanism regulating the actomyosin interaction in nonmuscle cells was obscure prior to the demonstration that MLC_{20} phosphorylation stimulated the AM ATPase of platelet myosin (Adelstein and Conti, 1975). Moreover, the subsequent demonstration that MLC_{20} phosphorylation also stimulated the actin-activated ATPase activity of macrophage myosin (Trotter and Adelstein, 1979) suggested that macrophage function may also be regulated by this reaction. Therefore, we have investigated the role of myosin phosphorylation in regulating chemotaxis by rat alveolar macrophages.

Studying the regulation of macrophage motility has proven to be technically more complex than studying the regulation of smooth muscle contraction. For instance, correlative experiments of the sort described above are not fruitful because it is difficult to get a population of macrophage to respond in unison. Moreover, it is difficult to quantitate partial responses and subtle changes in the behavior of an entire population of cells. In light of these problems, the most productive approach appeared to be one in which we manipulated intracellular MLCK activity and then determined the effect of increasing or decreasing MLCK activity on MLC_{20} phosphorylation and macrophage function. We judged MLCK and antibodies to MLCK (MK-Ab) to be the optimal agents for manipulating intracellular MLCK activity. On the other hand, using these proteins required a method for introducing them into macrophages without compromising the physiological responsiveness of the cells.

Our initial approach for introducing proteins into macrophages was to pressure inject proteins using a micropipette and to monitor the injected cells using video time-lapse photomicroscopy (Fig. 4). Of the cells shown in Fig. 4, the macrophage in panel A was injected with a drop of oil so that it could be identified easily. This cell continued to ruffle and migrate at will on the cover slip for about 2 h following the injection. The macrophage shown in panels B-F was injected with a solution containing MK-Ab. MK-Ab are affinity-purified rabbit antibodies to turkey gizzard MLCK that interact monospecifically with MLCK in a wide variety of smooth muscle and nonmuscle cells (de Lanerolle et al., 1981). These antibodies inhibit the catalytic activity of smooth muscle

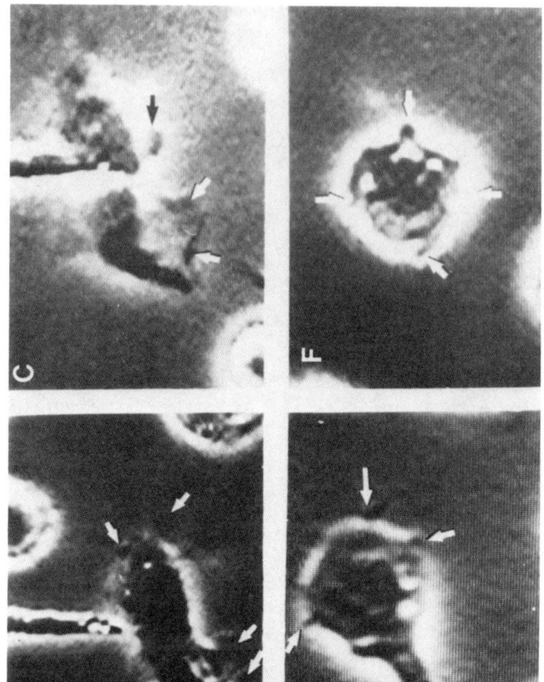

Figure 4 Microinjection of macrophages. Macrophages were injected with mineral oil (panel A) or MK-Ab (panels B-F) using a micropipette as described by Kiehart (1981). Microinjected cells were observed using time-lapse video microscopy for 2 h. The oil droplet injected into the cell in panel A is identified with a black arrowhead. This cell continued to ruffle and move about the coverslip for 2 h. Panel B shows the cell injected with MK-Ab just prior to injection. Injecting MK-Ab caused the membrane to pouch out (panel C, black arrow). This cell rounded up soon after the MK-Ab was injected into the cell. Although this cell remained attached to the same spot, the cell membrane continued to ruffle (white arrows) for 2 h.

and nonmuscle MLCK's (de Lanerolle et al., 1987), including MLCK activity in extracts prepared from macrophages (Wilson et al., 1991a). Macrophages pressure injected with MK-Ab round up and stop all translation motility in about 10 to 15 min but continue to ruffle for as long as the cells were observed (2 h).

The experiment shown in Fig. 4 established the feasiblity of using antibodies to manipulate macrophage function. Nevertheless, one has to be cautious in interpreting the results from such experiments because it is technically difficult to perform this experiment, particularly with respect to introducing the same amount of control and experimental antibody into cells, and the implications of data from experiments performed on a limited number of cells can be misleading. Therefore, we performed all of our subsequent experiments using a voltage discharge technique, known as electroinjection (Zimmermann, 1986), that can be used to introduce protein into 2 to 5 million cells at a time. Electroinjection is a variation of the electroporation technique that has been used extensively to introduce DNA into cells (Wong and Neumann, 1982). In both techniques a capacitor is charged to some level and then discharged through a solution containing cells and the material to be introduced into the cells. The major difference between the two techniques is that electroporation is usually performed using conditions that result in 50% cell death whereas electroinjection is performed using conditions that yield >90% cell viability. We have characterized this technique extensively (Wilson et al., 1991a, 1991b) and found, using precisely defined conditions, that (1) chemotaxis is essentially the same in electroinjected and noninjected macrophages, (2) micromolar concentrations of proteins can be introduced into individual cells, (3) there is a monodisperse distribution of cell loading as determined by fluorescence-activated cell sorting, and (4) there is a direct relationship between the extracellular protein concentration and the amount of protein loading.

The effect of injecting control antibodies (C-Ab) or MK-Ab on the chemotactic response of rat alveolar macrophages is shown in Fig. 5. Macrophages electroinjected with C-Ab display a characteristic chemotactic response that is virtually identical to uninjected cells. In contrast, electroinjecting cells with MK-Ab decreased motility by as much as 44 ±6% at the optimal concentration of chemoattractant. We also investigated the effect of increasing intracellular MLCK activity on macrophage motility. In order to do so, we produced an unregulated form of MLCK (MK^-) by using trypsin to digest MLCK purified from chicken gizzard smooth muscle. Trypsinization produces a constitutively active form of MLCK that does not require calcium/calmodulin for activity but still retains its specificity for MLC_{20} (Ikebe et al., 1987). Experiments with MK^- demonstrated that the number of macrophages that migrated decreased as function of the MK^- concentration in the electroinjection buffer (Fig. 6).

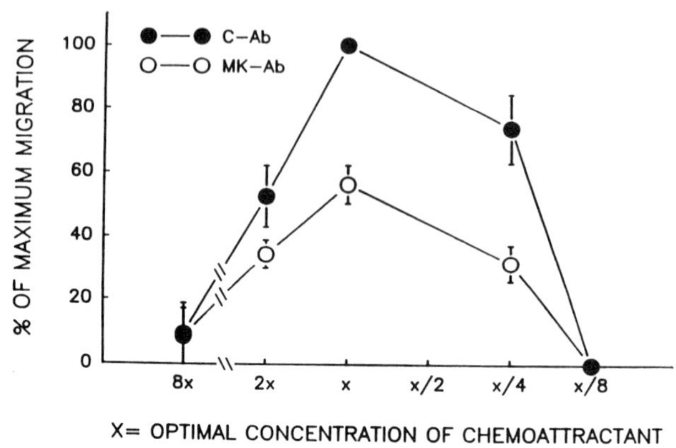

Figure 5 Effect of MK-Ab (○) and C-Ab (●) on macrophage motility as a function of chemoattractant concentration. Macrophages obtained by lung lavage were electroinjected at 700 V/cm in the presence of 2.5 mg/mL of C-Ab or MK-Ab. The chemotaxis assay was performed using varying concentrations of endotoxin-activated rat serum as described previously (Falk et al., 1980). Percentage migration was calculated based on the number of cells electroinjected with C-Ab that migrated at the optimum concentration of endotoxin-activated rat serum as 100%. There was no difference in the migration of untreated cells and cells electroinjected with C-Ab. However, inhibition of migration due to MK-Ab incorporation was significantly decreased ($p<0.001$) as determined by analysis of variance ($n = 3$). (Reproduced from Wilson et al., 1991a, by copyright permission from the Rockefeller University Press.)

The data in Figs. 5 and 6 suggest that both a decrease and an increase in MLCK activity are detrimental to the chemotactic response of macrophages. If this is truly due to changes in intracellular MLCK activity, this should be reflected in comparable changes in the levels of MLC_{20} phosphorylation. Therefore, we determined whether the electroinjection of MK-Ab or MK^- resulted in the appropriate changes in MLC_{20} phosphorylation. This was investigated by immunoprecipitating myosin from macrophages that had been electroinjected with C-Ab, MK-Ab, MK^-, or trypsinized BSA and then incubated with ^{32}P to label intracellular ATP pools. These experiments demonstrated that MK-Ab decreased and MK^- increased MLC_{20} phosphorylation relative to their respective controls (Wilson et al., 1991a).

The observation that an increase in MLC_{20} phosphorylation inhibited motility was unexpected. In order to investigate the mechanistic basis of this inhibition further, we studied the effect of okadaic acid on macrophage chemotaxis. Okadaic acid, a toxin purified from marine sponges, is a potent inhibitor of phosphoprotein phosphatases (Takai, 1988), including the phosphatases (type 2A and type 1) that are thought to dephosphorylate MLC_{20}. Phosphatase

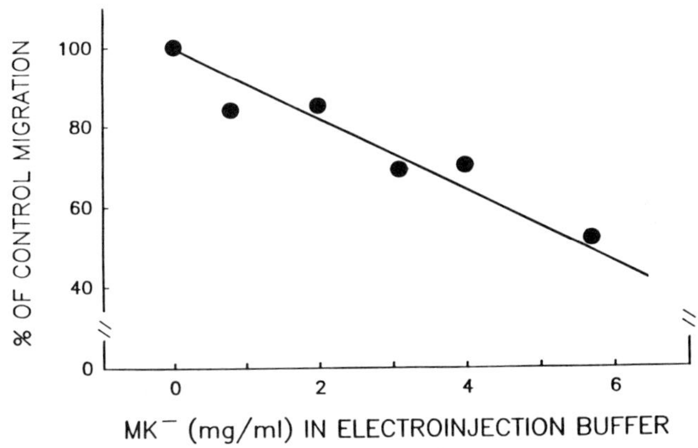

Figure 6 Effect of MK^- incorporation on macrophage motility. Macrophages were electroinjected at 700 V/cm in the presence of equivalent concentrations of trypsin-treated BSA (control cells) or MK^-. The protein concentration in the electroinjection buffer was varied as shown because there is a direct relationship between the protein concentration in the electroinjection buffer and incorporation into the cells (Wilson et al., 1991b). Chemotaxis was then performed using the optimal concentration of endotoxin-activated rat serum as defined in Fig. 5. Migration of cells electroinjected with MK^- are reported as the percent of control cell migration at each protein concentration. The migration of control cells (i.e., cells electroinjected with trypsinized BSA) was virtually identical to the migration of cells incubated, but not electroinjected, with BSA. Each data point represents one experiment. (Reproduced from Wilson et al., 1991a, by copyright permission from the Rockefeller University Press.)

inhibition results in an increase in protein phosphorylation and okadaic acid gave us a means to increase MLC_{20} phosphorylation more easily, albeit less specifically, than by electroinjecting MK^-. As predicted, incubating rat alveolar macrophages with okadaic acid results in a dose-dependent increase in MLC_{20} phosphorylation that correlates with a decrease in chemotaxis (Fig. 7).

We also investigated the effect of okadaic acid on the distribution of myosin in macrophages. We did this because it is well known that MLC_{20} phosphorylation also regulates the ability of smooth muscle and nonmuscle myosins to form filaments (Craig et al., 1983). Unphosphorylated smooth muscle and nonmuscle myosins have a folded conformation and a sedimentation coefficient of 10S. In this folded conformation, myosin does not aggregate into filaments. Thus, 10S myosin cannot perform work because, being unphosphorylated, it does not hydrolyze ATP; and it cannot be part of a sliding filament

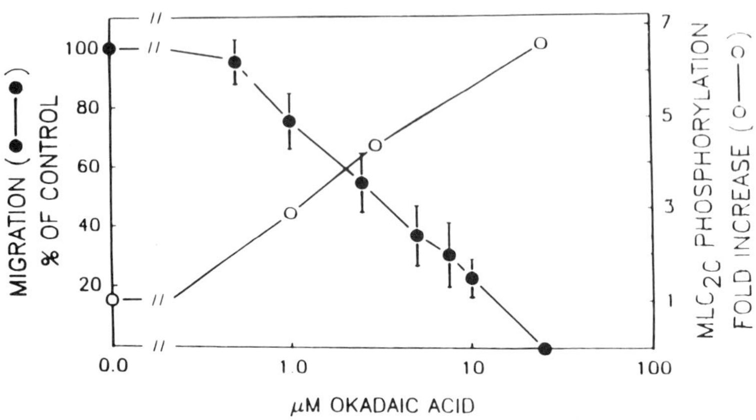

Figure 7 Effects of okadaic acid on MLC_{20} phosphorylation and macrophage motility. Data on migration are normalized to number of cells that migrated in the absence of okadaic acid (100%), whereas MLC_{20} phosphorylation data are reported as fold increase after okadaic acid treatment compared with cells not treated with okadaic acid (1.0). Washout of 3 and 25 μM okadaic acid resulted in restoration of 100% and 50% of control migration, respectively. Chemotaxis data were obtained from four separate experiments and are presented as means ± S.E. MLC_{20} phosphorylation data are presented as means of two to four separate experiments. (From Wilson et al., 1991c.)

mechanism because it does not form filaments. In contrast, phosphorylated myosin has a sedimentation coefficient of 6S, hydrolyzes ATP, and has an extended conformation that permits phosphorylated myosin to aggregate into filaments. Therefore, we prepared cytoskeletal and cytoplasmic fractions (Fox and Phillips, 1982) from control cells and cells treated with 25 μM okadaic acid and determined the amount of myosin associated with the cytoskeleton. When expressed as a function of the total cellular myosin, cytoskeletal myosin increased from 47.0 ±7.4% in control cells to 96.7 ±1% in okadaic acid-treated cells (Wilson et al., 1991c). Similar but less profound changes were observed following electroinjection of MK^- into macrophages (Wilson et al., 1991a).

Finally, we postulated that an increase in MLC_{20} phosphorylation, with accompanying increases in ATPase activity and myosin filaments, should be accompanied by morphological changes following okadaic treatment. Therefore, we investigated the distribution of myosin and F-actin in J774 cells, a cell line derived from mouse peritoneal macrophages that migrate and phagocytose particles, using an antibody to myosin purified from these cells. Figure 8 shows that treating J774 cells plated on cover slips with 5 μM okadaic acid results in the aggregation of myosin and a generalized retraction of the borders of the cell.

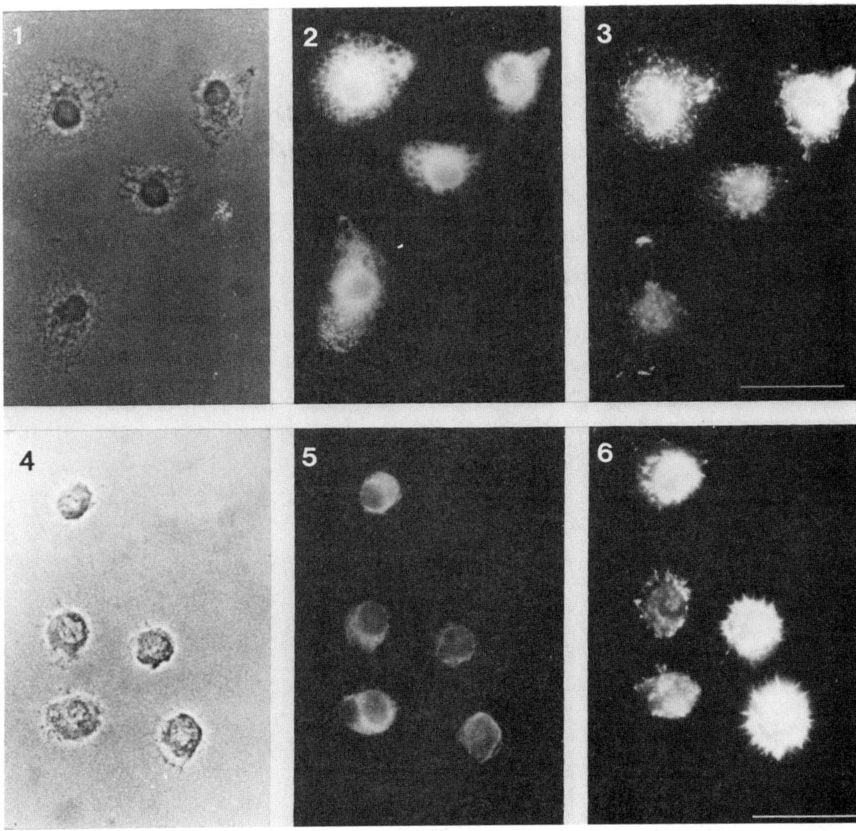

Figure 8 Phase-contrast microscopy (1, 4), myosin distribution (2, 5) and F-actin localization (3, 6) in cells treated with DMSO (1–3) or 5 μM okadaic acid (4–6) for 90 min. Cells grown on cover slips were fixed and permeabilized by treating the cells with 3.7% formalin in PBS containing 0.2% Triton X-100 for 7 min. The cells were then incubated with bodipy-labeled phalloidin to visualize F-action. Myosin localization was visualized using a rabbit antibody to J774 myosin and rhodamine-labeled secondary antibody. Panels 1 and 4 show that okadaic acid (4) causes the cells to contract. Panel 2 demonstrates a lacy, filamentous myosin network in control cells that is lost following okadaic acid treatment (5). F-actin is more granular in appearance and is seen in microspikes and around the nucleus in control cells (3). The microspikes and the perinuclear localization is more obvious following okadaic acid treatment (6). The bars represent 10 μm.

Taken together, these data demonstrate that MLCK activity must be tightly regulated and that MLC_{20} phosphorylation must be maintained within narrow limits to support motility. The observation that a decrease in MLC_{20} phosphorylation, following electroinjection of MK-Ab, inhibits chemotaxis was predictable based on the central role of MLCK in regulating the actin-myosin interaction in smooth muscle and nonmuscle cells. That is, a decrease in MLC_{20} phosphorylation should result in a predictable inhibition of motility due to the inability of cells injected with MK-Ab to convert chemical energy into mechanical work. On the other hand, the observation that an increase in MLC_{20} phosphorylation, following the electroinjection of MK^- or when cells were treated with okadaic acid, was unexpected. In fact, based on previous work (Fechheimer and Zigmond, 1983), we had anticipated that an increase in MLC_{20} phosphorylation would stimulate motility.

What, then, is the explanation for the fact that an increase in MLC_{20} phosphorylation inhibits motility? One explanation is that an increase in MLC_{20} phosphorylation inhibits motility due to the too much ATP hydrolysis, particularly in spatially inappropriate locations and/or due to the presence of a hyperabundance of myosin filaments. Cell motility is an extremely complex process in which the spatial distribution of the underlying regulatory reactions is as important as the temporal sequence in which they take place. For instance, a motile cell must be able to redistribute and continually reorganize the cytoskeleton in order to move in a particular direction. In addition, it must extend a pseudopod only in the direction it wants to migrate and withdraw all other cellular projections. Tail retraction is thought to be a MLCK-MLC_{20} phosphorylation-mediated process (Holzapfel et al., 1983), while extension of the leading edge is thought to depend on the polymerization of actin (Cande and Ezzell, 1986). A global increase in MLC_{20} phosphorylation could inhibit motility by decreasing the fluidity of the cytoskeleton, by increasing the number of myosin filaments in the cell, and/or by stimulating ATP hydrolysis by actin and myosin throughout the cell rather than confining this reaction to the trailing edge.

Finally, it is important to point out that another myosin molecule, known as myosin I, has been postulated to play an important role in cell motility in the slime mold *Dictyostelium discoideum. Dictyostelium* contains two myosin molecules, known as myosin I and myosin II (Korn and Hammer, 1988). Myosin I has ATPase activity that is activated by actin. However, in contrast to myosin II, it has a truncated heavy chain that cannot form filaments. Dictyostelium myosin II is very similar to mammalian smooth muscle and nonmuscle myosins structurally and in that it is primarily regulated by MLC_{20} phosphorylation (Korn and Hammer, 1988).

Experiments on *Dictyostelium* in which the expression of myosin II was genetically manipulated have demonstrated that these cells are still capable of cell motility (De Lozanne and Spudich, 1987; Knecht and Loomis, 1987). These data have suggested that myosin II is not essential for cell motility in *Dictyostelium* or, by extension, in mammalian cells. Although myosin I has not been proven to be present in macrophages, we cannot exclude the possibility that myosin I plays a role in macrophage motility. Nevertheless, the specificity of MK-Ab and MK^- clearly establishes that myosin II and its regulation by MLC_{20} phosphorylation are important in mammalian cell motility.

V. Conclusion and Summary

It is now clear that MLCK and MLC_{20} phosphorylation play important roles in signal transduction in smooth muscles and macrophages. There is a consensus of opinion that MLC_{20} phosphorylation is essential for initiating smooth muscle contraction, but the possible existence of alternative regulatory mechanisms or mechanisms that may modulate the contractile properties of smooth muscles has been raised. One interesting possibility is that there may be mechanisms that are specific to individual smooth muscles that are responsible for the unique contractile properties of these muscles (e.g., tonic versus phasic smooth muscles). The presence of multiple myosin isoforms in smooth muscles (Kawamoto and Adelstein, 1987) has also been demonstrated. Another interesting question is whether the differential expression of these myosin isoforms plays a role in airway smooth muscle hyperreactivity or hypertrophy. Finally, it would be of considerable interest to determine whether myosin I exists in smooth muscle and how such a molecule affects the physiological properties of a smooth muscle.

Our understanding of the role of MLCK and MLC_{20} phosphorylation in regulating the physiological responses of macrophages is at a more basic level. One of the most important questions in this regard is to establish whether myosin I exists in macrophages and how it is involved in macrophage function. Another important question relates to the possibility that there are other calcium-dependent processes (e.g., caldesmon, calponin, alpha-actinin) that are important, along with MLC_{20} phosphorylation, in regulating cell motility. This is almost surely true, because of the complexity of regulating the various components of the cytoskeleton. Finally, it is important to determine whether myosin is involved in regulating other macrophage responses, such as phagocytosis and cell division. One distinct advantage of macrophages over smooth muscle cells is that established cell lines such as the J774 cells permit genetic manipulation of macrophage function. We are currently performing such experiments and hope that they will increase our understanding of the molecular mechanisms that regulate macrophage function.

Acknowledgments

The authors thank Dan Kiehart and Tom Pollard, Johns Hopkins Medical School, for assistance in performing the microinjection experiments described in Fig. 4, and Janice Gentry for typing the manuscript. This work was supported in part by grants from the National Institutes of Health (HL35808 and HL02411) and a grant from the National Science Foundation (BNS 8720429) to P. de Lanerolle. P. de Lanerolle is also the recipient of the Florence and Arthur Brock Cancer Investigator Award from the Chicago Lung Association.

References

Adelstein RS, Conti MA. Phosphorylation of platelet myosin increases actin-activated myosin ATPase activity. Nature (London) 1975; 256:597–8.

Cande WZ, Ezzell RM. Evidence for regulation of lamellipodial and tail contraction of glycerinated chicken embryonic fibroblasts by myosin light chain kinase. Cell Motil Cytoskeleton 1986; 6:640–8.

Craig R, Smith R, Kendrick-Jones J. Light-chain phosphorylation controls the conformation of vertebrate non-muscle and smooth muscle myosin molecules. Nature (London) 1983; 302:436–9.

Dabrowska R, Sherry JMF, Aromatorio D, Hartshorne DJ. Modulator protein as a component of myosin light chain kinase from chicken gizzard. Biochemistry 1978; 17:253–8.

de Lanerolle P. cAMP, myosin dephosphorylation, and isometric relaxation of airway smooth muscle. J Appl Physiol 1988; 64:705–709.

de Lanerolle P, Paul RJ. Myosin phosphorylation/dephosphorylation and regulation of airway smooth muscle contractility. Am J Physiol 1991; 261:L1–14.

de Lanerolle P, Adelstein RS, Feramisco JR, Burridge K. Characterization of antibodies to smooth muscle myosin kinase and their use in localizing myosin kinase in nonmuscle cells. Proc Natl Acad Sci USA 1981; 78:4738–42.

de Lanerolle P, Condit JR Jr, Tanenbaum M, Adelstein RS. Myosin phosphorylation, agonist concentration and contraction of tracheal smooth muscle. Nature (London) 1982; 298:871–2.

de Lanerolle P, Nishikawa M, Felsen R, Adelstein RS. Immunological properties of myosin light-chain kinases. Biochim Biophys Acta 1987; 914:74–82.

de Lanerolle P, Strauss JD, Felsen R, Doerman GE, Paul RJ. Effects of antibodies to myosin light chain kinase on contractility and myosin phosphorylation in chemically permeabilized smooth muscle. Circ Res 1991; 68:457–65.

De Lozanne A, Spudich JA. Disruption of the *Dictyostelium* myosin heavy chain gene by homologous recombination. Science 1987; 236:1086–91.

Ebashi S, Endo M. Calcium ion and muscle contraction. Prog Biophys Mol Biol 1968; 18:123–83.

Edman KAP. The velocity of unloaded shortening and its relation to sarcomere length and isometric force in vertebrate muscle fibres. J Physiol (London) 1979; 291:143–59.

Falk W, Goodwin RH, Leonard EJ. A 48-well microchemotaxis assembly for rapid and accurate measurement of leukocyte migration. J Immunol Methods 1980; 33:239–47.

Fechheimer M, Zigmond SH. Changes in cytoskeletal proteins of polymorphonuclear leukocytes induced by chemotactic peptides. Cell Motility 1983; 3:349–61.

Fox JEB, Phillips DR. Role of phosphorylation in mediating the association of myosin with the cytoskeletal structures of human platelets. J Biol Chem 1982; 257:4120–6.

Gerthoffer WT. Regulation of the contractile element of airway smooth muscle. Am J Physiol 1991; 261:L15–28.

Haeberle JR, Hathaway DR, De Paoli-Roach AA. Dephosphorylation of myosin by the catalytic subunit of a type-2 phosphatase produces relaxation of a chemically skinned uterine smooth muscle. J Biol Chem 1985; 260:9965–8.

Hartshorne DJ. Biochemistry of the contractile process in smooth muscle. In: Johnson LR, ed. Physiology of the gastrointestinal tract, 2d ed. New York: Raven Press, 1987; 423–82.

Holzapfel G, Wehland J, Weber K. Calcium control of actin-myosin based contraction in triton models of mouse 3T3 fibroblasts is mediated by the myosin light chain kinase (MLCK)-calmodulin complex. Exp Cell Res 1983; 148:117–26.

Ikebe M, Stepinska M, Kemp BE, Means AR, Hartshorne DJ. Proteolysis of smooth muscle myosin light chain kinase. J Biol Chem 1987; 260:13828–34.

Kawamoto S, Adelstein RS. Characterization of myosin heavy chains in cultured aorta smooth muscle cells. J Biol Chem 1987; 262:7282–8.

Kiehart DP. Studies on the in vivo sensitivity of spindle microtubules to calcium ions and evidence for a vesicular calcium sequestering system. J Cell Biol 1981; 88:604–17.

Knecht DA, Loomis WF. Antisense RNA inactivation of myosin heavy chain gene expression in Dictyostelium discoideum. Science 1987; 236:1081–6.

Korn ED. Biochemistry of actomyosin-dependent cell motility. Proc Natl Acad Sci USA 1978; 75:588–99.

Korn ED, Hammer JA. Myosins of nonmuscle cells. Ann Rev Biophys Biophys Chem 1988; 17:23–45.

Pfitzer G, Merkel L, Ruegg JC, Hofmann F. Cyclic GMP-dependent protein kinase relaxes skinned fibers from guinea pig taenia coli but not from chicken gizzard. Pfluegers Archiv 1986; 407:87–91.

Ruegg JC, Paul RJ. Vascular smooth muscle. Calmodulin and cyclic AMP-dependent protein kinase alter calcium sensitivity in porcine carotid skinned fibers. Circ Res 1982; 50:394–9.

Sobieszek A. Vertebrae smooth muscle myosin. Enzymatic and structural properties. In: Stephens NL, ed. The biochemistry of smooth muscle. Baltimore: University Park Press, 1977; 413–43.

Sobue K, Muramoto Y, Fujita M, Kakiuchi S. Purification of a calmodulin-binding protein from chicken gizzard that interacts with F-actin. Proc Natl Acad Sci USA 1981; 78:5652–5.

Takai A. Okadaic acid: protein phosphatase inhibition and muscle contractile effects. J Muscle Res and Cell Motility 1988; 9:563–5.

Trotter JA, Adelstein RS. Macrophage myosin. Regulation of actin-activated ATPase activity by phosphorylation of the 20,000-dalton light chain. J Biol Chem 1979; 254:8781–5.

Wagner PD, Vu N. Regulation of the actin activated ATPase or aorta smooth muscle myosin. J Biol Chem 1986; 261:7778–3.

Wagner PD, Vu N. Actin-activation of unphosphorylated gizzard myosin. J Biol Chem 1987; 262:15556–62.

Wilson AK, Gorgas G, Claypool WD, de Lanerolle P. An increase or a decrease in myosin II phosphorylation inhibits macrophage motility. J Cell Biol 1991a; 114:277–83.

Wilson AK, Horwitz J, de Lanerolle P. Evaluation of the electroinjection method for introducing proteins into living cells. Am J Physiol 1991b; 260:C355–63.

Wilson AK, Takai A, Ruegg JC, de Lanerolle P. Okadaic acid, a phosphatase inhibitor, decreases macrophage motility. Am J Physiol 1991c; 260:L105–12.

Winder SJ, Walsh MP. Smooth muscle calponin. J Biol Chem 1990; 265:10148–55.

Wong T-K, Neumann E. Electric field mediated gene transfer. Biochem Biophys Res Commun 1982; 107:584–7.

Zimmermann U. Electrical breakdown, electropermeabilization and electrofusion. Rev Physiol Biochem Pharmacol 1986; 105:175–256.

17

Growth Factors and Signal Transduction in Smooth Muscle Systems

MORLEY D. HOLLENBERG

Faculty of Medicine
University of Calgary
Calgary, Alberta, Canada

I. Introduction

A. Growth Factors: What Are They and What Do They Do?

It has become evident, particularly over the past 10 years, that an enlarging group of highly potent hormonelike polypeptides can play a key role in the growth and differentiated function of a variety of somatic cells (Table 1). For some of these agents (e.g., platelet-derived growth factor (PDGF) or epidermal growth factor-urogastrone (EGF-URO), the isolation protocols that led to their purification depended on their ability to stimulate cell division, either in vivo (e.g., the ability of EGF-URO to stimulate epidermal cell growth and thereby cause precocious eyelid opening in mice: Cohen, 1962) or in vitro (e.g., the stimulation by PDGF of the proliferation of cultured fibroblasts or arterial smooth muscle cells: Ross et al., 1974; Antoniades et al., 1979). Thus, the term "growth factor" has been commonly used to refer to such agents. Nonetheless, apart from growth hormone itself, the physiological role that such compounds play in terms of the processes of growth and development remains an open question; and it is most likely that such agents, which, like EGF-URO and insulin, can cause a myriad of cellular effects apart from stimu-

Table 1 Putative Growth Factors[a]

Factor	Comments
Activin	Related to TGF-β
Adrenocorticotropic hormone (ACTH)	Trophic (but not mitogenic) for adrenal cortex
Amphiregulin	Structurally related to EGF-URO
Angiotensin-II*	Trophic, but not mitogenic for cultured aortic smooth muscle cells
Endothelin*	Both mitogenic and vasoactive
Epidermal growth factor-urogastrone (EGF-URO)*	One of a family of polypeptides including amphiregulin and TGF-α; causes inhibition of acid secretion as well as mitogenesis; related to *lin-* and *notch* developmental genes
Erythropoietin	Major factor responsible for red blood cell development
Fibroblast growth factor (FGF)	Two forms known (acidic and basic); related to heparin-binding growth factors
Gastrin	Proliferative action on stomach and duodenum; also promotes acid secretion
Glycine-histidine-lysine	A growth-modulating tripeptide with diverse effects ranging from stimulation to inhibition of cell division
Growth hormone	Action mediated via somatomedins (or insulinlike growth factors, IGFs); can act directly on target cells; essential for growth in humans and other vertebrates
Inhibin	Related to TGF-β
Insulin	Growth factor activity is distinct from insulin's traditional hormonal role
Insulinlike growth factor-I	Somatomedin family, thought to mediate the action of growth hormone; also referred to as somatomedin-C basic-somatomedin, or multiplication stimulating activity (MSA)
Interferons	A large family of polypeptides with antiviral and antiproliferative actions
Interleukins	A large family of distinct polypeptides that not only regulate leukocyte proliferation and function but also modulate the function of other cell types
Mullerian-inhibiting substance	Related to TGF-β
Platelet-derived growth factor*	One of several growth-modulating factors present in platelets, formed as homo- and heterodimers of PDGF-A and -B chains (PDGF-A,A; -A,B; and -B,B)
Secretin	In addition to regulating insulin secretion, can counteract the proliferative actions of gastrin; may have intrinsic antiproliferative action on stomach and duodenum

Table 1 Continued

Factor	Comments
Serotonin (5-hydroxytryptamine)*	In addition to regulating smooth muscle contractility, can regulate growth factor receptors in cultured smooth muscle cells
Thrombin*, trypsin, and other proteases	Proteolytic enzymes known to stimulate growth
Transforming growth factor-α*	Structurally related to EGF-URO and amphiregulin; possibly a fetal form of EGF-URO
Transforming growth factor-β	Structurally distinct from TGF-α, comprising an enlarging family of homo- and heterodimeric polypeptides typified by TGF-β1,1; -β1,2; -β2,2.
Vasopressin*	Stimulates cell growth and smooth muscle contraction via a V_1 receptor

[a] The structures of all of the factors listed are known. These agents can exert either a trophic (increase in size and/or cell number) or a growth-inhibitory action on their target cells. Those factors designated with an asterisk (*) also regulate smooth muscle contractility.

lating cell division (Table 2), may act as localized paracrine/autocrine cell regulators in a variety of organ systems. In this chapter, the focus is on the action of EGF-URO in smooth muscle systems, not in terms of its ability to stimulate DNA synthesis, but rather in terms of its ability to regulate acutely (i.e., within minutes) either tissue contractility or membrane enzyme activity (e.g., tyrosine kinase or phospholipase).

B. Mechanism of Growth Factor Action

Like other polypeptides, growth factors such as EGF-URO act via specific high-affinity cell membrane receptors. The intriguing property of the transmembrane receptors for many growth factors (e.g., EGF-URO, PDGF) is the presence, in the cytoplasmic portion, of a tyrosin kinase domain. It is presumably the substrate specificity of the tyrosine kinase domain in each unique receptor that results in both the common and selective actions of different agents, each of which may have their own receptors present in the same target tissue. It should be pointed out, however, that many agents that can act as "growth factors" to stimulate either cell division or cellular hypertrophy (e.g., vasopressin, angiotensin-II, or bradykinin) do not work via tyrosine kinase receptors, but act rather via a G-protein-linked mechanism (e.g., see Kaziro et al., 1991). In this respect, the ability of G-protein-linked agonists such as vasopressin and angiotensin-II to enhance cellular tyrosine phosphory-

Table 2 Diverse Actions of Epidermal Growth Factor-Urogastrone (EGF-URO)[a]

In vivo	In vitro
Cell proliferation	Increased cell proliferation
Keratinization	Enhanced keratinocyte growth
Accelerated newborn eyelid opening	Regulation of smooth muscle contractility
Incisor eruption	Enhanced metabolite transport (uridine, aminoisobutyrate, glucose)
Inhibition of gastric acid secretion	Increased synthesis of DNA, RNA, and protein
Healing of ulcers	Increased membrane phosphorylation
Enhanced ornithine and histidine decarboxylase	Increased activity of ornithine decarboxylase
Hepatic triglyceride accumulation	Increased cyclic nucleotide levels (both cAMP cGMP)
Regulation of blood flow	Enhanced hyaluronic acid synthesis
Regulation of hair growth	Enhanced fibronectin synthesis
	Increased cell binding to insolubilized concanavalin A
	Altered viral growth (cytomegalovirus, reduced; herpes simplex type I increased)
	Inhibition of palate fusion
	Inhibition of acid secretion
	Bone resorption

[a] This list, although long, is by no means exhaustive; for brevity, a number of other interesting actions of EGF-URO in vitro have been omitted. Qualitatively, both EGF-URO and TGF-α have similar actions.

lation (Force et al., 1991) may be of importance in terms of the trophic action of these polypeptides; it is to be noted, in the context of this chapter, that these G-protein-linked "growth factors" are much more widely recognized for their acute vasoactive properties than for their cytotrophic activities.

Despite the essential role that intrinsic receptor tyrosine kinase plays in the action of agents such as EGF-URO and PDGF, the exact receptor substrates that are involved directly in a specific cell response (e.g., mitogenesis) have not yet been identified with precision. Nonetheless, three interesting tyrosine kinase substrates, which have been found to be phosphorylated by the EGF-URO receptor kinase, could potentially play a role in the acute actions (i.e., seconds to tens of minutes) of this polypeptide: phospholipase C-γ1, calpactin-II/lipocortin-I, and the GTP-ase-activating protein, termed GAP. Of these three substrates, only phospholipase C-γ1 has been observed to have its function clearly regulated by tyrosine phosphorylation (Nishibe et al., 1990). The putative impact of tyrosine phosphorylation in terms of the ability of calpactin-II/lipocortin-I to modulate phospholipase A_2 activity or the ability of GAP to regulate the GTP-ase activity of Ras p21 still remains a matter for speculation.

C. Growth Factors and Smooth Muscle Contractility

As indicated in the above section, some agents well known for their vasoactive properties (e.g., vasopressin, angiotensin-II) are now known to be trophic (but not necessarily mitogenic) for a variety of cell types. However, it has not been widely recognized that at least two growth factors, EGF-URO and PDGF, widely known for their mitogenic actions, are also able to regulate acutely smooth muscle contractility. Our own work in this area was sparked by a report that EGF-URO administered in vivo to adrenalectomized sheep causes a fall in total peripheral resistance and an increase in cardiac output (Scoggins et al., 1984). Based on that report, we reasoned that EGF-URO would probably exhibit direct vasoactive properties, and in our preliminary work we went on to make the apparently contradictory observations that EGF-URO caused vasodilatation in canine arterial beds, when administered intraarterially in vivo (Gan et al., 1987), but elicited a contractile response in a rat ileocolic helical arterial strip in vitro (Muramatsu et al., 1985). During the course of our studies, others also reported growth factor-mediated contractile responses caused not only by EGF-URO but also by PDGF (Berk et al., 1985, 1986). It is the object of the sections that follow to elaborate upon the actions of EGF-URO in a variety of smooth systems that we have studied over the past five years, with a particular focus on growth factor-regulated contractility.

II. Multiple Responses to EGF-URO in Smooth Muscle Assays

A. Three Distinct Response Paradigms

As mentioned in Section I.C, we were initially puzzled by our observation that EGF-URO caused contraction of a rat helical arterial strip in vitro, because such an action would appear to be inconsistent with the ability of EGF-URO to act as a vasodilator in dogs in vivo. The clue to understanding these apparently conflicting data came from a comparative study of the actions of EGF-URO in similar vascular tissues derived from different species. In a nutshell, it became quickly apparent that the type of vascular resopnse to EGF-URO exhibited marked species differences. Thus, in canine arterial tissue, we discovered that, in contrast to the rat preparation, EGF-URO caused neither contraction nor relaxation (Gan et al., 1987; Muramatsu et al., 1986). Rather, pretreatment of the tissue with EGF-URO reduced the sensitivity of the preparation to contractile agonists such as noradrenaline or KCl, so as to shift their concentration-effect curves to the right (Fig. 1) (Muramatsu et al., 1986). The consequence of this action of EGF-URO in canine arterial tissue is to reduce the contractile impact of autonomically released neurotransmitter, as mimicked in our experiments by nerve stimulation (Fig. 2). Thus, the action observed in vitro provides a rationale for the vasodilatory action of EGF-URO administered in vivo in canine arterial beds. In contrast, in arterial preparations derived from other species (rodent, porcine, bovine), EGF-URO causes a contractile response (Fig. 3). The contractile response can be either transient, as illustrated by the rat and bovine preparations (Figs. 3 and 4), or tonic, as is the case for the porcine coronary artery (Gan and Hollenberg, 1990) and guinea pig tracheal smooth muscle preparation (Patel et al., 1988) (Fig. 4, lower tracing). We soon discovered that the contractile responses caused by EGF-URO in the various preparations were not generated by the same signal transduction mechanism, since in some tissues (e.g., rat ileocolic artery, porcine coronary artery, and guinea pig tracheal strip) the EGF-URO-mediated contraction could be completely blocked by indomethacin (Fig. 3), whereas in other tis-

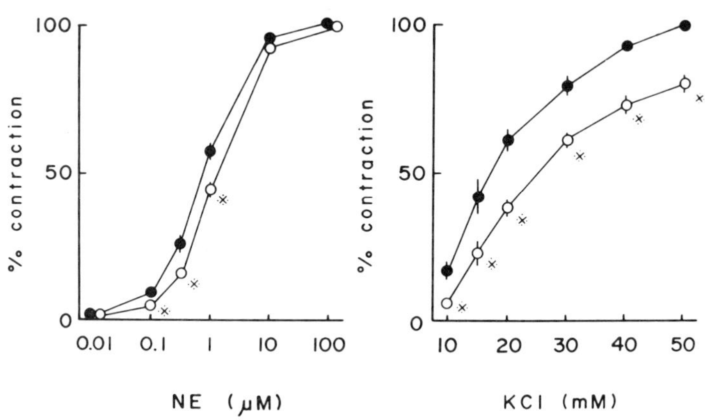

Figure 1 Effect of EGF-URO on canine arterial response to nonrepinephrine (NE) and potassium chloride (KCl). Contractile concentration-effect curves were measured for NE and KCl in helical canine mesenteric arterial strips either before (closed symbols) or after (open symbols) exposure of the tissue to EGF-URO (30 ng/mL). *Denotes responses significantly reduced, compared with the corresponding control response ($p < 0.005$). (From Muramatsu et al., 1986.)

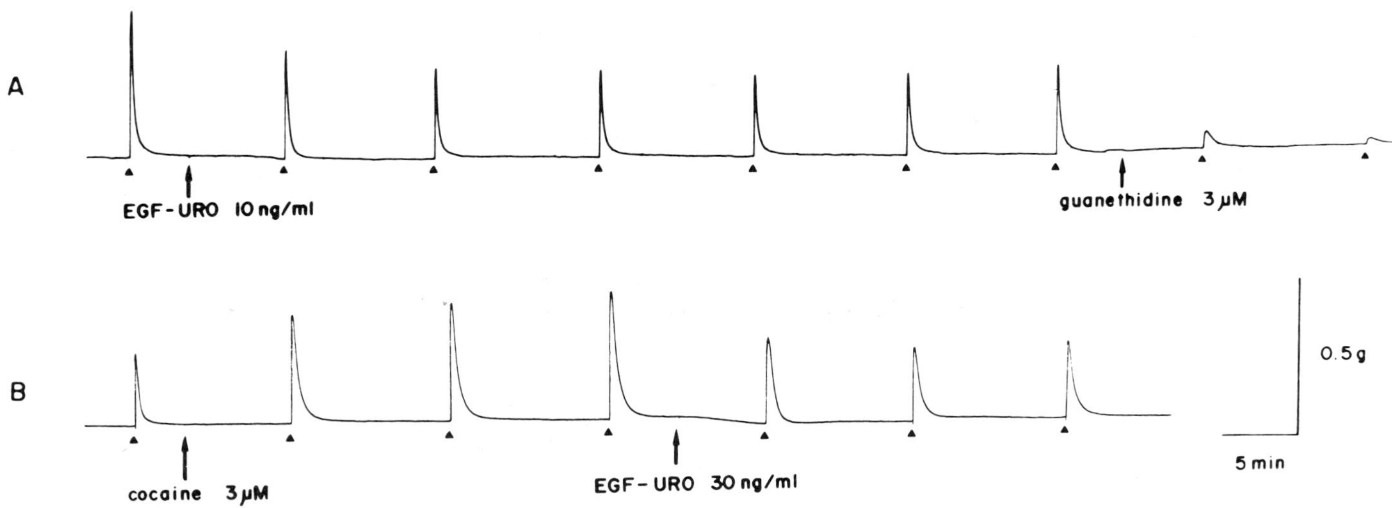

Figure 2 Reduction due to EGF-URO of contractions induced by transmural electrical stimulation of canine mesenteric arterial strips. The reduction of the electrically stimulated contractile response observed in the presence of EGF-URO was seen either under control conditions (tracing A) or in preparations treated with cocaine to enhance the action of electrically released neurotransmitter (tracing B). The blockade of contraction caused by guanethidine (tracing A) indicates that the electrically induced contractions were caused by the release of neurotransmitter. (From Muramatsu et al., 1986.)

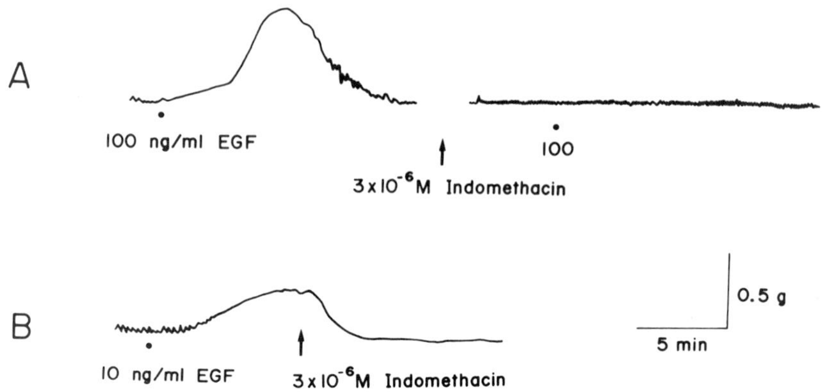

Figure 3 EGF-URO-stimulated contraction of rat ileocolic arterial strips: inhibition by indomethacin. Tissue was exposed to EGF-URO before or after treatment with indomethacin. (A) response before and after pretreatment with indomethacin. (B) treatment with indomethacin during EGF-URO-mediated contraction accelerated relaxation. (From Muramatsu et al., 1985.)

sues (e.g., bovine coronary artery) the contractile action of EGF-URO was not abolished by cyclooxygenase inhibitors (Fig. 4).

In summary, our work indicates that, in terms of EGF-URO action in vascular smooth muscle, there appear to be three distinct response systems. In two of these systems (type A and type B), EGF-URO causes contraction. However, in one system (type A), the contractile response is caused by the production of contractile prostanoids. In the second system (type B), arachidonate metabolites do not appear to be involved in the EGF-URO-mediated contraction, which, nonetheless, requires the presence of extracellular calcium (Gan and Hollenberg, 1990). One property of the type B system not seen in a prostanoid-dependent type A system is an intense desensitization to the contractile action of EGF-URO. For instance, in the bovine coronary artery preparation, even a single exposure of this tissue to EGF-URO renders the tissue refractory to a second exposure (Fig. 4). The third type of system (type C) is typified by the canine mesenteric artery preparation (Fig. 1), in which EGF-URO has no effect on tension on its own, but reduces the sensitivity of the tissue to a contractile agonist.

B. Diversity of Smooth Muscle Preparations Responding to EGF-URO

Although our early work was focused on vascular tissue as a smooth muscle target for the action of EGF-URO, we soon began to study other smooth mus-

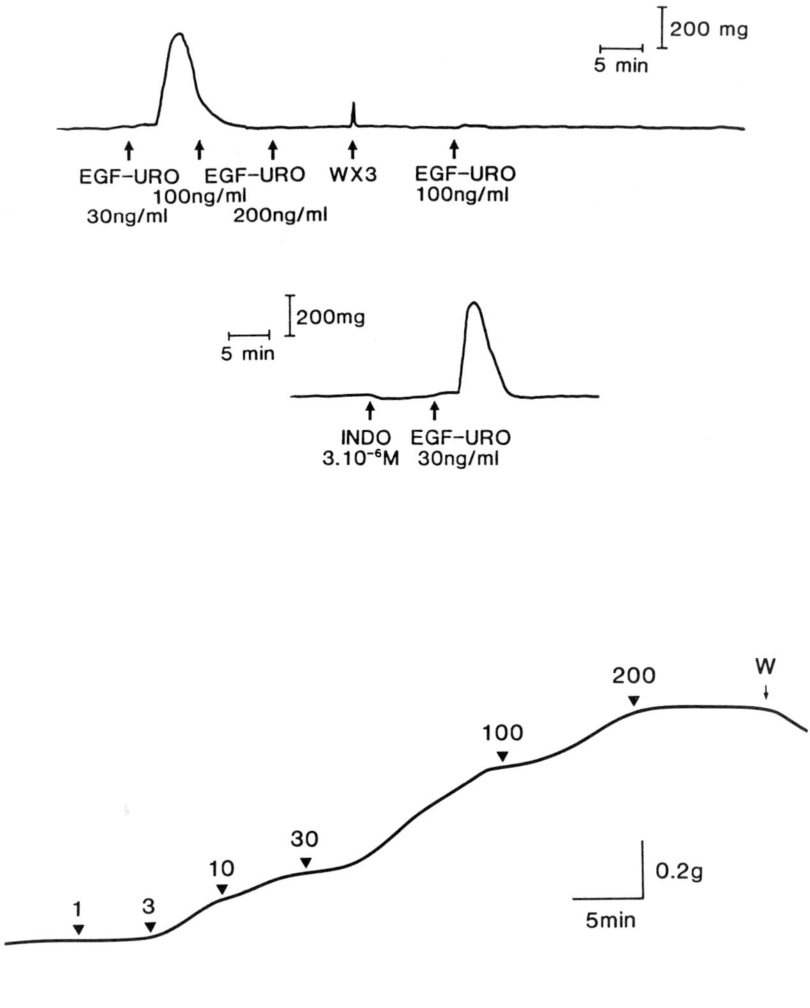

Figure 4 Contractile response of bovine coronary artery strips and guinea pig tracheal strips. The upper trace shows the first contractile response of bovine coronary artery strips to EGF-URO, followed by a complete refractoriness of the tissue to further stimulation by EGF-URO either before or after washing the tissue three times (W × 3). The middle trace shows that indomethacin does not block the contractile action of EGF-URO in the bovine coronary tissue. (From Gan and Hollenberg, 1990.) The lower trace shows the tonic contraction developed by increasing concentrations of EGF-URO in a guinea pig tracheal strip preparation. (Data from Gan and Hollenberg, 1990, and Patel et al., 1988.)

cle preparations. Work by us (Muramatsu et al., 1988; Patel et al., 1988; Yamaguchi et al., 1990) and by others (Mukku and Stancel, 1985) has established that EGF-URO can cause contractile responses in smooth muscle preparations obtained from tissues as diverse as the trachea, stomach, uterus, and the ciliary body of the eye (Table 3). What has also become evident from our work is that although a wide variety of smooth muscle preparations can respond to EGF-URO, there can be a marked species variation in responsiveness, even for the same type of preparation (e.g., arterial strip) (Table 4).

C. Distinct Response Systems for EGF-URO Within a Single Tissue: Gastric Longitudinal and Circular Muscle

In the course of our work, a gastric smooth muscle preparation obtained from guinea pig stomach has proved of particular value in analyzing the contractile actions of EGF-URO (Muramatsu et al., 1988). In this single strip of gastric smooth muscle, obtained from the gastric body, the direction of the dissecting cut (either along or at right angles against the circular muscle bundles) yields

Table 3 Smooth Muscle Preparations Responding to EGF-URO[a]

Species	Preparation	Response	System type
Rat	Ileocolic artery	Phasic contraction	A
Rat	Uterus	Phasic contraction, prolonged time course (1–4 h)	N.D.
Pig	Coronary artery	Phasic contraction	A
Pig	Ocular ciliary body	Tonic contraction	B
Cow	Coronary artery	Phasic contraction	B
Dog	Coronary artery, mesenteric artery	N/C and reduced agonist action	C
Guinea pig	Gastric longitudinal muscle	Phasic contraction	A
Guinea pig	Gastric circular muscle	Phasic contraction	B
Guinea pig	Aorta	N/C	N.D.
Guinea pig	Tracheal strip	Tonic contraction	A
Rabbit	Aorta and mesenteric artery	N/C and decrease of agonist action	N.D.
Ground squirrel	Aorta and mesenteric artery	N/C	N.D.

[a] The table summarizes responses obtained by us or by others to EGF-URO in a variety of smooth muscle response systems. Type A, contractile response blocked by indomethacin. Type B, contractile response resistant to indomethacin. Type C, reduction of action of second agonist after exposure to EGF-URO. N/C = no change in tension observed upon exposure to EGF-URO. N.D. = not determined.

Table 4 Species Differences in the Response of Arterial Preparations to EGF-URO[a]

Species	Preparation	Response
Rat	Mesenteric (ileocolic branch)	Contraction, type A
Dog	Mesenteric	N/C, reduced action of second agonist, type C
Cow	Mesenteric	Contraction, type B
Pig	Coronary	Contraction, type A
Cow	Coronary	Contraction, type B
Dog	Coronary	N/C, reduced action of second agonist, type C

[a] The table summarizes the distinct responses to EGF-URO observed for the same preparations (either from coronary or mesenteric arteries) obtained from different species. N/C = no change in tension. The type of response of each tissue is categorized as A (blocked by indomethacin), B (resistant to indomethacin), or C (reduced action of second agonist), as discussed in the text.

from the same tissue a preparation in which the contraction of either the circular (CM preparation) or longitudinal (LM preparation) muscle components can be monitored. Using the guinea pig gastric tissue, we observed that the LM preparation belongs to the type A response category described above (i.e., the EGF-URO-mediated contraction is blocked by indomethacin), whereas the CM preparation belongs to the type B category (the EGF-URO-mediated contractions are resistant to indomethacin) (Fig. 5). The guinea pig gastric LM and CM preparations have been used in conjunction with structure-activity studies with analogs of EGF-URO and its related peptide, transforming growth factor-α (TGF-α), to define two distinct receptor systems for EGF-URO in guinea pig tissue (Hollenberg et al., 1989). In one receptor system, represented by the LM preparation, the order of potencies for EGF-URO and TGF-α was EGF-URO > TGF-α, whereas in the second system (CM preparation), the order of potencies was reversed, i.e., TGF-α > EGF-URO. Pharmacologically, the data point to two receptor subtypes for EGF-URO in the guinea pig tissue, both in terms of concentration-response curves (Hollenberg et al., 1989) and in terms of ligand-binding measurements (Yang and Hollenberg, 1991). From the data we have obtained so far with tissues other than the LM and CM preparations, it would appear that the type A response systems (e.g., porcine coronary artery) possess the EGF-URO preferring receptor subtypes (i.e., potency of EGF-URO > TGF-α), whereas the type B

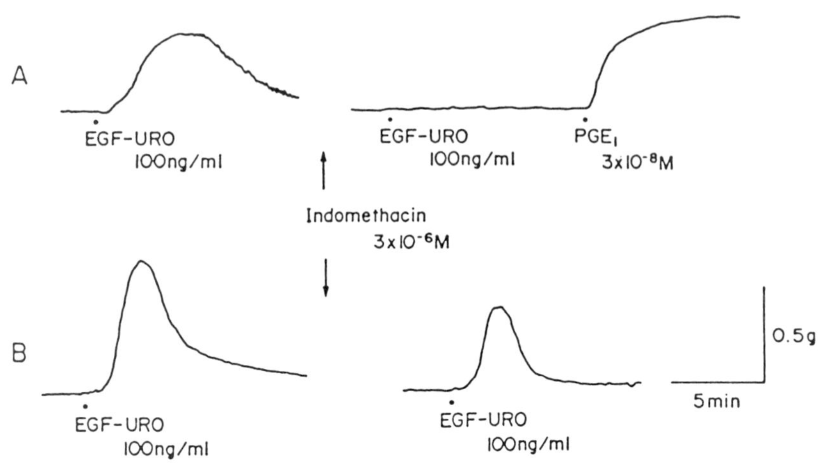

Figure 5 Contractile action of EGF-URO on guinea pig gastric longitudinal (LM) and circular (CM) preparations: effects of indomethacin. Tissue was exposed to EGF-URO either before (left) or after (right) pretreatment with indomethacin (3 μM). A. longitudinal muscle (LM) preparation. B. circular muscle (CM) preparation.

response systems (e.g., bovine coronary artery) possess the TGF-α-preferring receptor subtype (TGF-α > EGF-URO) (Gan and Hollenberg, 1990). We do not yet know if the two EGF-URO receptor subtypes that we have characterized in the guinea pig tissue result from distinct EGF-URO receptor sequences or are a consequence of differential posttranslational processing of the same receptor sequence in the different tissues. It remains to be seen if the contractile action of other growth factors in smooth muscle systems (e.g., PDGF) will mirror the heterogeneity of response systems that we have observed for EGF-URO, both in terms of interspecies variations and in terms of distinct signaling pathways.

III. Growth Factor-Triggered Signal Transduction Pathways

A. Receptors with Intrinsic Tyrosine Kinase Activity

As alluded to in Section I.B, it is believed that the first step in the action of an agent such as insulin or EGF-URO, for which agonists the transmembrane receptors possess intrinsic tyrosine kinase activity, is the phosphorylation on tyrosine of a key substrate that then transmits the signal to the cytoplasmic milieu so as to generate a cellular response. Because so many proteins have been identified as potential substrates for growth factor receptors (e.g., for the insulin receptor, see Hollenberg, 1990), it has proved difficult to determine which of the many substrates might be the critical one involved in any cellular response. In order to identify such a substrate, it is necessary to observe its phosphorylation in the context of the time course of a ligand-stimulated cellular response. This paradigm has been used to identify a 15-kDa substrate phosphorylated on tyrosine in adipocytes during the course of insulin-stimulated glucose uptake (Bernier et al., 1987). In terms of EGF-URO action, there are two substrates that are of particular interest in terms of playing a prominent role in EGF-URO action. One substrate for the EGF-URO receptor kinase, variously referred to as calpactin-II or lipocortin-I, has been identified in a variety of tissues and cell types (Fava and Cohen, 1984; Valentine-Braun et al., 1986) and has been observed to become phosphorylated on tyrosine within minutes of exposure of a receptor-bearing cell or vesicle preparation to EGF-URO (Blay et al., 1989). Although controversial, it has been suggested that calpactin-II/lipocortin-I phosphorylation may play some role in regulating phospholipase A_2 activity, as it relates to the inflammatory response (Flower, 1985). A second protein that becomes phosphorylated on tyrosine within minutes of treating a cell with EGF-URO is phospholipase C-γ1 (Wahl et al., 1988; Nishibe et al., 1990). The activation of this enzyme by EGF-URO-mediated tyrosine phosphorylation would be expected to lead to the formation of inositol 1,4,5-trisphosphate and diacylglycerol in intact cells. Both of these well-recognized intracellular mediators could readily generate a cellular response, either via the activation of protein kinase C or via an elevation of intracellular calcium (Berrige, 1987). Alternatively, as will be elaborated upon below, the diacylglycerol produced can be metabolized further to yield vasoactive prostanoids. In the work to be described in the following section, we have attempted to determine (1) whether or not tyrosine kinase activity plays a key role in the action of EGF-URO in the smooth muscle response systems with which we are familiar; (2) whether or not calpactin-II becomes phosphorylated on tyrosine during the course of a contractile response to EGF-URO; and (3) whether phospholipase C or phospholipase A_2-generated metabolites are involved in the contractile action of EGF-URO in the type A response systems. In the course of our work, we also sought to determine if tyrosine kinase activity per se might play a role in the contractile actions of growth factors such as vasopressin and angiotensin-II, which act via G-protein-coupled receptors rather than via receptors with intrinsic tyrosine kinase activity.

B. Tyrosine Kinase and the Contractile Action of EGF-URO

To evaluate the role of tyrosine kinase in the contractile actions of EGF-URO, we turned to the use of the enzyme inhibitors, genistein (GS) (Akiyama et al., 1987) and tyrphostin (TP) (Gazit et al., 1989), which have been found to be quite specific for the inhibition of growth factor receptor-related tyrosine kinase activity. In a number of cultured cell test systems, GS and TP have been observed to block the action of EGF-URO (Yaish et al., 1988; Linassier et al., 1989). Since the guinea pig gastric tissue comprises both a type A (LM preparation) and a type B (CM preparation) response system, we first examined the effects of GS and TP in the gastric contractile assays. As illustrated in Fig. 6, both GS and TP were able to block selectively the contractile actions of EGF-URO or TGF-α in both the LM and CM preparations, without affecting contractions caused by other agonists such as bradykinin. The data thus supported the conclusion that tyrosine kinase activity is a key element for the con-

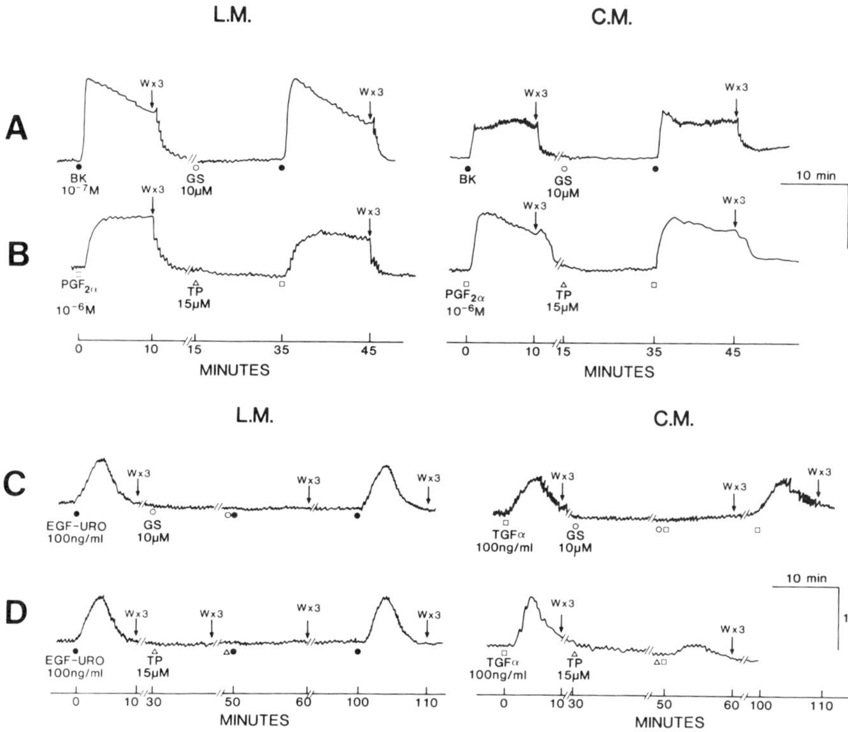

Figure 6 Effects of genistein and tyrphostin on contractile responses in guinea pig gastric longitudinal (LM) and circular (CM) muscle preparations. Tissue strips [left, longitudinal muscle (LM); right, circular muscle (CM)] were exposed to the various agonists (bradykinin, BK; prostaglandin $F_{2\alpha}$, $PGF_{2\alpha}$; EGF-URO; and TGF-α) either before or after pretreatment with either genistein (GS, 10 μM: ○) or tyrphostin (TP, 15 μM: △). Contractions caused by bradykinin (A) and $PGF_{2\alpha}$ (B) were not blocked by the tyrosine kinase inhibitors, whereas contractions caused by EGF-URO and TGF-α (C and D) were completely abolished. W = Wash. (From Yang et al., 1992.)

tractile action of EGF-URO in both the type A (LM) and type B (CM) response systems. While this finding was perhaps not surprising, in view of the critical importance of the tyrosine kinase activity of the EGF-URO receptor for its biological activity, it was of particular significance to us that the relative IC_{50}'s of GS and TP for the inhibition of contraction in the LM and CM preparations differed appreciably (Yang et al., 1992). In the LM preparation the IC_{50}'s for GS (1.1 μM) and TP (3.6 μM) yielded a relative inhibitor ratio GS:TP of 1:3. In contrast, in the CM tissue, the IC_{50}'s for GS (3.0 μM) and TP (2.4 μM) yielded a relative inhibitor ratio of 1:0.8. The difference between the inhibitor ratios in the LM and CM preparations implies that different substrates and/or different tyrosine kinases may be involved in the response to EGF-URO in the two preparations. The identification of such putative tyrosine kinase substrates is the focus of further ongoing work.

C. Calpactin-II/Lipocortin-I Phosphorylation and EGF-URO Mediated Contraction

In view of our wish to identify tyrosine kinase substrates that might play a role in EGF-URO-mediated contraction, we have used monoclonal antiphosphotyrosine antibodies and Western blot procedures to detect proteins in smooth muscle tissue both before and after exposure to EGF-URO. To this end, replicate guinea pig gastric LM preparations were set up in bioassay chambers either as control tissue or as experimental samples exposed to EGF-URO in the presence or absence of the tyrosine kinase inhibitor TP (at concentrations that selectively block contraction). At a time corresponding to the peak of contraction caused by EGF-URO (upper portion of Fig. 7), the tissues were harvested, quick-frozen, and processed for electrophoresis and Western blot analysis (lower portion, Fig. 7). Monoclonal antiphosphotyrosine (anti-PY) antibodies detected constituents in the 30- to 40-kDa range both in control and EGF-URO-treated samples (lanes 1a and 2a, Fig. 7). However, EGF-URO treatment caused an appreciable increase in immunostaining of these constituents by the anti-PY antibody and revealed the presence of a new band (arrow, Fig. 7, lane 2a) in the 35- to 40-kDa region. In the presence of TP, the signal yielded by the anti-PY antibody returned to control levels (lane 3a, Fig. 7). When the Western blot was counterstained with anticalpactin-II antibody, the blue color reaction overlaid precisely two of the constituents also detected by the anti-PY antibody (lanes 1b, 2b, and 3b, Fig. 7). Significantly, the color overlay yielded by the anticalpactin-II antibody for one of the constituents corresponded to the component for which phosphorylation (as detected by the anti-PY monoclonal) was apparent only during exposure to EGF-URO (arrow, Fig. 7, lanes 2a and 2b). Our data, although not yet entirely conclusive, thus indicate that phosphorylation of a calpactin-II-related protein occurs during the

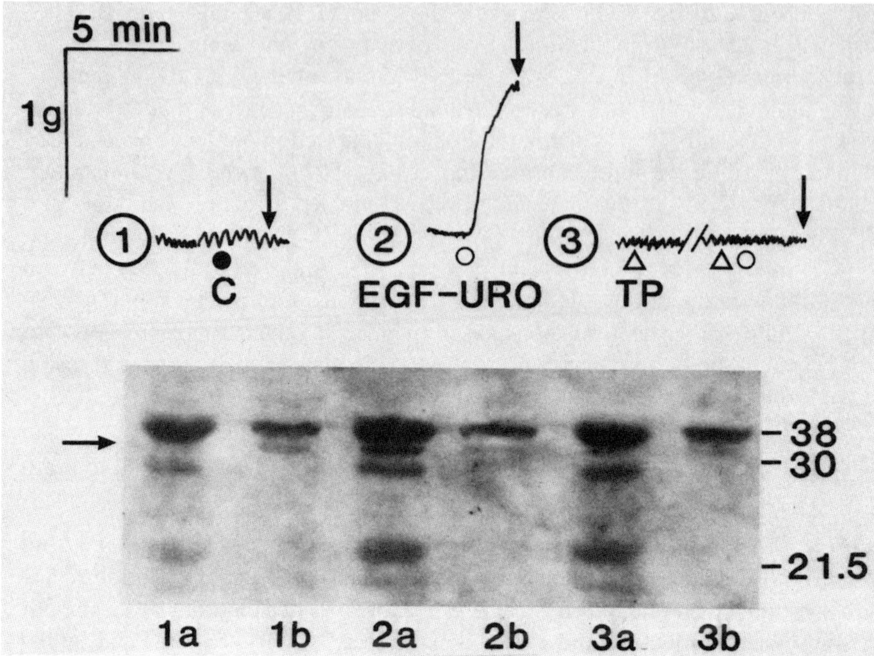

Figure 7 Western blot analysis of guinea pig gastric muscle preparations obtained before or after exposure to EGF-URO. Longitudinal muscle preparations were mounted in organ baths as for a bioassay and were either untreated (left) or were treated with EGF-URO either in the absence (middle) or presence (right) or tyrphostin (20 μM). At a time coincident with the peak of contraction (arrow, upper panel) tissues were harvested and processed for electrophoresis and Western blot analysis (lower panel) using either monoclonal antiphosphotyrosine antibody (lanes 1a, 2a, 3a) or polyclonal anticalpactin-II antibody (lanes 1b, 2b, 3b). For clarity, the phosphoprotein profile (lanes 1a, 2a, 3a) is shown alongside the counterstain that detects calpactin-II-related proteins (lanes 1b, 2b, 3b). The contractile measurements are shown in the upper panel; molecular mass positions, in kilodaltons, are shown on the right. The left-hand arrow denotes the position of the calpactin-II-related constituent for which phosphorylation was dramatically increased by EGF-URO treatment (middle panel, lane 2a), and for which the increase in phosphorylation was blocked by tyrphostin (TP). C, control ●; ○, EGF-URO; △, tyrphostin (TP). The scale for time and developed tension in the bioassay is shown in the upper left quadrant.

course of EGF-URO-stimulated contraction (Yang et al., 1992). Because of the nature of the tissue strip, it is not possible to determine the precise component (LM or CM) in which the increased phosphorylation occurs.

D. Diacylglycerol Lipase and the Contractile Action of EGF-URO in a Type A Response System

Although we had determined that in the type A, LM preparation, EGF-URO caused contraction via the production of a cyclooxygenase product, we wished to determine the source of the arachidonic acid. In principle, the arachidonate might come either directly via the activation of phospholipase A_2 or indirectly from the metabolism of diacylglycerol by diaclyglycerol lipase. The data obtained with indomethacin suggested that diacylglycerol was probably not causing contraction simply by activating protein kinase C, and would thus be a good candidate as a source for arachidonate, as mentioned above. Diacylglycerol could, in principle, come from the activation of phospholipase C by EGF-URO. Alternatively, diacylglycerol could also arise from the action of phospholipase D followed by phosphatidate phosphatase. To determine the role that diacylglycerol might play in the contractile response of the LM preparation to EGF-URO, we made use of the diacylglycerol lipase inhibitor, U57,908, which selectively blocks the sequential hydrolysis of diacylglycerol but does not affect phospholipase A_2. As shown in Fig. 8, the diacylglycerol lipase inhibitor completely and selectively blocked the contractile action of EGF-URO (or TGF-α) in the LM preparation, but had no effect on the contractile action of TGF-α in the CM preparation (Yang et al., 1991). We were thus able to conclude that phospholipase A_2 does not play a role in the contractile action of EGF-URO in the type A LM system; and we were able to highlight the importance of diacylglycerol as a metabolic substrate rather than as a kinase C activator in terms of the action of EGF-URO. Conversely, it would appear that neither diacylglycerol lipase nor phospholipase A_2 is of importance for the contractile action of EGF-URO in the type B, CM system, which nonetheless requires the presence of extracellular calcium for the generation of an EGF-URO-mediated response. The nature of the signal transduction pathway used in the type B, CM system, thus remains a most interesting topic for further study.

E. Tyrosine Kinase and the Contractile Action of G-Protein-Linked Growth Factors

During our work studying the inhibitory actions of the tyrosine kinase inhibitors, genistein (GS) and tyrphostin (TP), on the contractile effect of EGF-URO in the LM type A response system, we noticed that the actions of some G-protein-linked contractile agonists were also affected. In particular, the con-

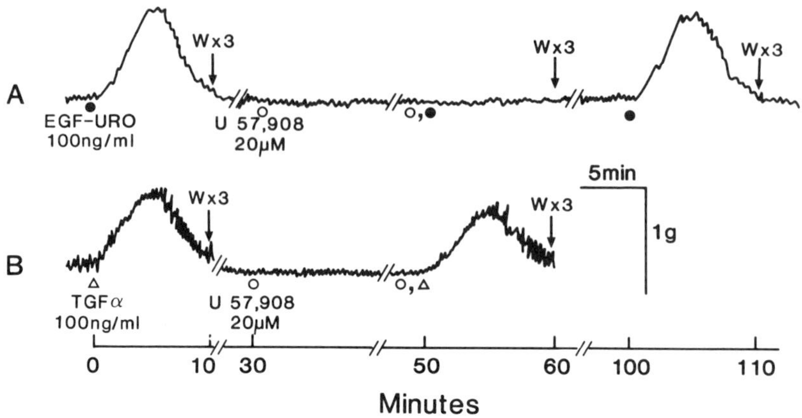

Figure 8 Effect of inhibition of diacylglycerol lipase on the contractile action of EGF-URO in guinea pig gastric longitudinal (A, LM) and circular (B, CM) muscle preparations. LM (A) and CM (B) muscle preparations were exposed to EGF-URO (●) or TGF-α (△) either before (left) or after (middle) treatment with the diacylglycerol lipase inhibitor, U57,908 (○). Tracing A shows the inhibition by U57,908 of EGF-URO-mediated contraction and the reversal of the inhibition after washing the tissue three times (W × 3). Tracing B shows that the contraction of the CM preparation was unaffected by U57,908.

tractile action of angiotensin-II was markedly attenuated by GS and TP in both the LM and CM guinea pig gastric preparations (Fig. 9). The contractile action of prostaglandin $F_{2\alpha}$ was also diminished, whereas the action of bradykinin was unaffected (not shown). In preliminary work we have also observed that the action of vasopressin in vascular preparations is abolished by the tyrosine kinase inhibitor (Laniyonu et al., 1992). We were thus able to conclude (Yang et al., 1992) that in addition to growth factors such as EGF-URO, which act via receptors with intrinsic tyrosine kinase activity, a tyrosine kinase pathway also probably plays a role in the contractile action of G-protein-linked agonists such as angiotensin and vasopressin. As noted in section I.B, the ability of angiotensin-II and vasopressin to stimulate tyrosine phosphorylation in certain target cells has already been documented (Force et al., 1991), as has the ability of these polypeptides to stimulate the expression of the protooncogene c-fos, which is associated with growth-promoting cellular signals. Our work with the tyrosine kinase inhibitors in the contractile systems provides a functional context in which to evaluate tyrosine phosphorylation stimulated by G-protein-linked agonists for comparison with the work we have begun to do with EGF-URO (Section III.C, Yang et al., 1992).

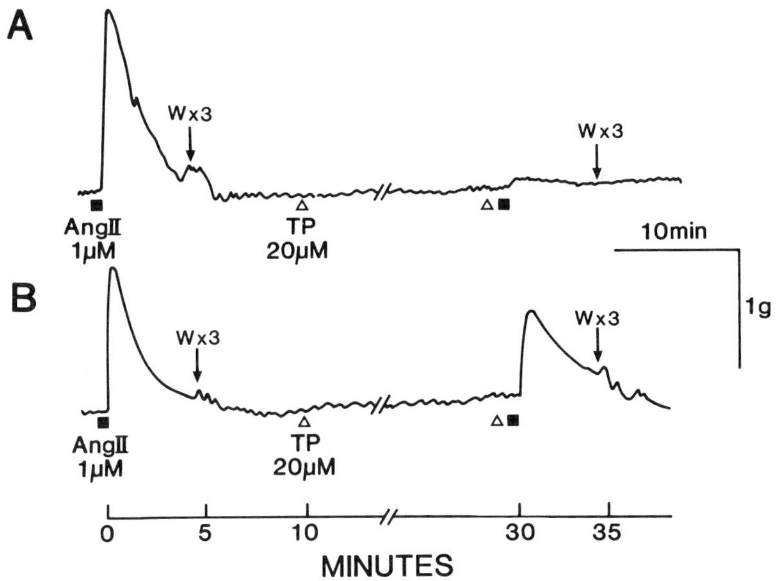

Figure 9 Effects of tyrphostin on the contractile action of angiotensin-II guinea pig gastric longitudinal (A) and circular (B) muscle preparations. Tracing A shows that 20 μM tyrphostin (△) abolished the LM contractile response to angiotensin-II (Ang II; ■, 1 μM); tracing B shows that 20 μM tyrphostin partially inhibited the CM contractile response to Ang II (1 μM). W = Wash. (From Yang et al., 1992.)

IV. Summary

As outlined in Sections I to III, work by us and by others over the past five to six years has established that mitogenic polypeptides, commonly referred to as "growth factors," can have acute regulatory actions (i.e., within minutes) in a variety of smooth muscle systems. Using EGF-URO as a prototype example, it is possible to describe three distinct smooth muscle response paradigms, two of which (type A and type B) are contractile and one of which (type C) comprises a reduced sensitivity to contractile agonists. Studies of the two EGF-URO-triggered contractile response systems, represented by guinea pig gastric longitudinal (type A, LM) and circular (type B, CM) muscle preparations, have revealed two quite distinct receptor subtypes and two distinct signal transduction pathways, which nonetheless both require intact tyrosine kinase activity and the presence of extracellular calcium. In one response system (type A, LM), diacylglycerol, via diacylglycerol lipase metabolism, rather than via kinase C activation, plays an important role in the EGF-URO-mediated contractile response; this response is caused by the production of cyclooxygenase products subsequent to the diacylglycerol lipase-catalyzed release of arachidonate from diacylglycerol. The signal transduction pathway for EGF-URO in the type B, CM system would appear not to require either phospholipase A_2 or phospholipase C activation and would appear to employ an as yet unidentified pathway for EGF-URO action. Our work has also revealed that vasoactive agonists such as angiotensin-II and vasopressin, which in cultured cells can act as trophic growth factors, may also work via a tyrosine kinase pathway to elicit contraction in some of the same systems that contract in response to growth factors such as EGF-URO. The elucidation of the common and distinct tyrosine kinase-modulated pathways that are involved in the control of smooth muscle contractility elicited by G-protein-linked agonists on the one hand and by growth factor-regulated tyrosine kinase receptors on the other hand remains an intriguing topic for further study.

Acknowledgments

Work in the author's laboratory is supported in large part by grants from the Medical Research Council of Canada and The Alberta Heart and Stroke Foundation.

References

Akiyama T, Ishida J, Nakagawa S, Ogawara H, Watanabe S, Itoh N, Shibuya M, Fukami Y. Genistein, a specific inhibitor of tyrosine-specific protein kinases. J Biol Chem 1987; 262:5592–5.

Antoniades HN, Scher CD, Stiles CD. Purification of human platelet-derived growth factor. Proc Natl Acad Sci USA 1979; 76:1809–13.

Berk BC, Alexander RW, Brock TA, Gimbrone MA Jr, Webb RC. Vasoconstriction, a new activity for platelet-derived growth factor. Science 1986; 323:87–9.

Berk BC, Brock TA, Webb RC, Taubman MB, Atkinson WJ, Gimbrone MA Jr, Alexander RW. Epidermal growth factor, a vascular smooth muscle mitogen, induces rat aortic contraction. J Clin Invest 1985; 75:1083–6.

Bernier M, Laird DM, Lane MD. Insulin-activated tyrosine phosphorylation of a 15-kilodalton protein in intact 3T3-L1 adipocytes. Proc Natl Acad Sci USA 1987; 84:1844–8.

Berridge MJ. Inositol trisphosphate and diacylglycerol: two interacting second messengers. Ann Rev Biochem 1987; 56:159.

Blay J, Valentine-Braun KA, Northup JK, Hollenberg MD. Epidermal-growth-factor-stimulated phosphorylation of calpactin II in membrane vesicles shed from cultured A-431 cells. Biochem J 1989; 259:577–83.

Cohen S. Isolation of a mouse submaxillary gland protein accelerating incisor eruption and eyelid opening in the newborn animal. J Biol Chem 1962; 237:1555–62.

Fava RA, Cohen S. Isolation of a calcium-dependent 35-kilodalton substrate for the epidermal growth factor receptor/kinase from A-431 cells. J Biol Chem 1984; 259:2636–45.

Flower RJ. Background and discovery of lipocortins. Agents and Actions 1985; 17:255–62.

Force T, Kyriakis JM, Avruch J, Bonventre JV. Endothelin, vasopressin, and angiotensin II enhance tyrosine phosphorylation by protein kinase C-dependent and -independent pathways in glomerular mesangial cells. J Biol Chem 1991; 266:6650–6.

Gan BS, Hollenberg MD. Distinct coronary artery receptor systems for epidermal growth factor-urogastrone. J Pharmacol Exp Ther. 1990; 252:1277–82.

Gan BS, MacCannell KL, Hollenberg MD. Epidermal growth factor-urogastrone causes vasodilatation in the anesthetized dog. J Clin Invest 1987; 80:199–206.

Gazit A, Yaish P, Gilon C, Levitzki A. Tyrphostins I: synthesis and biological activity of protein tyrosine kinase inhibitors. J Med Chem 1989; 32:2344–52.

Hollenberg MD. Insulin receptor-mediated transmembrane signalling. Handbook Exp Pharmacol 1990; 92:183–207.

Hollenberg MD, Muramatsu I, Itoh H, Patel P, Yang S-G, Lederis K. Contractile actions of epidermal growth factor-urogastrone in isolated smooth muscle preparations from guinea pig stomach: structure-activity relationships and comparison with the effects of human transforming growth factor-*alpha*. J Pharmacol Exp Ther 1989; 248:384–90.

Kaziro Y, Itoh H, Kozasa T, Nakafuku M, Satoh T. Structure and function of signal-transducing GTP-binding proteins. Ann Rev Biochem 1991; 60:349–400.

Laniyonu A, Saifeddine M, Yang S-G, Hollenberg M D. Tyrosine kinase inhibitors and the contractile action of G-protein linked growth promoting agonists. FASEB J., 1992; 6:A1862.

Linassier C, Pierre M, LePecq J-B, Pierre J. Mechanisms of action in NIH-3T3 cells of genistein, an inhibitor of EGF receptor tyrosine kinase activity. Biochem Pharmacol 1989; 39:187–93.

Mukku VR, Stancel GM. Receptors for epidermal growth factor in the rat uterus. Endocrinology (Baltimore) 1985; 117:149–54.

Muramatsu L, Hollenberg MD, Lederis K. Vascular actions of epidermal growth factor-urogastrone: possible relationship to prostaglandin production. Can J Physiol Pharmacol 1985; 63:994–9.

Muramatsu I, Hollenberg MD, Lederis K. Modulation by epidermal growth factor-urogastrone of contraction in isolated canine helical mesenteric arterial strips. Can J Physiol Pharamacol 1986; 64:1561–5.

Muramatsu I, Itoh H, Lederis K, Hollenberg MD. Distinctive actions of epidermal growth factor-urogastrone in isolated smooth muscle preparations from guinea pig stomach; differential inhibition by indomethacin. J Pharmacol Exp Ther 1988; 245:625–31.

Nishibe S, Wahl MI, Hernández-Sotomayor SMT, Tonks NK, Rhee SG, Carpenter G. Increase of the catalytic activity of phospholipase C-γ1 by tyrosine phosphorylation. Science 1990; 250:1253–6.

Patel P, Itoh H, Lederis K, Hollenberg MD. Contraction of guinea pig trachea by epidermal growth factor-urogastrone. Can J Physiol Pharmacol 1988; 66:1308–12.

Ross R, Glomset J, Kariya B, Harker L. A platelet-dependent serum factor that stimulates the proliferation of arterial smooth muscle cells in vitro. Proc Natl Acad Sci USA 1974; 71:1207–10.

Scoggins BA, Butkus A, Coghlan JP, Fei DTW, McDougall JG, Niall HD, Wang X-M. *In vivo* cardiovascular, renal and endocrine effects of epidermal growth factor in sheep. In: Abstracts of the 7th International Conference of Endocrinology. Amsterdam: Excerpta Medica, 1984; 124.

Valentine-Braun KA, Northup JK, Hollenberg MD. Epidermal growth factor (urogastrone)-mediated phosphorylation of a 35-kDa substrate in human placental membranes: Relationship to the β subunit of the guanine nucleotide regulatory complex. Proc Natl Acad Sci USA 1986; 83:236–40.

Wahl MI, Daniel TO, Carpenter G. Antiphosphotyrosine recovery of phospholipase C activity after EGF treatment of A-431 cells. Science 1988; 241:968–70.

Yaish P, Gazit A, Gilon C, Levitzki A. Blocking of EGF-dependent cell proliferation by EGF receptor kinase inhibitors. Science 1988; 242:933–45.

Yamaguchi K, Lederis K, Hollenberg MD. Contraction of porcine ocular ciliary muscle by epidermal growth factor-urogastrone. Eur J Pharmacol 1990; 191:245–51.

Yang S-G, Hollenberg MD. Distinct receptors for epidermal growth factor-urogastrone in cultured gastric smooth muscle cells. Am J Physiol (Gastrointest Liver Physiol) 1991; 260:G827–34.

Yang S-G, Saifeddine M, Chuang M, Severson DL, Hollenberg MD. Diacylglycerol lipase and the contractile action of epidermal growth factor-urogastrone: evidence for distinct signal pathways in a single strip of gastric smooth muscle. Eur J Pharmacol Mol Pharmacol Sec 1991; 207:225–30.

Yang S-G, Saifeddine M, Hollenberg MD. Tyrosine kinase inhibitors and the contractile action of epidermal growth factor-urogastrone and other agonists in gastric smooth muscle. Can J Physiol Pharmacol 1992; 70:85–93.

Part Four

ENDOTHELIAL CELLS

18

Ionic Currents in the Membranes of Cultured Endothelial Cells

P. BREGESTOVSKI, A. BAKHRAMOV, and V. MOKH

Cardiology Research Center
Moscow, Russia

I. Introduction

Regulation of endothelial cell function is performed in part by ionic channels whose gating properties depend on the membrane potential or action of drugs on specific receptors (see review by Bregestovski and Ryan, 1989). Recently, using the patch-clamp technique, voltage-gated inwardly rectifying ionic channels (Takeda et al., 1987; Johns et al., 1987; Bregestovski et al., 1988b), transient outward A-type currents (Takeda et al., 1987), stretch-activated ionic channels (Lansman et al., 1987; Olesen et al., 1988a), and agonist-activated ionic currents (Adams et al., 1987, 1989; see review by Bregestovski and Ryan, 1989; Takeda and Klepper, 1990) have been described in endothelial cells.

It is known that endothelium is a heterogeneous tissue and its physiological functions in various parts of the vascular system are different (Simionescu et al., 1982). However, the relationships between functional heterogeneity and membrane properties are not yet clear; it is unknown whether different types of endothelium possess different types of ionic channels. Using contemporary electrophysiological techniques in combination with cell culture methods could provide information about this and other questions concerning the physiology of endothelial cells.

A number of morphological ultrastructural data suggest the existence of communications between endothelial cells (Venter et al., 1975; Simionescu et al., 1975, 1978; Bottaro et al., 1985; Killakey et al., 1986). Functional gap junctions in endothelial cells have been demonstrated using electrophysiological methods (Larson et al., 1983; Bakhramov and Bregestovski, 1988) and dye transfer (Larson and Sheridan, 1985), but the properties of these gap junctions have not been studied in detail.

The results of pharmacological and biochemical studies show the existence in the membrane of endothelial cells of different receptors (Buonassisi and Venter, 1976; Freudenberg et al., 1983) which participate in the regulation of many functions concerning metabolic and secretory activities (Brotherton and Hoak, 1987; Furchgott, 1983, 1984; Feletou and Vanhoutte, 1988). However, the manner of receptor distribution in endothelial monolayers is not known. Theoretically, the membrane of each cell can possess a wide number of receptors. Alternatively, receptors can be distributed heterogeneously and each cell can possess only a few types of receptors. In the latter case the intercellular communication would play an important role for transportation of hormonal signal to neighboring cells and provide homogenous drug responses of the endothelial monolayer.

With the aim of clearing up some of the above discussed problems, we present here the results of:

1. Comparative study of electrophysiological characteristics of cultured endothelial cells from several sources.
2. Analysis of endothelial cell responses to the hormone histamine.
3. Study of gap junction properties in endothelial cell membranes.

Some data has been described previously (Bregestovski and Bakhramov, 1987; Bregestovski et al., 1988a, 1988b; Bakhramov et al., 1988; Mokh and Bregestovski, 1991).

II. Materials and Methods

A. Cell Culture

Experiments were conducted on both preconfluent and confluent endothelial cells isolated from several sources: bovine pulmonary artery (BPA), rabbit aorta (RA), bovine microvascular (BMV), pig aorta (PA), and human umbilical vein (HUV). Methods of isolation and cultivation has been described elsewhere (Ryan and Maxwell, 1986). For electrophysiological experiments, cells from the first to the tenth passages were mechanically harvested and seeded on glass cover slips with density 1–5 × 10^4 cells/cm^2 in 35-mm plastic Petri dishes. Patch-clamp recording was carried out on cells 2 to 7 days after seeding.

For junctional conductance measurements, pairs of cells were isolated by mechanical dispersion from a confluent monolayer. The cells were seeded in Petri dishes at densities of 0.5–1.0 × 10^4 and 0.5–1.0 × 10^5 cells/cm^2. On day 2–6, the cells were mechanically harvested, pipette-dispersed, and placed in a bathing solution for electrophysiological measurement. In some experiments, recording was done from cultured pairs of cells not isolated from the surface of the dish.

B. Electrophysiology

Conventional patch-clamp technique was used for whole-cell and single-channel current recordings. Currents were recorded on EPC-5 and EPC-7 patch-clamp amplifiers (List-Electronic, FRG) and stored on a 7006 FM tape recorder (Bruel and Kjaer, Denmark) or on a videorecorder (Sony, U.R. to check in the lab. P. Adams). Resistance of patch pipettes was 3–10 MΩ. The normal external solution contained: NaCl, 140 mM; KCl, 2.8 mM; $MgCl_2$, 2 mM; $CaCl_2$, 1 mM; HEPES-NaOH, 10 mM; pH 7.4. Micropipettes were filled with normal internal solution: KCl, 140 mM; $MgCl_2$, 2 mM; $CaCl_2$, 0.628 mM; EGTA, 1.1 mM; pCa7; HEPES-KOH 10; pH 7.4. Input resistance was calculated from plateau current responses to either depolarizing or hyperpolarizing voltage steps at the region of resting potential.

Properties of gap junctions were studied on cell pairs with the double patch-clamp technique described in detail elsewhere (Neyton and Trautmann, 1985). In brief, micropipettes with resistance 3–10 MΩ were approached to each cell and after establishing a gigaohm seal, the membrane under the patch was disrupted by applying gentle suction. This allowed the estimation of resting potentials and whole-cell currents for both cells. Holding potential was usually −20 or −40 mV. To measure gap junction conductance, voltage steps (ΔV) of from +10 to +90 mV in amplitude were applied to one cell and current responses (ΔI) were recorded from the other. Junctional conductance (G_j) was calculated as I/V. In some experiments at low G_j, constant continuous voltage gradients were set between cells, which allowed observation of current jumps probably corresponding to opening and closing of unitary gap junctional channels.

Experiments were performed at room temperature (20–22°C). Data were filtered at 1 or 3 kHz, digitized at 10–25 kHz, and analyzed on a computer (Labtam 3015, Australia, or IBM PC-XT). Drugs were applied either to the bath solution or from micropipettes using a pressure injector (Picosprizer, USA). Drugs and other chemicals were obtained from Sigma (USA).

III. Results and Discussion

A. Voltage-Gated Ionic Currents

General Electrophysiological Parameters

Table 1 represents comparative properties of different endothelial cell types. We took into consideration resting potential (RP), input conductance of isolated cells at the level of RP, and slope conductance of the cell at the membrane potential −140 mV, which describes the presence of inward rectifier currents.

Table 1

Type of cell	Parameters			
	RP (mV)	R_{in} (GΩ)	I_{-140} (pA)	$G_{I.R.}$ (nS)
Bovine pulmonary (BPA)	72±21 (n = 64)	1.9±0.3 (n = 21)	624±428 (n = 42)	9.08±3.57 (n = 34)
Rabbit aorta (RA)	56±21 (n = 13)	1.9±1.4 (n = 9)	741±163 (n = 8)	14.1±0.4 (n = 8)
Bovine microvascular (BMV)	46±15 (n = 8)	1.6±0.6 (n = 3)	184±79 (n = 7)	3.07±1.31 (n = 7)
Human umbilical vein (HUV) (isolated cells)	35±23 (n = 22)	5±2 (n = 20)		
HUV (confluent cells)	27±17 (n = 163)	87±11 (n = 72)		

It is seen that resting potentials of cells from arteries are higher in comparison with cells from HUV. This difference can be attributed to interspecies variations or, as was mentioned earlier (Bregestovski and Ryan, 1989), to differences in culture conditions. At measurement on single cells in preconfluent conditions, input resistance of all cells is rather high and varied on average from 2 to 5 GΩ; no systematic difference was observed.

B. Absence of Voltage-Gated Ca Channels in the Membranes of Endothelial Cells

Elevation of Ca^{2+} in the external solution to 20 mM, as well as substitution of Ca^{2+} by barium, did not lead to appearance of voltage-activated inward currents whose existence had been suggested earlier by some authors (Singer and Peach, 1982; Rubanyi et al., 1985). Electrophysiological evidence of absence of voltage-dependent Ca^{2+} currents in endothelial cells was also presented in reviews (Bregestovski and Ryan, 1989; Takeda and Klepper, 1990). None of these cells showed the existence of "voltage-gated" inward currents. These results coincide with studies on bovine aorta (Takeda et al., 1987; Olesen et al., 1988b), pig aorta (Lansman et al., 1987), and guinea pig coronary artery (Daut et al., 1988). No inward voltage-dependent Ca^{2+} or Na^{+} current has been found in these preparations. Whitmer et al., (1988) suggested that voltage-dependent Ca^{2+} channels are present in endothelial cells in situ and are lost during enzymatic dissociation. However, in our study, cells were isolated mechanically, without any enzymatic treatment. Nevertheless, one cannot exclude that Ca^{2+} channels can be lost during the period of cultivation of isolated cells.

Inward Rectifier Ionic Currents in Endothelial Cells

All types of cells (with the exception of umbilical vein) show prominent inward rectification. Typical currents are shown on Fig. 1. Upon application of

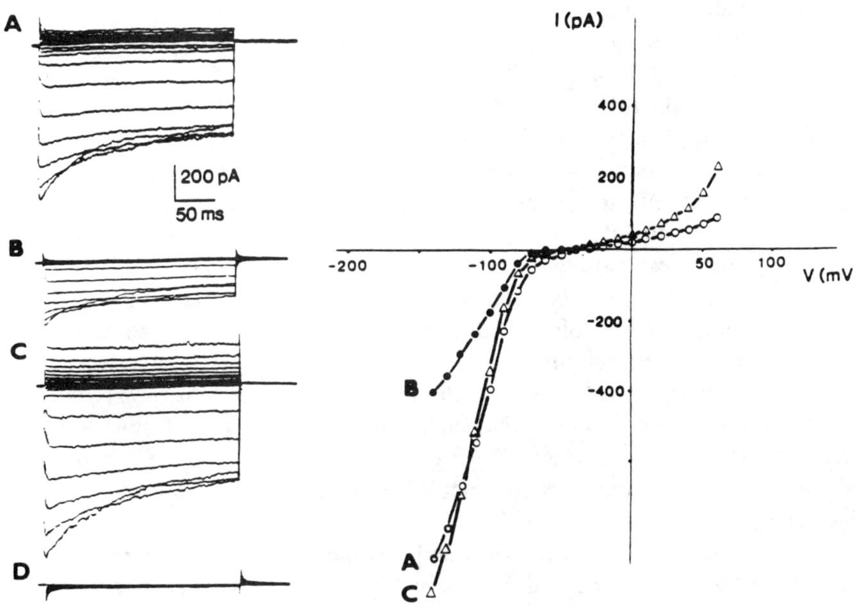

Figure 1 Inward rectifier currents in the membranes of endothelial cells from different sources. (Left) Examples of whole-cell current responses on voltage pulses with increments of 10 mV to maximum voltage step +100 mV from holding potential −40 mV for A, BPA; B, BMV; C, RA; D, HUV. (Right) Voltage-current relationships of currents presented in left. Data for D are not plotted.

hyperpolarized pulses from the level of the resting potential (−40, −60, or −80 mV), the development of transient inward currents was observed. At physiological conditions the threshold of inward current activation was from −70 to −80 mV for all cells studied. It was found that the average value of conductance for inwardly rectifying current is systematically different for cells from different preparations. It is seen from Table 1 that whole-cell conductance induced by inward rectifier channels ($G_{I.R.}$) estimated from the linear part of I-V relations was 3–5 times smaller for cells from BMV in comparison with BPA and RA.

Whole-cell inward rectifier currents for all cell types were found to have similar features. At membrane potentials more negative than −100 mV, current inactivation was observed. Its speed accelerated with hyperpolarization (Fig. 1A). This effect presumably results from a block of inward rectifier channels by sodium ions (Takeda et al., 1987). Indeed, upon substitution of Na^+ ions by K^+ in the external solution, current inactivation disappeared. These currents were inhibited in a voltage-dependent manner by Cs^+ ions (100 μM). Tenfold elevation of external potassium resulted in about 50 mV shift of reversal potential and elevation of slope conductance, suggesting a potassium nature for recorded currents.

These features suggest that endothelial cells from BPA, BMV, and RA possess a system of inward rectifier K channels.

The question arises on what level the difference in the amplitude of whole-cell inward rectifier current is developed. At given membrane potential, the whole-cell current (I) is proportional to the number of functional channels in the membrane (N), to the single-channel conductance (γ), and to the probability of open state (P_o): $I = f(N, \gamma, P_o)$. On the level of single-channel analysis, one can estimate two of these parameters (γ and P_o).

In the cell-attached configuration, in about 90% of cells from BPA, RA, and BMV, the activity of one or several inward rectifying channels was observed. The typical activity of currents recorded from BPA endothelial cells is presented on Fig. 2A. Current amplitude increased with hyperpolarization of the membrane. Average slope conductance was 23 pS for cells from BPA (Fig. 2B) and varied from 22 to 25 pS for other cell types. Thus, the difference in whole-cell current amplitudes does not arise on the level of single channel conductance.

Upon kinetic analysis it was found that the frequency of current fluctuation in the burst increased with membrane hyperpolarization (Fig. 2A), which reflected open-channel blocking action of external Na^+. Kinetic behavior for all cells studied was similar. Thus, we suggest that the difference in the whole-cell current amplitude of inward rectifier currents for endothelial cells from different sources arises due to the difference in the average amount of functional channels per cell. However, this suggestion requires further analysis.

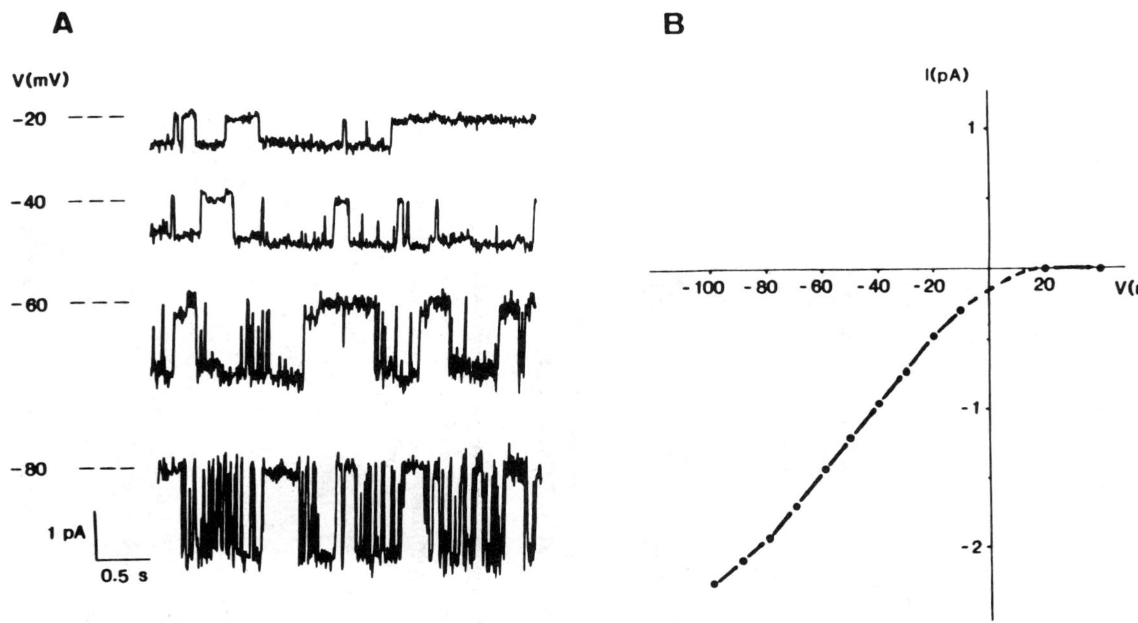

Figure 2 Inward rectifier single-channel currents. Cell-attached configuration. (A) Examples of single-channel currents at different membrane potentials. Numbers on the left (V mV) show amplitudes of pipette potentials for membrane hyperpolarization from the level of resting potential. (B) Voltage-current relationship for single-channel currents. O mV = value of resting potential.

The physiological role of inward rectifier channels is not clear. It has been suggested that these channels are involved in maintaining resting potential (Lindau and Fernandez, 1986). In excitable tissues these channels modulate cell repolarization and pacemaker activity (Di Francesco, 1985). Takeda et al. (1987) suggested that voltage-gated K^+-selective inward rectifier and A-currents serve for recovery of resting potential after agonist action, and thus play an important role in EDRF release.

Inward rectifier currents were not observed on endothelial cells from HUV (Fig. 1). This leads us to suggest that I.R. channels are absent in this type of cell. In this case the understanding of functional differences between venous and arterial endothelium could be a key to understanding the physiological meaning of this type of channel. On the other hand, inward rectifier channels in HUV may disappear more quickly during cell cultivation. In primary cultures from more than 50 HUV endothelial cells studied, typical inward rectifier currents were observed on two of them. However, in this case, a contamination of culture material by other cells could not be excluded.

IV. Hormone-Induced Ionic Currents in the Membrane of Cultured Endothelial Cells

The plasma membrane of endothelial cells contains receptors for a variety of hormones and neurotransmitters: histamine, thrombine, bradykinin, platelet-activating factor, serotonine, acetylcholine, etc. (Richardson and Beaulnes, 1971; Buonassisi and Venter, 1976; Makarski, 1981).

Mechanisms of drugs effects have been only poorly investigated. For instance, it was shown that histamine depolarizes the membrane of endothelial cells (Northover, 1975, 1980) and results in an increase in intracellular calcium concentration (D'Amore and Shepro, 1977). However, no relationship between the elevation of internal calcium and changes of membrane permeability was established. Here we present results of histamine effects in cultured endothelial cells from HUV. Some of these data has been published earlier (Bregestovski and Bakhramov, 1987; Bakhramov et al., 1988; Bregestovski et al., 1988a, 1988b).

A. General Characteristics of Histamine Responses

The effect of histamine was studied in whole-cell voltage clamp mode. The cells were maintained at holding potentials of either −20 or −40 mV and the cell resting potential was measured periodically under the current clamp. In approximately 20% of the cells, responses to applied histamine were seen and three classes of current responses were distinguished (Fig. 3): outward (hyperpolarizing), inward (depolarizing), and biphasic.

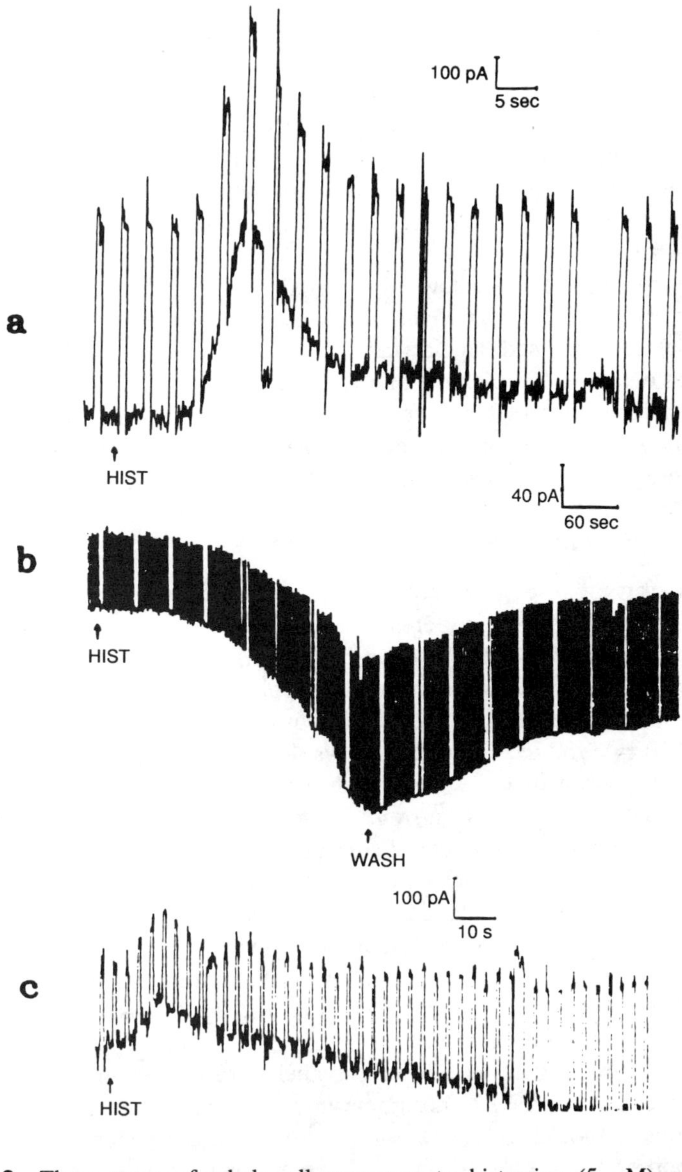

Figure 3 Three types of whole-cell responses to histamine (5 μM) of cultured endothelial cells from umbilical vein. Depolarizing pulses of 10 mV were given to estimate input conductance of the membrane. Resting potential was checked several times during each recording. (a) Hyperpolarizing histamine response. Note absence in changes of current pulses amplitude during histamine response. (b) Depolarizing histamine response. Note increase amplitude of depolarizing pulses, which shows elevation of membrane conductance. Note also different time scale. (c) Biphasic type of response to histamine. Note absence of changes in current pulse amplitude during development of the first phase of the response and slow increase during the second.

Outward response onset developed in 5–10 s after application of histamine. Peak responses occured within 10–20 s after onset, with a desensitization to essentially baseline levels following 1–2 min. Recovery from desensitization was slow, requiring at least 10–20 min of washout. During response, the cell membrane potential increased to 15–30 mV; i.e., the cell became hyperpolarized.

Inward response was slower compared to the hyperpolarizing one, with typical latencies of 1–2 min. This current was associated with a decrease in membrane resistance and depolarization of the cell of 2–15 mV. No desensitization was observed for this type of response.

A biphasic type of response was observed for a small number of cells. After histamine application, both outward and inward currents were observed. In all cases, the outward current preceded the maintained inward current.

B. Hyperpolarizing Type of Response

In the majority of responding cells, the drug induced outward currents of 50 pA-1.5 nA, which caused an increase of resting potential of 5–30 mV with no change of the input resistance of the cell (Fig. 3a). For instance, in 29 cells the average input resistance (R_{in}) before drug application was 82 MΩ and the MP was -23 mV. At the maximum of the response to 10 μM histamine, $R_{in} = 85$ MΩ and MP $= -32$ mV. Figure 4 shows the I-V relationship for this type of response. It is seen that the net I-V dependence of the histamine response (after subtraction of leakage current) is nearly parallel to the voltage axis. Average histamine-induced outward current was about 260 pA.

On four cells, the response to histamine was studied in symmetric KCl solutions, with a holding potential of -20 mV (resting potential of the cells was 0 to -4 mV). Under these conditions histamine produced a slight hyperpolarizing response on three cells of 2 to 7 mV, corresponding to an outward current of 10 to 80 pA, without any effect on the input resistance. In one cell, no obvious response was observed.

Presumably, histamine affects endothelial cells through H1 receptors. Figure 5 shows a response to histamine in control and 5 min after application of pirilamine (a blocker of H1 receptor). Pirilamine completely blocked histamine response. These data are in good agreement with previous observations (Rotrosen and Gallin, 1986), where the authors suggested the presence of H1 receptors in the membrane of endothelial cells.

The histamine response was nearly completely blocked by ouabain (10 μM).

This type of histamine response required the presence of Ca^{2+} in the external solution. In Ca-free external solution (less than 10^{-6} M), responses were reversibly reduced by three to five times.

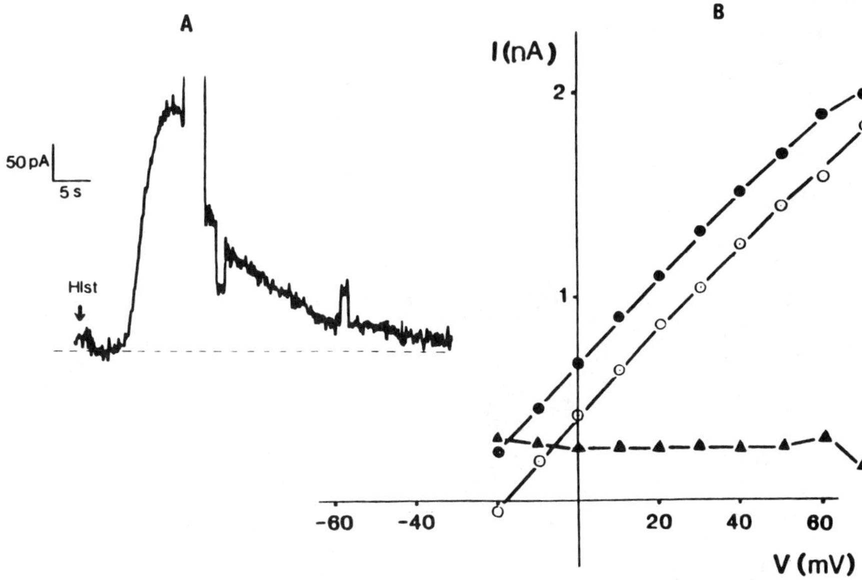

Figure 4 Hyperpolarized type of confluent endothelial cell response to histamine. Holding potential −20 mV. During the first break in the current trace, I-V relations were recorded; the two other breaks show measurement of resting potential in current-clamp mode. (A) Example of outward current. (B) I-V relationships in control (○), on the maximum of response (●), and after subtraction of leakage current (▲).

Two possibilities can explain the histamine hyperpolarizing type of response. First, a pump is activated. Two lines of evidence support this suggestion: (a) histamine increases the resting potential without elevating the membrane conductivity; (b) the histamine-induced currents are inhibited by ouabain. In addition, in symmetric potassium solution, the responses were of small amplitude but also of hyperpolarizing direction. Second, histamine application leads to closing of junctional contacts after increasing of $[Ca^{2+}]_{in}$ (Neyton and Trautmann, 1986). In this case the resting potential can increase due to diminishing "shunt" conductance.

To check these possibilities, experiments with a double patch-clamp were conducted. The protocol of one such experiment is shown on Fig. 6: On confluent cells the cell attached configuration was obtained on two neighboring cells using two pipettes and two patch amplifiers and then a whole-cell configuration was obtained. Both cells were held at −10 mV. Every 3 s, the membrane potential of one cell (top trace) was displaced by 10 mV during 0.5 s; the resulting current I_1 was 320 pA (R_{in} = 31.25 MΩ). The other cell responded on these pulses by currents I_2 of amplitude 137.6 pA. With accu-

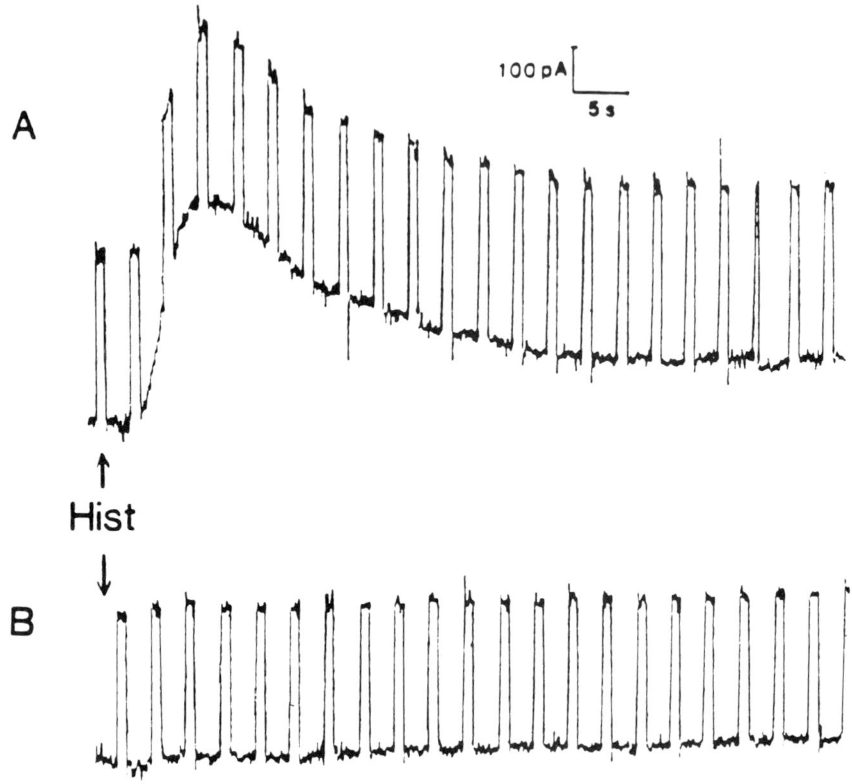

Figure 5 Effect of H1 receptors inhibitor pirilamine on whole-cell histamine response. Vh = −20 mV. (A) Control response to 10 μM histamine. (B) After 1 min pretreatment by 1 μM pirilamine.

racy to 10% (we did not take into consideration the relation of series resistance of the pipette to R_{in2}; see Neyton and Trautmann, 1985), I_2 represents junctional current, so junctional resistance was roughly 72.5 MΩ. After application of 10 μM histamine, both cells hyperpolarized 7 and 12 mV, respectively. During the first 7–10 s after the beginning of hyperpolarization, I_2 even increased slightly (see bottom trace) and then suddenly decreased to nearly 0 and later began slowly to recover. This experiment demonstrates at least two important facts:

1. Application of histamine leads to uncoupling between cells.
2. Uncoupling takes place with a delay of several seconds after achievement of the maximum response to histamine.

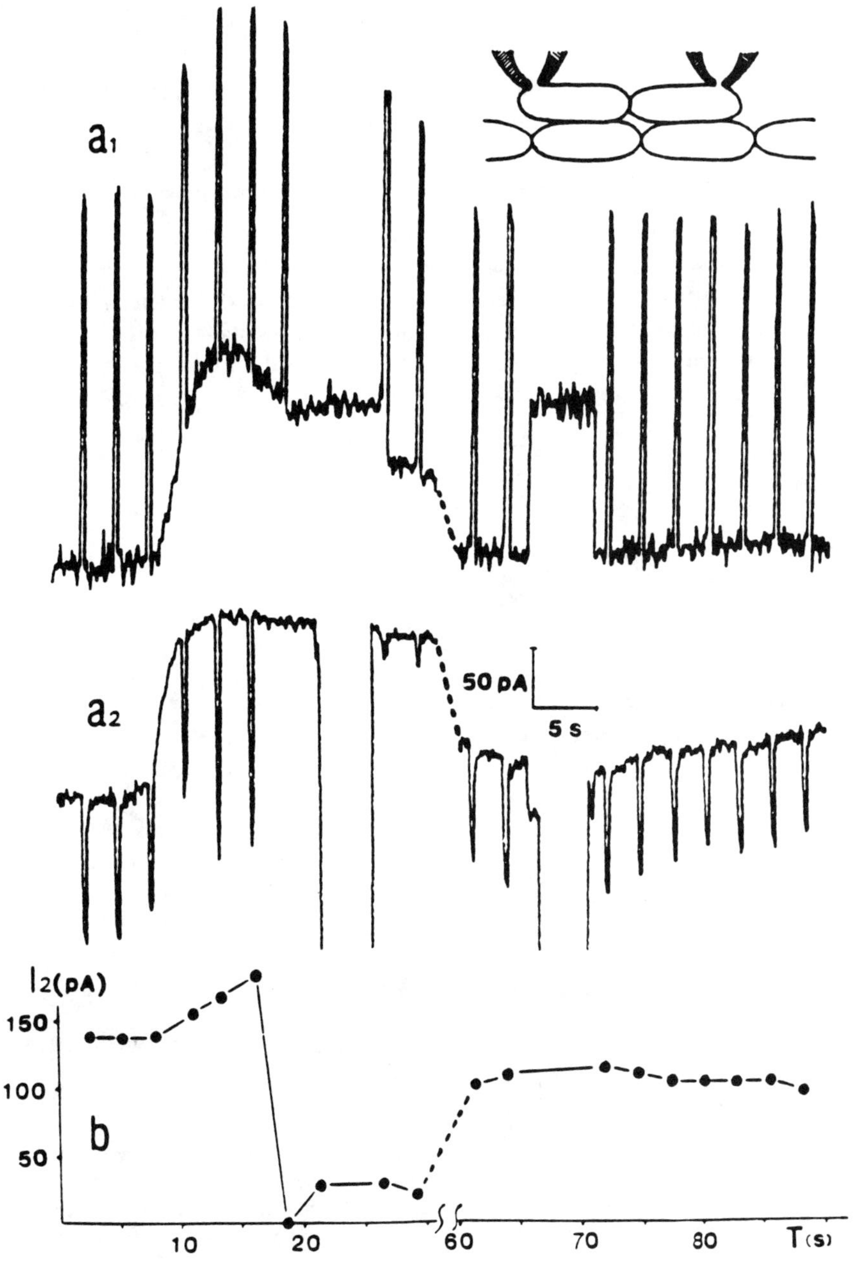

Figure 6 Double patch-clamp recording of intercellular junctions between endothelial cells during histamine response (see text).

Therefore it is strongly suggested that histamine-induced outward current is not a result of closing of junctional contacts.

The mechanism of the hyperpolarizing type of histamine-induced response requires further investigation.

C. Depolarizing Type of Histamine-Induced Response

The depolarizing type of histamine-induced response was described in detail earlier (Bregestovski et al., 1988a). Here we will review general features.

In about 9% of the cells tested, histamine produced depolarizing inward current associated with a decrease in membrane resistance (Fig. 3, middle trace). Calcium ionophore A23187 (1 μM) produced similar effects as histamine. In both cases, reversal potential was between -10 and 0 mV (Fig. 7).

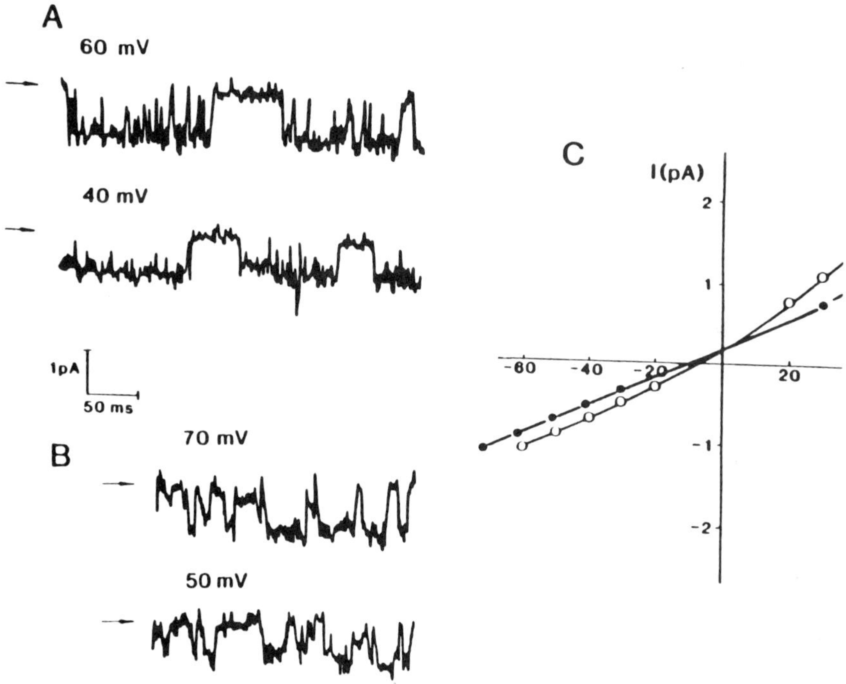

Figure 7 Single-channel currents activated by (A) 10 μM histamine and (B) 1 μM A23187. Cell attached configuration, two membrane patches. Close states are denoted by arrows on the left. (C) Current-voltage relationship for histamine (○)- and A23187 (●)-induced single-channel currents. Slope conductance is similar for histamine and A23187 and gives about 20 pS. Slightly different slope conductance may be due to different resting potentials of the cells and different intracellular ion concentrations.

Responses to A23187 were observed only in those cells which responded to histamine. These observations allow one to suggest that this type of response is due to the activation of a Ca^{2+}-dependent system of ionic channels.

In the cell attached configuration, in 1–2 min after addition to the external solution, histamine induced activation of ionic channels (Fig. 7A). Slope conductance of these channels was about 20 pS (Fig. 7C), average time of the open state was about 5 ms. The Ca^{2+} ionophore A23187 activated similar types of channels (Figs. 7B and 7C). Thus one can suggest that these channels may play a role in generation of the second phase of histamine response, which corresponds with an increase of membrane conductance.

Thus, histamine induced two types of Ca^{2+}-dependent responses. The depolarizing type was associated with ionic channel activation, while the hyperpolarizing type was not associated with changes in membrane conductance. This type of response did not involve activation of ionic channels, was inhibited by ouabain, and probably arose due to pump activation. It was shown recently on the dog coronary artery (Rubanyi and Vanhoutte, 1987) that ouabain inhibited relaxation caused by arachidonic acid in an endothelium-dependent manner. The authors suggested that one of the pathways of EDRF release involves activity of the Na^+/K^+ pump in endothelial cells.

V. Gap Junctional Ionic Currents Between Endothelial Cells

In the present study, two ways of investigation of junctional communication between endothelial cells were used. In the first case, we studied currents between cells in control and during response to histamine on a confluent monolayer. These results, described in previous paragraphs, gave us qualitative information about properties of junctional conductance between cells under "natural" conditions and its regulation by hormone. In the second case, currents between isolated pairs of cells were examined. This gives us a quantitative estimation of gap junction conductance.

A. Quantitative Description of Integral Junctional Conductance

Junctional conductance could be altered by applied potential in some systems but could not be altered in others. In the chironomus salivary gland cells, junctional conductance varied with membrane potential (Obaid et al., 1983); in fish and amphibian embryos, transjunctional potential affects junctional conductance (Spray et al., 1985). Electrophysiological properties of the gap junction between endothelial cells are not yet characterized in detail. In the present study we have examined the dependence of junctional conductance on transjunctional potential and membrane potential in endothelial cell pairs.

The influence of transjunctional voltage on the gap junction conductance between endothelial cells was studied in 52 cell pairs. Among the cell pairs tested, junctional conductance varied greatly from one pair to another. Junctional conductance measured after both membrane patches had been disrupted ranged from 0.18 to 30 nS (with mean value 6.0 ± 1.2 nS, n = 52). Two different types of the influence of transjunctional voltage on contact conductance were observed.

It has been found that macroscopic junctional conductance of contacts with conductance more than 4 ns showed no indications of voltage-dependent decay. Figure 8A shows the current responses recorded from both cells under voltage-clamp conditions, when one of the cells was stimulated by voltage pulses from 10 to 90 mV, with steps of 10 mV, from the holding level. The absence of voltage dependence in this junction is demonstrated in Fig. 8B, where the relationship between junctional current and transjunctional potential is shown. This relationship was linear in a wide range of transjunctional potentials.

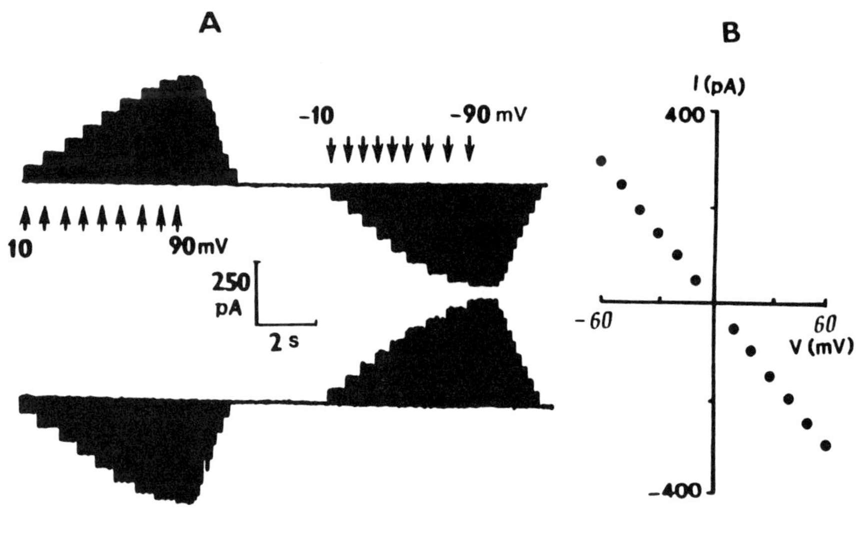

Figure 8 Current-voltage relationship of high-conductance gap junction between endothelial cells. (A) Current traces recorded with double patch-clamp technique from an isolated pair of cells. Holding potential was −20 mV for both cells. Upper trace: current responses of the cell which was stimulated by voltage pulses from holding potential with an increment of 10 mV; onsets of pulses are marked by arrows. Bottom trace: current responses of the cell which was not stimulated (junctional currents). (B) Corresponding current-voltage relationship of the junctional current.

The other type of voltage dependence was observed in contacts with conductance less than 4 nS. In some contacts of this group we have seen a decrease of junctional conductance when transjunctional potential was more than 20–50 mV. Figure 9A shows typical current responses recorded under voltage-clamp conditions from both cells of a cell pair with a voltage-dependent contact conductance.

Figure 9B illustrates a nonlinear relationship between junctional current and transjunctional potential, showing the asymetric character of this relationship: junctional conductance decreases for voltage pulses of one polarity and does not decrease for the other. In this case it is possible to speak about rectified behavior. Such behavior was observed in 11 of 22 cell pairs in which junctional conductance was less then 4 nS. In two cases we have seen symmetric behavior of the relationship between junctional current and transjunctional potential.

In four cell pairs the appearance of voltage dependence of initially transjunctional potential-independent contacs was observed during uncoupling. Figure 9C represents the results of one such experiment. Just after disruption of both patches, the contact conductance was 2.1 nS and the voltage-current relationship was linear; in 10 min, conductance decreased to 0.5 nS and the contact began to show rectified properties.

The influence of the membrane potential on the conductance of contacts between endothelial cells was examined on 15 cell pairs. In these experiments the holding potential was altered simultaneously for both cells from −60 to +40 mV, and one of the cells was stimulated by voltage pulses ranging from −90 to +90 mV. We did not find any influence of holding potential on junctional conductance. Figures 10A and 10B show that both the type of current-voltage relationship and conductance values remain the same for different levels of holding potential.

Thus, junctional conductance of well-coupled endothelial cells is voltage independent. When the cells are coupled more weakly, the contact shows sensitivity to transjunctional potentials. Neyton and Trautmann (1985) also observed the appearance of asymmetrical sensitivity to transjunctional potential in rat lacrimal cells before their irreversible uncoupling. The situation seems to be similar in neonatal rat heart cells (Rook et al., 1988). The authors found a disappearance of voltage dependence with the increase of junctional conductance, but it was not clear about the asymetry of the low-conductance state of these cells. Rook et al. (1988) suggested that voltage dependence could be a consequence of conformational changes in the junctional channels or result from different influences of cytosol access resistance on actual channel voltage drops for high- and low-conducting contacts, and, therefore, voltage-dependence of high-conducting contacts was not observed. This hypothesis cannot explain the asymmetric voltage dependence. It may be that the same

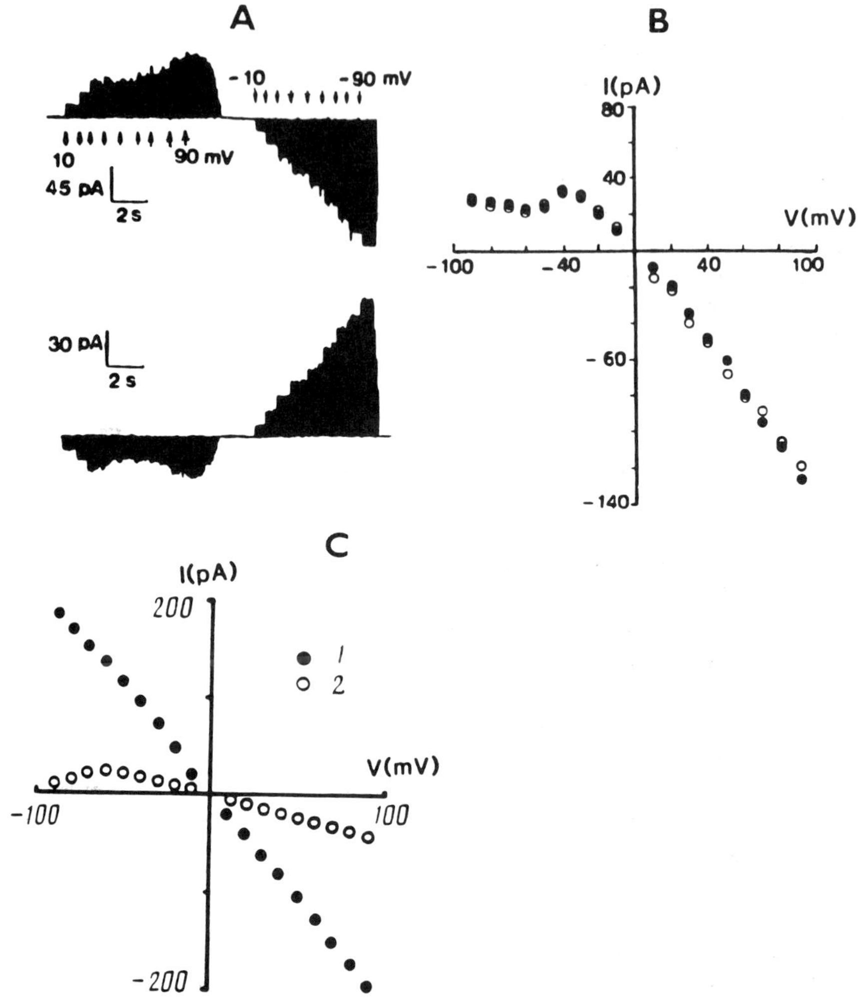

Figure 9 Current-voltage relationship of low-conductance gap junction between endothelial cells. (A) Current traces recorded with double patch-clamp technique from an isolated pair of cells. Holding potential was −20 mV for both cells. Upper trace: current responses of the cell which was stimulated by voltage pulses from holding potential with an increment of 10 mV; onsets of pulses are marked by arrows. Bottom trace: current responses of the cell which was not stimulated (junctional currents). (B) Corresponding current-voltage relationship of the junctional current. V (mv) indicates the transjunctional potential, $V_1 - V_2$. ○, the first cell was stimulated, junctional currents were recorded from the second cell; ●, the second cell was stimulated, junctional currents were recorded from the first cell. (C) Current-voltage relationships of intercellular contact at the beginning of recording (●) and 10 min later (○).

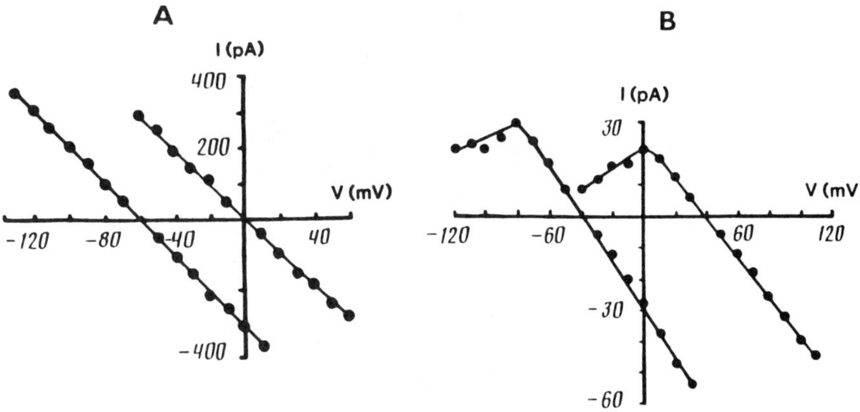

Figure 10 Dependence of junctional currents on the holding potential level. (A) Current-voltage relationships of a high-conductance contact for two levels of holding potential, −60 mV and 0 mV. (B) Current-voltage relationships of low-conductance contact for two levels of holding potential, −40 mV and +40 mV.

factors are responsible both for the voltage dependence and for the closure of the junction, and in most cases these factors do not appear symmetrically in both cells.

The nonlinearity of integral junction currents observed in these experiments may be a consequence of different influence of transjunctional potential on the number of functioning channels, on single-channel conductance, or on the probability of open state. To clear up this question, we studied properties of single-junction ionic channels.

B. Elementary Conductance of Single Gap Junction Channel

In the cell pairs, where junctional conductance was initially low or decreased during uncoupling, it was possible to observe the current flowing through the remaining open channels of the junction. These elementary events for constant transjunctional potential were recognized as symmetrical current fluctuations of opposite polarity.

Figure 11A shows an example of the simultaneous current recording from neighboring cells during applications of constant transjunctional potential. The equal and opposite conductance changes indicate openings and closings of single junctional channels. Three current steps are revealed in the corresponding amplitude histogram (Fig. 11B). Their values measured as distances between neighboring peaks are 2.6 pA, 2.5 pA, and 0.8 pA, which

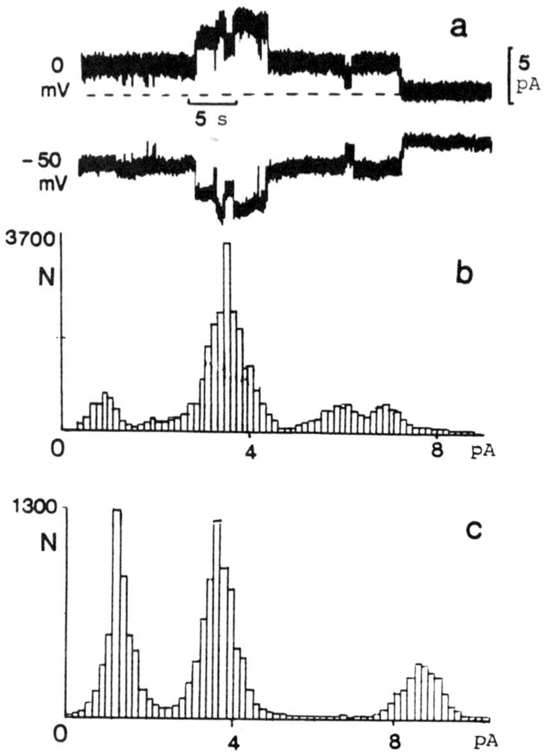

Figure 11 Characteristics of elementary conductance of junctional channel. (a) Currents recorded from a strongly uncoupled cell pair. Transjunctional potential −50 mV. Dotted line indicates the absence of junctional conductance. Channel openings are seen as upward deflections on the upper current trace and as downward deflections on the bottom trace. (b) Amplitude histogram of the same current record (upper trace). Sampling rate 1.5 ms, time of analysis 53 s. Peak amplitudes 1.0, 3.6, 6.1, 6.9 pA. (c) Amplitude histogram of elementary junctional current obtained in another experiment. Transjunctional potential 40 mV. Sampling rate 5 ms, time of analysis 75 s. Peak amplitudes 8.6, 3.8, 1.4 pA.

correspond to conductances 50 pS, 50 pS, and 15 pS. The preferential level had a conductance of 50 pS.

Conductance values close to 50 pS (range 40 to 60 pS) were observed in six experiments. In some experiments, conductances of smaller amplitudes could be discriminated. The minimal elementary conductance found in the present study was 10 pS.

In some experiments, in addition to the conductance of 50 pS, another conductance approximately twice as large was observed. Figure 11C demonstrates an amplitude histogram of junctional current obtained during one such experiment. Two distinct levels, corresponding to 59 pS and 120 pS, could be clearly identified. The elementary conductance of 80–120 pS was observed in three experiments.

Thus, the elementary conductance of junctional channels between endothelial cells varied over a wide range. It is quite possible that the value of about 100 pS reflects a conductance of a single junctional channel and other values correspond to different levels of subconductance states. Taking into account that in the present study the elementary conductance was determined when the contacts were strongly uncoupled, it cannot be excluded that in such conditions only subconductance states might be observed and the real conductance of a fully open channel exceeds 100–120 pS. This point of view is confirmed by the observation of junctional current fluctuations between three quantal levels of conductance with the distinguishable quanta of 170–180 pS in one of the experiments, where integral conductance was 0.8 nS.

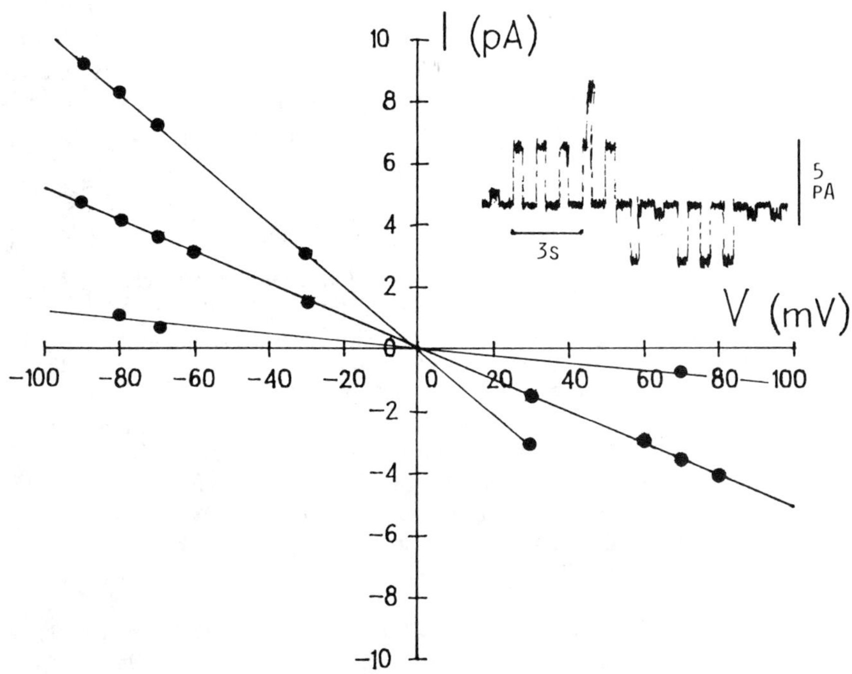

Figure 12 Current-voltage relationships of elementary junctional conductance. Holding potential was 0 mV for both cells, one cell was stimulated by voltage pulses with an increment of 10 mV, current responses were recorded from the second cell. Insert: an example of junctional current responses to voltage pulses of −70 mV (upward deflections) and +70 mV (downward deflections).

The influence of transjunctional potential on elementary conductance has been studied in strongly uncoupled cell pairs. Initially the junctional conductance of cell pairs demonstrated the typical asymmetrical voltage dependence for potentials above +30 mV. In 6 min, only one perferential level of conductance could be observed (Fig. 12, insert). As shown in Fig. 12, the current-voltage relationship at this time was linear, and three levels of conductance could be discriminated; 10 pS, 51 pS, and 100 pS. The level of conductance of 100 pS was not observed for potentials above +20 mV. These data suggest that asymmetry of low-conducting contacts is not linked to nonlinearity of elementary junctional conductance and that the alteration of kinetic characteristics of single junctional channels (or their sublevels) could be responsible for the asymmetric behavior.

VI. Acknowledgments

Part of the electrophysiological data have been obtained by P. Bregestovski at the Medical School, University of Miami (Florida). We thank B. Heyes and U. Ryan for kindly providing cultures of endothelial cells used in these experiments. D. Adams for offering the electrophysiological setup and helpful discussions, and the National Institute of Health for financial support.

References

Adams DJ, Baraken J, Laskey R, Van Breemen C. Ion channels and regulation of intracellular calcium in vascular endothelial cells. FASEB J 1989; 2389–400.

Adams DJ, Lategan TW, Lodge NJ, van Breemen C. Inward rectifying K^+ channels and thrombin-activated cation channels in cultured endothelial cells from bovine pulmonary artery. J Physiol (London) 1987; 394:45P.

Bakhramov A, Bregestovski P. Histamine action on the intercellular junctions of endothelial cells. Biol Membranes 1988; 5:950–9 (in Russian).

Bakhramov A, Bregestovski P, Takeda K. Histamine-dependent currents in endothelial cells isolated from human umbilical vein. J Physiol Abstr 1988; 406:90P.

Bottaro D, Shepro D, Peterson S, Hechtman HB. Serotonin, histamine and norepinephrine mediation of endothelial and vascular smooth muscle cell movement. Am J Physiol 1985; 248:C252–7.

Bregestovski P, Bachramov A. Activation by histamine ionic channels in the membrane of human endothelial cells. In: Hucho F, ed. Receptors and ionic channels. Berlin-New York: Walter de Gruyter, 1987:163–9.

Bregestovski P, Ryan U. Voltage-gated and receptor-activated ionic channels in endothelial cells. J Mol Cell Cardiol 1989; 21(Suppl. 1):103–8.

Bregestovski P, Bakchramov A, Danilov S, Moldobaeva A, Takeda K. Histamine-induced inward currents in cultured endothelial cells from human umbilical vein. Br J Pharmacol 1988a; 95:429–36.

Bregestovski P, Bakhramov A, Ryan U. Inward rectifier currents in bovine pulmonary artery endothelial cells. Proc Acad Sci USSR 1988b; 302:1247–50 (in Russian).

Bregestovski P, Ryan U. Voltage-gated and receptor-activated ionic channels in endothelial cells. J Mol Cell Cardiol 1989; 21(suppl 1):103–8.

Brotherton AFA, Hoak JC. Role of Ca^{2+} and cyclic AMP in the regulation of the production of prostacyclin by the vascular endothelium. Proc Natl Acad Sci USA 1982; 79:495–9.

Buonassisi V, Venter JC. Hormone and neurotransmitter receptors in an established vascular endothelial line. Proc Natl Acad Sci USA 1976; 73:1612–6.

D'Amore PA, Shepro D. Stimulation of growth and calcium influx in cultured bovine, aortic endothelial cells by platelets and vasoactive substances. J Cell Physiol 1977; 92:177.

Daut J, Mehrke S, Nees S, Newman WN. Passive electrical properties and electrogenic sodium transport of cultured guinea-pig coronary endothelial cells. J Physiol (London) 1988; 402:237–54.

Di Francesco D. The cardiac hyperpolarizing-activated current. In: Origins and developments. Progr Biophys Molec Biol 1985; 46:163–83.

Freudenberg N, Riese K, Freudenberg MA. The vascular endothelial system. Stuttgart, New York: Gustav Fisher Verlag, 1983.

Furchgott RF. Role of endothelium in responses of vascular smooth muscle. Circ Res 1983; 53:557–73.

Furchgott RF. The role of endothelium in the responses of vascular smooth muscle to drugs. Ann Rev Pharmac Tox 1984; 24:175–97.

Johns A, Lategan TW, Lodge NJ, Ryan US, Van Breemen C, Adams DJ. Calcium entry through receptor-operated channels in bovine pulmonary artery endothelial cells. Tissue & Cell 1987; 19:733–45.

Killakey JJF, Johnston MG, Movat HZ. Increased permeability of microcarrier-cultured endothelial monolayers in response to histamine and thrombine. A model for the in vitro study of increased vasopermeability. Am J Pathol 1986; 122:50–61.

Lansman JB, Hallam TS, Rink TJ. Single stretch-activated ion channels in vascular endothelial cells as mechanotransducers. Nature 1987; 325:811–3.

Larson DM, Sheridan JD. Junctional transfer in cultured vascular endothelium:II. Dye and nucleotide transfer. J Membrane Biol 1985; 83:157–67.

Larson DM, Kam EY, Sheridan JD. Junctional transfer in cultured vascular endothelium: I. Electrical coupling. J Membr Biol 1983; 74:103–13.

Lindau M, Fernandez JM. A patch-clamp study of histamine secreting cells. J Gen Physiol 1986; 88:349–68.

Makarski JS. Stimulation of cyclic AMP production by vasoactive agents in cultured bovine aortic and pulmonary artery endothelial cells. In Vitro 1981; 17:450–8.

Mokh VP, Bregestovski PD. Properties of intercellular junctions between endothelial cells. In Biol Membrany 1991; 394–401.

Neyton J, Trautmann A. Single-channel currents of an intercellular junction. Nature 1985; 317:331–5 (in Russian).

Neyton J, Trautmann A. Acetylcholine modulation of the conductance of intercellular junctions between rat lacrimal cells. J Physiol 1986; 377:283–95.

Northover AM. Action of histamine in endothelial cells of guinea-pig isolated hepatic portal vein and its modification by indomethacin or removal of calcium. Br J Exp Pathol 1975; 56:52–61.

Northover BJ. The membrane potential of vascular endothelial cells. Adv Microcirc 1980; 9:135–60.

Obaid AL, Socolar SJ, Rose B. Cell-to-cell channels with two independently regulated gates in series: Analysis of junctional conductance modulation by membrane potential, calcium, and pH. J Membrane Biol 1983; 73:69–89.

Olesen SP, Clapham DE, Davies PF. Haemodynamic shear stress activates a K^+ current in vascular endothelial cells. Nature 1988a; 331:168–70.

Olesen SP, Davies PF, Clapham DE. Muscarinic-activated K-current in bovine aortic endothelial cells. Circ Res 1988b; 62:1059–64.

Richardson JB, Beaulnes A. The cellular site of action of angiotensin. J Cell Biol 1971; 51:419-532.

Rook MB, Jongsma HJ, van Ginneken ACG. Properties of single gap junctional channels between isolated neonatal rat heart cells. Am J Physiol 1988; 255:H770–82.

Rotrosen D, Gallin JI. Histamine type I receptor occupancy increases endothelial cytosolic calcium, reduces F-actin, and promotes albumin diffusion across cultured endothelial monolayers. J Cell Biol 1986; 103:2379–87.

Rubanyi GM, Vanhoutte PM. Nature of endothelium-derived relaxing factor: are there two relaxing mediators? Circ Res (suppl. II) 1987; 61:II61–7.

Rubanyi GM, Schwartz A, Vanhoutte PM. The calcium agonists Bay K 8644 and (+)202,791 stimulate the release of endothelial relaxing factor from canine femoral arteries. Eur J Pharmacol 1985; 117:143–4.

Ryan U, Maxwell GJ. Isolation, culture and subculture of bovine pulmonary artery endothelial cells: mechanical methods. J Tissue Cult Methods 1986; 10:3–5.

Simionescu N, Heltianu C, Antohe F, Simionescu M. Endothelial cell receptors for histamine. In: Endothelium, annals of the New York Academy of Sciences, 1982; 401:132–40.

Simionescu M, Simionescu N, Palade GE. Segmental differentiation of cell junctions in the vascular endothelium. The microvasculature. J Cell Biol 1975; 67:863–86.

Simionescu M, Simionescu N, Palade GE. Structural basis of permeability in sequential segments of the microvasculature of the diaphragm. II. Pathways followed by microperoxidase across the endothelium. Microvasc Res 1978; 15:17–36.

Singer HA, Peach MJ. Calcium- and endothelial-mediated vascular smooth muscle relaxation in rabbit aorta. Hypertension 4 (suppl. II) 1982:19–25.

Spray DC, Bennett MVL. Physiology and pharmacology of gap junctions. Ann Rev Physiol 1985; 47:281–303.

Takeda K, Klepper M. Voltage-dependent and agonist-activated ionic currents in vascular endothelial cells: a review. Blood Vessels 1990; 27:169–83.

Takeda K, Shini V, Stoeckel H. Voltage-activated potassium, but not calcium currents in cultured bovine aortic endothelial cells. Pflugers Arch 1987; 410:385–93.

Venter JC, Buonassisi V, Bevan S, Heinemann S, Bevan JA. Hormone and neurotransmitter receptors on the intimal endothelium. Blood Vessels 1975; 12:381–2.

Whitmer KR, Williams-Lawson JS, Highsmith RF, Shcwartz A. Effect of calcium channel modulators on isolated endothelial cells. Biochem Biophys Res Commun 1988; 154:591–605.

19

Receptor-Dependent Regulation of $[Ca^{2+}]_i$ and Phospholipase C in Vascular Endothelial Cells

PAVEL AVDONIN

Institute of Experimental Cardiology
Cardiology Research Center
Moscow, Russia

UNA S. RYAN

Monsanto Company
and Washington University School of Medicine
St. Louis, Missouri

BRENDAN HAYES

Washington University School of Medicine
St. Louis, Missouri

I. Introduction

Relaxation of blood vessels is caused by many vasoactive substances, such as histamine, bradykinin, ATP, and substance P. The binding of these agents to specific receptors on vascular endothelial cells causes the release of two active substances: prostacyclin (PGI_2) (Moncada and Vane, 1978; Weksler et al., 1978); and endothelium-derived relaxing factor (EDRF) (Furchgott et al., 1981), which has been identified as nitric oxide (NO) (Palmer et al., 1987). These substances cause the relaxation of vascular smooth muscle cells and inhibit platelet aggregation (Moncada and Vane, 1978; Furchgott, 1984; Murray et al., 1986). It has been suggested that release of PGI_2 and NO is triggered by an increase in free cytoplasmic calcium concentration ($[Ca^{2+}]_i$) in endothelial cells, since it is potentiated by calcium ionophores (Weksler et al., 1978) and inhibited by the removal of extracellular calcium (Singer and Peach, 1982; Edwards et al., 1985; Long and Stone, 1985). To test the hypothesis that calcium ions play a triggering role in endothelial cell activation, it was necessary to determine the influence of vasodilators on $[Ca^{2+}]_i$. Progress in this field occurred after development of the second-generation fluorescent probes,

Fura-2 and Indo-1 (Grynkiewicz et al., 1985), which permit direct measurement of $[Ca^{2+}]_i$ in vascular endothelial cell monolayers and in individual cells. Several vasodilating agents were demonstrated to increase $[Ca^{2+}]_i$ in endothelial cells (Hallam and Pearson, 1986; Luckoff and Buss, 1986; Rotrosen and Gallin, 1986; Ryan, 1987; Ryan et al., 1988; Avdonin et al., 1989).

This chapter reviews the data obtained during the last 3 to 4 years on receptor-dependent $[Ca^{2+}]_i$ regulation in vascular endothelial cells and discusses both how agonists increase $[Ca^{2+}]_i$ and the mechanism of calcium regulation by protein kinase C. It is well known that stimulation of endogenous protein kinase C by phorbol esters reduces inhibition of the agonist-induced $[Ca^{2+}]_i$ rise in many cell types, including endothelial cells (Ryan et al., 1988). Inhibition of calcium response is accompanied by a reduction in phosphoinositide metabolism (MacIntyre et al., 1985). Thus it might be concluded that the agonist-induced $[Ca^{2+}]_i$ rise is inhibited by suppression of the first step in signal transduction at the level of phosphoinositide-specific phospholipase C (PLC). We have examined endothelial PLC properties and the effect of phorbol ester on its activities to test this hypothesis and clarify how protein kinase C regulates this enzyme.

II. Influence of Vasoactive Agonists on $[Ca^{2+}]_i$ in Endothelial Cells from Different Blood Vessels

Our experiments have studied receptor-dependent $[Ca^{2+}]_i$ regulation in vascular endothelial cells from bovine pulmonary artery and human umbilical vein, aorta, and pulmonary artery. The measurement of $[Ca^{2+}]_i$ was carried out in medium M199 in the presence of 2 mM added $CaCl_2$. The Indo-1 fluorescence from several dozen cells in the monolayer was registered simultaneously, and $[Ca^{2+}]_i$ traces recorded the mean Ca^{2+} levels obtained from these cells. Under these conditions the basal level of $[Ca^{2+}]_i$ in vascular endothelial cells ranges from 100 to 150 nM (Avdonin et al., 1989). Figure 1 shows the influence of vasoactive agents on $[Ca^{2+}]_i$ in cultured endothelial cells from the human umbilical vein (A), pulmonary artery (B), and aorta (C). The data show that in these human endothelial cells, the $[Ca^{2+}]_i$ level increases in response to histamine and thrombin. Human aortic endothelial cells also respond to ATP and bradykinin. The responses to histamine and thrombin were consistently observed in all other preparations of human endothelial cells that we tested, whereas the responses to bradykinin and ATP in the human aortic cells were found to be variable in that they were observed in only 60–80% of the preparations studied. We found that carbachol (100 μM), angiotensin II (40 μM), prostaglandin endoperoxide H_2 analog U46619 (0.5 μM), or

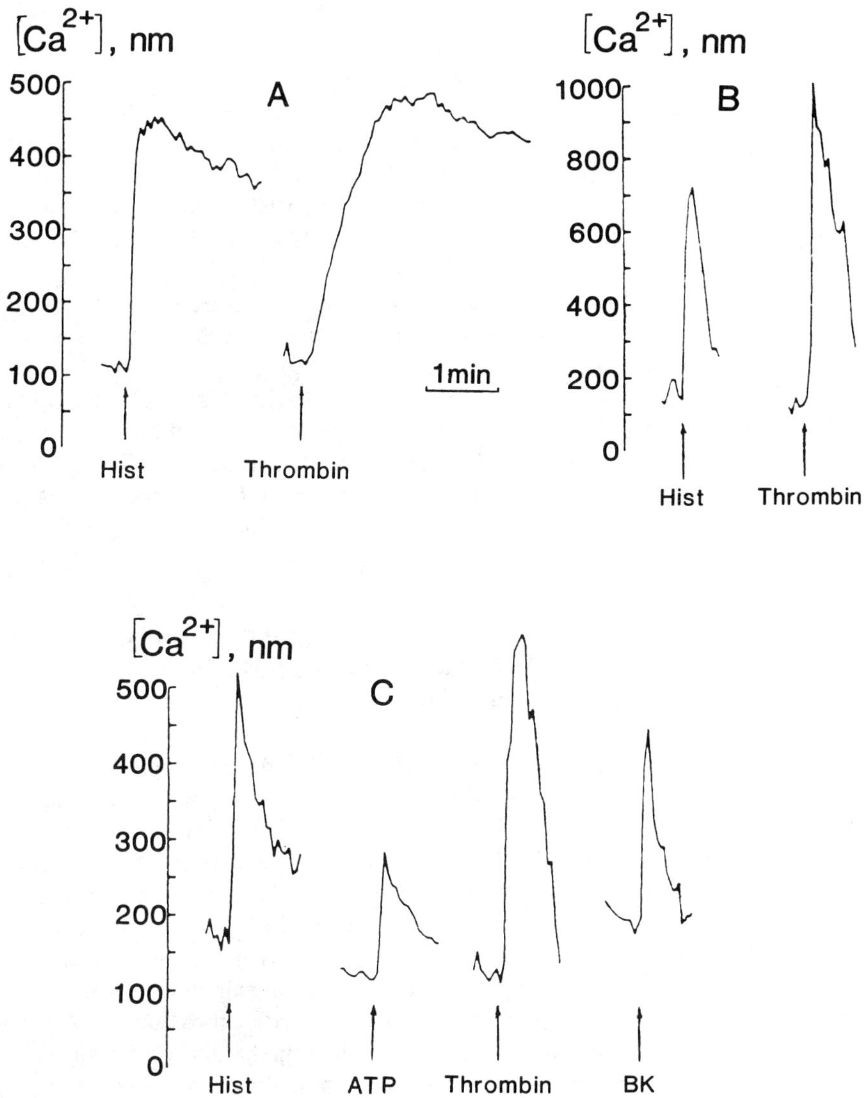

Figure 1 The effect of histamine (100 μM), thrombin (1 U/mL), ATP (200 μM), or bradykinin (10 μg/mL) on $[Ca^{2+}]_i$ in cultured endothelial cells from the human umbilical vein (A), pulmonary artery (B), or aorta (C). The cells were incubated in medium 199 with 2 mM additional exogenous Ca^{2+} at 37°C. Traces are representative of several measurements. (From Ryan et al., 1988, by permission.)

atrial natriuretic factor (1 mM) were without effect on $[Ca^{2+}]_i$ levels in cultured human endothelial cells.

Bovine pulmonary artery endothelial cells were found to behave similarly to human aorta endothelial cells in that they responded to thrombin, ATP, and bradykinin (Fig. 2). However, bovine endothelial cells never responded to histamine. In some preparations we also observed $[Ca^{2+}]_i$ rises in response to platelet activating factor (Ryan, 1987; Ryan et al., 1988). Thus, it is suggested that the sensitivity of bovine endothelial cells to PAF may depend on the conditions under which the cells were cultured.

As shown in Fig. 1, both histamine and thrombin produce marked and reproducible increases in $[Ca^{2+}]_i$ in human endothelial cells. Figure 3 shows the effect of different histamine concentrations ranging from 0.5 to 50 μM on $[Ca^{2+}]_i$ in human umbilical vein endothelial cells. Increasing concentrations of histamine resulted in a rise in both the peak level of $[Ca^{2+}]_i$ and an increase in the initial rates of $[Ca^{2+}]_i$ elevation. The histamine effect is mediated by H_1 receptors, since pyrilamine, a blocker of H_1 receptors, progressively decreased the response to histamine, while cimetidine, a specific H_2-receptor antagonist, had no effect.

III. Relative Contribution of Extracellular Ca^{2+} and Ca^{2+} from Intracellular Stores in Agonist-Induced $[Ca^{2+}]_i$ Elevations in Vascular Endothelial Cells

To determine the relative contribution of extra-/ and intracellular Ca^{2+}, we used bovine pulmonary artery endothelial cells (Fig. 2). After the addition of the agonist, a sharp increase in $[Ca^{2+}]_i$ to the maximum level was observed. With bradykinin and ATP, this process lasted 8–10 s, whereas the thrombin-induced $[Ca^{2+}]_i$ increase was slower and lasted 17 $\pm$6 s ($n = 6$). With thrombin there was an initial lag period lasting for 24 $\pm$15 s ($n = 6$), during which time $[Ca^{2+}]_i$ remained at the basal level. When measurements were performed in a Ca^{2+}-containing medium, after the maximum level had been reached, a slow decrease in $[Ca^{2+}]_i$ was observed. After thrombin administration, the $[Ca^{2+}]_i$ decrease was linear and reached the basal level within 1.5–2 min. Following exposure of cells to either bradykinin or ATP, $[Ca^{2+}]_i$ remained elevated for a period of about 3 min. The kinetics of the falling phase of the bradykinin-induced $[Ca^{2+}]_i$ transient were biphasic. Immediately after the maximum had been achieved, $[Ca^{2+}]_i$ rapidly decreased to an intermediate level (234 $\pm$55 nM, $n = 3$), then the rate of $[Ca^{2+}]_i$ decrease was considerably reduced. In contrast, the ATP-induced $[Ca^{2+}]_i$ rise was mainly followed by a linear decrease. However, in some experiments, the kinetics were biphasic, like those of bradykinin.

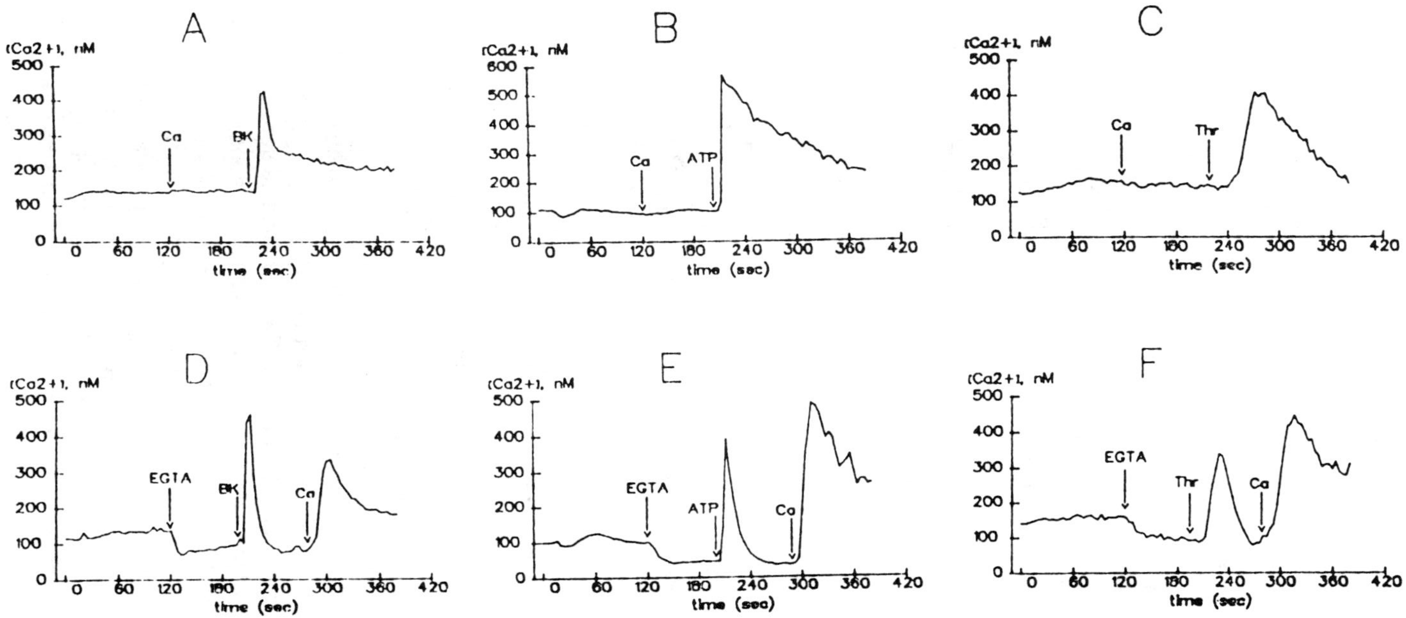

Figure 2 Effects of 1 μM bradykinin, 20 μM ATP, and 1 U/mL thrombin on $[Ca^{2+}]_i$ in bovine pulmonary artery endothelial cells in the presence of Ca^{2+} (A, B, C) or 2 mM EGTA (D, E, F). The second agonist-stimulated responses were induced by the addition of excess $CaCl_2$ after EGTA. (From Avdonin et al., 1989, by permission.)

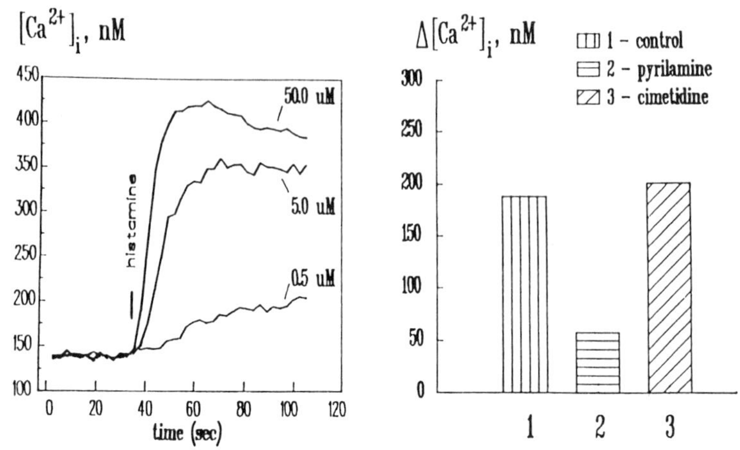

Figure 3 (Left) The influence of different histamine concentrations on $[Ca^{2+}]_i$ in EC from human umbilical vein. Each concentration was tested in two separate preparations. Cells were incubated in phosphate-buffered saline in the presence of 1 mM $MgCl_2$ and 2 mM $CaCl_2$ at 20°C. (Right) The influence of pyrilamine (5 μM) and cimetidine (5 μM) on histamine-induced $[Ca^{2+}]_i$ elevations in human umbilical vein endothelial cells. Histamine concentration was 5 μM. Incubations were performed (in replicates of 3) in medium 199 at 20°C in the presence of 2 mM additional exogenous Ca^{2+}. (From Ryan et al., 1988, by permission.)

The occurrence of the second phase of the response (Ca^{2+}-lowering part of the response) of endothelial cells to the agonist may be explained by either a reduction in Ca^{2+}-ATPase activity or by continued Ca^{2+} flow into the cytoplasm, although the rate of this flow is lower. The agonist-induced $[Ca^{2+}]_i$ increase may be due to Ca^{2+} release from intracellular stores and entry of Ca^{2+} from the extracellular medium. To elucidate the contribution of each source of Ca^{2+} in the response of endothelial cells to the agonist, 2 mM EGTA was added to the extracellular medium (Figs. 2D, 2E, 2F). In the presence of EGTA, the concentration of free Ca^{2+} in the medium ($[Ca^{2+}]_{out}$) was 1.1 μM. $[Ca^{2+}]_{out}$ was determined by measuring the fluorescence of hydrolyzed Indo-1 added to the medium. After chelation of the extracellular Ca^{2+} by EGTA, the baseline cellular level of $[Ca^{2+}]_i$ dropped from 146 $\pm$33 nM (n = 9) to 80 $\pm$133 nM (n = 9). At this low extracellular Ca^{2+} concentration, Ca^{2+} entry in response to agonists is virtually absent. Under these conditions, both the rate (Figs. 2D, 2E, 2F) and the extent (Table 1) of the increase in $[Ca^{2+}]_i$ in response to bradykinin, ATP, and thrombin remained unchanged. In contrast, a difference was observed in the kinetics of $[Ca^{2+}]_i$

Table 1 Agonist-Induced Increase in $[Ca^{2+}]_i$ in the Absence and Presence of Extracellular Ca^{2+}

	$[Ca^{2+}]_i$, nm		
	Bradykinin 1 μM	ATP 20 μM	Thrombin 1 U/mL
In the presence of Ca^{2+}	302 ± 34 (n = 3)	426 ± 33 (n = 3)	238 ± 39 (n = 3)
In the presence of 2 mM EGTA	299 ± 68 (n = 4)	338 ± 17 (n = 2)	243 ± 37 (n = 2)
After administration of 2 mM $CaCl_2$	287 ± 47 (n = 3)	512 ± 129 (n = 3)	512 ± 131 (n = 2)

lowering; i.e., there was a more rapid (less than 0.5 min) return to baseline after the initial peak. Thus, in Ca^{2+}-free medium, the maximum increase of $[Ca^{2+}]_i$ in response to the agonists was maintained, however the duration of the endothelial cell response was reduced severalfold. When excess Ca^{2+} was added to the extracellular medium after EGTA, a second $[Ca^{2+}]_i$ rise occurred. It should be noted that the decrease in $[Ca^{2+}]_i$ in this case was slower than the decrease observed in Ca^{2+}-free medium.

An explanation of these results is that bradykinin, ATP, and thrombin induce activation of Ca^{2+} entry from the extracellular medium in addition to Ca^{2+} mobilization from the intracellular stores. The extracellular Ca^{2+} that enters the cells provides a prolonged response to the agonists. Similar results were obtained with Fura 2-loaded cells (Avdonin et al., 1988).

IV. Fluctuations of Free Cytoplasmic Calcium and Magnesium Concentration in Vascular Endothelial Cells

As already noted, we measured the receptor-dependent increase in $[Ca^{2+}]_i$ occurring in hundreds of endothelial cells. Therefore the resultant picture (Figs. 1 and 2) is the mean of a population of individual cellular responses. In 1986, Woods et al. showed that, in individual hepatocytes, agonist-induced increases in $[Ca^{2+}]_i$ occur as periodic oscillations and not as a smooth increase. This phenomenon was later documented in many other types of cells, and its ubiquitous nature was demonstrated. Jacob et al. (1988) showed similar findings in blood vessel endothelial cells. In human umbilical vein endothelial cells loaded with Fura-2, $[Ca^{2+}]_i$ oscillations were found to occur in response to histamine at a frequency of 1 to 2 oscillations per min. The duration of individual oscillations was about 20 s. The frequency of the oscillations increased concomitantly with the increase in hormone concentration. It was suggested that oscillations may occur due to calcium mobilization from intracellular stores. However, extracellular Ca^{2+} was also necessary, since in its absence the oscillations soon disappeared. Further insight into this phenomenon was revealed by the study of the spatial changes in $[Ca^{2+}]_i$ in an individual endothelial cell as detected by Fura-2 fluorescence using the method of digital image analysis (Jacob, 1990). It appeared that the spreading of Ca^{2+} in the cytoplasm occurred as a "calcium wave." Initially, a local increase in $[Ca^{2+}]_i$ took place which induced a spreading of Ca^{2+} throughout the whole cell body at a rate of 50 μm/s. The movement of the calcium front could not be explained by simple diffusion. Berridge and Irvine (1989) suggested that Ca^{2+}-induced Ca^{2+} release from the internal stores gives rise to Ca^{2+} oscillations; at present, however, there is no direct evidence for the existence of this process in nonmuscular cells.

Is it possible to correlate the data on $[Ca^{2+}]_i$ regulation in an individual endothelial cell with that obtained from a cell monolayer? It is clear that the response of receptor-dependent increases in $[Ca^{2+}]_i$ in the cell population is an average of individual cell response, and an interesting peculiarity was revealed in a study of receptor-dependent $[Ca^{2+}]_i$ regulation in vascular endothelial cells. It appeared that changes in $[Ca^{2+}]_i$ in a confluent monolayer may mutually synchronize, since all the cells are interconnected through gap junctions. Due to these intercellular contacts, periodic oscillations of $[Ca^{2+}]_i$ develop in cell monolayers, which are characteristic of individual cells (Sage et al., 1989; Neylon and Irvine, 1990). In our experiments we usually failed to detect these calcium fluctuations (Figs. 1 and 2). However, in some experiments, as shown in Fig. 4, we observed one or two additional increases in $[Ca^{2+}]_i$. Similar data were obtained in the study of free cytosplasmic Mg^{2+} ($[Mg^{2+}]_i$) concentration in endothelial cells. $[Mg^{2+}]_i$ was measured using the fluorescent probe Magfura-2. Figure 5 shows that an increase in extracellular Mg^{2+} concentration from zero up to 10 mM in a cuvette results in a concomitant increase in $[Mg^{2+}]_i$ in the cells. It appeared that fluctuations similar to those observed with calcium develop in the process of this increase (Fig. 5). The mechanisms underlying the regulation of $[Mg^{2+}]_i$ have not been elucidated. Overall, these data concerning fluctuations in $[Ca^{2+}]_i$ and $[Mg^{2+}]_i$ indicate that under these conditions the layer of endothelial cells appears to act as it might behave in a blood vessel.

In endothelial cells, as in other animal cells, the mobilization of Ca^{2+} from internal depots is induced by inositol-1,4,5-trisphosphate (IP_3). Four receptors for IP_3, with a molecular mass of 313 kDa each, form a Ca^{2+} channel. These receptors have been isolated, and their structure and functional properties have been described elsewhere (see Mignery and Sudhof, 1990). Little data exist on the structure or regulation of plasma membrane Ca^{2+} chan-

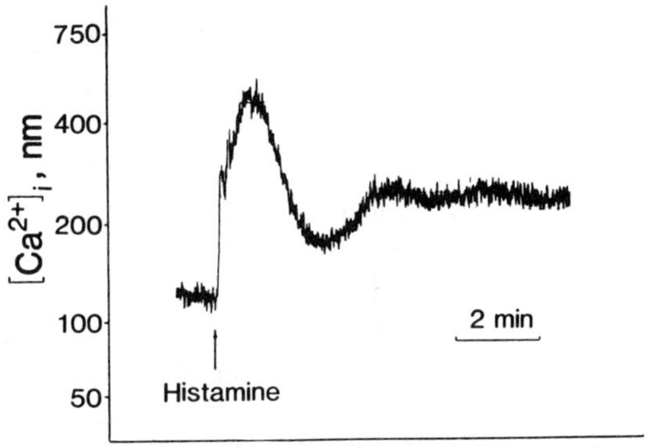

Figure 4 Histamine-induced fluctuations of $[Ca^{2+}]_i$ in human umbilical vein endothelial cells. The cells were loaded with Indo-1 and incubated in medium 199 without phenol red. Histamine concentration (0.1 mM). Fluorescence was measured at excitation wavelength 340 nm, emission wavelength 500 nm.

nels in endothelial or other cell types, and how these receptor-operated Ca^{2+} channels are activated is unknown. It has been suggested by some workers that plasma membrane receptor-operated channels are opened by inositol-1,4,5-trisphosphate (Kuno and Gardner, 1986; Mozshaeva et al., 1990). If this is the case, then the study of phospholipase C is necessary to understand the mechanisms underlying receptor-operated Ca^{2+}-channel regulation.

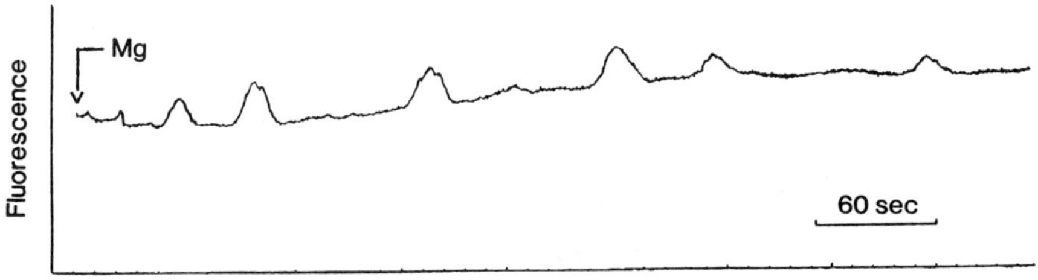

Figure 5 Spontaneous fluctuations of free cytoplasmic magnesium concentration ($[Mg^{2+}]_i$) in bovine pulmonary artery endothelial cells. The cells were grown until they formed a confluent monolayer and were loaded with magfura-2. The rise in fluorescence reflects elevation in ($[Mg^{2+}]_i$. The excitation and emission wavelengths were 340 nm and 500 nm, respectively. The cells were incubated in a medium M199 without phenol red, Ca^{2+}, and Mg^{2+}. The arrow designates addition of 10 mM $MgCl_2$.

V. Properties and Regulation of Polyphosphoinositide-Specific Phospholipase C in Vascular Endothelial Cells; Inhibition of Phospholipase C by Endogenous Protein Kinase C

While there is considerable understanding of how the calcium response is triggered, less is known about how the Ca^{2+} signal is terminated. It has been suggested in several laboratories that the Ca^{2+} signal is terminated by a negative feedback mechanism dependent on protein kinase C. This conclusion was based on the demonstration that activation of endogenous protein kinase C by diacylglycerol, or its analog, phorbol myristate acetate (PMA), results in suppression of inositol phosphate accumulation and blockage of agoinist-induced $[Ca^{2+}]_i$ rise. In vascular endothelial cells we also observed an inhibition of both the agonist-induced $[Ca^{2+}]_i$ rise (Fig. 6) and PI turnover (Ryan, 1987; Ryan et al., 1988) by phorbol myristate acetate. Figure 6 also shows that an elevation of the cAMP level in endothelial cells induced by treatment with isobutylmethylxanthine and forskolin, or a rise in the cGMP level by the guanylate cyclase activator nitroprusside, failed to exert a substantial effect on histamine-induced $[Ca^{2+}]_i$ elevation.

There are at least two possible explanations for the effect of protein kinase C (Fig. 6). According to Connolly et al. (1986), protein kinase C phosphorylates 5′-phosphomonoesterase (PME), the enzyme which hydrolyzes inositol-1,4,5-trisphosphate. Phosphorylation of PME causes substantial elevation in enzymatic activity. As a result, it prevents accumulation of IP_3 and a

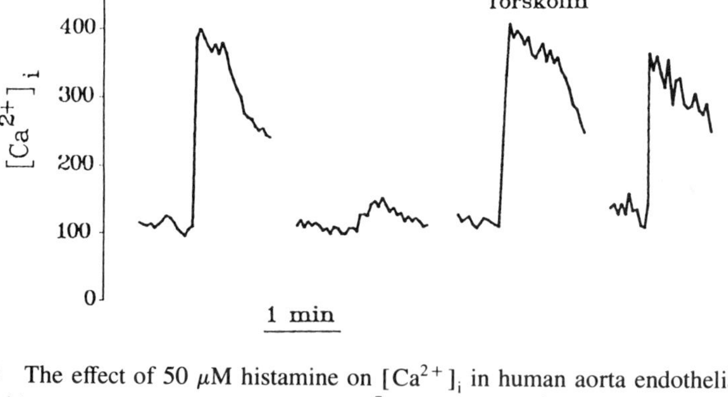

Figure 6 The effect of 50 μM histamine on $[Ca^{2+}]_i$ in human aorta endothelial cells with or without a 5-min incubation with 10^{-7} M PMA, 10^{-3} M IMBX, and 10^{-4} M forskolin or 5×10^{-5} M nitroprusside. (From Ryan et al., 1988, by permission.)

rise in intracellular Ca^{2+}. Alternatively, it is possible that protein kinase C inhibits phospholipase C, and this inhibition may be due to G-protein phosphorylation. Hence, to understand the mechanism of cellular Ca^{2+} regulation, the influence of protein kinase C on the biochemical reactions underlying signal transduction processes must be clarified and the role of PLC, the key enzyme in this system, must be elucidated.

While measurement of the enzymatic activity of PLC is difficult, two principal assay methods are used. One is based on the utilization by phospholipase C of endogenous ^3H-inositol-labeled PIP_2 in membranes isolated from cells incubated with radioactive inositol. However, in the case of endothelial cells, this method failed because during cell homogenization and membrane isolation most of the endogenous ^3H-PIP_2 was found to degrade to phosphatidylinositol. Therefore, we measured endothelial cell phospholipase C activity using exogenous ^3H-inositol-labeled phosphatidylinositol-4,5-bisphosphate (^3H-PIP_2). This assay was performed according to the method described by Rittenouse-Simmons (1977) with some modifications. Incubation was conducted for 10 min at 37°C in medium containing 25 mM Tris-maleate (pH 6.0), 1 mM $MgC1_2$, 100 mM phosphatidylinositol-4,5-bisphosphate, 0.01 mCi ^3H-phosphatidylinositol-4,5-bisphosphate (about 10,000 counts/min), 0.5 mM EGTA, and 0.3 mM $CaC1_2$. During this period the kinetics of PIP_2 hydrolysis were linear. In some experiments, pH and added $CaC1_2$ were varied. The reaction was started by the addition of cell protein (2–5 mg per sample). The incubation medium also contained a detergent necessary to support phospholipase C activity, and in most experiments sodium cholate was used for this purpose at a concentration of 1-1.5mM. Phospholipase C substrate was added to the incubation medium as a liposome suspension prepared by dissolving labeled and nonlabeled phosphatidylinositol-4,5-bisphosphate dissolved in $CHCl_3$ which was evaporated under a current of dry nitrogen. Tris-maleate buffer (10 mM) was added to the pellet to give a lipid concentration fivefold in excess of the final incubation medium. The mixture was sonicated four times for 20 s. To stop the reaction, 190 μL of an extraction mixture containing chloroform:methanol:concentrated HCl (100:200:3) was added to each sample with subsequent vigorous stirring and the samples were allowed to stand at room temperature for 15–30 min. To separate the aqueous and organic phases, 50 μL of chloroform and 50 μL of 1.8 M KCI were added to the samples, and stirring was again performed with subsequent centrifugation at 250 g for 5–10 min. Aliquots of the upper aqueous phase containing the water-soluble products of the phospholipase reaction were assayed for radioactivity. Products were separated and identified by column chromatography (Dowex AG1-X8, 0.5 g) by a previously described method (Berridge et al., 1983). Data showed that the ^3H-phosphatidylinostiol-4,5-bisphosphate hydrolysis product coeluted with the inositol trisphosphate fraction. Therefore, in most of the experiments, we limited ourselves to the measurement of the radioactivity in the aqueous phase rather than performing chromatography on anion-exchange resin for each sample. Data presented in all figures are means of two to three experiments.

Figure 7 shows the subcellular distribution of PIP_2-specific phospholipase C activity in bovine pulmonary artery endothelial cells. The enzyme activity in the homogenate is taken to be 100%. As seen from these data, most of the phospholipase C activity in endothelial cells is localized in the cytosol, and only 10–15% is present in the membrane fraction. This suggests that phospholipase C in these cells is heterogenous and comprised of two forms—soluble and membrane-bound. These results are in accordance with the available published data about the existence of several isoforms of this enzyme in mammalian cells. At present, five main isoforms of phospholipase C have been documented. They are designated as alpha, beta, gamma, delta, and epsilon (for review, see Meldrum et al., 1991). The intracellular distribution and regulation of these isoforms is not clearly established. It was demonstrated that isoform gamma-1 is stimulated by growth factors EGF and PDGF by means of phosphorylation of the tyrosine residue (Nishibe et al., 1990), whereas phospholipase C β^1 is stimulated by some agonists via a pertussis toxin-insensitive G_q protein (Taylor et al., 1991). The roles of other isoforms in hormonal regu-

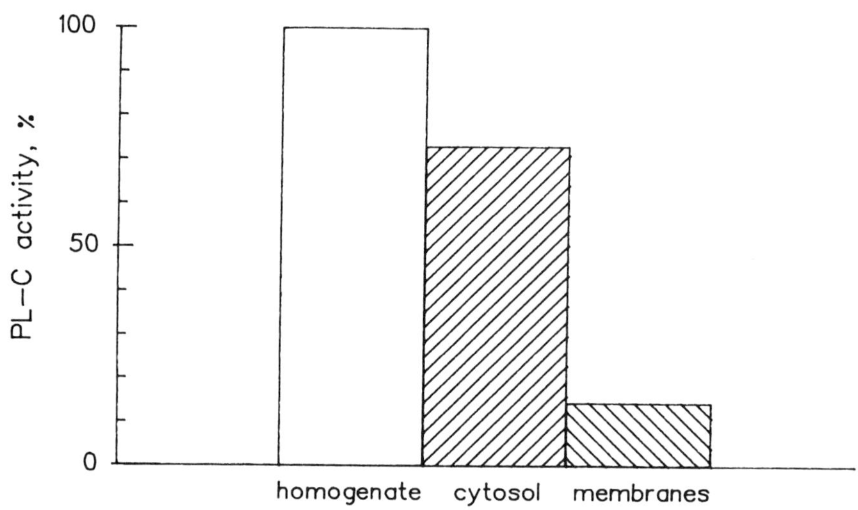

Figure 7 Subcellular distribution of phospholipase C activity in bovine pulmonary artery endothelial cells. Phospholipase C activity was measured in the presence of 1 mM EGTA, 1 mM $CaCl_2$, 0.03% Triton X-100 at pH 6.0.

lation of phosphoinositide turnover are unknown. No information has been reported concerning phospholipase C isozyme composition in vascular endothelial cells. In the experiments which will be presented here, only the membrane fraction was used.

The initial task in the investigation of membrane-associated phospholipase C of bovine endothelial cells was to characterize its general properties and to examine whether it participates in agonist-dependent regulation of PI-turnover. In Fig. 8 the pH dependence of ^{3}H-PIP_2 hydrolysis by membrane-bound phospholipase C is demonstrated. The optimum pH for phospholipase C activity is acidic, with maximum activity observed at pH 5–5.5. The same pH dependence was observed for phospholipase C activity in homogenates of endothelial cells. A similar pH dependence for phospholipase C activity has been demonstrated in other cell types (Crooke and Bennett, 1989). It is probable that the anomalous pH dependence is due to the artificial conditions at which phospholipase C activity was measured. It might be suggested that an

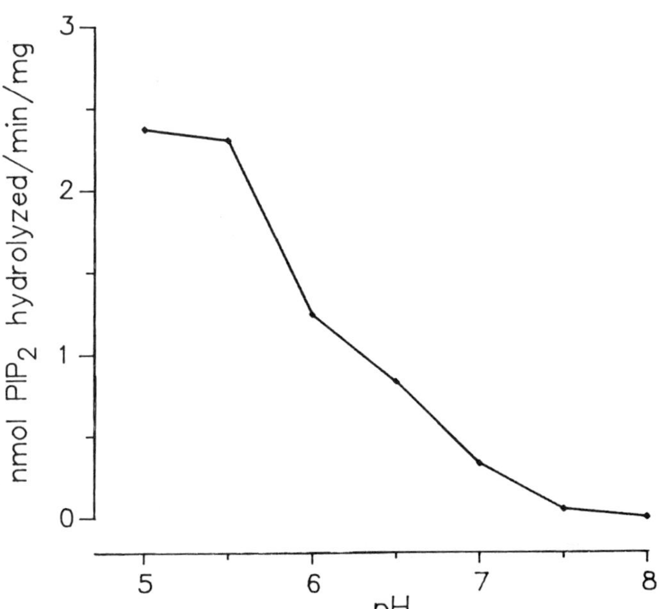

Figure 8 The effect of pH on phosphatidylinositol-4,5-bisphosphate hydrolysis by phospholipase C in bovine pulmonary artery endothelial cell membranes. The medium contained 1 mM EGTA, 0.6 mM $CaCl_2$, 1 mM sodium cholate, 3.5 μg of membrane protein. Each point is a mean of triplicate determination.

acidic pH makes exogenous PIP_2 more available to the enzyme. At neutral pH, its activity is substantially lowered. Since the phospholipase C activity assay method is not sensitive enough to measure very low levels of PIP_2 hydrolysis at neutral pH (7–7.5), we were obliged to study the pH properties of phospholipase C at pH 6–6.5.

Since it is known that detergent is necessary to support the activity of phospholipase C (Crooke and Bennett, 1989), sodium cholate was used for this purpose. Figure 9 shows the effect of different cholate concentrations on the activity of phospholipase C. The graph shows that basal phospholipase C activity increased with increasing concentrations of cholate ranging from zero to 0.5 mM, after which the effect reached a plateau. In the same experiment the regulatory properties of phospholipase C were tested. Phospholipase C in bovine pulmonary artery endothelial cell membranes is activated by GTPgammaS and thrombin (Fig. 9). In the absence of cholate, or at low detergent concentrations, the stimulatory effects of these compounds are additive. Increase in sodium cholate concentration resulted in inhibition of GTPgammaS-dependent phospholipase C activity. However, the combination of thrombin and guanyl nucleotide produced a synergistic stimulation of phosphatidylinositol-4,5-bisphosphate hydrolysis (Figs. 9 and 10).

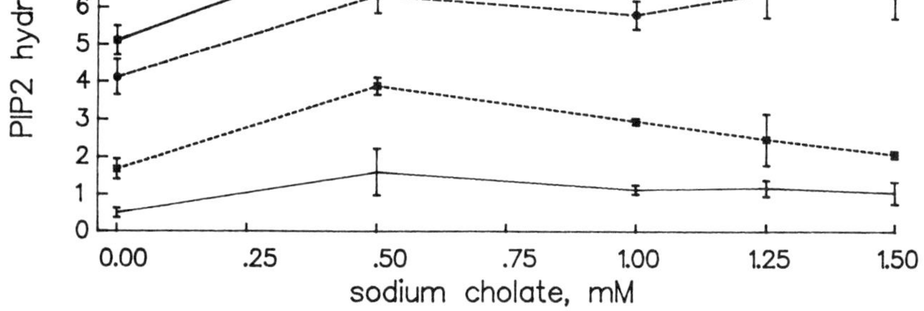

Figure 9 The effect of sodium cholate on phosphatidylinositol-4,5-bisphosphate hydrolysis by phospholipase C in bovine pulmonary artery endothelial cell membranes. The medium contained 1 mM EGTA, 0.6 mM $CaCl_2$, pH 6.0. Activity was measured in the presence of 100 μM GTP S, 2 U thrombin.

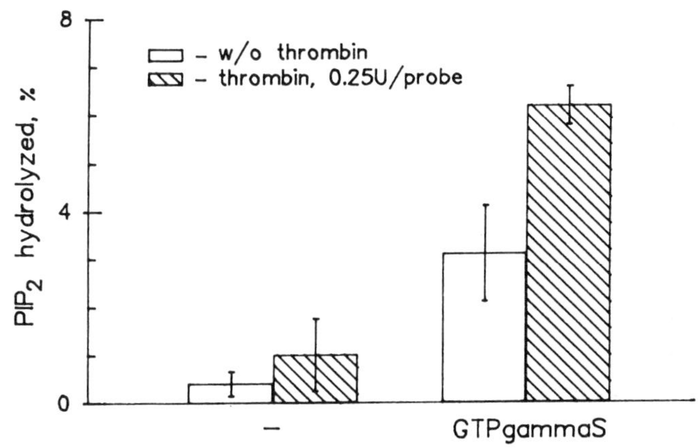

Figure 10 The synergistic stimulation of phosphatidylinositol-4,5-bisphosphate hydrolysis by thrombin (0.25 U/probe) and GTP S (100 μM).

The effects of different thrombin concentrations on phospholipase C activity in the absence and presence of GTPgammaS are shown on Fig. 11. It is seen from these data that GTPgammaS is necessary for thrombin-dependent stimulation of phospholipase C activity when low doses of thrombin are used. At higher concentrations (>5 U/mL), thrombin is able to stimulate PL-C activity without exogenous guanyl nucleotides (Fig. 11). The effect of high thrombin concentrations may be explained by its proteolytic action (Lazarowski and Lapetina, 1990).

It is concluded from these data that phospholipase C in endothelial cell membranes is stimulated by GTPgammaS and thrombin. These data indicate that the membrane-bound form of phospholipase C participates in agonist-induced signal transduction. The activity of endothelial cell phospholipase C depends on the Ca^{2+} concentration in the medium. Ca^{2+} ions stimulate phospholipase C activity at micromolar concentrations (data not shown).

The next question to be resolved was whether the suppression of agonist-stimulated PI turnover and $[Ca^{2+}]_i$ rise by protein kinase was due to direct inhibition of phospholipase C activity. In order to test this hypothesis, we incubated bovine pulmonary artery endothelial cells in culture with the protein kinase C stimulator phorbol-myristate-acetate (PMA) for 5 min. Control cells were incubated under the same conditions without PMA. Membranes from these cells were isolated, and phospholipase C was stimulated by GTPgammaS by more than 100% (Fig. 19.12). In contrast, in membranes isolated from cells treated with PMA, the stimulation of phospholipase C by

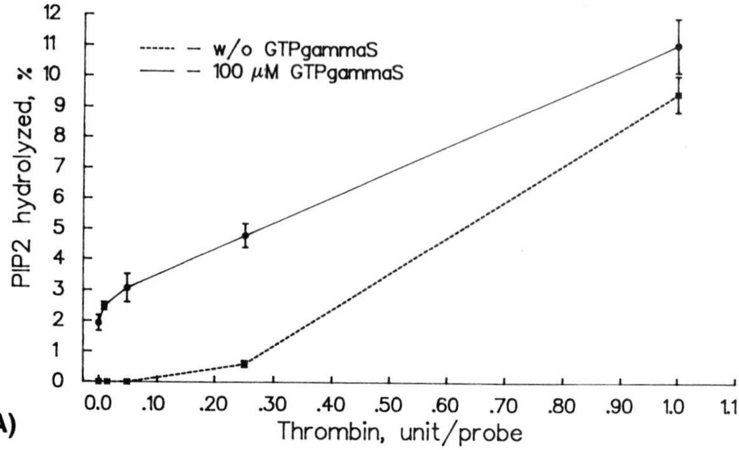

(A)

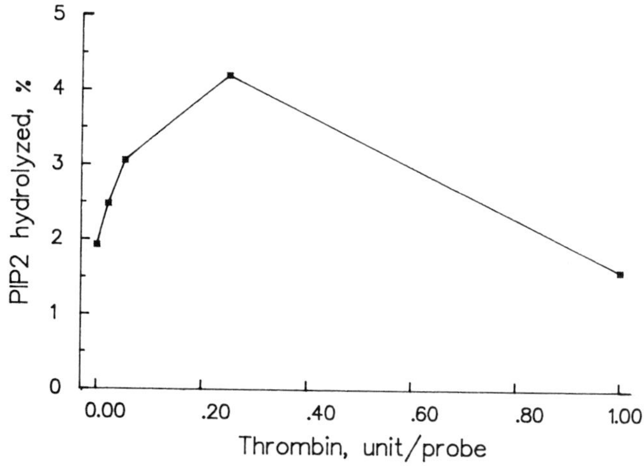

(B)

Figure 11 (A) The effect of different thrombin concentrations on phospholipase C activity in the absence (■—■) and presence of 100 μM GTP S (●—●). (B) GTPgammaS-dependent stimulation of phospholipase C by thrombin. The curve was obtained by subtraction of phospholipase C activity measured in the absence of GTPgammaS from this enzyme's activity measured in the presence of this nucleotide.

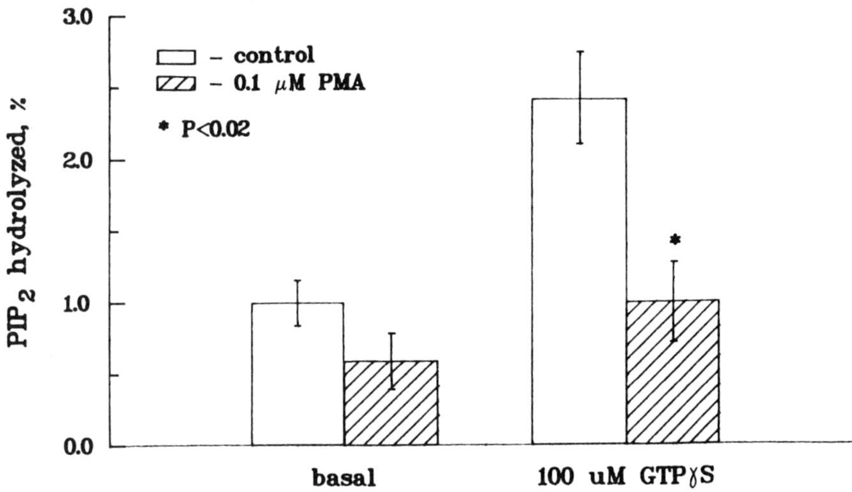

Figure 12 Inhibition of phospholipase C activity in bovine pulmonary artery endothelial cell membranes by endogenous protein kinase C. Membrane fractions were isolated from cells preincubated for 5 min with 1 μM PMA and in control conditions in the presence of DMSO.

GTPgammaS was abolished. Basal enzyme activity was also lower after treatment with PMA.

These results lead us to suggest the following scheme for PI turnover and $[Ca^{2+}]_i$ regulation in endothelial cells (Fig. 13). The initial event is the agonist-induced stimulation of membrane-bound phospholipase C by a GTP-dependent mechanism. Inositol-1,4,5-trisphosphate, formed in this reaction, triggers a $[Ca^{2+}]_i$ rise due to activation of Ca^{2+} channels on the endoplasmic reticulum. Simultaneously, extracellular Ca^{2+} enters the cells via plasma membrane Ca^{2+} channels. Mechanism of their activation by agonists is unclear. They may be opened directly by a receptor-G-protein complex or by IP_3 as observed in some cell types (Kuno and Gardner, 1986; Mozshaeva et al., 1990). The increase in intracellular Ca^{2+} concentration probably causes additional activation of phospholipase C. At the same time, calcium ions, together with the other product of the phospholipase C-catalyzed reaction, diacylglycerol, stimulate protein kinase C. This enzyme, according to our data, suppresses activation of membrane-bound phospholipase C by GTP-binding protein. As a result, it causes the termination of the endothelial cell response to calcium-mobilizing agonists. The available data do not permit us to explain how protein kinase C abolishes the stimulation of phospholipase C by G proteins. The data obtained in other cell types suggest that protein kinase C might phosphorylate the G protein directly (Katada et al., 1985; Carlson et al.,

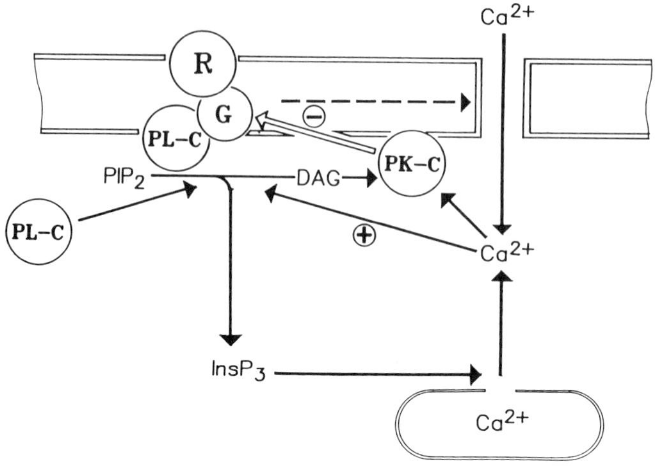

Figure 13 A model of phospholipase C and calcium channel regulation by endogenous protein kinase C.

1989). In addition to the membrane-bound phospholipase C, a soluble enzyme is demonstrated. The physiological role for this form of phospholipase C is unclear. It is possible that soluble phospholipase C also interacts with G protein(s) and mediates the agonist action on the cells. Soluble phospholipase C might represent isoform gamma, which is stimulated by the growth factors independently of a G protein.

VI. Acknowledgments

The authors are grateful to Z. Shumilina for technical assistance during preparation of the manuscript.

References

Avdonin PV, Hayes BA, Pozin Ey, Popov EG, Gavrilov Iy, Tkachuk VA, Ryan US. Dual-phase response of bovine pulmonary artery endothelial cells to agonists which increase free cytoplasmic calcium concentration. Tissue and Cell 1989; 21:171–8.

Avdonin PV, Hayes BA, Van Breeman C, Ryan US. Two phase regulation of cytosolic free calcium in endothelial cell response to agonists. FASEB J 1988; 2:A810.

Berridge MJ, Irvine RF. Inositol phosphates and cell signalling. Nature 1989; 341:197–205.

Berridge MJ, Dawson RMC, Downes CP, Heslop JP, Irvine RF. Changes in the levels of inositol phosphates after agonist-dependent hydrolysis of membrane phosphoinositides. Biochem J 1983; 212:473–82.

Carlson KE, Brass LF, Manning DR. Thrombin and phorbol ester cause the selective phosphorylation of a G protein other than G_i in human platelets. J Biol Chem 1989; 264:13298–305.

Connolly TM, Lawing WJ Jr, Majerus PW. Protein kinase C phosphorylates human platelets inositol phosphate 5′-phosphomonoesterase, increasing the phosphatase activity. Cell 1986; 46:951–8.

Crooke ST, Bennett CF. Mammalian phosphoinositide-specific phospholipase C isoenzymes. Cell Calcium 1989; 10:309–23.

Edwards DF, Griffith TM, Henderson AH, Lewis MG, Newby AC. Production of endothelium derived relaxant factor is both ATP and calcium dependent. Br J Pharmacol 1985; 85:344P.

Furchgott RF. The role of endothelium in the response of vascular smooth muscle to drugs. Ann Rev Pharmacol Toxicol 1984; 24:175–97.

Furchgott RF, Zawadski JV, Cherry PD. Role of endothelium in the vasodilator response to acetylchonline. In: Vanhoutte P, Leusen I, eds. Vasodilation. New York: Raven Press 1981; 49–66.

Grynkiewicz G, Poenie M, Tsien RY. A new generation of Ca^{2+} indicators with greatly improved fluorescence properties. J Biol Chem 1985; 260:3444–50.

Hallam TJ, Pearson JD. Exogenous ATP raises cytoplasmic free calcium in Fura-2 loaded piglet aortic endothelial cells. FEBS Lett 1986; 207:95–9.

Jacob R, Imagining cytoplasmic free calcium in histamine stimulated endothelial cells and in fMet-Leu-Phe stimulated neutrophils. Cell Calcium 1990; 11:241–9.

Jacob R, Merrit JE, Hallam TJ, Rink TJ. Repetitive spikes in cytoplasmic calcium evoked by histamine in human endothelial cells. Nature 1988; 335:40–5.

Katada T, Gilman AG, Watanabe Y, Bauer S, Jakobs KH. Protein kinase C phosphorylates the inhibitory guanine-nucleotide-binding regulatory component and apparently suppresses its function in hormonal inhibition of adenylate cyclase. Eur J Biochem 1985; 151:431–7.

Kuno M, Gardner P. Ion channels activated by inositol 1,4,5-triphosphate in plasma membrane of human T-lymphocytes. Nature 1987; 326:301–4.

Lazarowski ER, Lapetina EG. Persistent activation of platelet membrane phospholipase C by proteolytic action of trypsin and thrombin. Arch Biochem Biophys 1990; 276:265–9.

Long CJ, Stone T. The release of endothelium-derived relaxant factor is calcium dependent. Blood Vessels 1985; 22:205–8.

Luckhoff A, Busse R. Increased free calcium in endothelial cells under stimulation with adenine nucleotides. J Cell Physiol 1986; 126:414–20.

MacIntyre DE, McNicol A, Drummond AH. Tumor promoting phorbol esters inhibit agonist-induced phosphatidate formation and Ca^{2+} flux in human platelets. FEBS Lett 1985; 180:160–4.

Meldrum E, Parker PJ, Carozzi A. The PtdIns-PLC superfamily and signal transduction. Biochem Biophys Acta 1991; 1092:49–71.

Mignery GA, Sudhof TS. The ligand binding site and transduction mechanism in inositol-1,4,5-trisphosphate receptor. EMBO J 1990; 9:3893–8.

Moncada S, Vane JR. Prostacyclin (PGI_2), the vascular wall and vasodilation. In: Vanhoutte PM, Leusen I, eds. Mechanism of vasodilation. Basel: S. Karger 1978; 107–21.

Mozshaeva GN, Naumov AP, Kuryshev YA. Inositol 1,4,5-trisphosphate activates two types of Ca^{2+}-permeable channels in human carcinoma cells. FEBS Lett 1990; 277:233–4.

Murray JJ, Fridovich RG, Makhoul RG, Hagen P. Stabilization and partial characterization of endothelium derived relaxing factor from cultured bovine aortic endothelial cells. Biochem Biophys Res Commun 1986; 141:689–96.

Neylon CB, Irvine RF. Synchronized repetitive spikes in cytoplasmic calcium in confluent monolayers of human umbilical vein endothelial cells. FEBS Lett 1990; 275:173–6.

Nishibe S, Wahl MJ, Hermandez-Sotomayor SMT, Tonks NM, Rhee SG, Carpenter G. Increase of the catalytic activity of phospholipase C-gamma1 by tyrosine phosphorylation. Science 1990; 250:1253–6.

Palmer RMJ, Ferrige AG, Moncada S. Nitric oxide release accounts for the biological activity of endothelium-derived relaxing factor. Nature 1987; 327:524–6.

Rittenhouse-Simmons S. Production of diglyceride from phosphatidylinositol in activated human platelets. J Clin Invest 1979; 63:580–7.

Rotrosen D, Gallin JI. Histamine type I receptor occupancy increases endothelial cytosolic calcium, reduces F-action, and promotes albumin diffusion across cultured endothelial monolayers. J Cell Biol 1986; 103:2376–9.

Ryan US. Endothelial cell activation responses. In: Ryan US, ed. Lung biology in health and disease, Vol. 32, Pulmonary endothelium in health and disease. New York-Basel: Marcel Dekker, 1987; 3–62.

Ryan US, Avdonin PV, Posin E Ya, Popov EG, Danilov SM, Tkachuk VA. Influence of vasoactive agents on cytoplasmic free calcium in vascular endothelial cells. J Appl Physiol 1988; 65:2221–7.

Sage SO, Adams DJ, van Breeman C. Synchronized oscillations in cytoplasmic free calcium concentration in confluent bradykinin-stimulated pulmonary artery endothelial cell monolayers. J Biol Chem 1989; 264:6–9.

Singer HA, Peach MH. Calcium and endothelial-mediated vascular smooth muscle relaxation in rabbit aorta. Hypertension 1982; 4(Suppl.II):II-9-II-25.

Taylor SJ, Chae HZ, Rhee SG, Exton JH. Activation of the beta1 isozyme of phospholipase C alpha subunits of the Gg class of G protein. Nature 1991; 350:516–8.

Weksler BB, Ley CW, Jaffe EA. Stimulation of endothelial prostacyclin production by thrombin, trypsin, and the ionophore A23187. J Clin Invest 1978; 62:923–30.

Woods NM, Curthbertson KSR, Cobbold PH. Repetitive transient rises in cytoplasmic free calcium in hormone-stimulated hepatocytes. Nature 1986; 319:600–2.

20

Secretion Profiles of Vasoactive Amine-Induced Leukocyte Chemoattractants: Evidence for Endothelial Cell Heterogeneity

DENNIS J. BEER and HARRISON W. FARBER

Boston University School of Medicine
Boston, Massachusetts

I. Introduction

Vascular endothelial cells share certain common morphologic, metabolic, and physiologic features; however, there is striking heterogeneity among endothelial cells lining different vascular beds (Gerritsen, 1987; Zetter, 1988; Fajardo, 1989). This heterogeneity extends to cell surface antigenic and lectin-binding determinants, cellular adhesion molecules, receptor expression, bioenergetic enzyme activity, eicosanoid synthesis, cytokine secretion, and uptake, metabolism, and synthesis of vasoactive substances.

As part of our studies on functional differences amongst endothelial cells of varied origins, we have identified both lipid neutrophil-specific and protein lymphocyte-specific leukocyte chemoattractant factors derived from vasoactive amine-stimulated endothelial cells. There is heterogeneity of endothelial cell cytokine secretion that extends not only to endothelial cells obtained from different vascular beds of the same species, but also from endothelial cells harvested from different areas of the same vascular bed. In this chapter, we will describe different endothelial cell secretion profiles of vasoactive amine-induced leukocyte chemoattractants and relate these differences to potential pathophysiologic mechanisms of disease.

II. Endothelial Cell-Derived Lymphocyte Chemoattractant Activity

The accumulation of T lymphocytes at extralymphatic inflammatory foci represents the sequential effects of lymphocyte adhesion to vascular endothelium followed by chemoattractant-induced cell migration and growth factor-induced cell proliferation (Freemont, 1983; Jalkanen et al., 1986; Woodruff et al., 1987; Pober, 1988; Berman et al., 1990). Because many T-lymphocyte tissue responses are characterized by a lower percentage of antigen-specific cells (McCluskey et al., 1963), some element of amplification, generally attributed to secretion of multiple cytokines by mononuclear cells and by endothelial cells, is necessary for the full expression of a T-cell-dependent inflammatory response (Pober, 1988; Ryan, 1988; Berman et al., 1990). Of the numerous endogenous ligands capable of stimulating endothelial cell elaboration of pro-inflammatory cytokines, the vasoactive amines, originating from platelet and metachromatic cell degranulation or enzymatic cleavage of circulating proteins, are unique in their direct access to vascular endothelium.

A. Identification and Characterization

As assessed in a modified Boyden chamber assay, supernatants from bovine aortic endothelial cells (BAEC) incubated with 10^{-7} to 10^{-4} M serotonin enhanced human T cell migration (Fig. 1). The serotonin-stimulated secretion of this endothelial cell-derived lymphocyte chemotactic activity, ED-LCA, was blocked by serotonin type two ($5HT_2$) receptor antagonists. ED-LCA was not secreted following exposure of BAEC to histamine or angiotensin II, nor was it secreted by either serotonin-stimulated bovine pulmonary arterial (BPAEC) or human umbilical vein (HUVEC) endothelial cells (Figs. 1 and 2). Physiochemical characterization of ED-LCA demonstrated that it was a trypsin-sensitive protein with an apparent MW of 13–15 kDa (Fig. 3). Preparative isoelectric focusing demonstrated pIs of 6.0 and 7.5. Further investigation demonstrated that partially purified ED-LCA was specifically chemotactic for CD4+ and CD8+ T-lymphocyte subsets and did not enhance the migration of neutrophils or monocytes.

These in-vitro studies of bovine endothelial cell (Farber and Beer, 1991) were extended to human vascular endothelium (Farber et al., 1992a). As in the bovine endothelial cells, secretion of ED-LCA by human endothelial cells was greatest in response to 10^{-5} M serotonin; was inhibited by $5HT_2$ receptor antagonists; and was not induced by other vasoactive amines, such as histamine or angiotensin II. In addition, human ED-LCA had no effect on neutrophil or monocyte migration. Using human aortic endothelial cells (HAEC) and human

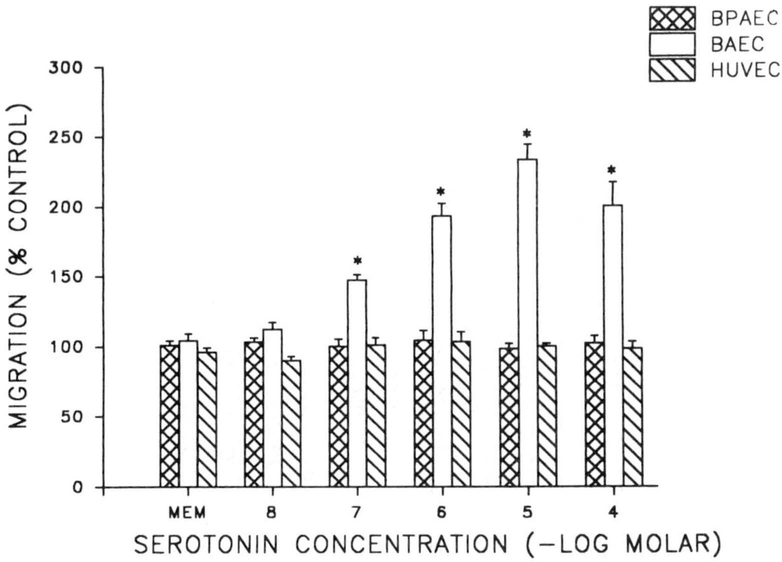

Figure 1 Bar graph showing secretion of endothelial cell-derived lymphocyte chemotactic activity by bovine aortic endothelial cells (BAEC). Confluent BAEC, bovine pulmonary arterial endothelial cells (BPAEC), and human umbilical vein (HUVEC) were incubated with minimal essential medium (MEM) alone or with varying concentrations of serotonin for 4 h, and the cell-free supernatants were assayed for T-lymphocyte chemoattractant activity. Data are mean ± S.E.M. for five experiments; * $p < 0.05$ compared with MEM. (Reproduced from Farber and Beer, 1991, with permission from the publisher.)

pulmonary arterial (HPAEC) endothelial cells from the same individual, we found that serotonin-stimulated HAEC, but not HPAEC, secreted ED-LCA.

Because of the demonstrated association between T-lymphocyte chemotactic and growth factor activity (Miossec et al., 1984; Kornfeld et al., 1985; Cruikshank et al., 1987; Akira et al., 1990; Heinrich et al., 1990), we (Katz et al., 1991) assessed the capability of ED-LCA to induce resting (G_0) human T cells to enter into the cell cycle. Incubation of G_0 freshly isolated human T lymphocytes for 48 h with chromatographically resolved, partially purified ED-LCA resulted in a threefold increase in cell surface expression of interleukin (IL)-2 receptors (IL-2R) and transferrin receptors and a twofold increase in HLA-DR. Evaluation of CD4+ and CD8+ T-cell subsets by double antibody labeling demonstrated that the IL-2R was induced in both T-cell subsets. This transition of resting T cells into the G_1 phase of the cell cycle was confirmed by cell cycle analysis demonstrating ED-LCA induction of

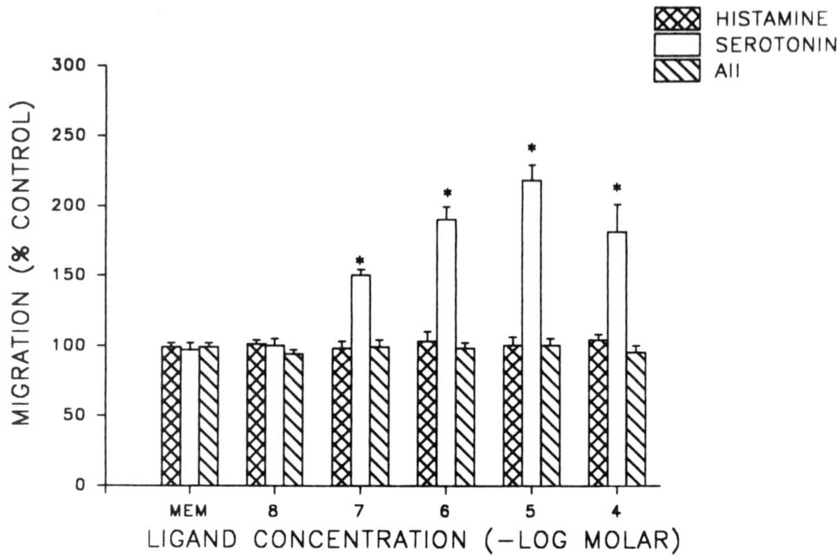

Figure 2 Bar graph showing induction of secretion of endothelial cell-derived lymphocyte chemotactic activity from bovine aortic endothelial cells (BAEC) by vasoactive amines. BAEC were incubated with minimal essential medium (MEM) alone or with varying concentrations of serotonin, histamine, angiotensin II (AII) for 4 h, and the cell-free supernatants were assayed for T-lymphocyte chemoattractant activity. Data are mean ± S.E.M. for five experiments; $*p < 0.05$ compared with MEM. (Reproduced from Farber and Beer, 1991, with permission from the publisher.)

increased cellular RNA but not DNA synthesis. Although incubation of human T cells with ED-LCA alone did not induce proliferation, addition of exogenous IL-2 to T cells pulsed with ED-LCA for 24 h caused a marked proliferative response (Fig. 4). Thus, by up-regulating functional cell surface receptors for IL-2, ED-LCA has been shown to be both a T-cell chemotaxin and a competence growth factor.

Although endothelial cells elaborate cytokines that affect many cell types (Gadjusek et al., 1980; DiCorleto et al., 1983; Jaye et al., 1985; Pober, 1988; Ryan, 1988), IL-1, IL-6, IL-8, and platelet-activating factor (PAF) are the only previously described endothelial cell-derived chemoattractants for T cells (Goetzel et al., 1980; Camussi et al., 1983; Libby et al., 1986; Wong and Clark, 1988; Shalaby et al., 1989; Larsen et al., 1989; Strieter et al., 1989; Gimbrone et al., 1989). ED-LCA, however, is distinct from these factors by both functional and physicochemical characteristics. ED-LCA is a protein with an apparent molecular weight of 13–15 kDa and is chemotactic for resting

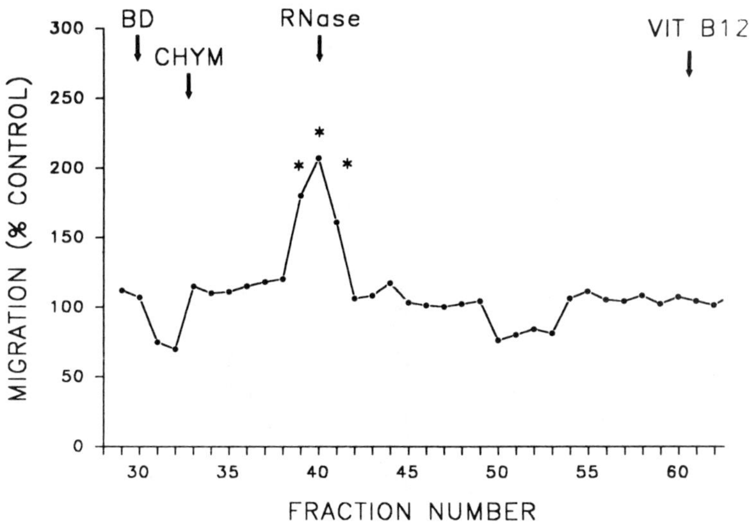

Figure 3 Graph showing representative G-50 molecular sieve column chromatography of endothelial cell-derived lymphocyte chemotactic activity. Pooled, dialyzed supernatants from serotonin-stimulated bovine aortic endothelial cells were eluted from a G-50 molecular sieve column, and each fraction was assayed for T-lymphocyte chemoattractant activity. Molecular mass markers included blue dextran (BD, 2000 kDa), chymotrypsin (CHYM, 25 kDa), ribonuclease (RNase, 13.7 kDa), and vitamin B_{12} (VIT B12, 1.2 kDa). *$p < 0.05$ compared with control. (Reproduced from Farber and Beer, 1991, with permission from the publisher.)

(G_0) CD4+ and CD8+ T lymphocytes, but not other leukocyte types. In contrast, although IL-1 has a similar molecular weight to ED-LCA, it is chemotactic for CD4+ but not CD8+ lymphocytes and is also chemotactic for neutrophils and monocytes (Miosecc et al. 1984; Oppenheim et al., 1986; Hunninghake et al., 1987). IL-6, a 26kD protein, is a T cell activator and chemoattractant, and also augments mitogen-induced proliferation of murine thymocytes (Heinrich et al., 1990; Akira et al., 1990). Crude serotonin-conditioned aortic endothelial cell supernatants and partially purified ED-LCA failed to augment PHA-stimulated proliferation of murine thymocytes or D10.G4.1 helper cells, thereby providing further functional evidence that ED-LCA is distinct from IL-1 and IL-6. IL-8, an 8.4-kDa protein, is chemotactic for both T lymphocytes and neutrophils (Larsen et al., 1989; Strieter et al., 1989; Gimbrone et al., 1989; Leonard and Yoshimura, 1990); however, the concentration necessary for chemoattraction of T cells is 2- to 10-fold less than that for chemoattraction of neutrophils (Larsen et al., 1989). We have recently

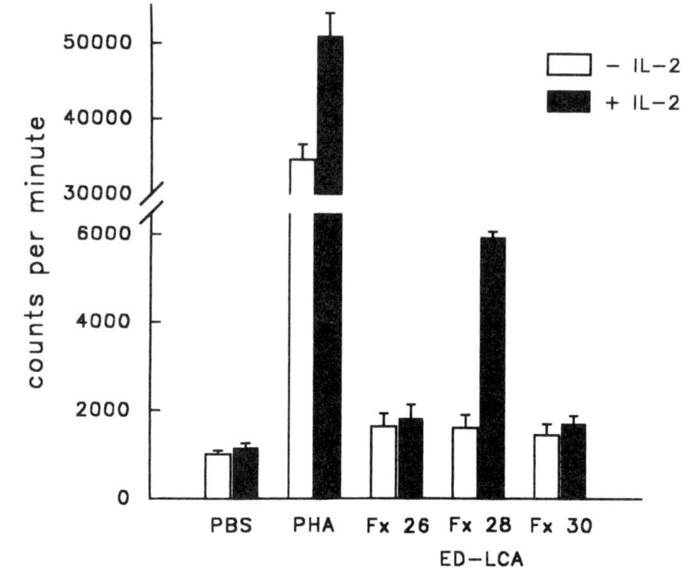

Figure 4 Interleukin (IL)-2-induced proliferation of human T cells pulsed with endothelial cell-derived lymphocyte chemotactic activity (ED-LCA). After 24 h of incubation with phosphate-buffered saline (PBS), 1 μg/mL phytohemagglutinin (PHA), 200 μL/mL of the two chemotactically inactive column fractions eluting just before or just after ED-LCA, or 200 μL/mL of partially purified ED-LCA, exogenous IL-2 (10^{-10} M final concentration) or control media was added to the T-cell cultures. After an additional 30 h of incubation, ^{3}H-thymidine was added and the samples were harvested at the end of a 72-h total incubation. T-lymphocyte proliferation is quantitated on the y axis s counts per minute (cpm).

observed that 10- to 100-fold dilutions of partially purified ED-LCA retained T-cell chemoattractant activity but contained no neutrophil chemoattractant activity (Farber HW, Beer DJ, unpublished observations). Lastly, anti-IL-1 and anti-IL-8 antibodies have no effect on ED-LCA-induced lymphocyte migration. Together, these findings suggest that ED-LCA is distinct from IL-1, IL-6, and IL-8. ED-LCA is physicochemically distinct from PAF, a phospholipid that is chemotactic for several different leukocyte types (Sturk et al., 1989).

B. Restricted Endothelial Cell Secretion Profile and Relationship to Atherosclerosis

The secretion of ED-LCA appears to be specific for both the ligand and the vascular bed from which the endothelial cells were isolated (Figs. 1 and 2).

Serotonin induced the secretion of ED-LCA from cultured aortic endothelial cells via interaction with $5HT_2$ cell surface receptors. The secretion of ED-LCA was not induced by the other vasoactive amines, histamine or angiotensin II, nor was its secretion blocked by either H_1 or H_2 receptor antagonists. Thus, despite the fact that aortic endothelial cells have been shown, by ligand binding and/or functional studies, to possess receptors for serotonin, histamine, and angiotensin II (Bottaro et al., 1986; Antohe et al., 1986; Alexander et al., 1987; Patel et al., 1989), only serotonin was capable of inducing secretion of ED-LCA from these cells.

A clinical consequence of ED-LCA production and its restriction to aortic endothelial cells might relate to atherosclerosis, a complex pathologic process involving predominantly large and medium-sized systemic arterial vessels. Although the initiating and perpetuating mechanisms which result in atherosclerotic lesions are not clearly understood, recent studies suggest that this process is similar to the inflammatory response in other tissues and may represent a response of the vascular wall to specific or nonspecific injury (Munro and Cotran, 1988).

In human and animal models of atherosclerosis, the fatty streak is the earliest recognized pathologic lesion. Recent studies have demonstrated that lymphocytes, along with macrophages, are an integral component of these fatty streaks and that lymphocyte accumulation is apparent even prior to the histologic appearance of fatty streaks (Munro et al., 1987; Emerson and Robertson, 1988). The stimulus for the appearance of lymphocytes in early atherosclerosis as well as their function in these early lesions is currently unknown. Lymphocytes, however, secrete cytokines which can activate macrophages (Steeg et al., 1982; Schreiber et al., 1983) and can induce expression of class II MHC (Ia) antigens in smooth muscle cells (Hansson et al., 1988). Thus, once present in early atherosclerotic lesions, lymphocytes, through interactions with macrophages and smooth muscle cells, may be important in the development of more extensive lesions in the arterial wall.

In some models of atherosclerosis, platelet adherence to vascular endothelium is thought to play an early and important role (Ross, 1986). The serotonin released by adhering platelets might stimulate the vascular endothelium to elaborate a T-cell chemotaxin, such as ED-LCA, leading to accumulation of T lymphocytes in the vascular wall. In this context, it is interesting that ED-LCA production is restricted to aortic endothelial cells, since it is the aorta, and not the pulmonary artery or umbilical vein, that is the usual site of atherosclerosis.

Both in vivo and in vitro, morphologic and phenotypic differences exist among endothelial cells from different areas of aorta depending on the degree of atherosclerotic involvement. In vivo, compared to endothelium from normal areas, endothelium from areas of vessels affected by atherosclerosis demonstrate differences in cellular morphology (Lewis et al., 1982; Repin et al., 1984), surface glycoproteins and glycolipids (Prokazova et al., 1989; Robbins et al., 1989), localization of apolipoproteins (Bedossa et al., 1989), and stress fibers (Guyton et al., 1989). In vitro, in animal models of atherosclerosis and in atherosclerotic human coronary arteries, impairment of endothelial-dependent vascular wall relaxation to a variety of agents has been demonstrated (Luscher, 1990). Likewise, in vitro, despite similarities in staining for Factor VIII, appearance of Weibel-Palade bodies, and lack of staining for anti-myosin antibodies, morphologic differences are evident in endothelial cells cultured from areas of human aorta involved differentially with atherosclerosis (Antonov et al., 1986). Compared to endothelial cells derived from normal areas of human aorta, endothelial cells derived from areas of aortas affected by either fatty streaks or plaques contain a greater number of multinucleated cells, more cells with a higher protein content and lipid inclusions, and a greater number of cells with increased DNA content (Antonov et al., 1986).

Interestingly, serotonin-stimulated ED-LCA secretion was related to the degree of atherosclerotic involvement of a segment of an individual vessel from which endothelial cells were harvested (Fig. 5) (Farber et al., 1992a).

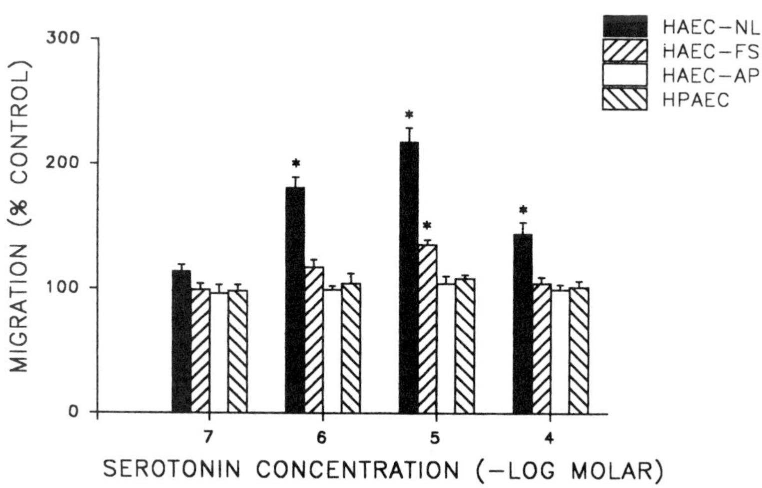

Figure 5 Secretion pattern of endothelial cell-derived lymphocyte chemoattractant activity by human endothelial cells. Human aortic endothelial cells (HAEC) derived from areas of aorta that were normal (NL), contained fatty streaks (FS), or contained atherosclerotic plaques (AP) and HPAEC were incubated with minimal essential medium (MEM) or with varying concentrations of serotonin for 4 h and the supernatants assayed for T-lymphocyte chemoattractant activity. Data are mean ± S.E.M. for six experiments; $*p < 0.05$ compared with MEM. (Reproduced from Farber et al., 1992a, with permission from the publisher.)

Endothelial cells harvested from normal areas of aorta secreted ED-LCA, cells from areas affected by fatty streaks secreted less ED-LCA, and cells from areas affected by plaques did not secrete ED-LCA. Diminished or absent ED-LCA in crude serotonin-stimulated supernatants from aortic endothelial cells derived from areas of fatty streaks or plaques was not a consequence of a shift in the dose response to serotonin. Furthermore, Sephadex G-50 column chromatography of crude serotonin-stimulated supernatants from endothelial cells derived from areas of aorta that were normal or contained fatty streaks demonstrated that ED-LCA eluted in a similar pattern. Additionally, no inhibitor of T-cell migration was found in column fractions or serotonin-stimulated supernatants from endothelial cells derived from areas of aorta containing fatty streaks or plaques.

The precise mechanism by which secretion of ED-LCA is altered in endothelial cells derived from areas of aorta affected by atherosclerosis is not clear. There was no difference in either baseline or serotonin-stimulated total protein synthesis among endothelial cells derived from any area of the aorta (Farber et al., 1992a). Thus, active protein synthesis is maintained within atherosclerotic endothelial cells; however, there may be a selective block of ED-LCA production which represents a small fraction of total celluar protein production. In addition, the absence of ED-LCA in lysates of serotonin-stimulated endothelial cells derived from areas containing fatty streaks or plaques suggests that this difference is not due to retention of active cytokine. Since secretion of ED-LCA from serotonin-stimulated normal aortic endothelial cells required interaction with $5HT_2$ receptors, it is possible that there is an alteration in these receptors or their signaling pathways in atherosclerotic endothelial cells. In endothelial cells derived from normal areas of aorta, serotonin-induced ED-LCA secretion was also blocked by imipramine, suggesting that uptake of serotonin is required for the secretion of ED-LCA. In this regard, the uptake of serotonin decreased as the atherosclerotic involvement of the area of aorta from which the HAEC were derived progressed. Further evaluation of this difference in endothelial cell secretory function will require a greater understanding of serotonin-induced signaling mechanisms as well as the development of specific probes for ED-LCA.

III. Endothelial Cell-Derived Lipid Neutrophil Chemoattractant Activity

Egress of neutrophils to a site of inflammation occurs in response to a chemoattractant gradient and requires interaction between neutrophils and vascular endothelial cells. Recent investigations have demonstrated that inflammatory cytokines and vasoactive amines are capable of modulating this interaction both by inducing expression of surface adhesion molecules (Bevilaqua et al., 1987; Tonnensen, 1989; Cotran and Pober, 1989; Mantovani and Dejana, 1989; Bevilaqua et al., 1989) and by production of neutrophil chemotaxins (Camussi et al., 1983; Farber et al., 1986, 1990; Gimbrone et al., 1989; Carveth et al., 1989; Streiter et al., 1989; Charles et al., 1991). The vasoactive amines, angiotensin II, histamine, and serotonin have all been implicated in the development of mononuclear cell inflammatory responses (Askenase et al., 1982; Beer et al., 1984; Weinstock, 1986; Weinstock et al., 1987; Owen, 1987; Silverman and Karnovsky, 1989; Sedgewick and Busse, 1990), and, more recently, their potential role in neutrophil-predominant inflammation has become clearer.

A. Angiotensin System

As assessed in a modified Boyden chamber assay, angiotensin I induced the secretion of neutrophil chemoattractant activity (NCA) from BAEC, BPAEC, and HUVEC in a concentration-dependent manner peaking at 10^{-5} M (Farber et al., 1985, 1990). Addition of either the angiotensin converting enzyme inhibitor, captopril, or the angiotensin II antagonist, saralasin, to endothelial cell monolayers blocked production of angiotensin I-stimulated NCA. Angiotensin II, in a concentration-dependent manner also peaking at 10^{-5} M, stimulated the secretion of NCA from BAEC, BPAEC, HUVEC, HAEC, and HPAEC (Fig. 6) (Farber et al., 1990). Taken together, these findings demonstrate that conversion of angiotensin I to angiotensin II is necessary for induction of NCA. Incubation of endothelial cells with bradykinin, another vasoactive amine metabolized by angiotensin converting enzyme, either alone or with captopril, did not cause secretion of NCA (Farber et al., 1985).

Further evaluation of this NCA defined it as a group of chemoattractant lipids (Farber et al., 1990, 1992b). NCA appeared in endothelial cell supernatants within 1 min of incubation with angiotensin II; maximal activity was present after 60 min. Coincubation with cycloheximide, an inhibitor of protein synthesis, did not block appearance of NCA. Addition of cyclooxygenase inhibitor, indomethacin, had no effect on angiotensin II-induced NCA, whereas the addition of either lipoxygenase inhibitor, diethylcarbamazine (DEC), or 5,8,11,14 eicosatetraynoic acid (ETYA) blocked its generation. Following extraction with diethyl ether, all NCA appeared in the organic phase; none remained in the aqueous phase. Reverse-phase high-pressure liquid chromatography (HPLC) resolution of the organic phases of extracted angiotensin II-stimulated endothelial cell supernatants demonstrated two major peaks of chemoattractant activity from HUVEC, HAEC, or HPAEC; three peaks of chemoattractant activity were obtained from BAEC or BPAEC, two of which appear similar to those observed from human endothelial cells (Fig. 7). Each of these HPLC-resolved peaks demonstrated gradient-dependent chemotactic

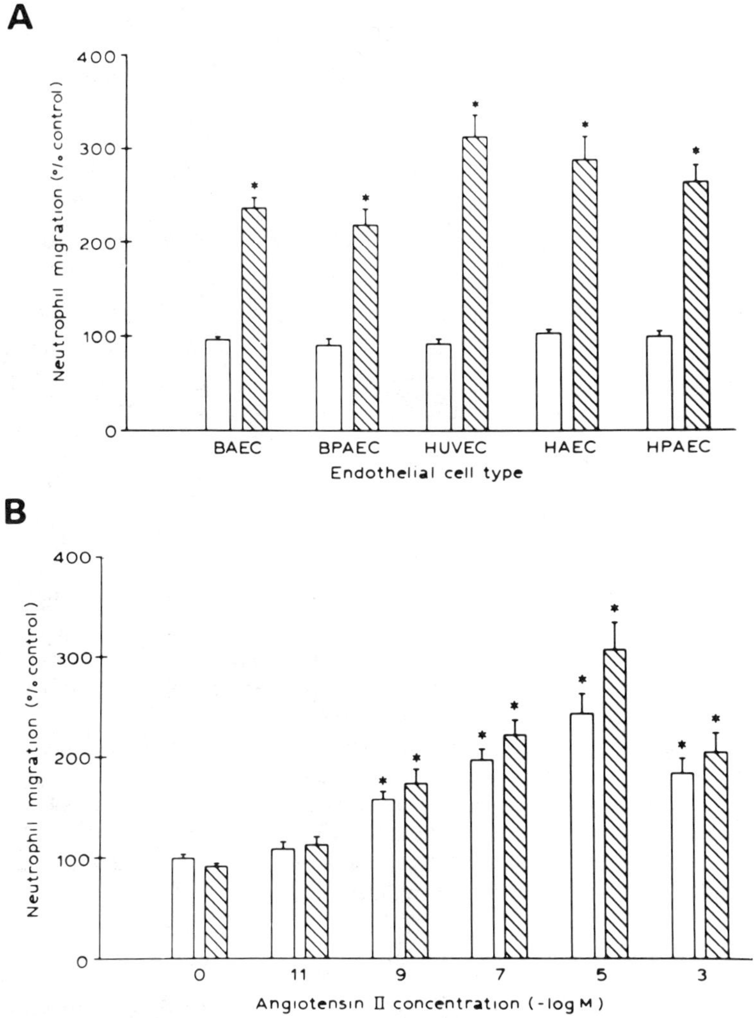

Figure 6 Bar graphs showing neutrophil chemoattractant activity production by the different endothelial cell types (A) and the concentration-dependent response to angiotensin II (B). In (A), bovine aortic (BAEC) and pulmonary arterial (BPAEC) and human umbilical vein (HUVEC), aortic (HAEC), and pulmonary arterial (HPAEC) endothelial cells were incubated for 4 h with buffer alone (open bars) or angiotensin II (10^{-5} M) (closed bars); and in (B), BAEC (open bars) and HUVEC (hatched bars) were incubated for 4 h with varying concentrations of angiotensin II. The cell-free supernatants were assessed for neutrophil chemoattractant activity and results expressed as a percentage of the neutrophil response to buffer alone (mean ± S.E.M.; monolayers tested = 4; *$p < 0.05$ compared to buffer alone). (Reproduced from Farber et al., 1990, with permission from the publisher.)

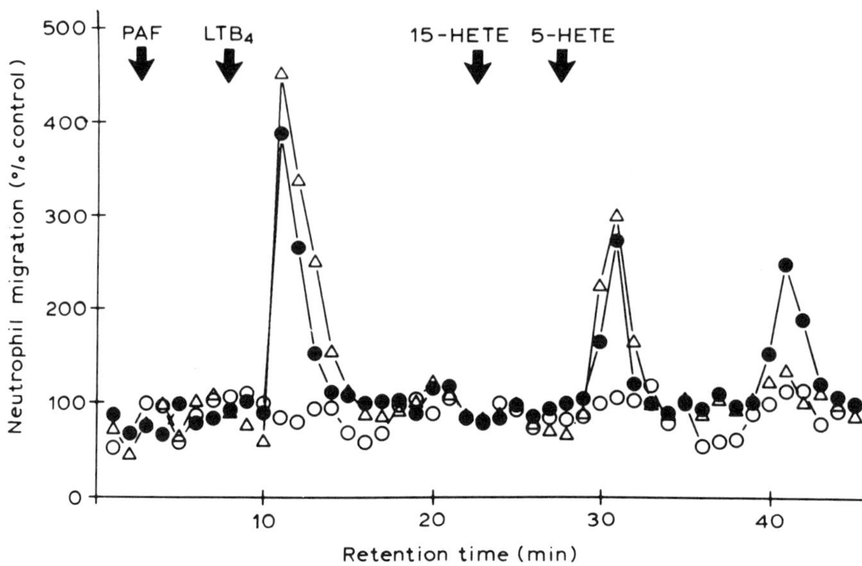

Figure 7 Representative HPLC resolution of supernatants from angiotensin II (AII)-stimulated bovine aortic (BAEC; closed circles) and human pulmonary arterial (HPAEC; open triangles) endothelial cells and supernatants from unstimulated (minimal essential medium; MEM) endothelial cells (open circles). After extraction, the organic phase was resolved by reverse-phase HPLC utilizing a methanol/water/acetic acid (78:22:0.01) solvent system. One-milliliter fractions were collected and assessed for neutrophil chemoattractant activity. Results are expressed as a percentage of the neutrophil response found in the initial fractions eluted from the HPLC column. The arrows represent the retention time of standard ^{3}H-PAF, ^{3}H-LTB$_4$, ^{3}H-15-HETE, and ^{3}H-5-HETE determined in replicate reverse-phase HPLC analysis. (Reproduced from Farber et al., 1990, with permission from the publisher.)

activity for neutrophils, but not for other leukocyte types. None of the chemotactic activities had a retention time corresponding to the retention time of synthetic ^{3}H-PAF, ^{3}H-leukotriene (LT) B$_4$, ^{3}H-15-hydroxyeicosatetranoic acid (HETE), or ^{3}H-5-HETE determined in replicate reverse-phase HPLC analyses.

B. Histamine

Similar to angiotensin, histamine stimulated both bovine and human endothelial cells to elaborate chemoattractant activity for human neutrophils in a concentration-dependent manner peaking at 10^{-5} M (Farber et al., 1986, 1992b). Inclusion of diphenhydramine, a histamine-type 1 (H_1) receptor antagonist, did not diminish the production of NCA, whereas the addition of the H_2 receptor antagonist, cimetidine, blocked NCA production.

Histamine stimulated endothelial cells to elaborate NCA within 1 min; maximal activity was present by 45 min. Cycloheximide did not inhibit the appearance of histamine-induced NCA. Addition of indomethacin did not inhibit NCA production from bovine or human endothelial cells, whereas histamine-induced NCA production was completely blocked by ETYA or DEC. Following extraction with diethyl ether, all NCA appeared in the organic phase; none remained in the aqueous phase. HPLC resolution demonstrated two major peaks of chemotactic activity from bovine and human endothelial cells (Fig. 8). A third peak of chemotactic activity was noted on HPLC-resolution of extracted supernatants from BAEC, BPAEC, and HUVEC. Each peak of chemotactic activity affected neutrophil, but not lymphocyte or monocyte, migration, but none had a retention time corresponding to the retention time of synthetic ^{3}H-PAF, ^{3}H-LTB4, ^{3}H-15-HETE, or ^{3}H-5-HETE determined in replicate reverse-phase HPLC analyses.

C. Serotonin

Similar to both angiotensin and histamine, serotonin caused both bovine and human endothelial cells to secrete chemoattractant activity for human neutrophils in a concentration-dependent manner peaking at 10^{-5} M (Charles et al., 1991; Farber et al., 1992b) (Fig. 9). Although each endothelial cell type produced NCA in response to incubation with serotonin, the mechanism of induction was variable depending on the origin of the endothelial cells. NCA production was blocked by the nonspecific $5HT_1/5HT_2$ receptor antagonist, methiothepin, in the two cell types examined (HAEC and HPAEC), by the $5HT_1$ receptor antagonist, buspirone, in BAEC and HUVEC, and by either $5HT_2$ receptor antagonist, ketanserin or LY53587, in BPAEC or HPAEC. These data suggest that NCA production occurs by interaction with $5HT_1$ receptors in BAEC, HAEC, and HUVEC and by interaction with $5HT_2$ receptors in BPAEC and HPAEC. Despite the difference in the serotonin receptor involved in the production of NCA, receptor interaction was specific, since neither diphenhydramine nor cimetidine blocked NCA production in any of the endothelial cell types. In those cell types in which NCA production was mediated by $5HT_2$ receptors, imipramine, an inhibitor of serotonin uptake, also blocked NCA production, whereas verapamil, a calcium-channel blocker that also acts as a competitive inhibitor of serotonin uptake by non-calcium-dependent mechanisms, blocked its production in all cell types. Further investigation demonstrated that serotonin itself, and not a metabolite, is important in the production of NCA. Co-incubation with iproniazid, an inhibitor of the intracellular metabolism of serotonin by monoamine oxidase, did not affect

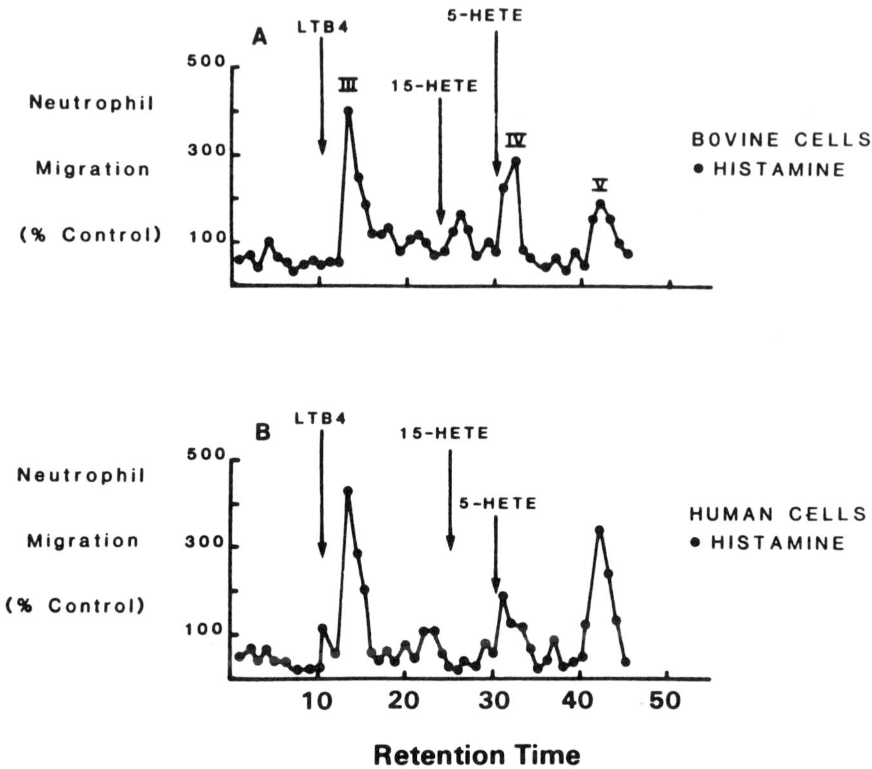

Figure 8 Representative HPLC resolution of supernatants from histamine-stimulated bovine aortic endothelial cells (BAEC) (A) and human umbilical vein endothelial cells (HUVEC) (B). After extraction, the organic phase was resolved by reverse-phase HPLC utilizing a methanol/water/acetic acid (78:22:0.01) solvent system. One-milliliter fractions were collected and assessed for neutrophil chemoattractant activity. The results are expressed as a percentage of the neutrophil response found in the initial two fractions eluted from the HPLC column. The arrows in each panel represent the retention time of standard ^{3}H-LTB$_4$, ^{3}H-15-HETE, and ^{3}H-5-HETE, determined in replicate reverse-phase HPLC analysis. (Reproduced from Farber et al., 1986, with permission from the publisher.)

production of NCA; in addition, the product of this enzyme, 5-hydroxyindolacetic acid, had no direct effect on neutrophil migration.

In all cell types, NCA appeared within 5 min of incubation with serotonin, and maximal activity was present by 30 min. Induction of NCA by serotonin, as opposed to induction by angiotensin or histamine, apparently required de-novo protein synthesis, since both cycloheximide and actinomycin D

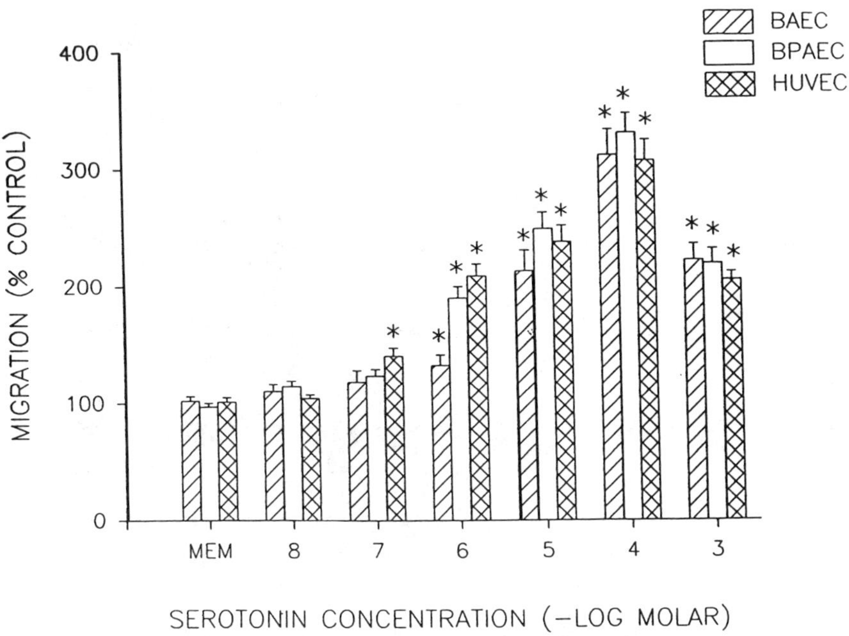

Figure 9 Bar graph showing production of neutrophil chemoattractant activity by serotonin-stimulated bovine and human endothelial cells. Bovine aortic (BAEC) and pulmonary arterial (BPAEC), and human umbilical vein (HUVEC) endothelial cells were incubated with minimal essential medium (MEM) or with varying concentrations of serotonin for 4 h and the cell-free supernatants assayed for neutrophil migration. Results are mean ± S.E.M.; n = 5 or more for each point. *$p < 0.05$ compared to MEM. (Reproduced from Charles et al., 1991, with permission from the publisher.)

blocked its production. As before, addition of indomethacin had no effect on NCA production, whereas addition of DEC, ETYA, or a third lipoxygenase inhibitor, nordihydroguaiaretic acid, completely blocked its production. Following extraction with diethyl ether, all NCA partitioned into the organic phase; none remained in the aqueous phase. HPLC resolution of the organic phase of extracted serotonin-stimulated HUVEC, HAEC, or HPAEC supernatants yielded two major peaks of chemotactic activity; three peaks of chemotactic activity were obtained from serotonin-stimulated BAEC or BPAEC (Fig. 10) As before, these peaks of chemotactic activity affected neutrophil, but not lymphocyte or monocyte, migration, and none had a retention time corresponding to the retention time of synthetic ^{3}H-PAF, ^{3}H-LTB4, ^{3}H-15-HETE, or ^{3}H-5-HETE determined in replicate reverse-phase HPLC analyses.

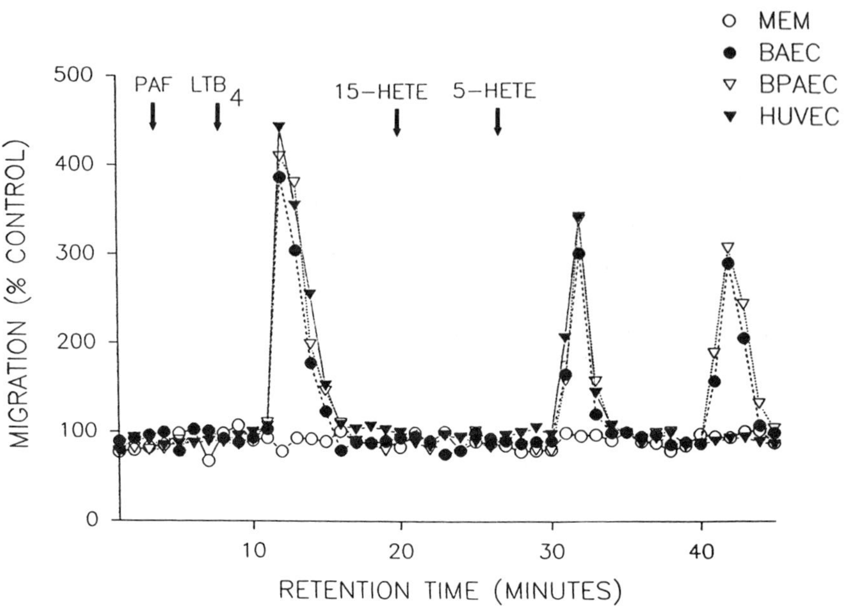

Figure 10 Representative HPLC resolution of supernatants from serotonin-stimulated bovine aortic (BAEC) and pulmonary arterial (BPAEC) and human umbilical vein (HUVEC) endothelial cells and supernatants from unstimulated (minimal essential medium; MEM) endothelial cells. After extraction, the organic phase was resolved by reverse-phase HPLC utilizing a methanol/water/acetic acid (78:22:0.01) solvent system. One-milliliter fractions were collected and assessed for neutrophil chemoattractant activity. The results are expressed as a percentage of the neutrophil response found in the initial fractions eluted from the HPLC column. The arrows represent the retention time of standard ^{3}H-PAF, ^{3}H-LTB_4, ^{3}H-15-HETE, and ^{3}H-5-HETE determined in replicate reverse-phase HPLC analysis. (Reproduced from Charles et al., 1991, with permission of the publisher.)

D. Heterogeneity in Endothelial Cell Production of Neutrophil Chemoattractant Activity

The endothelial cell-derived NCA induced by each of these vasoactive amines has several similar characteristics, including the concentration of amine that induces its maximal production, the time course of its appearance, its lipophilic nature, its resolution on HPLC, and its specificity for neutrophils. In addition, in each case, HPLC analysis demonstrated that NCA is a group of several chemotactic lipids, each of which is different from previously defined eicosanoid or phospholipid neutrophil chemoattractants. Moreover, despite the fact that these neutrophil-specific lipids have not been conclusively identified, they are similar in that each has an ultraviolet spectrum with both peaks and shoulders less than 235 nm and in each the existence of double bonds and carboxyl and/or hydroxyl groups is required for bioactivity (Farber, unpublished observations).

Despite these numerous similarities, depending on the species, site, or ligand, there is heterogeneity in vasoactive amine induction of NCA. Induction of NCA by serotonin, but not by angiotensin or histamine, was blocked by the addition of cycloheximide, suggesting that serotonin induction of NCA requires de-novo protein synthesis. The reason for this difference is not clear, but might represent the need for newly synthesized protein in elaboration, packaging, or release of the serotonin-stimulated NCA. In addition, induction of NCA by serotonin is mediated by different mechanisms depending on the vascular bed of origin of the endothelial cells. In BAEC, HAEC, and HUVEC, NCA production was mediated by interaction with $5HT_1$ receptors; in BPAEC and HPAEC, production of NCA occurred via $5HT_2$ receptors. Experiments with imipramine apparently demonstrated that the transmembrane transport of serotonin was important in NCA production only in pulmonary arterial endothelial cells; however, interpretation of these results is likely more complex. Serotonin, at the concentrations used in these experiments, will enter endothelial cells in significant amounts without involving the high-affinity carrier-mediated transport that is blocked by imipramine; yet imipramine was able to block NCA production in pulmonary arterial endothelial cells. When serotonin was used at the minimal concentration that would induce NCA production, imipramine still blocked NCA production only in bovine pulmonary arterial endothelial cells (Charles, Rounds, and Farber, unpublished observations). These findings suggest that imipramine may have inhibitory actions in bovine pulmonary arterial endothelial cells other than its effect on serotonin transport. Although the exact mechanisms by which interaction with serotonin results in the production of NCA are not yet known, these studies demonstrate that there is heterogeneity in some aspects of this interaction, depending on the vascular bed of origin of the endothelial cells.

These studies also demonstrate that there may be ligand, site, and species specificity in production of chemotactic lipids. Following resolution by HPLC, the two initial peaks of chemotactic activity from amine-stimulated BAEC, BPAEC, HUVEC, HAEC, and HPAEC eluted in a similar pattern compared to reference standards and, in each case, were specific for neutrophils but did not co-elute with other described lipid or phospholipid neutrophil chemoattractants, such as PAF, LTB_4, 5-HETE, or 15-HETE. A third peak of chemotactic activity, consistently produced by BAEC or BPAEC incubated with any of the three amines (Farber et al., 1986, 1990; Charles et al., 1991) or HUVEC incubated with histamine (Farber et al., 1986), was never elaborated by HAEC or HPAEC. These latter differences imply that BAEC, BPAEC, and possibly

Table 1 Vasoactive Amine-Induced Leukocyte Chemotactic Activities

		Vascular endothelial cells					
Chemotactic activity	Ligand	BAEC	BPAEC	HAEC	HPAEC	HUVEC	Characteristics
Lymphocyte chemotactic activity	Serotonin	+	−	+	−	−	13- to 15-kDa secreted protein with pI's 6.0 and 7.5; chemotactic for CD4+ and CD8+ lymphocytes; not chemotactic for neutrophils or monocytes; possesses T-cell growth factor activity; diminished secretion from endothelial cells affected by atherosclerosis
	Histamine	−	−	−	−	−	
	Angiotensin II	−	−	−	−	−	
Neutrophil chemotactic activity	Serotonin	+	+	+	+	+	Secreted family of de-novo synthesized lipids distinct from previously defined eicosanoids and phospholipids; for BAEC and BPAEC, three peaks resolved by HPLC whereas for HAEC and HPAEC, two peaks resolved by HPLC; in BAEC, HAEC, and HUVEC, serotonin type-1 receptor mediated; in BPAEC and HPAEC, serotonin type-2 receptor mediated
	Histamine	+	+	+	+	+	Similar to serotonin, except three peaks with HUVEC; histamine type-2 receptor mediated
	Angiotensin II	+	+	+	+	+	Similar to serotonin

BAEC = bovine aortic endothelial cells; BPAEC = bovine pulmonary artery endothelial cells; HAEC = human aortic endothelial cells; HPAEC = human pulmonary artery endothelial cells; HUVEC = human umbilical vein endothelial cells; HPLC = high-pressure liquid chromatography.

HUVEC, have lipid synthetic or degradative enzymes or a lipid precursor pool different from those of HAEC or HPAEC.

IV. Conclusion

Based on the secretion profiles of vasoactive amine-induced leukocyte chemoattractants, our investigations define intercellular signaling networks involved in leukocyte-specific inflammatory responses and heterogeneity of endothelial cell responses (summarized in Table 1). Although endothelial cells from varied arterial beds possess receptors for multiple vasoactive amines, the restricted secretion of LCA by serotonin-stimulated aortic, but not by pulmonary arterial, endothelial cells suggests the existence of heterogeneity of cell responsiveness depending on vessel of origin. The diminished secretion of LCA from aortic endothelial cells affected by atherosclerosis suggests that the ability of this differentiated cell type to synthesize or secrete cytokines may be lost as a consequence of disease. Further, although both aortic and pulmonary arterial endothelial cells possess both serotonergic type-1 and type-2 receptors, the secretion of NCA from aortic endothelial cells is mediated via a type-1 receptor whereas secretion of NCA from pulmonary artery endothelial cells is mediated via type-2 receptors, implying different mechanisms of intracellular signaling for the secretion of the same family of molecules, depending on organ of origin. Last, the ability of HUVEC, BAEC, and BPAEC to synthesize three chemotactic lipids while HAEC and HPAEC synthesize only two of these suggests that there are other differentially expressed functions of endothelial cells depending on both anatomic site of harvest and species of origin.

Acknowledgments

This work was supported by grant HL-32802 from the National Institutes of Health (D.J.B.) and by a grant from the Massachusetts Affiliate of the American Heart Association (H.W.F.). D.J. Beer is a recipient of a Career Investigator Award from the American Lung Association.

References

Akira S, Hirano T, Taga T, Kishimoto T. Biology of multifunctional cytokines: IL-6 and related molecules (IL-1 and TNF). FASEB J 1990; 4:2860–7.

Alexander JS, Hechtman HB, Shepro D. Serotonin induced actin polymerization and association with cytoskeletal elements in cultured bovine aortic endothelium. Biochem Biophys Res Commun 1987; 143:152–8.

Antohe F, Heltianu C, Simionescu N. Further evidence for the distribution and nature of histamine receptors on microvascular endothelium. Microcirculation, Endothelium, and Lymphatics. 1986; 3:163–85.

Antonov AS, Nikolaeva MA, Klueva TS, Romanov YA, Babaev VR, Bystrevskaya VB, Perov NA, Repin VS, Smirnov VN. Primary culture of endothelial cells from atherosclerotic human aorta. Part 1. Identification, morphological and ultrastructural characteristics of two endothelial cell populations. Atherosclerosis 1986; 59:1–19.

Askenase PW, Metzler CM, Gershon RK. Localization of leucocytes in sites of delayed-type hypersensitivity and in lymph nodes: dependence on vasoactive amines. Immunology 1982; 47:239–46.

Bedossa P, Poynard T, Abella A, Paraf F, Lemaigre G, Martin E. Localization of apolipoprotein A-I and apolipoprotein A-II in human atherosclerotic arteries. Arch Pathol Lab Med 1989; 113:777–80.

Beer DJ, Matloff SM, Rocklin RE. The influence of histamine on immune and inflammatory responses. In: Dixon FJ, Kunkel HG, eds. Advances in immunology, Vol 35. New York: Academic Press 1984; 206–68.

Berman JS, Beer DJ, Theodore AC, Kornfeld H, Bernardo J, Center DM. Lymphocyte recruitment to the lung. Am Rev Respir Dis 1990; 142:238–57.

Bevilacqua MP, Pober JS, Mendrick DL, Cotran RS, Gimbrone MA Jr. Identification of an inducible endothelial-leukocyte adhesion molecule. Proc Natl Acad Sci USA 1987; 84:9238–42.

Bevilacqua MP, Stengelin S, Gimbrone MA Jr, Seed B. Endothelial leukocyte adhesion molecule 1: An inducible receptor for neutrophils related to complement regulatory proteins and lectins. Science 1989; 243:1160–5.

Bottaro D, Shepro D, Peterson S, Hechtman HB. Serotonin, norepinephrine, and histamine mediation of endothelial cell barrier function in vitro. J Cell Physiol 1986; 128:189–94.

Camussi G, Aglietta M, Malavasi F, Tetta C, Piacibello W, Sanavio F, Bussolino F. The release of platelet-activating factor from human endothelial cells in culture. J Immunol 1983; 131:2397–403.

Carveth HJ, Bohnsack JF, McIntyre TM, Baggioloini M, Prescott SM, Zimmerman GA. Neutrophil activating factor (NAF) induces polymorphonuclear leukocyte adherence to endothelial cells and to subendothelial matrix proteins. Biochem Biophys Res Commun 1989; 162:387–93.

Charles A, Rounds S, Farber HW. Production of a neutrophil cytokine by serotonin-stimulated cultured bovine and human endothelial cells. Am J Physiol 1991; 261:L133–9.

Cotran RS, Pober JS. Effects of cytokines on vascular endothelium: Their role in vascular and immune injury. Kidney Int 1989; 35:969–75.

Cruikshank WW, Berman JS, Theodore AC, Bernardo J, Center DM. Lymphokine activation of $T4^+$ lymphocytes and monocytes. J Immunol 1987; 138:3817–23.

DiCorleto PE, Gadjusek CM, Schwartz SM, Ross R. Biochemical properties of the endothelium-derived growth factor: Comparison to other growth factors. J Cell Physiol 1983; 114:339–45.

Emerson EE, Robertson AL Jr. T lymphocytes in aortic and coronary intimas: Their role in atherogenesis. Am J Pathol 1988; 130:826–32.

Fajardo LF. The complexity of endothelial cells. Am J Clin Pathol 1989; 92:241–50.

Farber HW, Beer DJ. Restricted secretion of a T-lymphocyte chemotactic cytokine by serotonin-stimulated cultured aortic endothelial cells. Circ Res 1991; 69:257–65.

Farber HW, Antonov AS, Romanov YA, Smirnov VN, Scarfo LM, Beer DJ. Cytokine secretion by human aortic endothelial cells is related to the degree of atherosclerosis. Am J Physiol 1992a; 262:H1088–95.

Farber HW, Center DM, Rounds S. Bovine and human endothelial cell production of neutrophil chemoattractant activity in response to components of the angiotensin system. Circ Res 1985; 57:898–902.

Farber HW, Center DM, Rounds S, Danilov SM. Components of the angiotensin system cause release of a neutrophil chemoattractant from bovine and human endothelial cells. Eur Heart J 1990; 11 (supp. B):100–7.

Farber HW, Kozlova M, Moldobalva A, Charles A, Danilov SM. Production of neutrophil-specific lipid chemoattractant activity by cultured endothelial cells: Heterogeneity dependent on species, ligand, or endothelial cell site of origin. Tissue & Cell 1992b; 24:355–66..

Farber HW, Weller PF, Rounds S, Beer DJ, Center DM. Generation of a lipid neutrophil chemoattractant activity by histamine-stimulated cultured endothelial cells. J Immunol 1986; 137:2918–24.

Freemont AJ. A possible route for lymphocyte migration into diseased tissues. Clin Pathol 1983; 36:161–6.

Gadjusek C, DiCorleto P, Ross R, Schwartz SM. An endothelial cell-derived growth factor. J Cell Biol 1980; 85:467–72.

Gerritsen ME. Functional heterogeneity of vascular endothelial cells. Biochem Pharmacol 1987; 36:2701–11.

Gimbrone MA Jr, Obin MS, Brock AF, Luis EA, Hass PE, Hebert CA, Yip YK, Leung DW, Lowe DG, Kohr WJ, Darbonne WC, Bechtol KB, Baker JB. Endothelial interleukin-8: A novel inhibitor of leukocyte-endothelial interactions. Science 1989; 246:1601–3.

Goetzl EJ, Derian DK, Tauber AI, Valone FH. Novel effects of 1-O-hexadecyl-2-acyl-sn-glycero-3-phosphorylcholine mediators on human leukocyte function: Delineation of the specific roles of the acyl substituents. Biochem Biophys Res Commun 1980; 94:881–8.

Guyton JR, Shaffer DR, Henry PD. Stress fibers in endothelial cells overlying atherosclerotic lesions in rabbit aorta. Am J Med Sci 1989; 298:79–82.

Hansson GK, Jonassen L, Holm J, Clowes MM, Clowes AW. γ-Interferon regulates vascular smooth muscle proliferation and Ia antigen expression in vivo and in vitro. Circ Res 1988; 63:712–9.

Heinrich PC, Castell JV, Andus T. Interleukin-6 and the acute phase response. Biochem J 1990; 265:621–36.

Hunninghake GW, Glazier AJ, Monick MM, Dinarello CA. Interleukin-1 is a chemotactic factor for human T-lymphocytes. Am Rev Respir Dis 1987; 135:66–71.

Jalkanen S, Steere AC, Fox RI, Butcher EC. A distinct endothelial cell recognition system that controls lymphocyte traffic into inflamed synovium. Science 1986; 233:556–8.

Jaye M, Howk R, Burgess W, Ricca GA, Chiu IM, Ravera MW, O'Brien SJ, Modi WS, Maciag T, Drohan WN. Human endothelial cell growth factor: cloning, nucleotide sequence, and chromosome location. Science 1985; 233:541–5.

Katz MF, Farber MW, Cruikshank WW, Beer DJ. Serotonin-stimulated cultured aortic endothelial cells secrete a novel T-lymphocyte chemotactic cytokine possessing growth factor activity, 1991; submitted.

Kornfeld H, Berman JS, Beer DJ, Center DM. Induction of human T lymphocyte motility by interleukin 2. J Immunol 1985; 134:3887–90.

Larsen CG, Anderson AO, Appella E, Oppenheim JJ, Matsushima K. The neutrophil-activating protein (NAP-1) is also chemotactic for T lymphocytes. Science 1989; 243:1464–6.

Leonard EJ, Yoshimura T. Neutrophil attractant/activation protein-1 (NAP-1 [interleukin-8]). Am J Respir Cell Mol Biol 1990; 2:479–86.

Lewis JC, Taylor RG, Jones ND, St. Clair RW, Cornhill JF. Endothelial surface characteristics in pigeon coronary artery atherosclerosis, part 1 (cellular alterations during the initial stages of dietary cholesterol challenge). Lab Invest 1982; 46:123–38.

Libby P, Ordovas JM, Auger KR, Robbins AH, Birinyi LK, Dinarello CA. Endotoxin and tumor necrosis factor induce interleukin-1 gene expression in adult human vascular endothelial cells. Am J Pathol 1986; 124:179–85.

Luscher TF. Functional abnormalities of the vascular endothelium in hypertension and atherosclerosis. Scand J Clin Lab Invest 1990; 50(Suppl. 199):28–32.

Mantovani A, Dejana E. Cytokines as communication signals between leukocytes and endothelial cells. Immunol Today 1989; 10:370–5.

McClusky RT, Benacerraf B, McClusky JW. Studies on the specificity of the cellular infiltrate in delayed hypersensitivity reactions. J Immunol 1963; 90:466–77.

Miossec P, Yu C, Ziff M. Lymphocyte chemotactic activity of human interleukin 1. J Immunol 1983; 133:2007–11.

Munro JM, Cotran RS. The pathogenesis of atherosclerosis: atherogenesis and inflammation. Lab Invest 1988; 58:249–61.

Munro JM, van der Walt JD, Munro CS, Chalmers JAC, Cox EL. An immunohistochemical analysis of human aortic fatty streaks. Human Pathol 1987; 18:375–80.

Oppenheim JJ, Kovacs EJ, Matsushima K, Durum SK. There is more than one interleukin 1. Immunol Today 1986; 7:45–56.

Owen DAA. Inflammation—histamine and 5-hydroxytryptamine. Br Med Bull 1987; 43:256–69.

Patel JM, Yarid FR, Block ER, Raizada MK. Angiotensin receptors in pulmonary arterial and aortic endothelial cells. Am J Physiol 1989; 256:C977–83.

Pober JS. Cytokin-mediated activation of vascular endothelium. Am J Pathol 1988; 133:426–33.

Prokazova NV, Mukhin DN, Orekhov AN, Gladkaya EM, Vasilevskaya VV, Mikhailenko IA, Sadovskaya VL, Bushuev VN, Bergelson LD. Neutral glycolipids of atherosclerotic plaques and unaffected human aorta tissue. Eur J Biochem 1989; 180:167–71.

Repin VS, Dolgov VV, Zaikina OE, Novikov ID, Antonov AS, Nikolaeva MA, Smirnov VN. Heterogeneity of endothelium in human aorta—a quantitative analysis of scanning electron microscopy. Atherosclerosis 1984; 50:35–52.

Robbins RA, Wagner WD, Sawyer LM, Caterson B. Immunolocalization of proteoglycan types in aortas of pigeons with spontaneous or diet-induced atherosclerosis. Am J Pathol 1989; 134:615–26.

Ross R. The pathogenesis of atherosclerosis—an update. N Engl J Med 1986; 314:488–500.

Ryan US. Endothelial cells. Boca Raton, FL: CRC Press, 1988.

Schreiber RD, Pace JL, Russell SW, Altman A. Katz DH. Macrophage-activating factor produced by a T cell hybridoma: physicochemical and biosynthetic resemblance to γ-interferon. J Immunol 1983; 131:826–32.

Sedgewick JB, Busse WW. Histamine regulation of granulocyte function. In: Rocklin RE, ed. Histamine and H_2 antagonists in inflammation and immunodeficiency. New York: Marcel Dekker, 1990; 107–38.

Shalaby MR, Waage A, Espevik T. Cytokine regulation of interleukin 6 production by human endothelial cells. Cell Immunol 1989; 121:372–82.

Silverman DHS, Karnovsky ML. Serotonin and peptide immunoneuromodulators: recent discoveries and new ideas. Adv Enzymol 1989; 62:203–26.

Steeg PS, Moore RN, Johnson HM, Oppenheim JJ. Regulation of murine macrophage Ia antigen expression by a lymphokine with immune interferon activity. J Exp Med 1982; 156:1780–93.

Streiter RM, Kunkel SL, Showell HJ, Remick DG, Phan SH, Ward PA, Marks RM. Endothelial cell gene expression of a neutrophil chemotactic factor by TNF-α, LPS, and IL-1β. Science 1989; 243:1467–9.

Sturk A, Cate JWT, Hosford D, Mencia-Huerta J-M, Braquet P. The synthesis, catabolism, and pathophysiological role platelet-activating factor. In: Paoletti R, Kritchevsky D, eds. Advances in lipid research, Vol. 23. San Diego: Academic Press, 1989; 219–76.

Tonnensen MG. Neutrophil-endothelial cell interactions: mechanisms of neutrophil adherence to vascular endothelium. J Invest Dermatol 1989; 93:53S–8S.

Weinstock JV. The significance of angiotensin I converting enzyme in granulomatous inflammation: functions of ACE in granulomas. Sarcoidosis 1986; 3:19–26.

Weinstock JV, Blum AM, Kassab JT. Angiotensin II is chemotactic for a T-cell subset which can express migration inhibition factor activity in murine schistosomiasis mansoni. Cell Immunol 1987; 107:180–7.

Wong GG, Clark SC. Multiple actions of interleukin 6 within a cytokine network. Immunol Today 1988; 9:137–9.

Woodruff JJ, Clarke LM, Chin YH. Specific cell-adhesion mechanisms determining migration pathways of recirculating lymphocytes. In: Paul WE, Fathman CG, Metzger H, eds. Annual review of immunology, Vol. 5. Palo Alto, CA: Annual Reviews, 1987; 201–22.

Zetter BR. Endothelial heterogeneity: influence of vessel size, organ localization, and species specificity on the properties of cultured endothelial cells. In: Ryan US, ed. Endothelial cells, Vol. II. Boca Raton, FL: CRC Press 1988; 63–80.

21

High-Altitude Pulmonary Hypertension and Signal Transduction in the Cardiovascular System

A. A. ALDASHEV

Kirghiz Institute of Cardiology
Bishkek, Kirghizstan, Russia

I. Introduction

The pulmonary circulation has its peculiar features; the most prominent among them are a low level of vascular resistance, the possibility of depositing significant blood volumes, and characteristic blood flow regulation and redistribution. Studies of pulmonary circulation are important not only for physiology, but for clinical medicine as well, because pulmonary hypertension, either alone or associated with many pathologic states, occurs rather often, inducing right heart overload and decompensation. The data available show that 30% of patients dying of heart failure have cor pulmonale; such a high incidence proves the clinical importance of this problem (Zamotaev, 1978).

This chapter is a review of the recent data on the role of neurohumoral factors and secondary messengers in the development of high-altitude pulmonary arterial hypertension.

About 45 years have passed since it was experimentally shown that breathing a hypoxic gas mixture resulted in increases in pulmonary artery pressure (von Euler and Lilljestrand, 1946). It was suggested that a reflex mechanism underlies these increases. Hypoxic pulmonary vasoconstriction was attributed a beneficial value because of a more adequate relationship between ventilation and blood flow and better oxygenation of arterial blood. The significance of hypoxia in the genesis of pulmonary arterial hypertension was thoroughly studied in the model of hypoxic pulmonary arterial hypertension (PH) (breathing of a gas mixture with a low oxygen content) and in high-altitude PH. High-altitude PH as a separate entity was discussed by Mirrakhimov (1971).

Several stages can be distinguished in the development of hypoxic or high-altitude PH: (1) sustained vasoconstriction of pulmonary arterioles in response to hypoxia; (2) morphologic changes or remodeling of pulmonary vessels; (3) the development of right ventricular hypertrophy.

The first stage is the mysterious one: It is not clear why the decrease in oxygen tension is associated with pulmonary vasoconstriction. There are at least three questions in this puzzle: (1) What cellular system is a sensor of hypoxia? (2) What is the way of transforming information about a decrease in oxygen tension in contraction of smooth muscle cells in pulmonary vessels? (3) What is the advantage of pulmonary blood pressure elevation during hypoxia for the whole body? Two hypotheses try to explain these questions: (1) the indirect effect—alveolar hypoxia induces the release of certain vasoactive mediators, causing smooth muscle cell contraction from some hypoxia-sensitive pulmonary cells; (2) the direct effect—hypoxia induces the direct contraction of pulmonary arteriolar smooth muscle cells (Fishman, 1976). Up to now, neither of these two hypotheses has been given preference.

II. The Mechanism of Indirect Hypoxic Vasoconstriction

In the indirect hypothesis of hypoxic pulmonary vasoconstriction there has to be a sequence of cellular responses. First, specific cells have "oxygen sensors" which respond to decrease of oxygen tension below some threshold; they might be endothelial cells (Vender et al., 1987), alveolar cells, or carotide body cells (Fishman, 1976). Second, these sensor cells release factors which constrict smooth muscle cells in pulmonary vessels. Histamine was one of the first neurotransmitters suggested for the role of such a mediator (Houge, 1968). It was found, however, that histamine is released in response not to hypoxia but to vasoconstriction (Kay et al., 1974).

Pulmonary endothelium is known to contain significant quantities of angiotensin converting enzyme (ACE), and its activity in the lungs is very high (Ryan et al., 1985). Angiotensin II is a potent vasoconstrictor, hence its role in hypoxic vasoconstriction can be suggested. It has already been shown that ACE activity is significantly decreased in pulmonary cells during hypoxia ($pO_2 < 30$ mm Hg) (Stulcup et al., 1979). Endothelial cells up-regulate synthesis of ACE during hypoxia, but its activity does not change (King et al., 1989). There is no evidence yet that angiotensin II and ACE are involved in hypoxic vasoconstriction.

The most probable candidate for the role of hypoxic vasoconstrictor is endothelin—one of the most potent vasoconstrictors of biological origin (Yanagisawa et al., 1988). Endothelin, as one of the endothelium-derived vasoconstriction factors (EDCFs), will be discussed also.

The arachidonic acid metabolites, prostaglandins, thromboxane, and leukotrienes, play an important role in the regulation of pulmonary hemodynamics and in the genesis of pulmonary hypertension (Stenmark et al., 1989). This is related to the apparent local vasoactivity of these compounds (Schror, 1985). Increases in pulmonary vascular resistance and blood pressure in sheep during hypoxia was shown to be due to the efflux of leukotrienes (Ahmed and Oliver, 1983). The study of leukotriene content in alveolar lavage revealed high levels of LTC_4 during acute alveolar hypoxia (Morganroth et al., 1984). Injection of leukotriene inhibitors and antogonists relieved hypoxic vasoconstriction in rat lungs (Morganroth et al., 1984). However, it was pointed out that the leukotriene vasoconstrictor effect developed very slowly and was maintained for a prolonged time, whereas pulmonary vasoconstriction was relieved very quickly by normoxia (Morganroth et al., 1984).

The role of catecholamines in pulmonary vasoconstriction during hypoxia and high-altitude PH remains unclear. It was demonstrated in men that after ascent from sea level to high altitude (5100–6300 m above sea level), the catecholamine blood output increases dramatically and their level remains high for a longer period of time (Blume, 1984). It should be noted that mainly the norepinephrine level is increased. Similar increases in blood norepinephrine were observed in rats during high-altitude hypoxia (Myles and Drucker, 1971). Hypoxia increased the β-adrenergic activity in pulmonary vascular smooth muscles (Lock et al., 1979), and injections of isoproterenol prevented hypoxic pulmonary vasoconstriction (Freid and Reid, 1985). It has also been shown that reabsorption of norepinephrine in pulmonary vessels is altered in patients with primary and secondary pulmonary hypertension (Sole et al., 1979). However, there is a contradiction consisting of the mismatch between the onset of contractile response to hypoxia and the catecholamine release into the blood. Besides, the sympathetic nervous system is the main source of norepinephrine, but innervation was not significant for the development of hypoxic vasoconstriction (McMurtry et al., 1976).

The indirect mechanism hypothesis of hypoxic vasoconstriction became very attractive after Furchgott (Furchgott and Zavadski, 1980) found that endothelium could regulate the tone of underlying smooth muscle cells in the vessel by releasing endothelium—derived relaxing and constricting factors (EDRF and EDCF). Endothelium can release EDCF in some cases, as well as during hypoxia (Rubanyi, 1991). Endothelium—dependent vasoconstriction can be triggered by various biologically active substances (acetycholine, arachidonic acid, norepinephrine, prostaglandins, thromboxane), drugs (calcium ionophores, nicotine, chemicals which increase intracellular potassium), physical factors (stress, pressure) and hypoxia (Rubanyi, 1991). There are three main types of EDCF: (1) arachidonic acid metabolites and free radicals (e.g., superoxide anion); (2) an EDCF whose nature has not been determined that is released during severe hypoxia; and (3) endothelin, which causes slow, long-term vasoconstriction. It is obvious that all of them can play significant roles in hypoxic vasoconstriction. Recently, it was shown that, during hypoxia, smooth muscle cells increase synthesis and release of endothelin in the media (Langleben et al., 1991). This autocrine mechanism might be important in developing hypoxic vasoconstriction and high-altitude PH.

Various substances have been suggested for the role of mediator of hypoxic vasoconstriction: histamine, serotonine, prostaglandins, catecholamines, and peptides. It is hard to suggest which one of these physiologically active substances is the mediator of hypoxic vasoconstriction. At the same time, the absence of a proven mediator of indirect hypoxic vasoconstriction does not exclude the possibility of its existence.

III. Direct Mechanism of Hypoxic Pulmonary Vasoconstriction

As in the case of indirect mechanism, the hypothesis of the direct mechanism of hypoxic vasoconstriction includes the identification of a cellular system sensitive to decreases in oxygen concentration. But in this case the sensor system must be connected with the pulmonary vascular smooth muscle cell contractility. It was suggested that hypoxic vasoconstriction is related to the inhibition of oxidative phosphorylation (Lloyd, 1965). This suggestion resulted from the fact that uncouplers of oxidative phosphorylation induce hypoxic vasoconstriction in normoxic conditions. However, this hypothesis does not explain how this system can work if the level of oxidative phosphorylation does not change significantly until a decrease in oxygen tension down to 7 mm Hg (Fisher and Dodia, 1981). From the point of view of the chemist, it is clear that the hypoxic sensor triggering vasoconstriction should be one of redox reactions, but it is unclear which one.

According to one model (Archer et al., 1989), either oxygen or its partially reduced form oxidizes some key sulfhydryl groups in pulmonary vessel cells and this regulates the calcium flow in smooth muscle cells and induces the vasoconstriction. Hence, according to this model, the degree of oxidation of key sulfhydryl groups acts as the sensor and calcium ions as mediators of the hypoxic vasoconstriction. The above-mentioned review (Archer et al., 1989) presents the following proofs of this model: sulfoxide anions are formed in the lung proportionally to the oxygen tension; glutathione oxidants prevent hypoxic vasoconstriction; pulmonary oxygen changes the ratio between reduced and oxidized glutathiones; diamide is the sulfhydryl oxidant and rapidly reverses hypoxic vasoconstriction. This model does not exclude that this key sulfhydryl group can be found in calcium channel proteins. Recently it was shown that the redox status of pyridine nucleotides in mitochondria and thiol residues of membrane-bound proteins can reversibly open and close potassium channels and affect the calcium ion concentration in the cell (Mallery et al., 1991). The thiol residue redox state correlates in the cell with mobilization of intracellular calcium (Mallery et al., 1991).

Despite the attractiveness of this model, it has the same limitations as the previous one, if not more. At present there is no sufficiently convincing experimental data in favor of this hypothesis. It can also be contended that a cell really in some way is informed about the changes in oxygen concentration in the environment and responds to them by alterations in intracellular calcium concentration. The problem of cellular responses to decreased oxygen concentration in the environment not only concerns vasoconstriction, but also changes in cell metabolism, processes of transcription and translation, and cell division. Calcium is the putative intracellular messenger of the mitogen-induced responses (Berridge and Irvine, 1984). Now it is obvious that vasoconstriction is only one of many cellular adaptative responses to hypoxia and other stresses. It was known that cells cultured in unfavorable conditions can synthesize factors which increase their viability, such as heat shock proteins (HSPs). HSP is the family of proteins up-regulated in cells by different stress inductors, such as heat (41–43°C), heavy metal ions, ethanol, and others (Browder et al., 1989). It was suggested that HSPs play a very important role during adaptation of eukaryotes to environmental conditions (Browder et al., 1989). Injection of antibodies against HSP 70 in fibroblasts causes cell death even during a short increase of temperature to 45°C (Riabowol et al., 1988).

We suggest that, similar to HSP synthesis, hypoxia can stimulate the synthesis of factors in the cells which increase their viability, release factors contracting smooth muscle cells, or stimulate cell proliferation. Early it was shown that bovine pulmonary artery endothelial cells, cultured under hypoxic conditions, released polypeptide factors with molecular weights of less than 10 kDa or more than 60 kDa (Vender et al., 1987). These factors, when added to a preparation of smooth muscle cells, stimulated their division. A similar study on smooth muscle cells showed that, in response to hypoxia, they synthesize a peptide factor with molecular weight below 15 kDa, which results in a fourfold increase in elastin synthesis in pulmonary vessels (Mecham et al., 1987). Up-regulation of several proteins during hypoxia was shown in fibroblasts

(Anderson et al., 1989) and bovine endothelial cells (Zimmerman et al., 1991). We demonstrated that, when exposed to hypoxia, a number of proteins are up-regulated in cultivated human lymphocytes (Agibetov et al., 1990).

These proteins with molecular weights of 36 kDa, 38 kDa, 52 kDa, and 74–76 kDa are also synthesized under normoxic conditions, but in negligible amounts (Fig. 1). Increase of the cultivation temperature to 42°C for 30 min resulted in synthesis of classic HSPs of 70, 74, and 86 kDa and did not induce hypoxic proteins (Fig. 1). Cold stress (4°C, 1 h), like heat shock, induced synthesis of classic HSPs (Fig. 1). Thus the classic HSP inducers do not enhance

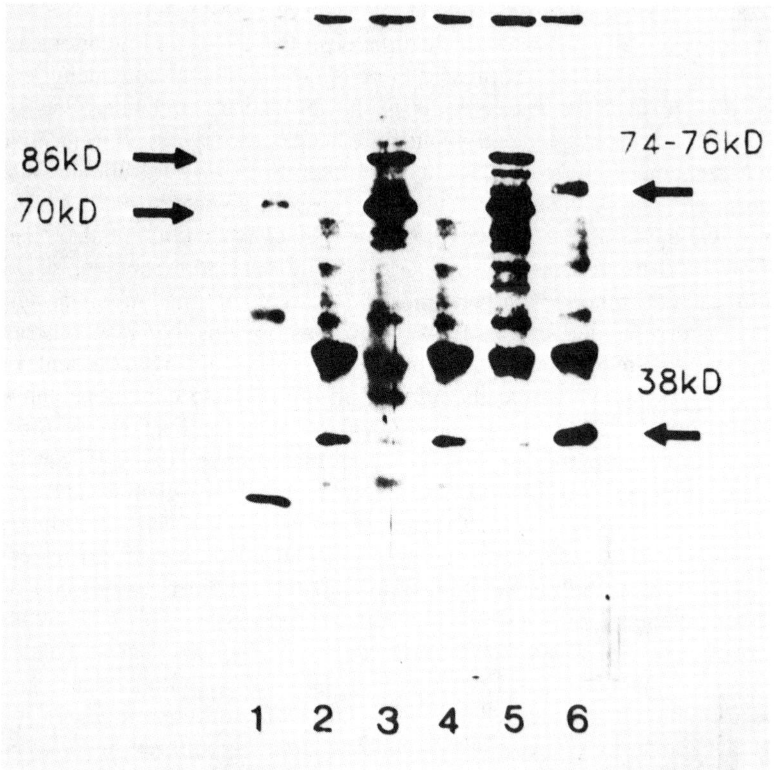

Figure 1 Effect of different stresses on protein synthesis in human lymphocytes. Lane 1, control (normoxia) lymphocyte proteins; lanes 2,3,4, lymphocyte proteins after exposure to cold stress (24 h, 4°C), Zn^{2+} (0.5 mM, 1 h), heat shock (42°C, 30 min), respectively. Lane 5, hypoxic lymphocytes (48 h, 4% O_2, 5% CO_2, 91% N_2). The arrows indicate molecular weights of proteins up-regulated by stresses (in kDa). Molecular weight standards are shown on the left side.

p38, p52, and p76 hypoxin synthesis. Cells fractioning into cytosol and membrane fractions revealed that p36, p38, and p52 are localized in cytosol, which p76 is membrane bound (Fig. 2). An increase in intracellular calcium level during hypoxia is known to be one of the first and principal responses of the cell to hypoxia (Archer et al., 1989). We demonstrated that the cultivation of human lymphocytes in a medium with a high concentration of Ca^{2+} (25 mM) under normoxic conditions (21% O_2, 5% CO_2, 37°C, 48 h) also effectively induced the synthesis of the hypoxins (Aldashev et al., 1991). But this elevated

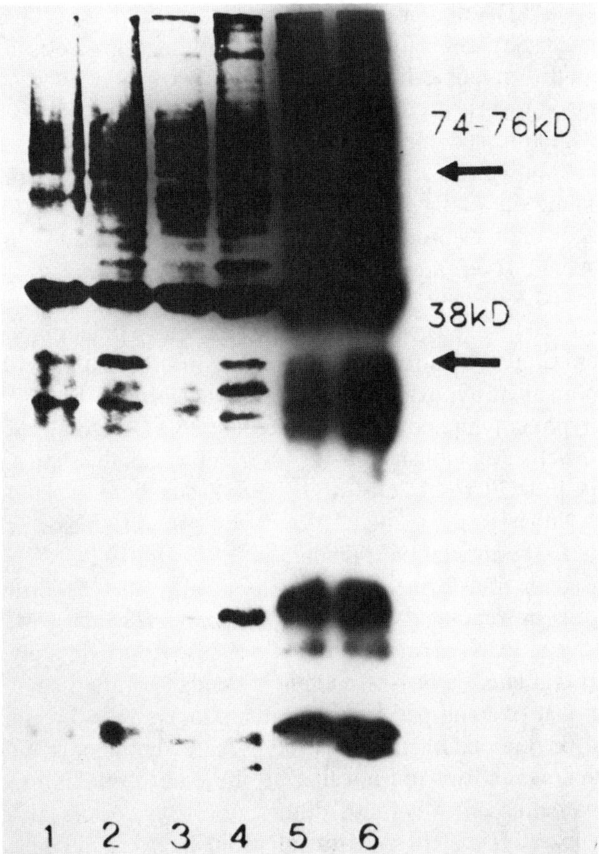

Figure 2 The spectrum of [^{35}S]-methionine-labeled proteins of human lymphocytes subcellular fractions after normoxia (lanes 1,3,5) and hypoxia (4% O_2, lanes 2, 4, 6). Lanes 1 and 2, total cell lysate; lanes 3 and 4, cytosol; lanes 5 and 6, membrane fraction. The arrows indicate molecular weights of up-regulated proteins.

calcium concentration increased only p21, p36, and p38 (Fig. 3). Two-dimensional gel electrophoresis clearly demonstrated that calcium induced synthesis of the same 38-kDa protein as hypoxia, and this up-regulation by calcium was more pronounced than by hypoxia (Figs. 4b and 4c). EGTA, a Ca^{2+} scavenger, when added to the medium, inhibited the hypoxia-induced up-regulation of p38 (Fig. 5). Thus it is obvious that p38 hypoxin expression, in human lymphocytes is a calcium-regulated process. Recently it was shown (Hoeffler et al., 1988; Sheng et al., 1991) that an increase of cAMP and calcium up-regulates cell transcription factors and cAMP-responsive transcriptional enhancer binding proteins (CREB and CREB B). CREB B has a molecular weight of 38 kDa and a structure similar to some oncogene proteins (Hoeffler et al., 1988). But our two-dimensional electrophoresis data (Fig. 4) revealed that p38 of hypoxic lymphocytes is PCNA/cyclin. PCNA (proliferating cell nuclear antigen), according to Celis (Celis et al., 1984), is a nuclear protein which is expressed during cell transition from G_1 to S phase of the cell cycle. Now it is known that PCNA is a cofactor of DNA polymerase δ, which builds the binding site of DNA polymerase δ in the replication fork; in cells lacking PCNA or another protein factor A1, DNA polymerase δ does not synthesize a DNA copy (Lee and Hurwitz, 1990).

The reported molecular weight of PCNA is 36 kDa (Celis et al., 1984) or 37 kDa (Tan et al., 1986); the difference from our p38 can be explained by the difference in protein electrophoretic mobility in different experimental conditions, or modified in hypoxic lymphocytes. Recently we found that isolated mRNA from hypoxic lymphocytes translated the 36-kDa protein in vitro while mainly p38 in cells (unpublished data). Normally, nondividing lymphocytes synthesize very little PCNA (Celis et al., 1986), but there is six- to sevenfold increase in synthesis after platelet-derived growth factor (PDGF) or interleukin-2 (IL-2) stimulation (Travali et al., 1989). PCNA synthesis is greatly increased in transformed cells (Celis et al., 1984). Fibroblasts under anoxic conditions up-regulated p34, p36, p51, p56, p78, and p94, and it was shown that p36 was PCNA (Anderson et al., 1989). Approximately all of these proteins are up-regulated in bovine endothelial cells (Zimmerman et al., 1991). We can expect that p52 and p76 in human lymphocytes are p51 and p78 in rat fibroblasts and bovine endothelial cells. Hypoxia in lymphocytes up-regulated only a few proteins, and we did not find all the other proteins found in fibroblasts and endothelial cells. Perhaps different cell types up-regulate different proteins in hypoxia? Thus, in endothelial cells, hypoxia did not up-regulate glucose-regulated proteins (Zimmerman et al., 1991) which are up-regulated in hypoxic fibroblasts (Anderson et al., 1989). It was reported that lymphocytes during hypoxia release IL-1 in the media (Pennathur-Das et al., 1989), so we cannot exclude the possibility of an autocrine mechanism.

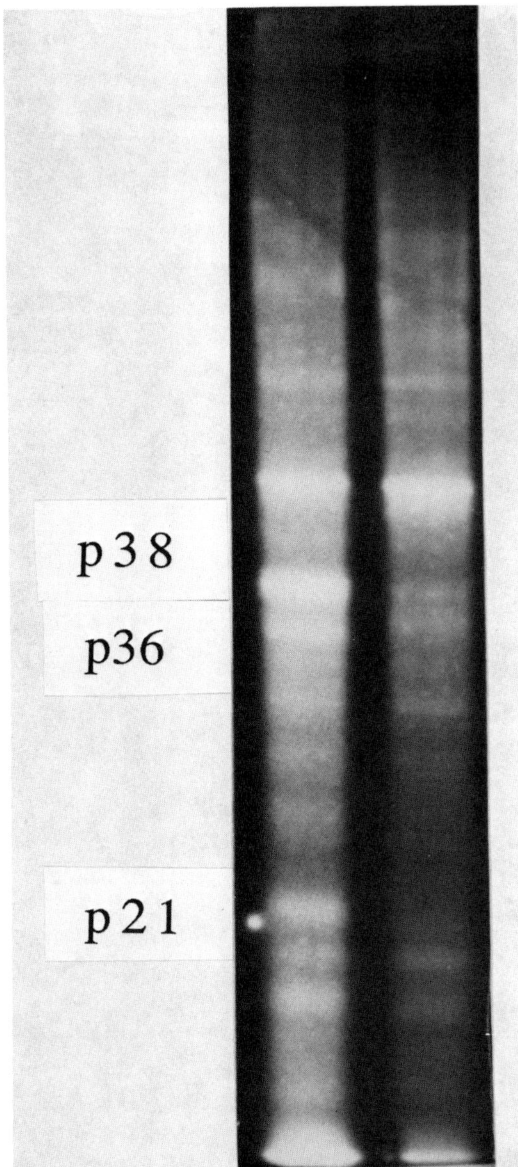

Figure 3 Effect of Ca^{2+} ions on protein synthesis in human lymphocytes. Left lane, cell lysate of lymphocytes cultivated in media with 25 mM $CaCl_2$; right lane, control lymphocytes (2 mM $CaCl_2$).

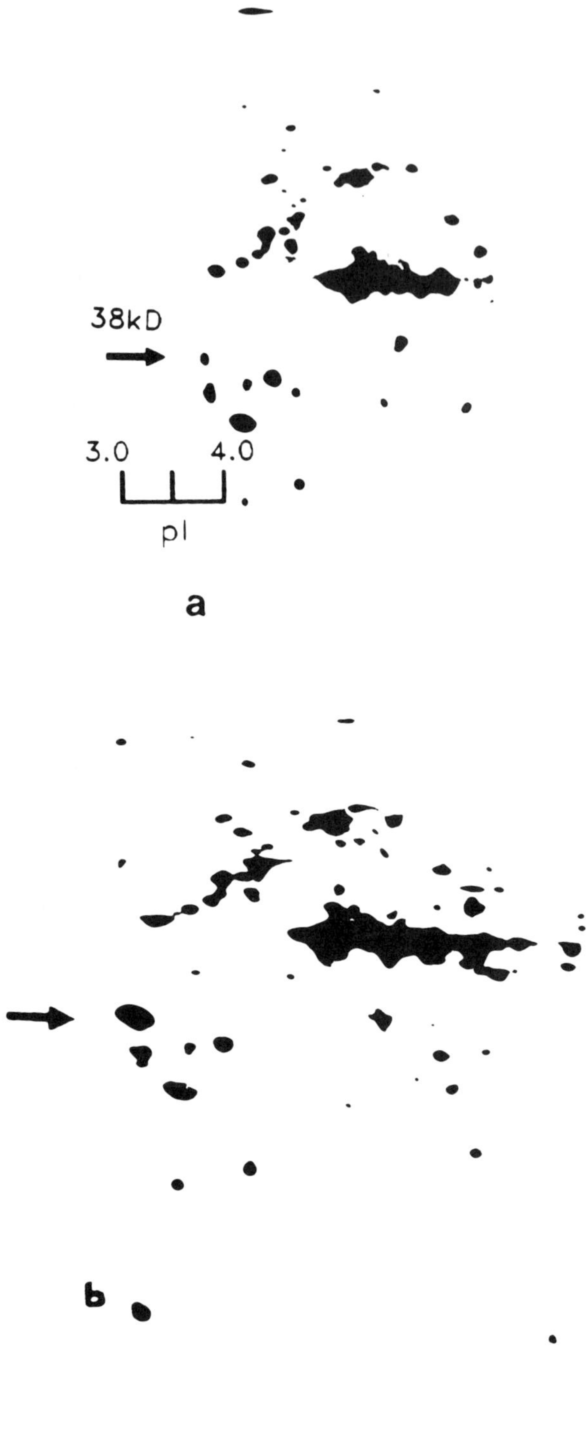
38kD
3.0
4.0
pI
a
b

Figure 4 Two-dimensional electrophoresis fluogram of [^{35}S]-methionine-labeled proteins from human lymphocytes cultivated in normoxic (a), hypoxic (b), and elevated Ca^{2+} conditions. The x direction is IEF, y is SDS gel electrophoresis. The arrow indicates the protein with a molecular weight of 38 kDa and pl of 3.5–4.0.

The finding of PCNA up-regulation by hypoxia suggests that hypoxia is a mitogen or mitogen-inducing factor. This is a very crucial point, because in this case hypoxia itself may play a primary role in the remodeling of pulmonary vessels, right ventricular hypertrophy at HAPH, and regeneration of tissues during wound healing. But it is still unknown what the role of PCNA is in the cellular adaptation to hypoxia.

IV. Effect of Hypoxia on Ca^{2+} Channels and Intracellular Calcium Level

Elevated intracellular calcium during hypoxia affects not only vasoconstriction, but, given the role of calcium as a secondary messenger and its relationship with cAMP and phosphoinositide turnover, significantly affects cellular metabolism. Increases in intracellular calcium and in metabolism of phosphoinisitides are known to result in cell division. It was shown that hypoxia enhances phosphoinositide turnover in endothelial cells and affects cell growth (Mirza-

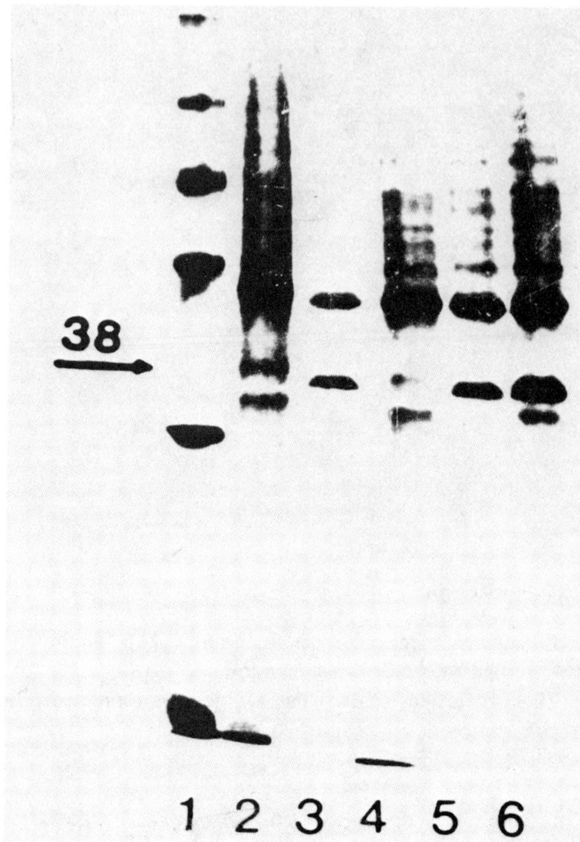

Figure 5 Effect of Ca^{2+} and EGTA, on hypoxic stress proteins. Fluorogram of SDS electrophoresis gel: lane 1, molecular-weight standards 14, 30, 46, 68, and 96 kDa beginning from the bottom; lane 2, normoxia; lane 3, hypoxia; lane 4, hypoxia in the presence of 2 mM EGTA; lanes 5 and 6, normoxia in the presence of 25 mM and 50 mM $CaCl_2$.

poyazova et al., 1988). Increase of intracellular calcium and phosphoinositide turnover were found to alter beta-adrenoceptor activity, resulting in decrease (Su et al., 1980) or increase of adenylate cyclase activity (Sibley and Lefkowitz, 1988).

Hypoxic constriction of pulmonary arterial smooth muscle requires Ca^{2+} influx via voltage-dependent and receptor-regulated Ca^{2+} channels. We studied the state of Ca^{2+} channels and calcium levels in human platelets during hypoxia as compared to platelets of patients with HAPH. The activity of receptor-regulated Ca^{2+} channels was assessed by an increase in Quin 2 probe fluorescence in response to increase of intracellular calcium. In high-altitude residents of Kirghizia who have variable pulmonary vascular responses to hypoxia, we hypothesized that increased Ca^{2+} channel activity relates to the increased vascular responses. Our approach in hyporesponders, hyperresponders, and patients with high-altitude pulmonary hypertension (HAPH) was to measure Ca^{2+} influx in platelets in response to known Ca^{2+} channel activator ADP (1 μM), because it was reported that activation of platelet Ca^{2+} channels reflects channel activation in vascular smooth muscle (Menshikov et al., 1986). Breathing 10% oxygen for 30 min increased mean pulmonary arterial pressure from 13 to 21 mm Hg in hypo-, and from 16 to 35 mm Hg in hyperresponders, and from 19 to 32 mm Hg in HAPH patients. Baseline platelet Ca^{2+} concentration (nM), the increase of influx with ADP stimulation, and the concentration of Nifedipine (nM) required to block 50% of the ADP-stimulated Ca^{2+} influx were all greater in hyperresponders and HAPH than in hyporesponders, as shown in Table 1.

The activity of Ca^{2+} channels in healthy lowlanders was similar to that in hyporesponders, but we suggest that hyperresponders can be found among them too. We propose that increased calcium channel activity increases pulmonary vascular responsiveness to hypoxia and predisposes to high-altitude pulmonary hypertension or secondary pulmonary hypertension in patients with obstractive lung diseases.

V. The Role of Adrenoceptors

As mentioned above, when humans (Blume, 1984) or rats (Myles and Drucker, 1971) are exposed to high altitude, their blood catecholamine levels increase and remain high over prolonged periods. Increases in norepinephrine were found to account for the elevated catecholamine levels, with no significant changes in epinephrine. In patients with pulmonary arterial hypertension, lung metabolism of norepinephrine is impaired (Sole et al., 1979). At

Table 1 Ca^{2+} Channel Activity in Highlanders with Different Responsiveness of Pulmonary Circulation to Hypoxia

	Hyporesponders	Hyperresponders	HAPH
Baseline Ca^{2+}	98 ± 6	115 ± 7	130 ± 18
Influx (ADP)	164 ± 6	299 ± 12	308 ± 15
ED_{50} (Nifedipine)	5	100	1000

All data mean ± S.D., $p < 0.05$.

the same time, hypoxia resulted in increased beta-adrenergic activity in pulmonary vascular smooth muscle and a decrease of vascular constrictive responses to epinephrine infusion (Lock et al., 1979). Infusion of the agonist isoproterenol prevented hypoxic pulmonary vasoconstriction in rats (Freid and Reid, 1985) and induced cardiac hypertrophy during normoxia (Nanoff et al., 1989). All these findings suggest altered adrenergic activity in pulmonary vessels during hypoxia.

The increased norepinephrine level accompanying chronic hypobaric hypoxia causes dramatic changes in beta-adrenoceptors located in rat heart (Voelkel et al., 1981). It is well known that increases in catecholamine and beta-adenergic agonist concentrations leads to the development of beta-adrenoceptor desensitization as reflected by the diminished density of surface receptors and their uncoupling from adenylate cyclase (Lefkowitz and Caron, 1990). Catecholamines and adrenoreceptors are not likely to play a role in vasoconstrictive response to hypoxia; however, such phenomena as desensitization due to chronic elevation of catecholamines, and the variety of effects of adrenoreceptor activation on cellular metabolism, may play an important role in the development of the pathological alterations seen in HAPH. Voelkel et al. (1981) observed a decrease in beta-adrenoceptor density with development of right ventricular hypertrophy in rats during hypobaric hypoxia. In similar experiments we showed that propranolol prevented development of right ventricular hypertrophy in hypoxic rats. Propranolol injections also prevented down-regulation of heart beta-adrenoreceptors. Hypoxia affected not only heart receptors but also those in the lung. Beta-adrenoceptor density (B_{max}) in hypoxic rat lung was 201 $\pm$45 fmol/mg protein, while in normoxic rats it was 308 $\pm$35. Propranolol increased B_{max} of both hypoxic and normoxic rats (384 $\pm$76 versus 414 $\pm$58 fmol/mg protein, respectively). Thus, hypoxia lead to the desensitization of beta-adrenoceptors in various tissues. High doses of beta-blockers prevented the development of right ventricular hypertrophy and beta-adrenoceptor desensitization, but did not prevent left ventricular mass from diminishing (Voelkel et al., 1981).

The problem of studying the role of receptors and secondary messengers in the hypoxic pulmonary vasoconstriction response is the lack of an adequate animal model. We performed studies of cardiac adrenoceptors in animals highly resistant to hypoxia—yaks—and homologous species sensitive to hypoxia—cows (Wier et al., 1974). The yak (*Bos grunniens*) is an animal well adapted to extreme altitudes; it is closely related to cattle and readily cross-breeds with them, producing hybrids fully capable of reproducing. Our data (Aldashev et al., 1988) show the decrease in adrenoceptor B_{max} and K_d in yak heart as compared to sea-level domestic cow (*Bos taurus*) (Table 2).

A similar decrease was observed both for beta- and alpha$_1$-adrenoceptors of yaks. Thus, adaptation to high altitude results in a decrease of adrenoceptor

Table 2 Adrenoceptor Density (B_{max}, fmol/mg protein) and K_d (nM) in the Right Atria of Bulls and Yaks

	β-Adrenoceptors		a_1-Adrenoceptors	
	B_{max}	K_d	B_{max}	K_d
	(^{3}H-dihydroalprenalol)		(^{3}H-Prasozin)	
Bulls	53.7 ± 7.8	1.0 ± 0.1	20.7 ± 1.5	1.3 ± 0.2
Yaks	19.3 ± 7.3	0.3 ± 0.1	13.1 ± 1.5	0.6 ± 0.1

All data mean ± S.D., $p < 0.05$.

density and an increase of their affinity. This decrease of adrenoceptor density suggests their desensitization, but beta-adrenoceptor-dependent activation of adenylate cyclase by isoproterenol was higher for yaks than bulls (217% and 149%, respectively).

Thus, evolution of the cardivascular system in high-altitude animals apparently involves selection in favor of high-affinity and highly effective adrenoceptors. Probably, in heart tissue, part of the adrenoreceptors are not coupled to G proteins and adenylate cyclase; the adaptation to high altitude abolished these low-affinity uncoupled receptors.

It is well known that activity of beta-adrenoreceptors is closely related to cholinergic receptors. We studied the state of muscarinic receptors in the right heart of yaks and cows. The density of M_2 cholinergic receptors in yak heart was decreased as compared to domestic cow right atrial tissue, but without any changes in affinity (Table 3). The main cellular effect of M_2 receptors is inhibition of adenylate cyclase (Birnbaumer et al., 1989), so we speculate that decrease of inhibitor receptors prevent the down-regulation of adenylate cyclase in yaks.

In collaboration with K. Stenmark and N. Voelkel (CVP Lab., Univ. Colorado Hlth Sci Ctr., Denver), we studied the effect of hypoxia on beta-adrenoceptors and adenylate cyclase (Stenmark et al., 1991), using an animal model of newborn calves which were in a hypobaric chamber at a simulated

Table 3 The State of M2-Cholinergic Receptor in Cows and Yak Right Atria

	B_{max} (fmol/mg protein)	K_d (nM)
Bulls	69.8 ± 10.0	0.43 ± 0.07
Yaks	50.2 ± 6.5	0.45 ± 0.06

All data mean ± S.D.

altitude of 4300 m above sea level. The newborn calves exposed to hypoxia developed severe pulmonary hypertension associated with many of the same functional and structural abnormalities observed in human infants with pulmonary hypertension (Stenmark et al., 1989). The development of PH in the newborn is associated with abnormalities of vessel tone and control of cell growth. Decrease in cAMP production inhibits vascular smooth muscle contraction by decrease of the concentration of intracellular calcium (Karaki, 1989). We investigated the changes in adenylate cyclase activity and beta-adrenoceptor density in right atria of hypoxic and control calves. We showed an age-dependent decrease in basal adenylate cyclase activity after birth (Table 4).

Chronic hypoxia interfered with the normal development of adenylate cyclase regulatory mechanisms. The 7-day hypoxia significantly decreased the basal adenylate cyclase activity, and this decrease was associated with an increase in G_i protein activity (Table 4). Pertussis toxin (PTX) treatment inactivates $G_i\alpha$ subunits and abolishes the G_i-related effects (Gilman, 1990). After PTX treatment of right atria membranes, the basal activity increased both in control and 7-day hypoxic calves and reached the same level; 6.8 pmol/mg.min (Table 4). So the decrease of basal activity in 7-day hypoxic calves was due to increased G_i activity. But continued hypoxia (14 days) lead to uncoupling of adenylate cyclase from G proteins and beta-adrenoceptors. The basal activity increased in hypoxic calves and became higher than in control calves, whereas intrinsic activity (stimulated by 100 μM of forskoline) was 40% lower (Table 4). And inhibition of adenylate cyclase by guanylyl imidodiphosphate (GuppNp) was also lower in 14-day hypoxic versus 14-day control calves. Thus the increase in basal adenylate cyclase activity was due mainly to uncoupling of enzyme from G_i protein. Chronic hypoxia also decreased the beta-adrenoceptor density in hypoxic calf right atria (Fig. 6). We speculated

Table 4 Effect of Hypoxia on Adenylate Cyclase Activity in Calf Right Atria ($\times 10^{-3}$ pmol/mg protein · min)

	Fetus	Newborn	Control, 7 days	Hypoxic, 7 days	Control, 14 days	Hypoxic 14 days
Basal	7.9 ± 1.7	7.0 ± 0.8	4.5 ± 0.7	1.9 ± 0.4	3.7 ± 0.9	5.6 ± 1.1
Iso	12.9 ± 1.9	7.6 ± 1.0	5.3 ± 1.6	1.2 ± 0.4	12.8 ± 1.0	6.7 ± 0.8
Forsk	127 ± 22				64.2 ± 0.8	39.2 ± 5.2
GuppNp	122 ± 15				48.2 ± 5.8	34.2 ± 5.8
PTX			6.7 ± 0.3	6.8 ± 1.4		

All data mean ± S.D.

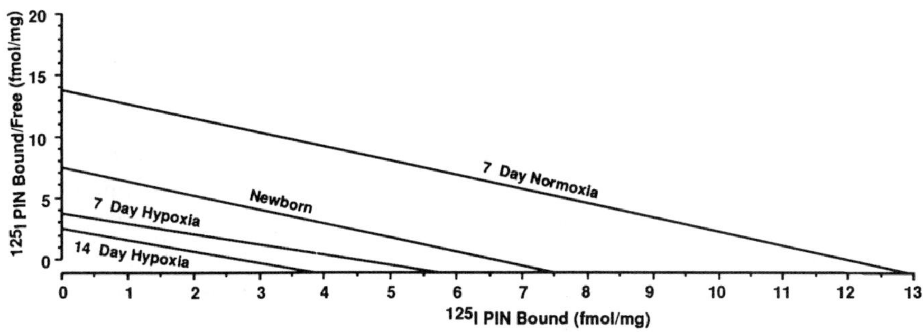

Figure 6 Scatchard plot of [^{125}I]-iodopindolol binding to hypoxic and control calf right atrium membranes. Data for newborn, 7-day control, 7-day hypoxia, 14-day hypoxic calves are shown.

that this decrease could be on the level of receptor mRNA. Recently, it was shown that chronic stimulation by beta-agonists decreased beta-adrenoceptor mRNA in the cells (Collins et al., 1989).

So we can see several effects of chronic hypoxia on adenylate cyclase: Heterologous desensitization of beta-adrenoceptors with uncoupling from adenylate cyclase and G proteins may be due to phosphorylation of G proteins by cAMP-dependent protein kinase (Lefkowtiz and Caron, 1990); beta-receptor density decreased during chronic hypoxia, perhaps due to decrease of their mRNA synthesis; the intrinsic activity of adenylate cyclase was down-regulated by chronic hypoxia. All these phenomena lead to decrease of cAMP production in hypoxic cells, which could cause increased contractility of smooth muscle cells (Karaki, 1989).

We naturally related these changes in beta-adrenoceptors and adenylate cyclase in the right heart of hypoxic calves to an increase in blood norepinephrine levels. But this is not the case, because we showed that there was no correlation between norepinephrine levels, and beta-adrenoceptors. Recent reports indicate that hypoxia induces a reversible 250% decrease in beta-adrenoceptor density in cardiomyocytes in cell culture (Marsh and Sweeney, 1989). Besides, a 2-h exposure to hypoxia decreased the extent of adenylate cyclase activation by isoproterenol and had no effect on basal cAMP levels. That is, hypoxia alone can cause beta-adrenoceptor desensitization without catecholamine involvement, but these changes are readily reversible by reoxygenation. The mechanism of this desensitization apparently involves phosphorylation of the receptor by protein kinase, which is probably activated by products of phosphoinositide turnover (Lefkowitz and Caron, 1990). In the case of the effect of hypoxia on the whole body, cardiac beta-adrenoceptors undergo desensitization during which changes occur not only in receptor density, but also in receptor affinity, accompanied by alternation in G-protein regulation of adenylate cyclase, and even adenylate cyclase activity itself (Lefkowitz and Caron, 1990). All these changes were noted by us in calves that, after birth, were chronically exposed to hypoxia in an altitude chamber and in comparative studies of high-altitude animals (yaks) and low-altitude domestic cows.

Thus, if we compare control calves, hypoxic calves, and yaks, it appears that evolution of signal transduction by adrenoceptors in the heart during adaptation to hypobaric hypoxia involves decreases in the density of receptors, with increases of their affinity; decrease of receptor density does not affect activity of adenylate cyclase due to decrease of some inhibitor receptors such as M_2 cholinergic receptors, and amplification of adenylate cyclase itself.

VI. Role of Adrenoceptors in the Genesis of High-Altitude Pulmonary Hypertension

Adenoceptors can play an important role in the adaptation to altitude hypoxia. We have already mentioned that blood norepinephrine levels rise in people ascending to 5400–6300 m above sea level (Blume, 1984). Moreover, norepinephrine metabolism in patients with pulmonary hypertension is impaired (Sole et al., 1979). It was shown that after 8 days at altitude, the beta-adrenoceptor density on lymphocytes from normal subjects decreased (Richalet et al., 1989). A few hours after descent to sea level, B_{max} returned to initial values. We studied the state of beta-adrenoceptors in residents of high altitudes (3600–4200 m above sea level). The desensitization of beta-adrenoceptors on lymphocytes from patients with high-altitude PH with severe right ventricular hypertrophy (RVH) was demonstrated (Aldashev et al., 1988; Aldashev et al., 1989a, 1989b). B_{max} was considerably lower in patients with severe RVH than in normal highlanders (1.6 ±1.4 and 6.9 ±3.2 fmol/mln cells, respectively). Assessment of beta-adrenoceptor-dependent activation of adenylate cyclase revealed a decrease in isoproterenol activation in patients with RVH (126%) versus 186% in healthy highlanders (Table 5). The G-protein stimulation of

Table 5 Activation of Adenylate Cyclase in Lymphocytes from Highlanders

	Basal (pmol/mln cells 10 min)	Activation (% of basal activity)	
		Iso	Forskoline
Controls	6.7 ± 3.3	186 ± 44	221 ± 27
Patients with grade III RVH	2.0 ± 0.5	126 ± 30	136 ± 35

All data mean ± S.D., $p < 0.05$.

adenylate cyclase by GuppNp up-regulates it in healthy highlanders, whereas inhibition prevails in patients with severe HAPH (Aldashev et al., 1989b). Direct activation of adenylate cyclase by forskoline is lower in highlanders with HAPH than in healthy highlanders (Table 5).

The existence of two types of receptor desensitization has been shown: a homologous and a heterologous type (Lefkowitz and Caron, 1990). Homologous desensitization is a fast process when receptors occupied by agonists are sequestered away from plasma membrane. Heterologous desensitization is the process by which receptors become less functional due to phosphorylation of receptor by cAMP-dependent protein kinase and uncoupling of it from G protein. The desensitization of beta-adrenoceptors on lymphocytes was shown in studies of patients with pheochromocytoma and cardiomyopathia (Maki et al., 1990). Thus, characteristic signs of beta-adrenoceptor desensitization are also observed in highlanders with HAPH.

When we studied the reactivity of pulmonary vessels to hypoxia in healthy highlanders at altitude 760 m above sea level, it appears that inhalation of 10% O_2 mixture in part of them (hyperresponders) highly increased the pulmonary arterial blood pressure (from 16.2 $\pm$1.6 to 34.8 $\pm$2.7 mm Hg), while others (nonresponders) only slightly increased it (from 12.8 $\pm$21.1 mm Hg). Hyperresponders had lower beta-adrenoceptor density, but even after 30 min inhalation of hypoxic gas mixture, increased to the level of nonresponders (Table 6).

According to Lefkowitz and Caron (1990), the rapid exposure of beta-adrenoceptors on plasma membrane can be only from internalized receptors, which were removed from the cell surface by high catecholamine levels. So approximately half of the beta-adrenoceptors on lymphocytes of hyperresponders easily externalized upon reexposure to hypoxia. The increase of beta-adrenoceptor density was not followed by adenylate cyclase activation (Table 6). This means that these reexposed receptors are not coupled to G pro-

Table 6 Effect of Hypoxia (10% O_2 for 30 min) on Lymphocyte β-Adrenoceptors from Control Highlanders

	B_{max} (fmol/mln cells)		% Activation Isoproterenol	
	Normoxia	Hypoxia	Normoxia	Hypoxia
Nonresponders	6.8 ± 3.2	7.2 ± 1.9*	119 ± 6	173 ± 27
Hyperresponders	3.7 ± 1.4	6.9 ± 1.8	130 ± 10	145 ± 8

All data mean ± S.D.
*Nonsignificant change.

teins, or desensitized. We speculated that these highlanders would develop severe HAPH with RVH. The beta-adrenoceptors of nonresponders are not desensitized, and these highlanders will resist HAPH.

Thus, we suggest that the development of HAPH is correlated with desensitization of beta-adrenoceptors.

VII. Summary

As can be seen, there are many levels at which hypoxia affects the organism—cells, organs, tissues, and the whole body. Its direct action on cells up-regulates synthesis of hypoxic stress proteins and mainly PCNA/cyclin, so hypoxia itself acts as a mitogen and could play a role in vascular wall remodeling.

One of the main manifestations of hypoxia in the cells is increase of intracellular Ca^{2+} (Archer et al., 1989). Any residue in the cell that is readily oxidized or reduced can be the oxygen sensor; it is not necessarily a sulfhydril group, but may be found in the ion channels. We showed that some disorders in regulation of calcium channel activity lead to hyperreactivity of pulmonary arterial vessels and hypoxic pulmonary arterial hypertension.

Hypoxia can cause the beta-adrenoceptor desensitization, and predisposition of highlanders to high-altitude pulmonary arterial hypertension is due to heterologous desensitization of their beta-adrenoceptors. The predisposition of beta-adrenoceptors to desensitization correlates with hyperreactivity of pulmonary blood vessels to hypoxia.

The evolution of the cardiovascular system during the adaptation to hypoxia in high-altitude animals involved decreases in the number of adrenoceptors with increases of their affinity and effectiveness of signal transduction into the cell.

References

Ahmed T, Oliver W. Does slow-reacting substance of anaphylaxis mediate hypoxic pulmonary vasoconstriction. Am Rev Resp Dis, 1983; 127:566–71.

Agibetov KA, Aldashev AA, Yugai AA. Identification of specific hypoxic protein in human lymphocytes. Dokl Akad Nauk SSSR (Proc Natl Acad Sci USSA), 1990; 311:989–92 (in Russian).

Aldashev AA, Borbugulov UM, Davletov BA, Titov VO. The state of beta-adrenoceptor system in the development of high-altitude pulmonary hypertension. In: Mirrakhimov MM, ed. Pulmonary arterial hypertension. Moscow: Frunze 1988; 4–7 (in Russian).

Aldashev AA, Borbugulov UM, Davletov BA, Mirrakhimov MM. Human adrenoreceptor system response to the development of high altitude pulmonary arterial hypertension. J Mol Cell Cardiol 1989; 21(Suppl.1):175–9.

Aldashev AA, Borbugulov UM, Davletov BA, Meimanaliev TS. Variation of the state of adrenoceptors in highlanders with right ventricular hypertrophy. Physiologiya Tcheloveka (Human Physiology), 1989b; 6:170–2 (in Russian).

Aldashev AA, Agibetov KA, Yugai AA, Shamshiev AT. Specific proteins synthesized in human lymphocytes during hypoxia. Am J Physiol., 1991; 261:92–6.

Anderson GR, Stoler DL, Scarcello LA. Normal fibroblasts responding to anoxia exhibit features of the malignant phenotype. J Biol Chem 1989; 265:14885-92.

Archer SL, McMurtry IF, Weir EK. Mechanisms of acute hypoxic and hyperoxic changes in pulmonary vascular reactivity. In: Weir EK, Reeves JT, eds. Pulmonary vascular physiology and pathophysiology. New York: Marcel Dekker, 1989; 241–90.

Berridge MG, Irvine RF. Inositol trisphosphate: a novel second messenger in cellular signal transduction. Nature (London), 1984; 312:315–21.

Birnbaumer L, Codina J, Yatani A, Mattera R, Graf R, Olate J, Themmen A, Liao C-F, Sanford J, Okabe K, Imoto Y, Zhou Z, Abramowitz J, Suki W, Hamm H, Iyengar R, Birnbaumer M, Brown A. Molecular basis of regulation of ionic channels by G proteins. In: Recent progress in hormone research, Vol. 45. New York: Academic Press, 1989; 121–70.

Blume F. Metabolic and endocrine changes at altitude. In: West J, Lahiri S, eds. High altitude and man. Bethesda, MD: Waverly Press, 1984; 37–45.

Browder LW, Pollock M, Nickells RW, Heikkila JJ, Winning RS. Developmental regulation of the heat shock response. In: DiBernardino M, Etkin LD, eds. Genomic adaptability in cell specialization. New York: Plenum Press, 1989; 97–147.

Celis JE, Bravo R, Larsen PM, Fey SJ. Cyclin: a nuclear protein whose level correlates directly with the proliferative state of normal as well as transformed cells. Leukemia Res 1984; 8:143–57.

Celis JE, Madsen P, Nielsen S, Celis A. Nuclear patterns of cyclin (PCNA) antigen distribution subdivide S-phase in cultured cells—some applications of PCNA antibodies. Leukemia Res 1986; 10:237–49.

Collins S, Bouvier M, Bolanowski MA, Caron MG, Lefkowitz RJ. cAMP stimulates transcription of the β2-adrenergic receptor gene in response to short-term agonist exposure. Proc Natl Acad Sci USA, 1989; 86:4853–7.

Fisher AB, Dodia C. Lung as a model for evaluation of critical intracellular PO_2 and PCO_2. Am J Physiol 1981; 241:E47–50.

Fishman AP. Hypoxia on the pulmonary circulation. How and where it acts. Circ Res 1976; 38:221–31.

Freid RE, Reid L. The effect of isoproterenol on the development and recovery of hypoxic pulmonary hypertension. A structural and hemodynamic study. Am J Pathol 1985; 121:102–11.

Furchgott FR, Zavadski JV. The obligatory role endothelial cells in the relaxation of arterial smooth muscle by acetycholine. Nature (London) 1980; 288:373–6.

Gilman AG. Regulation of adenylyl cyclase by G proteins. In: Nishizuka Y, et al., eds.

The biology and medicine of signal transduction. New York: Raven Press, 1990; 51–7.

Hoeffler JP, Meyer TE, Yun Y, Jameson JL, Habener JF. Cyclic AMP-responsive DNA-binding protein: structure based on a cloned placental cDNA. Science 1988; 242:1430–3.

Houge A. Role of histamine in hypoxic pulmonary hypertension in the rat. 1. Blockade or potentiation of endogenous amines kinins and ATP. Circ Res 1968; 22:371–83.

Karaki H. Calcium localization and sensitivity in vascular smooth muscle. Trends Pharmacol Sci 1989; 10:320–5.

Kay D, Waymire J, Grover RF. Lung mast cell hyperplasia and pulmonary histamine-forming capacity in hypoxic rats. Am J Physiol 1974; 226:178–84.

King S, Booyse F, Lin P-H, Traylor M, Narkates A, Oparil S. Hypoxia stimulates endothelial cell angiotensin-converting enzyme antigen synthesis. Am J Physiol 1989; 256:C1231–8.

Langleben D, Serban L, Blais D, Mohamed F, Stewart D. Hypoxia stimulates endothelin-1 release from pulmonary artery smooth muscle cells in-vitro. Abstracts 64th AHA Scientific Sessions, Anaheim, CA, 1991; A0672.

Lee S-H, Hurwitz J. Mechanism of elongation of primed DNA by DNA polymerase δ, proliferating cell nuclear antigen, and activator 1. Proc Natl Acad Sci USA 1990; 87:5672–6.

Lefkowitz RJ, Caron MG. The adrenergic receptors. In: Nishizuku Y, et al., ed. The biology and medicine of signal transduction. New York: Raven Press, 1990; 1–8.

Lloyd TC. Pulmonary vasoconstriction during histotoxic hypoxia. J Appl Physiol 1965; 20:488–90.

Lock JE, Olley PM, Coceani F. Hypoxia enhances pulmonary vascular beta-receptor reactivity. Circulation 1979; 60:5–16.

Maki T, Kontula K, Harkonen M. The beta-adrenergic system in man: physiological and pathophysiological response. Scand J Clin Lab Invest 1990; 50:25–43.

Mallery SR, Laufman HB, Solt CW, Stephens RE. Association of cellular thiol redox status with mitogen-induced calcium mobilization and cell cycle progression in human fibroblasts. J Cell Biochem 1991; 45:82–2.

Marsh J, Sweeney K. β-Adrenergic receptor regulation during hypoxia in intact cultured heart cells. Am J Physiol 1989; 256:H257–81.

McMurtry IF, Davidson AB, Reeves JT, Grover RF. Inhibition of hypoxic pulmonary vasoconstriction by calcium antagonists in isolated rat lungs. Circ Res 1976; 38:99–104.

Mecham RP, Whitehouse LA, Wrenn DS, Parks W, Griffin G, Senior R, Crouch E, Stenmark K, Voelkel N. Smooth muscle-mediated connective tissue remodelling in pulmonary hypertension. Science 1987; 237:423–6.

Menshikov M, Borbugulov U, Mazaev AV, Tkachuk VA. Calcium-blocking effect of nitro-compounds on human blood platelets. Proc Natl Acad Sci USSR (Dokl Akad Nauk SSSR) 1986; 4:1003–6 (in Russian).

Mirrakhimov MM. Heart disease and the mountains. Kirghizstan, Frunze, 1971 (in Russian).

Mirzapoyazova TY, Buravkova LV, Farber H, Grigorian GY. Cultivated vascular endothelium of human subjects and animals: chemoreceptor control in normoxia and hypoxia. In: Mirrakhimov, MM, ed. Pulmonary arterial hypertension. Moscow: Frunze, 1988; 52–4 (in Russian).

Morganroth ML, Reeves JT, Murphy R, Voelkel W. Leukotriene synthesis and receptor blockers block hypoxic pulmonary vasoconstriction. J Appl Physiol 1984; 56:1340–6.

Myles W, Drucker A. The excretion of catecholamines in rats during acute and chronic exposure to altitude. Can J Physiol Pharmacol 1971; 49:721–6.

Nanoff C, Freismith M, Tsuisl E, Schultz W. A different desensitization pattern of cardiac adrenoceptor subtypes by prolonged in vivo infusion of isoprenaline. J Cardiovasc Pharmacol 1989; 13:198–203.

Pennathur-Das R, Fung K, Larrick JW. Cytokine production by leukocytes exposed to hypoxic stress. Proc Hypoxia '89, Lake Louis, Canada, 1989; 34.

Riabowol KT, Mizzen LA, Welch WJ. Heat shock is lethal to fibroblasts microinjected with antibodies against hsp70. Science 1988; 242:433–5.

Richalet J-P, Delavier C, Keromes A, Herry J-P, Dubray C. Human lymphocytes beta-adrenoceptors in altitude hypoxia. Proc Hypoxia, '89, Lake Louis, Canada, 1989, 34.

Rubanyi GM. Endothelium-derived relaxing and contracting factors. J Cell Biochem 1991; 46:27–36.

Ryan U, Ryan J, Grutchley DJ. The pulmonary endothelial surface. Fed Proc 1985; 44:2603–9.

Schror K. Prostaglandins, other eicosanoids and endothelial cells. Basic Res Cardiol 1985; 80:502–14.

Sheng M, Thompson MA, Greenberg ME. CREB: a Ca^{2+}-regulated transcription factor phosphorylated by calmodulin-dependent kinases. Science 1991; 252:1427–30.

Sibley DR, Lefkowitz RJ. Biochemical mechanisms of β-adrenergic receptor regulation. ISI atlas sci, Pharmacol 2, 1988; 66–70.

Sole M, Drobac M, Schwartz I, Hussain M, Vanghain-Weil F. The extraction of circulating catecholamines by the lungs in normal man and in patients with pulmonary hypertension. Circulation 1979; 60:160–3.

Stenmark K, Abman SH, Accurso FJ. Ethiologic mechanism in persistent pulmonary hypertension of the newborn. In: Weir EK, Reeves JT, eds. Pulmonary vascular physiology and pathophysiology. New York: Marcel Dekker, 1989; 355–402.

Stenmark KR, Aldashev AA, Orton EC, Durmowicz AG, Badesh DB, Parks WC, Voelkel NF, Reeves JT. Cellular adaptation during chronic neonatal hypoxic pulmonary hypertension. Am J Physiol, 1991; 261:97–4.

Stulcup SA, Lipset J, Woan J, Lenenberger P, Mellins, RB. Converting enzyme activity in cultured endothelial cells by hypoxia. J Clin Invest 1979; 63:966–76.

Su YF, Harden TK, Perkins JP. Catecholamine-specific desensitization of adenylate cyclase. J Biol Chem 1980; 255:7410–9.

Tan CK, Castillo C, So AG, Downey KM. An auxiliary protein for DNA polymerase-delta from fetal calf thymus. J Biol Chem 1986; 261:12310–16.

Travali S, Ku D-H, Rizzo MG, Ottavio L, Baserga R, Calabretta B. Structure of human gene for proliferating cell nuclear antigen. J Biol Chem 1989; 264:7466–72.

Vender RL, Clemmons DR, Kwock L, Freidman M. Reduced oxygen tension induces pulmonary endothelium to release a smooth muscle cell mitogen(s). Am Rev Respir Dis 1987; 135:622–7.

Voelkel NF, Hegstrand J, Reeves JT, McMurtry IF, Molinoff PB. Effects of hypoxia on density of b-adrenergic receptors. J Appl Physiol 1981; 50:363–6.

von Euler US, Liljestrand G. Observations on the pulmonary arterial blood pressure in the cat. Acta Physiol Scand 1946; 12:301–20.

Weir EK, Tucker A, Reeves JT, Will DH, Grover RF. The genetic factor influencing pulmonary hypertension in cattle at high altitude. Cardiovasc Res 1974; 8:745–9.

Yanagisawa M, Kurihara H, Kimura S, Tomote Z, Kaboyashi M, Yazaki Y, Guot K, Masaki T. A novel potent vasoconstrictor peptide produced by vascular endothelial cells. Nature (London) 1988; 322:411–2.

Zamotaev IP. Lung and heart failure. Moscow: Medicine, 1978 (in Russian).

Zimmerman LH, Levine RA, Farber HW. Hypoxia induces a specific set of stress proteins in cultured endothelial cells. J Clin Invest 1991; 87:908–14.

Part Five

CELLS OF THE IMMUNE SYSTEM

22

Alveolar Macrophage Growth-Factor Signaling Systems

JAMES C. BONNER and ARNOLD R. BRODY

National Institute of Environmental Health Sciences
Research Triangle Park, North Carolina

I. Introduction

Several broad categories of inflammatory lung disease are characterized by abnormal patterns of cell growth and matrix production. Emphysema involves a continual destruction of alveolar walls (Demedts and Aumann, 1988), while fibrosis is associated with increased numbers of cells and connective tissue production at the alveolar level (Chang et al., 1988; Crystal et al., 1984; Keogh and Crystal, 1982). Although numerous etiological agents have been identified, and there are many good clues as to how these disease processes are mediated, the precise biochemical and molecular mechanisms which mediate these diseases remain undefined. The lung macrophage is central to many of the current hypotheses designed to explain inflammation and its sequelae. This large pleiomorphic cell originates from blood-borne monocytes which mature in the lung interstitium or on alveolar surfaces (Hance et al., 1985; Spurzem et al., 1987). Mature lung macrophages are not considered a highly replicative cell population under normal conditions (Van Oud Ablas and van Furth, 1979), although macrophages can replicate when activated at sites of inflammation (Bitterman et al., 1984). The reason macrophages are believed to play a central role in inflammatory processes in the lung and elsewhere (e.g., in wound healing, atherosclerosis, and arthritis) is their huge armamentarium of secreted enzymes, oxygen radicals, arachidonic acid metabolites and growth factors (reviewed in Nathan, 1987). Even though macrophages are associated with the disease processes noted above, there is no clear view of which of the many secreted products or their potential combinations actually mediates lung disease. This obviously is a fruitful area of research, but even more fundamental than establishing which factors are mediating disease is an understanding of the mechanisms which control their chemotaxis to the site of tissue damage, their subsequent activation by various foreign substances at this site, and finally their release of cytokines. Thus, this brings us to two major considerations of this chapter on macrophage signaling systems: (1) intracellular signaling (signal transduction) pathways induced by growth factors, chemoattractants, and stimulators of macrophage activation; and (2) extracellular paracrine signaling of various pulmonary cells by growth factors released by macrophages.

We divide this chapter into two sections only for convenience. The two, intra- and extracellular signaling, are inextricably linked, since the macrophage itself must be perturbed or somehow signaled extracellularly before activation of intracellular second messenger pathways that lead to the release of its own potent "signals" or growth factors, which, in turn, may bind to the specific receptors on target cells with their own signal transduction mechanisms. Through this cycle, the lung macrophage is likely to be a significant contributor to a wide variety of disease processes. Among the products proposed to play a major role in this regard are the peptide growth factors such as platelet-derived growth factor, the transforming growth factors, and insulinlike growth factors. These and others are known to influence the biochemistry and biology of other cell types. Following is a review of macrophage signal transduction pathways that are turned on by specific extracellular signals (activators or suppressors) and the control of these intracellular signals by growth factors which bind to macrophage cell surface receptors. Because several activators of macrophage signal transduction up-regulate the gene expression and secretion of growth factors, a wide variety of these macrophage-derived, growth-promoting cytokines will be reviewed which act as extracellular (paracrine) signals for pulmonary epithelial, mesenchymal, and endothelial cells.

II. Intracellular Signaling

A number of external stimuli are capable of activating signal transduction pathways within the alveolar macrophage. These agents may be growth factors, lymphokines, proteinase inhibitors, or particles, and their effect on macrophage function may either be stimulatory, suppressive, or both, depending on the nature of the external signal (Table 1). It has been proposed that the regulation of intracellular signaling by extracellular signals is likely controlled at at least three levels: (1) the nature of the molecule that interacts with the cell surface of the macrophages; (2) the specific biological function of the macrophage that is being studied; and (3) the presence of other signals (stimulatory or inhibitory) that may modulate the response to the first signal given (Adams et al., 1989). Activators of macrophage signal transduction include bacterial lipopolysaccharide (LPS), phorbol esters, oxidants, and inhaled particles (Fig. 1). Activation of intracellular second messenger cascades trigger specific mRNA expression, and the end result is some biological function of the macrophage, such as phagocytosis, chemotaxis, growth-factor secretion, expression of surface molecules, antibody-dependent cytotoxicity, or macrophage-mediated tumor cytotoxicity. The better-established second messenger pathways involving activation leading to macrophage function are those involving macrophage chemotaxis following the binding of specific

Table 1 Examples of Agents that Modulate Macrophage Function: Activators and/or Suppressors

Activators
Interferon-γ
Tumor necrosis factor
Interleukin-2
Interleukin-3
Interleukin-4
Granulocyte-macrophage colony-stimulating factor
Suppressors
Prostaglandin E2
α_2-Macroglobulin-protease complex
Transforming growth factors (TGF-β1 and β2)
Tumor-derived protein (TDP)
Activator/suppressors[a]
Bacterial lipopolysaccharide (LPS)
Tumor necrosis factor
Interferon-β
Interferon-α

[a] Agents that may be stimulatory or inhibitory to macrophage function, depending on the precise function studied or the circumstance of the experimental design. Reviewed in Uhing and Adams (1989).

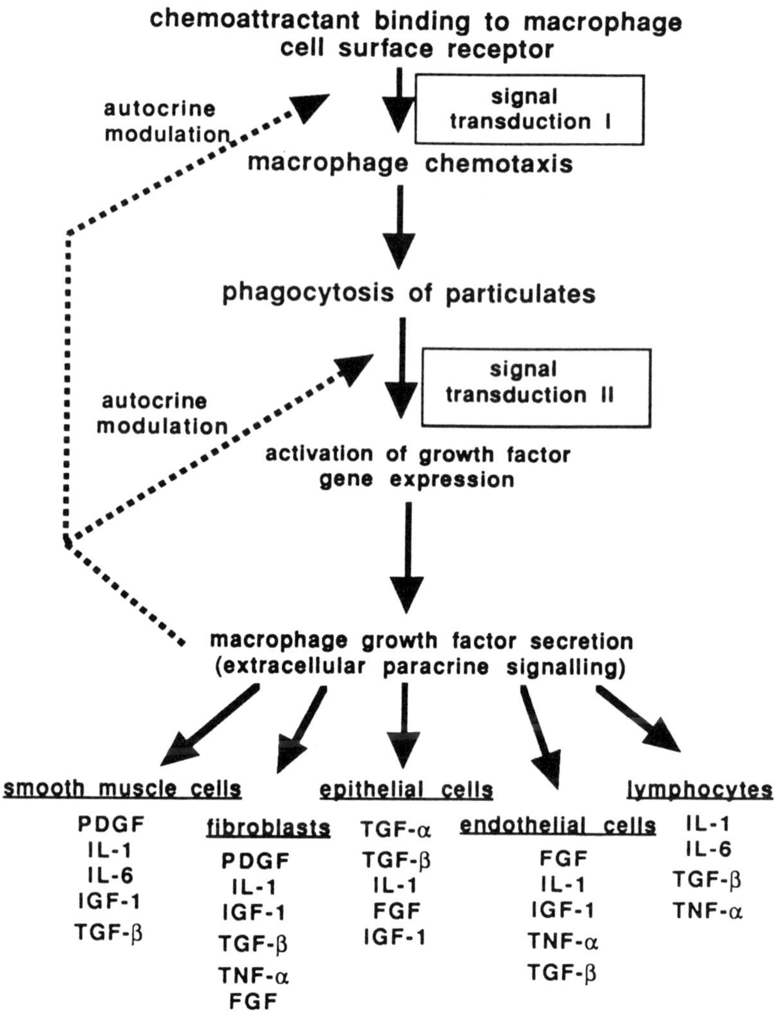

Figure 1 Intra- and extracellular signaling by alveolar macrophages via growth factors. Several stimuli, including fMet-Leu-Phe, C5a, and LTB4, generated at sites of tissue damage, bind to macrophage cell surface chemoattractant receptors and induce intracellular signal transduction pathways (I), resulting in chemotaxis to sites of inflammation, where macrophages ingest organic and/or inorganic insults inhaled and deposited within the airways. Phagocytosis or cell surface interaction with these particulates results in the activation of other signal transduction pathways (II), leading to the activation and up-regulation of specific growth-factor genes. Growth factors secreted by activated macrophages impact on other pulmonary cells in a paracrine fashion to influence mitogenesis, chemotaxis, immune function, and extracellular matrix production. Some growth factors secreted by macrophages (e.g., TNF-α) modulate macrophage signal transduction (dashed lines).

agents to chemoattractant cell surface receptors, and the tumoricidal activity of macrophages following IFN-γ priming and LPS activation. As shown in Fig. 1, activation of alveolar macrophages by LPS and inorganic particulates up-regulates the gene expression and subsequent secretion of a number of growth-promoting cytokines. The signaling events, both intra- and extracellular, will be discussed in order according to this flow scheme, which was developed in attempting to ease our comprehension of this complex topic.

A. Signal Transduction Pathways Stimulated by Chemoattractants

Alveolar macrophages are capable of phagocytizing a wide variety of inhaled organic and inorganic particles that deposit along the airways of the lung. Macrophages generally are attracted to these particles so they can be phagocytized, and the cells subsequently become activated to secrete growth factors. In this way, the alveolar macrophages are the primary effector cells in the lung that participate in a repair or disease process. For example, alveolar macrophages migrate in a directed pattern to sites of asbestos deposition at alveolar duct bifurcations within 24 h following asbestos inhalation (Warheit et al., 1984). The primary chemoattractant involved in this migration is the complement component C5a, produced by asbestos-mediated cleavage of C5 at the deposition site (Warheit et al., 1985). Other chemoattractants for macrophages include the arachidonic acid metabolite leukotriene B4 (LTB4), platelet-activating factor (PAF), and the formyl peptide N-formyl-methionyl-leucyl-phenylalanine (fMet-Leu-Phe) (Verghese et al., 1986b). These chemoattractants may originate from cells, bacteria, or from humoral sources (Table 2). Any of these chemoattractants could be central to the migration of macrophages to sites of inflammation, where the macrophage may become further activated and perform extracellular paracrine signaling with other pulmonary cell types via the release of growth factors.

With regard to intracellular signaling induced by a chemoattractant, perhaps the best-studied molecule is fMet-Leu-Phe (for review, see Snyderman et al., 1986). This chemoattractant, among others, activates a cascade of second messengers that ultimately lead to cytoskeletal reorganization, resulting in cell locomotion (Fig. 2). fMet-Leu-Phe binds to a specific cell surface receptor on the macrophage—i.e., an example of an organic stimulus that does not cross the plasma membrane. Although specific chemoattractant receptors have been identified, all of these receptor types for chemoattractants appear to operate through the same signal transduction pathways (Verghase et al., 1986b). G proteins have been shown to be complexed to chemotaxis receptors, ion channels, and enzymes of phosphotidylinositol metabolism (Stryer and Bourne, 1986). Exposure to chemoattractants leads to one of the earliest measurable responses: a rapid, transient increase in intracellular Ca^{2+} levels

Table 2 Selected Chemoattractants for Macrophages

Chemoattractant	Source
Endogenous humoral factors	
c5a	Proteolytic cleavage product from complement activation
Thrombin	Clotting cascade
Cell-derived chemoattractants	
Leukotriene B4 (LTB4)	Arachidonic acid metabolite
Platelet-activating factor (PAF)	Produced by PMNs
Lymphocyte-derived chemotactic factor (LDCF)	Produced by lymphocytes following antigen stimulation
Bacterial products	
N-formyl-methionyl-leucyl-phenylalanine (fMet-Leu-Phe)	product of bacterial protein synthesis

Reviewed in Verghese et al. (1986b).

(Verghese et al., 1986a). This increase in Ca^{2+} is mediated by a cleavage product of phosphatidylinositol 4,5-bisphosphate (PIP2), referred to as myo-inositol-1,4,5-triphosphate (IP3) (Berridge and Irvine, 1984; Verghese et al., 1986c; Shears, 1989). 1,2-Diacylglycerol is also produced by phospholipid hydrolysis and functions as a mediator of protein kinase C activation (Nishizuka, 1984).

After a chemoattractant binds to its receptor on the cell surface, there is a rapid hydrolysis of PIP_2 by membrane-associated phospholipase C, which leads to an intracellular release of IP_3 (Dougherty et al., 1984). This IP_3 accumulation occurs with chemoattractants other than fMet-Leu-Phe, including C5a, PAF, and LTB4 (Verghese et al., 1986b). The intracellular increase in IP_3 causes an increase in intracellular Ca^{2+}, that is derived either by the opening of ion channels to the cell surface or by releasing Ca^{2+} stored in intracellular organelles (Verghese et al., 1987). The increase in cytosolic Ca^{2+} is a signal that increases the level of cAMP and the activity of protein kinase C (PKC). The chemoattractant-induced increase in intracellular cAMP and PKC are apparently termination signals for the signal transduction pathways that drive the chemotactic machinery (Snyderman et al., 1986). This is because cAMP-dependent protein kinase may promote the inactivation of the phospholipase C which hydrolyzes PIP_2 (Fig. 2). Thus, Ca^{2+} and 1,2-diacylglycerol are stimulatory signals for chemotaxis, and their actions can be mimicked by Ca^{2+} ionophores and PKC activators (Snyderman et al., 1986). Mobilization of Ca^{2+} by chemoattractants also increases cAMP, and this elevated cAMP is

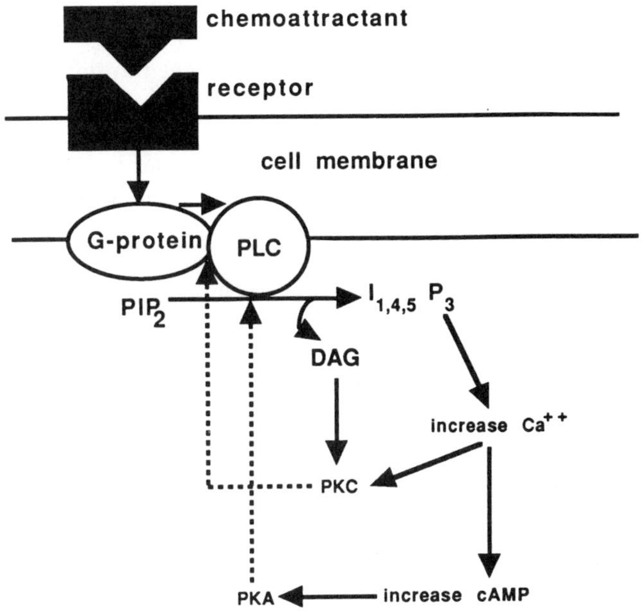

Figure 2 Signal transduction pathways following the binding of chemoattractants to cell surface receptors on macrophages. Second messenger pathways (solid lines) involved in signal transduction following the binding of a chemoattractant molecule to its cell surface receptor on macrophages. Specific chemoattractants (e.g., PAF and C5a) bind specific receptors, but activate a common signal transduction pathway via a receptor-linked G protein that activates phospholipase C (PLC) to hydrolyze polyphosphoinositol 4,5-bisphosphate (PIP2) to inositol 1,4,5-trisphosphate (IP3) and 1,2-sn-diacylglycerol (DAG). The formation of IP3 increases cytosolic Ca^{2+}, which in turn increases cAMP. Increased Ca^{2+} and DAG activate protein kinase C (PKC) and increased cAMP activates protein kinase A (PKA), and these protein kinases in turn modulate growth-factor gene expression and receptor function. PKC and PKA also serve to terminate the PLC-mediated conversion of PIP2 to IP3 and DAG. These negative regulatory pathways are indicated by dashed arrows. Many of the signal transduction pathways involved in macrophage chemotaxis are also utilized in macrophage activation for phagocytosis, and chemotaxis of macrophages is essential to positioning these effector cells to sites of tissue damage and inflammation, where macrophages can signal other cells in paracrine manner via growth factors (see Fig. 1). (Modified with permission from Snyderman et al., 1986.)

an inhibitory intracellular signal (Verghese et al., 1987). Importantly, chemoattractants neither stimulate nor inhibit membrane-bound adenylate cyclase, suggesting a novel role for G proteins in chemoattractant-induced signal transduction. PKC may also terminate PIP_2 hydrolysis by inhibiting the activation of phospholipase C. This type of intracellular signaling induced by chemoattractants could also serve as a model for understanding the actions of other stimuli, including PAF, C5a, and growth factors (Snyderman et al., 1986). Much of the work described above relates to monocytes and/or peritoneal macrophages. Macrophages, while derived from monocytes, possess some differences with regard to their cell surface receptors, and the growth factors that they secrete (Helmke et al., 1987). In addition, subpopulations of macrophages may alter their phenotype under circumstances of inflammation (Hance et al., 1985). For this reason, one must be cautious in interpreting data on macrophage-derived growth factors and correlating this information with a disease state. A recent study reports that alveolar macrophage signaling following the binding of fMet-Leu-Phe is similar to that found in polymorphonuclear cells, except that alveolar macrophages possess fewer chemoattractant receptors and G proteins and thus have a reduced response to this chemoattractant (Beaty et al., 1991).

B. Signal Transduction Leading to Growth-Factor Production

LPS Activation

As depicted in Fig. 1, alveolar macrophages are first activated to undergo a directed migration subsequent to the binding of chemoattractants to cell surface receptors. Once these phagocytes arrive at the site of chemoattractant generation, they encounter foreign substances that have been inhaled and deposited in the lung. The macrophages then actively phagocytize these foreign particulates, triggering another cascade of signal transduction, resulting in growth-factor gene expression and protein secretion. Inorganic particulates and LPS both up-regulate a number of growth factors produced by macrophages (Table 3). Little is known concerning the second messenger pathways following activation by inorganic particles such as asbestos and silica, but a large body of literature has been published on macrophage signal transduction following LPS activation.

LPS is a potent inducer of several biological responses of mononuclear phagocytes, including stimulation of the secretion of the growth factors TNF (Manogue and Cerami, 1988), IL-1 (Dinarello, 1988), PDGF (Shimokado et al., 1985), TGF-β (Assoian et al., 1987), IL-6 (Tosato et al., 1988), and IGF-1 (Rom et al., 1988). Other effects of LPS include priming for eicosanoid secretion, tumoricidal activity, and inhibition of interferon-γ (IFN-γ) induction of surface class II major histocompatibility complex immune-associated antigens

Table 3 Examples of Activators of Macrophage Function That Increase Growth-Factor Secretion

Activator	Growth factor studied	Reference
Organic		
LPS	PDGF	Shimokado et al., 1985
	TGF-β	Assoian et al., 1987
	IGF-1	Rom et al., 1988
	IL-1	Dinarello, 1988; Lepe-Zuniga and Gery, 1984
	IL-6	Tosato et al., 1988
	TNF-α	Manogue and Cerami, 1988
	TGF-α	Madtes et al., 1988
Thioglycollate	FGF	Baird et al., 1985
Bleomycin	IL-1, IL-6	Jordana et al., 1988
Inorganic particulates		
Asbestos	PDGF	Bonner et al., 1991
Silica	TNF-α	Piquet et al., 1990
Carbonyl iron	PDGF	Bonner et al., 1991;
Cell-derived factors		
IFN-γ	TNF, IL-1	Collart et al., 1986
	PDGF	Shaw et al., 1991[a]
TNF	IL-1, IL-6, IL-8, TNF, PDGF, M-CSF, GM-CSF	Vilcek and Lee, 1991[a]

[a] Review of growth factors stimulated by TNF.

(Reviewed in Adams and Hamilton, 1988). In murine macrophages, LPS stimulates phosphatidylinositol metabolism (Prpic et al., 1987) and protein phosphorylation (Rosen et al., 1989), while reducing prostaglandin-E2-stimulated cAMP accumulation (Okonog et al., 1991). With regard to the latter study, some responses of the macrophages that are stimulated by LPS are inhibited by cAMP elevation and thus cAMP is a negative modulator of macrophage function. Accordingly, prostaglandin E2 inhibits macrophage phagocytosis and growth-factor secretion, likely by increasing intracellular cAMP (Figueiredo et al., 1990). LPS inhibits the ability of PGE2 to stimulate cAMP accumulation via the activation of phosphodiesterase and thus LPS counteracts these negative transduction signals (Okonog et al., 1991).

The production of growth factors by macrophages can be primed by the lymphokine, interferon-γ (IFN-γ). Following IFN-γ treatment, macrophages can be fully activated for growth-factor secretion by adding activators such as

LPS, indicating that the transmembrane signaling pathways induced by IFN-γ and LPS are linked. For example, macrophages primed with IFN-γ prior to activation with inorganic particulates increase their production of PDGF-B chain dimers when compared to stimulation of the macrophages with inorganic particles alone (our unpublished observations). Resting macrophages, when primed with IFN-γ without subsequent LPS stimulation, increase their production of PDGF-B mRNA (Shaw et al., 1991a). Thus while IFN-γ can prime macrophages for LPS activation, IFN-γ alone can also stimulate growth-factor secretion. Furthermore, an increase in monocyte PDGF-B mRNA has been associated with increases in mRNA for c-fos, c-jun, and the early gene EGR-2 (Shaw et al., 1991b). Other effects of IFN-γ on macrophage function include alterations in cell surface markers, e.g., increases in cell surface antigens and down-regulation of the transferrin receptor (Uhing and Adams, 1989). Collectively, these studies indicate that IFN-γ and LPS are useful tools for reaching various stages of macrophage activation, the stage of activation defined in terms of both secreted protein and cell surface receptors.

Several studies have reported an increase in TNF production following priming with IFN-γ and subsequent activation with LPS (reviewed in Jaattela, 1991). TNF-α is particularly important in modulating the effects of other growth factors on pulmonary cells and, because much is known regarding the signal transduction pathways involved in its production by macrophages, TNF research may give insights into the production of other growth factors by macrophages. IFN-γ enhances the LPS-induced accumulation of mRNA for TNF in macrophages (Koerner et al., 1987), and several second messengers are generated which result in the up-regulation of TNF mRNA (Fig. 3). Following the binding of LPS to the cell surface, a G-protein-linked receptor complex is activated that in turn activates phospholipase C (PLC), which mediates the hydrolysis of polyphosphoinositol 4,5-bisphosphate (PIP_2) to diacylglycerol (DAG) and inositol 1,4,5-trisphosphate (IP_3). DAG activates protein kinase C (PKC), and PKC, along with increased Ca^{2+} caused by IP_3 stimulation, positively regulate the phosphorylation of specific proteins leading to the expression of early genes such as JE and causes transcription of the protooncogenes c-fos and c-myc. This transcription up-regulates the transcription of TNF. PKC and PKA modulate TNF production and receptor expression antagonistically (see Fig. 4). It is also of interest that the IFN-γ primed secretion of TNF and superoxide anion decline with age in rats (Davila et al., 1990).

Overall, LPS is regarded as a positive extracellular signal for stimulating macrophages to synthesize and secrete growth factors. Other agents that up-regulate growth factor production, such as bleomycin, asbestos, and silica (Table 3), could turn on intracellular signaling by similar mechanisms. Unlike chemoattractants, which initiate transmembrane signaling by binding to cell

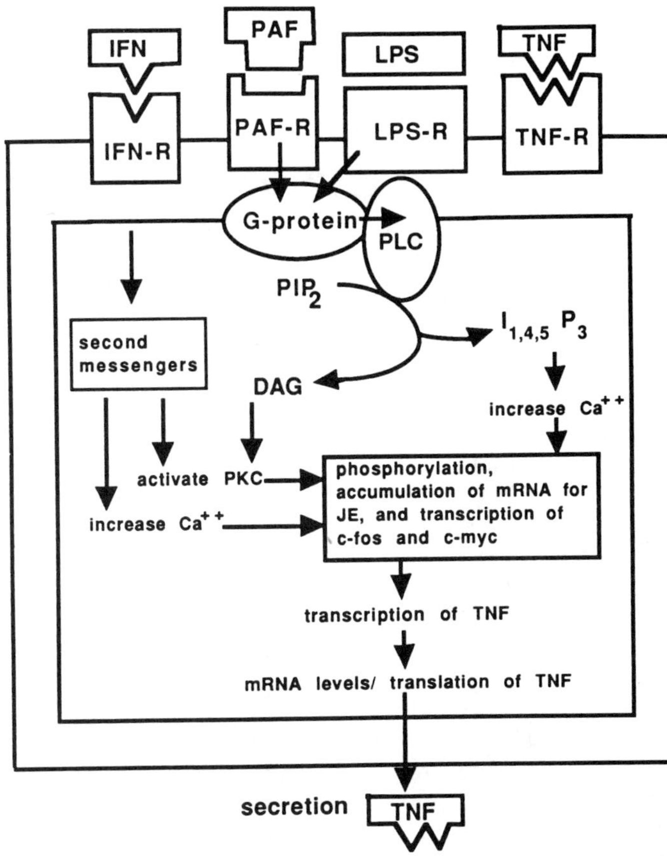

Figure 3 Signal transduction pathways involved in the activation of the macrophage to produce tumor necrosis factor (TNF) following the binding of bacterial lipopolysaccharide (LPS) to the cell surface. The additional effect of interferon-γ (IFN) on activation is also considered. As with chemoattractants (see Fig. 2), the LPS receptor is linked to a G protein that activates phospholipase C (PLC), which mediates the hydrolysis of polyphosphoinositol 4,5-bisphosphate (PIP2) to diacylglycerol (DAG) and inositol 1,4,5-triphosphate (IP3). DAG activates protein kinase C (PKC), and PKC, along with increased Ca^{2+}, caused by IP3 stimulation, positively regulate the phosphorylation of specific proteins and leads to the expression of early genes such as JE and causes transcription of c-fos and c-myc. This transcription up-regulates the transcription of TNF. PKC and PKA (not shown) modulate TNF production and receptor expression antagonistically (see Fig. 4). (Modified with permission from Uhing and Adams, 1989.)

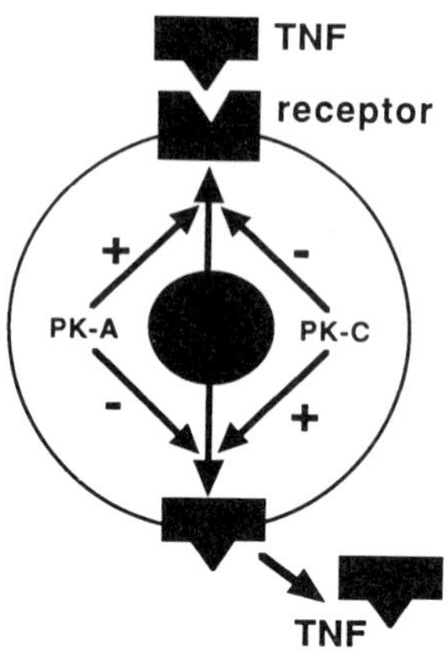

Figure 4 Modulation of macrophage signal transduction pathways by TNF-α and the Antagonistic action of protein kinases A and C on TNF-receptor binding and TNF production. The activation of PKC increases the production of TNF mRNA and stimulates the cleavage of a 17-kDa TNF from a 26-kDa membrane-bound precursor. PKA counteracts these positive signals and inhibits TNF production and release. The activation of PKA increases the responsiveness of TNF receptors, while PKA decreases TNF responsiveness (Modified with permission from Kronke et al., 1990.)

surface receptors, some activators of macrophage growth-factor secretion do not act via specific cell surface receptors. A variety of different stimuli may activate intracellular signaling, from LPS, which binds to the cell surface and activates G protein as a first step, to inorganic particulates such as asbestos and silica, which are phagocytized by the macrophage. In the latter instance, it is unclear which second messengers are activated as the initiating step(s) of the signaling cascade (Fig. 5), and a possible mechanism for activation by inorganic particulates is discussed below. It is clear that a number of very different types of activators (Table 3) can all lead to an up-regulation of several growth-factor genes by either entering the cell via phagocytosis or by binding to the cell surface.

Activation by Inorganic Particulates

As mentioned above, organic stimulators such as LPS and IFN-γ act to up-regulate growth-factor production, including TNF-α and PDGF. In addition to LPS, several inorganic particles, including asbestos (a fibrogenic particulate) and carbonyl iron (a nonfibrogenic particle), are also capable of increasing PDGF production. Macrophage-derived growth factors have been postulated as mediators in the progression of asbestosis and silicosis, and an enhanced

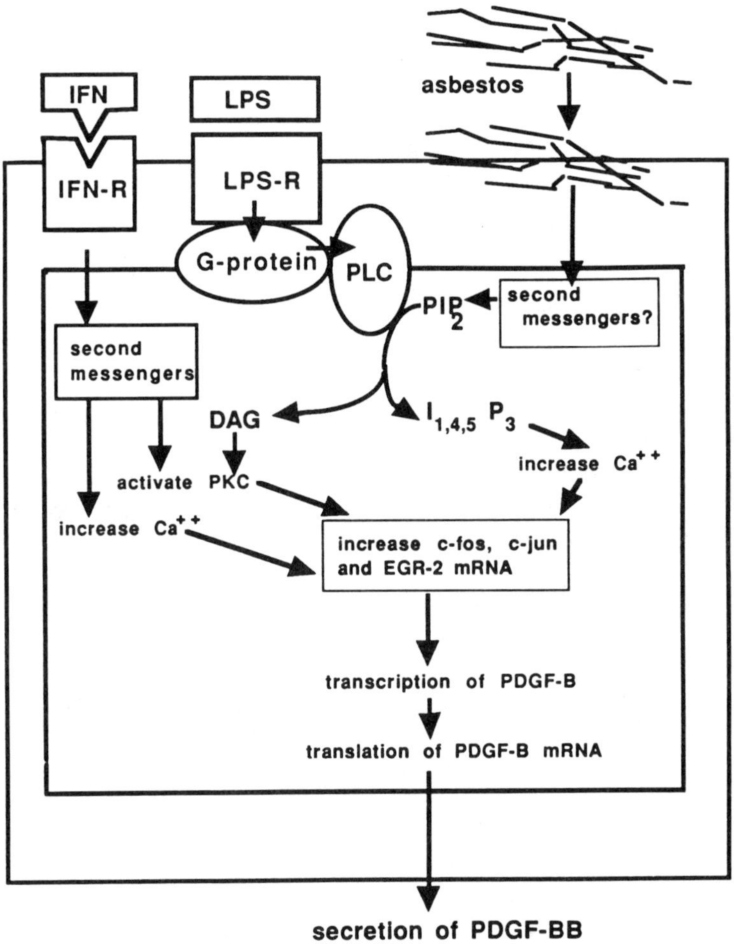

Figure 5 Postulated signal transduction pathways involved in the activation of the macrophage to produce PDGF-BB following the binding of bacterial lipopolysaccharide (LPS) to the cell surface or activation by the phagocytosis of asbestos. With regard to the macrophage, these pathways are speculative because asbestos activation of phosphoinositide hydrolysis, DAG accumulation, and PKC activation have been demonstrated so far in airway epithelial cells (Sesko et al., 1990; Perderiset et al., 1991). The pathways following IFN-γ and LPS activation are taken from studies of TNF production (Fig. 3). The additional effect of interferon-γ (IFN) on activation is also considered. As with chemoattractants (see Fig. 2), the LPS receptor is linked to a G protein that activates phospholipase C (PLC), which mediates the hydrolysis of polyphosphoinositol 4,5-bisphosphate (PIP2) to diacylglycerol (DAG) and inositol 1,4,5-triphosphate (IP3). DAG activates protein kinase C (PKC), and PKC along with increase Ca^{2+}, caused by IP3 stimulation, up-regulate c-fos, c-jun and EGR-2 mRNA. The expression of early gene EGR-2 and the protooncogenes c-fos and c-jun following asbestos exposure precede PDGF mRNA expression.

macrophage-fibroblast interaction has been reported to increase silica-induced fibrosis (Adamson et al., 1989). Macrophage-derived PDGF and TGF-β have been postulated as growth factors that could act in coordination in causing fibrosis (Brody, 1991), since PDGF is a potent stimulator of fibroblast proliferation and TGF-β promotes the increased synthesis of extracellular matrix proteins. The signaling for up-regulation of PDGF secretion by asbestos could be similar to the pathways involved in TNF up-regulation by LPS, and a postulated mechanism for asbestos-induced activation of PDGF production is presented in Fig. 5.

While second messenger activation following asbestos activation has not been investigated in mononuclear phagocytes, recent studies on asbestos-induced signal transduction in tracheal epithelial cells (Perderiset et al., 1991; Sesko et al., 1990) may give clues to the mechanism(s) by which asbestos up-regulates gene expression and secretion of growth factors such as PDGF by macrophages. Asbestos causes a rapid increase in accumulation of DAG and hydrolysis of phosphoinositides that precedes asbestos-induced proliferation of tracheal epithelial cells (Sesko et al., 1990). Further studies showed that asbestos-stimulated proliferation is mediated through PKC (Perderiset et al., 1991). Because asbestos stimulates the gene expression and secretion of PDGF by alveolar macrophages, it will be of interest to learn if activation for growth-factor production by asbestos follows the same signal transduction pathways involved in asbestos-induced epithelial cell proliferation.

C. Growth-Factor Modulation of Macrophage Signal Transduction

Several growth-factor receptors are present on the cell surface of alveolar macrophages. These include receptors for TNF-α, CSF, IL-2, and IGF-1. These growth factors may act via their cell surface receptors to modulate macrophage signal transduction pathways involved in activation and chemotaxis, cause the clonal growth of macrophages, or affect the production of other macrophage growth factors. In the latter instance, growth factors produced by activated macrophages could modulate their own activity in an autocrine fashion, or growth factors from other pulmonary cells could effect their function in a paracrine manner.

TNF is perhaps the best studied with regard to macrophage signal transduction. In HL60 macrophagelike cells, cAMP, which activates PKA, was found to up-regulate TNF-binding capacity, and TNF treatment of these cells results in arachidonic acid and PGE-2 production (Kronke et al., 1990). TNF treatment also increases PKC (Vilcek and Lee, 1991), and the protein kinases A and C have antagonist effects on TNF receptor binding and secretion (Fig. 4). Early transmembrane events following TNF binding include activation of G proteins, activation of phospholipase A2 activity, an increase in

Ca^{2+} from internal stores, and subsequent increases in cAMP and PKC levels (Kronke et al., 1991). Interestingly, there is no increase in IP_3 levels, and it is possible that DAG generated from phosphatidylcholine, rather than phosphatidylinositol-4,5-bisphosphate, is responsible for PKC activation (Vilcek and Lee, 1991). TNF induces the synthesis of several other growth factors, including interleukins, PDGF, colony-stimulating factors (CSFs), and TNF itself (Table 3).

CSF appears to play critical roles in both macrophage maturation and activation (Clark and Kamen, 1987), as well as macrophage proliferation (Chen et al., 1988). Exposure of macrophages to M-CSF in vitro induces the production of plasminogen activator (Hamilton et al., 1980), TNF, and IFN-γ (Warren and Ralph, 1986). The mechanisms responsible for alveolar macrophage activation or proliferation within the lung are not well defined (Nathan, 1986). However, alveolar macrophages are known to proliferate in response to CSFs (Akagawa et al., 1988) and CSF induces the expression of the nuclear protooncogenes c-fos and c-myc. Thus, CSF secreted by macrophages or other pulmonary cells could be important to macrophage activation and proliferation.

Interestingly, approximately 13% of cultured alveolar macrophages that have been activated or stimulated with LPS or chrysotile asbestos express receptors for IGF-1, while unstimulated macrophages do not possess these growth-factor receptors (Rom, 1991). It has been suggested that these receptors may function to initate the proliferation of alveolar macrophages during inflammation, and thus increase of the local macrophage population. While alveolar macrophages have been described as nondividing, end-stage cells under normal conditions, they have been reported to increase in number during inflammation (Bitterman et al., 1984). Rat alveolar macrophages, either resting or activated with asbestos, do not possess receptors for PDGF (unpublished observation). Thus, IGFs would not be expected to promote macrophage growth in a coordinated fashion with PDGF, but other factors could be operative in a competence-type manner to cause macrophage proliferation in the presence of IGF.

III. Extracellular Paracrine Signaling

A. Growth Factors Secreted by Activated Macrophages

A number of different growth factors secreted by alveolar macrophages likely contribute to tissue remodeling and repair in the lung. When these factors are not appropriately expressed or modulated, they may become mediators of several pulmonary diseases. A list of these macrophage-derived growth factors and their target cells is provided in Table 4. Macrophage-derived mitogenic

Table 4 Selected Growth Factors Secreted by Activated Macrophages Correlated with Pulmonary Cells That Possess Cell Surface Receptors for These Cytokines

Growth factor	Target cell type	Mitogenic activity	Reference
PDGF	Fibroblast	Promitogenic	Shimokado et al., 1985; Bonner et al., 1991
	Smooth muscle	Promitogenic	Mornex et al., 1986
TGF-β	Fibroblast	Pro- and anti-	Fine and Goldstein, 1987; Moses et al., 1990
	Epithelial, endothelial	Antimitogenic	Moses et al., 1990
FGF	Fibroblast	Promitogenic	Baird et al., 1985
	Endothelial	Promitogenic	Burgess and Maciag, 1989
	Smooth muscle cells	Promitogenic	Burgess and Maciag, 1989
IGF-1	Fibroblast	Promitogenic	Rom et al., 1988
	Epithelial	Promitogenic	Retsch-Bogart et al., 1991
	Endothelial, smooth Muscle cells	Promitogenic	Czech, 1989
TGF-α	Epithelial,	Promitogenic	Madtes et al., 1988
IL-6	T lymphocyte	Promitogenic	Hirano and Kishimoto, 1990
IL-1β	T lymphocyte	Promitogenic	Dinarello, 1988
	Fibroblast	Promitogenic	Libby et al., 1985; Elias et al., 1988
	Smooth muscle	Promitogenic	Libby et al., 1985
TNF-α	Fibroblast	Promitogenic	Elias et al., 1988
TNF-α + IL-1β	Fibroblast	Antimitogenic	Elias et al., 1988

activity for fibroblasts was first described as macrophage-derived growth factor in studies of wound repair where fibroblast infiltration and proliferation were observed (Leibovich and Ross, 1975). In these studies, the macrophage population at the wound site was greatly reduced by treatment with antimacrophage serum, and this treatment decreased fibroblast infiltration and proliferation. This fibroblast-stimulating factor, termed macrophage-derived growth factor (MDGF), was also shown to be secreted by alveolar macrophages (Bitterman et al., 1982), and was also demonstrated as a growth factor for smooth muscle cells (Greenburg and Hunt, 1978; Glenn and Ross, 1981; Martin et al., 1981) and endothelial cells (Greenburg and Hunt, 1978; Martin et al., 1981). However, at this time it was unclear whether or not MDGF was distinct from other known growth factors or whether it consisted of a mixture of distinct mitogenic factors. Later studies demonstrated that at least part of the mitogenic activity attributed to MDGF from peritoneal macrophages was immunologically and biochemically identical to brain and pituitary fibroblast-derived growth factor (Baird et al., 1985). Some of the mitogenic activity was also attributed to interleukin-1, a 13- to 18-kDa mitogen for fibroblasts and lymphocytes, but not for smooth muscle cells and endothelial cells (Libby et al., 1985). A significant portion of mitogenic activity due to MDGF from alveolar and peritoneal macrophages was identified as PDGF, and the majority of the mitogenic activity in macrophage-conditioned medium was blocked by antibodies to PDGF (Shimokado et al., 1985). It is noteworthy that while growth factors are likely important mediators, other macrophage-derived agents such as fibronectin could contribute to fibroblast growth, chemotaxis, and extracellular matrix production (Bitterman et al., 1984, 1986). Furthermore, mediators such as arachidonic acid metabolites can regulate growth-factor release by macrophages (Phan et al., 1987).

Platelet-Derived Growth Factor (PDGF)

The originally described PDGF released from human platelets is a 30-kDa dimeric polypeptide consisting of two monomeric chains that are linked by sulfhydral bridges (Ross et al., 1986). Two different types of monomeric chains (A and B) give rise to three possible dimers (AA, AB, and BB), and these dimers recognize two types of PDGF receptor subunits (α and β) that may also dimerize to produce three possible receptor types ($\alpha\alpha$, $\alpha\beta$, and $\beta\beta$). PDGF-BB recognizes all three receptor types, while the -AB dimer binds $\alpha\beta$ and $\alpha\alpha$ receptors and -AA binds only the $\alpha\alpha$ receptor (Seifert et al., 1989).

PDGF-like molecules constitute the major portion of macrophage-derived mitogenic activity in vitro (Shimokado et al., 1985). At least two molecular weight forms of PDGF are produced by human alveolar and peritoneal macrophages, possessing apparent molecular weights of 37–39 kDa and 12–17 kDa (Shimokado et al., 1985). Both of these human PDGF-like molecules recognized antibodies against PDGF and competed for PDGF binding to cell surface receptors on dermal fibroblasts. The human alveolar macrophages were also found to express the sis gene product, which codes for the PDGF-B chain. Recently, rat alveolar macrophages were shown to secrete two different molecular weight forms of PDGF (16–18 kDa and 30–34 kDa), and each of these forms was a mixture of all three PDGF dimers (AA, AB, and BB) (Bonner et al., 1991). Like human macrophage-derived PDGF, rat alveolar marcrophage-derived PDGF recognizes anti-human PDGF antibodies, stimulates fibroblast growth, and binds to cell surface PDGF receptors on lung fibroblasts (Fig. 6). The major portion of this macrophage PDGF mixture (~75%) was either -AB or -BB, and both molecular weight forms competed for PDGF-BB binding to cell surface receptors on early-passage rat lung fibroblasts and stimulated the growth of these same cells (Bonner et al., 1991).

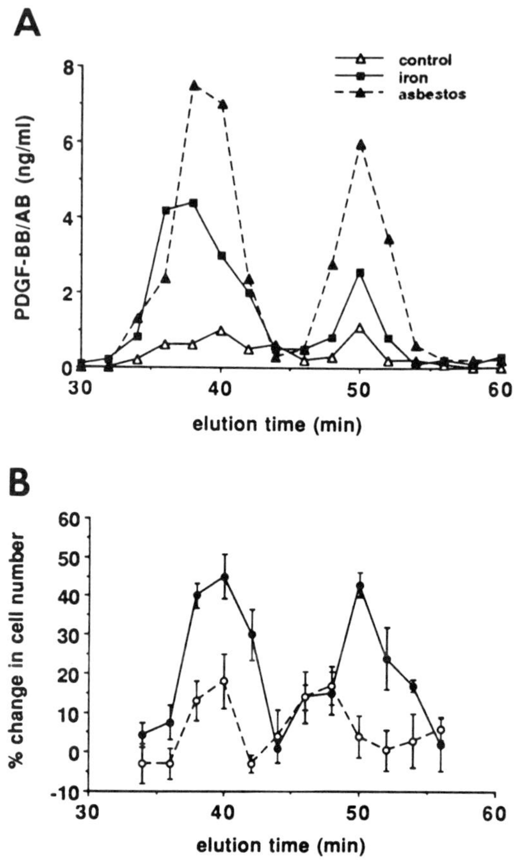

Figure 6 PDGF-like factors secreted by rat alveolar macrophages detected by immunoassay that stimulate the proliferation of rat lung fibroblasts. Media conditioned by 35–37 × 10^6 macrophages was collected after a 24-h incubation period. Concentrated media was fractionated by gel filtration chromatography in 1 M acetic acid. (A) An enzyme immunoassay specific for the PDGF-B-chain dimers showed that alveolar macrophages secrete two molecular weight forms (16–18 kDa and 30–34 kDa) of PDGF following activation with carbonyl iron (0.5 mg/cm^2) or chrysotile asbestos (15 $\mu g/cm^2$). (B) Both of these PDGF-like factor are signals for rat lung fibroblast proliferation in a cell growth assay, and the proliferation induced by these factors was inhibited by anti-PDGF (dashed line). These data demonstrate a paracrine-type signaling between the alveolar macrophage and lung fibroblast. (Reproduced with permission from Bonner et al., 1991.)

Interestingly, these rat lung fibroblasts proliferated in response to the -BB and -AB dimers, but only weakly to the -AA dimer. Macrophage-derived PDGF is chemotactic for lung fibroblasts (Osornio-Vargas et al., 1990), and a differential chemotactic response to the different PDGF isoforms mirrors the pattern observed for isoform-stimulated fibroblast proliferation (i.e., BB > AB > AA) (Osornio-Vargas et al., 1991).

PDGF produced by macrophages has been implicated as a key mediator in several lung disorders, including asbestosis (Bonner et al., 1991) and idiopathic pulmonary fibrosis (Mornex et al., 1986; Martinet et al., 1986; Antoniades et al., 1990; Nagaoka et al., 1990b). Alveolar macrophages lavaged from normal individuals, and patients with idiopathic pulmonary fibrosis (IPF) express the PDGF-B-chain mRNA and secrete PDGF (Mornex et al., 1986), and the IPF macrophages spontaneously release exaggerated amounts of PDGF (Martinet et al., 1986). More recent studies have shown that normal alveolar macrophages constitutively transcribe both PDGF-A and PDGF-B genes, and LPS stimulation causes more than a threefold increase in the gene expression for both of these chains (Nagaoka et al., 1990b). These workers found that the IPF alveolar macrophages transcribed the genes for both chains at an approximately equal rate, but the mRNA for the PDGF-B chain was 10-fold more abundant than the mRNA for the PDGF-A chain. Another study using in-situ hybridization of lung tissues from normal individuals and IPF patients demonstrated that IPF is related to the in-vivo expression of PDGF-2/c-sis (PDGF-B chain) expression in alveolar macrophages and epithelial cells (Antoniades et al., 1990). These studies have suggested that the increase in PDGF from alveolar macrophages obtained from patients with fibrotic lung disease could be causally related to increased fibroblast proliferation and collagen synthesis, which are hallmarks of pulmonary fibrosis.

Novel PDGF-like factors

PDGF isolated from platelets and at least some of the macrophage-derived PDGF are disulfide-linked dimers and can be cleaved by treatment with reducing agents, resulting in the loss of receptor-binding, mitogenic, and chemotactic activity (reviewed in Ross, 1985). However, some studies have revealed that peritoneal wound macrophages and activated human peripheral blood monocytes produce a novel PDGF-like factor that recognizes antibodies to PDGF and binds the PDGF receptor, but is a biologically active, 16-kDa monomer (Pencev and Grotendorst, 1988; Matsuoka and Grotendorst, 1989). This factor, like authentic PDGF, is a potent chemoattractant and mitogen for fibroblasts (Matsuoka and Grotendorst, 1989). Apparently, this molecule does not recognize antibodies raised against the amino terminals of A and B chains of PDGF, and thus may possess a different primary structure compared to PDGF. Because at least two different molecular weight forms of PDGF are present in human and rat alveolar macrophage supernatants, the lower-molecular-weight form (12–18 kDa) of the macrophage-derived PDGF could represent a molecule that is similar to the novel PDGF-like factor described in wound fluid. Further biochemical and molecular evaluation of the low-molecular-weight alveolar macrophage-derived PDGF (Shimokado et al., 1985; Bonner et al., 1991) are needed to determine whether or not it is more similar to the higher-molecular-weight macrophage-derived PDGF (30–38 kDa) or the novel 16-kDa macrophage/monocyte-derived PDGF-like monomer (Matsuoka and Grotendorst, 1989).

Fibroblast Growth Factor (FGF)

Fibroblast growth factors (FGFs) are monomeric proteins that are divided into two groups, basic FGF and acidic FGF, based on their relative affinities for heparin, and both classes of FGF bind to a common cell surface receptor (Neufeld and Gospodarowicz, 1986). The FGF family has recently been expanded to include proteins encoded by at least seven different genes (see Goldfarb, 1990, for review). As their name implies, the FGFs are mitogens for fibroblasts, but other cell types, such as endothelial cells, smooth muscle cells, and keratinocytes, possess FGF receptors (Burgess and Maciag, 1989). These growth factors may play a critical role in neovascularization because they elicit several key biological responses of endothelial cells, including mitogenesis, chemotaxis, plasminogen activator synthesis, and collagen synthesis (reviewed in Thomas, 1987).

As mentioned earlier in this chapter, part of the growth-promoting activity in the originally described macropage-derived growth factor was attributed to fibroblast growth factor (Baird et al., 1985). This factor was identified by immunoassay and was shown to be mitogenic for vascular endothelial cells. Characterization of this factor was carried out by using peritoneal exudate cells that are mostly (75%) macrophages. Thus, to date no direct evidence is available that demonstrates FGF expression and secretion by pulmonary macrophages. Further characterization of this growth factor from alveolar macrophages is needed to determine the relative mitogenic contribution of FGF compared to other macrophage-derived growth factors. It is conceivable that FGFs secreted by alveolar and interstitial macrophages could play important roles in tissue repair and remodeling in the lung.

Transforming Growth Factor alpha (TGF-α)

The alveolar macrophage is regarded as an important effector cell in the paracrine signaling of mesenchymal cell chemotaxis and proliferation. However, macrophages can also secrete growth factors that stimulate the proliferation of the pulmonary epithelium. Bovine alveolar macrophages secrete a growth factor for bronchial epithelial cells that is pepsin and acid-labile and lipid inextractable (Takizawa et al., 1990), and it has also been demonstrated that lung macrophages stimulate the proliferation of type II epithelial cells (Leslie et al., 1985). Perhaps the best candidate for an epithelial cell growth factor that is produced by activated alveolar macrophages is transforming growth factor-α (TGF-α). This macrophage-derived TGF-α is recognized by anti-TGF-α antibodies following reduction and denaturation. The macrophages secreted two forms of TGF-α (8.5 to 12 kDa and 28.5 kDa) that possessed TGF-α/EGF receptor-binding activity. Apparently, alveolar macrophages do not produce epidermal growth factor (EGF), since none of the macrophage-derived TGF-α/EGF receptor-binding activity can be attributed to EGF mature form by antibody recognition of EGF residues.

TGF-α is synthesized and secreted by human alveolar macrophages that have been activated with LPS, indicating that the macrophages are induced to transcribe the gene for TGF-α (Madtes et al., 1988). Macrophage production of TGF-α, like IL-1, TNF-α (Collart et al., 1986), and PDGF synthesis (Shimokado et al., 1985), is regulated at the level of transcription. In contrast, TGF-β secretion is controlled at the posttranscriptional level, and monocytes/macrophages constitutively transcribe the TGF-β gene and when activated become induced to secrete the growth factor (Assoian et al., 1987).

Transforming Growth Factor beta (TGF-β)

As mentioned in the preceding section of this chapter, TGF-β, unlike most other growth factors, is constitutively produced by the alveolar macrophage. Resting macrophages maintain high steady-state TGF-β mRNA levels and will rapidly secrete the growth factor upon activation (Assoian et al., 1987). TGF-β serves several functions in the lung. It is a chemoattractant for monocytes and activates monocytes to secrete IL-1 (Wahl et al., 1987). TGF-β also possesses chemotactic activity for fibroblasts (Postlethwaite et al., 1987), although this issue remains controversial (unpublished observations). It is clear that TGF-β causes lung fibroblasts to up-regulate their production of extracellular matrix proteins such as collagen and fibronectin (Fine and Goldstein, 1987; Ignotz and Massague, 1986). Rat alveolar macrophages secrete TGF-β (Kalter et al.,

1989) and rat lung fibroblasts possess high-affinity receptors for TGF-β1, while TGF-β2 binds with a lesser affinity to receptors on these cells (Kalter and Brody, 1991). Proteinase secretion is suppressed by TGF-β (Edwards et al., 1987). The proliferation of immature fibroblasts is stimulated by TGF-β, and it has been suggested that TGF-β could recruit fibroblasts from surrounding tissues and serve as a mitgen for these cells (Fine and Goldstein, 1987).

Alveolar macrophage-derived TGF-β has been implicated as a mediator in chronic pulmonary fibrosis induced by bleomycin (Khalil et al., 1989). This is based on the observation of a rapid, sustained up-regulation in TGF-β production by alveolar macrophages that was temporally related to an increase in collagen synthesis in lung fibroblasts. Additionally, an intense immunohistochemical staining of TGF-β in alveolar macrophages after bleomycin coincided with a peak of extractable total lung TGF-β activity and mRNA, indicating that the major source of this growth factor in the lung was the alveolar macrophage (Khalil et al., 1989).

Insulinlike Growth Factor (IGF-1)

A growing body of evidence suggest that the insulinlike growth factors (IGF-I and IGF-II) play a role in lung and airway growth and differentiation (see Stiles and D'Ercole, 1990, for review). IGFs are required for the optimal proliferation of cells by acting as a progression-type factor for cell replication (Czech, 1989). Human alveolar macrophages, when activated, express IGF-1 mRNA and release an IGF-type molecule that serves as a progression factor in that it signals "competent" lung fibroblasts to replicate (Rom et al., 1988). Once bound to its receptor on human lung fibroblasts, the macrophage IGF stimulates the phosphorylation of a tyrosine-containing substrate. This macrophage-derived IGF is similar to other tissue IGFs that possess a molecular weight of 26 kDa, and it has been suggested that they may play a role in acute and chronic inflammatory disorders (Rom et al., 1988).

Once released by activated macrophages, IGFs have the potential to signal other pulmonary cells such as fibroblasts (Stiles and D'Ercole, 1990) and airway epithelial cells (Retsch-Bogart et al., 1991) in a paracrine manner. Recently, alveolar macrophages that have been activated with LPS or asbestos, but not resting cells, have been reported to possess receptors for IGF-I and -II (Rom and Pääkkö, 1991; Rom, 1991), leading to the suggestion that IGF receptors on activated alveolar macrophages could lead to the autocrine-induced proliferation of at least a subpopulation of these cells during acute and chronic pulmonary disorders.

The information on signal transduction pathways linking IGF receptor internalization to cell proliferation is fragmentary for any cell type and generally lacking with regard to pulmonary cells. For this reason, it may be useful to draw on information on some nonpulmonary cell types. For example, in BALB/c 3T3 cells, calcium influx is a major determinant of IGF-induced DNA synthesis; and when the progression factor is removed in the middle of G1 phase, this prevents the cell from entering S phase, even in the presence of a "competence" signal such as PDGF (Kojima et al., 1988). Furthermore, in neutrophils it has been shown that ligand-induced calcium entry enhances the sustained production of diacylglycerol and protein kinase C (Truett et al., 1988). A complete elucidation of these second messengers, and probably other signals, will be necessary to the understanding of how pulmonary cells respond to a progression-type signal that results in cell replication.

IGF-binding proteins (IGFBPs) produced by lung mesencymal cells (i.e., fibroblasts and smooth muscle cells) may be important in the modulation and delivery of IGF action for a number of different pulmonary cells. Since IGF-1 is expressed and secreted by activated alveolar macrophages (Rom et al., 1988), it will be of interest to learn whether or not the biological activity of this progression-type growth factor is modulated locally by IGFBPs. It has been demonstrated by immunohistochemical localization (Hill et al., 1989) that IGFBPs result in the accumulation of IGF on pulmonary epithelium. Presumably, this is because these cells possess cell surface receptors for IGFBP. As discussed above, the cell surface association of some IGFBPs is related to the capacity of the IGFBP to potentiate IGF-stimulated cell growth. For this reason, IGFBPs may serve important functions for IGF in the lung under normal and pathological conditions. In either case, these binding proteins could serve to modulate both extracellular and intracellular signaling induced by IGFs.

Interleukin-1β

The activated macrophage is a major source of interleukin-1β (IL-1β). Transcription of IL-1β can be induced by adherence of macrophages to plastic tissue culture surfaces, but little translation and secretion of IL-1β follows (Dinarello, 1988). Activation of signal transduction pathways following the addition of LPS increases the transcription of IL-1β severalfold, and also results in the translation and secretion of IL-1β precursor. This secreted precursor is then processed to mature IL-1β by the action of serine proteinases (Lepe-Zuniga and Gery, 1984). While IL-1β is predominately secreted by macrophages, another form, referred to as IL-1α, remains membrane associated yet possesses biological activity (Oppenheim et al., 1986). Because macrophages secrete elastase and other serine proteinases, they are important not only as a primary source of this cytokine, but also because they secrete proteinases that are essential to the processing of the IL-1β precursor to its mature form.

IL-1β is considered a growth-promoting cytokine and stimulates the proliferation of T lymphocytes, smooth muscle cells, and fibroblasts, and IL-1β binding to receptors on the surface of these cells initiates sodium and potassium ion fluxes, and increases in protein kinase C and cytosolic calcium (reviewed in Dinarello, 1988). While IL-1β can promote the growth of fibroblasts, it can also act in synergy with TNF-α to inhibit fibroblast growth (Elias et al., 1988). Some of these effects induced by IL-1β are similar to those observed after stimulation with phorbol esters, which directly up-regulate protein kinase C. IL-1β binding to cell surface receptors results in the production of PGE-2 by the target cell, or the secreted IL-1β can stimulate the macrophage to increase its own PGE-2. PGE-2-induced cAMP production is a negative signal for IL-1β translation. IL-1β translation can also be postively modulated, and one such example is the increase in IL-1β mRNA following treatment of macrophages with calcium ionophore or leukotrienes. IL-1β increases the production and secretion of IL-1β by macrophages, and this up-regulation causes the directed migration (chemotaxis) of other macrophages and lymphocytes (Oppenheim et al., 1986; Dinarello, 1988). Macrophage-derived IL-1β stimulates several cell types, including fibroblasts, endothelial cells, lymphocytes, and macrophages themselves, to increase their production of other cytokines such as interferons and colony-stimulating factors (CSFs).

Interleukin-6

Interleukin-6 (interferon-β) is produced by alveolar macrophages (Jordana et al., 1988), but is also synthesized by lung fibroblasts, and monocytes (Tosato et al., 1988). In monocytes, LPS activates the production of this cytokine (Tosato et al., 1988). The mechanism by which IL-6 gene expression is regulated is unknown in any cell type. Like TNF, phorbol ester can activate PKC to stimulate gene expression in lymphocytes and fibroblasts (Seghal et al., 1987), which indicates that IL-6 up-regulation is a PKC-dependent process. A number of growth factors produced by macrophages enhance the production of IL-6 in fibroblasts, including TNF-α, IL-1β, and PDGF (reviewed in Hirano and Kishimoto, 1990). Another study has suggested that IL-6 gene expression is a cAMP-dependent (Zhang et al., 1988). IL-6 has multiple biological activities, including induction of macrophage differentiation and T-lymphocyte proliferation. While it has been reported that IL-6, along with IL-1 and PDGF, are three of the major cytokines release during bleomycin-induced pulmonary injury (Kelley, 1990), little information is available on the production and biology of IL-6 when compared to a vast amount of literature on PDGF and IL-1. Future studies should address some very basic questions relating to IL-6 production by macrophages and how this cytokine affects the secretion of other growth factors by alveolar macrophages.

Tumor Necrosis Factor-α (TNF-α)

As with the interleukins, macrophages are a primary source of TNF-α, and the quantity of this factor released by other cell types is extremely small when compared to the amounts released by activated macrophages. TNF-α is synthesized as a 26-kDa precursor which is cleaved to a 17-kDa mature polypeptide. Exposure of macrophages to activating substances, especially LPS, increases severalfold the production TNF-α. The LPS-induced transcription of TNF-α mediated by protein kinase C has been elaborated earlier in this chapter in Section II. Apparently, calmodulin-dependent kinases do not play a role in LPS-induced expression of TNF-α, but this kinase does inhibit the expression of IL-1 (Jäättelä, 1991). As with some other cytokines, the presence of mRNA for TNF-α does not always result in translation and secretion of the protein. For example, interferon-γ induces TNF-α gene expression, but other stimuli are required for the translation of the TNF-α precursor (Vilcek and Lee, 1991).

TNF-α is chemotactic for monocytes (Ming et al., 1987) and thus may play a role in the recruitment of these cells to sites of inflammation. Other cell types, such as T-lymphocytes, do not possess TNF-α receptors in a resting state, but activation of these cells leads to the production of cell surface receptors (Jäättelä, 1991). TNF-α enhances several biological activities of responding PMNs, including increased phagocytosis, antibody-dependent cytotoxicity, and production of superoxide anion (Vilcek and Lee, 1991). Like IL-1β, TNF-α alone causes growth stimulation of fibroblasts in vitro (Sugarman et al., 1985), but in combination with IL-1β it inhibits fibroblast growth (Elias et al., 1988).

Colony-Stimulating Factors (CSFs)

CSFs are glycoproteins that have been divided into two groups, granulocyte-macrophage (GM-CSF) and macrophage CSF (M-CSF). As with some other growth factors, macrophages possess high-affinity receptors for CSFs in normal and disease states (Kriepe et al., 1990), and the CSFs have several modulatory effects on macrophage function. Monocytes and macrophages differ in their response to CSF, possibly due to quantitiative differences in CSF receptor expression on these cells (Thomassen et al., 1989). M-CSF stimulates monocytes to produce IFN, TNF, and CSF itself (Warren and Ralph, 1986), and M-CSF induces macrophages to produce plasminogen activator (Hamilton et al., 1980), TNF and IFN (Warren and Ralph, 1986), and oxygen radicals (Wing et al., 1985). Both CSFs, as well as IL-3 or multipotential CSF (Suwabe et al., 1988), induce early c-fos and c-myc gene expression (Schaberg and Filderman, 1990). As shown in Fig. 3, other inducers of c-fos and c-myc, such as LPS, lead to the up-regulation of growth-factor secretion. Thus, stimulators of these early genes, whether they are growth factors binding to specific receptors or activators that do not possess specific receptors (e.g., LPS), appear to share some common signal transduction pathways with regard to growth-factor gene expression and growth-factor secretion.

B. Extracellular Signaling via Mixtures of Growth Factors

Activated macrophages secrete a wide variety of growth factors (Table 4), and the biological activity of these factors as studied in vitro may not reflect how the cytokine operates in vivo. One reason for this is that other growth factors may inhibit or stimulate the effect of the first growth factor, and the possible modulatory potential of additional cytokines depends on the concentration and temporal fashion of growth-factor release by the activated macrophage. For example, alveolar macrophage inhibition of lung fibroblast growth, a prostaglandin-dependent process, is due to synergy between TNF-α and IL-1β (Elias et al., 1985, 1988). In the case of PDGF, the different isoforms possess varying chemotactic potential for fibroblasts and monocytes (Seigbahn et al., 1990), and PDGF-AA is inhibitory to the PDGF-BB stimulated chemotaxis of rat lung fibroblasts (unpublished observation). The growth-promoting effect of PDGF for fibroblasts is also inhibited by TGF-β (Moses et al., 1990). Interestingly, the growth-promoting effect of IL-1 for fibroblasts is due to its ability to increase PDGF-AA (Raines et al., 1989). These examples illustrate the possibility of interactions between macrophage-derived growth factors, and the importance of investigating the complex network of cytokines both alone and in combination with other growth factors.

C. Modulation of Growth-Factor Signaling by α_2-Macroglobulin

α_2-Macroglobulin (α_2M) is a high-molecular-weight (725-kDa) proteinase inhibitor secreted by alveolar macrophages (White et al., 1980) that could be necessary to the normal defense of the lung, because of its function as a proteinase inhibitor and modulator of immune function (see Sottrup-Jensen, 1989; James, 1980). This macromolecule could be important both in modulating the growth-factor activity secreted by macrophages and in affecting the production of growth factor by macrophages, because α_2M is a growth-factor-binding protein and macrophages possess high-affinity cell surface receptors for α_2M-proteinase complex (Kaplan and Nielson, 1979). α_2M binds and modulates the in vitro activity of several growth factors that are known to be secreted by alveolar macrophages, including PDGF (Huang et al., 1984; Bonner et al., 1989), TGF-β (Danielpour and Sporn, 1990), basic FGF (Dennis et al., 1989), IL-1β (Borth and Luger, 1989), IL-6 (Matsuda et al., 1989), and TNF-α (Wollenberg et al., 1991). The biological activity of these growth factors, when complexed to α_2M, is modulated either positively or negatively in vitro, depending on the specific growth factor in question and the conformation of α_2M. For example, ~50% of rat alveolar macrophage-derived PDGF is secreted as a high-molecular-weight complex with a homolog of human α_2M (Bonner et al., 1989), and the biological activity of PDGF for rat lung fibroblasts is either inhibited or synergistically enhanced by α_2M, depending on the conformational state (receptor-recognized or non-receptor-recognized form) of the binding protein (Bonner et al., 1990). Additionally, the secretion of IL-6 and α_2M by alveolar macrophages is up-regulated from patients with sarcoidosis (Sibille et al., 1989), and inorganic particulates up-regulate the secretion of α-macroglobulins by rat alveolar macrophages (our unpublished observation). Thus, increases and/or decreases in the secretion of α_2M by alveolar macrophage could influence macrophage paracrine signaling of other pulmonary cells via growth factors.

It is not known if α_2M modulates growth-factor gene expression and secretion by macrophages. Because α_2M binds to receptors on macrophages and suppresses immune function, it would be of interest to investigate whether or not α_2M modulates the signal transduction pathways involved in macrophage activation for growth-factor secretion. Furthermore, it is conceivable that α_2M could serve to scavenge growth factors and inactivate them via internalization through the macrophage α_2M receptor.

IV. Macrophage-Derived Growth Factors in Pulmonary Disease

The overexpression and production of several growth factors by activated macrophages have been implicated in various pulmonary diseases both from clinical studies and in experimental animals (Table 5). With regard to clinical disorders, alveolar macrophages obtained from individuals with active pulmonary sarcoidosis, a granulomatous lung disorder (Hunninghake et al., 1984), spontaneously express the genes for and secrete elevated levels of IL-1β (Wewers et al., 1987; Hunninghake et al., 1984), IL-6 (Sibille et al., 1989), CSF (Itoh et al., 1990), and TNF-α (Bachwich et al., 1986). Alveolar macrophages from patients with IPF spontaneously express the gene for PDGF-BB (Nagaoka et al., 1990b), and these PDGF levels are severalfold that of normal individuals. Additionally, TGF-β production by macrophages is increased in IPF (Khalil et al., 1991).

Models of pulmonary fibrosis are invaluable for elucidating the mechanisms of growth-factor regulation and provide a basis of understanding that

Table 5 Selected Clinical Pulmonary Diseases Correlated with Overexpression and/or Increased Secretion of Macrophage-Derived Growth Factors[a]

Disorder	Growth factor	Reference
Clinical		
idiopathic pulmonary fibrosis	PDGF	Nagaoka et al., 1991b;
	TGF-β	Khalil et al., 1991
Familial pulmonary fibrosis	IGF	Bitterman et al., 1983
Sarcoidosis	TNF-α	Bachwich et al., 1986
	IL-1	Hunninghake et al., 1984
	IGF-1	Bitterman et al., 1983
Pneumoconionses	IGF-1	Rom et al., 1988
Hyperoxia	IGF-1	Davis et al., 1983
Animal model		
Bleomycin-induced fibrosis	TGF-β	Khalil et al., 1989
	IL-1, IL-6	Jordana et al., 1988
Asbestosis	PDGF	Kumar et al., 1988; Bonner et al., 1991
Silicosis	TNF-α	Piguet et al., 1990
Hyperoxia	PDGF	Fabisiak et al., 1989

[a] Some studies are not related directly to the macrophage-derived growth factor, but pertain to growth factors that can be secreted by activated macrophages.

could lead to a therapeutic intervention of the disease process. Several macrophage-derived growth factors have been implicated in fibrogenesis. For example, it has been reported that TNF is a necessary requirement for the development of silica-induced pulmonary fibrosis (Piguet et al., 1990). Asbestosis induced in rats is associated with increases in macrophage-derived PDGF (Bonner and Brody, 1991), and asbestos activates PDGF production by alveolar macrophages in vitro (Bonner et al., 1991). Bleomycin-induced pulmonary injury is associated with increases in macrophage PDGF (Kovacs and Kelley, 1985), TGF-β (Khalil et al., 1989; Hoyt and Lazo, 1988), IL-1β, and IL-6 (Jordana et al., 1988). While PDGF may be the most potent promoter of mesenchymal cell growth (Shimokado et al., 1985; Nagaoka et al., 1990b), TGF-β is likely the major contributor to increases in fibroblast collagen synthesis (Khalil et al., 1989). Thus, these two macrophage-derived factors (PDGF and TGF-β) could be responsible for the increased number of fibroblasts and the overproduction of extracellular matrix proteins that characterize the fibrotic lesion. Growth factors are essential for tissue repair following injury (Rennard et al., 1983), yet their prolonged overexpression appears to be detrimental to pulmonary health, and these agents could contribute to many different disorders.

V. Summary

The alveolar macrophage can be generally regarded as an effector cell that communicates with other pulmonary cell types through paracrine growth-factor signaling networks. These target cells are stimulated by macrophage-derived growth factors to undergo chemotaxis, proliferation, synthesis of extracellular matrix components, and production of their own growth factors and inflammatory mediators. A major avenue of research on growth-factor-induced signal transduction pathways is aimed at cell replication following the addition of growth factors to these cells in vitro, and much of this research is discussed in this volume. In contrast to fibroblasts, smooth muscle cells, endothelial cells, and epithelial cells, the mature macrophage is generally a nondividing cell. For this reason, the literature on signal transduction in macrophages generally reflects the biology of the macrophage; most of the available information relates to macrophage activation and its functions as an effector cell (e.g., growth-factor production). Because macrophages possess a variety of growth-factor receptors, there is growing interest in the modulatory effects of growth factors on macrophage activation for chemotaxis, phagocytosis, or the production of inflammatory mediators. Elucidating the pathways of transmembrane signaling involved in macrophage activation may give clues to therapeutic strategies for controlling macrophage function in pulmonary disease.

References

Adams DO, Hamilton TA. In: Gallin JI, Goldstein IM, Snyderman R, eds. Inflammation: basic principles and clinical correlates. New York: Raven Press, 1988; 471–492.

Adamson IY, Letourneau HL, Bowden DH. Enhanced macrophage-fibroblast interactions in the pulmonary interstitium increases fibrosis after silica injection to monocyte-depleted mice. Am J Pathol 1989; 134:411–8.

Akagawa KS, Kamoshita K, Tokunaga T. Effects of granulocyte-macrophage colony stimulating factor and colony stimulating factor-1 on the proliferation and differentiation of murine alveolar macrophages. J Immunol 1988; 141:3383–90.

Antoniades HN, Bravo MA, Avila RE, Galanopoulos T, Neville-Golden J, Maxwell M, Selman M. Platelet-derived growth factor in idiopathic pulmonary fibrosis. J Clin Invest 1990; 86:1055–64.

Assoian RK, Fleurdelys BE, Stevenson HC, Miller PJ, Madtes DK, Raines EW, Ross R, Sporn MB. Expression and secretion of type β transforming growth factor by activated human macrophages. Proc Natl Acad Sci USA 1987; 84:6020–4.

Bachwich PR, Chenuse SW, Larrick JW, Kunkel SL. Tumor necrosis factor stimulates interleukin-1 and prostaglandin E2 production in resting macrophages. Biochem Biophys Res Commun 1986; 136:94–101.

Baird A, Mormede P, Bohlen P. Immunoreactive fibroblast growth factor in cells of peritoneal exudate suggests its identity with macrophage-derived growth factor. Biochem Biophys Res Commun 1985; 126:358–64.

Beaty CD, Martin TR, Wilson CB. Signal transduction in human alveolar macrophages: diminished chemotactic response to FMLP correlates with a diminished density of Gi proteins and FMLP receptors. Am J Respir Cell Mol Biol 1991; 5:87–92.

Berridge MJ, Irvine RF. Inositol trisphosphate, a novel second messenger in cellular signal transduction. Nature (London) 1984; 312:315.

Bitterman PB, Rennard SI, Adelberg S, Crystal RG. Role of fibronectin as a growth factor for fibroblasts. J Cell Biol 1983; 97:1925–32.

Bitterman PB, Rennard SI, Hunninghake GW, Crystal RG. Human alveolar macrophage growth factor for fibroblasts. J Clin Invest 1982; 70:806–22.

Bitterman PB, Saltzman LI, Adelberg S, Ferrans VJ, Crystal RG. Alveolar macrophage replication. One mechanism for the expansion of the mononuclear phagocyte population in the chronically inflamed lung. J Clin Invest 1984; 74:460–9.

Bitterman PB, Wewers MD, Rennard SI, Adelberg S, Crystal RG. Modulation of alveolar macrophage-driven fibroblast proliferation by alternative mediators. J Clin Invest 1986; 77:700–8.

Bonner JC, and Brody AR. Asbestos-induced alveolar injury: evidence for macrophage-derived PDGF as a mediator of the fibrogenic response. CHEST 1991; 99:545–555.

Bonner JC, Badgett A, Osornio-Vargas AR, Hoffman M, Brody AR. PDGF-stimulated fibroblast proliferation is enhanced synergistically by receptor-recognized α_2-macroglobulin. J Cell Physiol 1990; 145:1–8.

Bonner JC, Hoffman M, Brody AR. Alpha-macroglobulin secreted by alveolar macrophages serves as a binding protein for a macrophage-derived homologue of platelet-derived growth factor. Am J Respir Cell Mol Biol 1989; 1:171–9.

Bonner JC, Osornio-Vargas AR, Badgett A, Brody AR. Differential proliferation of rat lung fibroblasts induced by the platelet-derived growth factor (PDGF)-AA, AB and BB isoforms secreted by rat alveolar macrophages. Am J Respir Cell Mol Biol 1991; 5:539–47 .

Borth W, Luger TA. Identification of α_2-macroglobulin as a cytokine binding protein: binding of interleukin-1β to "F" α_2-macroglobulin. J Biol Chem 1989; 264(10):5818–25.

Brandes ME, Finkelstein JN. Stimulated rabbit alveolar macrophages secrete a growth factor for type II pneumocytes. Am J Respir Cell Mol Biol 1989; 101–9.

Brody AR. Production of cytokines by particle-exposed lung macrophages. In: Cellular and molecular aspects of fiber carcinogenesis. Cold Spring Harbor Laboratory Press, 1991; 83–102.

Burgess WH, Maciag T. The heparin-binding (fibroblast) growth factor family of proteins. Ann Rev Biochem 1989; 58:575–606.

Chang L, Overby LH, Brody AR, Crapo JD. Progressive lung cell reactions and extracellular matrix production after a brief exposure to asbestos. Am J Pathol 1988; 131(1):156–70.

Chen BD, Mueller M, Chou TH. Role of granulocyte/macrophage colony-stimulating factor in the regulation of murine alveolar macrophage proliferation and differentiation. J Immunol 1988; 141:139–44.

Clark SC, Kamen R. The human hematopoietic colony stimulating factors. Science 1987; 236:1229–37.

Collart MA, Belin H, Vassalli J-D, de Kossodo S, Vassalli P. Gamma interferon enhances macrophage transcription of the tumor necrosis factor/cachectin, interleukin-1, and urokinase genes, which are controlled by short-lived repressors. J Exp Med 1986; 164:2133–18.

Crystal RG, Bitterman PB, Rennard SI, Hance AJ, Keogh BA. Interstitial lung diseases of unknown cause: disorders characterized by chronic inflammation of the lower respiratory tract. N Engl J Med 1984; 310:154–66.

Czech MP. Signal transmission by the insulin-like growth factors. Cell 1989; 59:235–8.

Danielpour D, Sporn MB. Differential inhibition of transforming growth factor $\beta1$ and $\beta2$ activity by α_2-macroglobulin. J Biol Chem 1990; 265:6973–7.

Davila DR, Edwards CK 3rd, Arkins S, Simon J, Kelley KW. Interferon-γ-induced priming for secretion of superoxide anion and tumor necrosis factor-α declines in macrophages from aged rats. FASEB J 1990; 4:2906–11.

Demedts M, Auman J. Early emphysema. Ten years' evolution. Chest 1988; 94:337–42.

Dennis PA, Saksela O, Harpel P, Rifkin DB. α_2-Macroglobulin is a binding protein for basic fibroblast growth factor. J Biol Chem 1989; 264 (13):7210–6.

Dinarello CA. Biology of interleukin-1. FASEB J 1988; 2:108–15.

Dougherty RW, Godfrey PP, Hoyle PC, Putney JW, Freer RJ. Secretagogue-induced phosphoinositide metabolism in human leukocytes. Biochem J 1984; 222:307.

Edwards DR, Murphy G, Reynolds JJ, Whitman SE, Docherty AJP, Angel P, Heath JK. Transforming growth factor beta modulates the expression of collagenase and metalloproteinase inhibitor. EMBO J 1987; 1899–1903.

Elias JA, Gustilo K, Freundlich B. Human alveolar macrophage and blood monocyte inhibition of fibroblast proliferation: evidence for synergy between interleukin-1 and tumor necrosis factor. Am Rev Respir Dis 1988; 138:1595–603.

Elias JA, Rossman MD, Zurier RB, Daniele RP. Human alveolar macrophage inhibition of lung fibroblast growth. A prostaglandin-dependent process. Am Rev Respir Dis 1985; 131:94–9.

Fine A, Goldstein RH. The effect of transforming growth factor b on cell proliferation and collagen formation by lung fibroblasts. J Biol Chem 1987; 262:3897–901.

Figueíredo F, Uhing RJ, Okonogi K, Gettys TW, Johnson SP, Adams DO, Prpic V. Activation of the cAMP cascade inhibits an early event involved in murine macrophage Ia expression J Biol Chem 1990; 265:12317–23.

Glenn K, Ross R. Human monocyte-derived growth factor(s) for mesenchymal cells: activation of secretion by endotoxin and concavalin A. Cell 1981; 25:603–15.

Goldfarb M. The fibroblast growth factor family. Cell Growth and Differentiation 1990; 1:439–45.

Greenburg GB, Hunt TK. Proliferative response in vitro of vascular endothelial cells and smooth muscle cells exposed to wound fluids and macrophages. J Cell Physiol 1978; 97:353–60.

Hamilton JA, Stanley ER, Burgess AW. Stimulation of macrophage plasminogen activator activity by colony-stimulating factors. J Cell Physiol 1980; 103:435–45.

Hance AJ, Douches S, Winchester RJ, Ferrans VJ, Crystal RG. Characterization of mononuclear phagocyte subpopulations in the human lung by using monoclonal antibodies: changes in alveolar macrophage phenotype associated with pulmonary sarcoidosiss. J Immunol 1985; 134:284–97.

Helmke RJ, Boyd L, German VF, Mangos JA. From growth factor dependence to growth factor responsiveness: the genesis of an alveolar macrophage cell line. In Vitro Cell Dev Biol 1987; 23:567–74.

Hill DJ, Clemmons DR, Wilson S, Khan VK, Strain AJ. Immunolocalization of insulin-like (IGF)-binding protein and IGF peptides in human fetal tissues. J Mol Endocrinol 1989; 2:31–8.

Hirano T, Kishimoto T. Interleukin-6. In: Sporn MB, Roberts AB, eds. Peptide growth factors and their receptors. 1990; 633–65. Springer-Verlag, New York.

Hoyt DG, Lazo JS. Alterations in pulmonary mRNA encoding procollagens, fibronectin, and transforming growth factor beta precede bleomycin induced pulmonary fibrosis in mice. J Pharmacol Exp Ther 1988; 246:765–71.

Huang JS, Huang SS, Deuel TS. Specific covalent binding of platelet-derived growth factor to human plasma α_2-macroglobulin. Proc Natl Acad Sci USA 1984; 81:342–6.

Hunninghake G, Garrett KC, Richerson HB, Fantone JC, Ward PA, Rennard SI, Bitterman PB, Crystal RG. Pathogenesis of granulomatous lung diseases. Am Rev Respir Dis 1984; 130:476–96.

Ignotz R, Massague J. Transforming growth factor beta stimulates the expression of fibronectin and collagen and their incorporation into the extracellular matrix. J Biol Chem 1986; 261-5.

Itoh A, Yamaguchi E, Kuzumaki N, Okazaki N, Furuya K, Abe S, Kawakami Y. Expression of granulocyte-macrophage colony stimulating factor mRNA by inflammatory cells in the sarcoid lung. Am J Respir Cell Mol Biol 1990; 3:245–9.

Jäättelä M. Biologic activities and mechanisms of action of tumor necrosis factor-α/cachectin. Lab Invest 1991; 64:724–42.

James K. Alpha-2-macroglobulin and its possible importance in immune systems. Trends Biochem Sci 1980; 5:43.

Jordana M, Richards C, Irving LB, Gauldie J. Spontaneous in vitro release of alveolar macrophage cytokines after the intratracheal instillation of bleomycin in rats. Am Rev Respir Dis 1988; 137:1135–40.

Kalter VK, Brody AR. Receptors for transforming growth factor β (TGF-β) on rat lung fibroblasts have higher affinity for TGF-β1 than for TGF-β2. Am J Respir Cell Mol Biol 1991; 4:397–407.

Kalter VK, Bonner JC, Brody AR. Secretion of TGF-β by rat alveolar macrophages and characterization of TGF-β receptors on rat lung fibroblasts. Cytokine 1989; 1:76.

Kaplan J, Nielsen ML. Analysis of macrophage receptors: I. binding of α-macroglobulin-protease complexes to rabbit alveolar macrophages. J Biol Chem 1979; 254(15):7323–8.

Kelley J. Cytokines of the lung. Am Rev Respir Dis 1990; 141:765–88.

Keogh BA, Crystal RG. Alveolitis: the key to understanding the interstitial lung disorders. Thorax 1982; 32:1–10.

Khalil N, Bereznay O, Sporn M, Greenburg AH. Macrophage production of transforming growth factor β and fibroblast collagen synthesis in chronic pulmonary inflammation. J Exp Med 1989; 170:727–37.

Khalil N, O'Conner RN, Unruh HW, Warren PW, Flanders KC, Kemp A, Bereznay OH, Greenberg GP. Increased production and immmunohistochemical localization of transforming growth factor-β in idiopathic pulmonary fibrosis. Am J Respir Cell Mol Biol 1991; 5:155–62.

Koerner TJ, Adams D, Hamilton TA. Regulation of tumor necrosis factor (TNF) expression: interferon-γ enhances the accumulation of mRNA for TNF induced by lipopolysaccharide in murine peritoneal macrophages. Cell Immunol 1987; 109:437–43.

Kojima I, Matsunaga H, Kurokawa K, Ogata E, Nishimoto I. Calcium influx: an intracellular message of the mitogenic action of insulin-like growth factor-I. J Biol Chem 1988; 263:16561–7.

Kovacs EJ, Kelley J. Intra-alveolar release of a competence-type growth factor after lung injury. Am Rev Respir Dis 1985; 133:68–72.

Kreipe H, Radzun HJ, Heidorn K, Barth J, Kiemle-Kallee J, Petermann W, Gerdes J, Parwaresch MR. Proliferation, macrophage-colony stimulating factor, and macrophage colony stimulating factor-receptor expression of alveolar macrophages in active sarcoidosis. Lab Invest 1990; 62:697–703.

Kronke M, Schutze S, Scheurich P, Meichle A, Hensel G, Thoma B, Kruppa G, Pfizenmaier K. Tumour necrosis factor signal transduction. Cell Signaling 1990; 2:1–8.

Leibovich SJ, Ross R. The role of the macrophage in wound repair. Am J Pathol 1975; 78:71–100.

Lepe-Zuniga B, Gery I. Production of intracellular and extracellular interleukin-1 (IL-1) by human monocytes. Clin Immunol Immunopathol 1984; 31:222–30.

Leslie CC, McCormick-Shannon MA, Cook JL, Mason R. Macrophages stimulate DNA synthesis in rat alveolar type II cells. Am J Respir Dis 1985; 132:1246–52.

Libby P, Wyler DJ, Janicka MW, Dinarello CA. Differential effects of human interleukin-1 on growth of human fibroblasts and vascular smooth muscle cells. Arteriosclerosis 1985; 5:186–91.

Madtes DK, Raines EW, Sakariassen KS, Assoian RK, Sporn MB, Bell GI, Ross R.

Induction of transforming growth factor α by activated human alveolar macrophages. Cell 1988; 53:255–93.

Manogue KR, Cerami A. In: Poste G, Crooke ST, eds. Cellular and molecular aspects of inflammation. New York: Plenum Press, 1988; 123–50.

Martin BM, Gimbrone MA, Unanue ER, Cotran RS. Stimulation of nonlymphoid mesenchymal cell proliferation by a macrophage-derived growth factor. J Immunol 1981; 126:1510–5.

Martinet Y, Rom WN, Grotendorst GR, Martin GR, Crystal RG. Exaggerated spontaneous release of platelet-derived growth factor by alveolar macrophages from patients with idiopathic pulmonary fibrosis. N Engl J Med 1986; 317:202–9.

Matsuda T, Hirano T, Nagasawa S, Kishimoto T. Identification of α_2-macroglobulin as a carrier protein for interleukin 6. J Immunol 1989; 142:148–52.

Matsuoka J, Grotendorst GR. Two peptides related to platelet-derived growth factor are present in human wound fluid. Proc Natl Acad Sci USA 1989; 86:4416–20.

Ming WJ, Bersan L, Mantovani A. Tumor necrosis factor is chemotactic for monocytes and polymorphonuclear leukocytes. J Immunol 1987; 138:1469–73.

Mornex J, Martinet Y, Yamaguchi K, Bitterman PB, Grotendorst A, Chytil-Weir A, Martin GR, Crystal RG. Spontaneous expression of the cis-gene and release of a platelet-derived growth factorlike molecule by human alveolar macrophages. J Clin Invest 1986; 78:61–6.

Moses HL, Yang EY, Pietenpol JA. TGF-β stimulation and inhibition of cell proliferation: new mechanistic insights. Cell 1990; 63:245–7.

Nagaoka I, Trapnell BC, Crystal RG. Regulation of insulin-like growth factor I gene expression in the human macrophage-like cell line U937. J Clin Invest 1990a; 85:448–55.

Nagaoka I, Trapnell BC, Crystal RG. Upregulation of platelet-derived growth factor-A and -B gene expression in alveolar macrophages of individuals with idiopathic pulmonary fibrosis. J Clin Invest 1990b; 85:2023–7.

Nathan CF. Macrophage activation: some questions. Ann Inst Pasteur Immunol 1986; 137C:345–51.

Nathan C. Secretory products of macrophages. J Clin Invest 1987; 79:319–26.

Neufeld G, Gospodarowicz D. Basic and acidic fibroblast growth factors interact with the same cell surface receptors. J Biol Chem 1986; 261:5631–7.

Nishizuka Y. The role of protein kinase C in cell surface signal transduction and tumour production. Nature (London) 1984; 308:693.

Okonog K, Gettys TW, Uhing RJ, Tarry WC, Adams DO, Prpic V. Inhibition of prostaglandin E2-stimulated cAMP accumulation by lipopolysaccharide in murine peritoneal macrophages. J Biol Chem 1991; 266:10305–12.

Oppenheim JJ, Kovacs EJ, Matsushima AK, Durum SK. There is more than one interleukin 1. Immunol Today 1986; 7:45–56.

Osornio-Vargas AR, Bonner JC, Badgett A, Brody AR. Rat alveolar macrophage-derived platelet-derived growth factor is chemotactic for rat lung fibroblasts. Am J Respir Cell Mol Biol 1990; 3:595–602.

Osornio-Vargas AR, Bonner JC, Brody AR. Rat lung fibroblast chemotaxis to PDGF-AA, AB and BB dimers. Am J Respir Dis 1991; 143:A524.

Pencev D, Grotendorst GR. Human peripheral blood monocytes secrete a unique form of PDGF. Oncogene Res 1988; 3:333–42.

Perderiset M, Marsh JP, Mossman BT. Activation of protein kinase C by crocidolite asbestos in hamster tracheal epithelial cells. Carcinogenesis 1991; 12:1499–502.

Phan SH, McGarry BM, Loeffler KM, Kunkel SL. Regulation of macrophage-derived growth factor release by arachidonate metabolites. J Leukoc Biol 1987; 42:106–13.

Piguet PF, Collart MA, Grau GE, Sappino A-P, Vassalli P. Requirement for tumour necrosis factor for development of silica-induced pulmonary fibrosis. Nature (London) 1990; 344:245–7.

Postlethwaite AE, Keski-Oja J, Moses HL, Kang AH. Stimulation of the chemotactic migration of human fibroblasts by transforming growth factor β. J Exp Med 1987; 165:251–9.

Prpic V, Weiel JE, Somers SD, Gonias SL, Pizzo SV, Hamilton TA, Herman B, Adams DO. Effects of bacterial lipopolysaccharide on the hydrolysis of phosphatidylinositol-4,5-bisphosphate in murine peritoneal macrophages. J Immunol 1987; 139:526–33.

Raines EW, Dower SK, Ross R. Interleukin-1 mitogenic activity for fibroblasts and smooth muscle cells is due to PDGF-AA. Science 1989; 243:393–6.

Rennard SI, Bitterman PB, Crystal RG. Response of the lower respiratory tract to injury: mechanisms of repair of the parenchymal cells of the alveolar wall. Chest 1983; 84:735–9.

Retsch-Bogart GZ, Stiles AD, Moats-Staat BM, Van Scott MR, Boucher RC, D'Ercole AJ. Tracheal epithelial cells express the insulin-like growth factor receptor and proliferate in response to insulin-like growth factor I. Am J Respir Cell Mol Biol 1991; in press.

Rom WN. Human mononuclear phagocytes express the insulin-like growth factor-II/mannonse-6-phosphate receptor. Am J Respir Cell Mol Biol 1991; 4:555–9.

Rom WN, Paakko P. Activated alveolar macrophages express the insulin-like growth factor I receptor. Am J Respir Cell Mol Biol 1991; 4:432–9.

Rom WN, Bassett P, Fells GA, Nukiwa T, Trapnell BC, Crystal RG. Alveolar macrophages release an insulin-like growth factor I-type molecule. J Clin Invest 1988; 82:1685–93.

Rosen A, Nairn AC, Greengard P, Cohn ZA, Aderem A. J Biol Chem 1989; 264:9118–21.

Ross R, Raines EW, Bowen-Pope DF. The biology of platelet-derived growth factor. Cell 1986; 46:155–69.

Sara VR, Hall K. Insulin-like growth factors and their binding proteins. Physiol Rev 1990; 70:591–614.

Schaberg T, Filderman AE. Colony-stimulating factor induction of protooncogene expression in rat alveolar macrophages. Am J Physiol 1990; 259 (Lung Cell Mol Physiol 3):L144–50.

Seghal PB, Walther Z, Tamm I. Rapid enhancement of β2-interferon/B cell differentiation factor BSF-2 gene expression in human fibroblasts by diacylglycerols and the calcium ionophore A23187. Proc Natl Acad Sci USA 1987; 84:3633–67.

Seifert RA, Hart CE, Phillips PE, Forstrom JW, Ross R, Murray MJ, Bowen-Pope DF. Two different subunits associate to create isoform-specific platelet-derived growth factor receptors. J Biol Chem 1989; 264:8771–8778.

Seigbahn A, Hammacher A, Westermark B, Heldin C-H. Differential chemotactic effects of the various isoforms of platelet-derived growth factor on chemotaxis of fibroblasts, monocytes and granulocytes. J Clin Invest 1990; 85:916–20.

Sesko A, Cabot M, Mossman B. Hydrolysis of inositol phospholipids precedes cellular proliferation in asbestos-stimulated tracheobronchial epithelial cells. Proc Natl Acad Sci USA 1990; 87:7385–9.

Shaw RJ, Bendict SH, Clark RAF, King TE Jr. Pathogenesis of pulmonary fibrosis in interstitial lung disease: alveolar macrophage PDGF (B) gene activation and up-regulation by interferon-gamma. Am Rev Respir Dis 1991a; 143:167–73.

Shaw RJ, Doherty DE, Ritter AG, Benedict SH, Clark RAF. Adherence-dependent increase in human monocyte PDGF (B) mRNA is associated with increases in c-fos, c-jun and EGR2 mRNA. J Cell Biol 1991b; in press.

Shears SB. Metabolism of the inositol phosphates produced upon receptor activation. Biochem J 1989; 260:313–24.

Shimokado K, Raines EW, Madtes DK, Barrett TB, Benditt EP, Ross R. A significant part of macrophage-derived growth factor consists of at least two forms of PDGF. Cell 1985; 43:277–86.

Sibille Y, Houssiau F, Pochet JM, Staquet P, Van Snick J, Van Leuven F. α_2-Macroglobulin (α_2M) and interleukin-6 (IL6) release by human alveolar macrophages (AM) from normals and sarcoidosis (SA) patients. Am Rev Respir Dis 1989; 141:abstract.

Snyderman R, Smith CD, Verghese MW. Model for leukocyte regulation of chemoattractant receptors: role of a guanine nucleotide regulatory protein and polyphosphoinositide metabolism. J Leukocyte Biol 1986; 40:785–800.

Sottrup-Jensen L. a-Macroglobulins: structure, shape, and mechanism of proteinase complex formation. J Biol Chem 1989; 264 (20):11539–42.

Spurzem JR, Saltini C, Rom W, Winchester RJ, Crystal RG. Mechanisms of macrophage accumulation in the lungs of asbestos-exposed subjects. Am Rev Respir Dis 1987; 136:276–80.

Stiles AD, D'Ercole AJ. The insulin-like growth factors and the lung. Am J Respir Cell Mol Biol 1990; 3:93–100.

Stryer L, Bourne HR. G proteins: a family of signal transducers. Ann Rev Cell Biol 1986; 2:391–419.

Sugarman BJ, Aggarwall BB, Hass PE, Figari IS, Palladine MA, Sheppard HM. Recombinant human tumor necrosis-α: effects on proliferation of normal and transformed cells in vitro. Science 1985; 230:943.

Suwabe A, Takahashi K, Yasui S, Arai S, Sendo F. Interleukin-3 (IL-3) stimulates the clonal growth of pulmonary alveolar macrophage of the mouse: role of IL-3 in the regulation of macrophage production outside the bone marrow. Blood 1988; 72:685–90.

Takizawa H, Beckmann JD, Shoji S, Claasen LR, Ertl RF, Linder J, Rennard SI. Pulmonary macrophages can stimulate cell growth of bovine bronchial epithelial cells. Am J Respir Cell Mol Biol 1990; 2:245–55.

Thomas KA. Fibroblast growth factors. FASEB J 1987; 1:434–40.

Thomassen MJ, Barna BP, Rankin D, Weidemann HP, Ahmad M. Differential effect of recombinant granulocyte macrophage colony stimulating factor on human monocytes and alveolar macrophages. Cancer Res 1989; 49:4086–9.

Tosato G, Seamon KB, Goldman ND. Monocyte-derived B cell growth factor identified as interferon $\beta 2$ (BSF-2; IL-6). Science 1988; 239:502–4.

Truett AP, Verghase MW, Dillon SB, Snyderman R. Calcium influx stimulates a second pathway for sustained diacylglycerol production in leukocytes activated by chemoattractants. Proc Natl Acad Sci USA 1988; 85:1549–53.

Uhing RJ, Adams DO. Molecular events in the activation of murine macrophages. Agents Actions 1989; 26:9–14.

Van Oud Ablas AB, van Furth R. Origin, kinetics and characteristics of pulmonary macrophages in the normal steady state. J Exp Med 1979; 149:1504–18.

Verghese MW, Charles L, Jakoi L, Dillon SB, Snyderman R. Role of guanine nucleotide regulatory protein in the activation of phospholipase C by different chemoattractants. J Immunol 1987; 138:4374–80.

Verghese MW, Smith CD, Charles LA, Jakoi L, Snyderman R. A guanine nucleotide regulatory protein controls polyphospoinositide metabolism, Ca+2 mobilization and cellular responses to chemoattractants in human monocytes. J Immunol 1986a; 137:271–5.

Verghese MW, Smith CD, Charles LA, Snyderman R. A common transduction pathway for leukocyte chemoattractant receptors: phospholipase C activation by a guanine nucleotide regulatory protein. Clin Res 1986b; 34:679A.

Verghese MW, Smith CD, Snyderman R. Role of guanine nucleotide regulatory protein in polyphosphoinositide degradation and activation of phagocytic leukocytes by chemoattractants. J Cell Biochem 1986c; 32:59–69.

Vilcek J, Lee TH. Tumor necrosis factor: new insights into the molecular mechanisms ot its multiple actions. J Biol Chem 1991; 266:7313–6.

Wahl SM, Hunt DA, Wakefield LM, McCartney-Francis N, Wahl LM, Roberts AB, Sporn MB. Transforming growth factor β induces monocyte chemotaxis and growth factor production. Proc Natl Acad Sci USA 1987; 84:5788–92.

Warheit DB, Chang LY, Hill LH, Hook GER, Brody AR. Pulmonary macrophage accumulation and asbestos-induced lesions at sites of fiber deposition. Am Rev Respir Dis 1984; 129:301–10.

Warheit DB, George G, Hill LH, Snyderman R, Brody AR. Inhaled asbestos activates a complement-dependent chemoattractant for macrophages. Lab Invest 1985; 52:505–14.

Warren MK, Ralph P. Macrophage growth factor CSF-1 stimulates human monocyte production of interferon, tumor necrosis factor and colony stimulating activity. J Immunol 1986; 137:2281–5.

Wewers MD, Saltini C, Sellers S, Tocci MJ, Bayne ER, Schmidt JA, Crystal RG. Evaluation of alveolar macrophages in normals and individuals with active pulmonary sarcoidosis for the spontaneous expression of the interleukin-1β gene. Cell Immunol 1987; 107:479–88.

White R, Janoff A, Godfrey HP. Secretion of α_2-macroglobulin by human alveolar macrophages. Lung 1980; 158:9–14.

Wing EJ, Ampel NM, Waheed A, Shadduck RK. Macrophage colony-stimulating factor (M-CSF) enhances the capacity of murine macrophages to secrete oxygen reduction products. J Immunol 1985; 135:2052–6.

Wollenberg GK, LaMarre J, Rosendal S, Gonias SL, Hayes MA. Binding of tumor necrosis factor alpha to activated forms of human plasma alpha-2-macroglobulin. Am J Pathol 1991; 138:265–72.

Zhang Y, Lin J-X, Vilcek J. Synthesis of interleukin 6 (interferon-β2/B cell stimulatory factor 2) in human fibroblasts is triggered by an increase in intracellular cyclic AMP. J Biol Chem 1988; 263:6177–82.

23

Role of the GTPase Superfamily in T-Lymphocyte Signal Transduction

THOMAS C. RYAN and WILLIAM W. CRUIKSHANK

Boston University School of Medicine
Boston, Massachusetts

I. T-Lymphocyte Activation

During an immune response, the thymus-derived T lymphocyte is activated in the presence of antigen and induced to proliferate, resulting in the generation of a population of cells which is capable of selectively regulating host defense. This activation of T cells progresses in two stages. In the initial stage, foreign antigen is internalized by phagocytic, antigen-presenting cells (APC), e.g., macrophages, processed, and transported to the cell surface bound to the antigen-presentation major histocompatability complex (MHC) (Unanue and Cerottini, 1989). The T-cell receptor (TCR), on the surface of the lymphocyte, binds both the processed antigen and the MHC molecule, resulting in an increase in both the production of cellular proteins (cytokines) and the expression of growth-factor surface receptors (Weiss, 1990). In addition to antigen-mediated activation, a number of accessory (CD4, CD8, CD45) and adhesion (CD2, LFA-1) molecules on the surface of the T cell act to increase both the avidity of cell-cell contact and the affinity of the TCR:antigen/MHC complex formed between the APC and the T cell (Springer et al., 1987; Ledbetter et al., 1988; Bierer et al., 1989). Integration between complementary sets of these molecules also transmits transmembrane signals necessary for lymphocyte activation (Springer et al., 1987; Ledbetter et al., 1988; Bierer et al., 1989). During the second stage of T-cell activation, the lymphocyte-specific cytokines, or lymphokines; produced during stage one, primarily interleukin-2 (IL2), bind to newly expressed high-affinity receptors on the surface of the T cell, resulting in the induction of cell proliferation (Smith, 1988). The bimodal activation of the T lymphocyte, therefore, requires the coordinated activity of a number of cell surface receptors which initiate carefully orchestrated intracellular events such as protein and DNA synthesis and mitosis (Weiss, 1989).

In order to better understand the mechanisms of T-lymphocyte activation, and thus the role of the T cell in immune function, an understanding of the systems which regulate the initiation and activity of these intracellular events is necessary. A number of signaling pathways have been identified which couple the activation of membrane receptors with intracellular effectors (Rao, 1991). These signaling pathways generate an activation cascade within the cell composed of second messenger and signaling intermediate molecules which interact to control cellular growth and differentiation. In this chapter we will examine the recent direct and indirect evidence which is suggestive of the involvement of a particular family of signaling intermediates, the members of the GTPase superfamily of proteins (Bourne et al., 1990), in the activity of a number of T-lymphocyte surface receptors.

II. The GTPase Superfamily

The GTPase superfamily of proteins is a group of intracellular proteins which regulate the activation and activity of a variety of intracellular transport processes and signal transduction pathways. The members of the GTPase superfamily include the protein synthesis-related GTPases and translational factors, the microtubule-associated proteins, the heterotrimeric "G proteins," and the monomeric low-molecular-mass or "ras-related" GTPases (Gilman, 1987; Bourne et al., 1990, 1991; Kaziro et al., 1991). The proteins in the GTPase superfamily are characterized by their ability to bind and hydrolyze GTP. Their role as intracellular mediators is regulated by the state of bound guanine nucleotide, with the GTP bound state being the active conformation, and the GDP bound state the resting. The mechanism for the initiation of the active state involves the exchange of the GDP on the resting molecule for GTP. Following activation, an intrinsic GTPase is induced which returns the molecule to the resting state. There are, however, a number of extrinsic factors which affect both the activation and deactivation of these proteins (Gilman, 1987; Bourne et al., 1990, 1991; Kaziro et al., 1991). In this chapter we will focus on the activity of the two GTPase families best studied in T lymphocytes, the heterotrimeric G proteins and the monomeric ras-related GTPases, and examine their relationship to T lymphocyte activation.

Signal transduction GTP-binding or "G" proteins form a group of plasma membrane-associated heterotrimers which function as signaling intermediates between cell surface receptors and intracellular second messenger and effector systems (Neer and Clapman, 1988; Freissmuth et al., 1989; Taylor, 1990; Birnbaumer, 1990). They are composed of an α (39-52 kDa), β (35-36 kDa), and a γ (5-10 kDa) subunit. While the β/γ subunit may play a role in the activation of some systems (Jelsema and Axelrod, 1987; Logothetis et al., 1987; Katada et al., 1986), the α subunit is the principal signal transducer (Gilman, 1987). The mechanism for the activation of G proteins involves first the binding of an extracellular factor or hormone to a receptor on the plasma membrane of the target cell. The newly formed ligand-receptor complex operates as a nucleotide exchange/release factor, inducing the release of GDP and the binding of GTP. GTP binding induces the dissociation of the heterotrimeric complex from the receptor and the uncoupling of the now active Gα from the β/γ subunits. Active Gα interacts with the effector, resulting in the amplification of the signal. The intrinsic GTPase activity hydrolyses GTP to GDP, resulting in the dissociation of Gα from the effector, the coupling of the heterotrimer, and reassociation of the complex with the receptor. Approximately 13 G proteins have been identified to date (Kaziro et al., 1991). They may be characterized by a number of factors, including cell specificity, receptor and effector system association, and sensitivity to bacterial toxins.

The low-molecule-mass or ras-related GTPases are a family of monomers, ranging in molecular weight from 20–35 kDa, which demonstrate a high degree of strucural and amino acid sequence homology (Barbacid, 1987; Hall, 1990; Downward, 1990; Macara, 1991). There are three principal groups within the family. The first of these is the ras and ras-related (50% ras homology) (Drivas et al., 1991) ral and rap proteins. Ras, the founding member of the family, was first described in association with the rat sarcoma virus and is the oncogene product responsible for the transformation of cells associated with the sarcoma (Harvey, 1964; Barbacid, 1987). It has also been suggested that ras may function as a signal transducer in normal cell growth and activation (Johnston et al., 1991). The second group of proteins is the rho family (35% ras homology), which have a possible role in the regulation of cytoskeletal events in the cell (Chardin et al., 1989; Paterson et al., 1990). The third group, the rab family (30% ras homology), is involved in the regulation of vesicular transport during both endo- and exocytosis (Balch, 1990). The structural homology of these proteins includes the GTP-binding pocket and the enzymatic GTPase site, common to GTPases in general (Bourne et al., 1991; Kaziro et al., 1991). In ras, this region includes the highly conserved glycine 12,13, and 61 amino acids which regulate the binding of guanine nucleotide and the oncogenic potential of the protein (Reddy et al., 1982; Tabin et al., 1982; Barbacid, 1987). There are also regions specific to this family. These include the effector-binding domain and the C-terminal consensus sequence, which mediates lipid modification and ultimately the cellular localization of the protein. Lipid modification of the C terminus of the ras-related proteins is required for membrane association (Grand and Owens, 1991). Modification involves the cleavage of a conserved cysteine sequence, generally CAAX, i.e., cysteine-aliphatic-aliphatic-X amino acid, followed by the addition of various isoprenyl groups (farnysyl, gerenyl-gerynyl, etc.) (Gutierrez et al., 1989). Other C-terminal modifications include carboxymethylation and palmitotylation (Buss and Sefton, 1986; Clarke et al., 1988). Ras is also modified by phosphorylation (Kamata and Feramisco, 1984). While the phosphorylation site is a threonine, and not a tyrosine, ras activity has been closely linked to tyrosine kinase activation (Smith et al., 1986).

The activation pathway of the proteins in the ras-related GTPase family differs from that of the G proteins. The exchange of nucleotides is regulated by a group of specific cytoplasmic nucleotide-exchange factors (NEFs), which includes both stimulatory and inhibitory proteins (Bollag and McCormick, 1991). The intrinsic rate of GTPase activity of the ras-related proteins is very low and is also regulated by extrinsic factors. The activation of hydrolysis is mediated by a GTPase-activating protein, GAP (Trahey and McCormick, 1987). A number of GAP proteins have been identified, including ras GAP, rho GAP and the neurofibromatosis gene product. NF-1 (Gibbs et al., 1988; McCormick 1989; Garret et al., 1989; Ballester et al., 1990; Martin et al., 1990; Xu et al., 1990). The mechanism of activation and activity of GAP has yet to be determined. However, GAP has been shown to be a substrate for the tyrosine kinases. The phosphorylated form of GAP is associated with the platelet-derived growth-factor (PDGF) receptor in a large multisubunit complex along with the signal transduction-associated phospholipase C gamma, the PI_3 kinase, and the raf serine/threonine kinase (Kaplan et al., 1990). This suggests a possible link between the activity of GAP and ras with transmembrane signaling.

Evidence has accumulated which points toward the association of both heterotrimeric and ras-related GTPases with a number of T lymphocyte receptors. These include the TCR-CD3 receptor complex, the accessory/adhesion molecules CD4, CD8, CD2, and select lymphokine receptors including IL2, which will be reviewed in the following section.

III. T-Lymphocyte Surface Receptors and the GTPase Superfamily

A. T-Cell Receptor (Ti-CD3)

The T-cell receptor (TCR) is a celi surface glycoprotein complex composed of a clonotypic, antigen-recognition component coupled to a multisubunit signaling complex. The antigen-recognition component "Ti" is a disulfide-linked heterodimer composed of an α (47–54 kDa) and a β (40–44 kDa) subunit (Clevers et al., 1988). Ti α and β both contain immunoglobulin like regions which selectively bind both peptide antigen and MHC. These components therefore regulate the antigen specificity of the TCR-mediated activation of the lymphocyte. γ/δ and γ/γ dimers of Ti have also been identified, but at present are restricted to developing thymocytes (Brenner et al., 1988) and a subpopulation of peripheral blood lymphocytes which may play a distinct role in immune surveilance (Goodman and Lefrancois, 1988). The Ti heterodimer is noncovalently coupled to the multimeric CD3 complex. CD3 is composed of at least six subunits, a 25- to 28-kDa glycoprotein γ and a 20-kDa glycoprotein δ chain, two 20-kDa nonglycoprotein ϵ chains present as γ–ϵ and δ–ϵ dimers, and a pair of disulfide linked ζ–ζ or ζ–η chains (de la Hera et al., 1991; Koning, 1991; Rojo and Portoles, 1991). Due to the large cytoplasmic domains of the proteins in the CD3 complex, as compared to the short cytoplasmic tail of Ti α and β, it has been suggested that CD3 is the signal transduction unit for the TCR (Rao, 1991). This is supported by the observation that, in response to both antigen and monoclonal antibodies directed against CD3, both tyrosine phosphorylation and phosphatidylinositol hydrolysis occur.

Activation of CD3 has been demonstrated to result in the generation of a number of serine-and tyrosine-specific products, including CD3γ (serine), the ζ chain of the complex, and phospholipase Cγ (tyrosine) (Cantrell et al., 1985; Samelson et al., 1986). The zeta chain has recently been demonstrated to transmit transmembrane signals as a CD8–ζ chimera (Irving and Weiss, 1991). The phosphorylation of the ζ subunit and PLCγ therefore suggests a regulatory role for tyrosine kinases in signal transduction. None of the CD3 components, however, contains tyrosine kinase domains. Recent studies have suggested the association of the nonreceptor tyrosine kinase $p59^{fyn}$ kinase with the complex (Samelson et al., 1990). The fyn kinase belongs to the src oncogene tyrosine kinase family (Bolen et al., 1991). These kinases have been demonstrated to be involved in the transformation of a number of cell types.

Following exposure to mitogen, anti-TCR, and CD3 monoclonal antibodies and pharmacological agents, the T lymphocyte experiences a rapid transient increase in cytosolic calcium as well as an increase in plasma membrane calcium permeability (Imboden and Stobo, 1985; Kuno and Gardner, 1987). The initial rise in cytoplasmic calcium concentration is dependent primarily on the release of intracellular stores of calcium and not on the influx of extracellular calcium (Imboden and Stobo, 1985). The initial mobilization of intracellular calcium stores is mediated by the cleavage of phosphatidylinositol 4,5-bisphosphate (PIP_2), in the inner leaflet of the plasma membrane, by the membrane-associated enzyme phospholipase C (PLC), into the highly charged, water-soluble, 1,4,5-trisphosphate (IP_3) and the hydrophobic phospholipid backbone diacylglycerol (DAG) (Berridge 1987; Rhee et al., 1989). Inositol phosphate released into the cytosol binds to organellar receptors, resulting in the release of calcium (Ferris et al., 1989). Both the calcium and the DAG released from this reaction bind to and act as a co-activator for the serine/threonine-specific kinase, protein kinase C (PKC), (Ferrera 85) which in turn phosphorylates a number of lymphocyte proteins, including the CD3γ subunit (Cantrell et al., 1985), and acts as a key activator in T-lymphocyte response. It is the activation of the phosphatidylinositol pathway by the TCR that is most suggestive of the involvement of a heterotrimeric G protein in CD3-mediated signal transduction (Gilman, 1987).

During the past several years, much work has been performed investigating the importance of GTP-binding proteins in TCR-mediated lymphocyte signaling and function. Historically, the earlier studies determined GTP-binding protein involvement through the use of bacterial toxins, pertussis toxin and cholera toxin, which act to irreversibly ADP ribosylate the catalytic (alpha) subunits of the Gi, Go, and Gs subclasses, respectively, and thereby prevent signal transduction. The actual involvement of GTPases in TCR activation is difficult to substantiate based solely on these findings, since the effects of both pertussis and cholera toxin are not restricted solely to interactions with GTPases in T cells. For example, pertussis toxin alone is a potent mitogen for T (Rosoff et al., 1987), as well as B, lymphocytes (Kolb et al., 1990) and platelets (Banga et al., 1987). High concentrations of pertussis toxin treatment results in IL2 secretion from human T-cell clones and rodent T lymphocytes (Vistica 86, Gilmore and Weiner, 1988; Aussel et al., 1988), and induces rises in intracellular calcium and inositol phosphates in human T lymphocytes and Jurkat cells (Gray et al., 1989; Stewart et al., 1989), a human lymphocyte cell line. However, low concentrations of PT act to suppress activation by several lymphocyte mitogens (Vistica 86, Gilmore and Weiner, 1988). In fact, while PT has been shown to interact directly with GTP-binding proteins, many of the effects of PT may be completely independent of this interaction and result primarily from association with membrane-expressed surface protein. Studies by Rogers et al. (1990) have sugested that activating effects of PT can be found only in cells expressing a cell surface PT receptor; cells such as neutrophils, which lack this membrane protein, are unresponsive. Other complicating issues that must be addressed to determine GTP-binding protein involvement in cell signaling are the dose response and timing of ADP ribosylation as they relate to the inhibitory effects. Inhibition of GTPase activity is accomplished by ADP ribosylation, yet in some assay systems it appears the PT blocking of signal transduction occurs at different doses or different times apparently required for ADP ribosylation (Modesto et al., 1991). Given all these caveats, much of the evidence indicating G-protein involvement in TCR signaling originated using either CT or PT to alter normal Gs or Gi function. Initially, Imboden (1986) demonstrated that TCR signaling in Jurkat cells was blocked by CT treatment, suggesting Gs involvement. Other studies have demonstrated that PT treatment depresses TCR signaling, implicating Gi involvement (Stanley et al., 1989; Ledbetter et al., 1987; Gilmore and Weiner, 1988; Aussel et al., 1988).

An alternative, indirect approach to investigate GTPase involvement in signal transduction pathways is to activate the G protein directly by using guanosine 5′-O-[3-thiotrisphosphate] (GTPγS), a nonhydrolyzable analog of GTP. GTPγS irreversibly binds to the catalytic subunit of the heterotrimeric GTP-binding protein, resulting in a perpetually activated conformation. Bonvini et al. (1991) report that GTPγS stimulation of murine T-helper cells results in an increase in inositol phosphate generation. This is further supported by the results that guanosine 5′ [B-thio]diphosphate (GDPβS), a guanine nucleotide analog that inhibits G-protein function, inhibits TCR-induced InsPL formation (Bonvini et al., 1991; Graves and Cantrell, 1991). This type of approach using permeablized cells and direct Gi and Gs activation has indicated G-protein involvement in several other functional assay systems (Davies et al., 1988; Schrezenmeier et al., 1988; Ledbetter et al., 1987; O'Shea et al., 1987) of TCR activation.

While these studies give indirect evidence to suggest G-protein participation of TCR signaling, other studies indicate that coupling of TCR to PLC is not directly G-protein mediated, or at least not mediated by substrate of either CT or PT. Augustine et al., (1991) demonstrated that insPL formation induced by anti-CD3 antibodies was inhibited by the protein tyrosine kinase inhibitor, staurosporin, as well as CT. However, neither CT nor staurosporin was able to block insPL generation induced by GTPγS. The implication of their data is that TCR signaling is transduced by a protein tyrosine kinase. To address the possible role of the Gi subunit in T-cell signaling, Chaffin et al. (1990) expressed the S1 subunit of PT in the thymocytes of transgenic mice. While they found no change in signaling responsiveness to several mitogens, they did report that S1 containing thymocytes accumulated in the thymus but didi not appear in secondary lymph nodes nor in the circulation. They suggest therefore that a PT substrate functions to faciliate thymocyte emigration rather than cell signaling.

Thus, the question concerning heterotrimeric G-protein involvement in T-cell signaling remains for the most part unanswered. However, recent data are suggestive of the association of ras-related GTPases with TCR signaling events. In human T lymphocytes, both mitogen and monoclonal antibodies directed against CD3 (UCHL-1) induce an increase in the GTP-bound state of immunoprecipitated ras (Downward et al., 1990). The mechanism which couples ligand-receptor binding with the increased activation state of ras is not known. However, using alpha ^{32}P-labeled nucleotides, Downward et al. were able to demonstrate that the exchange of GDP for GTP was very rapid, while the accumulation of GTP-unstimulated samples was negligible, suggesting both a rapid exchange rate and rapid hydrolysis. The intrinsic GTPase activity of ras, however, is known to be low. This suggested that an active GAP was present in the cells, which was rendered quiescent following exposure of the cells to CD3 specific ligands. To determine the effect of CD3 signaling on GAP activity, one of the downstream effector molecules of CD3 activation, PKC, was activated directly with the phorbol ester, phorbol dibutyrate (PDBu). By comparing GAP activity in resting and PDBu-stimulated Triton X-100 cell lysates, Downward et al. were able to demonstrate the PDBu was capable of inhibiting GAP activity, suggesting that PKC is a regulator of the activation state of ras in CD3-stimulated lymphocytes. Further investigations are required at this time to determine if the observed change in activation state of ras-related GTPases following TCR stimulation is involved directly in antigen/TCR receptor signaling or is an epiphenomenon of the signaling pathway.

B. CD4/CD8

CD4 (55 kDa) and CD8 (31-34 kDa dimer) were originally identified as differentiation antigens for T lymphocytes and have been demonstrated to play significant roles in T-cell activation (Reinherz et al., 1979; Reinhertz and Schlossman, 1980; Parnes, 1989). As surface receptors for the nonpolymorphic region of the major histocompatability complex, both CD4 (MHC type II) and CD8 (MHC type I) act to increase cell-cell contact, in the context of the Ti/CD3 complex (Sleckman, 1989, Gay et al., 1989; Doyle and Strominger, 1987; Rosenstein et al., 1989). Both CD4 and CD8 also regulate antigen-mediated T-lymphocyte proliferation (Gay et al., 1987; Landegren et al., 1982). CD4/CD8-mediated regulation of T-cell activity may involve the activation of the lymphocyte-specific, nonreceptor tyrosine kinase $p56^{lck}$, which is noncovalently associated with the cytoplasmic tails of the molecules. The complexing of CD4 and CD8 with $p56^{lck}$ (Marth et al., 1985; Veillette et al., 1988), member of the src family of protein tyrosine kinases, presents a structure similar to several growth-factor receptors, such as epidermal growth-factor receptor, insulin receptor, platelet-derived growth-factor receptor, and fibroblast growth-factor receptor (Ullrich and Schlessinger, 1990; White, 1991). CD4 and CD8 could, therefore, function in a manner similar to these receptors and have the capability to transmit intracellular signals which affect cell growth. Recent evidence has in fact indicated that CD4 may function independently as a growth-factor receptor (Kornfeld et al., 1988; Neudorf et al., 1990; Cruikshank et al., 1991; Acres et al., 1986). While there are no known soluble CD8 ligands, there are specific ligands for CD4 which induce signaling. These are HIV-1 surface glycoprotein, gp120, and CD4-binding lymphokine, lymphocyte chemoattractant factor (LCF), as well as a number of CD4-specific antibodies (Kornfeld et al., 1988; Neudorf et al., 1990; Cruikshank et al., 1987, 1991; Carrel et al., 1988). Exposure of T lymphocytes to these CD4 ligands results in a rise in intracellular calcium and the generation of inositol phosphates (Kornfeld et al., 1988; Neudorf et al., 1990; Cruikshank et al., 1991). The generation of these products from the phosphatidylinositol signal transduction pathway is suggestive of the involvement of GTPases.

Incubation of T cells with most anti-CD4 antibodies does not results in second messenger generation but does induce subsequent blocking of second messenger generation and cell responsiveness by specific antigen or anti-CD3 antibodies (Bank 85, Biddison 86). However, two reported antibodies appear to have the capacity to induce generation of second messengers (Neudorf et al., 1990; Carrel et al., 1988). CD4 is also bound by two other reported ligands, the envelope glycoprotein from HIV-1 virus, and a lymphokine generated from CD8+ lymphocytes, LCF (Cruikshank et al., 1991). Binding to CD4 by antibodies and HIV-1 gp120 has been extensively described, while the association of LCF with CD4 has only recently been reported. There are several lines of evidence to suggest that LCF binds to CD4. First, all LCF-induced cellular responses, chemotaxis and cell cycle progression, are blocked by Fab components of anti-CD4 antibodies. Second, cellular responses to LCF are in many cases proportional to CD4 expression on the cell surface. Third, unresponsive cells become responsive to LCF stimulation following transfection of human CD4 antigen, and the function is inhibited by anti-CD4 Fab fragments. Cells expressing truncated CD4 (residues 403-439 of the cytoplasmic tail deleted) are not responsive to LCF stimulation. And fourth, LCF binds specifically to matrices of recombinant soluble CD4. Studies using these ligands to CD4 have demonstrated that binding to CD4 results in a transient rise in Ca^{2+} and inositol-trisphosphate (IP_3) formation. While the signal pathway for CD4 has not been elucidated, some evidence does exist to indicate that a G protein may be involved. Our laboratory has reported that lymphocytes pretreated with PT, but not CT, demonstrate a marked unresponsiveness to CD4-stimulated migration (Cruikshank 89). In addition, we have preliminary data which indicates that there is an increase in GTPase enzyme activity and GTPγS binding following stimulation with LCF. Also, we have found that CD4 immunoprecipitates demonstrate the capacity to exchange GDP for labeled GTP, indicating that a physical association exists between CD4 and a form of GTP-binding protein.

In strong support of this ligand selective data, Tefler et. al., (1991) have reported that CD4 immunoprecipitates have the capacity to facilitate GTP/GDP exchange. All attempts thus far to demonstrate high-molecular-mass GTPases associated with CD4 have not been successful. However, Tefler has shown the presence of a GTPase in CD4 immunoprecipitates with a molecular weight of 34 kDa. This GTPase was identified using an antibody directed against a conserved region of Gi-alpha subunits. The presence of a novel alpha subunit or LMM GTPase associated directly with CD4 could help to explain many of the previous findings.

C. CD2

In addition to the antigen-recognition complex, the interaction of a variety of cell surface adhesion molecules, including members of both the immunoglobulin and the integrin family (ICAM-1,2; LFA-1,3; etc.) on both the APC and the T cell act to increase the affinity of the cell-cell interaction (Springer, 1990). One of the molecules involved in cell adhesion is the CD2 molecule. CD2, a 50-kDa single-chain cell surface glycoprotein, is one of the earliest of the T-lymphocyte differentiation antigens expressed during T-lymphocyte ontogeny (Moingeon et al., 1987). In conjunction with the formation of the TCR:antigen/MHC complex, CD2 binds to its natural ligand LFA-3 (55–70 kDa), aiding the stabilization of cell contact (Makagoba et al., 1989).

CD2 is also the target for an alternative T-lymphocyte activation pathway and has been suggested to play a role in thymic selection of lymphocytes prior to the expression of the TCR-CD3 antigen-recognition complex (Kabelitz, 1990). Anti-CD2 antibodies have been demonstrated to induce the activation of a number of signal transduction events. These include both an increase in PIP_2 hydrolysis and intracellular calcium concentration and the translocation of PKC from the cytosolic to the membrane fraction of the cell, resulting in T-cell activation and proliferation (Pantaleo et al., 1987; Ledbetter et al., 1988a; Banasco et al., 1989). CD2 also induces the tyrosine-specific phosphorylation of a number of proteins, including CD3 ζ (Ley et al., 1991). Recent studies examining the mechanism of CD2 activation have demonstrated the association of the low-molecular-mass GTPase ras with the receptor (Graves et al., 1991). In response to CD2 trivalent antibody conjugates (GT2-

OKT11), there is an increase in the GTP bound state of ras in peripheral blood lymphoblasts. As with the CD3 receptor complex, the exchange rate of GDP for GTP is unaffected by the agonist. Triton X-100 lysates of CD2-stimulated cells, however, demonstrated a decreased GAP activity similar in nature, but smaller in magnitude, to that observed following PDBu stimulation of lymphocytes. This suggests a common GAP-related mechanism for both CD3- and CD2-mediated ras activation.

D. Interleukin-2 Receptor (IL2R), IL3, GM-CSF

In T lymphocytes, the binding of IL2 to its high-affinity receptor appears to be the primary mediator of cell proliferation and progression. IL2 receptor (IL2R) is composed of two independent binding proteins, the low-affinity (10^{-8}) IL2 R alpha (p55) and the moderate-affinity (10^{-9}) IL2R beta (p70/75), which complex to form a single high-affinity (10^{-11}) binding site for IL2 (Waldmann, 1991). Recent studies examining the possible signal transduction mechanism of IL2 receptor activation have revealed the association of $p56^{lck}$ with the beta unit of the receptor complex (Hatakeyama et al., 1991). Following stimulation of the IL2 receptor, there is both the serine/threonine phosphorylation of lck and the induction of the tyrosine kinase activity of the molecule (Horak et al., 1991). The role of the $p56^{lck}$ in T-cell signaling has yet to be determined. In addition to $p56^{lck}$, the activation of IL2 and other select lymphokine receptors has been shown to involve an increase in the GTP bound state of monomeric ras (Satoh et al., 1991).

Using a series of lymphokine-specific cell lines, Satoh et al. were able to demonstrate the activation of the ras in response to IL2, the pleitropic growth factor interleukin-3 (IL3), and the neutrophil/macrophage growth factor granulocyte/monocyte-colony stimulating factor (GM-CSF), but not multifunctional growth factor interleukin-4 (IL4). While each of these is capable of inducing tyrosine phosphorylation, the mechanism which differentiates their signaling is unknown. An early report by Farrar et al., suggested that a high-molecular-mass, PT-sensitive, GTP-binding protein was associated with the IL2R. They demonstrated decreased IL2 binding to IL2R in the presence of GTPγS and increased GTPγS binding in the presence of IL2. Continued investigations are required to further delineate GTPases involvement in this system.

IV. Conclusion

While there is now a growing body of information that strongly suggests the presence of GTPases in T-cell signaling via TCR, CD4, CD2, IL2R, and IL3R, there are as yet no direct functional links to any of these receptors that would satisfy all stringent criteria. Of greatest interest is the suggestive evidence that potentially links low-molecular-mass GTPase of the ras family to receptor-coupled signal transduction; data would expand our understanding of the functions of these proteins.

V. Acknowledgments

Grant support for this work was provided by NIH individual grant NRSA HL08260, American Lung Association Research Training Fellowship, AmFAR Research Scholarship, RO1 A128736-03, 1 P50 HL46563-01.

References

Acres RB, Conlon PJ, Mochizuki DY, Gallis B. Rapid phosphorylation and modulation of the T4 antigen on cloned helper T cells induced by phorbol myristate acetate or antigen. J Biol Chem 1986; 261:16210–4.

Anderson D, Tsoukas C. Cholera toxin inhibits resting human T cell activation via a cAMP-independent pathway. J Immunol 1989; 143:3647–52.

Augustine J, Secrist JP, Daniels J, Leibson P, Abraham R. Signal transduction through the T cell antigen receptor: activation of phospholipase C though a G protein-independent coupling mechanism. J Immunol 1991; 146:2889–97.

Aussel C, Mary D, Peyron JF, Pelassy C, Ferrua B, Fehlmann M. Inhibition and activation of interleukin 2 synthesis by direct modification of guanosine triphosphate-binding proteins. J Immuno 1988; 140:215–20.

Balch WE. Small GTP-binding proteins in vesicular transport. TIBS 1990; 15:473–7.

Ballester R, Murchuk D, Boguski M, Saulino A, Letchter R, Wigler M, Collins F. The NF1 locus encodes a protein functionlly related to mammalian GAP and yeast IRA proteins. Cell 1990; 63:851–9.

Banasco M, Nunes J, Lopez M, Cerdan C, Pierres A, Mawas C, Olive D. T cell activation via CD2 molecule is associated with protein kinase C translocation from the cytosol to the plasma membrane. Eur J Immunol 1989; 19:823–7.

Banga HS, Walker RK, Winberry LK, Rittenhouse SE. Pertussis toxin can activate human platelets. J Biol Chem 1987; 262:14871–4.

Barbacid M. ras genes. Ann Rev Biochem 1987: 56:779–827.

Berridge MJ. Inositol trisphosphates and diacylglycerol: two interacting second messengers. Ann Rev Biochem 1987; 56:159–93.

Bierer BE, Sleckman BP, Ratnofsky SE, Burakoff SJ. The biologic role of CD2, CD4, and CD8 in T-cell activation. Ann Rev Immunol 1989; 7:579–600.

Birnbaumer L. G proteins in signal transduction. Ann Rev Pharmacol Toxicol 1990; 30:675–705.

Bolen JB, Thompson PA, Eiseman E, Horak ID. Expression and interaction of the src family of tyrosine protein kinases in T lymphocytes. Adv Cancer Res 1991; 57:103–49.

Bollag G, McCormick F. Regulators and effectors of ras proteins. Ann Rev Cell Biol 1991; 7:601–32.

Bonvini E, Debell K, Taplits M, Brando C, Laurenza A, Seamon K, Hoffman T. A role for guanine-nucleotide-binding proteins in mediating T-cell-receptor coupling to inositol phospholipid hydrolysis in a murine T-helper (type II) lymphocyte clone. Biochem J 1991; 275:689–96.

Bourne HR, Sanders DA, McCormick F. The GTPase superfamily: a conserved switch for diverse cell functions. Nature 1990; 348:125–32.

Bourne HR, Sanders DA, McCormick F. The GTPase superfamily: conserved structure and molecular mechanism. Nature 1991; 349:117–27.

Brenner MB, Strominger JL, Krangel MS. The $\gamma\delta$ T cell receptor. Adv Immunol 1988; 43:133–91.

Buss JE, Sefton BM. Direct identificaiton of palmitic acid as the lipid attached to p21ras. Mol Cell Biol 1986; 6:116–22.

Cantrell DA, Davies AA, Crumpton MJ. Activations of protein kinase C down regulate and phosphorylate the T3-T-cell antigen receptor complex of human T lymphocytes. Proc Natl Acad Sci USA 1985; 82:8158–62.

Carrel S, Moretta A, Pantaleo G, Tambussi G, Isler P, Perussia B, Cerottini JC. Stimulation and proliferation of CD4+ peripheral blood T lymphocytes induced by an anti-CD4 monoclonl antibody. Eur J Immunol 1988; 18:333–9.

Chaffin K, Beals C, Wilkie T, Forbush K, Simon M, Perlmutter R. Dissection of thymocyte signaling pathways by invivo expression of pertussis toxin ADP-ribosyltransferase. EMBO J 1990; 9:3821–9.

Chardin P, Boquet P, Madaule P, Popoff MR, Rubin EJ, Gill DM. The mammalian G protein rho C is ADP-ribosylated by Clostridium botulinum exoenzyme C3 and affect action microfilaments in Vero cells. EMBO J 1989; 8:1087–92.

Clarke S, Vogel JP, Deschenes RH, Stock J. Post-translational modification of the H-ras oncogene protein: evidence for a third class of protein carboxy methyltransferase Proc Natl Acad Sci USA 1988; 85:4643–7.

Clevers H, Alacron B, Wilesman T, Terhorst C. The T cell receptor/CD3 complex: a dynamic protein ensemble. 1988; 6:629–62.

Cruikshank WW, Berman JS, Theodore AC, Bernardo J, Center DM. Lymphokine activation of T4+ lymphocytes and monocytes. J Immunol 1987; 138:3817–23.

Cruikshank WW, Greenstein JL, Theodore AC, Center DM. Lymphocyte chemoattractant factor induces CD4-dependent intracytoplasmic signaling in lymphocytes. J Immuno 1991; 146:2935–43.

Davies A, Hexham M, Cantrell D, Crumpton M. Evidence that a GTP binding protein regulates phosphorylation of the CD3 antigen in human T lymphocytes. BBRC 1988; 156:328–34.

de la Hera A, Muller U, Olsson C, Isaaz S, Tunnacliffe A. Structure of the T cell antigen receptor (TCR): two CD3ϵ subunits in a functional TCR/CD3 complex. J Exp Med 1991; 173:7–17.

Downward J. The ras superfamily of small GTP-binding proteins. TIBS 1990; 15:469–72.

Downward J, Graves JD, Warne PH, Rayter S, Cantrell DA. Stimulation of p21ras upon T-cell activation. Nature 1990; 346:719–23.

Doyle C, Strominger JL. Interaction between CD4 and Class II MHC molecules mediates cell adhesion. Nature 1987; 330:256–9.

Drivas GT, Palmieri S, D'Eustachio P, Rush MG. Evolutionary grouping of the ras-protein family. Biochem Biophys Res Commun 1991; 176:1130–5.

Ferris CD, Huganir RL, Supattapone S, Snyder SH. Purified inositol 1,4,5-trisphosphophate receptor mediates calcium flux in reconstituted lipid vesicles. Nature 1989; 342:87–9.

Freissmuth M, Casey PJ, Gilman AF. G proteins control diverse pathways of transmembrane signaling. FASEB J 1989; 3:2125–31.

Garrett MD, Self AJ, van Oers C, Hall A. Identification of distinct cytoplasmic targets for ras/R-ras and rho regulatory proteins.

Gay D, Maddon P, Sckaly R, Talle MA, Godfrey M, Long E.

Gerardy-Schahn R, Mittrucker HW, Schultze U, Fleischer B. Selective loss of pertussis toxin-sensitive G-protein form the plasma membrane after antibody-induced internalization of T-cell surface molecules. J Biol Chem 1991; 266:6942–7.

Gibbs JB, Schaber MD, Allard WJ, Sigal IS, Scolnick EM. Purification of ras GTPase activating protein from bovine brain. Proc Natl Acad Sci USA 1988; 85:5026–30.

Gilman AF. G proteins: transducers of receptor-generated signals. Ann Rev Biochem 1987; 56:615–49.

Gilmore W, Weiner L. The effects of pertussis toxin and cholera toxin on mitogen-induced Interleukin-2-production: evidence for G protein involvement in signal transduction. Cell Immunol 1988; 113:235–50.

Goldstein G, Chess L, Axel R, Kappler L, Marrack P. Functional interaction between human T-cell prteoin CD4 and the major histocompatibility complex HLA-DR antigen. Nature 1987; 328:626–9.

Goodman T, Lefrancois. Expression of the $\gamma\delta$ T-cell receptor on intestinal CD8+ intraepithelial lymphocytes. Nature 1988; 333:855–7.

Grand RJA, Owens D. The biochemistry of ras p21. Biochem J 1991; 279:609–31.

Graves J, Cantrell D. An analysis of the role of guanine nucleotide binding proteins in antigen receptor/CD3 antigen coupling to phospholipase C. J Immunol 1991; 146:2102–7.

Graves JD, Downward J, Rayter S, Warne P, Tutt AL, Glennie M, Cantrell DA. CD2 antigen mediated activation of the guanine nucleotide binding proteins p21ras in human T lymphocytes. J Immunol 1991; 146:3709–12.

Gray LS, Huber KS, Gray MC, Hewlett EL, Englehard VH. Pertussis toxin effects on T lymphocytes are mediated through CD3 and not by pertussis toxin catalyzed modification of a G protein. J Immunol 1989; 142:1631–8.

Gutierrez L, Magee AI, Marshall CL, Hancock JF. Post-translational processing of p21ras is two-step and involves carboxyl-methylation and carboxy-terminal proteolysis. EMBO J 1989; 8:1093–8.

Hall A. The cellular functions of small GTP-binding proteins. Science 1990; 249:636–40.

Hatakeyama M, Kono T, Kobayashi N, Kawahara A, Levin SD, Perlmutter RM,

Tanaguchi T. Interaction of the IL-2 receptor with the src-family kinase p56lck: identificaiton of novel intermolecular association. Science 1991; 252:1523–8.

He Y, Hewlett E, Temeles D, Qusenberry P. Inhibition of interleukin 3 and colony-stimulating factor 1-stimulated marrow cell proliferation by pertussis toxin. Blood 1988; 71:1187–95.

Horak ID, Gress RE, Lucas PJ, Horak EM, Waldmann TA, Bolen JB. T-lymphocyte interleukin 2-dependent tyrosine protein kinase signal transduction involves the activation of p56lck. Proc natl Acad Sci USA 1991; 88:1996–2000.

Imboden JB, Stobo JD. Transmembrane signalling by the T cell antigen receptor: perturbation of the T3-antigen receptor complex generates inositol phosphates and releases calcium ions from intracelular stores. J Exp Med 1985; 161:446–56.

Irving BA, Weiss A. The cytoplasmic domain of the T cell receptor ζ chain is sufficient to couple to receptor-associated signal transduction pathway. Cell 1991; 64:891–901.

Jelsema Cl, Acelrod J. Stimulation of phospholipase A2 activity in bovine rod outer segments by the $\beta\gamma$ subunits of transducin and its inhibition by the α subunit. Proc Natl Acad Sci USA 1987; 84:3623–7.

Johnston C, Morris J, Hall A. The role of ras in signal transduction. Biochem Soc Trans 1991; 19:269–99.

Kabelitz D. Do CD2 and CD3-TCR T-cell activation pathways function independently? Immunol Today 1990; 11:44–7.

Kamata T, Feramisco JR. Epidermal growth factor stimulates guanine nucleotide binding activity and phosphorylation of ras oncogene proteins. Nature 1984; 310:147–50.

Kaplan DR, Morrison DK, Wong G, McCormick F, Williams LT. PDGF β-receptor stimulates tyrosine phosphorylation of GAPand association of GAP with a signaling complex. Cell 1990; 61:125–33.

Katada T, Oinuma M, Ui M. Mechanisms for inhibition of the catalyytic activity of adenylate cyclase by the guanine nucleotide binding proteins serving as the substrate of islet-activating protein, pertussis toxin. J Biol Chem 1986; 261:5215–21.

Kaziro Y, Itoh H, Kozasa T, Nakafuku M, Satoh T. Structure and function of signal-transducing GTP-binding proteins. Ann Rev Biochem 1991; 60: 349–400.

Kelvin D, Schreeve M, McAuley C, McLeod DL, Simard G, Connolly JA. Interleukin 3-stimulated proliferation is sensitive to pertussis toxin: evidence for a guanyl nucleotide regulatory protein-mediated signal transduction mechanism. J Cell Physiol 1989; 138:273–80.

Kolb JP, Genot E, Petit-Koskas E, Paul-Eugene N, Dugas D. Effect of bacterial toxins on human B cell activation. I. Mitogenic activity of pertussis toxin. Eur J Immunol 1990; 20:969–76.

Koning F. Lymphocyte antigen receptors: a common design? Immunol Today 1991; 12:100–5.

Kornfeld H, Cruikshank WW, Pyle SW, Berman JS, Center DM. Lymphocyte activiation by HIV-1 envelope glycoprotein. Nature 1988; 335:445–8.

Kuno M, Gardner P. Ion channels acticated by inositol 1,4,5-trisphosphate in plasma membrane of human T-lymphocytes. Nature 1987; 326:301–4.

Kvanta A, Nordstedt C, van der Ploeg I, Jondal M, Fredholm B. CD3/T-cell receptor coupling to a pertussis and cholera toxin-insensitive G-protein. FEB 1989; 250:536–40.

Landegren U, Ramstedt U, Axberg, I, Ullberg M, Jondal M, Wigzell H. Selective inhibition of human T cell cytotoxicity at levels of target recognition or initiation of lysis by monoclonal OKT3 and Leu-2a antibodies. J Exp Med 1982; 155:1579–84.

Ledbetter JA, June CH, Grosmaire LS, Rabinovitch PS. Crosslinking of surface antigens causes mobilization of intracellular ionized calcium in T lymphocytes. Proc Natl Acad Sci USA 1987; 84:1384–8.

Ledbetter JA, Rabinovitch PS, Hellstrom I, Hellstrom KE, Grosmarie LS, June CH. Role of CD2 crosslinking in cytoplasmic calcium responses and T cell activation. Eur J Immunol 1988a; 18:1601–8.

Ledbetter JA, Tonks NK, Fischer EH, Clark EA. CD45 regulates signal transduction and lymphocyte activation by specific association with recepor molecules on T and B cells. Proc Natl Acad Sci USA 1988b; 85:8628–32.

Ley SC, Davies AA, Druker B, Crumpton MJ. The T cell receptor/CD3 complex and CD2 stimulate the tyrosine phosphorylation of indistinguishable patterns of polypeptides in the human T leukemic cell line Jurkat. Eur J Immunol 1991; 21:2203–9.

Logothetis DE, Kurachi Y, Galper J, Neer EJ, Clapman DE. The $\beta\gamma$ subunits of GTP-binding proteins activate the muscarinic K+ channel in heart. Nature 1987; 325:321–6.

Macara IG. The ras superfamily of molecular switches. Cell Signaling 1991; 3:179–87.

Makagoba MW, Sanders ME, Shaw S. The CD2-LFA-3 and LFA-1-ICAM pathways: relevence to T-cell recognition. Immunol Today 1989; 10:417–22.

Marth JD, Peet R, Krebs EG, Perlmutter RM. A lymphocyte-specific protein-tyrosine kinase gene is rearranged and overexpressed in the murine T cell lymphoma LSTRA. Cell 1985; 43: 393–404.

Martin GA, Viskochil C, Boolag G, McCabe PC, Crosier WJ, Haubruck H, Conroy L, Clark R, O'Connell P, Cawthon RM, Innis MA, McCormick F. The GAP-related domain of the neurofibromatosis type 1 gene product interacts with ras p21. 1990; 63,943–849.

McCormick F. Ras GTPase activating protein: signal transmitter and signal terminator. Cell 1989; 56:5–8.

Modesto J, Breittmayer JP, Grenier-Brossette N, Fehlmann M, Cousin JL. Pertussis toxin-sensitive G-proteins are not involved in activation of T-lymphocytes. Cell Signaling 1991; 3:25–33.

Moingeon P, Chang H-C, Sayre PH, Clayton LK, Alcover A, Gardner P, Reinherz EL. The structural biology of CD2. Immunol Rev 1987; 111:111–44.

Neer EJ, Clapman DE. Role of G protein subunits in transmembrane signaling. Nature 1988; 333:129–34.

Neudorf SML, Jones MM, McCarthy BM, Harmony JA, Choi EM. Cell Immunol 1990; 125:301–14.

O'Shea J, Urdahl K, Luong HT, Chused TM, Samelson LE, Klausner RD. Aluminum fluoride induces phosphatidylinositol turnover, elevation of cytoplasmic free calcium, and phosphorylation of the T cell antigen receptor in murine T cells. J Immunol 1987; 139:3463–9.

Pantaleo G, Olive D, Poggi A, Kozumbo W, Moretta L, Moretta A. Tranmembrane signalling via T11-dependent pathway of human T cell activation. Evidence for the involvement of 1,2 diacylglycerol and inositol phosphates. Eur J Immunol 1987; 17:55–60.

Parnes JR. Moelcular biology and function of CD4 and CD8. 1989; 44:265–311.

Paterson HF, Self AJ, Garrett MD, Just I, Aktories K, Hall A. Microinjection of recombinant p21 rho induces rapid changes in cell morphology. J Cell Biol 1990; 111:1001–7.

Pessa-Morikawa T, Mustelin T, Andersson L. Functional maturation of human T lymphocytes is accompanied by changes in the G-protein pattern. J Immunol 1990; 144:2690–5.

Rao A. Signaling mechanisms in T cells. Crit Rev Immunol 1991; 10:495–519.

Reddy EP, Reynolds RK, Santos E, Barbacid M. A point mutation is responsible for the acquistion of the transforming properties by the T24 human bladder carcinoma oncogene. Nature 1982; 300:149–52.

Reinherz EL, Kung PC, Goldstein G, Schlossmann SF. Separation of functional subsets of human T cells by amonoclonal antibody. Proc Natl Acad Sci USA 1979; 76:4061–5.

Reinherz EL, Schlossman SF. The differentiation and function of human T lymphocytes. Cell 1980; 19:821–7.

Rhee SG, Suh P-G, Ryu S-H, Lee SY. Studies of inositol phospholipid-specific phospholipase C. Science 1989; 244:546–50.

Rogers T, Corey S, Rosoff P. Identification of a 43-kilodalton human T lymphocyte membrane protein as a receptor for pertussis toxin. J Immunol 1990; 145:678–83.

Rojo JM, Portoles P. A symmetrical view of the T-cell receptor-CD3 complex. Immunol Today 1991; 12:377–8.

Rosenstein Y, Ratnofsky S, Burakoff SJ, Herrmann SH. Direct evidence for the binding of CD8 to HLA class I antigens. J Exp Med 1989; 169:149–60.

Rosoff PM, Walker R, Winberry L. Pertussis toxin triggers rapid second messenger production in human T lymphocytes. J Immunol 1987; 139:2419–23.

Rudd CE. CD4, CD8 and the TCR-CD3 complex: a novel class of protein-tyrosine kinase receptor. Immunol Today 1990; 11:400–6.

Samelson LE, Patel MD, Weissman AM, Hartford JB, Klausner RD. antigen activation of murine T cells induces tyrosine phosphorylation of a polypeptide associated with the T cell antigen receptor. Cell 1986; 46:1083–90.

Samelson LE, Phillips AF, Luong ET, Klausner RD. Association of the fyn protein-tyrosine kinase with the T-cell antigen receptor. Proc Natl Acad Sci USA 1990; 87:4358–62.

Satoh T, Nakafuku M, Miyajima A, Kaziro Y. Involvement of ras p21 protein in signal transduction pathways from interleukin 2, interleukin 3 and granulocyte/macrophage colony stimulating factor, but not from interleukin 4. Proc Natl Acad Sci USA 1991; 88:3314–8.

Schrezenmeier H, Ahnert-Hilger G, Fleischer B. A T cell receptor-associated GTP-binding protein triggers T cell receptor-mediated granule exocytosis in cytotoxic T lymphocytes. J Immunol 1988a; 141:3785–90.

Schrezenmeier J, Ahnert-Hilger G, Fleischer B. Inactivation of a T cell receptor-associated GTP-binding protein by antibody-induced modulation of the T cell receptor/CD3 complex. J Exp Med 1988b; 168:817–22.

Smith K. Interleukin-2: inception, impact, and implications. Science 1988; 240:1169–76.

Smith MR, DeGudicibus SJ, Stacey DW. Requirement for c-ras proteins during viral oncogene transformation. Nature 1986; 320:540–3.

Sommermeyer H, Resch K. Pertussis toxin B-subunit-induced Ca^{2+} fluxes in Jurkat human lymphoma cells: the action of long-term pre-treatment with cholera and pertussis holotoxins. Cell Signalling 1990; 2:115–28.

Springer TA. Adhesion receptors of the immune system. Nature 1990; 346:425–34.

Springer TA, Dustin ML, Kishimoto TK, Marlin SD. The lymphocyte function-associated LFA-1, CD2 and LFA-3 molecules: cell adhesion receptors of the immune system. Ann Rev Immmunol 1987; 5:223–52.

Stanley JB, Gorczynski RM, Delovitch TL, Mills GB. IL-2 secretion is pertussis toxin sensitive in a T lymphocyte hybridoma. J Immunol 1989; 142:3546–52.

Stewart SJ, Prpic V, Johns JA, Powers FS, Graber SE. Bacterial toxins affect early events of T lymphocyte activation. J Clin Invest 1989; 83:234–42.

Tabin CJ, Bradley SM, Bargmann CI, Weinberg RA, Papageorge AG, Scolnick EM, Dhar R, Lowry DR, Chang EH. Mechanism of activation of a human oncogene. Nature 1982; 300:143–9.

Taylor CW. The role of G proteins in transmembrane signalling. Biochemical J 1990; 272:1–13.

Telfer JC, Rudd CE. A 32kD GTP binding protein associated with the CD4-$p56^{lck}$ and CD8-$p56^{lck}$ T cell receptor complexes. Science 1991; 254:439–41.

Too C, Murphy P, Friesen H. G-proteins modulate prolactin- and interleukin-2-stimulated mitogenesis in rat Nb2 lymphoma cells. Endocrinology 1989; 124:2185–92.

Trahey M, McCormick F. A cytoplasmic protien stimulates normal N-ras p21 GTPase, but does not effect oncogenic mutants. Science 1987; 238:542–5.

Ullrich A, Schlessinger J. Signal transduction by receptors with tyrosine kinase activity. Cell 1990; 61:203–12.

Unanue E, Cerottini J-C. Antigen presentation. FASEB J 1989; 3:2496–502.

Veillette A, Bookman MA, Horak EM, Bolen JB. The CD4 and CD8 T cell surface antigens are associated with the internal membrane tyrosine-protein kinase p56lck. Cell 1988; 55:301–8.

Waldmann TA. The interleukin-2 receptor. J Biol Chem 1991; 266:2681–84.

Weiss A. T lymphocyte activation. In Paul WE, ed. *Fundamental immunology*, 2d ed. New York: Raven Press, 1989; 359–84.

Weiss A. Structure and function of the antigen receptor. J Clin Invest 1990; 86:1015–22.

White MF. Structure and function of tyrosine kinase receptors. J Bioenerg Biomem 1991; 23:63–82.

Xu G, Lin B, Tanaka K, Dunn D, Wood D, Gesteland R, White R, Weiss R, Tamano F. The catalytic domain of the neurofiromatosis type 1 gene product stimulates ras GTPase and complements ira mutants of *S. cerevisiae*. Cell 1990; 63:835–41.

24

Mechanisms of Mast Cell Signaling

STEPHEN I. WASSERMAN, DIANA L. MARQUARDT, and KIM E. BARRETT

University of California, San Diego
San Diego, California

I. Introduction

Allergic disorders owe their characteristic pathophysiologic features to the release of potent biologically active molecules from tissue mast cells. These molecules, termed mediators (Wasserman), include preformed and generated vasoactive compounds, such as histamine, platelet-activating factor, leukotriene C4 and prostaglandin D2, chemoattractant molecules, active enzymes, particularly tryptase; and in the case of connective tissue mast cells, chymase, and a panel of cytokines (Plaut et al.), including tumor necrosis factor, and interleukins 3 and 4. Due to the generation and release of these mediators, increases in vascular permeability, smooth muscle contraction, mucus secretion, and a characteristic infiltration of tissue by CD4 positive, activated T lymphocytes and eosinophils occur. These events contribute to the pathobiology of asthma, allergic rhinitis, atopic eczema, and urticaria and anaphylaxis. The purpose of this review is to describe some of the mechanisms whereby mast cells can be triggered to generate and release these potent biologically active molecules.

II. The Mast Cell

Mast cells are positioned at portals of entry of noxious materials to the host throughout the body, including the skin, the upper and lower respiratory tracts, and the gastrointestinal tract. Two types of mast cell, the mucosal and connective tissue types, have been identified (Enerbach, 1966). The mucosal type, predominant primarily in the superficial layers of the gastrointestinal tract and in the lower respiratory tract, can be differentiated from the connective tissue mast cell in its lack of content of the enzyme chymase (Irani et al., 1986) and its more exquisite sensitivity to proliferation under inflammatory stimuli. The mucosal mast cell appears to be controlled by CD4-positive lymphocytes, and it is notably diminished in tissue in advanced HIV infection and in severe combined immunodeficiency disorders (Irani et al., 1987). Recent advances indicate that mast cell growth and proliferation are regulated by a complex interaction of a variety of growth factors and cytokines, including interleukins 3, 4, 9, and 10, nerve growth factor, and a molecule termed stem cell factor or kit-ligand.

III. Mast Cell Activation

Mast cells can be activated by a variety of processes, including physical stimuli, enzymes (e.g., chymotryptic proteases), numerous highly charged basic compounds, and a variety of neuropeptides (e.g., substance P and ATP). However, the most important activators of the mast cell are IgE molecules, acting through mast cell membrane receptors for this immunoglobulin subclass, and a family of IgE-independent ligands termed "histamine-releasing factors."

Mast cells acquire the IgE receptor very early in their development (Thompson et al., 1990). It is possible to demonstrate the presence of IgE receptors linked to signal transduction pathways, notably calcium influx (Rossi et al., in press), prior to the full granulation of the cell, and prior to the ability of IgE-mediated mechanisms to activate the mast cell. The biology of the IgE immunoglobulin is beyond the scope of this review, but it is important to note that IgE synthesis is regulated positively by interleukin 4 and negatively by interferon-γ. The receptor for IgE binds this immunoglobulin with extraordinary affinity. The high affinity of binding is felt to be due to a slow dissociation of bound IgE from the receptor (Ishizaka et al., 1985). The receptor for IgE is composed of three separate proteins, termed alpha, beta, and gamma, and all are required to develop surface expression of this complex (Blank et al., 1989). The current concept of this receptor suggests that a single alpha subunit, two gamma subunits, and a single beta subunit comprise the intact receptor (Metzger 1991). This receptor resembles, in many ways, those which have been demonstrated to be important in G-protein binding, in that it possesses seven transmembrane sequences, one in the alpha, four in the beta chains, and two in the cross-linked gamma subunits. However, as noted below, it is by no means clear that G proteins participate in signal transduction mechanisms in mast cells. The intact IgE receptor binds a single molecule of IgE (Mendoza and Metzger, 1976), and it is a fragment of the $C\epsilon 2/C\epsilon 2$ domain of the IgE heavy chain which appears to be central to the ability of this immunoglobulin to bind to the receptor (Helm et al., 1988). It is to the extracellular portion of the alpha chain which IgE binds, a phenomenon confirmed by the fact that antibodies to the extracellular moieties of the alpha component of the IgE receptor can activate the cells by a mechanism mimicking that due to cross-linking of IgE antibody by antigen (Ishizaka et al., 1977). That receptor cross-linking is necessary for activation of mast cells has been known for many years (Ishizaka and Ishizaka, 1968). This finding was confirmed by the finding that polyvalent but not monovalent anti-IgE antibodies mimic antigen cross-linking of adjacent IgE molecules (Ishizaka et al., 1971). It is now known that dimeric cross-linking of IgE receptors is sufficient to activate the cell, whereas trimeric and higher oligomers may be somewhat more potent in this regard (Kagey-Sobotka et al., 1981). IgE receptors are found in large numbers on mast cells, and appear to be increased in patients with elevated serum IgE. Utilizing peripheral blood basophils, Kagey-Sobotka and colleagues have demonstrated a direct relationship between serum IgE concentrations and the number and proportional occupancy of IgE receptors (Malveaux et al., 1978). It is accepted, but not yet rigorously proven, that the effect of the cross-linking of IgE receptors is translated through the membrane to the intracellular domains of the receptor, including one alpha, two beta, and two gamma intracellular portions. It remains totally unknown as to what this particular transmission mechanism is and how the intracellular domains participate and link to subsequent intracellular events (see below).

A second major family of molecules capable of activating mast cells are termed histamine-releasing factors (HRFs). In contrast to IgE, virtually nothing is known of the mast cell receptor for these factors. While most have not been characterized, one, interleukin 3, is a known growth factor for mast cells, and possesses the ability to activate the cell to release histamine. In addition, the known growth factor GM-CSF and the inflammatory cytokine, interleukin 1 are also capable of activating mast cell secretion (Alem et al., 1989).

The best-characterized species of HRF is a 12-kDa protein which is generated by mononuclear leukocytes. It has been purified, and a partial amino acid sequence determined. Other, less well characterized HRFs have been described, some, interestingly, requiring mast cells to be sensitized with IgE from atopic, but not IgE from nonallergic, individuals to express their actions

(Baeza et al., 1990). HRFs have been identified in a variety of biological materials, including nasal secretions and bronchial lavage fluid (MacDonald et al., 1987). Unfortunately, to date, none of these HRFs has been cloned or convincingly purified to homogeneity in sufficient quantities to enable investigators to determine receptor identity or to elucidate any details of the signal transduction pathways they activate.

IV. Participation of G Proteins in Mast Cell Signaling Pathways

The importance of G proteins to mast cell secretory events has been an area of some speculation. Two approaches have been used to assess the role of G proteins in mast cell signal transduction: the addition of specific toxins that irreversibly alter G-protein function, and the introduction of GTP analogs into permeabilized cells.

Although initial reports were inconsistent (Nakamura and Ui, 1985; Şaito et al., 1987), it appears that pertussis toxin does not directly modulate the ability of IgE antibodies (and related stimuli) to induce the release of preformed or newly synthesized mast cell mediators (Saito et al., 1987), despite ADP-ribosylation of the alpha subunit of a 41-kDa G protein in the cell membrane. Thus, IgE-mediated mast cell secretory events are pertussis toxin-independent and do not require a functional G_i or G_i-related protein for activity. However, a small, yet significant, inhibition of thrombin-induced histamine release from mast cells preincubated with pertussis toxin has been described. Moreover, pertussis toxin can partially inhibit the ability of adenosine to potentiate the ongoing release of granule-associated mast cell mediators (Marquardt and Walker, 1988; Gilfillan et al., 1990), suggesting a secondary regulatory pathway in mast cell degranulation utilizing a pertussis toxin-sensitive G protein.

Similar studies were undertaken utilizing cholera toxin, an agent reported to ADP-ribosylate the alpha subunit of the stimulatory G protein, G_s, to modulate mast cell mediator release induced by specific antigen or thrombin (Saito et al., 1988). Although cholera toxin clearly increases intracellular cyclic AMP levels, it does not alter mediator release, suggesting that G_s is not involved in IgE-mediated mast cell signal transduction. However, thrombin-triggered mast cell histamine, but not arachidonate, release is enhanced by such treatment, suggesting a role for both G_i and G_s proteins in the mechanism of action of thrombin as a mast cell secretagogue.

To complement the indirect effect of toxins to modulate G-protein pathways, direct stimulation of G proteins (usually in permeabilized mast cells) has been attempted to identify cell transduction machinery. Rat mast cells permeabilized by streptolysin-o secrete histamine in response to nucleoside triphosphates and appropriate concentrations of Ca^{2+} (Howell et al., 1987). GTP and its stable analogs, GTP-gamma-S and GppNHp, are also effective secretagogues, suggesting that calcium and the activation of a GTP-regulatory protein are sufficient stimuli for mast cell exocytosis.

Other workers have provided additional evidence supporting the dependence of exocytosis from rat mast cells on Ca^{2+} and guanine nucleotides (Koopman and Jackson, 1990; Aridor et al., 1990). Thus, digitonin-permeabilized mast cells release histamine in response to GTP-gamma-S and Ca^{2+}. As expected when GTP analog and pertussis toxin are co-administered, no inhibition of GTP-induced mediator release was appreciated (Lindau and Nusse, 1987).

An additional approach to the study of G proteins is the use of a wasp venom peptide, mastoparan, which is thought to activate G proteins directly. Mastoparan induces histamine release from rat mast cells via a pertussis toxin-sensitive mechanism (Mousli et al., 1989). It also acts synergistically with GTP-gamma-S to potentiate exocytosis from mast cells and is thought to act downstream from a phospholipase C process in a receptor-independent manner (Aridor et al., 1990).

Another G-protein-dependent signal transduction system which has been quite thoroughly studied is that involving cyclic AMP as a second messenger. Dynamic changes in cyclic AMP concentrations occur during mast cell secretion and include both direct early and indirect late increases in cyclic AMP levels (Lewis et al., 1979). However, some agents which directly increase mast cell cyclic AMP, such as exogenous adenosine (Marquardt et al., 1986) and cholera toxin (Saito et al., 1988), enhance and have no effect on mast cell degranulation respectively. Forskolin, a direct activator of adenylate cyclase (Marone et al., 1987), and beta-adrenergic agonists (in human, not rat mast cells) inhibit IgE-mediated secretion in a concentration-effect relationship.

However, it is clear that elevations of intracellular cyclic AMP alone are not sufficient to induce mast cell mediator release.

The dynamic alterations in intracellular cyclic AMP concentrations observed during stimulation of mast cells by various secretagogues, including anti-IgE, the calcium ionophore A23187, or compound 48/80, could exert a number of secondary effects that might modulate the extent of mast cell secretion and the relative quantities of preformed and newly generated mediators that are released. It is equally likely that the activation or inhibition of adenylate cyclase and subsequent activation or inhibition of protein kinase A is not the pivotal part of mast cell secretion, as was once thought. Studies with specific inhibitors of protein kinase A or with genetically altered G proteins may help to clarify how agents that alter cyclic AMP levels influence secretion in mast cells, and identify which, if any, G proteins play a role in this process.

V. Role of Phospholipid Turnover

It is now widely accepted that the receptor-linked stimulation of membrane phospholipase C serves as the signal for the generation of a variety of intracellular second messengers in many cell types. In keeping with this paradigm, activation of mast cells has been correlated with dynamic changes in the cellular levels of a number of polyphosphates produced by the turnover of phosphatidylinositol. For example, antigen stimulation of rat basophil leukemia cells (RBL-2H3 subline) is accompanied by the release of three isomers of inositol monophosphate, three of inositol bisphosphate, two of inositol trisphosphate [Ins(1,4,5)P_3 and Ins(1,3,4)P_3], and inositol (1,3,4,5) tetrakisphosphate (Beaven et al., 1987b). Of these substances, Ins(1,4,5)P_3 has received the most attention because it releases calcium sequestered in intracellular stores (Putney, 1987). This is achieved by binding of the Ins(1,4,5)P_3 to intracellular receptors (Joseph and Williamson, 1989) on membrane sites which are most likely part of the endoplasmic reticulum. In some other secretory cell types, such as pancreatic acinar cells, discrete calcium storage vesicles designated "calciosomes" have also been described (Volpe et al., 1988). Subcellular fractionation and/or morphologic techniques have yet to be applied to determine whether structures analogous to calciosomes similarly contribute to mast cell calcium homeostasis.

In keeping with its ability to mobilize calcium, an intracellular ion which can exert toxic or even lethal effects on the cell if sustained at extremely elevated levels, Ins(1,4,5)P_3 is rapidly broken down in mast cells, thereby extinguishing its effects. A primary pathway for inactivation, as observed in other cell types (Irvine et al., 1986), is its rapid conversion to Ins(1,3,4,5)P_4, which is subsequently dephosphorylated to Ins(1,3,4)P_3 (Beaven et al., 1987b). The biologic activities of these derived molecules are not yet well understood, but Ins(1,3,4,5)P_4 has been reported to mobilize (with low efficiency) Ins(1,4,5)P_3-sensitive calcium stores to increase entry of calcium across the plasma membrane, to affect calcium pool reloading, to alter cellular phosphatase activities, or to modulate other ion channels, such as conductances for potassium or chloride (Putney, 1987; Joseph and Williamson, 1989). Any of these effects could contribute to signal transduction in mast cells.

The turnover of inositol phospholipids has a close link to the control of secretion, at least in RBL-2H3 cells. Manipulations which perturb phosphatidylinositol hydrolysis, either in a positive or a negative fashion, are accompanied by analogous changes in the extent of secretion (Beaven et al., 1987b; Maeyama et al., 1988). Similarly, even when cells are stimulated with a calcium ionophore, a mode of stimulation classically considered to bypass receptor-linked activation of phospholipase C, secretion occurs only if small effects on phosphatidylinositol hydrolysis ensue (Lo et al., 1987).

One note of caution should be raised regarding the interpretation of these findings. Of necessity, most studies of the role of PI turnover in mast cell secretion must employ cultured mast cells, since cells are required in large numbers and at high purity. Very little is known about the precise pathways of PI turnover, and dependence of secretion on such turnover, in tissue mast cells of human origin.

The bifurcating second messenger hypothesis promulgated by Berridge (1987) suggests that the release of Ins(1,4,5)P_3 in cells is accompanied by the production of diacylglycerol (DAG), a molecule whose chief role appears to be the activation of the phospholipid and calcium dependent-protein kinase, protein kinase C (PKC). In keeping with this scheme, the activation of mast cells has long been known to be associated with a marked increase in cellular levels of DAG, which precedes the onset of secretion. However, a more recent examination of this area reveals that in many cell types, including mast cells isolated from the rat peritoneal cavity, changes in DAG levels may differ either in extent or kinetics (or both) from changes in phosphatidylinositols stimulated by agonists. Kennerly (1987) has utilized the technique of molecular species analysis to demonstrate that the majority of DAG produced in stimulated rat mast cells is derived from phosphatidylcholine by phospholipase D action rather than from phosphatidylinositol. A similar pathway of DAG production may also be important in transducing other mast cell responses, such as proliferation (Duronio et al., 1989). DAG species derived from phosphatidylinositol may, however, have functional importance out of proportion to their abundance because of their relative enrichment in arachidonic acid.

VI. Calcium Signaling

The release of mediators from mast cells has long been recognized as intimately involving calcium. Early studies revealed that extracellular calcium was required for the full expression of IgE-mediated secretory responses. Over the last 10 years, the availability of fluorescent intracellular calcium indicators, as developed by Tsien and associates, has permitted the dynamic monitoring of intracellular concentrations of this cation, and consequently, an improved appreciation of its role in mast cell signaling. More recently, whole-cell patch-clamp experiments are enabling a clearer interpretation of the sources and flux routes of the calcium changes measured.

The linkage between the free cytosolic calcium concentration ($[Ca^{2+}]_i$) and secretion in suspensions of rat peritoneal mast cells was first demonstrated by White et al. (1984), using the calcium indicator quin-2. Manipulations that enhanced or inhibited antigen-stimulated mediator release from the cells were accompanied by corresponding changes in the degree of elevation in $[Ca^{2+}]_i$.

It has also been demonstrated that there is a close correlation between $[Ca^{2+}]_i$ and the rate of secretion of granule-associated mediators from RBL-2H3 cells (Beaven et al., 1984). If $[Ca^{2+}]_i$ returns to basal levels, secretion is promptly terminated (Maeyama et al., 1991). The sensitivity of $[Ca^{2+}]_i$ to IgE-mediated signaling is acquired very early in mast cell development, even before granule maturation or the ability of IgE to induce significant mediator release (Rossi et al., in preparation).

Increases in $[Ca^{2+}]_i$ represent a balance between the release of calcium from intracellular stores, the entry of calcium from the extracellular compartment, and elimination of calcium from the cytoplasm, either back into stores or across the plasma membrane (Rasmussen, 1989). It is widely accepted that Ins(1,4,5)P_3 mobilizes intracellular calcium, and this property is likely due to direct activation of a Ca^{2+}-selective channel (Meyer et al., 1988). As examined in RBL-2H3 cells, channel opening is highly cooperative, and requires the binding of at least three molecules of Ins(1,4,5)P_3. This may enable the cell to detect and amplify very small changes in the concentration of this messenger. However, there appears to be a consensus in the literature that intracellular pools are not the source of calcium that is primarily important for the promotion of antigen-induced secretion. For example, calcium can be released from intracellular sores to markedly elevate $[Ca^{2+}]_i$ in the absence of secretion (Marcotte et al., 1990; Penner, 1988; Neher and Almers, 1986). Rather, entry of calcium from the extracellular space appears to determine the extent of secretion, and this process, like the mobilization of intracellular calcium, may also represent an action of Ins(1,4,5)P_3 (Penner et al., 1988).

Using patch-clamp techniques, two conductance pathways for calcium entry into rat peritoneal cells have been identified (Penner et al., 1988). The first is a voltage-independent cation channel through which divalent cations permeate according to their thermodynamic driving force. The second is a hyperpolarization-driven calcium influx, which may be an ion channel highly specific for calcium and therefore capable of allowing large changes in $[Ca^{2+}]_i$ with relatively small changes in whole-cell currents. Activation of rat mast cells also activates a slow, outwardly rectifying chloride current, which is proposed to clamp the membrane potential of the cell to negative values, and provide a driving force for calcium entry via the two calcium conductances (Penner et al., 1988). This chloride channel may represent an amplification mechanism for secretion, as it may be activated by high levels of $[Ca^{2+}]_i$ in association with cAMP.

Although an increase in $[Ca^{2+}]_i$ appears to be necessary for secretion to proceed, it is not by itself a sufficient signal. Thus, treatment of cells with calcium ionophores at concentrations that are able to markedly elevate $[Ca^{2+}]_i$ does not cause mediator release unless there is also measurable turnover of phosphatidylinositol (Lo et al., 1987). This leads to the hypothesis that antigen, and possibly non-IgE-receptor-dependent stimuli, activate additional signal transduction mechanisms that modulate the sensitivity of the secretory machinery to calcium (Penner, 1988; Beaven et al., 1987a). Conversely, this second signal may be the primary activator of the secretory response, with calcium enhancing its effectiveness.

VII. Role of Protein Kinase C

Since mast cell activation is accompanied by increases in both $[Ca^{2+}]_i$ and DAG, the species required for stimulation of the ubiquitous protein kinase, PKC, it became logical to examine whether this enzyme is involved in the subsequent promotion of secretion. A number of investigators showed that direct activation of PKC with phorbol esters could lead to mast cell mediator release, though in most cases inefficiently and requiring extended periods of incubation to produce relatively modest effects (Chakravarty et al., 1990; Heiman and Crews, 1985; Sagi-Eisenberg et al., 1987, Cunnife and Pearce, 1988). More interestingly, mast cells treated with a phorbol ester and a low dose of calcium ionophore sufficient to increase $[Ca^{2+}]_i$ (each unable to cause secretion alone) rapidly secrete mediators (Katakami et al., 1984).

The hypothesis that PKC is central to mast cell secretion was further supported by its direct measurement. PKC is a cytosolic enzyme in resting cells, but upon cell stimulation is translocated to the plasma membrane, where it becomes enzymatically active. Antigen stimulation of cultured murine mast cells leads to a rapid increase in the PKC content associated with the plasma membrane, and a concommitant decrease in cytosol levels of this enzyme (White et al., 1985). Similarly, a variety of PKC inhibitors block IgE-dependent mediator release (Marquardt and Walker, 1990; White et al., 1990). Further, stabilization of DAG (via inhibition of DAG kinase) enhances IgE-dependent histamine release from rat peritoneal mast cells (Cantwell and Foreman, 1987), and secretion is associated with the phosphorylation of proteins which in broken cell preparations are known to be PKC substrates (White et al., 1990; Sagi-Eisenberg et al., 1987).

There is, however, experimental evidence that suggests that the role of PKC in mediator secretion is more complicated than is implied by the foregoing discussion. For example, higher concentrations of phorbol esters inhibit mast cell mediator release, and the ability of PKC inhibitors to block secretion has not been uniformly observed (Marquardt and Walker, 1990; Beaven et al., 1987; Chakravarty et al., 1990). In RBL-2H3 cells where PKC activity had been depleted by prolonged treatment with phorbol esters, a partial secretory response to antigen remained (White and Metzger, 1988). Koopman and Jackson (1990), working with permeablized mast cells, concluded that PKC could modulate but was not required for exocytosis induced by calcium and G-protein activation to occur. Similarly, a cultured murine mast cell line incapable of exocytosis actually displayed higher levels of PKC activity in the membrane fraction, and a greater number of high-affinity phorbol ester-binding sites, than a related cell line capable of secretion (Mazurek et al., 1987).

Part of the confusion in this area may relate to the fact that there is a family of PKC isozymes, which may vary widely in their catalytic properties, substrate specificities, requirement for calcium, and susceptibility to regulation by phorbol ester treatment (Nishizuka, 1988). Thus it is possible that one PKC isoform may be responsible for positive modulation of secretion, while another transduces negative signals. This hypothesis has yet to be tested directly. Additionally, the effects of PKC activation may be highly dependent on the type and/or species of origin of the mast cells being studied (Cunnife and Pearce, 1988), or the nature of the stimulus (Cohan et al., 1991).

It is apparent that PKC can potentially interact with several aspects of the proximal signal transduction pathway in mast cells, which are likely to be mechanisms whereby it exerts its effects on the overall extent of secretion. For example, PKC activity regulates the activity of the PI-specific phospholipase C, reducing the levels of Ins(1,4,5)P_3, and consequently affects calcium mobilization. Another important action is related to later events in degranulation. By phosphorylating myosin chains, PKC may alter the properties of the cytoskeleton so that granules may be extruded (Ludowyke et al., 1989). PKC may also be required for the ability of adenosine to potentiate mediator release (Marquardt and Walker, 1990). This potential for PKC to interact with several, potentially conflicting, signaling pathways has recently been reviewed.

Taken together, it is clear that the signal transduction processes utilized by mast cells are like those in other cells, complex, intertwined, and potentially redundant. The precise roles and interrelationships of PKA, PKC, cAMP, $[Ca^{2+}]_i$, inositol phosphates, DAG, the secretion-activating principles (IgE, HRF), and their receptors (i.e., FcERI) remain incompletely understood. These interactions continue to be the focus of intense investigation and provide promise for therapeutic interventions in allergic disease.

References

Alem R, Welter JB, Forsythe PA, et al. Comparative effect of recombinant interleukin 1,2,3,4,and 6, IFN-gamma, granulocyte-macrophage colony stimulating factor, tumor necrosis factor-α, and histamine-releasing factors on the secretion of histamine release from basophils. J Immunol 1989; 142:3431–5.

Aridor M, Traub LM, Sagi-Eisenberg R. Exocytosis in mast cells by basic secretagogues: evidence for direct activation of GTP-binding proteins. J Cell Biol 1990; 111:909–17.

Baeza ML, Reddigari SR, Kornfeld D, et al. Relationship of one form of human histamine-releasing factor to connective tissue activating peptide III. J Clin Invest 1990; 85:1516–21.

Beaven MA, Guthrie DF, Moore JP, et al. Synergistic signals in the mechanism of antigen-induced exocytosis in 2H3 cells: evidence for an unidentified signal required for histamine release. J Cell Biol 1987a; 105:1129–36.

Beaven MA, Maeyama K, Wolde Mussie E, et al. Mechanism of signal transduction in mast cell/s and basophils: studies with RBL-2H3 cells. Agents Actions 1987b; 20:137–45.

Beaven MA, Rogers J, Moore JP, et al. The mechanism of the calcium signal and correlation with histamine release in 2H3 cells. J Biol Chem 1984; 259:7129–36.

Berridge MJ. Inositol trisphosphate and diacylglyconol: two interacting second messengers. Ann Rev Biochem 1987; 56:159–93.

Blank U, Ra C, Miller L, et al. Complete structure and expression in transected cells of high affinity IgE receptor. Nature 1989; 337:187–9.

Broide DH, Smith CM, Wasserman SI. Mast cells and pulmonary fibrosis: identification of a histamine-releasing factor in bronchial lavage fluid. J Immunol 1990; 145:1838–44.

Cantwell ME, Foreman JC. The actions of inhibitors of diacyl-glycerol kinase and diacylglycerol lipase on histamine release from rat peritoneal mast cells. Br J Pharmacol 1987; 92:597–601.

Chakravarty N, Kjeldsen B, Hansen M, et al. The involvement of protein kinase C in exocytosis in mast cells. Exp Cell Res 1990; 186:245–9.

Cohan VL, MacGlashan DW, Warner JA, et al. Mechanisms of mediator release from human skin mast cells upon stimulation by the bradykinin analog, $[{}_{d}ARG^{\cdot} - Hyp^{3} - {}_{D}Phe^{7}]$ bradykinin. Biochem Pharmacol 1991; 41:293–300.

Cunniffe TM, Pearce FL. Effects of phorbol ester, serine phospholipids, and calcium on histamine release from mast cells and basophils. Agents Actions 1988; 23:170–2.

Duronio V, Nip L, Pelech SL. Interleukin 3 stimulates phosphatidylcholine turnover in a mast/megakaryocyte cell line. Biochem Biophys Res Commun 1989; 164:804–8.

Enerbach L. Mast cells in rat gastrointestinal mucosa. Acta Pathol Microbiol Scand 1966; 66:303–2.

Gilfillan AM, Wiggan GA, Welton AF. Pertussis toxin pretreatment reveals differential effects of adenosine analogs on IgE-dependent histamine and peptidoleukotriene release from RBL-2H3 cells. Biochim Biophys Acta 1990; 1052:467–74.

Heiman AS, Crews FT. Characterization of the effect of phorbol esters on rat mast cell secretion. J Immunol 1985; 134:548–55.

Helm B, March P, Ferchelli D, et al. The mast cell binding site on human IgE. Nature 1988; 331:180–3.

Houslay MD. "Crosstalk": a pivotal role for protein kinase C in modulatory relationships between signal transduction pathways. Em J Biochem 1991; 195:9–27.

Howell TW, Cockcroft S, Gomperts BD. Essential synergy between Ca^{2+} and guanine nucleotides in exocytotic secretion from permeabilized rat mast cells. J Cell Biol 1987; 105:191–7.

Irani AA, Schecter NM, Craig S, et al. Two human mast cell subsets with different neutral protease compositions. Proc Natl Acad Sci USA 1986; 83:4464–8.

Irani AA, Schecter NM, Schwartz LB. Deficiency of the tryptase positive, chymase negative mast cell type in gastrointestinal mucosa of patients with defective T lymphocyte function. J Immunol 1987; 138:4381–7.

Irvine RF, Legchen AJ, Heslop JP, et al. The inositol tris/tetrakisphosphate pathway-demonstration of Ins(1,4,5)P_3 3-kinase activity in animal tissues. Nature 1986; 320:631–4.

Ishizaka K, Ishizaka T. Induction of erythema/wheal reactions by soluble antigen/gamma E antibody complexes in humans. J Immunol 1968; 101:68–78.

Ishizaka T, Chang TH, Taggert M, et al. Histamine release from rat mast cells by antibodies against rat basoleukemia cell membrane. J Immunol 1977; 119:1589–96.

Ishizaka T, Dvorak AM, Conrad DH, et al. Morphologic and immunologic characterization of human basophils developed in cultures of core blood mononuclear cells. J Immunol 1985; 134:532–40.

Ishizaka T, Tamiyoka H, Ishizaka K. Degranulation of human basophil leukocytes by anti-gamma E antibody. J Immunol 1971; 106:705–10.

Joseph SK, Williamson JR. Inositol polyphosphates and intracellular calcium release. Arch Biochem Biophys 1989; 273:1–15.

Kagey-Sobotka, Dembo M, Goldstein B, et al. Qualitative characteristics of histamine release from human basophils by covalently cross-linked IgE dimers and trimers. J Immunol 1981; 127:2285–91.

Katakami Y, Kaibuchi K, Sawamura M, et al. Synergistic action of protein kinase C and calcium for histamine release from rat peritoneal mast cells. Biochem Biophys Res Commun 1984; 121:573–8.

Kennerly DA. Diacylglyconol metabolism in mast cells. Analysis of lipid metabolic pathways using molecular species analysis of intermediates. J Biol Chem 1987; 262:16305–13.

Koopman WR, Jackson RC. Calcium- and guanine-nucleotide-dependent exocytosis in permeabilized rat mast cells. Biochem J 1990; 265:265–373.

Lewis RA, Holgate ST, Roberts LJ, Maguire JF, Oates JA, Austen KF. Effects of indomethacin on cyclic nucleotide levels and histamine release from rat serosal mast cells. J Immunol 1979; 123:1663–9.

Lindau M, Nusse O. Pertussis toxin does not affect the time course of exocytosis in mast cells stimulated by intracellular application of GTP-gamma-S. FEBS Lett 1987; 222:317–21.

Lo TN, Saul W, Beaven MA. The actions of Ca^{2+} ionophores on rat basophilic (2H3) cells are dependent on cellular ATP and hydrolysis of inositol phospholipids. A comparison with antigen stimulation. J Biol Chem 1987; 262:4141–5.

Ludowyke RI, Peleg I, Beaven MA, et al. Antigen-induced secretion of histamine and the phosphorylation of myosin by protein kinase C in rat basophilic leukemia cell. J Biol Chem 1989; 264:12492–501.

MacDonald SM, Lichtenstein LM, Proud D, et al. Students of IgE-dependent histamine releasing factors: heterogeneity of IgE. J Immunol 1987; 139:506–12.

Maeyama K, Hohman R, Ali H, et al. Assessment of IgE receptor function through measurement of hydrolysis of membrane inositol phospholipids. New insights on the phenomena of biphasic antigen concentration-response curves and desensitization. J Immunol 1988; 140:3919–27.

Maeyama K, Sasaki M, Watanabe T. Simultaneous determination of intra-cellular calcium concentration and histamine secretion in rat basophilic leukemia cells (RBL-2H3). Anal Biochem 1991; 194:316–25.

Malveaux F, Conroy M, Adkinsin F. IgE receptors on human basophils: relationship to serum IgE concentration. J Clin Invest 1978; 62:176–81.

Marcotte GV, Millard PJ, Fewtrell C. Release of calcium from intracellular stores in rat basophilic leukemia cells monitored with the fluorescent probe chlortetracycline. J Cell Physiol 1990; 142:78–88.

Marone G, Columbo M, Triggiani, et al. Inhibition of IgE-mediated release of histamine and peptide leukotriene from human basophils and mast cells by forskolin. Biochem Pharmacol 1987; 36:13–20.

Marquardt DL, Walker LL. Alteration of mast cell responsiveness to adenosine by pertussis toxin. Biochem Pharmacol 1988; 37:4019–25.

Marquardt DL, Walker LL. Modulation of mast cell responses to adenosine by agents that alter protein kinase C activity. Biochem Pharmacol 1990; 39:1929–34.

Marquardt DL, Gruber HE, Wasserman SI. Aminophylline exposure alters mouse bone marrow-derived mast cell adenosine responsiveness. J Allergy Clin Immunol 1986; 78:462–9.

Mazurek N, Regazzi R, Borner C, et al. Altered protein kinase C in a mast cell variant defective in exocytosis. Proc Natl Acad Sci USA 1987; 84:1277–81.

Mendoza GR, Metzger, H. Distribution and valency of the receptor for IgE on rodent mast cells and related tumor cells. Nature 1976; 264:548–50.

Metzger H. The high affinity receptor for IgE on mast cells. Clin and Exp Allergy 1991; 21: 269–79.

Meyer T, Holowka D, Stryer L. Highly cooperative opening of calcium channels by inositol 1,4,5-trisphosphate. Science 1988; 240:653–6.

Mousli M, Bronner C, Bueb J-L, et al. Activation of rat peritoneal mast cells by substance P and mastoparan. J Pharmacol Exp Ther 1989; 250:329–35.

Nakamura T, Ui M. Simultaneous inhibitions of inositol phospholipid breakdown, arachidonic acid release, and histamine secretion in mast cells by islet-activating protein, pertussis toxin. J Biol Chem 1985; 260:3584–93.

Neher E, Almers W. Fast calcium transients in rat peritoneal mast cells are not sufficient to trigger exocytosis. EMBO J 1986; 5:51–3.

Nishizuka Y. The molecular heterogeneity of protein kinase C and its implications for cellular regulation. Nature 1988; 334:661–5.

Penner R. Multiple signaling pathways control stimulus-secretion coupling in rat peritoneal mast cells. Proc Natl Acad Sci USA 1988; 85:9856–60.

Penner R, Matthews G, Neher E. Regulation of calcium influx by second messengers in rat mast cells. Nature 1988; 334:499–504.

Plaut M, Pierce JH, Watson CJ, et al. Mast cell lines produce lymphokines in response to cross-linkage of FcϵRI or to calcium ionophores. Nature 19XX; 339:64–7.

Putney JW. Formation and actions of calcium-mobilizing messengers, inositol 1,4,5-trisphosphate. Am J Physiol 1987; 252:G149–57.

Rasmussen H. The cycling of calcium as an intracellular messenger. Sci Am 1989; 261:66–73.

Rossi GL, Yen A, Barrett KE. Ige and ATP receptors on immature murine mast cells are functionally linked to signal transduction mechanisms. J Allergy Clin Immunol, in press.

Sagi-Eisenberg R, Foreman JC, Raval PJ, et al. Progein and diacylglycerol phosphorylation in the stimulus-secretion coupling of rat mast cells. Immunology 1987; 61:201–6.

Saito H, Okajima F, Molski TFP, et al. Effects of ADP-ribosylation of GTP-binding protein by pertussis toxin on immunoglobulin E-dependent and -independent histamine release from mast cells and basophils. J Immunol 1987; 138:3927–34.

Saito H, Okajima F, Molski TFP, et al. Effect of cholera toxin on histamine release from bone marrow-derived mouse mast cells. Proc Natl Acad Sci USA 1988; 85:2504–8.

Thompson HL, Metcalfe DD, Kinet JP. Early expression of high-affinity receptor for immunoglobulin E (FcεRI) during differentiation of mouse mast cells and human basophils. J Clin Invest 1990;85:1227–88.

Volpe P, Kraus K-H, Hashimoto S, et al. "Calciosome," a cytoplasmic organelle: the inositol 1,4,5-trisphosphate-sensitive Ca^{2+} store of nonmuscle cells? Proc Natl Acad Sci USA 1988; 85:1091–5.

Wasserman SI. Mediators of immediate hypersensitivity. J Allergy Clin Immunol 19XX; 72:101–8.

White KN, Metzger H. Translocation of protein kinase C in rat basophilic leukemic cells induced by phobol ester of by aggregation of IgE receptors. J Immunol 1988; 141:942–7.

White JR, Ishizaka T, Ishizaka K, et al. Direct demonstration of increased intracellular concentration of free calcium as measured by quin-Z in stimulated rat peritoneal mast cell. Proc Natl Acad Aci USA 1984; 81:3978–82.

White JR, Pluznik DH, Ishizaka K. Antigen-induced increase in protein kinase C activity in plasma membrane of mast cells. Proc Natl Acad Sci USA 1985; 82:8193–7.

White JR, Zembryki D, Hanna H, et al. Differential inhibition of histamine release from mast cells by protein kinase C inhibitors: staurosporine and K-252a. Biochem Pharmacol 1990; 40:447–56.

25

Beta-Adrenergic Receptors on Alveolar Macrophages and Lymphocytes: Studies with Cells from Asthmatics

SERGEY A. RUDCHENKO, IRNA L. KORICHNEVA, VLADIMIR G. TUPYKIN, VSEVOLOD A. TKACHUK, and ALEXANDER G. CHUCHANLIN

Cardiology Research Center
Moscow, Russia

I. Bronchoalveolar Lavage Macrophage Study

A. Introduction

The autonomic nervous system regulates airway smooth muscle and blood vessel tone, submucosal gland secretion, mast cell mediator release, and surfactant production by cellular responses to specific hormones mediated through adrenergic and cholinergic receptors (Barnes, 1982; Barnes et al., 1983; Carstairs et al., 1985).

Receptors for beta-adrenergic agonists, linked to adenylate cyclase, are present on smooth muscle cells, epithelial cells, secretory mast cells, and basophils. The physiologic results of beta-agonist binding in these cells are bronchodilatation, inhibition of mast cell secretion, reduction of bronchial secretion, and decrease in mucosal edema. Because of these antiinflammatory effects of beta-agonist, primary or secondary changes in beta-adrenergic receptor number or affinity have been postulated as parts of the asthmatic diathesis.

Experimental evidence for the "blockade" theory came from Szentivanji (1968), who injected mice and rats with pertussis vaccine. The development of bronchial hyperresponsiveness to serotonin, and histamine, and eosinophilia was associated with a decreased response to catecholamines (Szentivanji, 1979; Szentivanji and Fishel, 1976). Some component of this theory appears to be applicable to human asthma, which can be initiated or exacerbated by beta-adrenergic receptor antagonists and cholinomimetic agents. This theory prompted studies of the cAMP response to beta-agonists of lung tissue and peripheral blood leukocytes from asthmatics compared to normals. A decreased receptor number and isoproterenol-elicited cAMP response was noted in lymphocytes from asthmatics (Cookson and Redd, 1963; Lockey et al., 1968; Smith and Parker, 1970), as was a decrease in beta-adrenergic number in lungs (removed for treatment of lung cancer) of patients with reversible airway obstruction compared to two other groups of patients (Barnes et al., 1983). In some cases, the decrease in beta-adrenergic receptor number could be a result of sympathomimetic therapy (Bruijnzeel, 1979; Pieter et al., 1982), as similar levels were found in peripheral blood lymphocytes of healthy people and patients with heart failure after treatment with beta-agonists (Baroms and Hills, 1983; Colucci et al., 1981), likely as a result of receptor desensitization (Jenne, 1982; Vermeire and Vanhaitte, 1979).

However, it is still debatable if there is a primary beta-receptor number decrease; that is, if a hereditary or acquired defect in beta-receptors exists that leads to bronchial asthma. Some more recent investigations have found no differences in beta-receptor number (Glant et al., 1980; Pamela et al., 1986), while others have found a decrease in beta-adrenergic receptor number on peripheral blood lymphocytes in asthmatic patients as compared to control groups (Alston et al., 1974; Sano et al., 1983). These contradictions may be due to use of different peripheral blood cells or the possibility that any defects are concentrated in lung cells. Since bronchoalveolar lavage (BAL) macrophages of the rat, guinea pig, monkey, and human express beta-adrenergic receptors (Abrass et al., 1985; Verghese and Snyderman, 1983; Loscalzo and Daniel, 1979; Liggett, 1989), we thought that BAL macrophages might be a more representative population of cells to study. There are no data in the literature on alveolar macrophage beta-adrenergic receptors in asthmatic patients. The aim of the first part of our study was to evaluate beta-adrenergic receptor numbers and binding affinity of BAL macrophages and compare these parameters with those of the peripheral blood cells in asthmatic patients.

B. Methods

The experimental group consisted of nine patients (Table 1). The first group consisted of four patients with minimal clinical and endoscopic bronchitis without any signs of bronchial obstruction. The second group consisted of five patients with asthma of mixed origin. These patients had developed asthma within 3 years, but were in a state of disease remission during the study. Four

Table 1 Characteristics of Patients

Patient	Sex	Age	Diagnosis	Treatment
1	f	62	Chronic bronchitis	Kefzol (cefazolin sodium salt), 2 × 1 g/day
2	m	38	Chronic bronchitis	Aethazolum, 4 × 1 g/day
3	m	58	Chronic bronchitis	—
4	f	41	Chronic bronchitis	—
5	m	48	Chronic bronchitis	Euphyllin (aminophylline), 2.4% 10 mL/day by intravenous
6	f	59	Bronchial asthma, chronic bronchitis	Euphyllin (aminophylline), 2.4%, 10 mL/day by intravenous, Vibriomycin (dihydrostreptomycin), 100 mg/day
7	f	37	Bronchial asthma, chronic bronchitis	Euphyllin (aminophylline), 2.4% 10 mL/day by intravenous
8	m	48	Bronchial asthma	Lasolvan, 3 × 1 tb/day
9	f	58	Bronchial asthma	Polcortolon (triamcinolone 16c, 21-diacetate), 5 × 4 mg/day, Berotec (fenoterol hydrobromide), 5–8 inhalations/day

of them did not use beta-agonists or glucocorticoids. One patient (#9) received beta-agonists (Berotec, 4–6 times a day) and glucocorticoids (Polcortoloni, 20 mg/day). Methylxanthine treatment was withdrawn 24 h before the beginning of the experiments.

All bronchoalveolar lavage procedures were performed with a fiber optic bronchoscope (model BF, type IO, Olympus Co., Japan) wedged in a segmental or subsegmental bronchus of the right middle lobe. The lung was lavaged with 150–200 mL of sterile 0.9% saline solution at 37°C. The cell suspension thus obtained was washed twice with Dulbecco's modified Eagle medium (DMEM, from GIBCO, United States) by centrifugation at 400g for 10 min. The final cell suspension contained 75–95% alveolar macrophages. The other cells were most frequently lymphocytes, with occasional erthrocytes and epithelial cells. The viability of cells was evaluated by trypan blue exclusion and was not less than 95%. Mononuclear lymphocytes were isolated from peripheral blood of patients on Ficoll-Paque gradient according to the method described by Boyum (1968).

For evaluating beta-adrenoreceptors on intact cells, we use flow cytometry (Steinkamp and Kraemer, 1979). In our experiments, the beta-adrenergic ligand 9-aminoacridine propranolol (9-AAP) (Atlas and Levitzki, 1977) was used as a fluorescent probe for beta-adrenoceptor testing by cytofluormetry. Using Scatchard plot analysis, the number of beta-adrenergic receptors per cell, as well as the dissociation constant, were determined.

Cells were incubated with 9-AAP in 1 mL of DMEM for 45 min at 37°C with periodic shaking. Each tube contained 2.5 10^5 cells. The 9-AAP concentration ranged from 2.5 to 100 nM. To determine nonspecific binding, cells were incubated in the presence of 10^{-5} M L-alprenolol. At the end of incubation, samples were centrifuged at 400g for 5 min, the supernatant was aspirated, and the cells were resuspended in a volume of 50 μL of incubation medium and stored at 2–4°C in an ice-water bath until analysis. Prior to the flourescence flow of cytometric assay, the samples were diluted with 2 mL of DMEM.

The analysis of fluorescence was performed using the FACS-II (Becton-Dickinson, United States). The ultraviolet wavelength of an argon laser (model 171–19, Spectra Physics, United States), operating at 500 mW power, was used for excitation of fluorescence. Emission of fluorescence was cut off with a BP 440 interference filter (Ditric Optics, United States), with maximum transparence 440 nm. The population of macrophages was selected for size by forward light scatter. All calculations were made using an IN-IIO computer (Intertechnique, France) interfaced with the FACS-II.

C. Results

Figure 1 demonstrates a representative example of specific and nonspecific binding (Fig. 1A) and a Scatchard plot (Fig. 1B) of 9-AAP interaction with bronchoalveolar macrophages from a bronchitic patient produced by flow cytometry analysis.

The results of our analysis have shown no differences between patients with bronchial asthma and chronic bronchitis in the number or affinity for 9-AAP of beta-adrenoceptors on the surface of leukocytes from peripheral blood

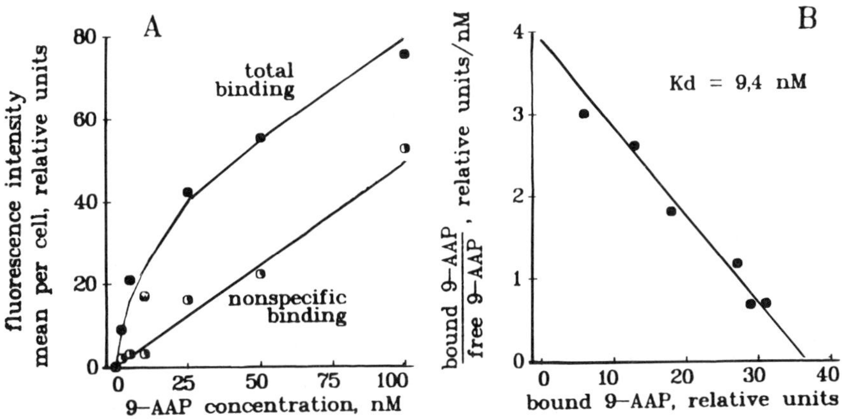

Figure 1 9-AAP binding (A) and Scatchard plot (B) of bronchoalveolar macrophages.

(Table 2). However, the dissociation constant of propranolol for beta-adrenoceptors on alveolar macrophages of patients with bronchial asthma appeared to be higher by an order of magnitude than those of patients with chronic bronchitis. In patients with bronchial asthma, the number of beta-adrenoceptors on macrophages from bronchoalveolar lavage fluid was 1.6 times higher than that from patients (n = 9) who received beta-agonist and corticosteroid therapy.

D. Discussion

These results suggest that differences in beta-adrenergic receptor affinity and number can be detected in alveolar macrophages when no differences are noted in peripheral blood mononuclear cells in asthmatics. The increased dissociation constant for propranolol beta-adrenergic receptors on the surface of macrophages from bronchoalveolar lavage fluid of bronchial asthma patients suggests that there may be decreased functional activity of beta-adrenoceptors in lung cells. The increased numbers of beta-adrenoceptors may reflect a compensatory reaction for that decreased functional activity or could be an independent event. The decreased numbers and the increased affinity for propranolol of beta-adrenoceptors on the surface of bronchoalveolar lavage macrophages of the beta-agonist-treated patient with bronchial asthma may be the consequence of beta-agonist and corticosteroid therapy. This may represent in-vivo desensitization to beta-agonists (Sibley and Lefkowitz, 1985). Thus, local decreases in beta-adrenoceptor sensitivity in lung cells may accompany bronchial hyperreactivity in bronchial asthma. Our results on peripheral blood

Table 2 Characteristics of Beta-Adrenoceptors from Patients

Chronic bronchitis					Bronchial asthma				
	Macrophages		Leukocytes			Macrophages		Leukocytes	
No.	K_d	B_{max}	K_d	B_{max}	No.	K_d	B_{max}	K_d	B_{max}
1	1.6	98	9.0	35	5	56.0	74	10.3	37
2	2.0	42	—	—	6	49.3	128	—	—
3	8.2	57	9.8	40	7	30.8	82	—	—
4	9.4	79	—	—	8	75.5	155	—	—
M ± m	5.3 ± 2.0	69 ± 12				53 ± 9[a]	110 ± 19		
					9	9.1	43	9.4	15

K_d = dissociation constant for 9-AAP, nM; B_{max} = number of beta-adrenoceptors per cell, relative units.
[a] $P < 0.01$ (statistical analyses were done with unpaired Student's test).

mononuclear cells are in agreement with previous studies using peripheral blood lymphocytes in asthmatics.

II. Association of Peripheral Blood T-Lymphocyte Beta-Adrenoreceptor Number and Affinity with Cellular Responses to IL-2 and PHA

A. Introduction

Activation of T and B cells is accompanied by phospholipase C hydrolysis of phosphotidyl inositol (PI) to form inositol-1,4,5-trisphosphate (IP3) and diacylglycerol (DAG). This results in Ca^{2+} mobilization from the endoplasmic reticulum and protein kinase C activation (Berridge, 1984). Stimulation of PI turnover occurs after challenge with mitogens (Taylor et al., 1984; Sugiura and Waku, 1984), growth factors (Nakamura and Ui, 1985), and antibodies (Coggeshall and Cambier, 1984). In the current studies we activated T lymphocytes by the T-cell mitogen phytohemagglutinin (PHA) and the T-cell growth factor interleukin-2 (IL-2) and determined cytoplasmic pH (which has been demonstrated to rise secondary to activation of sodium-proton exchange (Mills et al., 1985), which may be regulated by protein kinase C (Grinstein et al., 1985; Farrar and Anderson, 1985) and intracellular Ca^{2+} concentration (Grinstein and Cohen, 1987).

B. Methods

In these studies, mononuclear lymphocytes were isolated from peripheral blood of healthy donor-volunteers on Ficoll-Paque gradients according to the method described by Boyum (1968). Monocytes were removed by double adhesion on plastic dishes; B lymphocytes were separated by a panning technique using monoclonal antibodies to light chains of human immunoglobulins, which were the gift of O. V. Rokhlin (Institute of Experimental Cardiology, National Cardiology Research Center, Moscow). Phenotypic analysis of the cell populations were determined by flow cytometry using indirect immunofluorescence. T lymphocytes represented more than 85%, B-lymphocytes did not exceed 1%, and monocytes were less than 0.1%.

The T lymphocytes were divided into two parts, and both were cultured for 18 h in medium RPMI 1640, supplemented with 2 mM glutamine, 0.1% (w/v) human serum albumin, 50 μg/mL gentamycin, and 20 mM HEPES-NaOH, pH 7.4 (all reagents from Flow Laboratories, U.K., and Gibco, U.K.). The incubation medium of one of the two lymphocyte cultures was further supplemented with 2.5 μg of PHA/mL (Sigma, United States) and 20 U of IL-2/ml (Amersham Int., U.K.). These concentrations were earlier found to maximally stimulate ^{3}H-thymidine incorporation by T-lymphocytes. IL-2 receptor expression was evaluated by flow cytometry using indirect immunofluorescence with anti-Tac antibodies (Amersham Int., U.K.) to the IL-2 receptor (Dower et al., 1985).

The activity of phospholipase C was determined using a modification of a procedure previously described (Resink et al., 1987). After incubation with 100 μCi/mL myo-[2-^{3}H]-inositol (Amersham Int., U.K.) for 18 h, cells were washed by centrifugation, then preincubated with 10 mM LiCl. Aliquots containing 3×10^6 cells were then incubated with different agonists. Inositol phosphates were separated on columns containing 0.25g Dowex (1×4, 200–400 mesh) by increasing concentrations of ammonium formate as described previously (Berridge et al., 1983). Columns were calibrated using labeled standards of inositol phosphates (obtained from Amersham Int., U.K.).

Proliferative activity was measured by evaluating [^{3}H]-thymidine (Amersham Int. U.K.) incorporation. Cell cycle was determined by flow cytometry after treating cells with RNAse and then propidium iodide.

Beta-adrenoceptor density was determined in binding experiments at seven concentrations of ^{125}I-cyanopindolol (^{125}I-CYP) (Amersham Int., U.K.), as described previously (Krasnikova et al., 1989). Cells ($0.3\text{-}0.5 \times 10^6$) were incubated in the culture medium (300 μL) for 18 h at 4°C. Then the samples were filtered through GF/C filters. The filters were washed with an ice-cold buffer containing 20 mM Tris-HCl and 0.9% NaCl (pH 7.8). Nonspecific binding was that detected in the presence of 1 μM L-alprenolol (Sigma, United

States). Specific binding was calculated as the difference between total and nonspecific binding. The receptor density (B^{max}) and the antagonist dissociation constant (K_d) were calculated for Scatchard plots. Substitution of ^{125}I-CYP by L-isoproterenol (Sigma, United States) was done in intact T-lymphocytes at 4°C in three parallel experiments at 50–100 pM ^{125}I-CYP in the presence of 0.1 mM sodium ascorbate, at an L-isoproterenol concentration range of 10^{-9} to 10^{-5} M. Isoproterenol binding constant and the low- and high-affinity binding site ratios were calculated using a "Ligand" program.

Adenylate cyclase activity was determined in a freshly prepared lymphocyte homogenate, using alpha-^{32}P-ATP (Amersham Int., U.K.) as a substrate (Grigorian and Tkachuk, 1982). Incubation medium (50 μL) contained 50 mM Tris-HCl (pH 7.4), 0.1 mM cAMP, 1 mM isobutylmethylxanthine, 0.1 mM PMSV, 1 mM dithiotreitol, 0.1 mM ATP, alpha-^{32}P-ATP (1 μCi per probe), 2 mM $MgCl_2$, 10 mL creatine phosphate, 1 mg/mM creatine kinase, 0.4% bovine serum albumin (BSA). ^{32}P-cAMP was separated on a microcolumn packed with 1 g of aluminum oxide (neutral, by Brockman). All reagents were obtained from Sigma (United States), and Merck (Germany). Protein concentration was measured by the method of Lowry with Peterson's modification (Peterson, 1977) using BSA as a standard. Intracellular cAMP content was determined using standard kits (Amersham Int., U.K.).

C. Results

It is known that T-lymphocyte activation is accompanied by IL-2 receptor expression and an increase of proliferative activity. Using flow cytometry techniques, we have been able to demonstrate that antibodies to IL-2 receptors (anti-Tac) bind to T-lymphocytes preincubated with IL-2 and PHA together for 18 h. This binding did not occur to control T-lymphocyte cultures not exposed to IL-2 and/or PHA, corresponding well to the data of other authors (Larsson, 1981; Shirakawa et al., 1986).

Culture of T lymphocytes with IL-2 and PHA resulted in activation of the proliferative response of cells. The data on ^{3}H-thymidine incorporation by lymphocytes is presented in Table 3. The incorporation of ^{3}H-thymidine in T lymphocytes and the fraction of cells entering into S phase were greatest after co-culture with IL-2. Table 4 presents the data on stimulation of PI turnover in lymphocytes by PHA. After 18 h of cultivation with IL-2 and PHA, T cells accumulate the labeled products of PI turnover more actively than those under control conditions. PHA stimulates formation of inositol phosphates both in T-lymphocytes cultivated with PHA and IL-2, and in control cells. Table 2 shows representative data obtained on the cells from three donors. Similar effects were observed in all experiments (n = 12). However, activated T lymphocytes do not lose their phospholipase C sensitivity for PHA. This fact

Table 3 Effect of PHA and Interleukin 2 on Lymphocyte Proliferation

Additions	^{3}H-Thymidine incorporation[a]	Index of stimulation	Percent of cells in S phase of cell cycle
	T cells		
Control	824 ± 75	1.0	0.85%
PHA (5 μg/mL)	1073 ± 133	1.3	1.8%
IL-2 (20 U/mL)	1660 ± 81	2.0	4.7%
PHA + IL-2	2169 ± 379	2.6	4.2%
	PBMC		
Control	950 ± 90	1.0	—
PHA (5 μg/mL)	17730 ± 880	18	—

[a] ^{3}H-Thymidine incorporation was determined in quadruplicate by 6-h pulse with 5 μCi/mL ^{3}H-thymidine. Data are expressed as the mean ± S.D.

Table 4 PHA-Induced Formation of Inositol Phosphates in Intact and Activated Lymphocytes

	Intact lymphocytes		Activated lymphocytes[a]	
	Control	PHA[b]	Control	PHA[b]
Experiment 1: T cells				
IP	1290 ± 57	2214 ± 62	2447 ± 41	5028 ± 98
IP2	336 ± 30	431 ± 31	603 ± 47	1022 ± 49
IP3	178 ± 15	253 ± 18	337 ± 26	632 ± 38
Experiment 2: T cells				
IP	6196 ± 101	7654 ± 105	7237 ± 117	9523 ± 172
IP2	3352 ± 56	3783 ± 62	3952 ± 70	4542 ± 96
IP3	1650 ± 38	2108 ± 60	2138 ± 49	2647 ± 56
Experiment 3: PBMC				
IP	2151 ± 51	3690 ± 85	4077 ± 71	8379 ± 151
IP2	561 ± 36	717 ± 52	1005 ± 48	1704 ± 52
IP3	297 ± 27	423 ± 39	627 ± 38	1053 ± 91

Data (dpm/3 × 10^{6} cells) are means ± S.D. from representative experiments performed in duplicates.

[a] Activated T cells were obtained by cultivation for 18 h in the presence of PHA (2.5 μg/mL) and IL-2 (20 U/mL) in the culture medium, as decribed in the text. For activation of PBMC, only PHA was used.

[b] Incubation was performed with PHA (5 μg/mL) for 10 min at 37°C.

allowed us to compare the effect of PHA and IL-2 on the phosphoinositide turnover in activated cells.

Figure 2 represents typical time curves of inositol phosphate formation stimulated by PHA (5 μg/mL) and/or IL-2 (200 μL) in cells cultivated for 18 h with PHA and IL-2. Both agents activated breakdown of polyphosphoinositides. The most active formation of inositol phosphates occurred during the first 5 min after addition of stimulants. Accumulation of IP, IP_2, and IP_3 reached saturation within 15 min after addition of PHA as well as IL-2. However, by 15 min after addition of IL-2, the amount of IP and IP_2 decreased in comparison to the maximal effect. There was no significant accumulation of inositol phosphates in control cells.

The IL-2 and PHA-induced formation of inositol phosphates was dose-dependent. Figure 3 shows typical dose-response curves of IL-2 or PHA-stimulated breakdown of polyphosphoinositides for 10 min in T lymphocytes which had been cultured previously for 18 h with IL-2 and PHA. Maximal effect on the IP_3 formation was observed at 5 μg/mL PHA and 10μ/mL IL-2. In all cases PHA was more effective than IL-2 in stimulating formation of inositol phosphates. It should be noted that the variability of the IL-2 effect on PI turnover was greater than that of PHA.

Figure 4 represents the IL-2 production by T cells in response to IL-2 and PHA. IL-2 stimulated its own accumulation up to seven times, while PHA had only little effect during the 54 h of culture.

Having determined that PI turnover and cell cycle changes following IL-2 and PHA stimulation were similar in our study population to others in the literature, we next examined the properties of the beta-adrenergic-linked adenylate cyclase system in human peripheral blood T lymphocytes previously cultured with IL-2 and PHA and in control cells.

Culturing of human peripheral blood T lymphocytes with IL-2 and PHA for 18 h increased the density of surface beta$_2$-adrenoceptors. Table 5 summarizes beta$_2$-adrenoceptor densities (B_{max}) and binding constant for ^{125}I-CYP (K_d) measured in the T lymphocytes of four donors, isoproferenol-binding constants for high- and low-affinity receptors (Ki_H and Ki_L), and the proportions of the receptors in high- and low-affinity states determined on T lymphocytes from two of four donors. Though the initial density of beta$_2$-adrenoceptors varied from 0.71 fmol/10^6 cells to 2.25 fmol/10^6 cells, their number in activated T lymphocytes was about 40% greater than in control cells. Isoproterenol binding constants for high- and low-affinity receptors differed by two orders of magnitude. It should be mentioned that the proportion of high-affinity binding sites in control T lymphocytes was greater than in activated cells.

The increase in beta$_2$-adrenoceptor density on activated T lymphocytes coincided with a decrease in dissociation constants for isoproterenol and

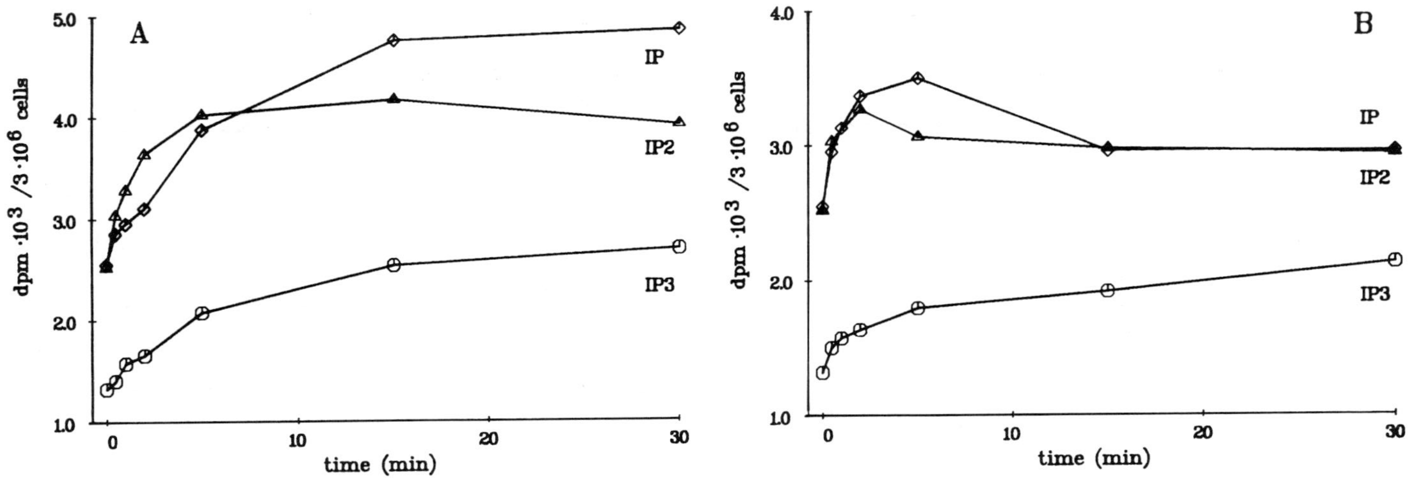

Figure 2 Kinetics of inositol phosphates accumulation in activated T lymphocytes in response to 5 μg/mL PHA (A) and 200 U/mL IL-2 (B). T lymphocytes were activated in the presence of 2.5 μg/mL PHA and 20 U/mL IL-2 during 18 h in an atmosphere of 5% CO_2 at 37°C. Each data point is the average of two separate determinations. Date varied by no more than 10% between duplicate samples. There was no significant accumulation of inositol phosphate without additions.

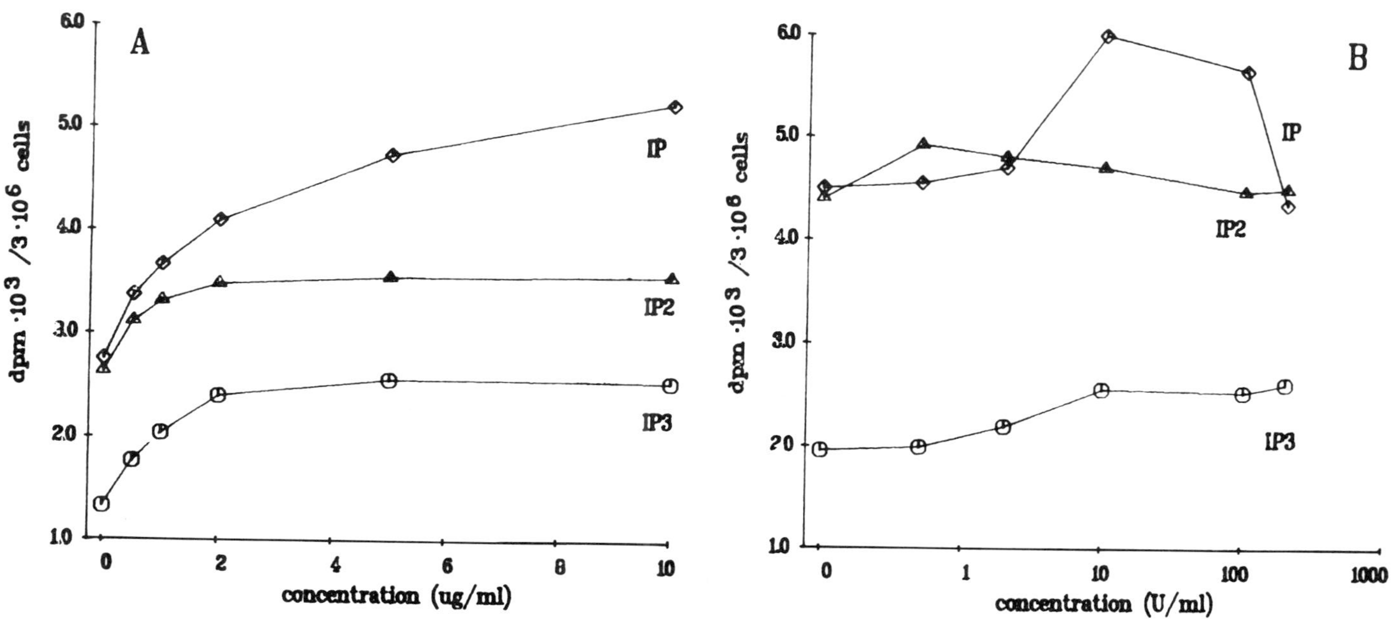

Figure 3 Dose dependencies for inositol phosphate accumulation in response to PHA (A) and IL-2 (B). Incubation time with different concentrations of agonists was 15 min. T lymphocytes were activated as described in the legend to Fig. 1. Values are the average of two samples. Date varied by no more than 10% between duplicates.

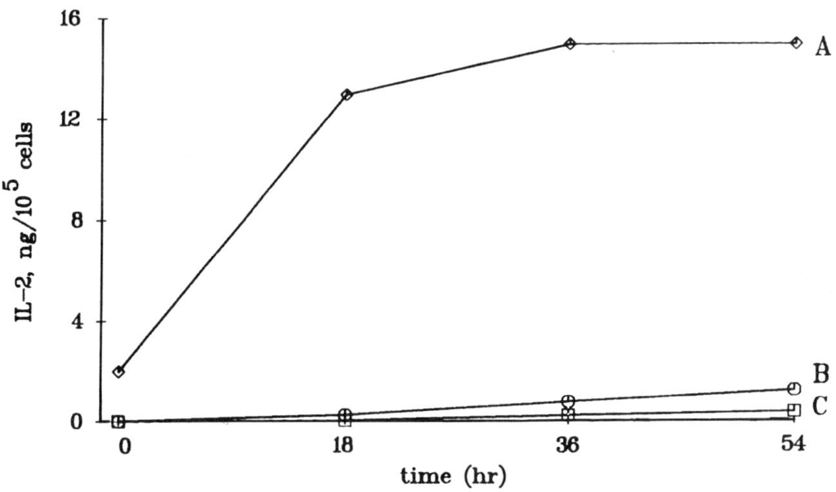

Figure 4 Effect of IL-2 (A) and 5 μg/mL PHA (B) on interleukin-2 production by culturing T lymphocytes. Control production (C) was measured without additions. Each data is the average of two samples. Date varied by no more than 10% between duplicates.

^{125}I-CYP. This reduced affinity for agonists could be explained by expressed receptors which are not completely coupled to regulatory G proteins. The possibility that the expressed beta-receptors are not coupled to G_s proteins was further suggested by the data on the regulatory properties of adenylate cyclase in control and activated T lymphocytes in Table 6, in which we illustrate the

Table 5 Properties and Density of Beta$_2$-Adrenoceptors on T Lymphocytes Activated with IL-2 and PHA

	Control cells				Activated cells			
	B_{max}	K_d	Ki_H/Ki_L	$\%B_{max_H}/\%B_{max_L}$	B_{max}	K_d	Ki_H/Ki_L	$\%B_{max_H}/\%B_{max_L}$
1	0.71	95	0.02/1.60	56/44	1.05	166	0.01/0.68	29/71
2	1.30	35	0.01/2.00	85/15	1.87	37	0.04/18.10	32/68
3	1.87	72	—	—	2.29	90	—	—
4	2.25	33	—	—	3.17	94	—	—

B_{max} = density of beta$_2$-adrenoceptors, fmol/10^6 cells; K_d = dissociation constant for ^{125}I-CYP, pM; Ki_H and Ki_L = isoproterenol binding constants for high- and low-affinity receptors, μM; $\%B_{max_H}$ and $\%B_{max_L}$ = proportions of high- and low-affinity binding sites.

Table 6 Regulatory Properties of Adenylate Cyclase in Control and Activated T Lymphocytes

	Specific activity (pmol cAMP/mg · min)	
	Control cells	Activated cells
Basal activity	6.1 ± 0.4	7.9 ± 1.0
Isoproterenol	9.9 ± 0.6	11.8 ± 0.4
GppNHp	12.3 ± 0.2	16.7 ± 0.6
GppNHp + isoproterenol	13.8 ± 0.8	19.5 ± 0.5
Forskolin	39.4 ± 1.8	53.3 ± 2.3

Data are the means of three parallel determinations with standard error. All agents were added at a final concentration of 10 μM.

effects of isoproterenol, guanylyl imidodiphosphate (GppNHp), and forskolin on adenylate cyclase activity in T lymphocytes cultured in the presence of IL-2 and PHA. An increase in basal and stimulated adenylate cyclase activities in the presence of IL-2 and PHA coincided with enhanced stimulatory effects of GppNHp and forskolin, while the degree of adenylate cyclase activation by isoproterenol remained unchanged.

Figure 5(1) shows the kinetics of the alterations in beta$_2$-adrenoceptor density on the surface of activated T lymphocytes. The receptor density increased by 41% after an 18-h incubation with PHA and IL-2 and then decreased to values lower than the initial ones. After 36-h and 54-h incubations with these agents, B_{max} in activated T lymphocytes decreased by 9% and 18%, respectively, compared with control.

A similar pattern of changes in cAMP content was observed [Fig. 5, (2)]. After an 18 h incubation in the presence of IL-2 and PHA, intracellular cAMP concentration increased more than 1.5-fold, and then decreased by 25% and 41% after 36 h and 54 h incubations, respectively, compared with controls. Isoproterenol produced similar effects on cAMP concentration in control and activated T lymphocytes [Fig. 5, (3)].

To elucidate the mechanisms underlying the alterations in beta$_2$-adrenoceptors on IL-2- and PHA-activated T lymphocytes, we determined beta-adrenoceptor density and cAMP content of lymphocytes incubated with phorbol myristate acetate (PMA), an analog of diacylglycerol. Figure 6 shows the intracellular cAMP content after 10 min of incubation with varied concentrations of PMA. At 0.1 μM PMA, cAMP concentration increased from 0.9 to 1.9 pmol/10^6 cells. At 0.001 and 0.01 μM PMA, similar stimulatory effects of isoproterenol were observed, while at 0.1 μM PMA this effect was markedly potentiated. The density of beta$_2$-adrenergic receptors was determined at given

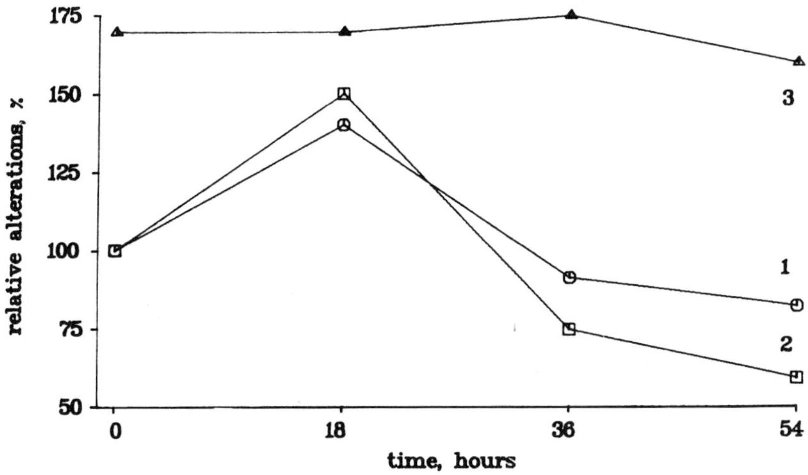

Figure 5 Kinetics of beta$_2$-adrenoceptor density (1), cAMP content (2), and isoproterenol effect (3) in T lymphocytes activated with IL-2 and PHA. The data are expressed as percent of the control. The effect of isoproterenol is expressed as percent of the changes in cAMP content after the action of 10 μM isoproterenol. Maximal cAMP content and beta$_2$-adrenoceptor density were 1.02 pmol/10^6 cells and 3.17 fmol/10^6 cells, respectively (mean values of two parallel determinations).

time periods during a 2 h incubation of T lymphocytes with 0.1 μM PMA (Fig. 7). After 15 min of incubation, B_{max} increased significantly, from 2.860.44 to 3.85 0.23 fmol/10^6 cells, and in 45 min decreased to 1.93 0.08 fmol/10^6 cells.

D. Discussion

Our results, indicating that IL-2 stimulates phosphoinositide turnover, correspond well with those performed by other investigators on the IL-2-dependent cell lines CT6 and CTB6 (Bonvini et al., 1987). On the line CTLL-2, IL-2 did not affect phospholipase C, but did induce a proliferative response (Mills et al., 1986). Differences in the effect of IL-2 on the phosphoinositide metabolism among cell lines may be attributed to their lineage or to transformation of their properties during long-term culture.

Comparison of time and dose dependencies of inositol phosphate accumulation in response to PHA and IL-2 revealed that PHA stimulates more phosphoinositide turnover than IL-2 in T lymphocytes. In contrast, IL-2 was more effective than PHA in stimulating the proliferative response. This difference cannot be explained by the time needed for PHA to stimulate IL-2

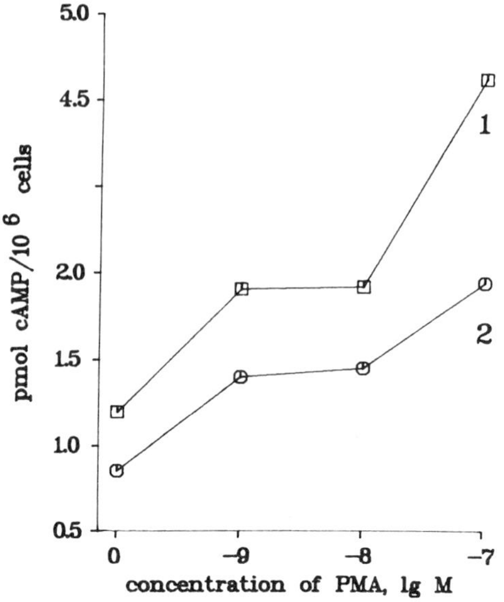

Figure 6 Effect of PMA on cAMP content in T lymphocytes after a 15 min preincubation. (1), 10 min incubation with 10 μM isoproterenol; (2), control cells. The data are the means of two parallel determinations.

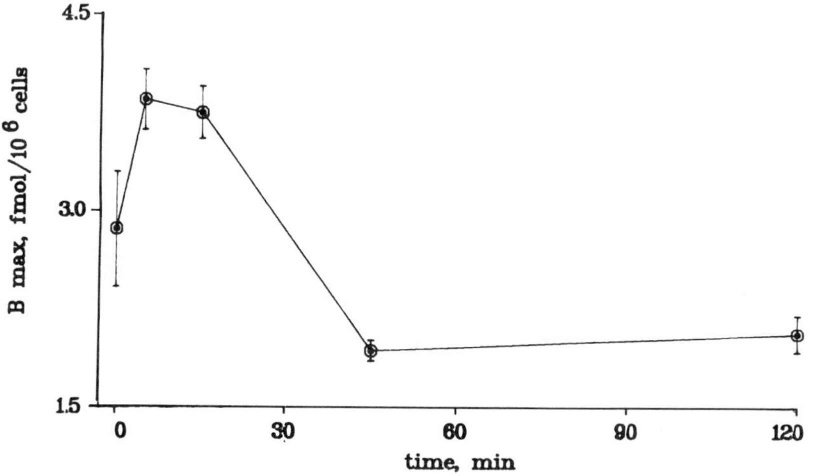

Figure 7 Kinetics of the changes in beta$_2$-receptor density on the surface of T lymphocytes under the action of 0.1 μM PMA. Data are the means of three determinations with standard error.

production by T cells, because PHA alone did not affect IL-2 production in our experiments.

Our data, like others, suggest that the proliferative response cannot be directly, quantitatively related only to phosphoinositide turnover (Katzen et al., 1985; Mills et al., 1985). Since the activation of phospholipase C and protein kinase C induces receptor expression, and PHA (Depper et al., 1984) and phorbol ester (Kumagai et al., 1987) induce expression of IL-2 receptors, a role of phosphoinositide turnover in the process of the proliferative response may be in the translocation of growth-factor receptors onto the lymphocyte surface. The ability of IL-2 to stimulate phosphoinositide turnover may explain its ability to activate expression of its own receptor (Harel-Bellan et al., 1986).

The results of our experiments suggest that protein kinase C is involved in the increase and then decrease with time in beta$_2$-adrenoceptor density followed by activation of T lymphocytes with PHA and IL-2. A decrease in the number of beta$_2$-adrenoceptors upon activation of protein kinase C has been observed in various cellular models, and may be mediated by heterologous receptor cross-desensitization (Sibley and Lefkowitz, 1985). However, the mechanisms responsible for the increase in beta$_2$-adrenoceptor density on T lymphocytes involving protein kinase C remain unclear.

III. Summary

In this chapter we have characterized beta-adrenergic receptors on macrophages from bronchoalveolar lavage in an attempt to reflect the local status of airway cells more precisely than utilizing mononuclear cells from peripheral blood. We used flow cytometry with the beta-adrenergic ligand 9-aminoacridine propranolol (9-AAP) for detecting beta-adrenoceptors on intact cells. No differences were noted in the number or in the affinity for 9-AAP of beta-adrenoceptors in mononuclear cells from peripheral blood between patients with bronchial asthma and patients with chronic bronchitis. The dissociation constant of 9-APP and the number of beta-adrenoceptors in alveolar macrophages of patients with bronchial asthma were significantly higher than those of patients with chronic bronchitis. Thus our results demonstrate local changes in receptor systems of bronchial tree cells which have been directly exposed to mediators associated with asthma.

Since antigens cause specific activation of T cells, we next explored the relationship of the beta-adrenergic adenylate cyclase system to the T-cell activation process. The results of our in-vitro experiments have shown that the rise of cytoplasmic inositol phosphates secondary to PI turnover following mitogen stimulation does not completely explain the proliferative response of T cells; and that protein kinase C must play some additive role. We demonstrated for the first time that IL-2 and PHA can regulate the state of the beta-adrenergic-linked adenylate cyclase system by stimulating phosphoinositide turnover and protein kinase C. These biochemical changes may suggest a mechanism for an association between adrenergic and immunological mechanisms of T-cell activation, perhaps explaining some of the differences in T-cell beta-adrenergic receptor function observed in the T cells of individuals with asthma.

References

Abrass DK, O'Connor SW, Scarpace PJ, Abrass IB. Characterization of the beta-adrenergic receptor of the rat peritoneal macrophage. J Immunol 1985; 135:1338–41.

Alston W, Patel K, Kerr J. Response of leukocyte adenylcyclase to isoprenaline and effect of alpha blocking drugs in extrinsic bronchial asthma. Br Med J 1974; 1:90–100.

Atlas D, Levitzki A. Probing of beta-adrenergic receptors by novel fluorescent beta-adrenergic blockers. Proc Natl Acad Sci USA 1977; 74:5290–4.

Baner S, Jakobs KH. Phorbol ester treatment impairs hormone—but not stable GTP analog-induced inhibition of adenylate cyclase. FEBS Lett 1986; 198(1):43–6.

Barnes P. Localization and function of autonomic receptors. In: Receptors and cold. Proceeding of an international meeting, Utrecht. 1982; 187–97.

Barnes P, Basbaum C, Nadel J. Autoadiographic localization of autonomic receptors in airway smooth muscle. Am Rev Respir Dis 1983a; 127:758–62.

Barnes P, Skoogh B-E, Brown et al. Activation of beta-adrenergic response ub tracheal smooth muscle, a postreceptor mechanism. J Appl Physiol 1983b; 54(6):1469–76.

Baroms R, Hills T. Decreased beta-adrenergic receptor density on human lymphocytes after chronic treatment with agonists. J Pharmacol Exp Ther 1983; 224(1):1–6.

Berridge MJ. Inositol trisphosphate and diacylglycerol as second messengers. Biochem J 1984; 220:345–60.

Berridge MS, Dawson RMC, Downes CP, Hesplop GP, Irvine RF. Changes in the level of inositol phosphates after agonist-dependent hydrolysis of membrane phosphoinositides. Biochem J 1983; 212:473–82.

Bonvini E, Ruscetti FW, Ponzoni M, Hoffman T, Farrar WL. Interleukin 2 rapidly stimulates synthesis and breakdown of polyphosphoinositides in interleukin 2-dependent, murine T-cell lines. J Biol Chem 1987; 262:4160–4.

Boyum A. Isolation of mononuclear cells and granulocytes from human blood. Scand J Clin Lab Invest 1968; 21(suppl. 97): 77–79.

Bruijnzeel P. Desensitization of the beta-adrenergic receptor on leukocytes after long-term oral use of beta-symphathucomimetic, its effect on the beta-adrenergic blockage hypothesis of Szentivanji. Ann Allerg 1979; 43:105–16.

Carstairs J, Nimmo S, Barnes P. Autoradiographic visualization of beta-adrenoreceptors subtypes in human lung. Am Rev Respir Dis 1985; 132:541–7.

Coggeshall KM, Cambier JC. B cell activation. VIII. Membrane immunoglobulins transduce signals via activation of phosphatidylinositol hydrolysis. J Immunol 1984; 133:3382–6.

Colucci W, Alexander R, Williams G, et al. Decreased lymphocyte beta-adrenergic receptor density in patients with heart failure and tolerance to the beta-adrenergic against pirbuterol. N Engl J Med 1981; 305:185–92.

Cookson D, Redd C. A comparison of the effects of isoproterenol in the normal and asthmatic subject. Am Rev Respir Dis 1963; 88:636–80.

Depper JM, Leonard VJ, Waldmann TA, Green WS. In: Pick E, ed. Lymphokines, Vol. 9. New York: Academic Press, 1984; 127–52.

Dower SK, Hefencider SH, Albert AR, Urda DL. Quantitative measurement of human interleukin 2 receptor levels with intact and detergent-solubilized human T-cells. Mol Immunol 1985; 22:937–47.

Engle G, Hoyer D, Berthold R, and Wagner H. [^{125}Iodo]cyanopindolol, a new ligand for beta-adrenoceptors: identification and quantitation of subclasses of beta-adrenoceptors in guinea pig. Naunn-Schmeideberg Arch Pharmacol 1981; 317:277–85.

Farrar WL, Anderson WB. Interleukin-2 stimulates association of protein kinase C with plasma membrane. Nature 1985; 315:233–5.

Fels AO, Cohn ZA. The alveolar macrophage. J Appl Physiol 1986; 60:352–69.

Glant S, Duriseti L, Underwood S., et al. Beta-adrenergic receptor of polymorphonuclear particulates in bronchial asthma. J Clin Invest 1980; 65:577–85.

Godard P. Cjhaintreuil J, Damon M., et al. Functional assessment of alveolar macrophages: comparison of cells from asthmatics and normal subjects. J Allergy Clin Immunol 1982; 70:88–93.

Grigorian GYu, Tkachuck VA. Tubulin restores the sensitivity of the rabbit hear adenylate cyclase catalytic subunit to guanyl nucleotides and NAF. Biochem Int 1982; 4(6):592–603.

Grinstein S, Cohen S. Cytoplasmic [Ca^{2+}] and intracellular pH in lymphocytes. J Gen Physiol 1987; 89:185–213.

Grinstein S, Cohen S, Goetz JD, Rothstein A, Gelfand EW. Characterization of the activation of Na^+/H^+ exchange in lymphocytes by phorbol esters: change in cytoplasmic pH dependence of the antiport. Proc Natl Acad Sci USA 1985; 82:1429–33.

Harel-Bellan A, Bertoglio J, Quillet A, Marchiol C, Wakasugi H, Mishall Z, Fradelizi D. Interleukin 2 (IL-2) up-regulates its own receptor on a subset of human unprimed peripheral blood lymphocytes and triggers their proliferation. J Immunol 1986; 136:2463–9.

Katada T, Gilman AG, Watanabe Y. Protein kinase C phosphorylates Ni and apparently supresses its function in normal inhibition of adenylate cyclase. Eur J Biochem 1985; 151:431–7.

Katzen D, Chu E, Terhost C, Leung DY, Gesner M, Miller RA, Geha RS. Mechanisms of human T cell response to mitogens: IL 2 induces IL 2 receptor expression and proliferation but not IL 2 synthesis in PHA-stimulated t cells. J Immunol 1985; 135(3):1840–5.

Korichneva IL, Grigorian GYu, Krasnikova TL, Rudchenko SA, Tkachuk VA. Interleukin 2 and phytohemagglutinin activated proliferation of human T-lymphocytes is accompanied by stimulation of phosphoinositide turnover. Biochim Biophys Acta 1989a; 1014(1):173–7.

Korichneva IL, Rudchenko SA, Krasnikova TL, Grigorian GYu, Smirnov VN. Stimulation of phosphoinositide turnover is an important but not a singular requirement for phytohemagglutinin- and interleukin 2-activated proliferation of human T-lymphocytes. Proc USSR Acad Sci 1989b; 304(1):221–4.

Krasniova TL, Korivhneva IL, Radyukhin VA. The dependence of the kinetic constants of beta$_2$-adrenoceptors of intact human lymphocytes on the nature of the radioligands applied. Biokhimia 1989; 54(2):235–43.

Kumagai N, Benedict SH, Mills GB, Gelfand EW. Requirement for the simultaneous presence of phorbol esters and calcium inophores in the expression of human T-lymphocyte proliferation-related genes. J Immunol 1987; 139:1393–9.

Larsson EL. Mechanism of T-cell activation. 11. Antigen-and lectin-dependent acquisition of responsiveness to TCGF is a nonmitogenic, active response of T cells. J Immunol 1981; 126:1323–6.

Liggett SB. Identification and characterization of a homogenous population of beta$_2$-adrenergic receptors on human alveolar macrophages. Am Rev Respir Dis 1989; 139:552–5.

Limas CJ, Limas C. Carbachol induces desensitization of cardiac beta-adrenergic receptors through muscarinic M_I receptors. Biochem Biophys Res Commun 1985; 128(2):699–704.

Lockey S, Ilennon S, Redd C. Comparison of some metabolic response in normal and asthmatic subjects to epinephrine and glucagon. J Allergy 1968; 42:288–91.

Loscalzo J, Daniel RP. Modulation of beta-adrenergic receptors in alveolar macrophage membranes the effect of phagocytosis and adherence (abstract). Am Rev Respir Dis 1979; 119(4, pt 2):79.

Lynch CJ, Charest R, Bocckino SB, Exton JH, Blackmore PF. Inhibition of hepatic alpha$_I$-adrenergic effects and binding by phorbol myristate acetate. J Biol Chem 1985; 260(5):2844–51.

Mills GB, Cheung RK, Grinstein S, Gelfand EW. Interleukin 2-induced lymphocyte proliferation is independent of increases in cytosolic-free calcium concentrations. J Immunol 1985a; 134:2431–5.

Mills GB, Cragoe EJ Jr. Gelfand EW, Grinstein S. Interleukin 2 induces a rapid increase in intracellular pH through activation of a Na^+/H^+ antiport. J Biol Chem 1985b; 260:12500–7.

Mills GB, Stewart DJ, Mellors A, Gelfand EW. Interleukin 2 does not induce phosphatidylinositol hydrolysis in activated T-cells. J Immunol 1986; 136:3019–24.

Nakamura T, Ui M. Simulataneous inhibitions of inositol phosphilipid breakdown, arachidonic acid release and histamine secretion in mast cells by islet-activation protein pertussis toxin. J Biol Chem 1985; 260:3584–93.

Pamela B, David M, Dale M, et al. Beta-adrenergic responses in drug-free subjects with asthma. J Allergy Clin Immunol 1986; 77(6):133–49.

Peterson GL. A simplification of the protein assay method of Lowry et al. which is more generally applicable. Anal Biochem 1977; 83(2):346–56.

Pieter L, Bruynzee Willen van Berg, Lefernik G. The bronchial tree and the beta-adrenergic system. Eur J Respir Dis 1982; 63(suppl.121):64–71.

Resink TJ, Grigorian GY, Moldabaeva AK, Danilov SM, Buhler FR. Histamine-induced phosphoinositide metabolism in cultured human umbilical vein endothelial cells. Association with thromboxane and prostacyclin release. Biochem Biophy Res Commun 1987; 144:438–46.

Sano Y, Gavin W, Towuley M. Decreased mononuclear cell beta-adrenergic receptors in bronchial asthma: parallel studies of lymphocytes and granulocyte desensitization. J Allergy Clin Immunol 1983; 72(5):495–507.

Sasaki T, Hasegawa-Sasaki H. Breakdown of phosphatidylinositol 4,5-biophosphate in T-cell leukemia line stimulated by phytoheagglutinin is not dependent on Ca^{2+} mobilization. Biochem J 1985; 227:971–9.

Shirakawa F, Tanaka Y, Eto S, Suzuki H, Yodol J, Jamashito U. Effect of interleukin I on the expression of interleukin 2 receptor (tac antigen) on human natural killer cells and natural killer-like cell line (yt cells). J Immunol 1986; 137:551–6.

Sibley DR, Lefkowitz RJ. Molecular mechanisms of receptor desensitization using the beta-adrenergic receptor-coupled adenylate cyclase system as a model. Nature 1985; 317:124–9.

Smith J, Parker C. The responsiveness of leukocyte cyclic adenosine monophosphate to adrenergic agents inpatients with asthma. J Lab Clin Med 1970; 76:993–4.

Steinkamp JA, Kraemer PM. Quantitation of lectin binding by cells. In: Melamed MR, et al., eds. Flow cytometry and sorting. New York: John Wiley, 1979; 497–504.

Stiles GL, Caron MG, Lefkowitz RJ. Beta-adrenergic receptors: biochemical mechanisms of physiological regulation. Physiocological regulation. Physiol REv 1984; 64(2):661–743.

Sugiura T, Waku K. Enhances turnover of arachidonic acid-containing species of phosphatidic acid of concanavalin A-stimulated lymphocytes. Biochem Biophys Acta 1984; 796:190–8.

Svedmyc N. Airway hyperreactivity. Cholinergic and adrenergic receptors. Eur J Respir Dis 1983; 64 (suppl.131):326–40.

Szentivanji A. The beta-adrenergic theory of the atopic abnormality in bronchial asthma. J Allergy 1968; 42:203–10.

Szentivanji A. The conformation flexibility of adrenoreceptors and the constitutional basis of atopy. Triangle 1979; 18(4):109–15.

Szentivanji A. Variations de structure des adrenorecepteurs et bases constitutionneles de L'atopie. Triangle (J. Sandoz des sciences medicals) 1980; 3:35–42.

Szentivanji A, Fishel C. The beta-adrenergic theory and cyclic AMP-mediated control mechanisms in human asthma. In: Weiss EB, Segal MS, eds. Bronchial asthma mechanisms and therapeutics. Boston: Little Brown, 1976;137.

Taylor MV, Metcalft JC, Hesketh TR, Smith GA, Moore JP. Mitogens increase phosphorilation of phosphoinositides in thymocytes. Nature 1984; 312:462–5.

Tonnel AB, Gosset P, Joseph M, Fourneir E, Capron A. Stimulation of alveolar macrophages in asthmatic patients after local provocation test. Lancet 1983; I:1406–9.

Verghese MW, Snyderman R. Hormonal activation of adenylate cyclase in macrophage membranes is regulated by guanine nucleotides. J Immunol 1983; 130:869–73.
Vermeire P, Vanhaitte P. Inhibitory effects of catecholamine sin isolated canine bronchial smooth muscle. J Appl Physiol 1979; 46:787–91.
Yamashita A, Kurokawa T, Dan'ura T, Higashi K, Ishibashi S. Induction of desensitization by phorbol ester to beta-adrenergic agonist stimulation in adenylate cyclase system of rat reticulocytes. Biochem Biophys Res Commun 1986; 138:125–30.

AUTHOR INDEX

Italic numbers give the page on which the complete reference is listed.

B

G

M

N

O

P

S

T

U

V

W

SUBJECT INDEX

A

B

C

D

E

H

K

L

M

T